The Nyström Method in Electromagnetics

The Nyström Method in Electromagnetics

Mei Song Tong
Tongji University, Shanghai, China

Weng Cho Chew
Purdue University, West Lafayette, USA

WILEY

IEEE PRESS

This edition first published 2020

Registered Offices
John Wiley & Sons, Inc., 111 River Street, Hoboken, NJ 07030, USA
John Wiley & Sons Singapore Pte. Ltd, 1 Fusionopolis Walk, #07-01 Solaris South Tower, Singapore 138628

Editorial Office
The Atrium, Southern Gate, Chichester, West Sussex, PO19 8SQ, UK

For details of our global editorial offices, customer services, and more information about Wiley products visit us at www.wiley.com.

Wiley also publishes its books in a variety of electronic formats and by print-on-demand. Some content that appears in standard print versions of this book may not be available in other formats.

Library of Congress Cataloging-in-Publication Data

Names: Tong, Mei Song, author. | Chew, Weng Cho, author.
Title: The Nyström method in electromagnetics / Mei Song Tong (Tongji University Shanghai, China), Weng Cho Chew (Purdue University, West Lafayette, USA).
Description: Hoboken, NJ, USA : Wiley-IEEE Press, 2020. | Includes bibliographical references and index. |
Identifiers: LCCN 2018060139 (print) | LCCN 2019002836 (ebook) | ISBN 9781119284888 (Adobe PDF) | ISBN 9781119284871 (ePub) | ISBN 9781119284840 | ISBN 9781119284840(hardcover) | ISBN 1119284848(hardcover)
Subjects: LCSH: Electromagnetism–Mathematics. | Integral equations–Numerical solutions.
Classification: LCC QC760 (ebook) | LCC QC760 .T59 2019 (print) | DDC 537.01/51–dc23
LC record available at https://lccn.loc.gov/2018060139

Cover Design: Wiley
Cover Image: © Mark Garlick / Science Photo Library/Getty Images

Set in 10/12pt WarnockPro by SPi Global, Chennai, India

10 9 8 7 6 5 4 3 2 1

To Mei Song Tong's Family
Wei Liu and Zikang Tong

To Weng Cho Chew's Family
Chew Chin, Huibin Amelia, Shinen Ethan, and Sharon Ling

Contents

About the Authors

Professor Mei Song Tong received the Ph. D. degree in electrical engineering from Arizona State University, Tempe, Arizona, USA, in 2004. He is currently the Distinguished Professor, Department Head of Electronic Science and Technology, and Vice Dean of College of Microelectronics, Tongji University, Shanghai, China. Before he joined Tongji University, he was a Research Scientist at the Center for Computational Electromagnetics and Electromagnetics Laboratory, Department of Electrical and Computer Engineering, University of Illinois at Urbana-Champaign, Urbana, Illinois, USA. He is currently an adjunct professor at the University of Illinois at Urbana-Champaign and an honorary professor at the University of Hong Kong, China. He has published more than 400 papers in refereed journals and conference proceedings, and co-authored six books. His research interests include electromagnetic field theory, antenna theory and design, simulation and design of RF/microwave circuits and devices, interconnect and packaging analysis, inverse electromagnetic scattering for imaging, and computational electromagnetics.

Professor Tong is a Fellow of Electromagnetics Academy, Fellow of Japan Society for the Promotion of Science (JSPS), Full Member (Commission B) of the U.S. National Committee for the International Union of Radio Science, and Member of the Applied Computational Electromagnetics Society and the Sigma Xi Honor Society. He has been the chair of Shanghai Chapter and the chair of SIGHT committee of IEEE Antennas and Propagation Society since 2014 and in 2018, respectively. He has served as an associate editor or guest editor for several well-known international journals, including IEEE Antennas and Propagation Magazine, IEEE Transactions on Antennas and Propagation, IEEE Transactions on Components, Packaging and Manufacturing Technology, International Journal of Numerical Modeling: Electronic Networks, Devices and Fields, Progress in Electromagnetics Research, and Journal of Electromagnetic Waves and Applications, etc. He also served as a session organizer, session chair, technical program committee chair or member, and general chair for some prestigious conferences in the electromagnetics community such as IEEE International Symposium on Antennas and Propagation (IEEE AP-S) and USNC/URSI National Radio Science Meeting, Progress in Electromagnetics Research Symposium (PIERS), and IEEE International Conference on Computational Electromagnetics (ICCEM). He was the recipient of a Visiting Professorship Award from Kyoto University, Japan, in 2012. He advised and coauthored ten papers that received the Best Student Paper Award from 2014 International Workshop on Finite Elements for Microwave Engineering, PIERS'2014, PIERS'2015, PIERS'2016, ICCEM'2018, ICCEM'2019, PIERS'2019, and 2019 International Applied Computational Electromagnetics Society Symposium, respectively. He was the recipient of the Travel Fellowship Award of USNC/URSI for the 31th General Assembly and Scientific Symposium (GASS) of URSI in 2014, Advance Award of Science and Technology of Shanghai Municipal Government in 2015, Fellowship Award of JSPS in 2016, Innovation Award of Universities' Achievements of Ministry of Education of China in 2017, and many teaching-type awards from Tongji University and Shanghai Municipal Government, respectively. In 2018, Professor Tong was selected as the distinguished lecturer by IEEE Antennas and Propagation Society for the 2019-2021 term.

Professor Weng Cho Chew received all his degrees from Massachusetts Institute of Technology (MIT). His research interests are in wave physics, specializing in fast algorithms for multiple scattering imaging and computational electromagnetics in the last 30 years. His recent research interest is in combining quantum theory with electromagnetics, and differential geometry with computational electromagnetics. After MIT, he joined Schlumberger-Doll Research in 1981. In 1985, he joined University of Illinois at Urbana-Champaign, and was then the director of the Electromagnetics Lab from 1995 to 2007. During 2000–2005, he was the Founder Professor, 2005–2009 the Y. T. Lo Chair Professor, and 2013–2017 the Fisher Distinguished Professor. During 2007–2011, he was the Dean of Engineering at the University of Hong Kong. He joined Purdue University in August 2017 as a Distinguished Professor. He has co-authored three books, many lecture notes, over 400 journal papers, and over 600 conference papers. He is a fellow of various societies, and an ISI's highly-cited author. In 2000, he received the IEEE Graduate Teaching Award, in 2008, he received the C. T. Tai Distinguished Educator Award of IEEE Antennas and Propagation Society, in 2013, he was elected to the National Academy of Engineering, and in 2015, he received the ACES Computational Electromagnetics Award. Also, he received the IEEE Electromagnetics Award in 2017 and served as the president of IEEE Antennas and Propagation Society in 2018.

Preface

When I was invited with multiple times by the representative of Wiley to write this book in 2015, I was very hesitant because I worried about whether or not there will be enough readers who want to read the book. This is because the method of moments (MoM) as the primary numerical method for solving electromagnetic (EM) integral equations has been so widely used and accepted, it is very difficult to attract readers' interest by presenting another similar numerical method without demonstrating its obvious advantages. Fortunately, the Nyström method as one of nontraditional numerical methods has received more and more attention and has been fast-increasingly applied to solve various EM problems since it was introduced to EM community in 1990. The Nyström method has shown certain merits which the MoM does not have and has become a strong competitor and alternative in many applications. Also, as a nontraditional and relatively-new approach, it is natural to be lately accepted since people need some time to recognize its value, but this should not become a reason to stop studying and promoting it. No new exploration and competition means no advance! The rule should be applicable to all situations, including the investigation on different numerical methods for solving EM problems. Therefore, we can confidently write this book now.

This book is a monograph instead of a textbook which is mainly used for research reference. Many authors of such type of books said that the reason why they wrote their books was that there were not related books on relevant topics or the existent books were too old, so they should present new books. This is of course a strong reason, but we think that more books on similar topics in some research areas are also needed and are very worthy to be written even though there have been other books which have not been very old, because a diversity of books should be presented to satisfy different readers' diverse needs. Also, different books can provide different views or emphases even on similar topics. More importantly, keeping a moderate competition can certainly improve the quality of books just like papers, tons of which can simultaneously address the same or similar topics.

The technical contents of books usually come from published papers which have gone through a peer review, but writing books is still necessary and is actually a very important thing even if readers can find similar contents in papers. This is because books and papers serve different groups of readers who could have very different needs. The papers mainly aim at those professional researchers who have enough background and knowledge on related topics and can easily understand what papers present. In the contrast, the books primarily serve the newbies or inexperienced researchers whose representatives could be graduate students who have not had enough background and knowledge on relevant topics and could feel hard to directly read papers. An investigation by our graduate students association shows that more than 80% graduate students strongly rely on reading books instead of papers in their research work. The books should re-narrate the sophisticated technical contents of papers by an easier-understanding style so that readers can more easily understand and accept them. This requires that the authors of books creatively compose, explain, clarify, or illustrate the technical contents of papers according to their understanding and experience, instead of just a simple and mechanical collection or repeat of papers' contents. Due to the significant difference of functions between the books and papers, we should not make the books become a collective copy of some papers as some books' authors did. Those authors often claimed that their books mainly wanted to provide a convenience for readers to read related papers. To avoid such kind of style, we have provided an extensive introduction to the background, basic principle or mechanism, implementation method, physical applications, etc. for the book's theme, i.e. the Nyström method, with a more easily-accepted manner. We believe that these non-paper contents are very necessary for inexperienced researchers and students although experienced researchers may not need them.

This book is a summary for the work that we have done on the Nyström method in the past sixteen years. Thanks to Professor Weng Cho Chew who initialized the study on the interesting method and gave a continuous and huge instruction and advice on later investigations, resulting in making a significant progress on this method. Compared with other similar books, we specially strengthen the introduction to the singularity treatment techniques which could be viewed as the life and core of the Nyström method and its great expansion on practical applications, including the incorporation with the multilevel fast multipole algorithm (MLFMA), solution of multiphysics problems, and solution of time-domain integral equations. These features have never or seldom been dealt with by other authors and could be the distinct highlights of this book.

Also, the Nyström method is an integral equation solver or belongs to an integral equation approach, so it is intimately related to integral equations and knowing the EM integral equations is a prerequisite for reading this book. We already wrote a book entitled "Integral Equation Methods for Electromagnetic and Elastic Waves", which was published by Morgan & Claypool, San Rafael, CA, in 2008. That book mainly focuses on addressing the mathematics and physics behind the EM integral equations although it also includes some numerical solution methods and numerical examples. As a comparison, this book can be thought of as an engineering book which emphasizes the implementations and applications of the Nyström method for solving EM integral equations for practical engineering problems. Therefore, it may be a good idea to jointly read or refer to these two books.

This book has ten chapters in total and is divided into two parts. The first part includes the first five chapters which mainly address the background of EM or computational EM (CEM), and the basic principle and necessary elements of implementation such as quadrature rules and singularity treatment in the Nyström method. The second part includes other five chapters which mainly deal with the applications and physical implementations of the Nyström method for solving various EM problems as we have emphasized in the above. The first chapter was written by Professor Chew and the other nine chapters were written by myself. The primary contents of each chapter are summarized as follows.

Chapter 1 can be thought of as Professor Chew's recent observation and thinking about the EM and CEM. He first recalls the history of classical EM theory and draws a picture describing its connection with other well-known physical theories like quantum electrodynamics and Yang-Mills theory to emphasize the mathematics and physics behind the EM theory. He then clarifies the relationship of the rising quantum optics and quantum EM to the classical EM which can be viewed as the descendent of the fluid physics. Immediately followed is his narration about the complete development of the Maxwell's equations and derivation of wave equations, which allows him to divide the wave into three regimes in terms of frequency, i.e. the circuit physics, wave physics, and ray physics, respectively, and the mechanism of plasmonic resonance is also explained by the way. With the physics, he then starts to address the solution methods for the wave problems. He first narrates the closed-form solutions and their asymptotic approximations at high frequencies. Next, he addresses the CEM or numerical methods and shows a procedure of converting an operator equation into a matrix equation which is the underpinning method behind the finite element method (FEM) and the MoM. After that, he further addresses fast algorithms and particularly introduces the MLFMA and domain decomposition method (DDM). Finally, he reviews some key components of high-frequency solutions, inverse problems, metamaterials, and small antennas. With those reviews, he predicts that EM will still remain an important area of study and a fundamental status impacting many other technologies, even if the onset of quantum mechanics could bring new possibilities to make a big change.

Chapter 2 also addresses the CEM as the background of applying the Nyström method, but it is in a more microscopic manner compared with the more macroscopic picture and wider range of Chapter 1. The CEM, which is the basis of solving EM problems by computer modeling and simulation, is first defined and categorized. The analytical solutions of spherical objects, which are Mie-series solutions and are usually used as exact solutions, are then presented. After that, a brief introduction to the three mainstream numerical methods, i.e. the finite-difference time-domain (FDTD) method, FEM, and MoM, is given. Finally, all EM integral equations for different cases are summarily presented and they serve the basis of applying the Nyström method to solve EM problems.

Chapter 3 presents a careful look on the theme of this book, i.e. the Nyström method. It introduces the history, basic principle, and implementation scheme for the Nyström method, but there is no demonstration and application which will be shown in later chapters. It also deals with the singularity treatment which is the key of the Nyström method and describes a higher-order scheme in principle. Finally, it makes a comparison between the Nyström method and the MoM or point-matching method, with which many readers could be concerned.

Chapter 4 is devoted to the introduction for numerical quadrature rules which also play a vital role in the Nyström method. The definition of quadrature rules is introduced first, and then different types of quadrature rule are presented in details. These rules are designed for segmental, surface, and volumetric meshes, respectively. For the segmental meshes, the Gaussian quadrature rules in different orders with different points are addressed. For the surface and volumetric meshes, the widely used quadrature rules defined over triangular patches and tetrahedral elements are presented, but the quadrature rules defined over square patches and cuboid elements are also considered because they could be needed in certain cases although they are less employed.

Chapter 5 deals with the singularity treatment techniques which are very essential to the Nyström method. The singularity subtraction method is presented first, which is indispensable for strongly singular and hypersingular integrals. It is based on the Tayor's series expansion of the scalar Green's function to distinguish the singular cores from singular integrands and then subtract them to regularize the original singular integrals. Next, the singularity cancellation method is addressed, which is preferable but can only be applied to weakly-singular integrals. Much effort is then paid to the derivation of closed-form formulations for evaluating strongly-singular and hypersingular integrals which are from the subtraction based on the Cauchy-principal-value (CPV) sense. Various cases including surface domain with triangular patches and volume domain with tetrahedral or cylindrical elements are considered. Finally, the near-singular integrals which are seldom considered are also addressed and a nearness factor (NF) is proposed to determine if the near singularity should be handled or not.

In Chapter 6, we narrate the application of the Nyström method to solve EM problems with conducting media. The two-dimensional (2D) structures with or without incorporating edge conditions are considered and the treatment of logarithm singularity is described as a supplementary. We then apply the Nyström method to solve the body-of-revolution (BOR) problems by using Fourier series expansion and the formulations of evaluating singular modal Green's function and its derivative are derived. Regarding regular three-dimensional (3D) conducting objects, we first present a higher-order Nyström scheme to solve the electric field integral equation (EFIE) and then particularly consider very-thin or super-thin conducting structures by using the magnetic field integral equation (MFIE) and combined field integral equation (CFIE) to formulate, respectively. Robust formulations for evaluating those near-singular integrals based on the Green's lemma or the Stokes' theorem are developed and they are particularly suitable for being used in thin structures.

Chapter 7 depicts the application of the Nyström method or Nyström-based mixed scheme to solve EM problems with penetrable media. The surface integral equations (SIEs), including the EFIEs, MFIEs and CFIEs, with an assumption of homogeneous and isotropic media are solved first and a comparison with the MoM is illustrated. The Nyström method is then applied to solve volume integral equations (VIEs) which allow the inclusion of inhomogeneous and anisotropic media by developing an efficient local correction scheme for singular or near singular integrals over tetrahedral elements. The method is also applied to the case with conductive media which are treated as penetrable media when the skin depth is large. Finally, the Nyström-based mixed scheme is developed for analyzing interconnect and packaging problems with both penetrable dielectrics and impenetrable conductors based on the volume-surface integral equations (VSIEs) and the comparison with the conventional MoM is also presented.

In Chapter 8, we recount the incorporation of the Nyström method with the MLFMA for solving electrically large problems. The basic principle and characteristics of MLFMA are introduced first and the Nyström-based MLFMA is developed for solving various problems. We consider to solve the SIEs with conducting objects, SIEs with penetrable media, VIEs with conductive media, and VSIEs with conducting-anisotropic media, respectively. Finally, a comparison between the Nyström-based MLFMA and the conventional MoM-based MLFMA is made and its distinct features are clarified.

Chapter 9 discusses the application of the Nyström method to solve multiphysics problems which are of interdisciplinary feature. Although there are many types of multiphysics problems, we select the coupled EM–elastodynamic problem as an example to illustrate the solution by the Nyström method. We begin from the introduction of elastic wave equations and acoustic wave equations, with which EM researchers may not be familiar, and then present their numerical solutions by the Nyström method. We also give a unified form of these two wave equations because the acoustic wave is actually the special case of elastic wave. Moreover, we develop the Nyström-based MLFMA for solving both elastic wave equations and acoustic wave equations of large problems, in particular, we develop a multiple-tree structure of MLFMA which does not exist in EM problems. Based on those prerequisite investigations, we then develop the coupled integral equations for the multiphysical coupling of EM wave and elastic wave and their Nyström solutions based on an alternative iterative scheme are illustrated.

In the last chapter, Chapter 10, we focus on the solutions of time-domain integral equations (TDIEs), including the time-domain version of EFIEs (TDEFIEs), MFIEs (TDMFIEs), CFIEs (TDCFIEs), and VIEs (TDVIEs), for transient EM problems. A corresponding hybrid scheme is developed by combining the Nyström method in space domain and Galerkin's method with Laguerre basis and testing functions in time domain, and the TDIEs are solved with a march-on in degree (MOD) manner. Based on the hybrid scheme, we solve the time-domain surface integral equations (TDSIEs) for conducting objects, TDSIEs for penetrable media, TDVIEs for penetrable media, and TDCFIEs for mixed media which include both conducting and penetrable media, respectively. A comparison with the conventional scheme, which uses the MoM to discretize the space domain and employs a march-on in time (MOT) scheme to discretize the time domain, is performed and its advantages or disadvantages are clarified.

The Nyström method as a nontraditional method is still growing rapidly and many relevant topics can be further studied or explored. For instance, the singularity treatment techniques still have a big space to be investigated as it is tightly related to the Green's function which is full of wave physics. Also, the applications of the Nyström method can be significantly expanded since it has not been applied to many areas such as those mentioned in Chapter 1. We hope that this book can stimulate the readers' strong interest on the interesting method and attract more researchers to study and innovate it.

Mei Song Tong
Tongji University
Shanghai, China

Acknowledgment

The book reports what we have done on the Nyström method in electromagnetics in the past sixteen years. It was impossible to do such an extensive study in such a long time without enough financial supports. Therefore, we are highly grateful to our research sponsors for their generous fund supports on the research work. Particularly, we would like to strongly appreciate U.S. Air Force Office of Scientific Research (AFOSR), Construction Engineering Research Laboratory (CERL) of U.S. Army, Semiconductor Research Corporation (SRC), Intel Corporation, IBM Corporation, National Natural Science Foundation of China (NSFC), Ministry of Science and Technology (MOST) of China, Ministry of Education (MOE) of China, Science and Technology Committee of Shanghai Municipality (STCSM) of China, and Shanghai Institute of Intelligent Science and Technology of Tongji University. Also, we would like to thank all colleagues and students who have been involved in the related research work. Finally, our gratitude would go to the related staffs of Wiley who have made the publication of this book become possible. Their hard work and tremendous support and help are highly appreciated.

1

Electromagnetics, Physics, and Mathematics

Mathematics is the mother of science, science is the mother of technology and father of innovation, and technology is a gift of God.

1.1 A Brief History of Electromagnetics

Electromagnetics is a subject that entails the study of electromagnetic theory, its physical interaction with environments and objects, and its use for design and engineering applications. Electromagnetic theory is governed by a set of equations that originated from many great scientists such as Gauss (1835), Amperes (1823), Faraday (1838), and Coulomb (1785) [1–4]. However, this set of equations is now commonly known as Maxwell's equations due to the work of James Clerk Maxwell who put on them in mathematical form in 1864 and added a term to Ampere's law that included displacement current density [5, 6, 33][1].

This set of equations is used to describe static electromagnetic fields, radio waves, microwaves, optical fields, as well as x-rays. It is valid from subatomic length scale to inter-galactic length scale. The Coulomb potential due to a point charge of a proton or an electron inside an atom can be derived from Maxwell's equations. Meanwhile, the propagation of radio waves and optical signals from outer galaxies can be described by the same set of equations [7].

Before the completion of Maxwell's theory, optical fields were thought to be different from electromagnetic fields. Snell (1621), Huygens (1660), Newton (1660), and Fresnel (1814) were studying optical fields along a different line. Maxwell's theory unified electromagnetics with optics. When quantum theory is added to electromagnetic theory, quantum electrodynamics (QED) ensues, as was done by Dirac (1920). When differential geometry was used, electromagnetic theory inspired Yang-Mills theory which has been regarded as a generalized electromagnetic theory (see Figure 1.1).

Moreover, the impact and importance of electromagnetics on modern technologies are unquestionable. This knowledge is pervasively used in engineering, and is used as a supporting technology for many aspects of scientific investigation. For instance, in high-energy physics, particles are accelerated by an electromagnetic field [8, 9], and

1 The theory was presented to the British Royal Society in 1864, published in 1865, and a treatise was written in 1873. The equations were later beautified by O. Heaviside.

The Nyström Method in Electromagnetics, First Edition. Mei Song Tong and Weng Cho Chew.

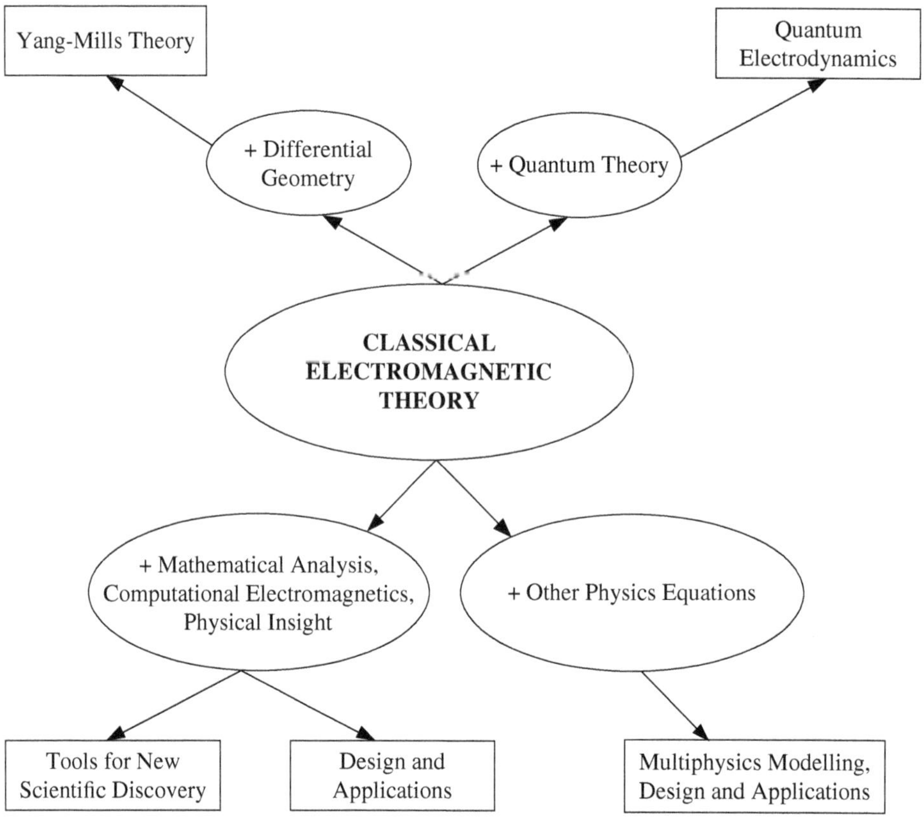

Figure 1.1 The relationship of classical electromagnetic theory with other knowledge areas.

hence electromagnetic engineering is needed. The measurement of electromagnetic fields and optical signals arriving to the earth from the outer universe is used in astronomy to enhance our understanding of the universe. Electromagnetic fields are used in electrophoresis in electrochemistry [10, 11] and biomolecular electrostatics [12, 13]. The electromagnetic force is a predominant force in nano-electronic transport [14–17]. The Poisson-Boltzmann equation is usually solved in these contexts [18–25]. More recently, electromagnetics is of fundamental importance in wireless power transfer and also in bioelectromagnetics.

Electromagnetics is also important in the design of rotating machineries, electric generators, energy conversion devices, and electric power distribution networks, as well as green energy such as solar cells and solar collectors [26–32]. It is important for antenna design for wireless communication, radar, and remote sensing. The maser was first invented at microwave frequency (1952) before its close cousin, laser, was invented (1958). As the field of optics grows, many technologies that were first proven at microwave frequencies are now being realized at optical frequencies, such as optical interferometric imaging, optical coherence tomography, and nano-antennas. Electromagnetics also becomes increasingly important in electromagnetic interference (EMI) and electromagnetic compatibility (EMC) as more electronic components are packed into smaller spaces in, e.g., the cellphone industry.

The recent advent in nano-fabrication technology, where structures on the order of tens of nanometers can be made, has spurred strong interest in nano-optics. The advent of single photon measurement, and the validation of the Bell's theorem in favor of the interpretation of Copenhagen school in the 1980s, spurred new interest in quantum optics and quantum information. Quantum electromagnetics is emerging. The validation of the Casimir force in 1997 also spurred new interest in this force.

1.2 Enduring Legacy of Electromagnetic Theory–Why?

Electromagnetic theory as completed by James Clerk Maxwell in 1865 [5] is just over 150 years old now. It has withstood the test of time, and the emergence of new physics theories such as relativity and quantum theory. This is unlike Newton's laws of mechanics that were quickly superseded by new theories.

Putatively, Maxwell's equations were difficult to understand [33]; distillation and cleaning of the equations were done by Heaviside and Hertz [34]; experimental confirmations of these equations were not done until some 20 years later in 1888 by Hertz [35].

As is the case with the emergence of new knowledge, it is often confusing at times, and inaccessible to many people. Numerous developments of electromagnetic theory have ensued since Maxwell's time and we shall discuss them in the next section.

Since the advent of Maxwell's equations in 1865, the enduring legacy of these equations has been pervasive in many fields. As aforementioned, Hertz confirmed the remote induction effect in 1888. And in 1893, Tesla [36] demonstrated the possibility of radio. In 1897, Marconi [37] demonstrated wireless transmission, followed by transatlantic transmission in 1901. Maxwell did not know the importance of the equations that he had completed. Much advanced understanding of electromagnetic theory in its modern form did not emerge until many years after his death; it will be interesting to recount these facts.

- Since Maxwell's equations unify the theories of electromagnetics and optics, they are valid over a vast length scale. Electromagnetic theory is valid for subatomic particle interaction, as well as being responsible for the propagation of light waves and radio waves across the galaxies.
- With the theory of special relativity developed by Einstein in 1905 [38], these equations were known to be relativistically invariant. In other words, Maxwell's equations remain the same in a spaceship irrespective of how fast it is moving. Electrostatic theory in one spaceship becomes electrodynamic theory in a moving spaceship relative to the first one.
- The development of QED by Dirac in 1927 [39] indicated that Maxwell's equations are valid in the quantum regime as well. Initially, QED was studied by Dyson, Feynman, Schwinger, and Tomonaga, mainly to understand the interparticle interactions in the context of quantum field theory to determine fine structure constants and their anomalies due to quantum electromagnetic fluctuations [40]. However, the recent increase in quantum information has spurred on the application of QED in optics, giving rise to the field of quantum optics [41–47].

- Later, with the development of differential forms by Cartan in 1945 [48, 49], it was found that electromagnetic theory is intimately related to differential geometry. Electromagnetic theory inspired the Yang-Mills theory, which was developed in 1954 [50, 51]; it is regarded as a generalized electromagnetic theory. In fact, as quoted by Misner, Thorne, and Wheeler, it is said that "differential forms illuminate electromagnetic theory, and electromagnetic theory illuminates differential forms" [52, 53].
- In 1985, Feynman wrote that quantum electrodynamics (a superset of electromagnetic theory) had been validated to be one of the most accurate equations to a few parts in a billion [54]. This is equivalent to an error of a few human hair widths compared to the distance from New York to Los Angeles. More recently, Styer wrote in 2012 [55] that the accuracy had been improved to a few parts in a trillion [56]; such an error is equivalent to a few human hair widths in the distance from the Earth to the Moon.
- More importantly, since Maxwell's equations have been around for over 150 years, they have pervasively influenced the development of a large number of scientific technologies. This impact is particularly profound in electrical engineering, ranging from rotating machinery, oil-gas exploration, and magnetic resonance imaging, to optics, wireless and optical communications, computers, remote sensing, bioelectromagnetics, etc.

Despite the cleaning up of Maxwell's equations by Heaviside, he has great admiration for Maxwell as seen from his following statement [34], "A part of us lives after us, diffused through all humanity – more or less – and through all nature. This is the immortality of the soul. There are large souls and small souls. The immoral soul of the 'scienticulists'[2] is a small affair, scarcely visible. Indeed its existence has been doubted. That of a Shakespeare or Newton is stupendously big. Such men live the bigger part of their lives after they are dead. Maxwell is one of these men. His soul will live and grow for long to come, and hundreds of years hence will shine as one of the bright stars of the past, whose light takes ages to reach us."

1.3 The Rise of Quantum Optics and Electromagnetics

The rise of the importance of quantum electromagnetics has been spurred on by technologies for single photon sources and measurements [57, 58], the validation of Bell's theorem [59], and the leaps and bounds progress in nanofabrication technologies.

The validation of Bell's theorem in favor of the interpretation of Copenhagen school opens up new possibilities for quantum information, computing, cryptography, and communication [60]. Nanofabrication techniques further allow the construction of artificial atoms such as quantum dots that are microscopic in scale. Moreover, the potential for using such artificial atoms to manipulate quantum information abounds. In this case, semi-classical calculations where the fields are treated classically and the atoms treated quantum mechanically [61, 62] do not suffice to support many of the emerging technologies when the number of photons is limited, such as single photon

2 This term was coined by Heaviside for his nemesis, Preece [34].

based devices and high sensitivity photo-detectors. Another interesting example is circuit QED at microwave frequencies where a superconducting quantum interference device (SQUID) based artificial atom is entangled with coplanar waveguide microwave resonators [63]. For these situations, full quantum field-artificial-atom calculations need to be undertaken [64].

The recent progress in nano-fabrication technology underscores the importance of quantum effects at nanoscale, first on electron transport [65], and now on the importance of photon-artificial-atom interaction at nanoscale. Moreover, nanofabrication emphasizes the importance of photons and the accompanying quantum effects in heat transfer. While phonons require material media for heat transfer, photons can account for near-field heat transfer through vacuum where the classical heat conduction equation and Kirchhoff's law of thermal radiation are invalid. Furthermore, the confirmation of the Casimir force in 1997 [66] revived it as an interesting research topic. Several experiments confirmed that the Casimir force is in fact real, and entirely quantum in origin: it can be only explained using quantum theory of electromagnetic fields in its quantized form [67, 68]. Also, the Casimir force cannot be explained by classic electromagnetics theory, which assumes null electromagnetic field in vacuum.

More importantly, the use of the ubiquitous Green's function is still present in many quantum calculations [69, 71]. Hence, the knowledge and effort in computational electromagnetics for computing the Green's functions of complicated systems have not become obsolete or in vain [72, 74–76]. Therefore, the development of computational electromagnetics, which has been important for decades for the development of many classical electromagnetics technologies all across the electromagnetic spectrum, will be equally important in the development of quantum technologies.

1.3.1 Connection of Quantum Electromagnetics to Classical Electromagnetics

Vacuum space consists of electron-positron (e–p) pairs that represent nothingness. However, when an electromagnetic wave passes through vacuum, the e–p pairs are polarized to form simple harmonic oscillators. The propagation of electromagnetic waves through vacuum is due to the coupling of these simple harmonic oscillators [79]. From this concept, the quantum Maxwell's equations are derived to be [80, 81]:

$$\nabla \times \hat{\mathbf{H}}(\mathbf{r}, t) - \partial_t \hat{\mathbf{D}}(\mathbf{r}, t) = \hat{\mathbf{J}}_{\text{ext}}(\mathbf{r}, t) \tag{1.1}$$

$$\nabla \times \hat{\mathbf{E}}(\mathbf{r}, t) + \partial_t \hat{\mathbf{B}}(\mathbf{r}, t) = 0 \tag{1.2}$$

$$\nabla \cdot \hat{\mathbf{D}}(\mathbf{r}, t) = \hat{\varrho}_{\text{ext}}(\mathbf{r}, t) \tag{1.3}$$

$$\nabla \cdot \hat{\mathbf{B}}(\mathbf{r}, t) = 0. \tag{1.4}$$

The Green's function technique applies when the quantum system is linearly time-invariant. Hence, past knowledge in classical computational electromagnetics can be invoked to arrive at these Green's functions. These quantum Maxwell's equations portend well for a better understanding of quantum effects that are observed in many branches of electromagnetics, as well as in quantum optics, quantum information, communication, computing, encryption and related fields. More details about this work can be found in [62, 80–83]. Hence, the combination of computational electromagnetics with quantum theory is cogent for the development of computational quantum optics.

1.4 The Early Days – Descendent from Fluid Physics

In the early days, mathematics was developed to describe different physical laws such as Newton's laws. Later, calculus was developed to study solid mechanics, elasticity, and fluid mechanics. In fluids and elastomechanics, the concepts of field and its conservative and rotational properties were introduced. Such concepts were later carried over to describe the physics of electromagnetic fields. The concepts of gradient, curl, and divergence operators have originated in fluid mechanics. The concepts of scalar field and vector field also occur in fluids. The gradient of a scalar field gives rise to a vector field that could be a velocity field. For instance, in acoustics, the gradient of the pressure, which is a scalar field, gives rise to a force field density, which is a vector field, namely, $\mathbf{F} = -\nabla\phi$.

The curl operator is a measure of the rotation of a vector field. Hence, it is a good measure of angular momentum that exists in a fluid field. It also reflects on the torsional field that exists in the force field in a solid when it is deformed. For instance, $\boldsymbol{\omega} = \nabla \times \mathbf{v}$ is a measure of rotation in the velocity field $\mathbf{v}$.

The divergence operator indicates how much field actually oozes from a point in space. It is useful for measuring the conservation of "things", such as conservation of mass and momentum in fluid mechanics. For example, the conservation of mass in fluid can be written as

$$\nabla \cdot (\mathbf{v}\varrho) + \frac{\partial \varrho}{\partial t} = 0 \tag{1.5}$$

where $\mathbf{v}$ is the velocity field, while ϱ is the mass density.

The divergence operator has been used to describe Coulomb's law that the total electric flux that emerges from a charge source is proportional to the charge [3], i.e.

$$\nabla \cdot \mathbf{D} = \varrho \tag{1.6}$$

where ϱ is the charge density. The concept of electric flux density $\mathbf{D}$ was introduced to relate $\mathbf{D}$ to the surface charge density. The above has a conservative concept in the sense that a fixed amount of flux emerges from a fixed amount of charge.

Similarly, the Gauss' law [4] was introduced for the conservation of magnetic flux, and the concept of magnetic flux density $\mathbf{B}$ was introduced, namely,

$$\nabla \cdot \mathbf{B} = 0 \tag{1.7}$$

which represents the conservation of magnetic flux density. It also points to the non-existence of magnetic charges.

Ampere's law [2] was introduced by experimental observation that a wire carrying an electric current produces a swirling magnetic field that "circulates" around the current, i.e.

$$\nabla \times \mathbf{H} = \mathbf{J}. \tag{1.8}$$

Taking the divergence of the above yields $\nabla \cdot \mathbf{J} = 0$, which implies the conservation of current flow. It also implies Kirchhoff's current law in circuit theory.

Faraday's law has been formulated following the observation that a time-varying magnetic flux induces a voltage in a loop of wire. Consequently,

$$\nabla \times \mathbf{E} = -\frac{\partial \mathbf{B}}{\partial t}. \tag{1.9}$$

The above equations can be written in an integral form, but the physics is entailed in the differential form as lucidly as they are in integral form. However, much of circuit theory is derived from the integral form. The integral form of Maxwell's equations brings out the fact that it is the topology that governs the laws of circuits. So if two circuits may look very different in shape, but their behavior is similar if they are topologically the same. The above equations (1.8) and (1.9) are pre-Maxwellian, and the development of telegraphy very much depends on them [34].

1.5 The Complete Development of Maxwell's Equations

Maxwell's equations are descended from Gauss', Coulomb's, Ampere's, and Faraday's laws. Maxwell's contribution was to add a displacement current term to Ampere's law making it into the generalized Ampere's law. The displacement current is produced by the time variation of the electric flux density **D** (also called the displacement field). The generalized Ampere's law can be written as

$$\nabla \times \mathbf{H} = \mathbf{J} + \frac{\partial \mathbf{D}}{\partial t}. \tag{1.10}$$

It is the addition of this displacement current term $\frac{\partial \mathbf{D}}{\partial t}$ that connects the laws of electricity and magnetism with the laws of optics. Prior to this, the group of four equations without the displacement current was used to describe the theory of electricity and magnetism, and did not predict the existence of wave phenomena.

The term $\frac{\partial \mathbf{D}}{\partial t}$ introduces "springiness" into the electromagnetic media like a capacitor. This displacement current, in turn, produces a magnetic field. The time-varying magnetic field produces an electric field that opposes the flow of the displacement current, or a reluctance (analogous to the inertial of a mass) as in Lenz's law. This inductance together with the capacitive effect yield a simple harmonic oscillator[3].

For material media,

$$\mathbf{D} = \epsilon_0 \mathbf{E} + \mathbf{P} \tag{1.11}$$

where **P** is the dipole density. The dipoles in material media are polarized by the electric field that exerts a force that pulls the positive and negative charge of a molecule or an atom apart, forming a dipole. For a time harmonic field, the time-varying charge produces a current, and hence a magnetic field like an inductor. The dipole together with the inductor forms a resonant circuit very much like an LC tank circuit. For a weak field, **P** is linearly related to **E**, and we can write $\mathbf{P} = \epsilon_0 \chi \mathbf{E}$. Therefore, for material media, we introduce $\epsilon = \epsilon_0(1 + \chi)$ so that $\mathbf{D} = \epsilon \mathbf{E}$.

For a vacuum, $\mathbf{D} = \epsilon_0 \mathbf{E}$, implying that even a vacuum has "springiness". It is this springiness of vacuum and material media that gives rise to electromagnetic waves

3 The equation for capacitance, $V = Q/C$ is similar to Hook's law in that $F = kx$ where k is the spring constant. The voltage across an inductor $V = L\dot{I} = L\ddot{Q}$ is similar to Newton's law in that $F = m\ddot{x}$. Hence, the harmonic oscillator of a mass on a spring is similar to the harmonic oscillator formed by an LC tank circuit.

in material and even in vacuum. Electron-positron pairs that lurk in vacuum, in their quiescent state, constitute nothingness. It has also been observed that when vacuum is bombarded with energetic photons (or energetic electromagnetic fields), electron-positron pairs are produced [79, 84, 85].

Electromagnetic theory in the classical sense is hence completely described by the four equations:

$$\nabla \times \mathbf{H} = \mathbf{J} + \frac{\partial \mathbf{D}}{\partial t} \tag{1.12}$$

$$\nabla \times \mathbf{E} = -\frac{\partial \mathbf{B}}{\partial t} \tag{1.13}$$

$$\nabla \cdot \mathbf{D} = \varrho \tag{1.14}$$

$$\nabla \cdot \mathbf{B} = 0. \tag{1.15}$$

When Maxwell's equations were first derived by Maxwell, they were not as elegant and succinct as the above. It was the work of Heaviside and Herz that cast them into the above form [34, 86]. It was deemed that the electric field and the magnetic field are the measurable quantities.

The above equations look deceptively simple, but they have very profound meanings. They are relativistically correct in all inertial frames of special relativity [9, 74]. They are also intimately related to geometry.

It is to be noted that for time-varying solutions, the second two equations above are derivable from the first two. For instance, if we take the divergence ($\nabla\cdot$) of the first equation (1.12), and make use of the fact that $\nabla \cdot (\nabla \times \mathbf{H}) = 0$, we obtain that

$$0 = \nabla \cdot \mathbf{J} + \frac{\partial(\nabla \cdot \mathbf{D})}{\partial t}. \tag{1.16}$$

The conservation of charge necessitates that

$$\nabla \cdot \mathbf{J} + \frac{\partial \varrho}{\partial t} = 0. \tag{1.17}$$

Using this fact in (1.16) yields

$$\frac{\partial(\nabla \cdot \mathbf{D} - \varrho)}{\partial t} = 0. \tag{1.18}$$

The above implies that

$$\nabla \cdot \mathbf{D} - \varrho = \varrho_{\text{const}} \tag{1.19}$$

where ϱ_{const} corresponds to a constant charge independent of time. We can assume such a charge to be zero, or that it produces a static field that does not contribute to our time-varying field. Then (1.14) follows from (1.12). Equation (1.15) can be similarly derived from (1.13). Therefore, for time-varying fields or electrodynamics, Equations (1.12) and (1.13) are sufficient. However, when a numerical code is written for a time-varying field, it often breaks down when the frequency is lowered, because at low frequencies, the first two of Maxwell's equations are not stable, as they do not reduce to the last two equations when the frequency is identically zero [73].

To preclude the low-frequency breakdown, the vector and scalar potential formulation has been suggested [73]. This formulation solves all four Maxwell's equations concurrently, and hence, does not have low-frequency breakdown. This formulation is in fact closer to the original formulation of Maxwell's [33].

However, there are four vector unknowns **E**, **H**, **D**, **B**, with two vector equations. To increase the number of equations, we rely on the constitutive relations. A set of general constitutive relations are

$$\mathbf{D} = \overline{\epsilon} \cdot \mathbf{E} + \overline{\xi} \cdot \mathbf{H} \tag{1.20}$$

$$\mathbf{B} = \overline{\zeta} \cdot \mathbf{E} + \overline{\mu} \cdot \mathbf{H} \tag{1.21}$$

where $\overline{\epsilon}$, $\overline{\xi}$, $\overline{\zeta}$, and $\overline{\mu}$ are tensors. The medium with the above relations is known as a bianisotropic medium [74]. When $\overline{\xi} = \overline{\zeta} = 0$, the medium is anisotropic. When the tensors are replaced with scalars, the medium is isotropic. Other non-local relations are possible such as a convolutional relation in space

$$\mathbf{D}(\mathbf{r}) = \int d\mathbf{r}' \epsilon(\mathbf{r} - \mathbf{r}')\mathbf{E}(\mathbf{r}'). \tag{1.22}$$

When Fourier transformed to the **k** space or spatial frequency space, the above becomes

$$\mathbf{D}(\mathbf{k}) = \epsilon(\mathbf{k})\mathbf{E}(\mathbf{k}) \tag{1.23}$$

which is a scalar relationship. Such a medium is known as spatially dispersive. Certain multiple scattering media that are translationally invariant in space can be described by such constitutive relations. Also, a relationship involving convolution in time is also possible. When transformed into the ω or the frequency space, the permittivity is a function of frequency. Such media are termed frequency dispersive. When the constitutive parameters are functions of space, the media are inhomogeneous.

The constitutive relations describe the physical properties of the media. Most media can be described by simpler constitutive relations

$$\mathbf{D} = \epsilon \mathbf{E} \tag{1.24}$$

$$\mathbf{B} = \mu \mathbf{H} \tag{1.25}$$

where ϵ is the permittivty and μ is the permeability. Then Equations (1.12) and (1.13) and the constitutive relations are sufficient to be solved for the unknowns **E** and **H** if the current **J** is given. In some media, the electric current **J** is induced. Then we divide $\mathbf{J} = \mathbf{J}_{\text{imp}} + \mathbf{J}_{\text{ind}}$, where $\mathbf{J}_{\text{imp}}$ is the impressed current into the equations, while $\mathbf{J}_{\text{ind}}$ is the induced current in the medium[4]. For instance, for a simple conductive medium,

$$\mathbf{J}_{\text{ind}} = \sigma \mathbf{E} \tag{1.26}$$

where σ is the conductivity.

1.5.1 Derivation of Wave Equation[5]

To show the existence of the wave solution, we take the curl of (1.13) and one gets

$$\begin{aligned} \nabla \times \nabla \times \mathbf{E} &= -\frac{\partial(\nabla \times \mathbf{B})}{\partial t} = -\mu \frac{\partial(\nabla \times \mathbf{H})}{\partial t} \\ &= -\mu \frac{\partial \left(\mathbf{J} + \frac{\partial \mathbf{D}}{\partial t}\right)}{\partial t} = -\mu \frac{\partial \mathbf{J}}{\partial t} - \mu\epsilon \frac{\partial^2 \mathbf{E}}{\partial t^2} \end{aligned} \tag{1.27}$$

---

4 An impressed current is a current put into Maxwell's equations that is immutable irrespective of what the environment is.

5 This can be skipped for the initiated.

or

$$\nabla \times \nabla \times \mathbf{E} + \mu\epsilon \frac{\partial^2 \mathbf{E}}{\partial t^2} = -\mu \frac{\partial \mathbf{J}}{\partial t}. \tag{1.28}$$

The above is the vector wave equation with a source term **J**. In the absence of a source, it reduces to

$$\nabla \times \nabla \times \mathbf{E} + \mu\epsilon \frac{\partial^2 \mathbf{E}}{\partial t^2} = 0 \tag{1.29}$$

whose solutions are the homogeneous solutions to (1.28). It can be simplified to the scalar wave equation, verifying the existence of a wave solution to Maxwell's equations. By using the vector identity that

$$\nabla \times \nabla \times \mathbf{E} = -\nabla \cdot \nabla \mathbf{E} + \nabla \nabla \cdot \mathbf{E} \tag{1.30}$$

and that for a source free region $\nabla \cdot \mathbf{E} = 0$, (1.29) becomes

$$\nabla^2 \mathbf{E} - \mu\epsilon \frac{\partial^2 \mathbf{E}}{\partial t^2} = 0 \tag{1.31}$$

which manifestly resembles the scalar wave equation. When written in Cartesian coordinates, the above becomes three scalar wave equations for E_x, E_y, E_z. However, one has to be mindful that these field components are not independent of each other since $\nabla \cdot \mathbf{E} = 0$.

1.6 Circuit Physics, Wave Physics, Ray Physics, and Plasmonic Resonances

Even though Maxwell's equations are deceptively simple, there are three regimes of physical phenomena described by them that are distinctly different. At low frequencies or long wavelengths, circuit physics prevails. When the object size is *on the order of* [6] the wavelength, wave physics prevails. However, when the wavelength is very short compared to the size of the object, ray physics predominates [87].

When the frequency is very low, or at statics, the **E** and **H** fields are decoupled in free space. This is the regime of circuit physics, where we have the world of the inductors and the world of the capacitors. The inductive world encompasses devices that store magnetic energy, while the capacitive world encompasses devices that store electric energy. Moreover, at low frequencies, devices in these two worlds are weakly coupled to each other.

1.6.1 Circuit Physics

When $\omega \to 0$ or $\partial/\partial t \to 0$, the four Maxwell's equations reduce to:

$$\nabla \times \mathbf{H} = \mathbf{J} \tag{1.32}$$

$$\nabla \times \mathbf{E} = 0 \tag{1.33}$$

$$\nabla \cdot \mathbf{D} = \varrho \tag{1.34}$$

$$\nabla \cdot \mathbf{B} = 0. \tag{1.35}$$

6 This is the mathematical parlance for "about".

In this case, the bottom two equations are not derivable from the top two. Hence, In the low frequency regime, all four of Maxwell's equations are important. Moreover, the electric field and the magnetic field are decoupled in free space. Namely, $\{\mathbf{E}, \mathbf{D}, \rho\}$ are decoupled from $\{\mathbf{H}, \mathbf{B}, \mathbf{J}\}$ in free space. This is the circuit physics regime where topology is more important than shape.

At very low frequencies,

$$\nabla \times \mathbf{E} = -\frac{\partial \mathbf{B}}{\partial t} \approx 0 \tag{1.36}$$

can be interpreted as Kirchhoff's voltage law. When written in integral form, it implies that the sum of voltages in a loop is zero. The above implies that the electric field $\mathbf{E}$ is a conservative field: The total work done by such a field in a loop is zero.

Similarly, by taking the divergence of the generalized Ampere's law (1.10), and making use of $\nabla \cdot \mathbf{D} = \rho$, we deduce that

$$\nabla \cdot \mathbf{J} = -\frac{\partial \varrho}{\partial t} \approx 0 \tag{1.37}$$

which is Kirchhoff's current law. It implies the conservation of current flow.

The above two laws, together with the voltage-current (V–I) relations of lumped elements such as resistors, capacitors, inductors, form the basic equations in circuit theory. Nonlinear elements such as diodes and transistors can be added to further enrich circuit physics. Notice that in the static or direct current (DC) limit, taking the divergence of (1.32) implies that $\nabla \cdot \mathbf{J} = 0$. However, the current, by Helmholtz decomposition, can be decomposed into curl-free (irrotational) and divergence-free (solenoidal) currents, namely, $\mathbf{J} = \mathbf{J}_{\text{irr}} + \mathbf{J}_{\text{sol}}$. It implies that $\nabla \cdot \mathbf{J}_{\text{irr}} = 0$ but $\mathbf{J}_{\text{irr}}$ need not be zero. The DC current $\mathbf{J}$ can be coupled to the magnetic field only, as in the super-conducting current loop case. A current can flow in a superconducting loop to generate a DC magnetic field without the need for an electric field to push the current along, viz., $\mathbf{E} = 0$.

On the other hand, the DC current $\mathbf{J}$ can be a conduction current. In this case, $\mathbf{J} = \sigma\mathbf{E}$ where σ is the conductivity of the medium in which the current is flowing. It couples the electric field to the magnetic field via (1.32). Since $\nabla \times \mathbf{E} = 0$, it follows that

$$\nabla \times \frac{\mathbf{J}}{\sigma} = 0. \tag{1.38}$$

By applying the above around a loop of resistors and using Ohm's law that $V = IR$, which follows from $\mathbf{E} = \sigma^{-1}\mathbf{J}$, we obtain

$$\sum_{i=1}^{N} I_i R_i = 0 \tag{1.39}$$

which is Kirchhoff's voltage law. Since by $\nabla \cdot \mathbf{J} = 0$, the current must be conserved at every node (Kirchhoff's current law), the above can be zero only if one or more of terms $I_i R_i$ is negative. One way that $I_i R_i$ can be negative is to have a negative resistor (see Figure 1.2). This can be understood by summing the above over a simple loop. In the negative resistor, the current flows in the opposite direction to the electric field. A battery can be modeled by a negative resistor. We can make the negative resistor dependent on current to model a voltage source, or make it dependent on voltage to model a current source.

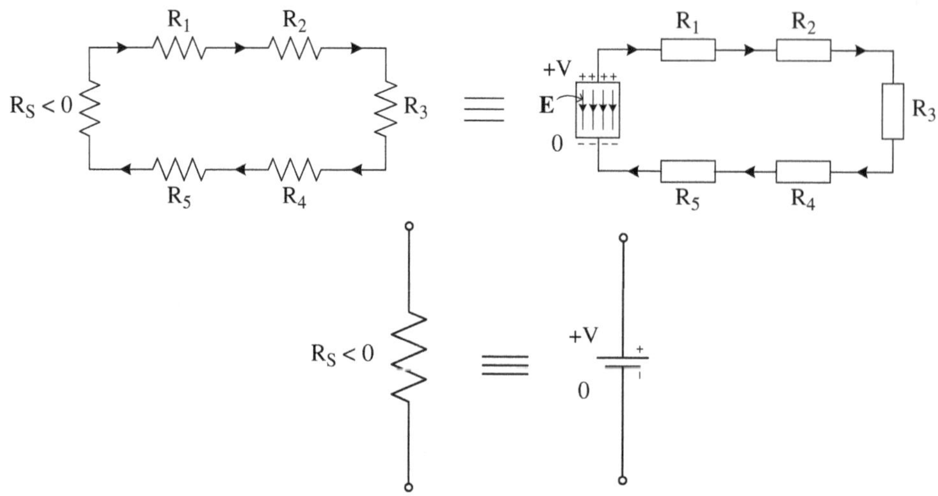

Figure 1.2 A negative resistor is needed in order for the total voltage drop in a loop to add up to zero. A battery can be modeled by a negative resistor.

For some devices such as an inductor, the $\frac{\partial \mathbf{B}}{\partial t}$ term can be made large by concentrating the magnetic field. In such a case, its V–I relation is

$$V = L\frac{\mathrm{d}I}{\mathrm{d}t} \tag{1.40}$$

which follows from Faraday's law that states $\nabla \times \mathbf{E} = -\partial \mathbf{B}/\partial t = -\mu \partial \mathbf{H}/\partial t$. By studying this law for a current loop, and using the fact that the magnetic field $\mathbf{H}$ can be related to the current I in the loop, one can derive (1.40).

From $\nabla \cdot \mathbf{J} = 0$, we derive Kirchoff's current law, which is

$$\sum_{i=1}^{N_j} I_i = 0. \tag{1.41}$$

Similarly, for a capacitor, its charge storage capability is enhanced and the right-hand side of (1.37) is non-negligible. In other words, the current is now coupled to the displacement current. Then its V–I relation is

$$I = C\frac{\mathrm{d}V}{\mathrm{d}t} \tag{1.42}$$

where the right-hand side is due to the displacement current $\mathbf{J}_D = \epsilon \partial \mathbf{E}/\partial t$.

From these circuit equations, one can also derive the telegrapher's equations, which describe the propagation of signals (or waves) on a telegraph line. These equations predate Maxwell's equations [88, 89]. By approximating a transmission line as a ladder of inductors and capacitors (see Figure 1.3), one can easily show that

$$\Delta V \approx -\Delta z L\frac{\mathrm{d}I}{\mathrm{d}t} \tag{1.43}$$

$$\Delta I \approx -\Delta z C\frac{\mathrm{d}V}{\mathrm{d}t}. \tag{1.44}$$

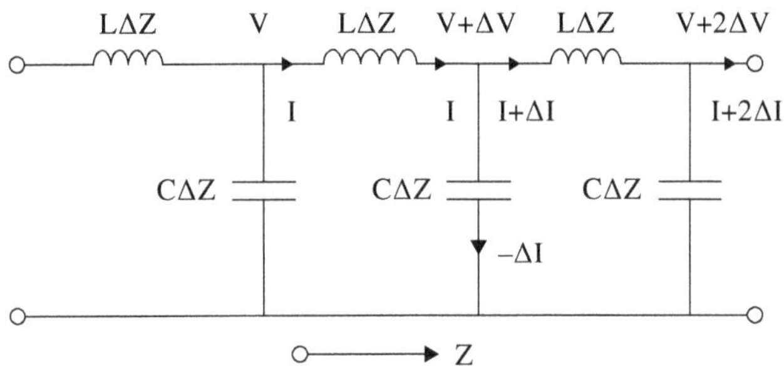

Figure 1.3 Two pieces of wire constituting a transmission line can be approximated with a ladder of inductors and capacitors.

In the limit when $\Delta z \to 0$, we obtain the telegrapher's equations:

$$\frac{\partial V}{\partial z} = -L\frac{\partial I}{\partial t} \tag{1.45}$$

and

$$\frac{\partial I}{\partial z} = -C\frac{\partial V}{\partial t}. \tag{1.46}$$

With this historical setting, it is inevitable that the missing displacement current in Ampere's law be discovered.

In general, in the low frequency regime, the electric field and the magnetic field are weakly coupled through space, except via energy storage devices such as capacitors and inductors where the electric field and magnetic field are enhanced, or via resistors where the current and electric field are coupled. It is through the lumped elements that the dynamic ($\partial/\partial t$) terms or the whole bounty of Maxwell's equations are retained. This gives rise to the richness of circuit physics even though the circuits may be confined to a domain that is much smaller than wavelength.

Moreover, when the tiny objects are coupled to electronic devices such as transistors and diodes, the physics of circuits is richly endowed, and wonderful electronic technology has emerged which allows the devices to be packed in miniaturized dimensions. This is almost like the Alice in Wonderland story, where when one descends to nanometer dimensions, one can see a whole different world of physics in action. This is also reflected in the singularity of the electromagnetic dyadic Green's function [78], which is more elaborate than that of the elastodynamic dyadic Green's function. There was an avid interest in the dyadic Green's function singularity [76, 77].

The electromagnetic field is related to the electric current by

$$\mathbf{E}(\mathbf{r}) = i\omega\mu \int_V \overline{\mathbf{G}}(\mathbf{r}, \mathbf{r}') \cdot \mathbf{J}(\mathbf{r}') \tag{1.47}$$

where

$$\overline{\mathbf{G}}(\mathbf{r}, \mathbf{r}') = \left(\overline{\mathbf{I}} g(\mathbf{r}, \mathbf{r}') + \frac{\nabla g(\mathbf{r}, \mathbf{r}')\nabla}{k_0^2} \right) \tag{1.48}$$

and

$$g(\mathbf{r}, \mathbf{r}') = \frac{e^{ik_0|\mathbf{r}-\mathbf{r}'|}}{4\pi|\mathbf{r} - \mathbf{r}'|}. \tag{1.49}$$

The second term is related to the electric field produced by charges in the system, which is very important in the near field to the source or when capacitance physics is important. The first term is related to the electric field produced by magnetic induction. However, if we take the curl of the above equation to obtain the magnetic field via $\mathbf{H} = \nabla \times \mathbf{E}/(i\omega\mu)$, the second term disappears completely, yielding

$$\mathbf{H}(\mathbf{r}) = \nabla \times \int_V g(\mathbf{r}, \mathbf{r}')\mathbf{J}(\mathbf{r}). \tag{1.50}$$

Moreover, if the dyadic Green's function acts on a divergence-free current, only the first term, or the induction term, is important. Hence, the rich interplay between inductance physics and capacitance physics occurs in the dyadic Green's function.

In contrast, the elastodynamic Green's function is

$$\overline{\mathbf{G}}(\mathbf{r}, \mathbf{r}') = \frac{1}{\mu}\left(\overline{\mathbf{I}} + \frac{\nabla\nabla}{k_s^2}\right)\frac{e^{ik_s|\mathbf{r}-\mathbf{r}'|}}{4\pi|\mathbf{r}-\mathbf{r}'|} - \frac{1}{\lambda + 2\mu}\frac{\nabla\nabla}{k_c^2}\frac{e^{ik_c|\mathbf{r}-\mathbf{r}'|}}{4\pi|\mathbf{r}-\mathbf{r}'|}. \tag{1.51}$$

It can be shown that the leading singularities produced by the $\nabla\nabla$ term will cancel each other, making its physics close to the source point quite different from that of electromagnetics.

1.6.2 Wave Physics

When the frequency is higher, the **E** and **H** fields are tightly coupled in space. As a consequence, oscillatory behavior through space is possible. The oscillation comes about due to the exchange of stored energies in the form of electric field and magnetic field. In the early days, this was possible in a telegraph line where inductors and capacitors coexist to enhance the magnetic field and electric field.

The wave physics regime is important for radio frequency (RF) and microwave when the wavelength is on the order of the dimension of the objects. In the past, the optical wavelength was usually much smaller than the size of the objects. However, wave physics has recently been important in optics and nano-optics due to the advent of nanofabrication. It is now possible to fabricate objects on the order of optical wavelength or smaller [90].

The wave physics regime is the regime in which the solutions to Maxwell's equations are the most difficult. The vector nature and the wave nature of the electromagnetic field have to be accounted for. It is also in this regime that computational electromagnetics plays an important role in seeking viable solutions to Maxwell's equations.

The physics of waves is quite different from the physics of static fields. Waves can be used to send information over a long distance, as in communication, remote sensing, and geophysical probing. However, a static field cannot be used to send information over a long distance [87]. Hence, when we look at the moon through an optical telescope, we can observe the fine details on the moon's surface. The gravitational field of the moon is a Coulombic or Laplacian field. If we observe the moon through its gravitational field, a big blob with no detail is observed [87, 91].

It turns out that the presence of $\epsilon\partial\mathbf{E}/\partial t$ implies the existence of "capacitance" in space, and the presence of $\mu\partial\mathbf{H}/\partial t$ implies the existence of "inductance" in space. Due to the "smallness" of this phenomenon (because ϵ and μ are small numbers), the velocity of the wave is tremendous, given by $c = 1/\sqrt{\mu\epsilon}$. In other words, the coupling becomes significant only when **E** and **H** are varying rapidly with respect to time. For instance, for the same wavelength, electromagnetic wave oscillates much more rapidly compared to an acoustic wave.

The coupling allows the existence of electromagnetic waves even in vacuum. In wave physics, both **E** and **H** are equally important since stored energies are exchanged between them. When this physics exists, the yardstick is the wavelength λ of the electromagnetic wave at the pertinent frequency, viz., $\lambda = c/f$ where f is the frequency of the field. When electromagnetic field interacts with objects that are on the order of wavelength, wave physics is important. On the other hand, when electromagnetic field interacts with objects that are much smaller than the wavelength, then circuit physics predominates.

The aforementioned fact can be appreciated by eye-balling the scalar wave equation

$$\nabla^2\phi - \mu\epsilon\frac{\partial^2\phi}{\partial t^2} = 0 \tag{1.52}$$

which is valid for a Cartesian component of the electromagnetic field. When the field is time harmonic, the above simplifies to

$$\nabla^2\phi + k^2\phi = 0 \tag{1.53}$$

where $k = \omega^2\mu\epsilon$. When an object is small, and is of length scale L, such that $kL \ll 1$, then the field has to vary rapidly on the length scale L around the object in order to satisfy the boundary condition on it. We can do a coordinate stretching transformation by letting $x = Lx'$, $y = Ly'$, and $z = Lz'$ [92]. In other words, x', y', and z' are of $\mathcal{O}(1)$ when x, y, and z are of $\mathcal{O}(L)$. Consequently, $\nabla^2 = \nabla'^2/L^2$ where ∇' is of $\mathcal{O}(1)$. For time-harmonic field, the above becomes

$$\nabla'^2\phi + k^2L^2\phi = 0. \tag{1.54}$$

Therefore, if $k^2L^2 \ll 1$, then the above equation becomes

$$\nabla'^2\phi = 0 \tag{1.55}$$

which is just the Laplace equation describing the static (circuit) physics of electromagnetics. On the other hand, if $k^2L^2 \approx 1$, then the full wave equation has to be solved [93].

1.6.3 Ray Physics

When the frequency becomes exceedingly high, electromagnetic physics morphs into another regime, making electromagnetic waves behave like particles. This is the ray physics regime. In this regime, since the wavelength is very short, most electromagnetic fields can be locally approximated by a plane wave. A plane wave can be thought of as a ray. A ray bounces off surfaces and penetrates media very much like a particle. Surprisingly, it is in this regime that the quantization of an electromagnetic field is important [94], and that electromagnetic energy propagating through space should be thought of as packets of energy [95–99]. In the ray physics regime, solutions are amenable to asymptotic approximations [100–105].

In the early days, optics was studied primarily with scalar wave equations: the polarization effect of light was de-emphasized in those studies [106]. It was because light, by its nature, is randomly polarized, hence washing out the importance of the vector nature of light. Due to the short wavelength of light, the concept of coherence comes into play [107, 108]. A light may not be able to maintain the purity of its polarization over long distances due to the short wavelength nature of light and the environmental perturbation of the wave.

Ray Equations

In the extremely-high frequency limit, the wave equation can be further simplified. In this regime, the wave that propagates through the medium is locally a plane wave. In other words, the magnitude variation of the field in one direction is much more rapid than the phase variation of the field in other directions. The dominant variation of the field is in the phase. Therefore, we let [76]

$$\phi(\mathbf{r}) = A e^{i\omega\tau(\mathbf{r})} \tag{1.56}$$

where A is a constant. Substituting the above into (1.53), we have

$$\nabla\tau(\mathbf{r}) = i\omega\phi(\mathbf{r})\nabla\tau(\mathbf{r}) \tag{1.57}$$

and

$$\nabla \cdot \nabla\phi(\mathbf{r}) = \left\{ i\omega\nabla^2\tau(\mathbf{r}) - \omega^2\left[\nabla\tau(\mathbf{r})\right]^2 \right\} \phi(\mathbf{r}). \tag{1.58}$$

Using the above in (1.53), we have

$$i\omega\nabla^2\tau(\mathbf{r}) - \omega^2(\nabla\tau)^2 + k^2(\mathbf{r}) = 0. \tag{1.59}$$

At this point, (1.59) is still exact but nonlinear. However, we can solve (1.59) perturbatively by letting $\omega \to \infty$. In this limit, we let

$$\tau(\mathbf{r}) = \tau_0(\mathbf{r}) + \omega^{-1}\tau_1(\mathbf{r}) + \cdots, \quad \omega \to \infty. \tag{1.60}$$

If the series is used in (1.59), the ω^2 terms in the equation balance each other, yielding a leading order approximation that

$$\left[\omega\nabla\psi_0(\mathbf{r})\right]^2 = k^2(\mathbf{r}) = \omega^2 s^2(\mathbf{r}) \tag{1.61}$$

where $s = k/\omega = \sqrt{\mu\epsilon}$ is the slowness of the wave. The above equation can be written as

$$\nabla\psi_0(\mathbf{r}) = \pm s(\mathbf{r})\hat{s} \tag{1.62}$$

where $\hat{s}$ is an arbitrary unit vector. The above equation can be integrated to yield the ray passing through a point. This is known as the eikonal equation. Along the ray direction or in one dimension, it can be written as

$$\tau_0'(z) = \pm s(z). \tag{1.63}$$

It can be easily integrated to yield

$$\tau_0(z) = \pm \int_{z_0}^{z} dz'\, s(z') + C_0. \tag{1.64}$$

By collecting the first order terms, we have

$$i\omega\nabla \cdot \nabla\tau_0(\mathbf{r}) - 2\omega\nabla\tau_0(\mathbf{r}) \cdot \nabla\tau_1(\mathbf{r}) = 0. \tag{1.65}$$

The above is the transport equation. In one dimension, or along the ray direction, it can be written as

$$i\tau_0''(z) - 2\tau_0'(z)\tau_1'(z) = 0. \tag{1.66}$$

The above can be solved to yield

$$\tau_1(z) = \frac{\mathrm{i}}{2}\ln \tau_0'(z) + C_1 = \frac{\mathrm{i}}{2}\ln s(z) + C_{1\pm}. \tag{1.67}$$

Consequently, along the ray direction, or in one dimension, the solution to the wave equation in a slowly varying inhomogeneous medium is

$$\phi(z) \sim \frac{A_+}{\sqrt{s}}\exp\left[\mathrm{i}\omega\int_{z_0}^{z} s(z')\,dz'\right] + \frac{A_-}{\sqrt{s}}\exp\left[-\mathrm{i}\omega\int_{z_0}^{z} s(z')\,\mathrm{d}z'\right]. \tag{1.68}$$

The eikonal equation contributes to describing the phase variation of the field, while the transport equation contributes to the $1/\sqrt{s}$ factor, which is important for energy conservation.

Importance of Polarization Charges

When a light impinges on a dielectric interface, the Fresnel reflection coefficients for TE (transverse electric) and TM (transverse magnetic) fields are quite different. This is because for TM fields, a component of electric field normal to the dielectric interface exists and gives rise to polarization charges. However, when the medium is slowly varying, and the frequency is high, these polarization charges are less important [76]. The inhomogeneity at optical frequencies usually is with the dielectric permittivity ϵ and we can assume that the permeability μ is a constant. We start with the vector wave equation for magnetic field **H**, or

$$\nabla\times\epsilon^{-1}\nabla\times\mathbf{H} - \omega^2\mu\mathbf{H} = 0 \tag{1.69}$$

where a time-harmonic field is assumed with frequency ω. Using the vector identity, we rewrite the above as

$$\epsilon^{-1}\nabla\times\nabla\times\mathbf{H} + (\nabla\epsilon^{-1})\times\nabla\times\mathbf{H} - \omega^2\mu\mathbf{H} = 0. \tag{1.70}$$

If the field is locally a plane wave, then $\nabla = \mathbf{ik}$. If the permittivity is slowly varying, then $\nabla\epsilon^{-1}$ is small. Only the first term and the last term are needed to balance the equation when $\omega \to \infty$. Hence, the vector wave equation for high frequency and slowly varying inhomogeneity simplifies to

$$\nabla\times\nabla\times\mathbf{H} - \omega^2\mu\epsilon\mathbf{H} = 0. \tag{1.71}$$

Again, the above reduces to the scalar Helmholtz wave equation

$$\nabla^2\phi + k^2\phi = 0 \tag{1.72}$$

where $k^2 = \omega^2\mu\epsilon$.

1.6.4 Plasmonic Resonance

Even though we have attributed the resonance/oscillation behavior of an electromagnetic field to energy exchange between **E** and **H** fields, there is an emerging technology that relies on another physics for resonances; this is the field of plasmonic resonances. In this system, the resonance is due to the exchange between the stored energy in the electric field and the kinetic energy of the electrons. In plasma, free electrons abound such as in the ionosphere. In certain metal, such as gold and silver, the sea of electrons in the metal behaves like plasma. It can acquire enough kinetic energy at high-enough frequencies (optical frequencies) to exchange energy with that stored in the electric field, giving rise to resonance behavior. In this regime, the permittivity of the material medium is negative [74, 109], as shall be shown below.

One can easily show by Newton's law that the acceleration of the electron due to an applied electric field is given by

$$m_e \ddot{x} = qE \tag{1.73}$$

where x is the displacement of the electron from the mean, m_e is the effective mass of the electron[7], q is the charge of the electron, and E is the applied electric field. By assuming a time-harmonic field, so that $\ddot{x} = -\omega^2 x$, one deduces that

$$x = -\frac{qE}{m_e \omega^2}. \tag{1.74}$$

Since the polarization density $P = n_e q x$, where n_e is the electron density, one deduces that

$$\mathbf{P} = -\frac{n_e q^2}{m_e \omega^2} \mathbf{E}. \tag{1.75}$$

Hence, using the fact that $\mathbf{D} = \epsilon_0 \mathbf{E} + \mathbf{P}$, one gets

$$\epsilon = \epsilon_0 \left(1 - \frac{\omega_p}{\omega^2}\right) \tag{1.76}$$

where $\omega_p = \sqrt{n_e q^2/(m_e \epsilon_0)}$. In the above, the plasma medium becomes ϵ-zero material when $\omega = \omega_p$.

Simple Drude-Lorentz-Sommerfeld Model

This is often just called the Drude model. Since Lorentz and Sommerfeld also made significant contributions to this model, we will call it the Drude-Lorentz-Sommerfeld model. To begin, we have

$$m\ddot{x} + \frac{m}{\tau}\dot{x} = qE \tag{1.77}$$

$$\dot{p} + \frac{p}{\tau} = qE \tag{1.78}$$

$$-i\omega p + \frac{p}{\tau} = qE \tag{1.79}$$

where $p = m\dot{x}$ and $I \sim p$. Using the fact that the kinetic energy (KE) density of the electrons is

$$\mathrm{KE} = \frac{1}{2} n_e m_e (\dot{x})^2 = \frac{1}{2} n_e m_e \omega^2 x^2 \tag{1.80}$$

---

7 The plasma effect is also observed in some metallic nano-particles. In this case, the electron moves in a lattice with an effective mass [98, 99].

and that the displacement x is given by (1.74), one can show that at $\omega = \omega_p$, the kinetic energy density is exactly equal to the electric field energy density [109], i.e.

$$\text{KE} = \frac{1}{2}\frac{\partial \omega \epsilon E^2}{\partial \omega}. \tag{1.81}$$

Hence, at the plasma frequency, the material becomes ϵ-zero, the velocity of the wave is infinite, and all the electrons are in a time-harmonic motion but in phase with each other. In other words, the particles move in unison and oscillate as a block; the medium is in bulk plasma resonance.

When the frequency is below the plasma frequency, the material is ϵ-negative, and the wave is evanescent in the medium. This is because the electrons are light and agile; they can shield the electromagnetic field by annulling its effect to induce positive polarization current in a medium. However, above the plasma frequency, the inertia of the electrons prevents them from following the rapidly-varying electric field; they lose their shielding effect, and the electromagnetic field penetrates the plasma medium.

Surface Plasmons

When the plasma medium is ϵ-negative, the kinetic energy storage is dominant. This kinetic energy can be exchanged with the stored electric field energy. At an air-plasma interface, when the electric field has a normal component to the interface, a strong electric field is produced. The stored energy in the electric field of the charges can be exchanged with the stored kinetic energy of the electrons, yielding surface plasmon resonance. A surface plasmon mode can propagate on an interface with this energy exchange.

The guidance condition for the surface plasmon can be derived by finding the pole of the Fresnel reflection coefficient

$$R^{TM} = \frac{\epsilon_1 k_{0z} - \epsilon_0 k_{1z}}{\epsilon_1 k_{0z} + \epsilon_0 k_{1z}} \tag{1.82}$$

where $k_{iz} = \sqrt{k_i^2 - k_x^2}$, $(i = 0, 1)$ and k_x is the wavenumber along the interface, and k_{iz} is the wavenumber normal to it. The pole of the reflection coefficient is obtained by solving the equation

$$\epsilon_1 k_{0z} = -\epsilon_0 k_{1z}. \tag{1.83}$$

Assuming that the medium is non-magnetic material, solving the above yields the guidance condition that

$$\epsilon_1 k_{0z} = -\epsilon_0 k_{1z} \tag{1.84}$$

which further gives

$$k_x = \omega\sqrt{\frac{\mu_0 \epsilon_0 \epsilon_1}{\epsilon_0 + \epsilon_1}}. \tag{1.85}$$

This is the same as the condition for the Brewster angle, except that they are on different Riemann sheets of the complex plane [76]. This is the rare case where an electromagnetic wave can be guided by a single interface, whereas in elastic waves, single interface guided waves exist as Rayleigh waves [110], and Stoneley waves [111].

Since the stored energy of electric field need not exchange with the stored energy of magnetic field, plasmonic resonance can occur in nanoparticles even when they are much smaller than the wavelength. When an ϵ-negative particle is immersed in an electric field, much electric field energy is stored outside the particle. Hence, resonance by exchanging its kinetic energy with the stored electric field energy can occur. The resonance occurs despite the fact that the particle is much smaller than a wavelength.

If a dielectric spherical particle with a radius a is immersed in a long wavelength electric field, it has a dipole moment given by

$$p = 4\pi\epsilon_0 a^3 E_0 \frac{\epsilon_s - \epsilon_0}{\epsilon_s + 2\epsilon_0} \tag{1.86}$$

where ϵ_s is the permittivity of the particle. For an ϵ-negative material, the denominator of the above can become zero, giving rise to plasmonic resonance of the nanoparticle. This is observed in gold and silver nanoparticles immersed in the optical field, where the particles can be in tens of nanometers, while the wavelength of the optical field is in hundreds of nanometers [90, 112, 113].

1.7 The Age of Closed Form Solutions

Maxwell's equations have extremely high predictive power compared to many other physical laws. The range of length scales and the range of frequencies over which they are valid make these equations become the bedrock of many technologies. Moreover, since many electromagnetic phenomena are linear, Maxwell's equations are relatively simple to solve compared to many other physics equations. By comparison, fluid equations are nonlinear, and often break into turbulence when loss or viscosity is considered [114].

Hence, there has been great interest in finding closed form solutions of Maxwell's equations. These closed form solutions offer researchers physical insights into the interactions of electromagnetic fields with simple shape objects [115–118].

The derivation of closed form solutions was preceded by mathematical knowledge in the field of fluids and acoustics that predated electromagnetics. In fluids, non-turbulent flows are amenable to closed form solutions, such as high viscosity flow (low Reynolds number) and Helmholtz flow. In fluids, the onset of turbulence is due to the existence of viscosity. In an ideal fluid with no viscosity, called inviscid fluid, turbulence cannot occur, and closed form solutions can be found [114]. Also, the acoustic wave equation, which is scalar and linear, is also amenable to closed form solutions. Many special functions and mathematical knowledge of ordinary differential equations were derived before Maxwell's equations [116]. Hence, they helped set the stage for deriving closed form solutions for electromagnetics.

1.7.1 Separable Coordinate Systems

The early closed form solutions were obtained by objects in separable coordinate systems for which the separation of variables technique applies. For example, scattering solution of a sphere can be obtained in the spherical coordinate system. Solutions to Maxwell's equations are derivable from solutions to the scalar wave equation. The solutions to the vector wave equation in spherical coordinates can be expressed by the derivatives of two scalar Debye potentials. The Debye potentials are in turn solutions to the scalar wave equation in spherical coordinates [113, 119].

The scalar wave equation in spherical coordinates can be expressed by separation of variables into solutions of the spherical Bessel equation, the Legendre equation, and cylindrical harmonics. These functions together are called spherical harmonics. When the solution is expressed in terms of a summation of series involving spherical harmonics, it is called the Mie-series solution [113].

When the sphere size is small compared to wavelength, the approximate solution, the Rayleigh scattering solution, can be obtained [120]. It provides insight that a wave experiences more scattering when the frequency increases, hence, explaining why the sky is blue and the sunset is red. When the scatterer is comparable to wavelength, the solution is computed via the Mie scattering solution. A typical Mie-series solution has Debye potentials that are of the form

$$\Phi_{\text{sca}}(\mathbf{r}) = \sum_{n=0}^{\infty} \sum_{m=-n}^{n} a_{n,m} h_n^{(1)}(kr) P_n^m(\cos\theta) \mathrm{e}^{\mathrm{i}m\phi} \tag{1.87}$$

where $a_{n,m}$ is selected to match boundary conditions. The above solution is given in the frequency domain, or the Fourier transform space where the time dependence is $\exp(-\mathrm{i}\omega t)$. The spherical Hankel function of the first kind, $h_n^{(1)}(kr)$, is chosen to correspond to an outgoing wave at infinity, $P_n^m(x)$ is a Legendre's polynomial, and r, θ, and ϕ are variables in the spherical coordinate system. For an incident wave, one would replace the spherical Hankel function above with spherical Bessel function $j_n(kr)$ [121]. These special functions are often discussed in standard textbooks in electromagnetics [74, 122–126].

The scattering solution of a cylinder can be obtained in the cylindrical coordinate system. Again, the solution to the vector wave equation in cylindrical coordinates can be derived from two scalar potentials representing the transverse[8] electric and transverse magnetic waves. The solution is expressed in terms of the cylindrical Hankel function $H_n^{(1)}(k\rho)$, and Bessel function $J_n(k\rho)$ and cylindrical harmonics $\mathrm{e}^{\mathrm{i}n\phi}$. Other closed form solutions are possible by similar reduction to scalar wave equations, when such solutions are available, e.g., in spherical coordinates, elliptical coordinates, etc.

In addition to the scattering problems, the inside-out solutions of cavity resonances in spherical coordinates, waveguide solutions of a cylinder can be derived by similar mathematical techniques. The above solutions can also be generalized to layered media.

A quirky problem is the case of the cavity resonances of a cuboid formed by perfect electric conducting wall or perfect magnetic conducting wall. While the inside-out problem is separable and can be solved in closed form, the outside-in scattering problem of a cuboid has no closed-form solution. The same case applies to a rectangular waveguide that has a closed form solution, while the outside-in problem has no closed form solution.

1.7.2 Integral Transform Solution

Other important problems with closed form solutions using integral transform techniques include the solutions of a point source on top of layered media, which were used to explain the propagation of radio wave over the layered earth. This problem was first solved by Sommerfeld [127]. The field due to a point source can be expanded in terms

8 Transverse here implies transverse to the axial direction of the cylinder.

of plane waves, which can be easily propagated through the planar layered media. An important mathematical identity to this end is the Weyl identity [128]

$$\frac{e^{ik_0 r}}{r} = \frac{i}{2\pi} \iint_{-\infty}^{\infty} dk_x dk_y \frac{e^{ik_x x + ik_y y + ik_z |z|}}{k_z} \tag{1.88}$$

where $k_x^2 + k_y^2 + k_z^2 = k_0^2$, or $k_z = (k_0^2 - k_x^2 - k_y^2)^{1/2}$. It expands the spherical wave due to a point source in terms of plane waves emanating from the $z = 0$ plane. Both propagating and evanescent waves are included in the above expansion. The double integral above can be reduced to a single integral using the integral representation for Bessel function, yielding the Sommerfeld identities [76]

$$\frac{e^{ik_0 r}}{r} = i \int_0^{\infty} dk_\rho \frac{k_\rho}{k_z} J_0(k_\rho \rho) e^{ik_z |z|} = \frac{i}{2} \int_{-\infty}^{\infty} dk_\rho \frac{k_\rho}{k_z} H_0^{(1)}(k_\rho \rho) e^{ik_z |z|} \tag{1.89}$$

where $\rho = \sqrt{x^2 + y^2}$ and $k_\rho = \sqrt{k_x^2 + k_y^2}$. The above identity was first used by Sommerfeld to solve for the solution of a point source over a half-space. The transform integrals thus obtained are known as Sommerfeld integrals. Various kinds of potential were defined to solve this problem. It was later generalized to layered media using the Kong formulation with just E_z and H_z components of the electromagnetic fields [129].

Similarly, when a point source field is expanded in terms of cylindrical harmonics, or spherical harmonics, they can be propagated through layered cylinders or layered spheres.

Another problem of great importance with a closed form solution is the Sommerfeld half-plane problem [130–132]. While this problem is deceptively simple, it resists a closed form solution unless the Fourier transform technique is applied. When the Fourier transform technique is applied with the Wiener-Hopf technique [133], a closed form solution in terms of an integral is obtained[9]. The Wiener-Hopf technique is used to deconvolve a function that has semi-infinite support and the convolved result is known only over semi-infinite support. A typical problem of this type can be written as

$$\Phi_s(x) = g(x) * j(x) = -\Phi_i(x), \quad x \geq 0, \tag{1.90}$$

$$j(x) = 0, \quad x < 0 \tag{1.91}$$

where $*$ implies a convolution. In the above, $\Phi_i(x)$, which represents the incident field, is known for $x \geq 0$ while $j(x)$ is unknown to be solved with a support for $x \geq 0$. The above integral equation is a special class of the integral equation that can be deconvolved using the Wiener-Hopf technique, which seems like a lost art. The final solution to the Sommerfeld half-plane problem can be expressed in terms of the Fourier integral as

$$\Phi_s(x, y) = -\frac{e_0 (k + k_x^i)^{1/2}}{2\pi i} \int_{-\infty}^{\infty} \frac{e^{i(k_x x + k_y y)}}{(k_x - k_x^i)(k_x + k)^{1/2}} dk_x \tag{1.92}$$

where the incident field is $e_0 e^{i(k_x^i x - k_y^i y)}$, $k = \omega/c$, and $k_x^2 + k_y^2 = k^2$.

When asymptotic expansion using the contour deformation technique and the method of steepest descent is performed on the Fourier integral, the physics of the

9 Sommerfeld obtained his own solution by inspection, which is quite different from the Wiener-Hopf technique, but his method cannot be generalized.

scattering of a plane wave by a half plane can be elucidated. One can clearly see the contribution of edge diffraction, specular reflection, and the existence of reflection boundary and the shadow boundary in the above solution. This problem forms the canonical solution that explains the high-frequency scattering physics of a half plane.

1.8 The Age of Approximations

Many closed form solutions are in terms of integral transforms, or a summation of series. For instance, the solution of the point source on top of layered media is in terms of the Sommerfeld integral, which offers little physical insight into the problem. However, when the frequency is high, asymptotic approximations can be derived so as to offer more physical insight. This was done for the Sommerfeld half-plane problem, but it can also be done for the Sommerfeld integrals.

1.8.1 Asymptotic Expansions

While the asymptotic expansion for the half-plane problem could be easily obtained due to the simplicity of the integrand, it was more difficult to obtain the asymptotic solution to the Sommerfeld integrals for layered media. This was due to the presence of poles corresponding to guided waves and leaky waves in the integrand. The presence of branch cuts can be related to lateral waves, or surface waves [76, 134, 135]. A typical Sommerfeld integral is

$$E_{1z} = \frac{-I\ell}{8\pi\omega\epsilon_1} \int_{-\infty}^{\infty} \mathrm{d}k_\rho \frac{k_\rho^3}{k_{1z}} H_0^{(1)}(k_\rho\rho) \left[\mathrm{e}^{\mathrm{i}k_{1z}|z|} + \tilde{R}_{12}^{TM} \mathrm{e}^{\mathrm{i}k_{1z}z + 2\mathrm{i}k_{1z}d_1} \right] \tag{1.93}$$

where $k_{1z} = (k_1^2 - k_\rho^2)^{\frac{1}{2}}$, and $k_1^2 = \omega^2\mu_1\epsilon_1$ which is the wave number in Region 1. In the above, $\tilde{R}_{12}^{TM}$ is the generalized reflection coefficient given by

$$\tilde{R}_{i,i+1} = R_{i,i+1} + \frac{T_{i,i+1}\tilde{R}_{i+1,i+2}\,T_{i+1,i}\mathrm{e}^{2\mathrm{i}k_{i+1,z}(d_{i+1}-d_i)}}{1 - R_{i+1,i}\tilde{R}_{i+1,i+2}\mathrm{e}^{2\mathrm{i}k_{i+1,z}(d_{i+1}-d_i)}} \tag{1.94}$$

where $R_{ij} = \frac{p_j k_{iz} - p_i k_{jz}}{p_j k_{iz} - p_i k_{jz}}$, $T_{ij} = 1 + R_{ij}$, $p_i = \epsilon_i$ for TM waves and $p_i = \mu_i$ for TE waves. Furthermore, $k_{iz} = \sqrt{k_i^2 - k_\rho^2}$. It appears that there is a branch point at $k_\rho = k_i$ for each of the layered media, but it can be proved that the branch point exists only for the first and the last layer [76, 135, 136]. It is a beautiful harmony between physics and mathematics as the branch points correspond to space waves that can only exist in the first and the last regions.

One important lesson to learn from the exercise of deriving asymptotic expansions for spectral integrals that follow from the Sommerfeld problems is that the approximations are associated with critical points on the complex plane: poles, branch points, saddle point (stationary phase point) [76, 137]. When the far field can be approximated by a simple critical point, simple ray-optics physics follows. However, when the critical points start to merge on the complex plane, the approximation of the spectral integral with simple functions becomes difficult and special functions have to be used [76, 137]. This usually corresponds to observation points near the shadow boundary, reflection boundary, or caustics, where simple ray optics does not work. Uniform asymptotic approximations have to be designed in the vicinity of such boundaries [101–103, 138]. Since integrals are amenable to asymptotic expansions, even when a series solution, such as a Mie-series solution, is obtained, attempts will be made to express it into an integral by Watson's transformation to facilitate its asymptotic approximations [135, 139–141]. Watson's transformation elucidates the physics of the creeping waves, as well as the radiation from the lit-shadow boundary, which can be corrected with the incremental length diffraction coefficient [142].

Approximate asymptotic solutions offer physical insight not available with many closed form solutions. They have also motivated other asymptotic-based approximate solutions to many problems [100]. This was because the cases of solvable geometries were quickly exhausted by researchers. However, there was still a quest for solutions of more complex geometries to increase the predictive power of theoretical modeling of physical problems using mathematics. Nevertheless, many problems have no closed form solutions and are amenable only to approximate solutions.

1.8.2 Matched Asymptotic Expansions

A notable approximate method is the method of matched asymptotic expansions [92]. It was used in fluids to solve many boundary layer problems, involving the problems of different length scales. For instance, when fluid flows near a surface, the fluid velocity varies rapidly normal to the surface compared to the tangential direction. For this type of problem, it can be divided into an inner solution and an outer solution. Coordinate stretching technique is used to emphasize the variation of the fluid normal to the surface when one seeks the inner solution. For the outer solution, the inner solution provides an effective "boundary condition" for it [143].

One can view the Rayleigh scattering solution of a small particle [120] as a simple matched asymptotic solution. When one is close to the particle, which is much smaller than a wavelength, electrostatic physics prevails. The field is rapidly varying in the vicinity of the particle, but an electrostatic solution is sufficient to capture the interaction of the field with the particle. Hence, one can solve a static problem in the vicinity of the particle, to derive the induced dipole moment by a field. But far away from the particle, the radiation field of the particle is important to produce wave physics results. Hence, one can match the far field solution to obtain the dipole scattering strength [74].

A place where matched asymptotics is useful is in the WKB (Wentzel, Kramers, Brillouin) [144–146] solution in an inhomogeneous profile by assuming that the profile is smooth, and that the frequency is high [76]. If the profile is becoming optically less dense (decreasing refractive index profile) as the wave penetrates into the medium, eventually, at one point called the turning point, the wave will be refracted back and totally reflected by total internal reflection (Figure 1.4). At the region before the turning point, it can be described by ray optics. At the region beyond the turning point, the wave can be described by the evanescent wave of total internal reflection. However, at the turning point region, only a special function can be used to describe the field. In this case, it is the Airy function [121]. Solutions can be written down in each region, and asymptotic matching is used to find the coefficients of the solution in each region. The turning point region is where ray optics breaks down, because of the convergence of the rays into a tiny region. This is a caustic region [76].

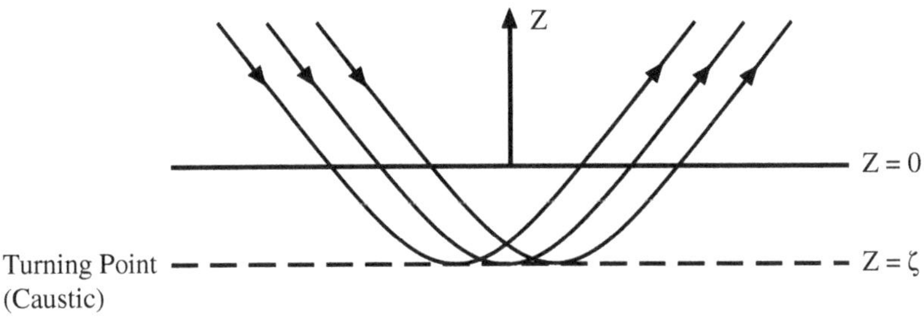

Figure 1.4 In an inhomogeneous profile with decreasing index, the ray will be continuously refracted until it turns around at the turning point.

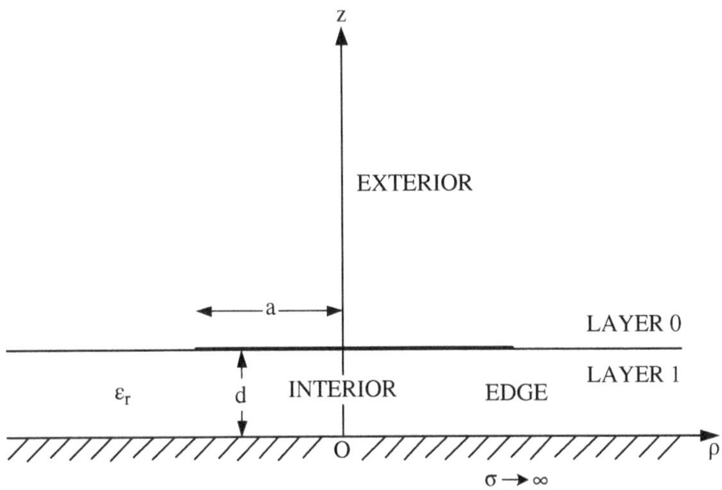

Figure 1.5 Matched asymptotics can be used to find the solution of the capacitance between two disks. The solution is divided into edge solution near the edge, inner solution between the plate, and outer solution outside the two plates.

Another place where matched asymptotics has been used is to find the capacitance of the two circular parallel plates. When the separation between the two plates is small, the solution near the edge, or the edge solution, is rapidly varying, and is quite different from the solution in the region between the two parallel plates, or the inner solution, and outside the plates, or the outer solution (see Figure 1.5). The solution near the edge can be obtained by coordinate stretching when $\delta = d/a$ is small. After coordinate stretching, the edge solution to leading order is similar to the solution of a straight edge which can be solved by conformal mapping or the Wiener-Hopf technique [133]. The outer solution can be easily sought. The first approximate solution to such a problem was obtained by Kirchhoff [147] heuristically, but was correct asymptotically:

$$C \sim \frac{\pi \epsilon_0 a^2}{2d} + \epsilon_0 a \log\left(\frac{a}{d}\right) + \epsilon_0 a\,(\log 8\pi - 1), \quad \frac{d}{a} \to 0 \tag{1.95}$$

where $2d$ is the separation of the disks, and a is their radii. It was generalized to the case of two disks separated by a dielectric slab by Chew and Kong [148]. The formula in this case is

$$C \sim \frac{\pi\epsilon_r\epsilon_0 a^2}{2d} + \epsilon_0 a \log\left(\frac{a}{d}\right)$$
$$+ \epsilon_0 a \left[\log 8 - 1 - 2\epsilon_r \sum_{n=1}^{\infty}\left(\frac{1-\epsilon_r}{1+\epsilon_r}\right)^n \log n + (\epsilon_r - 1)\log 2 + \epsilon_r \log\pi\right],$$
$$\frac{d}{a} \to 0. \tag{1.96}$$

The Wiener-Hopf method was used to find the edge solution in the above problem. The above reduces to Kirchhoff's formula when $\epsilon_r = 1$.

Leppington and Levine [149] obtained a correction to Kirchhoff's formula, and Shaw [150] further refined the correction given by

$$C \sim \frac{\pi\epsilon_0 a^2}{2d} + \epsilon_0 a \log\left(\frac{a}{d}\right) + \epsilon_0 a\,(\log 8\pi - 1) + \frac{\epsilon_0 d}{2\pi}\left[\log\left(\frac{d}{8\pi a}\right)\right]^2, \quad \frac{d}{a} \to 0. \tag{1.97}$$

The conformal mapping technique [151] was used to find the edge solution since a dielectric slab was absent.

A formula to the same order of approximation as Shaw's formula has also been given by Chew and Kong in [152] when a dielectric slab is present.

$$C \sim \frac{\pi\epsilon_r\epsilon_0 a^2}{2d} + \epsilon_0 a \log\left(\frac{a}{d}\right) + \epsilon_0 a\,(\log 8\pi - 2 + A)$$
$$+ \frac{\epsilon_0 d}{2\pi}\left\{\left[\log\left(\frac{8a}{d}\right) + A - 1\right]^2 - 2\right\}, \quad \frac{d}{a} \to 0 \tag{1.98}$$

where

$$A = -2\epsilon_r \sum_{n=1}^{\infty}\left(\frac{1-\epsilon_r}{1+\epsilon_r}\right)^n \log n + (\epsilon_r - 1)\log 2 + \epsilon_r \log\pi + 1. \tag{1.99}$$

When $\epsilon_r = 1$, the above reproduces Shaw's formula but with a correction term to her formula.

A similar formula for capacitance per unit length of two parallel strips, with width w separated by a dielectric slab of thickness d, has also been derived in [148] with an improvement in [153]. It is

$$C \sim \frac{\epsilon_r\epsilon_0 w}{2d} + \frac{\epsilon_0}{\pi}\left[\log\left(\frac{w}{d}\right) + A\right] + \frac{2\epsilon_0 d}{w\pi^2\epsilon_r}\left[\log\left(\frac{w}{d}\right) + A - \frac{1}{2}\right], \quad \frac{w}{b} \to 0. \tag{1.100}$$

When $\epsilon_r = 1$, the first two terms reduce to the result of Michell, Bromwich, and Love [154–156]. The higher order correction to the above is an improvement over Shaw's formula [150].

The matched asymptotic technique has also been generalized to find the resonance frequencies and guidance condition of a microstrip line where wave physics is included in the outer solution while static physics is involved in the edge solution [157, 158]. The matched asymptotic expansion method has also been applied to understand the interaction of electromagnetic field with electrical double layer in electrochemistry.

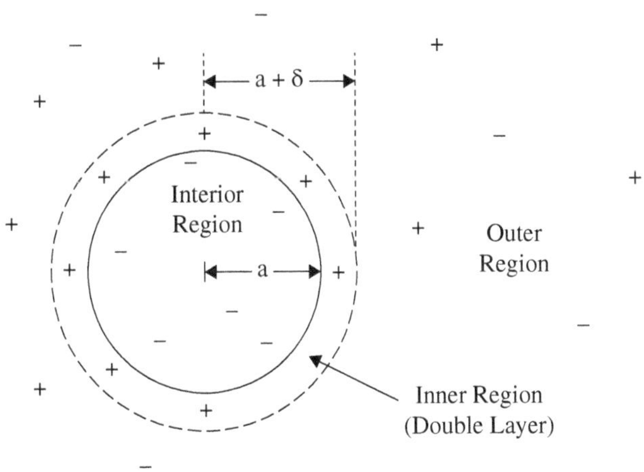

Figure 1.6 The occurrence of a thin double layer outside a charged particle makes the problem suitable for matched asymptotics analysis [10].

It helps explain the dielectric dispersion and electrophoretic properties of colloidal suspension of particles in ionic solutions [10, 159–161]. For instance, when a positively charged polystyrene particle is immersed in an ionic solution, it acquires negative counter-ion charges on its surface, forming a thin double layer. The thin double layer, where the ionic density varies rapidly, is the inner solution of a matched asymptotic analysis (see Figure 1.6).

1.8.3 Ansatz-Based Approximations

When a scatterer is very complex, a high frequency scattering solution can be motivated by physical insight derived from asymptotic approximations of closed form solutions. The approximate solutions can be derived based on an ansatz. It is assumed that at high frequencies the dominant contribution of a scatterer comes primarily from important critical points (see Figure 1.7). Moreover, the scattering phenomena are localized, meaning that the scattered field will come from localized points on surfaces, edges, vertices [100, 162–164]. At high frequencies, the specular reflection from an object is of leading order whereas the scattered field from edges and vertices are of increasingly higher order or weaker.

Since in ray physics the interaction of the field with the object is local, one can use canonical solutions such as the Sommerfeld half-plane solution to ascertain the amplitude of the scattered fields from the critical points. By studying Watson's transformation, one can understand the physics of the surface or creeping wave contribution on a curved surface. On a convex curved surface, the surface wave is radiative, meaning that it sheds energy as it creeps on the curved surface. Hence, it decays along its creeping direction, implying that in the high-frequency limit, their contributions are exponentially small, but in the twilight zone between wave physics and ray physics they may be important. In the high frequency limit, in the order of importance, the scattering events are: specular reflection, edge diffraction, vertex diffraction, and creeping wave.

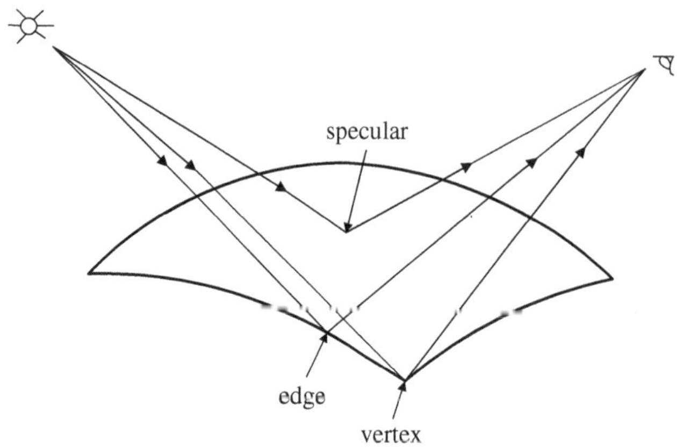

Figure 1.7 The essential physics of high-frequency scattering is controlled by critical point reflections and diffractions.

Moreover, many ray-optics approximations break down near shadow boundaries and caustics. Uniform solutions with the help of special functions and asymptotic matching are needed for solutions in this region [142].

1.9 The Age of Computations

The untenability of many problems in order to meet the demand of science and technology developments calls for new mathematical modeling methods of solving intricate equations with complex geometries. Numerical methods were the most promising general purpose methods to solve nonlinear differential equations involving elaborate geometries.

During the Second World War, a bevy of women were hired to do numerical computing by hand, but it was far too laborious. These women were actually called computers. Hence, there was a pressing need to use machines to replace humans as computers [165, 166]. Mechanical computing machines have been around for a while, but they were not feasible for high-speed computations [167]: machines with moving parts were slow and too unwieldy. Therefore, the electronic computer, with no moving mechanical parts, that promised high speed computation was developed in 1946 (Eniac) [168, 169] and an all-transistor computer was developed in 1951 (Illiac) [170].

The invention of the computer heralded in a new era and motivated many researchers to investigate alternative methods of solving sophisticated problems. In engineering, many researchers were using the finite element method (FEM) and the finite difference method (FDM) to solve complex engineering problems [171–178]. The final equation becomes a sparse matrix equation that is best solved by numerical methods and methodically by machines.

The basic equations describing many science and engineering problems involving fields, such as fluid dynamics and electromagnetics, are partial differential equations. These partial differential equations can be linear or nonlinear. The fields are functions of space as well as time. When the problems are nonlinear, they need to be solved in their original form in the time domain.

However, if the problem is linear, the Fourier transform technique can convert these equations from the time domain to the frequency domain. When the frequency is fixed, then the fields are functions of space alone, making them easier to solve. Furthermore, for linear problems, one can define a Green's function to further convert the differential equations into integral equations by using the principle of linear superposition [122, 123]. Consequently, numerical methods for solving integral equations have become important and popular [179, 181]. It is important to note that reduction to the frequency domain and integral equations is only possible for linear problems. It is, however, possible to solve a nonlinear problem in the frequency domain by considering multiple frequencies simultaneously, and to use the harmonic balancing method [182–185] to seek the solutions.

Numerical methods can be easily developed for linear problems. When a non-linear problem is encountered, it is linearized, and then an approximate solution is sought and iterated until convergence is reached. This is a Newton related method. In one dimension, this is called the Newton-Raphson method, while in higher dimensions, it is called the Gauss-Newton method or by other variants [186–188].

For inhomogeneous media, the electromagnetic equation reduces to

$$\nabla \times \mu^{-1} \nabla \times \mathbf{E} + \epsilon \frac{\partial^2 \mathbf{E}}{\partial t^2} = -\frac{\partial \mathbf{J}}{\partial t}. \tag{1.101}$$

Finite element or finite difference techniques can be applied directly to solve the above equation in the time domain. These techniques can be applied to nonlinear problems as well, for instance, when ϵ or μ are functions of **E** or **H**. However, if the problem is linear, then the above equation can be transformed to the frequency domain and it becomes

$$\nabla \times \mu^{-1} \nabla \times \mathbf{E} - \omega^2 \epsilon \mathbf{E} = i\omega \mathbf{J}. \tag{1.102}$$

The above can be rewritten as

$$\mathcal{L}\mathbf{E} = i\omega \mathbf{J} \tag{1.103}$$

where $\mathcal{L}$ is a linear operator. This equation can be solved with the finite-element frequency domain method or the finite-difference frequency domain method.

By rearranging terms, the above equation can be rewritten as [76, 123]

$$\nabla \times \mu_0^{-1} \nabla \times \mathbf{E} - \omega^2 \epsilon_0 \mathbf{E} = i\omega \mathbf{J} - \omega^2 (\epsilon_0 - \epsilon)\mathbf{E} + \nabla \times (\mu_0^{-1} - \mu^{-1}) \nabla \times \mathbf{E}. \tag{1.104}$$

Assuming that μ_0 is a constant, it can be multiplied through the equation to arrive at

$$\nabla \times \nabla \times \mathbf{E} - \omega^2 \mu_0 \epsilon_0 \mathbf{E} = i\omega \mu_0 \mathbf{J} - \omega^2 \mu_0 (\epsilon_0 - \epsilon)\mathbf{E} + \mu_0 \nabla \times (\mu_0^{-1} - \mu^{-1}) \nabla \times \mathbf{E}. \tag{1.105}$$

Now if we have a closed form solution to the following equation

$$\nabla \times \mu_0^{-1} \nabla \times \overline{\mathbf{G}}(\mathbf{r} - \mathbf{r}') - \omega^2 \epsilon_0 \overline{\mathbf{G}}(\mathbf{r} - \mathbf{r}') = \overline{\mathbf{I}} \delta(\mathbf{r} - \mathbf{r}') \tag{1.106}$$

the solution is the point source response to the vector wave equation [72, 77]. Using the principle of linear superposition, we can treat the right-hand side of (1.105) as sources, and write down the solution of it as

$$\begin{aligned} \mathbf{E}(\mathbf{r}) = {} & i\omega \mu_0 \int d\mathbf{r}' \overline{\mathbf{G}}(\mathbf{r} - \mathbf{r}') \cdot \mathbf{J}(\mathbf{r}') - \omega^2 \mu_0 \int d\mathbf{r}' \overline{\mathbf{G}}(\mathbf{r} - \mathbf{r}') \cdot \left[\epsilon_0 - \epsilon(\mathbf{r}')\right] \mathbf{E}(\mathbf{r}') \\ & + \mu_0 \int d\mathbf{r}' \overline{\mathbf{G}}(\mathbf{r} - \mathbf{r}') \cdot \nabla \times \left[\mu_0^{-1} - \mu(\mathbf{r}')^{-1}\right] \nabla \times \mathbf{E}(\mathbf{r}'). \end{aligned} \tag{1.107}$$

The above equation can be simplified further if needed. Assuming that **J** is known, the first term on the right-hand side can be regarded as the incident field on the inhomogeneity, and then the above is an integral equation for the unknown **E**. We can rewrite the above as

$$\mathbf{E} = \mathbf{E}_{\text{inc}} + \mathcal{L}_v \mathbf{E} \tag{1.108}$$

where $\mathcal{L}_v$ represents the integral operator associated with the second and the third terms on the right-hand side of (1.107).

Since the unknown is inside the integral as well as outside the integral, it is an integral equation of the second kind. Moreover, it is a volume integral equation, and the integration need only be over the domain where $\epsilon_0 - \epsilon(\mathbf{r}')$ and $\mu_0 - \mu(\mathbf{r}')$ are non-zero. So if the scatterer is of finite support, this domain is finite. The unknown **E** needs only to be solved over a smaller domain compared to the original differential equations [76, 93, 179, references therein]. The early concepts were formulated in 2D [180] and gradually evolved to 3D.

If the domains are piecewise homogenous, then the vector Green's theorem can be used to derive integral equations where only surface unknowns are needed [189]. The support of the unknown domain is even further reduced. These are surface (or boundary) integral equations. The number of unknowns is greatly reduced, beating the cruelty of dimensionality that shall be discussed later.

1.9.1 Computations and Mathematics

Mathematics is useful for classifying partial differential equations (PDEs) that are encountered in electromagnetics. For electromagneitcs, when the equations involve no derivative in time, the PDEs are considered elliptic. Such PDEs cannot propagate a singularity such as the Laplace's equation or the Poisson's equation. The field solutions of such equations are highly smooth. When the PDEs involve the first-order derivative in time only, such as the diffusion equation, they are classified as parabolic. These equations only evolve in one direction in time: the singularity can only propagate forward in time. When the PDEs involve the second-order derivatives in time, such as the wave equation, they are classified as hyperbolic. Such equations can propagate the field or singularity both forward and backward in time [190, 191].

Partial differential equations need to be solved with boundary conditions in order to guarantee unique solutions [192]. However, when internal resonance solutions exist, uniqueness is not guaranteed. Internal resonances can exist in wave-like solutions. Both hyperbolic and parabolic equations can have wave-like solutions. Internal resonance corresponds to a homogeneous solution without excitation, and hence, is equivalent to the existence of a null space solution. The solution is non-unique for PDEs when such resonance solutions exist.

Derivation of Integral Equations

When a problem is cast into an integral equation, the Green's function becomes the kernel of the integral equation. For example, in the integral equation

$$\int \mathrm{d}\mathbf{r}' g(\mathbf{r}, \mathbf{r}') f(\mathbf{r}') = h(\mathbf{r}) \tag{1.109}$$

where $g(\mathbf{r}, \mathbf{r}') = \exp(ik|\mathbf{r} - \mathbf{r}'|)$ is the Green's function and also the kernel. For wave problems, $k \neq 0$, and the Green's function is oscillatory. The oscillatory nature of the Green's function implies that it can send information over a long distance. For static problems or low-frequency problems, $k = 0$, and the Green's function is smooth (non-oscillatory); it cannot send information over a long distance [87]. These two distinct physics affect how these integral equations can be solved. Internal resonances can affect integral equations to give rise to non-uniqueness in their solutions, just as for differential equations [193–208]. An excellent review was given by Peterson [204].

Most problems are linear problems in electromagnetics. In the case when the problem is nonlinear, it is linearized and solved iteratively as a set of linear problems. For instance, differential equations and integral equations in electromagnetics are often linear. Hence, we shall discuss next the numerical methods to solve a linear equation [76, 93, 179].

Subspace Projection Methods

We can describe a general linear problem with a linear operator equation

$$\mathcal{L}f = h \tag{1.110}$$

$$\mathcal{L}^t f_a = h_a \tag{1.111}$$

where $\mathcal{L}$ represents a linear operator. The second equation is the auxiliary equation, which is discussed in conjunction with the first equation. It represents a differential operator in the case of differential equations, while it represents an integral operator for integral equations. In mathematical notation, one often writes $\mathcal{L} : V \to W$, meaning that it is a linear operator that maps elements of the vector space V to the elements of the vector space W, where V is known as the domain space and W the range space of the operator $\mathcal{L}$.

For a differential operator, an explicit form may be

$$(\nabla^2 + k^2)f(\mathbf{r}) = h(\mathbf{r}) \tag{1.112}$$

which is the Helmholtz wave equation or the frequency domain version of the wave equation. A boundary condition and a loss condition have to be stipulated in order to make the solution to the above equation unique. The Laplacian operator ∇^2 can be proved to be a negative definite operator because by using integration by parts or using the fact that $\nabla \cdot (f\nabla f) = (\nabla f)^2 + f\nabla^2 f$, we have

$$\langle f, \nabla^2 f\rangle = \int_V \mathrm{d}\mathbf{r} f(\mathbf{r})\nabla^2 f(\mathbf{r}) = -\int_V \mathrm{d}\mathbf{r}(\nabla f(\mathbf{r}))^2 + \int_S \mathrm{d}\mathbf{S} \cdot f(\mathbf{r})\nabla f(\mathbf{r}). \tag{1.113}$$

When $f(\mathbf{r})$ satisfies the appropriate boundary condition on the surface S, the second term on the right-hand side disappears and the first term is always negative for real $f(\mathbf{r})$. Hence, for real k^2, the operator $\nabla^2 + k^2$ is indefinite, implying that it can have a null space. This null space corresponds to the resonance solution of the wave equation.

For an integral equation, an explicit form is given in (1.109). These equations are not amenable to computation. They are equations in infinitely dimensional Hilbert space with uncountably infinite indices. To make them computable, we find a subspace in which an approximate solution of the above equations can be found, and then project the solution into this subspace. This process can be methodically described as follows.

We choose a basis set with N basis functions f_n, $n = 1, \cdots, N$ that spans the subspace that can approximate the domain space of the operator. We expand the unknown f in terms of the basis functions in this subspace, and the auxiliary unknown f_a in terms of the auxiliary basis functions, namely

$$f(\mathbf{r}) \doteq \sum_{n=1}^{N} a_n f_n(\mathbf{r}) \tag{1.114}$$

$$f_a(\mathbf{r}) \doteq \sum_{n=1}^{N} a'_n f'_n(\mathbf{r}) \tag{1.115}$$

The linear operator equation can then be written as

$$\sum_{n=1}^{N} a_n \mathcal{L} f_n \doteq h. \tag{1.116}$$

The above equation can only be approximately satisfied since we have picked a finite number of basis functions. Moreover, the solution to (1.116) is still untenable. To make it more easily solvable, we convert (1.116) into a matrix equation by weighting or testing the equation with $w_m(\mathbf{r}), m = 1, \cdots, N$. Consequently, we have

$$\sum_{n=1}^{N} a_n \langle w_m, \mathcal{L} f_n \rangle = \langle w_m, h \rangle, \quad m = 1, \cdots, N \tag{1.117}$$

where the inner product is defined by

$$\langle u, v \rangle = \int u(\mathbf{r}) v(\mathbf{r}). \tag{1.118}$$

The above equation (1.117) implies that it can be exactly satisfied if we project (1.116) onto a subspace spanned by $w_m(\mathbf{r}), m = 1, \cdots, N$. Recent research has shown that if the testing functions w_m are chosen from the dual space of the range space, the above can be a good approximation or matrix representation of the original operator equation [210]. The dual space can be larger or smaller than the range space.

Equation (1.117) now constitutes a matrix equation

$$\overline{\mathbf{L}} \cdot \mathbf{a} = \mathbf{h} \tag{1.119}$$

$$(\overline{\mathbf{L}})_{mn} = \langle w_m, \mathcal{L} f_n \rangle \tag{1.120}$$

$$(\mathbf{a})_n = a_n \tag{1.121}$$

$$(\mathbf{h})_m = \langle w_m, h \rangle \tag{1.122}$$

$$\overline{\mathbf{L}}^t \cdot \mathbf{a}' = \mathbf{h}_a \tag{1.123}$$

$$(\overline{\mathbf{L}})_{mn} = \langle w_m, \mathcal{L} f_n \rangle \tag{1.124}$$

$$(\mathbf{a}')_n = a'_n \tag{1.125}$$

$$(\mathbf{h}_a)_m = \langle w_m, g_a \rangle \tag{1.126}$$

In the above, the choice of the testing function is determined by the auxiliary equation [76]. The testing function should be chosen so that the auxiliary equation is solved accurately, whereas the choice of the expansion or basis function is to solve the primary equation accurately. The auxiliary equation is the adjoint or the transpose of the original equation where the range space and the domain space are swapped. It can be shown that the left domain space of an operator is the dual space of the range space. Hence, if the testing functions solve the auxiliary equation well, they can well approximate the left domain space of the original operator, which is the dual space of the range space of the original equation. This is in agreement with the aforementioned recent findings.

The above procedure of converting the operator equation into a matrix equation is the underpinning method behind the finite element method, or the method of moments. They are variously known as Galerkin's method, Petrov-Galerkin's method, method of weighted residuals, collocation method, and point matching method [212–214]. However, they are all subspace projection methods [215, 216].

When $\mathcal{L}$ is a differential operator, the matrix $\overline{\mathbf{L}}$ is sparse because a differential operator is a local operator. However, when $\mathcal{L}$ is an integral operator, the matrix $\overline{\mathbf{L}}$ is dense. In differential equations, the unknown is the field that permeates all of space. Hence, the unknown count is usually larger in differential equation solvers. However, they come with sparse matrix systems that are cheaper to solve and store.

For integral equations, the unknown is the induced current on the scatterer. Hence, the unknown count is smaller since the current resides only on or in a scatterer with finite support. However, the ensuing matrix system is dense and is hard to store. Moreover, the matrix assembly of forming the matrix elements

$$\langle w_m, \mathcal{L}f_n \rangle \tag{1.127}$$

is tedious since the operator involves singularities that have to be evaluated with care [217, 218].

1.9.2 Sobolev Space and Dual Space

Since many quantities (fields and currents) in electromagnetics correspond to finite energy, they are square integrable over a finite volume. In mathematical parlance, these quantities live in the $L2$ space. For instance,

$$\int_V |\mathbf{E}(\mathbf{r})|^2 \mathrm{d}V < \infty \tag{1.128}$$

for physical reasons. Hence, $\mathbf{E}(\mathbf{r})$ lives in the $L2$ space. However, the above is not specific enough to describe the space in which $\mathbf{E}$ lives. Since $\nabla \times \mathbf{E}$ produces the magnetic field $\mathbf{H}$, which must have finite energy and hence, the square integrable $\mathbf{E}$ lives in a space where itself plus its curl are square integrable. We can define such a space to be

$$H(\mathrm{curl}) = \{\mathbf{E} \in L2 \,:\, \nabla \times \mathbf{E} \in L2\}. \tag{1.129}$$

Spaces where the function is in Lp as well as its certain-order derivatives are also in Lp are known as Sobolev spaces. The particular Sobolev space above is also called curl conforming space. A Sobolev space imposes smoothness constraint on a function, and hence, it is a smaller space than $L2$.

What kind of space does $\mathbf{J}$ live in? If $\mathbf{J}$ is physical, then its divergence exists since from the continuity equation, $\nabla \cdot \mathbf{J} = -i\omega\varrho$ where ϱ is charge density. Since both the current density $\mathbf{J}$ and the associated charge density ϱ are finite energy functions that are square integrable, $\mathbf{J}$ lives in the space

$$H(\mathrm{div}) = \{\mathbf{J} \in L2 \,:\, \nabla \cdot \mathbf{J} \in L2\}. \tag{1.130}$$

Given an inner product $\langle f, h \rangle$ that exists where h is chosen from a vector space V, then f is said to live in the dual space of V. For instance, in electromagnetics, the reaction $\langle \mathbf{E}, \mathbf{J} \rangle$ always exists corresponding to a physical $\mathbf{E}$ field and a physical current $\mathbf{J}$. If all physical currents live in a space V, then $\mathbf{E}$ must live in the dual space of V.

Iterative Methods

When the unknown count is large, the matrix $\overline{\mathbf{L}}$ is never generated. Instead, iterative matrix-free methods in numerical linear algebra are used to solve the matrix equation. Iterative methods can be made matrix-free because one needs only to write a code to produce the result

$$\overline{\mathbf{L}} \cdot \mathbf{a} \tag{1.131}$$

namely, the result of the action of a matrix on a vector, but not the matrix itself. For sparse matrices, this action can be effected with $\mathcal{O}(N)$ operations. For dense matrices, it costs $\mathcal{O}(N^2)$ to effect this action. However, various fast solvers have been developed that allow this action to be effected in $\mathcal{O}(N)$ or $\mathcal{O}(N \log N)$ operations [115, 219–224].

When it comes to solve a matrix equation iteratively, numerical linear algebra plays an important role [225]. The condition number and the distribution of the eigenvalues of the system matrix determine the convergence rate of iterative methods. The Krylov subspace method [225–228] is a popular way to understand the convergence rate of iterative methods. In this method, one finds the solution to the matrix equation by finding the best fit solution in a subspace called Krylov subspace

$$\mathcal{K}^K(\overline{\mathbf{L}}, \mathbf{r}_0) = \left\{ \mathbf{r}_0, \overline{\mathbf{A}} \cdot \mathbf{r}_0, \overline{\mathbf{A}}^2 \cdot \mathbf{r}_0, \cdots, \overline{\mathbf{A}}^{K-1} \cdot \mathbf{r}_0 \right\} \tag{1.132}$$

where $\mathbf{r}_0 = \mathbf{h} - \overline{\mathbf{L}} \cdot \mathbf{a}_0$ where a_0 is the initial guess to the solution. Thus, $\mathbf{r}_0$ is the residual error in the first iteration in the method. The above Krylov subspace can be generated by performing a $K-1$ matrix-vector product with the matrix $\overline{\mathbf{L}}$ on the vector $\mathbf{r}_0$. The method finds the optimal solution at the Kth iteration, $\mathbf{a}_K$, by letting it be $\mathbf{a}_K = \mathbf{a}_0 + \mathbf{z}_K$, such that $\mathbf{z}_K \in \mathcal{K}^K(\overline{\mathbf{L}}, \mathbf{r}_0)$. By doing so, the residual at the Kth iteration is

$$\begin{aligned} \mathbf{r}_K &= \mathbf{h} - \overline{\mathbf{L}} \cdot \mathbf{a}_K \\ &= \mathbf{r}_0 - \overline{\mathbf{L}} \cdot \mathbf{z}_K \in \left\{ \mathbf{r}_0, \overline{\mathbf{L}} \cdot \mathbf{r}_0, \overline{\mathbf{L}}^2 \cdot \mathbf{r}_0, \cdots, \overline{\mathbf{L}}^K \cdot \mathbf{r}_0 \right\} \\ &\in \mathcal{K}^{K+1}(\overline{\mathbf{L}}, \mathbf{r}_0). \end{aligned} \tag{1.133}$$

In other words, the residual error that is to be minimized can be written as

$$\begin{aligned} \mathbf{r}_K &= \mathbf{r}_0 + \alpha_1 \overline{\mathbf{L}} \cdot \mathbf{r}_0 + \alpha_2 \overline{\mathbf{L}}^2 \cdot \mathbf{r}_0 + \cdots + \alpha_K \overline{\mathbf{L}}^K \cdot \mathbf{r}_0 \\ &= \sum_{k=0}^{K} \alpha_k \overline{\mathbf{L}}^k \cdot \mathbf{r}_0 = P_K^o(\overline{\mathbf{L}}) \cdot \mathbf{r}_0 \end{aligned} \tag{1.134}$$

where $P_K^o(x)$ is an optimal polynomial with $\alpha_0 = 1$, and the coefficients $\alpha_1, \cdots, \alpha_K$ are chosen to minimize $\mathbf{r}_K$.

To understand the convergence of the above matrix polynomial, we expand $\mathbf{r}_0$ in terms of the left and right eigenvectors of $\overline{\mathbf{L}}$, or [229–233]

$$\mathbf{r}_0 = \sum_{n=1}^{N} \mathbf{v}_n \left(\mathbf{w}_n^t \cdot \mathbf{r}_0 \right) \tag{1.135}$$

where $\mathbf{v}_n$ is the right eigenvector, while $\mathbf{w}_n$ is the left eigenvector. They can be shown to share the same set of eigenvalues, and are mutually orthogonal. That is $\mathbf{w}_n \cdot \mathbf{v}_{n'} = \delta_{nn'}$. Therefore,

$$\overline{\mathbf{L}} \cdot \mathbf{v}_n = \lambda_n \mathbf{v}_n \tag{1.136}$$

$$\mathbf{w}_n^t \cdot \overline{\mathbf{L}} = \lambda_n \mathbf{w}_n^t \tag{1.137}$$

where the eigenvalue λ_n can be complex. We can easily show that

$$\overline{\mathbf{L}}^k \cdot \mathbf{r}_0 = \sum_{n=1}^{N} \lambda_n^k \mathbf{v}_n (\mathbf{w}_n^t \cdot \mathbf{r}_0). \tag{1.138}$$

By substituting the above into (1.134), we have

$$\mathbf{r}_K = \sum_{n=1}^{N} \sum_{k=0}^{K} \alpha_k \lambda_n^k \mathbf{v}_n (\mathbf{w}_n^t \cdot \mathbf{r}_0) = \sum_{n=1}^{N} P_K^o(\lambda_n) \mathbf{v}_n (\mathbf{w}_n^t \cdot \mathbf{r}_0) \tag{1.139}$$

$$\mathbf{r}_K = \sum_{k=0}^{K} \overline{\mathbf{L}}^k \cdot \mathbf{r}_0 = \sum_{n=1}^{N} P_K^o(\lambda_n) \mathbf{v}_n (\mathbf{w}_n^t \cdot \mathbf{r}_0) \tag{1.140}$$

$$\mathbf{r}_0 = \mathbf{h} - \overline{\mathbf{L}} \cdot \mathbf{a}_0 \tag{1.141}$$

$$\mathbf{r}_1 = \mathbf{r}_0 + \alpha_1' \overline{\mathbf{L}} \cdot \mathbf{r}_0. \tag{1.142}$$

It is seen that the residual error in the Kth iteration is proportional to the optimal polynomial $P_K^o(\lambda_n)$ at the N eigenvalues λ_n. If $K = N$, a polynomial can be found such that $P_K^o(\lambda_n)$ is exactly zero at all these eigenvalues, meaning that the residual error is zero. When K is less than N, if the eigenvalues are clustered together on the complex plane, the residual error can still be made small. Since the polynomial has a value such that $P_K^0(0) = 1$, if there are many eigenvalues near the origin, it is difficult to fit the polynomial so that it is small close to the origin.

1.10 Fast Algorithms

Due to the increased workload of numerical methods when the scatterer becomes electrically large (large compared to wavelength), there has been much interest in fast methods to solve the ensuing matrix equations derived from Maxwell's equations.

Differential equation solvers naturally give rise to sparse matrices. The solution process can be made matrix-free easily. The down side is the existence of grid dispersion error whose deleterious effect increases with the size of the problem [234]. In this case, the effort has been in reducing the unknown counts as the unknowns are fields that permeate space pervasively. Also, accuracy improvement is necessary to mitigate grid dispersion errors. A higher-order method is used to reduce unknown counts, but with the peril of reducing the sparsity of the matrix. Also, the unknowns in differential equation solvers grow with the volume of the simulation region, and hence suffer from the cruelty of dimensionality as we shall discuss later.

Methods to invert the finite-element matrix directly have been studied extensively. Because of the sparsity of the FEM matrix for differential equations, it can be directly inverted in $\mathcal{O}(N^{1.5})$ in 2D and $\mathcal{O}(N^2)$ in 3D by the nested dissection ordering method

[235, 236]. Moreover, for domains where the shapes are oblong, frontal method proves popular in inverting the finite-element matrix [237, 238].

Another hot area of research in differential equation solvers is the design of absorbing boundary conditions (ABCs) [239–250]. Berenger's perfectly matched layers (PMLs) [251], and coordinate-stretching PMLs [252] have become highly popular among numericists in this area. Coordinate-stretching PMLs draw inspirations from the area of matched asymptotics where coordinate stretching is used to emphasize a certain physics of the problem [252]. Furthermore, an anisotropic medium PML has been developed [253]. This topic has spurred the interest of many researchers [254–261].

The coordinate stretching PML is quite general, as coordinate stretching can be applied to any derivative operators. So it can be applied to different types of PDEs, and hence, it has allowed the generalization of PMLs to curvilinear coordinates, and other equations of physics easily [261–264].

Most ABCs are not perfect; a rigorous, perfect ABC is actually a boundary integral equation truncation of a differential equation solution domain solved by FEM or FDM [174–178]. This has been avoided in the past because boundary integral equations give rise to dense matrix systems that are expensive to solve and store. However, the advent of fast algorithms has changed the landscape [115, 265]. Boundary integral equations accelerated by fast algorithms have been used to reduce the domain size of finite element methods since they act as absorbing boundary conditions.

1.10.1 Cruelty of Computational Complexity

Integral equations were difficult and expensive to solve in the past. They usually give rise to dense matrix system requiring $\mathcal{O}(N^3)$ computer time and $\mathcal{O}(N^2)$ memory to solve. These computational complexities are just too unwieldy for large problems. However, advances in fast methods have eliminated this bottleneck by reducing the size of α in the exponent in $\mathcal{O}(N^\alpha)$ in these computational complexity scalings.

First, direct solvers such as RATMA (recursive aggregate T matrix method) [266, 267] have been designed to solve integral equations in $\mathcal{O}(N^2)$ in 2D, and $\mathcal{O}(N^{7/3})$ in 3D, faster than the $\mathcal{O}(N^3)$ time needed to solve a dense matrix system. A method similar to the nested dissection ordering method, called the nested equivalence principle algorithm (NEPAL) [268, 269] can directly solve a matrix in $\mathcal{O}(N^{1.5})$ in 2D and $\mathcal{O}(N^2)$ in 3D. When the object is oblong, it can be reduced to $\mathcal{O}(N\log^2 N)$ [270, 271]. Recently, some direct solver work has been done to produce low-complexity solutions to scattering problems [272, 273].

For iterative solvers, a sleuthing of the method can reduce the computational time complexity to $\mathcal{O}(N\log N)$. Moreover, many of these methods can be made matrix-free so that only the unknown vector, and only the diagonal part of the matrix, need to be stored. Hence, the memory requirements can be reduced to $\mathcal{O}(N)$. There are essentially three popular methods to speed up the solutions of integral equations.

- Fast Fourier transform (FFT) based techniques: In this technique, the integral operator is cast into a form that resembles a convolutional integral as much as possible. Then FFT is used to expedite the convolution enabling its evaluation of a matrix-vector products in $\mathcal{O}(N\log N)$ operations. For surface scatterers with a lot of zero-padding in the FFT, the complexity is worse and is not optimal [274–285].
- Matrix compression based techniques: the method of moments matrix that follows from solving integral equations is of low rank. The reasons for low rank are two: over

discretization of mesh density; and far interactions between currents on the object. The Nyquist sampling theorem necessitates the discretization of at least two points per wavelength in order to capture the oscillatory nature of the currents on an object. However, oftentimes, discretization far above the Nyquist sampling rate is used. In this case, redundancies are created in the unknown counts, and the ensuing matrix system is of lower rank than the unknown counts. Such matrix systems can be easily compressed using matrix compression techniques, such as wavelets [286–295], adaptive cross approximations, simple fast multipole, etc. [219, 296–305]. Interaction matrices with low-ranks due to far interactions are harder to compress. They cannot be compressed beyond the Nyquist barrier [291], even though ray-physics based methods have been used to compress them further beyond the Nyquist barrier [306–310]. When wave physics is involved, the only viable way to compress the matrices for far interaction efficiently is the multipole based methods [222, 223]. However, recent advances in direct solvers [272, 273] herald new possibility for further matrix compression.

- Multipole-based methods: the simple fast multipole method can expedite the solution involving circuit physics or Laplace solutions very easily [219, 311, 312]. Extension of such an algorithm for wave physics cases has been proposed, but its verbatim use for wave physics cases does not work [313, 314]. The only viable method of expediting the wave physics case is the multilevel fast multipole algorithm [222, 223], where anterpolation and interpolation [315] between levels are added. The algorithm is a tree-based algorithm. The matrices for far interactions are analytically diagonalized [313, 316, 317] on paper rather than by algebraic or numerical means. The number of diagonal elements needed is proportional to the rank of the matrices. The reason is that the ranks of the matrices increase with the group size compared to wavelength. Near the upper levels of the inverted tree, the ranks of the matrices are higher, while near the lower levels, their ranks are lower.

  To be of the varying ranks of these matrices, their dimensions are different for different levels even if they are diagonalized. Hence, anterpolation and interpolation are needed between levels due to their different ranks, very much in the spirit of multilevel multigrid schemes [318]. With this augmentation, the multilevel fast multipole algorithm provides optimal complexity of performing a matrix-vector product with $\mathcal{O}(N \log N)$ complexity. It is with this algorithm that wave physics problems with tens of millions to hundreds of millions, and over a billion unknowns have been solved [115, 319–324].

The key to the multilevel fast multipole algorithm is the factorization of the the matrix element L_{ij} when elements i and j are far apart. Then one can express [115, 223]

$$L_{ij} = \tilde{\mathbf{V}}^t_{f,i,i_1} \cdot \overline{\mathbf{I}}^t_1 \cdot \tilde{\overline{\boldsymbol{\beta}}}_{i_1,i_2} \cdot \overline{\mathbf{I}}^t_2 \cdots \tilde{\overline{\boldsymbol{\beta}}}_{i_N,L} \cdot \tilde{\overline{\boldsymbol{\alpha}}}_{LL'} \cdot \tilde{\overline{\boldsymbol{\beta}}}_{L',j_N} \cdots \overline{\mathbf{I}}_2 \cdot \tilde{\overline{\boldsymbol{\beta}}}_{j_2,j_1} \cdot \overline{\mathbf{I}}_1 \cdot \tilde{\mathbf{V}}_{s,j_1,j}. \tag{1.143}$$

The factorization allows a matrix-vector product to emulate a multilevel telephone network connection, as shown in Figure 1.8, where the number of telephone lines can be greatly reduced compared to direct telephone connection. In the above, the $\tilde{\overline{\boldsymbol{\alpha}}}$ and $\tilde{\overline{\boldsymbol{\beta}}}$ matrices are diagonal, while the $\overline{\mathbf{I}}$ matrices are quasi-diagonal interpolation and anterpolation matrices that are not square. Hence, the storage requirement of the factorized matrices is small, and they can be reused in a tree algorithm. Other factorizations do not lead to $\mathcal{O}(N \log N)$ complexity. There has been much activity in this field as well [325–336].

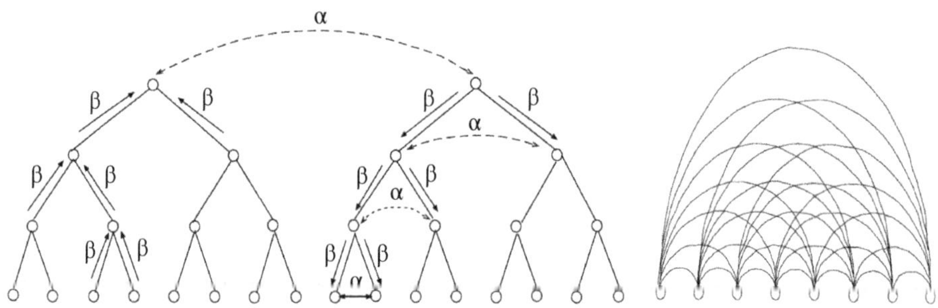

Figure 1.8 The factorization of the matrix L_{ij} in (1.143) allows the matrix-vector product to emulate a telephone network connection (left) [115]. The direct line connections between telephones (right) require a large number of lines.

1.10.2 Curse of Dimensionality

If the discretization density is in terms of n per wavelength, it implies that n is linearly proportional to $k = 2\pi/\lambda$. The fact that the total $N = n^d$ where d is the dimensionality of the problem can make N exorbitantly large if d is 3. For some problems, like the finite difference problem, the grid dispersion error makes the frequency scaling of n as $n = C_0k^{1.5}$ [234]. Consequently, in 3D,

$$N = Ck^{4.5} \tag{1.144}$$

which is a rather pathological scaling.

One way to reduce the grid dispersion error is to use higher-order methods. Another way to reduce the dimensionality of the problem is to combine it with analytic methods. Integral equation techniques generally can reduce a volume problem to a surface problem. The use of the closed-form Green's function as a propagator reduces the grid dispersion error, as it is an exact propagator.

Yet another way to reduce the dimensionality of the problem is to use the semi-analytic, or numerical mode-matching method [337–339]. In this method, the problem is decomposed into the propagation dimension and the transverse dimensions. The transverse dimension is usually one less than the original dimensions of the problem. The problem is solved numerically in the traverse dimension, and the solution is analytically propagated in the propagation dimension (see Figure 1.9). These methods are variously known as the R-matrix method, rigorous coupled wave analysis, numerical mode matching, the recursive Green's function method, and the contact block recursion approaches [337, 338, 340–344]. They are also pervasively used in nano-electronic simulation involving Schrödinger equations. The use of the FEM or FDM to solve the transverse dimension can give rise to an increase in speed of orders of magnitude compared to a full 3D method [338, 345].

1.10.3 Multiscale Problems

As the complexity of a geometry increases, there is a need for multiscale modeling of complex geometry. Certain parts of the geometry may need discretization sizes much smaller than wavelengths, while other parts may be modeled with much larger discretization sizes. This can happen in the modeling of complex antennas mounted on

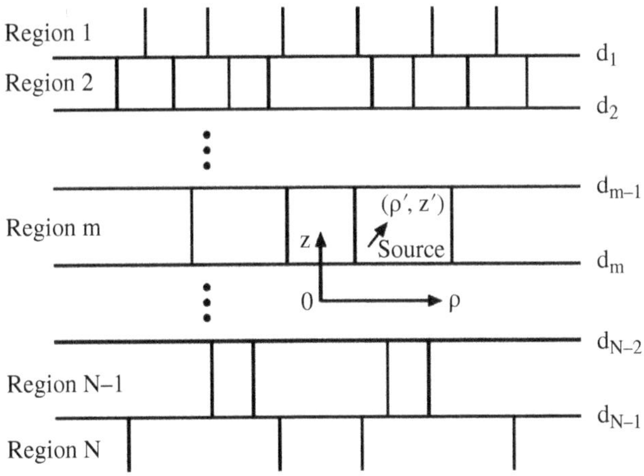

Figure 1.9 A multiregion problem can be solved semi-analytically using numerical mode matching, where the problem is solved numerically in the transverse dimension, but analytically in the longitudinal *z* direction [76].

a platform or inside a computer. The modeling of the fineries of an antenna structure is important in capturing the circuit physics necessary for the functioning of the antenna. This part may need fine meshes for its modeling. On the other hand, the interaction of the radiation field from the antenna with the rest of the platform does not involve circuit physics, and hence can be modeled with much coarser meshes.

The concurrent existence of disparate mesh sizes can be dealt with using a number of methods. In the case of multilevel FMM (fast multipole method), tree structures of different depths have been proposed for its solution, or different multiscale methods have been developed [210, 346–349].

However, the disparate mesh sizes often give rise to ill-conditioned matrix systems. Hence, a divide and defeat (DAD) scheme is more expedient to overcome such problems [265, 350]. For instance, the regions of fine meshes can be solved first with a different method, and later on sewn with the other regions. In the FEM, the domain decomposition method (DDM) [265] is best suited to solve such problems. For integral equation solvers, the equivalence principle algorithm (EPA) [350] can be used to overcome the ill-conditioning of the matrix system due to disparate mesh sizes. More recently, well-conditioned formulations such as the augmented electric field integral equation [351] and the Calderon multiplicative preconditioner have been designed to overcome such problems [209, 210].

1.10.4 Fast Algorithm for Multiscale Problems

When the discretization sizes are uneven, some parts of the object can be in the circuit-physics regime, while the long-range interactions between different parts of the object are in the wave-physics regime. A simple fast multipole algorithm that works only in the circuit physics (low frequency), or the fast multipole algorithm that captures only wave physics, is insufficient to model such problems. For these problems, the mixed-form fast multipole algorithm has been specially designed to overcome these problems [352–355]. More importantly, a recent advance has been made in developing a low-complexity fast multipole algorithm that is valid from the static to the wave-physics regime [356].

1.10.5 Domain Decomposition Methods

A popular way to attack large scale problem is via the DAD scheme by DDMs. In this scheme, a large problem is subdivided into smaller subproblems. Then each subproblem is solved using a method that requires less resource. DDM has been applied first to differential equation solvers. Examples of such solvers are the frontal algorithms [357], Schwarz algorithms [358], as well as the nested dissection ordering method [235].

The precursor to domain decomposition method is the diakoptic approach where a complex problem is broken into simpler subproblems [359]. The subproblems are easily solved and pieced together to obtain the solution to the larger problem [360–365]. This subsequently gives rise to the DDM. The domain of numerical simulation is systematically divided into smaller subdomains, and the problem is solved for each subdomain using a numerical method, and usually, the FEM. This allows for ease of parallel computing. Then the solution for the entire domain is sewn together by imposing boundary conditions at the interface of the domain. Overlapping domains are also used, called the overlapping Schwarz method [358]. When a condition is imposed at an interface, it is also called the finite element tearing interconnect method [265, 366–388]

Other DDM methods include the frontal algorithm [357], and the nested dissection ordering (NDO) algorithm [235]. Recently, an integral equation approach of domain decomposition has been proposed using Huygens' equivalence principle [268, 269, 389–397]. This is particularly useful when each domain contains "white space" due to free space or a homogeneous medium. A differential equation solver would have wasted many unknowns by having to discretize and represent such white space with many unknowns. An integral equation solver, however, would propagate the field through such white space using the closed-form Green's function. The DDM for an integral equation was first proposed in 1993 known as the NEPAL [268, 269]. It is similar to the NDO for differential equation solvers. Its advantage is that there is little grid dispersion error, as in differential equation solvers, and the radiation boundary condition is automatically satisfied due to the used Green's function. Subsequently, a simpler version of it was used to solve practical problems involving complex structures. The simple version is known as the EPA [389, 390].

While the NEPAL does a multilevel subdivision of a domain into smaller subdomains, the EPA does a single-level division of a domain into smaller subdomains. The EPA is particularly useful for solving multiscale problems. Interaction between subdomains is calculated using surface equivalence currents on equivalence surfaces. A scattering operator (also called equivalence principle operator) is defined to encapsulate the scattering physics of an object within the operator. The mesh for capturing the physics of surface equivalence current is usually coarser than the fine mesh needed to capture the physics of a multiscale structure. Hence, the system matrix equation is better conditioned, and easier to solve, by iterative methods.

The EPA has been applied to solve for the solution of a microstrip patch antenna mounted on a car or a large ground plane. The EPA allows for solutions of large problems when it is accelerated with the multilevel fast multipole algorithm [115, 223, 395]. Hence, a problem size with several million unknowns has been solved on a single CPU personal computer. More recently, a version of the EPA known as the augmented-EPA (A-EPA) has been developed to preclude low-frequency breakdown [391, 392]. Consequently, the A-EPA has been used to solve low-frequency micro-chip problems where the structures are much smaller than the wavelength. In this regime, both circuit physics and wave physics have to be captured concurrently. A more efficient version of the EPA using a single source rather than both electric and magnetic currents has also been developed [393]. Other methods that can be classified as an integral equation DDM are: (1) the self-box inclusion preconditioner method, where a parallel code has been used to solve problems with several million unknowns [115, 223, 394, 395]. (2) The generalized impedance boundary condition where the scattering solution is encapsulated within an impedance operator [396]. Some problems involving nano-optics have been solved using such a method [397].

1.11 High Frequency Solutions

When the frequency is extremely high, electromagnetic problems are not easily solved with numerical methods. Most numerical methods are based on the discretization of complex objects into smaller pieces, each of which is much smaller than the wavelength. Hence, the number of discretizations or unknowns grows exorbitantly as the frequency becomes high.

In the high frequency limit, the dominant contribution comes from specular reflections and diffractions off critical points of an object. Ray tracing technique can be used to find these dominant contributions. Moreover, as the wavelength becomes short, the phase of the wave becomes random and unimportant. In the infra-red (IR) and optical regime, the phase of the field is often neglected.

However, in the microwave regime, there is some interest in keeping track of the phase of the wave. The shooting-and-bouncing ray method is popular in seeking solutions in this regime [398]. For complex targets, the current on the object is approximated with the physical optics approximation. Then the radiation integral is evaluated to find the scattered field. But these integrals are highly oscillatory in nature. One pressing need is the efficient evaluation of highly oscillatory integrals making the workload frequency-independent. This has been a subject of avid study by a number of researchers [100–103, 105, 162–164, 398–414]. Also, efficient numerical methods have also been developed to enable them to solve higher frequency problems [115, 313, 315, 415].

Another approach is to solve the integral equation directly using techniques that are frequency-independent. This has also been a subject of avid study [416–419].

1.12 Inverse Problems

Since an electromagnetic field has been used in remote sensing, geophysical prospection, microwave imaging, and optical coherence tomography, inverse problems are an important aspect of this field. Inverse problems in 1D, surprisingly, are driven by incredible tour-de-force mathematics. Many inverse problems are not solvable, but the 1D inverse problems can be tackled with many mathematical techniques, such as the Gelfand-Levitan-Marchenko method [76, 420].

In multidimensions, there is no closed form solution for the inverse problems except when linear approximations are made. If only single scattering is assumed, which is also called the Born approximation, one can show that the scattered field data in the far field are related to the Fourier transform of the object. Hence, one can perform a Fourier inverse transform to obtain the object. This is used in synthetic aperture radar (SAR) with the production of marvelous pictures of the Earth surface from remotely sensed data. Diffraction tomography is another example of linear inverse problem that can be performed rapidly. Since FFT can be used in the reconstruction, the complexity of the image reconstruction is proportional to $O(N \log N)$ [421–424].

In general, inverse problems in multidimensions have no closed-form solutions, and are usually solved as an optimization problem, where the forward problem is solved with a scatterer. Then a scatterer that generates the best fit data is searched and selected as the solution. Such a method generally gives rise to non-uniqueness of solutions. The way to overcome the non-uniqueness is to arrive at the minimum norm solution [76, 425].

This is very much in the spirit of finding the Fourier inverse transform. If one knows the Fourier transform of an object up to a finite spectrum, the inverse Fourier transform of the object is indeterminate. This is because if we can add an object with spectrum outside the finite spectrum, it will produce the same Fourier transform within the same window. The way to remove the non-uniqueness is to define a minimum norm inverse that is equivalent to pad the data outside the finite window with zero values.

When multiple scattering is accounted for in inverse problems, the multiple scattering (nonlinear) solution can be solved by minimization of the cost functional. A cost functional can be set up and the minimum of the cost functional can be found by a gradient search approach. The gradient in multidimensional space, called the Frechet derivative, can be found by the distorted Born approximation, or other Gauss-Newton methods [426–441]. The Gauss-Newton method is the general name for a nonlinear problem solver, whereas the distorted Born approximation relates this multiple scattering (nonlinear) method directly to the inverse problems.

1.12.1 Distorted Born Iterative Method

By the above approaches, it can be shown that the calculation of the Frechet derivative is equivalent to the solving of the forward scattering problem. Hence, if the forward problem can be solved rapidly, it portends a rapid solution to the inverse problem. With the advent in computer technologies, there is hope that the multiple scattering (nonlinear) inverse problem can be solved rapidly and made practical [442].

Since inverse scattering involves a volume scatterer, we can write the volume integral equation of the Green's function as

$$\mathbf{E}(\mathbf{r}) = \mathbf{E}_{\text{inc}}(\mathbf{r}) + \int_V \mathrm{d}\mathbf{r}' \, \overline{\mathbf{G}}_0(\mathbf{r}, \mathbf{r}') \cdot \left[k^2(\mathbf{r}') - k_0^2\right] \mathbf{E}(\mathbf{r}'), \tag{1.145}$$

where k_0^2 is a constant wavenumber for the homogeneous background. Here, $\overline{\mathbf{G}}_0(\mathbf{r}, \mathbf{r}')$ is a homogeneous medium Green's function with wavenumber k_0. Hence, $\mathbf{E}_{\text{inc}}$ is the field in the absence of the scatterer, but $\mathbf{E}(\mathbf{r})$ is the total field in the presence of the scatterer. By letting

$$\mathbf{E}(\mathbf{r}) = \overline{\mathbf{G}}(\mathbf{r}, \mathbf{r}'') \cdot \mathbf{a} \tag{1.146}$$

$$\mathbf{E}_{\text{inc}}(\mathbf{r}) = \overline{\mathbf{G}}_0(\mathbf{r}, \mathbf{r}'') \cdot \mathbf{a} \tag{1.147}$$

(1.145) can be written as

$$\overline{\mathbf{G}}(\mathbf{r},\mathbf{r}'') = \overline{\mathbf{G}}_0(\mathbf{r},\mathbf{r}'') + \int_V \mathrm{d}\mathbf{r}'\, \overline{\mathbf{G}}_0(\mathbf{r},\mathbf{r}') \cdot \left[k^2(\mathbf{r}') - k_0^2\right] \overline{\mathbf{G}}(\mathbf{r}',\mathbf{r}''). \tag{1.148}$$

The above can be written as an operator equation

$$\overline{\mathcal{G}} = \overline{\mathcal{G}}_0 + \overline{\mathcal{G}}_0 \cdot \overline{\mathcal{O}} \cdot \overline{\mathcal{G}} \tag{1.149}$$

where $\overline{\mathcal{O}}$ is the operator representation of $k^2(\mathbf{r}') - k_0^2$. The above can be solved formally

$$\overline{\mathcal{G}} = (\overline{\mathcal{I}} - \overline{\mathcal{G}}_0 \cdot \overline{\mathcal{O}})^{-1} \cdot \overline{\mathcal{G}}_0 \tag{1.150}$$

where $\overline{\mathcal{G}}$ is manifestly nonlinear with respect to $\overline{\mathcal{O}}$. If one lets

$$\overline{\mathcal{O}} = \overline{\mathcal{O}}_b + \delta\overline{\mathcal{O}} \tag{1.151}$$

$$\overline{\mathcal{G}} = \overline{\mathcal{G}}_b + \delta\overline{\mathcal{G}} \tag{1.152}$$

where $\overline{\mathcal{G}}_b = (\overline{\mathcal{I}} - \overline{\mathcal{G}}_0 \cdot \overline{\mathcal{O}}_b)^{-1} \cdot \overline{\mathcal{G}}_0$, then it can be easily shown that

$$\delta\overline{\mathcal{G}} = \overline{\mathcal{G}}_b \cdot \delta\overline{\mathcal{O}} \cdot \overline{\mathcal{G}}_b. \tag{1.153}$$

The above is the functional or Frechet derivation that can be used in conjunction with a gradient search method such as the Gauss-Newton method. However, it is also a distorted Born approximation for finding the perturbed Green's operator $\delta\overline{\mathcal{G}}$. We have called the iterative method of solving the inverse problem the distorted Born iterative method (DBIM).

1.12.2 Super-Resolution Reconstruction

The use of the DBIM has been shown to yield super-resolution reconstruction of objects with subwavelength resolution (see Figure 1.10) [443–448]. The super-resolution can be ascribed to the fact that the data used for reconstruction have been embedded in the multiple scattering information. The high-resolution information of an object is buried in the evanescent wave spectrum of the scattered field. When single scattering is assumed, the evanescent wave remains localized to the vicinity of the scatterer. Hence, the far field has no super-resolution information of the object. However, when multiple scattering is accounted for, evanescent waves can be converted to propagating waves due to multiple scattering. The multiply-scattered propagating wave hence contains super-resolution information on the object. Thus, if a reconstruction algorithm includes the multiple-scattering physics in the model, super-resolution reconstruction of an object is possible. Super-resolution phenomena have also been observed in [449, 450].

1.12.3 Super-Resolution and the Weyl-Sommerfeld Identity

The physics of super-resolution can be gleaned from the Weyl (1919) identity given as

$$\frac{\mathrm{e}^{\mathrm{i}k_0 r}}{r} = \frac{i}{2\pi} \iint_{-\infty}^{\infty} \mathrm{d}k_x \mathrm{d}k_y \, \frac{\mathrm{e}^{\mathrm{i}k_x x + \mathrm{i}k_y y + \mathrm{i}k_z |z|}}{k_z} \tag{1.154}$$

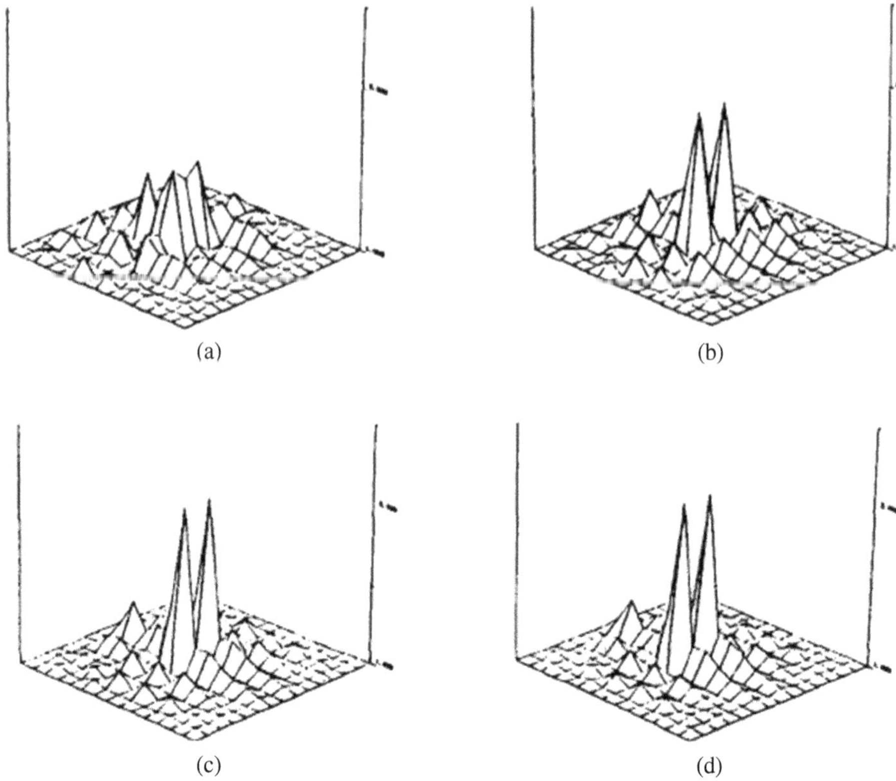

Figure 1.10 The use of multiple-scattering information in multiple scattering (nonlinear) inverse scattering algorithm allows the super-resolution reconstruction of two objects spaced 0.1λ apart [443].

where $k_x^2 + k_y^2 + k_z^2 = k_0^2$ or $k_z = (k_0^2 - k_x^2 - k_y^2)^{1/2}$. The above expression says that a point source generates a spherical wave that can be expressed as an integral summation of plane waves propagating in all directions. Notably, k_z becomes pure imaginary corresponding to an evanescent wave when $k_x^2 + k_y^2 > k_0^2$. If we view the above right-hand side as the Fourier transform of the left-hand side, the field is from a point source. It is clear that the Fourier data for $k_x^2 + k_y^2 > k_0^2$ cannot propagate to a large z location. For large z, the Fourier data that can propagate over long distances are within the Ewald circle $k_x^2 + k_y^2 < k_0^2$. Hence, a low-pass filtered version of the source field is accessible at long distances. This is the source of the Rayleigh limit for the resolution of a point source.

The bunches of plane waves only interfere constructively when they are almost parallel to each other. In other words, the plane wave bundles become rays in the long distance limit or high-frequency limit. If one is to collect the Fourier data with a lens of radius a and subtending an angle θ at the source location, the Fourier data are collected with a circular disk of radius $k_0 \sin\theta$, even smaller than the Eward circle aforementioned. Here, $\sin\theta$ is the numerical aperture of the lens. It is seen that the larger its numerical aperture, the more Fourier data the lens can collect, and hence, of the higher resolution the image that can be reconstructed.

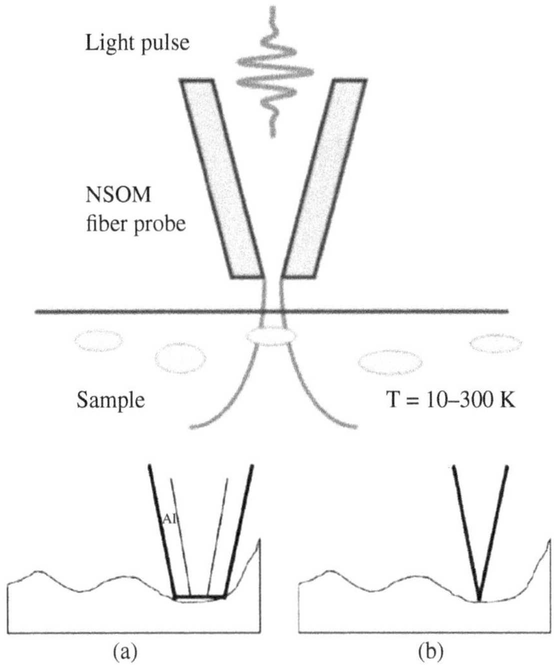

Figure 1.11 Since super-resolution information is buried in the evanescent field of the wave, near-field scanning optical microscopy (NSOM) exploits this fact by using subwavelength sharp tips to generate and sense evanescent field (courtesy of Wiki).

A space propagator for wave is a Fourier transformer, as indicated by the following formula

$$\phi(\mathbf{r}) = \int_V \mathrm{d}\mathbf{r}' g(\mathbf{r}, \mathbf{r}') s(\mathbf{r}') \approx \frac{\mathrm{e}^{\mathrm{i}k_0 r}}{4\pi r} \int_V \mathrm{d}\mathbf{r}' \mathrm{e}^{-\mathrm{i}\mathbf{k}\cdot\mathbf{r}'} s(\mathbf{r}') = \frac{\mathrm{e}^{\mathrm{i}k_0 r}}{4\pi r} \tilde{S}(\mathbf{k}) \tag{1.155}$$

where $\tilde{S}(\mathbf{k})$ is the Fourier transform of $s(\mathbf{r})$. A lens is a phase-shift provider so that after the wave field has passed through it, its Fourier transform by space creates the image of the original object.

If a lens has a numerical aperture of $\sin\theta$, the Fourier data are provided by a circle of radius $k_0 \sin\theta$ on the k_x-k_y Fourier space. A precise point source field cannot be created. It is interesting to note that

$$\left. \frac{\partial}{\partial z} \frac{\mathrm{e}^{\mathrm{i}k_0 r}}{r} \right|_{z=0} = \delta(x)\delta(y). \tag{1.156}$$

The above can be proved easily by using Weyl's identity. The space propagator takes the Fourier of the field at $z = 0$ and propagates to the lens location. The lens adds the appropriate phase shift to the wave field so that the second part of the space propagator essentially does a Fourier inverse transform of the wave field at $z = h$, which tries to reconstruct the wave field at $z = 0$. However, due to low-pass filtering and the finite numerical aperture of the lens, the point image reconstructed is

$$I(x, y) = \int_0^{k_0 \sin\theta} \mathrm{d}k_\rho k_\rho J_0(k_\rho \rho). \tag{1.157}$$

The above can be evaluated to

$$I(\rho) = \frac{J_1(k_0 \sin\theta)}{k_0 \sin\theta} \tag{1.158}$$

which is also called the Airy disk function. It is the point-spread function of the low-pass filter system that will propagate a point image at $z = 0$ to the focal point of a lens.

1.13 Metamaterials

The study of metamaterials has become a hot topic in recent years mainly due to two recent concepts proposed by Pendry [451, 452]. One was the use of double-negative materials to obtain super-resolution. The second was to engineer a reflectionless body by cloaking so that the object is invisible to a field [453, 454].

While super-resolution was possible on paper, it was never realized to satisfaction. The high-resolution information of a point source is embedded in the evanescent spectrum of the field. However, the evanescent spectrum decays rapidly away from the source point, and hence this information is lost as one moves away from the source point. However, if one can amplify this evanescent spectrum, it can potentially recover the high-resolution information and give rise to super-resolution. This is the gist of the concept of super-resolution in a double-negative slab [455].

When a double negative material with $\mu = -\mu_0$ and $\epsilon = -\epsilon_0$ interfaces with a vacuum, it can be shown that plasmonic-type resonances occur at the interface for all frequencies. For a double negative slab placed in vacuum, the plasmon resonances occur at both surfaces. The character of a surface plasmon mode is that it is exponentially decaying away from the surface. So when a point source is placed next to the slab, the evanescent spectrum of the point source will excite the surface plasmon resonance at the interface. The surface plasmon resonance at the first interface will excite the surface plasmon resonance at the second interface. The surface plasmon of the second interface, though exponentially decaying from the interface, is exponentially growing with respect to the first interface. Hence, within the slab, it will appear that the evanescent spectrum is exponentially growing, but it is actually due to resonance coupling between two surface plasmon modes. This amplification of the evanescent mode is the source of the proposed super-resolution in a double-negative slab.

However, this amplification process is a very fragile process. The evanescent field, by the time it reaches the first interface, could be in the noise floor of the system. When such an evanescent wave is amplified by resonance coupling, the noise content of the amplified field can be high. So in practice, even if such double-negative material exists in nature, the physics behind this amplification is highly fragile. Any system imperfection or loss could annul this physics [455, 456].

The second difficulty is to construct an artificial double negative material. This has been mimicked at microwave frequencies due to the longer wavelength and larger component sizes, but has been difficult to achieve at optical frequencies due to the shorter wavelengths, and potential loss.

Regarding the second cloaking phenomenon, it has been long observed that the material properties described by μ and ϵ are intimately related to the metric of space [457]. This is particularly clear when Maxwell's equations are written in differential forms [458]. The metric of space and material properties are inseparable.

In Cartesian coordinates in free space, a plane wave can propagate without reflections. However, one can change the metric of space. The wave will propagate through a distorted space with respect to the Cartesian coordinates, and yet it can propagate without reflections. Hence, space can be deformed. Since metric and material properties are multiplied together, one can distort space with changing material properties. For instance, a point can be morphed into a sphere. In the original space, a point is invisible to a plane wave. In the distorted space, the wave will travel around the sphere without reflection. Hence, such material can be used to make an object invisible. However, one needs to engineer materials with μ and ϵ ranging from zero to infinity in order to be able to distort space fully [459, 460].

The above two ideas have given rise to a flurry of activity in this area, generally called the metamaterial or transformation optics area [461–463]. The promise of super-resolution and the possibility of cloaking to make objects invisible have induced a large amount of funding in this arena. Even though the original goals may not have been achieved, there has been much new knowledge created on the making of artificial materials due to the large amount of resource being invested.

1.14 Small Antennas

Antennas are used to radiate electromagnetic energy to free space, mainly for communication purposes. The first famous radiation experiment was done by Hertz [35] to confirm the wave nature of an electromagnetic field. One can model a Hertzian dipole current as

$$\mathbf{J}(\mathbf{r}) = I\ell\delta(\mathbf{r})\hat{z} \tag{1.159}$$

where $\hat{z}$ is a unit vector pointing in the z direction. The field produced by this current source is

$$\mathbf{E}(\mathbf{r}) = \mathrm{i}\omega\mu\left(\bar{\mathbf{I}} + \frac{\nabla\nabla}{k^2}\right)\cdot\hat{z}I\ell\frac{\mathrm{e}^{\mathrm{i}k|\mathbf{r}|}}{4\pi|\mathbf{r}|}. \tag{1.160}$$

In the far field, when $kr \to \infty$, the above can be approximated by replacing $\nabla = ik\hat{r}$ where $\hat{r}$ is a unit vector pointing in the r direction. For a Hertzian dipole, $I = \mathrm{i}\omega Q$ where Q is the charge on the dipole. Consequently,

$$\mathbf{E}(\mathbf{r}) \approx -\omega^2\mu(\hat{\theta}\hat{\theta} + \hat{\phi}\hat{\phi})\cdot\hat{z}Q\ell\frac{\mathrm{e}^{\mathrm{i}k|\mathbf{r}|}}{4\pi|\mathbf{r}|}. \tag{1.161}$$

It is seen that the radiation field is proportional to ω^2 because it is proportional to the acceleration of the charge Q. Strangely enough, a DC current loop does not radiate even though the electrons carrying the current are traveling in a circle, and hence, experiencing centrapetal acceleration. The reason is because of charge neutrality: for every electron carrying the current, a positive charge is present to neutralize the electron charge, but an electron alone traveling in a circle will radiate.

The effective aperture of an antenna is given by

$$A_{eff} = G\frac{\lambda^2}{4\pi} \tag{1.162}$$

where G is the antenna gain, and λ is the wavelength of the operating frequency. For small antennas, the above formula gives the erroneous concept that the aperture of the antenna becomes larger without a bound as the wavelength increases. This is incorrect, because as the frequency lowers, the radiation resistance of the antenna becomes smaller. Finally, copper loss or material loss starts to dominate in the antenna, precluding the effective aperture from increasing without bound. In other words, the gain of the antenna drops as the frequency lowers.

A small antenna like a Hertzian dipole is a poor radiator because the dipole moment $Q\ell$ of the antenna is small. This dipole moment can be enlarged by enlarging the size of the dipole. Since a Hertzian dipole is capacitive, one can make the antenna resonate with an inductor to drive more current, and hence charge into the antenna to make it radiate more. Alternatively, one can make the dipole into a half-wave dipole so that it can self resonate to enhance the current on the antenna.

Due to the growth in the cellular phone industry, antennas have to be designed to be as inconspicuous as possible. Hence, the field of small antennas has been in vogue [464–472].

An antenna can be made into a resonating structure to enhance its radiation efficiency, such as in half-wave dipole, microstrip patch, etc. A Hertzian dipole is a small antenna, but with poor radiation efficiency because of its non-resonating nature. An inductor can be used to resonate the antenna and hence increase the current amplitude on the antenna, but at the peril of increasing loss coming from the inductor as well as increasing the Q (quality factor) of the antenna.

The Q of an antenna is defined to be [473]

$$Q = 2\pi \frac{\text{energy stored}}{\text{energy radiated per cycle}} = \frac{f_r}{\Delta f} \tag{1.163}$$

where f_r is the resonance frequency while Δf is the bandwidth. Hence, a high-Q antenna has a narrow bandwidth, which is undesirable. The Q of the antenna is also reflective of its bandwidth. A small dipole has high Q because it has much stored energy in the near field, and radiates little. The fundamental limit for the Q of a small antenna has been derived by Chu in [474]. That and others [469, 475] have improved on the tightness of the bound derived by Chu.

The alternative has been to build inductors into the antenna, which also contribute to the radiation of the antenna. This has been the concept of many small antenna designs. The resonating structure has been made smaller by using meandering line to reduce the size of the antenna. It has also exploited the inductor as radiating magnetic dipole in addition to having a radiating electric dipole. In the case of a microstrip patch, a simple microstrip patch resonator requires about half-wavelength for its resonance. Attempts have been made by using tortuous lines on a patch to increase the patch inductance so as to achieve a resonance at a lower frequency [470]. Also, one has used metamaterials to improve the efficiency for small antennas [476].

1.15 Conclusions

Electromagnetic theory and Maxwell's equations were completed in 1864. They are over 150 years old now, but their usefulness in science and technology has not diminished, as they are fundamental to many engineering technologies, especially electrical engineering technologies. From the very beginning, electrical technology grew because of the need for power transfer and communication. The importance of these two needs has not diminished in the modern world, but more technologies have emerged, such as wireless communication, remote sensing, and emerging applications in renewable energies and wireless power transfer.

The onset of quantum mechanics brings new possibilities to electromagnetic technologies in the area of nano-optics, quantum optics, quantum communications, cryptography, and computing. All these systems use electromagnetic technologies, old and new. Therefore, electromagnetics will remain an important area of study, impacting many technologies for many years to come.

Bibliography

1 Faraday, M. (1843). On static electrical inductive action. *Philos. Mag.* 22 (144): 200–204. Faraday, M. (1839). *Experimental Researches in Electricity and Magnetism*. vol. 1. London: Taylor & Francis; vol. 2, London: Richard & John E. Taylor, 1844; vol. 3, London: Taylor and Francis, 1855. Reprinted by Dover, 1965. Also see Faraday, M. (1858). Remarks on static induction. *Proc. Roy. Inst.* (12 February 1858).

2 Ampére, A.M. (1823). Mémoire sur la théorie des phénoménes électrodynamiques. *Mem. Acad. R. Sci. Inst. France* 6: 228–232.

3 Gillmore, C.S. (1971). *Charles Augustin Coulomb: Physics and Engineering in Eighteenth Century France*. Princeton, NJ: Princeton University Press.

4 Gauss, C.F. (1841). General theory of terrestrial magnetism. In: *Scientific Memoirs*, vol. 2 (ed. R. Taylor), 184–251. London: R & J. E. Taylor.

5 Maxwell, J.C. (1865). A dynamical theory of the electromagnetic field. *Philos. Trans. R. Soc. London* 155: 459–512.

6 Maxwell, J.C. (1873). *A Treatise of Electricity and Magnetism*. 2 vols. Oxford: Clarendon Press. Also see Harman, P. M. (ed.) (1995). The Scientific Letters and Papers of James Clerk Maxwell, Vol. II, 1862–1873. Cambridge, UK: Cambridge University Press.

7 Woodruff, T.S. III (1984). *The Early Years of Radio Astronomy*. Cambridge, UK: Cambridge University Press.

8 Carlsmith, D. (2013). Particle Physics. *Upper Saddle.* River, NJ: Pearson Education, Inc.

9 Jackson, J.D. (1962). *Classical Electrodynamics*. New York, NY: Wiley.

10 Chew, W.C. and Sen, P.N. (1982). Dielectric enhancement due to electrochemical double layer: thin double layer approximation. *J. Chem. Phys.* 77 (9): 4683–4693.

11 Altman, M.D., Bardhan, J.P., White, J.K., and Tidor, B. (2008). Accurate solution of multi-region continuum biomolecule electrostatic problem using linearized Poisson-Boltzmann equation with curved boundary elements. *J. Comput. Chem.* 30 (1): 132–153.

12 Yoon, B.J. and Lenhoff, A.M. (1990). A boundary element method for molecular electrostatics with electrolyte effects. *J. Comput. Chem.* 11 (9): 1080–1086.

13 Yokota, R., Barba, L., Narumi, T., and Yasuoka, K. (2012). Scaling fast multipole methods up to 4000 GPUs. *Proceedings of the ATIP/A*CRC Workshop on Accelerator Technologies for High-Performance Computing: Does Asia Lead the Way?*

14 Wang, F., Yam, C.Y., Chen, G.H. et al. (2007). Linear scaling time-dependent density-functional tight-binding method for absorption spectra of large systems. *Phys. Rev. B* 76 (4): 045114(1)–045114(6).

15 Ionescu, A.M. and Riel, H. (2011). Tunnel field-effect transistors as energy-efficient electronic switches. *Nature* 479: 329–337.

16 Ferain, I., Colinge, C.A., and Colinge, J.-P. (2011). Multigate transistors as the future of classical metal-oxide-semiconductor field-effect transistors. *Nature* 479: 310–316.

17 del Alamo, J.A. (2011). Nanometre-scale electronics with III-V compound semiconductors. *Nature* 479: 317–323.

18 Phillips, J.R. and White, J.K. (1997). A precorrected-FFT method for electrostatic analysis of complicated 3-D structures. *IEEE Trans. Comput. Aided Des.* 16 (10): 1059 1072.

19 Zauharand, R.J. and Morgan, R.S. (1988). The rigorous computation of the molecular electric potential. *J. Comput. Chem.* 9 (2): 171–187.

20 Juffer, A.H., Botta, E.F.F., Van Keulen, B.A.M. et al. (1991). The electric potential of a macromolecule in a solvent: a fundamental approach. *J. Comput. Phys.* 97 (1): 144–171.

21 Kuo, S.-H., Tidor, B., and White, J. (2008). A meshless, spectrally accurate, integral equation solver for molecular surface electrostatics. *ACM J. Emerging. Technol. Comput.* 4 (2, Art. 6).

22 Holst, M.J. and Said, F. (1995). Numerical solution of the non-linear Poisson-Boltzmann: developing more robust and efficient methods. *J. Comput. Chem.* 16 (3): 337–364.

23 Baker, N., Holst, M., and Wang, F. (2000). Adaptive multilevel finite element solution of the Poisson Boltzmann equations II: refinement at solvent accessible surfaces in biomolecular systems. *J. Comput. Chem.* 21 (15): 1343–1352.

24 Lu, B., Zhou, Y., Holst, M., and McCammon, J. (2008). Recent progress in numerical methods for the Poisson-Boltzmann equation in biophysical applications. *Commun. Comput. Phys.* 3 (5): 973–1009.

25 Cai, W. (2013). *Computational Methods for Electromagnetic Phenomena: Electrostatics in Solvation, Scattering, and Electron Transport.* Cambridge: Cambridge University Press.

26 Chau, K.T., Li, W.L., and Lee, C.H.T. (2012). Challenges and opportunities of electric machines for renewable energy. *Prog. Electromagn. Res. B* 42: 45–74.

27 Li, G., Zhu, R., and Yang, Y. (2012). Polymer solar cells. *Nat. Photonics* 6: 153–161.

28 Atwater, H.A. and Polman, A. (2010). Plasmonics for improved photovoltaic devices. *Nat. Mater.* 9: 205–213.

29 Yoon, J., Baca, A.J., Park, S.-I. et al. (2008). Ultrathin silicon solar microcells for semitransparent, mechanically flexible and microconcentrator module designs. *Nat. Mater.* 7: 907–915.

30 Shin, J.C., Kim, K.H., Yu, K.J. et al. (2011). $In_xGa_{1-x}As$ nanowires on silicon: one-dimensional heterogeneous epitaxy, bandgap engineering, and Photovoltaics. *Nano Lett.* 11 (11): 4831–4838.

31 Li, X.H., Choy, W.C.H., Huo, L. et al. (2012). Dual plasmonic nanostructures for high performance inverted organic solar cells. *Adv. Mater.* 24 (22): 3046–3052.

32 Zhang, D., Choy, W.C.H., Wang, C.C.D. et al. (2011). Polymer solar cells with gold nanoclusters decorated multi-layer graphene as transparent electrode. *Appl. Phys. Lett.* 99 (22): 223302(1)–223302(3).

33 Yaghjian, A.D. (2014). Reflections on Maxwell's treatise. *Prog. Electromagn. Res.* 149: 217–249.

34 Nahin, P.J. (2002). *Oliver Heaviside: The Life, Work, and Times of an Electrical Genius of the Victorian Age.* JHU Press.

35 Hertz, H.R. (1888). On the propagation velocity of electrodynamic effects (in German: Ueber die Ausbreitungsgeschwindigkeit der electrodynamischen Wirkungen). *Ann. Phys.* 270 (7): 551–560.

36 Tesla, N. (1982). *My Inventions.* (Parts I through V published in the *Electrical Experimenter* monthly magazine from February through June 1919. Part VI published October 1919.) Reprint edition with introductory notes by B Johnson, Barnes, and Noble.

37 Marconi, G. Transmitting electrical signals, (using Ruhmkorff coil and Morse code key). US Patent 586,193, filed December 1896, patented July 1897.

38 Einstein, A. (1905). On the electrodynamics of moving bodies (in German: Zur elektrodynamik bewegter körper). *Ann. Phys.* 17: 891.

39 Dirac, P.A.M. (1927). The quantum theory of the emission and absorption of radiation. *Proc. R. Soc. London, Ser. A* 114 (767): 243–265.

40 Schweber, S.S. (1994). *QED and the Men Who Made It: Dyson, Feynman, Schwinger, and Tomonaga.* Princeton, NJ: Princeton University Press.

41 Mandel, L. and Wolf, E. (1995). *Optical Coherence and Quantum Optics.* Cambridge, UK: Cambridge University Press.

42 Scully, M.O. and Zubairy, M.S. (1997). *Quantum Optics.* Cambridge, UK: Cambridge University Press.

43 Loudon, R. (2000). *The Quantum Theory of Light.* Oxford, UK: Oxford University Press.

44 Gerry, C. and Knight, P. (2005). *Introductory Quantum Optics.* Cambridge, UK: Cambridge University Press.

45 Fox, M. (2006). *Quantum Optics: An Introduction,* vol. 15. Oxford, UK: Oxford University Press.

46 Garrison, J. and Chiao, J. (2014). *Quantum Optics.* Oxford, UK: Oxford University Press.

47 Milonni, P.W. (1994). *The Quantum Vacuum: An Introduction to Quantum Electrodynamics.* San Diego, CA: Academic Press.

48 Cartan, É. (1945). *Exterior Differential Systems and Its Applications (in French: Les systemes différentiels extérieurs et leurs applications géométriques).* Hermann.

49 Warnick, K.F. and Russer, P.H. (2014). Differential forms and electromagnetic field theory. *Prog. Electromagn. Res.* 148: 83–112.

50 Yang, C.N. and Mills, R.L. (1954). Conservation of isotopic spin and isotopic gauge invariance. *Physiol. Rev.* 96 (1): 191–195.

51 Hooft, G.T. (ed.) (2005). *50 Years of Yang-Mills Theory.* Singapore: World Scientific.

52 Misner, C.W., Thorne, K.S., and Wheeler, J.A. (1973). *Gravitation.* New York, NY: Macmillan.

53 Teixeira, F.L. and Chew, W.C. (1999). Differential forms, metrics, and the reflectionless absorption of electromagnetic waves. *J. Electromagn. Wave* 13 (5): 665–686.

54 Feynman, R.P. (1985). *QED: The Strange Story of Light and Matter.* Princeton, NJ: Princeton Science.

55 Styer, D. (2012). Calculation of the anomalous magnetic moment of the electron, [Online]. Available: http://www.oberlin.edu/physics/dstyer/StrangeQM/Moment.pdf

56 Aoyama, T., Hayakawa, M., Kinoshita, T., and Nio, M. (2012). Tenth-order QED contribution to the electron g-2 and an improved value of the fine structure constant. *Phys. Rev. Lett.* 109 (11): 111807.
57 Clauser, J.F. (1974). Experimental distinction between the quantum and classical field-theoretic predictions for the photoelectric effect. *Phys. Rev. D: Part. Fields* 9 (4): 853–860.
58 Kimble, H.J., Dagaenais, M., and Mandel, L. (1977). Photon antibunching in resonance fluorescence. *Phys. Rev. Lett.* 39 (11): 691–695.
59 Aspect, A., Dalibard, J., and Roger, G. (1982). Experimental test of Bell's inequalities using time-varying analyzers. *Phys. Rev. Lett.* 49 (25): 1804–1807.
60 Griffiths, D.J. (2004). *Introduction to Quantum Mechanics.* Harlow, Essex: Pearson Education.
61 Miller, D.A.B. (2008). *Quantum Mechanics for Scientists and Engineers.* New York, NY: Cambridge University Press.
62 Ryu, C.J., Liu, A.Y., Sha, W.E.I., and Chew, W.C. (2016). Finite-difference time-domain simulation of the Maxwell-Schrodinger system. *IEEE J. Multiscale Multiphys. Comput. Tech.* 1: 40–47. http://dx.doi.org/10.1109/JMMCT.2016.2605378.
63 Xiang, Z.-L., Ashhab, S., You, J.Q., and Nori, F. (2013). Hybrid quantum circuits: superconducting circuits interacting with other quantum systems. *Rev. Mod. Phys.* 85 (2): 623–653.
64 Cohen-Tannoudji, C., Dupont-Roc, J., and Grynberg, G. (1992). *Atom-Photon Interactions: Basic Processes and Applications.* New York, NY: Wiley.
65 Datta, S. (2005). *Quantum Transport: Atom to Transistor.* New York, NY: Cambridge University Press.
66 Lamoreaux, S.K. (1997). Demonstration of the Casimir force in the 0.6 to 6 μm range. *Phys. Rev. Lett.* 78 (1): 5–8.
67 Casimir, H.B.G. (1948). On the attraction between two perfectly conducting plates. *Proc. K. Ned. Akad. Wet.* 51: 793–795.
68 Casimir, H.B.G. and Polder, D. (1948). The influence of retardation on the London-van der Waals forces. *Physiol. Rev.* 73: 360–372.
69 Novotny, L. and Hecht, B. (2008). *Principles of Nano-Optics.* New York, NY: Cambridge University Press.
70 Dung, H.T., Knöll, L., and Welsch, D.-G. (1998). Three-dimensional quantization of the electromagnetic field in dispersive and absorbing inhomogeneous dielectrics. *Phys. Rev. A* 57 (5): 3931–3942.
71 Scheel, S. and Buhmann, S.Y. (2008). Macroscopic quantum electrodynamics concepts and applications. *Acta Phys. Slov.* 58 (5): 675–809.
72 Tai, C.T. (1994). Dyadic Green Functions in Electromagnetic Theory, *2e*. Piscataway, NJ: IEEE Press.
73 Chew, W.C. (2014). Vector potential electromagnetics with generalized gauge for inhomogeneous media: formulation. *Prog. Electromagn. Res.* 149: 69–84.
74 Kong, J.A. (1986). *Theory of Electromagnetic Waves.* New York, NY: Wiley.
75 Ishimaru, A. (1991). *Electromagnetic Wave Propagation, Radiation, and Scattering.* Upper Saddle River, NJ: Prentice Hall.
76 Chew, W.C. (1995). *Waves and Fields in Inhomogeneous Media*, vol. 522. IEEE Press. First Printing 1990.

77 Yaghjian, A.D. (1980). Electric dyadic Green's functions in the source region. *Proc. IEEE* 68 (2): 248–263.
78 Green, G. (1828). *An Essay on the Application of Mathematical Analysis to the Theories of Electricity and Magnetism.* Nottingham: T. Wheelhouse. Also, see Challis, L. and Sheard, F. (2003). The green of Green functions. *Physics Today* (December 2003), pp. 41–46.
79 Lancaster, T. and Blundell, S.J. (2014). *Quantum Field Theory for the Gifted Amateur.* OUP Oxford.
80 Chew, W.C., Liu, A.Y., Salazar-Lazaro, C., and Sha, W.E.I. (2016). Quantum electromagnetics: a new look, Part I. *IEEE J. Multiscale Multiphys. Comput. Tech.* 1: 73–84.
81 Chew, W.C., Liu, A.Y., Salazar-Lazaro, C., and Sha, W.E.I. (2016). Quantum electromagnetics: a new look, Part II. *IEEE J. Multiscale Multiphys. Comput. Tech.* 1: 85–97.
82 Chen, Y.P., Sha, W.E.I., Choy, W.C.H. et al. (2012). Study on spontaneous emission in complex multilayered plasmonic system via surface integral equation approach with layered medium Green's function. *Opt. Express* 20 (18): 20210–20221.
83 Sha, W.E.I., Liu, A.Y., and Chew, W.C. Dissipative quantum electromagnetics: a Novel Approach, accepted by *JMMCT*.
84 Dirac, P.A.M. (1928). The quantum theory of the electron. *Proc. R. Soc. London, Ser. A* 117 (778): 610–624.
85 Anderson, C.D. (1932). The apparent existence of easily deflectable positives. *Science* 76 (1967): 238–239.
86 Heaviside, O. (1888). On electromagnetic waves, especially in relation to the vorticity of the impressed forces, and the forced vibration of electromagnetic systems. *Philos. Mag.* 25: 130–156.
87 Chew, W.C. (2004). Computational electromagnetics: the physics of smooth versus oscillatory fields. *Philos. Trans. R. Soc. London, Ser. A* 362 (1816): 579–602.
88 Morse, S.F.B. (1869). *Examination of the Telegraphic Apparatus and the Processes in Telegraphy.* Washington, WA: Philip and Solomons.
89 Pohl, R.W. (1924). *Einführung in die Physik*, vol. 3. Göttingen: Springer.
90 Novotny, L. and Hecht, B. (2006). *Principles of Nano-Optics.* Cambridge, UK: Cambridge University Press.
91 Rokhlin,V. Yale University, New Haven, CT, private communication.
92 Van Dyke, M.D. (1964). *Perturbation Methods in Fluid Mechanics.* New York, NY: Academic Press.
93 Chew, W.C., Tong, M.S., and Hu, B. (2008). *Integral Equations Methods for Electromagnetic and Elastic Waves.* Morgan & Claypool.
94 Dirac, P.A.M. (1927). The quantum theory of the emission and absorption of radiation. *Proc. R. Soc. London, Ser. A* 114 (767): 243–265.
95 Gerry, C. and Knight, P. (2004). *Introductory Quantum Optics.* Cambridge, UK: Cambridge University Press.
96 Fox, M. (2006). *Quantum Optics: An Introduction.* Oxford, UK: Oxford University Press.
97 Cohen-Tannoudji, C., Dupont-Roc, J., and Grynberg, G. (1997). *Photons and Atoms: Introduction to Quantum Electrodynamics.* New York, NY: Wiley.

98 Miller, D.A.B. (2008). *Quantum Mechanics for Scientists and Engineers*. New York, NY: Cambridge University Press.

99 Chew, W.C. (2016). Lecture Notes: Quantum Mechanics Made Simple. Internet: http://wcchew.ece.illinois.edu/chew/course/QMAll20161206.pdf (6 December 2016).

100 Keller, J.B. (1962). Geometrical theory of diffraction. *J. Opt. Soc. Am.* 52 (2): 116–130.

101 Kouyoumjian, R.G. (1965). Asymptotic high-frequency methods. *Proc. IEEE* 53 (8): 864–876.

102 Kouyoumjian, R.G. and Pathak, P.H. (1974). A uniform geometrical theory of diffraction for an edge in a perfectly conducting surface. *Proc. IEEE* 62 (11): 1448–1461.

103 Lee, S.W. and Deschamps, G.A. (1976). A uniform asymptotic theory of electromagnetic diffraction by a curved wedge. *IEEE Trans. Antennas Propag.* 24 (1): 25–34.

104 Hansen, R.C. (ed.) (1981). *Geometric Theory of Diffraction*. Piscataway, NJ: IEEE Press.

105 Pathak, P.H. (1992). High-frequency techniques for antenna analysis. *Proc. IEEE* 80 (1): 44–65.

106 Born, M. and Wolf, E. (1964). *Principles of Optics: Electromagnetic Theory of Propagation, Interference and Diffraction of Light*, 1e. London, UK: Macmillan. Cambridge, UK: Cambridge University Press, 1999.

107 Mandel, L. and Wolf, E. (1995). *Optical Coherence and Quantum Optics*. Cambridge, UK: Cambridge University Press.

108 Stroke, G.W. (1966). *An Introduction to Coherent Optics and Holography*. New York, NY: Academic Press.

109 Haus, H.A. (2000). *Electromagnetic Noise and Quantum Optical Measurements*. Berlin: Springer-Verlag.

110 Viktorov, I.A. (1967). *Rayleigh and Lamb Waves: Physical Theory and Applications*. New York, NY: Plenum Press.

111 Stoneley, R. (1928). The dispersion of waves in a double superficial layer. *Geophys. J. Int.* 1 (s10): 527–532.

112 Faraday, M. (1857). Experimental relations of gold (and other metals) to light. *Philos. Trans. R. Soc. London* 147: 145–181.

113 Mie, G. (1908). Beiträge zur optik trüber medien, speziell kolloidaler metallösungen. *Ann. Phys.* 330 (3): 377–445.

114 Choudhuri, A.R. (1998). *The Physics of Fluids and Plasma*. Cambridge, UK: Cambridge University Press.

115 Chew, W.C., Jin, J.M., Michielssen, E., and Song, J.M. (eds.) (2001). *Fast and Efficient Algorithms in Computational Electromagnetics*. Boston, MA: Artech House.

116 Strutt Rayleigh, J.W. (Lord Rayleigh) (1976). *Theory of Sound*. New York, NY: Dover Publications (originally published 1877).

117 Strutt Rayleigh, J.W. (Lord Rayleigh) (1897). On the passage of electric waves through tubes, or the vibrations of dielectric cylinders. *Philos. Mag.* 43: 125–132.

118 Bowman, J.J., Senior, T.B.A., Uslenghi, P.L.E., and Asvestas, J.S. (1970). *Electromagnetic and Acoustic Scattering by Simple Shapes*. Amsterdam: North-Holland Publishing Co.

119 Debye, P. (1909). Der lichtdruck auf kugeln von beliebigem material. *Ann. Phys.* 4 (30): 57.

120 Strutt Rayleigh, J.W. (Lord Rayleigh) (1871). On the scattering of light by small particles. *Philos. Mag.*, Series 4, 41: 447–454.

121 Abramowitz, M. and Stegun, I.A. (1965). *Handbook of Mathematical Functions.* New York, NY: Dover Publications.

122 Stratton, J.A. (1941). *Electromagnetic Theory.* New York, NY: McGraw-Hill.

123 Harrington, R.F. (1961). *Time-Harmonic Electromagnetic Fields.* New York, NY: McGraw-Hill.

124 Collin, R.E. (1991). *Field Theory of Guided Waves.* Piscataway, NJ: IEEE Press.

125 Balanis, C.A. (1989). *Engineering Electromagnetics.* New Jersey, NJ: Wiley.

126 Jin, J.M. (2010). *Theory and Computation of Electromagnetic Fields.* New York, NY: Wiley.

127 Sommerfeld, A. (1909). Über die ausbreitung der wellen in der drahtlosen telegraphie. *Ann. Phys.* 28 (25): 665–737.

128 Weyl, H. (1919). Ausbreitung electromagnetischer wellen über einem ebenen leiter. *Ann. Phys.* 365 (21): 481–500.

129 Kong, J.A. (1972). Electromagnetic field due to dipole antennas over stratified anisotropic media. *Geophysics* 37 (6): 985–996.

130 Sommerfeld, A. (1896). Mathematische Theorie der Diffraction. *Math. Ann.* 47 (s319): 317–374.

131 Heins, A.E. (1982). The Sommerfeld half-plane problem revisited I: the solution of a pair of coupled Wiener-Hopf integral equations. *Math. Methods Appl. Sci.* 4 (1): 74–90.

132 Ziolkowski, R.W. (1987). Exact solution of the Sommerfeld half-plane problem: a path integral approach without discretization. *J. Electromagn. Waves Appl.* 1 (4): 377–402.

133 Wiener, N. and Hopf, E. (1931). Ueber eine Klasses singulärer Integralgleichungen. *S.-B. Deutsch. Akad. Wiss. Berlin KI. Math. Phys. Tech.* 696–706.

134 Banos, A. (1966). *Dipole Radiation in the Presence of a Conducting Half-Space.* Oxford: Pergamon Press.

135 Felsen, L.B. and Marcuvitz, N. (1972). *Radiation and Scattering of Waves.* Englewood Cliffs, NJ: Prentice-Hall.

136 Chew, W.C. (1983). The singularities of a Fourier-type integral in a multicylindrical layer problem. *IEEE Trans. Antennas Propag.* 31 (4): 653–655.

137 Bleistein, N. and Handelsman, R.A. (1974). *Asymptotic Expansions of Integrals.* New York, NY: Holt Rinehart and Winston.

138 Lewis, R. M. and Keller, J. B. (1964). Asymptotic methods for partial differential equations: the reduced wave equation and Maxwell's equations. *N.Y.U. Research Report No. EM-194.*

139 Watson, G.N. (1918). The diffraction of electric waves by the earth. *Proc. R. Soc. London, Ser. A* 95 (666): 83–99.

140 Li, M.K. and Chew, W.C. (2004). A new Sommerfeld-Watson transformation in 3-D. *IEEE Antennas Wirel. Propag. Lett.* 3 (1): 75–78.

141 Sha, W.E.I. and Chew, W.C. (2009). High frequency scattering by an impenetrable sphere. *Prog. Electromagn. Res.* 97: 291–325.

142 Shore, R.A. and Yaghjian, A.D. (1988). Incremental diffraction coefficients for planar surfaces. *IEEE Trans. Antennas Propag.* 36 (1): 55–70.

143 Batchelor, G.K. (1967). *An Introduction to Fluid Dynamics*. New York, NY: Cambridge University Press.

144 Wentzel, G. (1926). A generalization of the quantum conditions for the purposes of wave mechanics. *Z. Angew. Phys.* 38 (6-7): 518–529.

145 Kramers, H.A. (1926). Wave mechanics and semi-numerical quantisation. *Z. Angew. Phys.* 30 (10-11): 828–840.

146 Brillouin, L. (1926). The undulatory mechanics of Schrödinger: a general method of solution by successive approximations. *C.R. Hebd. Seances Acad. Sci.* 183 (5): 24–26.

147 Kirchhoff, G. (1877). Zur Theorie des Condensators, *144–162*. Berlin: Akad. Monatsber.

148 Chew, W.C. and Kong, J.A. (1981). Asymptotic formula for the capacitance of two oppositely charged discs. *Math. Proc. Cambridge Philos. Soc.* 89: 373–384.

149 Leppington, F. and Levine, H. (1970). On the capacity of the circular disc condenser at small separation. *Cambridge Philos. Soc.* 68 (1): 235–254.

150 Spurling, K.L., Poitras, A.E., McGranaghan, M.F., and Shaw, J.H. (1987). Analysis and operating experience for back-to-back 115 kV capacitor banks. *IEEE Trans. Power Delivery* 2 (4): 1255–1263.

151 Hildebrand, F.B. (1976). *Advanced Calculus for Applications*. Prentice-Hall.

152 Chew, W.C. and Kong, J.A. (1982). Microstrip capacitance for a circular disk through matched asymptotic expansions. *SIAM J. Appl. Math.* 42 (2): 302–317.

153 Poh, S.Y., Chew, W.C., and Kong, J.A. (1981). Approximate formulas for line capacitance and characteristic impedance of microstrip line. *IEEE Trans. Microwave Theory Tech.* 29 (2): 135–142.

154 Michell, J.H. (1894). A map of the complex Z-function: a condenser problem. *Messenger. Math. Cambridge.* 23: 72–78.

155 I'a Bromwich, T.J. (1902). Note on a condenser problem. *Messenger Math. Cambridge* 31: 184–192.

156 Love, A.E.H. (1924). Some electrostatic distributions in two dimensions. *Proc. London Math. Soc.*, Series 2, 22 (1460): 337–369.

157 Chew, W.C. and Kong, J.A. (1981). Asymptotic formula for the resonant frequencies of a circular microstrip antenna. *J. Appl. Phys.* 52 (8): 5365–5369.

158 Chew, W.C. and Kong, J.A. (1981). Asymptotic eigenequations and analytic formulas for the dispersion characteristics of open, wide microstrip lines. *IEEE Trans. Microwave Theory Tech.* 29 (9): 933–941.

159 Dukhin, S. and Shilov, V.N. (1974). *Dielectric Phenomena and the Double Layer in Disperse Systems and Polyelectrolytes*. New York, NY: Wiley.

160 Fixman, M. (1980). Charged macromolecules in external fields. I. The sphere. *J. Chem. Phys.* 72 (9): 5177–5186.

161 Hinch, E.J., Sherwood, J.D., Chew, W.C., and Sen, P.N. (1984). The dielectric response of a dilute suspension of spheres with thin double layers in an asymmetric electrolyte. *J. Chem. Soc., Faraday Trans.* 2 (80): 535–551.

162 Conde, O.M., Pérez, J., and Cátedra, M.F. (2001). Stationary phase method application for the analysis of radiation of complex 3-D conducting structures. *IEEE Trans. Antennas Propag.* 49 (5): 724–731.

163 Carluccio, G., Albani, M., and Pathak, P.H. (2010). Uniform asymptotic evaluation of surface integrals with polygonal integration domains in terms of UTD transition functions. *IEEE Trans. Antennas Propag.* 58 (4): 1155–1163.

164 Ufimtsev, P.Y. (2007). *Fundamentals of the Physical Theory of Diffraction*. New York, NY: Wiley.

165 Wikipedia, Human Computer. Internet: http://en.wikipedia.org/wiki/Human_computer (4 December 2012).

166 Gumbrecht, J. (2011) Rediscovering WWII's female 'computers'. Internet: http://www.cnn.com/2011/TECH/innovation/02/08/women.rosies.math/index.html (8 February 2011).

167 Wikipedia, Mechanical Computer. Internet: http://en.wikipedia.org/wiki/Mechanical_computer (7 December 2012).

168 Ceruzzi, P.E. (1998). *A History of Modern Computing*. Cambridge, MA: MIT Press.

169 ENIA Wikipedia, ENIAC. Internet: http://en.wikipedia.org/wiki/ENIAC (22 December 2012).

170 ILLIAC Wikipedia, ILLIAC. Internet: http://en.wikipedia.org/wiki/ILLIAC (17 December 2012).

171 Zienkiewicz, O.C. (1977). The Finite Element Method, *3e*. London: McGraw-Hill.

172 Silvester, P.P. and Ferrari, R.L. (1983). *Finite Elements for Electrical Engineers*. Cambridge: Cambridge University Press.

173 Silvester, P.P., Itoh, T., and Pelosi, G. (1996). *Finite Element Software for Microwave Engineering*. New York, NY: Wiley.

174 Jin, J.M. (2002). The Finite Element Method in Electromagnetics, *2e*. New York, NY: Wiley-IEEE Press.

175 Volakis, J.L., Chaterjee, A., and Kempel, L.C. (1998). *Finite Element Method Electromagnetics: Antennas, Microwave Circuits, and Scattering Applications*. New York, NY: Wiley-IEEE Press.

176 Zhu, Y. and Cangellaris, A.C. (2006). *Multigrid Finite Element Methods for Electromagnetic Field Modeling*. Hoboken, NJ: Wiley-IEEE Press.

177 Yee, K.S. (1966). Numerical solution of initial boundary value problems involving Maxwell's equation in isotropic media. *IEEE Trans. Antennas Propag.* 14: 302–307.

178 Taflove, A. (1995). *Computational Electrodynamics: The Finite Difference Time Domain Method*. Boston: Artech House.

179 Harrington, R.F. (1968). *Field Computation by Moment Method*. New York, NY: Macmillan.

180 Richmond, J. (1965). Scattering by a dielectric cylinder of arbitrary cross section shape. *IEEE Trans. Antennas Propag.* 13 (3): 334–341.

181 Rao, S.M., Wilton, G.R., and Glisson, A.W. (1982). Electromagnetic scattering by surfaces of arbitrary shape. *IEEE Trans. Antennas Propag.* 30 (3): 409–418.

182 Nakhla, M.S. and Vlach, J. (1976). A piecewise harmonic balance technique for determination of periodic response of nonlinear systems. *IEEE Trans. Circuits Syst.* 23 (2): 85–91.

183 Gilmore, R.J. and Steer, M.B. (1991). Nonlinear circuit analysis using the method of harmonic balance-A review of the art. Part I. Introductory concepts. *Int. J. Microwave Millimeter-Waves Comput.-Aided Eng.* 1 (1): 22–37.

184 Yamada, S., Bessho, K., and Lu, J. (1989). Harmonic balance finite element method applied to nonlinear AC magnetic analysis. *IEEE Trans. Magn.* 25 (4): 2971–2973.

185 Parrón, J., Collado, C., Mateu, J. et al. (2001). General electromagnetic simulation tool to predict the microwave nonlinear response of planar, arbitrarily-shaped HTS structures. *IEEE Trans. Appl. Supercond.* 11 (1): 399–402.

186 Newton, I. *Methodus fluxionum et serierum infinitarum*, 1664–1671.

187 Raphson, J. *Analysis aequationum universalis*, London, 1690.

188 Fletcher, R. (1987). Practical Methods of Optimization, *2e*. New York, NY: Wiley.

189 Medgyesi-Mitschang, L.N., Putnam, J.M., and Gedera, M.B. (1994). Generalized method of moments for three-dimensional penetrable scatterers. *J. Opt. Soc. Am. A* 11 (4): 1383–1398.

190 Sommerfeld, A. (1949). *Partial Differential Equation*. New York, NY: Academic Press.

191 Garabedian, P.R. (1964). *Partial Differential Equations*. New York, NY: Wiley.

192 Courant, R. and Hilbert, D. (1953–1962). *Methods of Mathematical Physics*. New York, NY: Interscience Publishers.

193 Mautz, J.R. and Harrington, R.F. (1978). *H*-field, *E*-field, and combined-field solutions for conducting bodies of revolution. *A. E. Ü.* 32: 157–164.

194 Harrington, R.F. (1989). Boundary integral formulations for homogenous material bodies. *J. Electromagn. Waves Appl.* 3 (1): 1–15.

195 Müller, C. (1969). *Foundation of the Mathematical Theory of Electromagnetic Waves*. Berlin: Springer-Verlag.

196 Werner, P. (1963). On the exterior boundary value problem of the perfect reflection for the stationary electromagnetic wave fields. *J. Math. Anal. Appl.* 7: 348–396.

197 Schenck, H.A. (1968). Improved integral formulation for acoustic radiation problems. *J. Acoust. Soc. Am.* 44 (1): 41–58.

198 Mitzner, K.M. (1968). Numerical solution of the exterior scattering problem at the eigen-frequencies of the interior problem, *URSI Meeting Digest*, 75, Boston.

199 Burton, *A. J.* and Miller, G.F. (1971). The application of integral equation methods to the numerical solution of some exterior boundary value problems. *Proc. R. Soc. London, Ser. A* 323 (1553): 201–210.

200 Bolomey, J.C. and Tabbara, W. (1973). Numerical aspects on coupling between complementary boundary value problems. *IEEE Trans. Antennas Propag.* 21 (3): 356–363.

201 Mittra, R. and Klein, C.A. (1975). Stability and convergence of moment method solutions. *Numer. Asymp. Tech. Electromagn.* 3: 129–163.

202 Morita, N. (1979). Resonant solutions involved in the integral equation approach to scattering from conducting and dielectric cylinders. *IEEE Trans. Antennas Propag.* 27 (6): 869–871.

203 Yaghjian, A.D. (1981). Augmented electric- and magnetic-field integral equations. *Radio Sci.* 16: 987–1001.

204 Peterson, A.F. (1990). The interior resonance problem associated with surface integral equations of electromagnetics: numerical consequences and a survey of remedies. *Electromagnetics* 10 (3): 293–312.

205 Chew, W.C. and Song, J.M. (2007). Gedanken experiments to understand the internal resonance problems of electromagnetic scattering. *Electromagnetics* 27 (8): 457–471.

206 Song, J.M., Shu, W.W., and Chew, W.C. (2007). Numerical resonances in method of moments. *IEEE APS Int. Symp. Dig.* 4861–4864.

207 Wilton, D.R. (1992). Review of current status and trends in the use of integral equations in computational electromagnetics. *Electromagnetics* 12 (3-4): 287–341.

208 Mautz, J.R. and Harrington, R.F. (1979). A combined-source solution for radiation and scattering from a perfectly conducting body. *IEEE Trans. Antennas Propag.* 27 (4): 445–454.

209 Andriulli, F.P., Cools, K., Bagci, H. et al. (2008). A multiplicative Calderon preconditioner for the electric field integral equation. *IEEE Trans. Antennas Propag.* 56 (8): 2398–2412.

210 Cools, K., Andriulli, F.P., Olyslager, F., and Michielssen, E. (2009). Improving the MFIE's accuracy by using a mixed discretization. *IEEE Antennas Propag. Soc. Int. Symp.* 1–4.

211 Bagci, H., Andriulli, F.P., Cools, K. et al. (2010). A Calderon multiplicative preconditioner for coupled surface-volume electric field integral equations. *IEEE Trans. Antennas Propag.* 58 (8): 2680–2690.

212 Petrov, G. (1940). Application of Galerkin's method to a problem of the stability of the flow of a viscous liquid." (Russian). *Prikad. Matem. i Mekh* 4 (3): 3–12.

213 Belytschiko, T., Lu, Y.Y., and Gu, L. (1994). Element-free Galerkin methods. *Int. J. Numer. Methods Eng.* 37 (2): 229–256.

214 Brooks, A.N. and Hughes, T.J.R. (1982). Streamline upwind/Petrov-Galerkin formulations for convection dominated flows with particular emphasis on the incompressible Navier-Stokes equations. *Comput. Meth. Appl. Mech. Eng.* 32 (1–3): 199–259.

215 Saad, Y. (2003). *Iterative Methods for Sparse Linear Systems.* SIAM.

216 Chan, T.F. and Ng, M.K. (1999). Galerkin projection methods for solving multiple linear systems. *SIAM J. Sci. Comp.* 21 (3): 836–850.

217 Graglia, R.D. (1993). On the numerical integration of the linear shape functions times the 3D Green's function or its gradient on a plane triangle. *IEEE Trans. Antennas Propag.* 41 (10): 1448–1456.

218 Graglia, R.D., Wilton, D.R., and Peterson, A.F. (1997). Higher order interpolatory vector bases for computational electromagnetics. *IEEE Trans. Antennas Propag.* 45 (3): 329–342.

219 Greengard, L. and Rokhlin, V. (1987). A fast algorithm for particle simulations. *J. Comput. Phys.* 73: 325–348.

220 Willis, D.J., Peraire, J., and White, J.K. (1997). A combined pFFT-multipole tree code, unsteady panel method with vortex particle wakes. *Int. J. Numer. Methods Fluids* 53: 1399–1422.

221 Bleszynski, E., Bleszynski, M., and Jaroszewicz, T. (1996). AIM: Adaptive integral method for solving large-scale electromagnetic scattering and radiation problems. *Radio Sci.* 31 (5): 1225–1251.

222 Lu, C.C. and Chew, W.C. (1994). Multilevel algorithm for solving boundary-value scattering. *Microwave Opt. Technol. Lett.* 7 (10): 466–470.

223 Song, J.M. and Chew, W.C. (1995). Multilevel fast-multipole algorithm for solving combined field integral equations of electromagnetic scattering. *Microwave Opt. Technol. Lett.* 10 (1): 14–19.

224 Velamparambil, S., Chew, W.C., and Song, J.M. (2003). 10 million unknowns, is it that large. *IEEE Antennas Propag. Mag.* 45 (2): 43–58.

225 Trefethen, L.N. and Bau, D. (1997). *Numerical Linear Algebra,* No. 50. SIAM.

226 Saad, Y. (1989). Krylov subspace methods on supercomputers. *SIAM J. Sci. Stat. Comput.* 10: 1200–1232.

227 Saad, Y. (1981). Krylov subspace methods for solving large unsymmetric linear systems. *Math. Comput.* 37: 105–126.

228 Saad, L.Y. and Schultz, M.H. (1986). GMRES: a generalized minimal residual algorithm for solving nonsymmetric linear systems. *SIAM J. Sci. Stat. Comput.* 7 (3): 856–869.

229 Trefethen, L.N. (1990). *Approximation theory and numerical linear algebra*. In: *Algorithms for Approximation II*, Chapman, London (eds. J.C. Mason and M.G. Cox).

230 Nachtigal, N.M., Reddy, S.C., and Trefethen, L.N. (1992). How fast are nonsymmetric matrix iterations. *SIAM J. Matrix Anal. Appl.* 13 (3): 778–795.

231 Driscoll, T.A., Toh, K.-C., and Trefethen, L.N. (1998). From potential theory to matrix iterations in six steps. *SIAM Rev.* 40: 547–578.

232 Liesen, J. and Tichy, P. (2004). Convergence analysis of Krylov subspace methods. *GAMM-Mitteilungen* 27 (2): 153–173.

233 Siefert, C. and de Sturler, E. (2006). Preconditioners for generalized saddle-point problems. *SIAM J. Numer. Anal.* 44 (3): 1275–1296.

234 Lee, R. and Cangellaris, A.C. (1992). A study of discretization error in the finite element approximation of wave solution. *IEEE Trans. Antennas Propag.* 40 (5): 542–549.

235 George, J.A. (1973). Nested dissection of a regular finite element mesh. *SIAM J. Numer. Anal.* 10: 345–363.

236 George, A. and Liu, J.W.H. (1976). An algorithm for automatic nested dissection and its application to general finite element problems. In: *Proceedings of the Sixth Conference on Numerical Mathematics and Computing, Winnipeg, Manitoba*, 59–94.

237 Irons, B.M. (1970). A frontal solution scheme for finite element analysis. *Int. J. Numer. Methods Eng.* 2: 5–32.

238 Axelsson, O. and Barker, V.A. (1984). *Finite Element Solution of Boundary Value Problems: Theory and Computation*. Academic Press.

239 Bayliss, A. and Turkel, E. (1980). Radiation boundary conditions for wave-like equations. *Commun. Pure Appl. Math.* 33: 707–725.

240 Engquist, B. and Majda, A. (1977). Absorbing boundary conditions for the numerical simulation of waves. *Math. Comput.* 31: 629–651.

241 Fang, J. and Mei, K.K. (1988). A super-absorbing boundary algorithm for solving electromagnetic problems by time-domain finite-difference method. *IEEE AP-S Int. Symp. Dig.* 2, 472–474.

242 Liao, Z.P., Wong, H.L., Yang, B.P., and Yuan, Y.F. (1984). A transmitting boundary for transient wave analysis. *Sci. Sin., Ser. A* 27 (10): 1063–1076.

243 Lindman, E.L. (1975). Free space boundary conditions for the time dependent wave equation. *J. Comput. Phys.* 10: 66–70.

244 Mur, G. (1981). Absorbing boundary conditions for the finite-difference approximation of the time-domain electromagnetic-field equations. *IEEE Trans. Electromagn. Compat.* 23: 377–382.

245 Randall, C.L. (1988). Absorbing boundary conditions for the elastic wave equation. *Geophysics* 53 (5): 611–624.

246 Wagatha, L. (1983). Approximation of pseudodifferential operators in absorbing boundary conditions for hyperbolic equations. *Numer. Math.* 42: 1–64.

247 Higdon, R.L. (1987). Numerical absorbing boundary conditions for the wave equation. *Math. Comput.* 49: 65–90.

248 Mittra, R., Ramahi, O.O., Khebir, A. et al. (1989). A review of absorbing boundary conditions for two and three-dimensional electromagnetic scattering problems. *IEEE Trans. Magn.* 25 (4): 3034–3039.

249 Ziolkowski, R.W. (2008). FDTD absorbing boundary conditions: From global lookback schemes to metamaterial PMLs. *Antennas and Propagation Society International Symposium, 2008. AP-S 2008. IEEE*: 1–4 (5–11 July 2008).

250 Alavikia, B., Ramahi, O.M.(2010). Limitation of using absorbing boundary condition to solve the problem of scattering from a cavity in metallic screens. *Antennas and Propagation Society International Symposium (APSURSI), 2010 IEEE*: 1–4 (11–17 July 2010).

251 Berenger, J.-P. (1994). A perfectly matched layer for the absorption of electromagnetic waves. *J. Comput. Phys.* 114: 185–200.

252 Chew, W.C. and Weedon, W.H. (1994). A 3-D perfectly matched medium from modified Maxwell's equations with stretched coordinates. *Microwave Opt. Technol. Lett.* 7 (13): 599–604.

253 Sacks, Z.S., Kingsland, D.M., Lee, R., and Lee, J.-F. (1995). A perfectly matched anisotropic absorber for use as an absorbing boundary condition. *IEEE Trans. Antennas Propag.* 43 (12): 1460–1463.

254 Katz, D.S., Thiele, E.T., and Taflove, A. (1994). Validation and extension to three dimensions of the Berenger PML absorbing boundary condition for FD-TD meshes. *IEEE Microwave Guided Wave Lett.* 4 (8): 268–270.

255 Mittra, R. and Pekel, U. (1995). A new look at the perfectly matched layer (PML) concept for the reflectionless absorption of electromagnetic waves. *IEEE Microwave Guided Wave Lett.* 5 (3): 84–86.

256 Navarro, E.A., Wu, C., Chung, P.Y., and Litva, J. (1994). Application of PML superabsorbing boundary condition to non-orthogonal FDTD method. *Electron. Lett.* 30 (20): 1654–1656.

257 Gribbons, M., Lee, S.K., and Cangellaris, A.C. (1995). Modification of Berenger's perfectly matched layer for the absorption of electromagnetic waves in layered media. In: *11th Annual Review of Progress in ACES*, 498–503.

258 Andrew, W.V., Balanis, C.A., and Tirkas, P.A. (1995). A comparison of the Berenger perfectly matched layer and the Lindman higher-order ABC's for the FDTD method. *IEEE Microwave Guided Wave Lett.* 5 (6): 192–194.

259 Fang, J. and Wu, Z. (1995). Generalized perfectly matched layer—an extension of Berenger's perfectly matched layer boundary condition. *IEEE Microwave Guided Wave Lett.* 5 (12): 451–453.

260 Abarbanel, S. and Gottlieb, D. (1998). On the construction and analysis of absorbing layers in CEM. *Appl. Numer. Math.* 27: 331–340.

261 Teixeira, F.L. and Chew, W.C. (1997). PML-FDTD in cylindrical and spherical grids. *IEEE Micro. Guided Wave Lett.* 7 (9): 285–287.

262 Chew, W.C. and Liu, Q.H. (1996). Perfectly matched layers for elastodynamics: A new absorbing boundary condition. *J. Comput. Acoust.* 4 (4): 341–359.

263 Zeng, Y.Q., He, J., and Liu, Q.H. (2001). The application of the perfectly matched layer in numerical modeling of wave propagation in poroelastic media. *Geophysics* 66 (4): 1258–1266.

264 Cheng, C., Lee, J.-H., Lim, K.H. et al. (2007). 3D quantum transport solver based on the perfectly matched layer and spectral element methods for the simulation of semiconductor nanodevices. *J. Comput. Phys.* 227: 455–471. https://doi.org/10.1016/j.jcp.2007.07.028.

265 Peng, Z., Wang, X.C., and Lee, J.F. (2011). Integral equation based domain decomposition method for solving electromagnetic wave scattering from non penetrable objects. *IEEE Trans. Antennas Propag.* 59 (9): 3328–3338.

266 Chew, W.C. and Wang, Y.M. (1990). A fast algorithm for solution of a scattering problem using a recursive aggregate tau matrix method. *Microwave Opt. Technol. Lett.* 3 (5): 164–169.

267 Chew, W.C. and Lu, C.C. (1995). The recursive aggregated interaction matrix algorithm for multiple scatterers. *IEEE Trans. Antennas Propag.* 43 (12): 1483–1486.

268 Chew, W.C. and Lu, C.C. (1993). NEPAL–An algorithm for solving the volume integral equation. *Microwave Opt. Technol. Lett.* 6 (3): 185–188.

269 Lu, C.C. and Chew, W.C. (1995). The use of Huygens' equivalence principle for solving 3D volume integral equation of scattering. *IEEE Trans. Antennas Propag.* 43 (5): 500–507.

270 Chew, W.C. and Lu, C.C. (1993). A fast recursive algorithm to compute the wave scattering solution of a large strip. *J. Comput. Phys.* 107: 378–387.

271 Michielssen, E., Boag, A., and Chew, W.C. (1996). Scattering from elongated objects: direct solution in $O(N \log^2 N)$ operations. *IEEE Proc. Microwaves Antennas Propag.* 143 (4): 277–283. IET.

272 Chai, W. and Jiao, D. (2008). An H2-matrix-based integral-equation solver of linear-complexity for large-scale full-wave modeling of 3D circuits. In: *Electrical Performance of Electronic Packaging*, 283–286. IEEE-EPEP.

273 Guo, H., Liu, Y., Hu, J., and Michielssen, E. (2017). A butterfly-based direct integral-equation solver using hierarchical LU factorization for analyzing scattering from electrically large conducting objects. *IEEE Trans. Antennas Propag.* 65 (9): 4742–4750.

274 Borup, D.T. and Gandhi, O.P. (1984). Fast-Fourier transform method for calculation of SAR distributions in finely discretized inhomogeneous models of biological bodies. *IEEE Trans. Microwave Theory Tech.* 32 (4): 355–360.

275 Shen, C.Y., Glover, K.J., Sancer, M.I., and Varvatsis, A.D. (1989). The discrete Fourier transform method of solving differential-integral equations in scattering theory. *IEEE Trans. Antennas Propag.* 37 (8): 1032–1041.

276 Zwamborn, P. and van den Berg, P.M. (1992). The three-dimensional weak form of the conjugate gradient FFT method for solving scattering problems. *IEEE Trans. Microwave Theory Tech.* 40 (9): 1757–1766.

277 Catedra, M.F., Gago, E., and Nuno, L. (1989). A numerical scheme to obtain the RCS of three-dimensional bodies of size using the conjugate gradient method and the fast Fourier transform. *IEEE Trans. Antennas Propag.* 37 (5): 528–537.

278 Phillips, J.R. and White, J.K. (1994). Efficient capacitance computation of 3D structures using generalized pre-corrected FFT methods. In: Proceedings of the 3rd

Topical Meeting on Electric Performance of Electronic Packaging, *Nov.* 2–4, Monterey, CA.
279 Lumme, K. and Rahola, J. (1994). Light scattering by porous dust particles in the discrete-dipole approximation. *Astrophys. J.* 425: 653–667.
280 Bleszynski, E., Bleszynski, M., and Jaroszewicz, T. (1994). A fast integral-equation solver for electromagnetic scattering problems. *IEEE APS Int. Symp. Dig.* 416–419.
281 Chan, C.H. and Tsang, L. (1995). A sparse-matrix canonical-grid method for scattering by many scatterers. *Microwave Opt. Technol. Lett.* 8 (2): 114–118.
282 Gan, H. and Chew, W.C. (1995). A discrete BCG-FFT algorithm for solving 3D inhomogeneous scatterer problems. *J. Electromagn. Waves Appl.* 9 (10): 1339–1357.
283 Chew, W.C., Lin, J.H., and Yang, X.G. (1995). An FFT T-matrix method for 3D microwave scattering solution from random discrete scatterers. *Microwave Opt. Technol. Lett.* 9 (4): 194–196.
284 Lin, J.H. and Chew, W.C. (1996). BiCG-FFT T-matrix method for solving for the scattering solution from inhomogeneous bodies. *IEEE Trans. Microwave Theory Tech.* 44 (7): 1150–1155.
285 Seo, S.M. and Lee, J.F. (2005). A fast IE-FFT algorithm for solving PEC scattering problems. *IEEE Trans. Magn.* 41 (5): 1476–1479.
286 Kim, H. and Ling, H. (1993). On the application of fast wavelet transform to the integral- equation solution of electromagnetic scattering problems. *Microwave Opt. Technol. Lett.* 6 (3): 168–173.
287 Steinberg, B.Z. and Leviatan, Y. (1993). On the use of wavelet expansions in the method of moments. *IEEE Trans. Antennas Propag.* 41 (5): 610–619.
288 Wagner, R.L., Otto, G.P., and Chew, W.C. (1993). Fast waveguide mode computation using wavelet-like basis functions. *IEEE Microwave Guided Wave Lett.* 3 (7): 208–210.
289 Sabetfakhri, K. and Katehi, L.P.B. (1994). Analysis of integrated millimeter-wave and submillimeter-wave waveguides using orthonormal wavelet expansion. *IEEE Trans. Microwave Theory Tech.* 42: 2412–2422.
290 Wang, G., Pan, G., and Gilbert, B.K. (1995). A hybrid wavelet expansion and boundary element analysis for multiconductor transmission lines in multilayered dielectric media. *IEEE Trans. Microwave Theory Tech.* 43: 664–674.
291 Wagner, R.L. and Chew, W.C. (1995). A study of wavelets for the solution of electro- magnetic integral equations. *IEEE Trans. Antennas Propag.* 43 (8): 802–810.
292 Golik, W.L. (2000). Sparsity and conditioning of impedance matrices obtained with semi-orthogonal and bi-orthogonal wavelet bases. *IEEE Trans. Antennas Propag.* 48 (4): 473–481.
293 Deng, H. and Ling, H. (1999). On a class of predefined wavelet packet bases for efficient representation of electromagnetic integral equations. *IEEE Trans. Antennas Propag.* 47 (12): 1772–1779.
294 Jaffard, S. (1992). Wavelet methods for fast resolution of elliptic problems. *SIAM J. Numer. Anal.* 29 (4): 965–986.
295 Alpert, B., Beylkin, G., Coifman, R.R., and Rokhlin, V. (1993). Wavelet-like bases for the fast solution of second-kind integral equations. *SIAM J. Sci. Comput.* 14 (1): 159–184.
296 Hackbusch, W. (1999). Sparse matrix arithmetic based on H-matrices. Part I: introduction to H-matrices. *Computing* 62 (2): 89–108.

297 Bebendorf, M. (2000). Approximation of boundary element matrices. *Numer. Math.* 86 (4): 565–589.

298 Borm, S., Grasedyck, L., and Hackbusch, W. (2003). Hierarchical matrices. *Lecture Note 21 of the Max Planck Institute for Mathematics in the Sciences.*

299 Bebendorf, M. (2008). *Hierarchical Matrices: A Means to Efficiently Solve Elliptic Boundary Value Problems.* Berlin: Springer.

300 Baharav, Z. (1996). Impedance matrix compression with the use of wavelet expansions. *Microwave Opt. Technol. Lett.* 12 (2): 268–272.

301 Kurz, S., Rain, O., and Rjasanow, S. (2002). The adaptive cross-approximation technique for 3-D boundary-element method. *IEEE Trans. Magn.* 38 (2): 421–424.

302 Zhao, K., Vouvakis, M., and Lee, J.F. (2004). Application of the multilevel adaptive cross-approximation on ground plane designs. *IEEE Int. Symp. Electromagn. Compat.* 1: 124–127.

303 Bebendorf, M. and Grzhibovskis, R. (2006). Accelerating Galerkin BEM for linear elasticity using adaptive cross approximation. *Math. Methods Appl. Sci.* 29 (14): 1721–1747.

304 Guo, H., Hu, J., Shao, H., and Nie, Z. (2011). Multilevel sparse approximate inverse preconditioning for solving dynamic integral equation by H-matrix method. In: *Proc. IEEE Int. Workshop Antenna Tech.*, 124–127.

305 Liberty, E., Woolfe, F., Martinsson, P.G., and Tygert, M. (2007). Randomized algorithms for the low-rank approximation of matrices. *Proc. Natl. Acad. Sci. U.S.A.* 104 (51): 20167–20172.

306 Canning, F.X. (1990). Transformations that produce a sparse moment matrix. *J. Electromagn. Waves Appl.* 4 (9): 983–993.

307 Canning, F.X. (1991). Interaction matrix localization (IML) permits solution of larger scattering problems. *IEEE Trans. Magn.* 27 (5): 4275–4278.

308 Michielssen, E. and Boag, A. (1994). Multilevel evaluation of electromagnetic fields for the rapid solution of scattering problems. *Microwave Opt. Technol. Lett.* 7 (17): 790–795.

309 Michielssen, E. and Boag, A. (1996). A multilevel matrix decomposition algorithm for analyzing scattering from large structures. *IEEE Trans. Antennas Propag.* 44 (8): 1086–1093.

310 Rius, J., Parrón, J., Úbeda, E., and Mosig, J. (1999). Multilevel matrix decomposition algorithm for analysis of electrically large electromagnetic problems in 3-D. *Microwave Opt. Technol. Lett.* 22 (3): 177–182.

311 Appel, A.W. (1985). An efficient program for many-body simulation. *SIAM J. Sci. Stat. Comput.* 6 (1): 85–103.

312 Barnes, J. and Hut, P. (1986). A hierarchical $O(N \log N)$ force calculation algorithm. *Nature* 324 (4): 446–449.

313 Rokhlin, V. (1990). Rapid solution of integral equations of scattering theory in two dimensions. *J. Comput. Phys.* 86 (2): 414–439.

314 Coifman, R., Rokhlin, V., and Wandzura, S. (1993). The fast multipole method for the wave equation: a pedestrian prescription. *IEEE Antennas Propag. Mag.* 35 (3): 7–12.

315 Brandt, A. (1991). Multilevel computations of integral transforms and particle interactions with oscillatory kernels. *Comput. Phys. Commun.* 65 (1–3): 24–38.

316 Rokhlin, V. (1993). Diagonal forms of translation operators for the Helmholtz equation in three dimensions. *Appl. Comput. Harmon. Anal.* 1 (1): 82–93.

317 Chew, W.C., Koc, S., Song, J.M. et al. (1997). A succinct way to diagonalize the translation matrix in three dimensions. *Microwave Opt. Technol. Lett.* 15 (3): 144–147.

318 Brandt, A. (1982). Guide to multigrid development. In: *Multigrid methods*, vol. 96 (eds. W. Hackbusch and U. Trottenberg), 220–312. Berlin: Springer.

319 Song, J.M., Lu, C.C., Chew, W.C., and Lee, S.W. (1998). Fast Illinois solver code (FISC). *IEEE Antennas Propag. Mag.* 40 (3): 27–33.

320 Song, J.M. and Chew, W.C. (2000). Large scale computations using FISC. *IEEE Antennas Propag. Soc. Int. Symp., Salt Lake City, Utah (16–21 July)* 4: 1856–1859.

321 Hastriter, M.L. and Chew, W.C. (2004). Comparing Xpatch, FISC, and ScaleME using a cone-cylinder. *IEEE Antennas Propag. Soc. Int. Symp.* , Monterey, CA (20–25 June) 2: 2007–2010.

322 Taboada, J.M., Araujo, M.G., Bertolo, J.M. et al. (2010). MLFMA-FFT parallel algorithm for the solution of large-scale problems in electromagnetics. *Prog. Electromagn. Res.* 105: 15–30.

323 Pan, X.M., Pi, W.C., Yang, M.L. et al. (2012). Solving problems with over one billion unknowns by the MLFMA. *IEEE Trans. Antennas Propag.* 60 (5): 2571–2574.

324 Ergul, O. and Gurel, L. (2011). Rigorous solutions of electromagnetics problems involving hundreds of millions of unknowns. *IEEE Antennas Propag. Mag.* 53 (1): 18–26.

325 Lee, J., Zhang, J., and Lu, C.C. (2004). Sparse inverse preconditioning of multilevel fast multipole algorithm for hybrid integral equations in electromagnetics. *IEEE Trans. Antennas Propag.* 52 (9): 2277–2287.

326 Yang, M.L. and Sheng, X.Q. (2010). Parallel FE-BI-MLFMA for scattering by extremely large targets with cavities. *International Conference on Electromagnetics in Advanced Applications*, Sydney, Australia (20–24 September 2010).

327 Nie, Z.P., Ma, W.M., Ren, Y. et al. (2009). A wideband electromagnetic scattering analysis using MLFMA with higher order hierarchical vector basis functions. *IEEE Trans. Antennas Propag.* 57 (10): 3169–3178.

328 Shao, H.R., Hu, J., and Nie, Z.P. (2011). Hybrid tangential equivalence principle algorithm with MLFMA for analysis of array structures. *Prog. Electromagn. Res.* 113: 127–141.

329 Kishimoto, S., Ohnuki, S., Ashizawa, Y. et al. (2012). Time domain analysis of nanoscale electromagnetic problems by a boundary integral equation method with fast inverse Laplace transform. *J. Electromagn. Waves Appl.* 26: 997–1006.

330 Kim, B., Song, J.M., and Song, X. (2010). Calculations of the binding affinities of protein-protein complexes with the fast multipole method. *J. Chem. Phys.* 133, Paper 095101.

331 Cui, T.J., Chew, W.C., Chen, G., and Song, J.M. (2004). Efficient MLFMA, RPFMA and FAFFA algorithms for EM scattering by very large structures. *IEEE Trans. Antennas Propag.* 52 (3): 759–770.

332 Hu, J., Nie, Z.P., Wang, J. et al. (2004). Multilevel fast multipole algorithm for solving scattering from 3-D electrically large object. *Chin. J. Radio Sci.* 19 (5): 509–524.

333 Dembart, B. and Yip, E. (1995). A 3D fast multipole method for electromagnetics with multiple levels. *Ann. Rev. Prog. Appl. Comput. Electromagn.* 1: 621–628.

334 Ergin, A.A., Shanker, B., and Michielssen, E. (1999). The plane-wave time-domain algorithm for the fast analysis of transient wave phenomena. *IEEE Antennas Propag. Mag.* 41 (4): 39–52.

335 Darve, E. and Havé, P. (2004). A fast multipole method for Maxwell equations stable at all frequencies. *Philos. Trans. R. Soc. London, Ser. A* 362 (1816): 603–628.

336 Kishimoto, S., Ohnuki, S., Ashizawa, Y. et al. (2012). Time domain analysis of nanoscale electromagnetic problems by a boundary integral equation method with fast inverse Laplace transform. *J. Electromagn. Waves Appl.* 26: 997–1006.

337 Moharam, M.G. and Gaylord, T.K. (1981). Rigorous coupled-wave analysis of planar-grating diffraction. *J. Opt. Soc. Am. A* 71: 811–818.

338 Chew, W.C., Barone, S., Anderson, B., and Hennessy, C. (1984). Diffraction of axisymmetric waves in a borehole by bed boundary discontinuities. *Geophysics* 49 (10): 1586–1595.

339 Druskin, V.L. and Knizhnerman, L.A. (1988). A spectral semi-discrete method for the numerical solution of 3-D nonstationary problems in electrical prospecting. *Izv. Phys. Solid Earth* 24: 63–74.

340 Berrington, K.A., Burke, P.G., Le Dourneuf, M. et al. (1978). A new version of the general program to calculate atomic continuum processes using the R-matrix method. *Comput. Phys. Commun.* 14: 367–412.

341 Svizhenko, A., Anantram, M.P., Govindan, T.R., and Biegel, B. (2002). Two-dimensional quantum mechanical modeling of nanotransistors. *J. Appl. Phys.* 91 (4): 2343–2354.

342 Khan, H.R., Mamaluy, D., and Vasileska, D. (2007). Quantum transport simulation of experimentally fabricated nano-FinFET. *IEEE Trans. Electron Devices* 54 (4): 784–796.

343 Milnikov, G., Mori, N., Kamakura, Y., and Ezaki, T. (2008). R-matrix theory of quantum transport and recursive propagation method for device simulations. *J. Appl. Phys.* 104 (4): 044506.

344 Anantram, M.P., Lundstrom, M.S., and Nikonov, D.E. (2008). Modeling of nanoscale devices. *Proc. IEEE* 96 (9): 1511–1550.

345 Radhakrishnan, K. and Chew, W.C. (2001). Efficient analysis of waveguiding structures. In: *Fast and Efficient Algorithms in Computational Electromagnetics*, Chapter 10. Boston, MA: Artech House.

346 Greengard, L. and Gropp, W.D. (1990). A parallel version of the fast multipole method. *Comput. Math. Appl.* 20 (7): 63–71.

347 Ying, L., Biros, G., and Zorin, D. (2004). A kernel-independent adaptive fast multipole algorithm in two and three dimensions. *J. Comput. Phys.* 196 (2): 591–626.

348 Bagci, H., Yilmaz, A.E., Jin, J.-M., and Michielssen, E. (2008). Time domain adaptive integral method for surface integral equations. *Model. Comput. Electromagn.* 65–104.

349 Pirinoli, P., Matekovits, L., Vecchi, G. et al. (2003). Synthetic functions, multiscale MoM analysis of arrays. *IEEE Antennas Propag. Soc. Int. Symp.* 4: 799–802.

350 Li, M.K. and Chew, W.C. (2007). Wave-field interaction with complex structures using equivalence principle algorithm. *IEEE Trans. Antennas Propag.* 55 (1): 130–138.

351 Qian, Z.-G. and Chew, W.C. (2009). Fast full-wave surface integral equation solver for multiscale structure modeling. *IEEE Trans. Antennas Propag.* 57 (11): 3594–3601.

352 Zhao, J.S. and Chew, W.C. (2000). Integral equation solution of Maxwell's equations from zero frequency to microwave frequencies. *IEEE Trans. Antennas Propag.* James R. Wait Memorial Special Issue 48 (10): 1635–1645.

353 Jiang, L.J. and Chew, W.C. (2005). A mixed-form fast multipole algorithm. *IEEE Trans. Antennas Propag.* AP-53 (12): 4145–4156.

354 Cheng, H., Crutchfield, W.Y., Gimbutas, Z. et al. (2006). A wideband fast multipole method for the Helmholtz equation in three dimensions. *J. Comput. Phys.* 216 (1): 300–325.

355 Wallen, H., Jarvenpaa, S., and Yla-Oijala, P. (2006). Broadband multilevel fast multipole algorithm for acoustic scattering problems. *J. Comput. Acoust.* 14 (04): 507–526.

356 Xia, T., Meng, L.L., Liu, Q.S. et al. (2018). A low-frequency stable broadband multilevel fast multipole algorithm using plane wave multipole hybridization. *IEEE Trans. Antennas Propag.* 66 (11): 6137–6145.

357 Hood, P. (1976). Frontal solution program for unsymmetric matrices. *Int. J. Numer. Methods Eng.* 10 (2): 379–399.

358 Smith, B., Bjorstad, P., and Gropp, W. (2004). *Domain Decomposition: Parallel Multilevel Methods for Elliptic Partial Differential Equations.* Cambridge University Press.

359 Kron, G. (1963). *Diakoptics: The Piecewise Solution of Large Scale Systems.* MacDonald.

360 Lai, C.H. (1994). Diakoptics, domain decomposition, and parallel computing. *Comput. J.* 37 (10): 840–846.

361 Goubau, G., Puri, N.N., and Schwering, F.K. (1982). Diakoptic theory for multielement antennas. *IEEE Trans. Antennas Propag.* 30 (1): 15–26.

362 Schwering, F.K., Puri, N.N., and Butler, C.M. (1986). Modified diakoptic theory of antennas. *IEEE Trans. Antennas Propag.* 34 (11): 1273–1281.

363 Niver, E., Smith, H.H., and Whitman, G.M. (1992). Frequency characterization of a thin linear antenna using diakoptic antenna theory. *IEEE Trans. Antennas Propag.* 40 (3): 245–250.

364 Merugu, L.N. and Fusco, V.F. (1993). Concurrent network diakoptics for electromagnetic field problems. *IEEE Trans. Microwave Theory Tech.* 41 (4): 708–716.

365 Ooms, S. and Zutter, D.D. (1998). A new iterative diakoptics-based multilevel moments method for planar circuits. *IEEE Trans. Microwave Theory Tech.* 46 (3): 280–291.

366 Notaros, B.M. (2008). Higher order frequency-domain computational electromagnetics. Special Issue on Large and Multiscale Computational Electromagnetics. *IEEE Trans. Antennas Propag.* 56 (8): 2251–2276.

367 Olcan, D.I., Stevanovic, I.M., Mosig, J.R., and Djordjevic, A.R. (2007). Diakoptic surface integral-equation formulation applied to large 2-D scattering problems. In: *Proceedings of EuCAP 2007*, 11–16. Edinburgh, UK.

368 Stupfel, B. (1996). A fast-domain decomposition method for the solution of electromagnetic scattering by large objects. *IEEE Trans. Antennas Propag.* 44 (10): 1375–1385.

369 Stupfel, B. (1996). A fast-domain decomposition method for the solution of electromagnetic scattering by large objects. *IEEE Trans. Antennas Propag.* 44 (10): 1375–1385.

370 Stupfel, B. and Mognot, M. (2000). A domain decomposition method for the vector wave equation. *IEEE Trans. Antennas Propag.* 48 (5): 653–660.

371 Lu, Y. and Shen, C.Y. (1997). A domain decomposition finite-difference method for parallel numerical implementation of time-dependent Maxwell's equations. *IEEE Trans. Antennas Propag.* 45 (3): 556–562.

372 Wolfe, C.T., Navsariwala, U., and Gedney, S.D. (2000). Parallel finite element tearing and interconnecting algorithm for solution of the vector wave equation with PML absorbing medium. *IEEE Trans. Antennas Propag.* 48 (2): 278–284.

373 Boag, A. (2001). A fast multilevel domain decomposition algorithm for radar imaging. *IEEE Trans. Antennas Propag.* 49 (4): 666–671.

374 Xu, F. and Hong, W. (2004). Analysis of two dimensions sparse multicylinder scattering problem using DD-FDTD method. *IEEE Trans. Antennas Propag.* 52 (10): 2612–2617.

375 Liu, P. and Jin, Y.Q. (2005). Numerical simulation of the Doppler spectrum of a flying target above dynamic oceanic surface by using the FEM-DDM method. *IEEE Trans. Antennas Propag.* 53 (2): 825–832.

376 Vouvakis, M.N., Cendes, Z., and Lee, J.F. (2006). A FEM domain decomposition method for photonic and electromagnetic band gap structures. *IEEE Trans. Antennas Propag.* 54 (2): 721–733.

377 Yin, L., Wang, J., and Hong, W. (2002). A novel algorithm based on the domain-decomposition method for the full-wave analysis of 3-D electromagnetic problems. *IEEE Trans. Microwave Theory Tech.* 50 (8): 2011–2017.

378 Hamandi, L., Lee, R., and Ozgdner, F. (1995). Review of domain-decomposition methods for the implementation of FEM on massively parallel computers. *IEEE Antennas Propag. Mag.* 37 (1): 93–98.

379 Barka, A. and Caudrillier, P. (2007). Domain decomposition method based on generalized scattering matrix for installed performance of antennas on aircraft. *IEEE Trans. Antennas Propag.* 55 (6): 1833–1842.

380 Rubia, V. and Zapata, J. (2007). Microwave circuit design by means of direct decomposition in the finite-element method. *IEEE Trans. Microwave Theory Tech.* 55 (7): 1520–1530.

381 Zhao, K., Rawat, V., Lee, S.C., and Lee, J.F. (2007). A domain decomposition method with nonconformal meshes for finite periodic and semi-periodic structures. *IEEE Trans. Antennas Propag.* 55 (9): 2559–2570.

382 Li, Y.J. and Jin, J.M. (2006). A vector dual-primal finite element tearing and interconnecting method for solving 3-D large-scale electromagnetic problems. *IEEE Trans. Antennas Propag*. 54: 3000–3009. 55 (10): 2803–2810, 2007.

383 Collino, F., Ghanemi, S., and Joly, P. (2000). Domain decomposition method for harmonic wave propagation: a general presentation. *Comput. Meth. Appl. Mech. Eng*. 184: 171–211 . Elsevier.

384 Rebollo, T.C. and Vera, E.C. (2004). Study of a non-overlapping domain decomposition method: Poisson and Stokes problems. *Appl. Numer. Math.* 48: 169–194 . Elsevier.

385 Farhat, C. and Roux, F.-X. (1991). A method of finite element tearing and interconnecting and its parallel solution algorithms. *Int. J. Numer. Methods Eng.* 32: 1205–1227.

386 Lee, S.-C., Vouvakis, M.N., and Lee, J.-F. (2005). A nonoverlapping domain decomposition method with nonmatching grids for modeling large finite antenna arrays. *J. Comput. Phys.* 203: 1–21.

387 Zhao, K., Rawat, V., and Lee, J.-F. (2008). A domain decomposition method for electromagnetic radiation and scattering analysis of multi-target problems. *IEEE Trans. Antennas Propag.* 56: 2211–2221.

388 Ilic, M.M. and Notaros, B.M. (2009). Higher order FEM-MoM domain decomposition for 3-D electromagnetic analysis. *IEEE Antennas Wirel. Propag. Lett.* 8: 970–973.

389 Li, M.-K. and Chew, W.C. (2007). Wave-field interaction with complex structures using equivalence principle algorithm. *IEEE Trans. Antennas Propag.* 55 (1): 130–138.

390 Li, M.-K. and Chew, W.C. (2008). Multiscale simulation of complex structures using equivalence principle algorithm with high-order field point sampling scheme. *IEEE Trans. Antennas Propag.* 56 (8): 2389–2397.

391 Sun, L.E., Chew, W.C., and Jin, J.M. (2010). Augmented equivalence principle algorithm at low frequencies. *Microwave Opt. Technol. Lett.* 52 (10): 2274–2279.

392 Ma, Z.H., Jiang, L.J., and Chew, W.C. (2013). Loop-tree free augmented equivalence principle algorithm for low frequency problems. *Microwave Opt. Technol. Lett.* 55 (10): 2475–2479.

393 Shao, H.R., Hu, J., Chew, W.C. (2013). Single source EPA for scattering by array structures. AP-S/URSI *Symposium*, Orlando (July, 2013).

394 Chao, H.Y., Song, J.M., Michielssen, E., and Chew, W.C. (1998). The multilevel fast multipole algorithm for analyzing electromagnetic radiation from complex surface-wire structures. *Antenna Applications Symposium*, Monticello, IL (16–18 September, 1998).

395 Song, J.M., Lu, C.C., and Chew, W.C. (1997). Multilevel fast multipole algorithm for electromagnetic scattering by large complex objects. *IEEE Trans. Antennas Propag.* 45 (10): 1488–1493.

396 Hesford, A.J. and Chew, W.C. (2008). On preconditioning and the eigensystems of electromagnetic radiation problems. *IEEE Trans. Antennas Propag.* 56: 2413–2420.

397 He, S.Q., Sha, W.E.I., Jiang, L.J. et al. (2012). Finite element based generalized impedance boundary condition for modeling plasmonic nanostructures. *IEEE Trans. Nanotechnol.* https://doi.org/10.1109/TNANO.2011.2171987.

398 Ling, H., Chou, R.C., and Lee, S.W. (1989). Shooting and bouncing rays: calculating the RCS of an arbitrarily shaped cavity. *IEEE Trans. Antennas Propag.* 37: 194–205.

399 Engquist, B. and Runborg, O. (2003). Computational high frequency wave propagation. *Acta Numer.* 12: 181–266.

400 Chandler-Wilde, S.N., Graham, I.G., Langdon, S., and Spence, E.A. (2012). Numerical-asymptotic boundary integral methods in high-frequency acoustic scattering. *Acta Numer.* 21: 89–305.

401 Osher, S.J. and Sethian, J.A. (1988). Fronts propagating with curvature-dependent speed: algorithms based on Hamilton-Jacobi formulations. *J. Comput. Phys.* 79: 12–49.

402 Sethian, J.A. (1996). A fast marching level set method for monotonically advancing fronts. *Proc. Natl. Acad. Sci. U.S.A.* 93: 1591–1595.

403 Červený, V., Popov, M.M., and Pšenč′ lk, I. (1982). Computation of wave fields in inho-mogeneous media–Gaussian beam approach. *Geophys. J. R. Astron. Soc.* 70: 109–128.

404 Tanushev, N.M., Engquist, B., and Tsai, R. (2009). Gaussian beam decomposition of high frequency wave fields. *J. Comput. Phys.* 228: 8856–8871.

405 Macdonald, H.M. (1913). The effect produced by an obstacle on a train of electric waves. *Philos. Trans. R. Soc. London, Ser. A* 212: 299–337.

406 Gordon, W.B. (1975). Far-field approximations to the Kirchhoff-Helmholtz representation of scattered fields. *IEEE Trans. Antennas Propag.* 23: 590–592.

407 Ludwig, A.C. (1968). Computation of radiation patterns involving numerical double integration. *IEEE Trans. Antennas Propag.* 16: 767–769.

408 Burkholder, R.J. and Lee, T. (2005). Adaptive sampling for fast physical optics numerical integration. *IEEE Trans. Antennas Propag.* 53: 1843–1845.

409 Obelleiro-Basteiro, F., Rodriguez, J.L., and Burkholder, R.J. (1995). An iterative physical optics approach for analyzing the electromagnetic scattering by large open-ended cavities. *IEEE Trans. Antennas Propag.* 43: 356–361.

410 Burkholder, R.J. and Lundin, T. (2005). Forward-backward iterative physical optics algorithm for computing the RCS of open-ended cavities. *IEEE Trans. Antennas Propag.* 53: 793–799.

411 Wu, Y.M., Jiang, L.J., and Chew, W.C. (2012). An efficient method for computing highly oscillatory physical optics integral. *Prog. Electromagn. Res.* 127: 211–257.

412 Chandler-Wilde, S.N. and Graham, I.G. (2009). Boundary integral methods in high-frequency scattering. In: *Highly Oscillatory Problems* (eds. B. Engquist, T. Fokas, E. Hairer and A. Iserles), 154–193. Cambridge: Cambridge University Press.

413 Domlnguez, V., Graham, I.G., and Smyshlyaev, V.P. (2007). A hybrid numerical-asymptotic boundary integral method for high-frequency acoustic scattering. *Numer. Math.* 106: 471–510.

414 Chandler-Wilde, S.N. and Langdon, S. (2007). A Galerkin boundary element method for high frequency scattering by convex polygons. *SIAM J. Numer. Anal.* 45: 610–640.

415 Bruno, O.P. and Kunyansky, L.A. (2001). A fast, high-order algorithm for the solution of surface scattering problems: basic implementation, tests, and applications. *J. Comput. Phys.* 169 (1): 80–110.

416 Perrey-Debain, E., Trevelyan, J., and Bettess, P. (2003). Plane wave interpolation in direct collocation boundary element method for radiation and wave scattering: numerical aspects and applications. *J. Sound Vib.* 261 (5): 839–858.

417 Huybrechs, D. and Vandewalle, S. (2006). On the evaluation of highly oscillatory integrals by analytic continuation. *SIAM J. Numer. Anal.* 44 (3): 1026–1048.

418 Geuzaine, C., Bruno, O., and Reitich, F. (2005). On the O(1) solution of multiple-scattering problems. *IEEE Trans. Magn.* 41 (5). 1400–1491.

419 Davis, C.P. and Chew, W.C. (2008). Frequency-independent scattering from a flat strip with TE_z-polarized fields. *IEEE Trans. Antennas Propag.* 56 (4): 1008–1016.

420 Burridge, R. (1980, 1980). The Gelfand-Levitan, the Marchenko, and the Gopinath-Sondhi integral equations of inverse scattering theory, regarded in the context of inverse impulse-response problems. *Wave Motion* 2: 305–323.

421 Cutrona, L.J. (1970). Synthetic aperture radar , Chapter 23. In: *Radar Handbook* (ed. M.I. Skolnik). New York, NY: McGraw Hill.

422 McCandless, S.W. and Jackson, C.R. (2004). Principles of synthetic aperture radar," Chapter 1. In: *SAR Marine Users Manual* (eds. C.R. Jackson and J.R. Apel), 1–23. NOAA.

423 Devaney, A.J. (1982). A filtered backpropagation algorithm for diffraction tomography. *Ultrason. Imaging* 4 (6): 336–350.

424 Devaney, A.J. (1983). A computer simulation study of diffraction tomography. *IEEE Trans. Biomed. Eng.* 30 (7): 377–386.

425 Tarantola, A. (1984). Inversion of seismic reflection data in the acoustic approximation. *Geophysics* 49: 1259–1266.

426 Chew, W.C. and Wang, Y.M. (1990). Reconstruction of two-dimensional permittivity distribution using the distorted Born iterative method. *IEEE Trans. Med. Imaging* 9: 218–225.

427 Bond, E.J., Bond, E.J., Li, X. et al. (2003). Microwave imaging via space-time beamforming for early detection of breast cancer. *IEEE Trans. Antennas Propag.* 51: 1690–1705.

428 Cui, T.J., Chew, W.C., Aydiner, A.A., and Chen, S. (2001). Inverse scattering of two-dimensional dielectric objects buried in a lossy earth using the distorted Born iterative method. *IEEE Trans. Geosci. Remote Sens.* 39: 339–346.

429 Wang, G.L., Chew, W.C., Cui, T.J. et al. (2004). 3D near-to-surface conductivity reconstruction by inversion of VETEM data using the distorted Born iterative method. *Inverse Prob.* 20: S195–S216.

430 Mojabi, P. and LoVetri, J. (2009). Overview and classification of some regularization techniques for the Gauss-Newton inversion method applied to inverse scattering problems. *IEEE Trans. Antennas Propag.* 57: 2658–2665.

431 Liu, Q.H., Zhang, Z.Q., Wang, T.T. et al. (2002). Active microwave imaging I–2-D forward and inverse scattering methods. *IEEE Trans. Microwave Theory Tech.* 50: 123–133.

432 Cui, T.J., Chew, W.C., Yin, X.X., and Hong, W. (2004). Study of resolution and super resolution in electromagnetic imaging for half-space problems. *IEEE Trans. Antennas Propag.* 52: 1398–1411.

433 Deuflhard, P. (2004). *Newton Methods for Nonlinear Problems. Affine Invariance and Adaptive Algorithms*, Springer Series in Computational Mathematics, 1e, vol. 35. Berlin: Springer.

434 Franchois, A., Joisel, A., Pichot, C., and Bolomey, J.-C. (1998). Quantitative microwave imaging with a 2.45-GHz planar microwave camera. *IEEE Trans. Med. Imaging* 17 (4): 550–561.

435 Van Den Berg, P.M. and Kleinman, R.E. (1997). A contrast source inversion method. *Inverse Prob.* 13 (6): 1607–1620.

436 Abubakar, A., Habashy, T.M., Druskin, V.L. et al. (2008). 2.5D forward and inverse modeling for interpreting low-frequency electromagnetic measurements. *Geophysics* 73: F165–F177.

437 Zhong, Y. and Chen, X. (2007). MUSIC imaging and electromagnetic inverse scattering of multiple-scattering small anisotropic spheres. *IEEE Trans. Antennas Propag.* 55: 3542–3549.

438 Litman, A., Lesselier, D., and Santosa, F. (1998). Reconstruction of a two-dimensional binary obstacle by controlled evolution of a level-set. *Inverse Prob.* 14 (3): 685–706.

439 Torres-Verdín, C. and Habashy, T.M. (1994). Rapid 2.5-dimensional forward modeling and inversion via a new nonlinear scattering approximation. *Radio Sci.* 29: 1051–1079.

440 Cui, T.J. and Chew, W.C. (1999). Fast algorithm for electromagnetic scattering by buried 3-D dielectric objects of large size. *IEEE Trans. Geosci. Remote Sens.* 37: 2597–2608.

441 Cui, T.J. and Liang, C.H. Nonlinear differential equation for the reflection coefficient of an inhomogeneous lossy medium and its inverse scattering solutions. *IEEE Trans. Antennas Propag.* 42 (5): 621–626.

442 Hesford, A.J. and Chew, W.C. (2010). Fast inverse scattering solutions using the distorted Born iterative method and the multilevel fast multipole algorithm. *J. Acoust. Soc. Am.* 128: 679–690.

443 Moghaddam, M. and Chew, W.C. (1992). Nonlinear two-dimensional velocity profile inversion using time domain data. *IEEE Trans. Geosci. Remote Sens.* 30 (1): 147–156.

444 Wang, Y.M. and Chew, W.C. (1992). Accelerating the iterative inverse scattering algorithms by using the fast recursive aggregate T-matrix algorithm. *Radio Sci.* 27 (2): 109–116.

445 Weedon, W.H. and Chew, W.C. (1999). Time-domain inverse scattering using the local shape function (LSF) method. *Inverse Prob.* 9 (5): 551.

446 Chen, F.C. and Chew, W.C. (1998). Experimental verification of super resolution in nonlinear inverse scattering. *Appl. Phys. Lett.* 72 (23): 3080–3082.

447 Ali, M.A. and Moghaddam, M. (2010). 3D nonlinear super-resolution microwave inversion technique using time-domain data. *IEEE Trans. Antennas Propag.* 58 (7): 2327–2336.

448 Gilmore, C., Mojabi, P., Zakaria, A. et al. (2010). On super-resolution with an experimental microwave tomography system. *IEEE Antennas Wirel. Propag. Lett.* 9: 393–396.

449 Lehman, S.K. and Devaney, A.J. (2003). Transmission mode time-reversal super-resolution imaging. *J. Acoust. Soc. Am.* 113: 2742.

450 Lerosey, G., De Rosny, J., Tourin, A. et al. (2004). Time reversal of electromagnetic waves. *Phys. Rev. Lett.* 92 (19): 193904.

451 Pembry, J.B. (2000). Negative refraction makes a perfect lens. *Phys. Rev. Lett.* 85: 3966–3969.

452 Pendry, J.B., Schurig, D., and Smith, D.R. (2006). Controlling electromagnetic fields. *Science* 312: 1780–1782.

453 Schurig, D., Mock, J.J., Justice, B.J. et al. (2006). Metamaterial electromagnetic cloak at microwave frequencies. *Science* 314: 977–980.

454 Cummer, S.A., Popa, B.I., Schurig, D. et al. (2006). Full-wave simulations of electromagnetic cloaking structures. *Phys. Rev. E* 74: 036621.

455 Chew, W.C. (2005). Some reflections on double negative materials. *Prog. Electromagn. Res.* 51: 1–26.

456 Thomas, J.R. and Ishimaru, A. (2002). Transmission properties of material with relative permittivity and permeability close to -1. *Proc. SPIE* 4806: 167–175.

457 Dolin, L.S. (1961). *Izv. Vyssh. Uchebn. Zaved. Radiofizika* 4: 964.
458 Odabasi, H., Teixeira, F.L., and Chew, W.C. (2011). Impedance-matched absorbers and optical pseudo black holes. *J. Opt. Soc. Am. B: Opt. Phys.* 28 (5): 1317–1323.
459 Chen, H. and Chan, C.T. (2007). Acoustic cloaking in three dimensions using acoustic metamaterials. *Appl. Phys. Lett.* 91 (18).
460 Liu, R., Ji, C., Mock, J.J. et al. (2009). Broadband ground-plane cloak. *Science* 323: 366–369.
461 Shu, W.W. and Song, J.M. (2006). Complete mode spectrum of a grounded dielectric slab with double negative metamaterials. *Prog. Electromagn. Res.* 65: 103–123.
462 Cui, T.J., Smith, D.R., and Liu, R. (2009). *Metamaterials: Theory, Design, and Applications*. Springer.
463 Ma, H.F. and Cui, T.J. (2010). Three-dimensional broadband ground-plane cloak made of metamaterials. *Nat. Commun.* 1 (3): 21.
464 Choo, H.S., Rogers, R.L., and Ling, H. (2005). Design of electrically small wire antennas using a pareto genetic algorithm. *IEEE Trans. Antennas Propag.* 53: 1038–1046.
465 Adams, J.J. and Bernhard, J.T. (2009). Tuning method for a new electrically small antenna with low Q. *IEEE Antennas Wirel. Propag. Lett.* 8: 303–306.
466 Hansen, R. and Collin, R.E. (2011). *Small Antenna Handbook*. Wiley.
467 Best, S.R. (2007). A discussion on power factor, quality factor and efficiency of small antennas. *IEEE Antennas Propag. Soc. Int. Symp. Dig.* 2269–2272.
468 Yaghjian, A.D., O'Donnell, T.H., Altshuler, E.E., and Best, S.R. (2008). Electrically small supergain end-fire arrays. *Radio Sci.* 43: 13.
469 Thal, H.L. Jr. (2006). New radiation Q limits for spherical wire antennas. *IEEE Trans. Antennas Propag.* 54: 2757–2763.
470 Wong, H., Luk, K.-M., Chan, C.H. et al. (2012). Small antennas in wireless communications. *Proc. IEEE* 100 (7): 2109–2121.
471 Wong, K.L. (2003). *Planar Antennas for Wireless Communications*. New York, NY: Wiley.
472 Chen, Z.N., See, T.S.P., and Qing, X. (2007). Small printed ultrawideband antenna with reduced ground plane effect. *IEEE Trans. Antennas Propag.* 55: 383–388.
473 Jordan, E.C. and Balmain, K.G. (1968). *Electromagnetic Waves and Radiating Systems*. Upper Saddle River, NJ: Prentice-Hall.
474 Chu, L.J. (1948). Physical limitations of omni-directional antennas. *J. Appl. Phys.* 19: 1163–1175.
475 McLean, J.S. (1996). A re-examination of the fundamental limits on the radiation Q of electrically small antenna. *IEEE Trans. Antennas Propag.* 44: 672–676.
476 Erentok, A. and Ziolkowski, R.W. (2008). Metamaterial-inspired efficient electrically small antennas. *IEEE Trans. Antennas Propag.* 56: 691–707.

2

Computational Electromagnetics

2.1 Introduction

In the preceding chapter, we provided a macroscopic observation for electromagnetics (EM) by putting it into a larger picture with physics and mathematics as its background. Although the fundamental or root of EM resides in physics and mathematics [1], many EM problems arising from practical engineering applications have their own physical features and need a zoom-in examination from microscopic angles. As modern products, including EM devices and systems, have become more and more complicated and meticulous, such an examination or analysis has become indispensable for optimally designing them. Analyzing EM problems is a non-trivial thing in general due to the complexity of the involved physics and mathematics, and it is challenging to obtain a complete solution for the problems.

Generally, engineering problems, including EM problems, can be solved by either an analytical approach or experimental approach. The analytical approach can give exact or very accurate solutions in theory and is always preferred as a first choice, but unfortunately, most of problems cannot be solved by the approach due to their complexity. The approach can only be used to solve some simple problems whose structures usually have simple regular geometries and ideal material constitutions; therefore, the application of such an approach is very limited. The experimental approach, on the other hand, looks more powerful and can solve unlimited problems when combined with necessary theoretical analyses, so it has been the primary approach for a long time, and even today. However, the experimental approach requires many pieces of expensive equipment, a special site and space, and much human resource. Also, it is very inflexible, meaning that people cannot use the experimental approach when they leave their laboratories.

As the most powerful machine of human beings, the computer and its relevant technologies, which are continuously and rapidly developed or renewed, have completely changed the research method and means. Based on computers, a brand-new approach of solving problems, i.e. the computer modeling and simulation (CMS) approach [2], was proposed and has been significantly developed. The approach is so powerful that almost all research areas have been using it. The approach has been thought of as the third approach for research or solving problems, in addition to the first approach, which is the analytical or theoretical approach, and the second approach,

The Nyström Method in Electromagnetics, First Edition. Mei Song Tong and Weng Cho Chew.

which is the experimental approach as mentioned above. The CMS approach is a bridge between the first approach and second approach and can take the advantages from both approaches while overcome their drawbacks at the same time. With the CMS approach, human beings can dramatically expand their ability to solve problems. A vast number of problems, either theoretical or practical, were previously impossible to be solved, but they can now be easily solved by the CMS approach. In addition, it is very flexible in both time and space, and requires much less resource in both equipment and human power. Due to the huge contribution of this approach to the research in chemistry, three pioneering scientists who first used this approach to solve chemical problems in their research work even received the Nobel prize in chemistry in 2013. Although some researchers still doubt the reliability of results produced by the CMS approach, it can be expected that this approach will fully replace the first and second approaches in the future when the reliability is not problematic or the final experimental verification becomes unnecessary [3].

In EM, the CMS approach has also rapidly developed and has a wide application. The CMS approach uses computational methods that include both analytical and numerical methods (but mainly numerical methods) to solve EM problems in the platform of computers and we usually categorize the relevant work as computational electromagnetics (CEM). Since available analytical methods are quite few and numerical methods are the major ones in computational methods, the numerical methods with their relevant techniques and applications for solving EM problems have been the main content of research in CEM. More specifically, CEM is to study the accuracy, speed, cost, robustness, complexity, etc. of numerical methods in solving EM problems in addition to developing new or improving existing computational methods.

Classical EM problems are fully governed by Maxwell's equations, which can be expressed into first-order partial differential equations (PDEs) in mathematics. The equations describe the relationship between the fields (electric field and magnetic field) and sources (electric current density, magnetic current density, electric charge density, and magnetic charge density) in a microscope and can be thought of universal equations that govern all electric phenomena. Maxwell's equations can be mathematically manipulated and take different forms. For example, the original equations include a coupling between the electric field and magnetic field and we can obtain uncoupled second-order PDEs, which are vector wave equations or Helmholtz equations, by uncoupling the first two equations. By introducing the Green's function and representing fields in terms of potentials, we can further transform the equations into a form of integral equations. Aiming at the different forms of Maxwell's equations, different numerical methods and relevant techniques have been developed to solve them. The most well-known methods include the finite-difference time-domain (FDTD) method [4], the finite element method (FEM) [5], and the method of moments (MoM) [6], and they are used to solve the corresponding three forms of Maxwell's equations, respectively. We will give a brief introduction for each of them, but they are not our theme. The Nyström method, which is the theme of this book, is also an important numerical method to solve EM integral equations and has become a competitor to the widely-used MoM in recent years. We will give a detailed introduction to it and present its applications in the subsequent chapters. Although there are very few analytical methods, we also present the Mie-series method [7] for solving EM scattering problems by spherical objects in this chapter because its solutions are widely used to verify numerical solutions.

2.2 Analytical Methods

Although almost no analytical methods can be used to solve complicated problems in terms of today's standard, they did make a huge contribution to solving simple problems in early stages, and some of them are still being used for certain applications nowadays. For example, the separation-of-variables method is widely used to solve vector wave equations for waveguide problems with regular simple geometrical shapes, and the solutions of fields can be written in closed-form expressions [8]. Another analytical method is the Mie-series method of solving EM scattering problems by spherical objects [7], and the solutions are widely used to verify the solutions of numerical methods. In fact, the Mie-series solutions are not exactly analytical because their expressions are in infinite series or open form instead of purely-closed form. Therefore, more accurately speaking, the Mie-series method is semi-analytical, but its solutions are usually treated as "exact" solutions in comparison to numerical solutions.

The analytical methods are always desirable and should be the first choice no matter how complicated their derivation and formulas or expressions look. Some researchers could prefer numerical methods instead of analytical methods in certain applications, for instance, for integrations, because numerical quadratures are simpler in implementation while analytical formulas are hard to use sometimes even though they are available. However, analytical methods are of very high value, not only because they can be used to validate numerical solutions, but also because they are "continuous" methods that can cover all possibilities and possess a "completeness" from a mathematical point of view. In contrast, the numerical methods are "discrete" methods that can only give solutions at chosen discrete spatial and temporal points. For non-chosen points, one has to use interpolation methods based on those chosen points to calculate their values and this will introduce extra errors and much postprocessing inconvenience if compared with analytical methods.

Owing to the importance and extensiveness of the Mie-series method in CEM, we particularly illustrate this method as a representative of analytical methods. Suppose we have a spherical object embedded in free space with a permittivity ϵ_0 and a permeability μ_0. The object could be a perfectly electric conductor (PEC), a homogeneous dielectric body, or a mixture of both materials, and its center is located at the origin of a rectangular coordinate system $\{x, y, z\}$ and its corresponding spherical coordinate system $\{r, \theta, \phi\}$. A plane wave, which has an angular frequency ω and is polarized in the x direction, is propagating along the $-z$ direction and impinges upon the object to produce scattered fields at a spatial point $\mathbf{r}$ which is located at (r, θ, ϕ) in the spherical coordinate system. The incident fields can be expressed as

$$\mathbf{E}^{\rm inc}(\mathbf{r}) = E_0 \mathrm{e}^{-ik_0 z}\hat{x} \tag{2.1}$$

$$\mathbf{H}^{\rm inc}(\mathbf{r}) = -\sqrt{\frac{\epsilon_0}{\mu_0}} E_0 \mathrm{e}^{-ik_0 z}\hat{y} = -\frac{E_0}{\eta_0}\mathrm{e}^{-ik_0 z}\hat{y} \tag{2.2}$$

where $k_0 = \omega\sqrt{\epsilon_0\mu_0}$ is the wavenumber and $\eta_0 = \sqrt{\mu_0/\epsilon_0}$ is the wave impedance in free space, and E_0 is the magnitude of the incident electric field. Also, $\hat{x}$, $\hat{y}$, and $\hat{z}$ represent the three unit vectors of three coordinate axes in the rectangular coordinate system. The Mie-series solution of scattered fields at the point $\mathbf{r}$ can be found by using spherical vector wave functions as basis functions to represent the scattered fields and imposing boundary conditions at boundaries to find the expansion coefficients of spherical vector wave functions. The earliest derivation can be found in [9] and the scattered fields can be written as

$$\mathbf{E}^{\mathrm{sca}}(\mathbf{r}) = E_0 \sum_{n=1}^{\infty} \left[A_n \mathbf{M}_{1n}(\mathbf{r}) + B_n \mathbf{N}_{2n}(\mathbf{r})\right] \tag{2.3}$$

$$\mathbf{H}^{\mathrm{sca}}(\mathbf{r}) = -\mathrm{i}\sqrt{\frac{\epsilon_0}{\mu_0}} E_0 \sum_{n=1}^{\infty} \left[B_n \mathbf{M}_{2n}(\mathbf{r}) + A_n \mathbf{N}_{1n}(\mathbf{r})\right] \tag{2.4}$$

where $\mathbf{M}_{jn}$ and $\mathbf{N}_{jn}$ $(j = 1, 2)$ are the spherical vector wave functions which are only related to the position of field point $\mathbf{r}$. The constant expansion coefficients A_n and B_n are determined by enforcing boundary conditions and they are only related to the geometry and material of the object which are independent of the field point. The spherical vector wave functions are defined by [10]

$$\mathbf{M}_{\substack{1\\2}n}(\mathbf{r}) = \pm\frac{1}{\sin\theta} h_n^{(1)}(k_0 r) P_n^1(\cos\theta) \begin{bmatrix}\cos\phi\\ \sin\phi\end{bmatrix} \hat{\theta} - h_n^{(1)}(k_0 r)\frac{\mathrm{d}}{\mathrm{d}\theta} P_n^1(\cos\theta) \begin{bmatrix}\sin\phi\\ \cos\phi\end{bmatrix} \hat{\phi} \tag{2.5}$$

$$\begin{aligned}\mathbf{N}_{\substack{1\\2}n}(\mathbf{r}) = {} & \frac{n(n+1)}{k_0 r} h_n^{(1)}(k_0 r) P_n^1(\cos\theta) \begin{bmatrix}\sin\phi\\ \cos\phi\end{bmatrix} \hat{r} \\ & + \frac{1}{k_0 r}\frac{\mathrm{d}}{\mathrm{d}(k_0 r)}\left[k_0 r h_n^{(1)}(k_0 r)\right] \frac{\mathrm{d}}{\mathrm{d}\theta} P_n^1(\cos\theta) \begin{bmatrix}\sin\phi\\ \cos\phi\end{bmatrix} \hat{\theta} \\ & \pm \frac{1}{k_0 r \sin\theta}\frac{\mathrm{d}}{\mathrm{d}(k_0 r)}\left[k_0 r h_n^{(1)}(k_0 r)\right] P_n^1(\cos\theta) \begin{bmatrix}\cos\phi\\ \sin\phi\end{bmatrix} \hat{\phi}\end{aligned} \tag{2.6}$$

where $h_n^{(1)}(k_0 r)$ is the first kind of spherical Hankel function of order n while $P_n^1(\cos\theta)$ represents the first kind of associate Legendre function of order n. Also, $\hat{r}$, $\hat{\theta}$, and $\hat{\phi}$ are the three unit vectors of three coordinate axes in the spherical coordinate system. For calculating near-zone fields, the above formulas should be used without any approximation. However, for calculating far-zone fields or radar cross sections, the spherical vector wave functions can be greatly simplified by an approximation with $r \to \infty$ and the scattered fields can also be simplified accordingly. The simplified scattered fields are given by

$$\mathbf{E}^{\mathrm{sca}}(\mathbf{r}) = E_0 \frac{e^{\mathrm{i}k_0 r}}{k_0 r}\left[\cos\phi S_1(\theta)\hat{\theta} - \sin\phi S_2(\theta)\hat{\phi}\right] \tag{2.7}$$

$$\mathbf{H}^{\mathrm{sca}}(\mathbf{r}) = \sqrt{\frac{\epsilon_0}{\mu_0}}\mathbf{E}^{\mathrm{sca}}(\mathbf{r}) \tag{2.8}$$

where

$$S_1(\theta) = \sum_{n=1}^{\infty} (-\mathrm{i})^{n+1}\left[A_n \frac{P_n^1(\cos\theta)}{\sin\theta} + \mathrm{i}B_n \frac{\mathrm{d}}{\mathrm{d}\theta} P_n^1(\cos\theta)\right] \tag{2.9}$$

$$S_2(\theta) = \sum_{n=1}^{\infty} (-\mathrm{i})^{n+1}\left[A_n \frac{\mathrm{d}}{\mathrm{d}\theta} P_n^1(\cos\theta) + \mathrm{i}B_n \frac{P_n^1(\cos\theta)}{\sin\theta}\right]. \tag{2.10}$$

To determine the expansion coefficients A_n and B_n, we need to consider the physical properties of the object. If the object is a PEC sphere with a radius a, then we can find [10]

$$A_n = -(-\mathrm{i})^n \frac{2n+1}{n(n+1)} \frac{j_n(k_0 a)}{h_n(k_0 a)} \tag{2.11}$$

$$B_n = (-\mathrm{i})^{n+1} \frac{2n+1}{n(n+1)} \frac{\left[k_0 a j_n(k_0 a)\right]'}{\left[k_0 a h_n(k_0 a)\right]'} \tag{2.12}$$

where the prime denotes a derivative with respect to $k_0 a$. Also, $j_n(k_0 a)$ is the first kind of spherical Bessel function of order n and we have written $h_n^{(1)}$ into h_n for simplicity. If the object is a homogeneous dielectric sphere that could be lossy and it has a radius a, a permittivity $\epsilon_1 = \epsilon_0 \epsilon_{r1}$, and a permeability $\mu_1 = \mu_0 \mu_{r1}$, respectively, then we can find the coefficients as [10]

$$A_n = (-\mathrm{i})^n \frac{2n+1}{n(n+1)} \frac{j_n(k_0 a)\left[k_1 a j_n(k_1 a)\right]' - j_n(k_1 a)\left[k_0 a j_n(k_0 a)\right]'}{j_n(k_1 a)\left[k_0 a h_n(k_0 a)\right]' - h_n(k_0 a)\left[k_1 a j_n(k_1 a)\right]'} \tag{2.13}$$

$$B_n = (-\mathrm{i})^{n+1} \frac{2n+1}{n(n+1)} \frac{j_n(k_0 a)\left[k_1 a j_n(k_1 a)\right]' - m_1^2 j_n(k_1 a)\left[k_0 a j_n(k_0 a)\right]'}{h_n(k_0 a)\left[k_1 a j_n(k_1 a)\right]' - m_1^2 j_n(k_1 a)\left[k_0 a h_n(k_0 a)\right]'} \tag{2.14}$$

where $k_1 = \omega\sqrt{\epsilon_1 \mu_1} = k_0 m_1$ is the wavenumber related to the material of the object and $m_1 = \sqrt{\epsilon_{r1}\mu_{r1}}$ is called the refractive index of the object's material, which is a complex number when the material is lossy. Similarly, the prime denotes a derivative with respect to $k_0 a$ or $k_1 a$.

When the object is a homogeneous dielectric sphere coated with another homogeneous dielectric layer, as shown in Figure 2.1, where the geometrical parameters and property of each material have been labelled, it is a radially stratified or radially inhomogeneous object, which is also treated as a piecewise homogeneous object. The coefficients of such an object can be found as [10]

$$A_n = -(-\mathrm{i})^n \frac{2n+1}{n(n+1)} \frac{k_0 a_0 j_n(k_0 a_0) - \mathrm{i} Z_n(k_0 a_0)\left[k_0 a_0 j_n(k_0 a_0)\right]'}{k_0 a_0 h_n(k_0 a_0) - \mathrm{i} Z_n(k_0 a_0)\left[k_0 a_0 h_n(k_0 a_0)\right]'} \tag{2.15}$$

$$B_n = (-\mathrm{i})^{n+1} \frac{2n+1}{n(n+1)} \frac{k_0 a_0 j_n(k_0 a_0) - \mathrm{i} Y_n(k_0 a_0)\left[k_0 a_0 j_n(k_0 a_0)\right]'}{k_0 a_0 h_n(k_0 a_0) - \mathrm{i} Y_n(k_0 a_0)\left[k_0 a_0 h_n(k_0 a_0)\right]'} \tag{2.16}$$

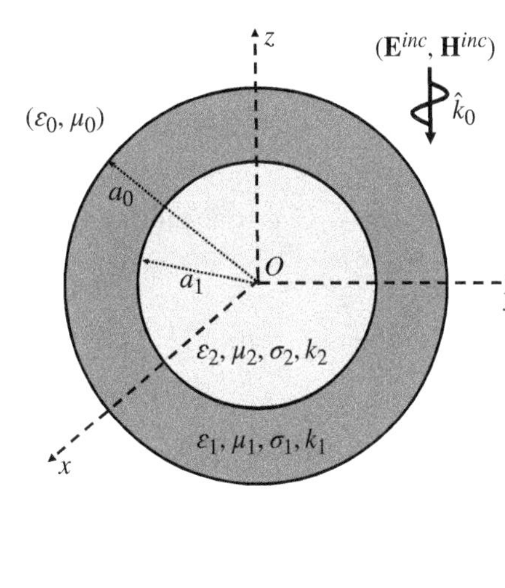

Figure 2.1 Scattering by an homogeneous dielectric spherical core coated with a one-layer homogeneous dielectric coating.

where $Z_n(k_0a_0)$ and $Y_n(k_0a_0)$ are called normalized nth-order modal surface impedance and admittance, respectively [11]. The surface impedance and admittance can be derived as

$$Z_n(k_0a_0) = ik_0a_0\left(\frac{f_2}{f_1}\right), \quad Y_n(k_0a_0) = ik_0a_0\left(\frac{f_4}{f_3}\right) \tag{2.17}$$

where

$$\begin{aligned} f_1 = {} & \frac{j_n(k_2a_1)}{\mu_1^2}\left\{[k_1a_0j_n(k_1a_0)]'[k_1a_1h_n(k_1a_1)]' - [k_1a_1j_n(k_1a_1)]'[k_1a_0h_n(k_1a_0)]'\right\} \\ & + \frac{[k_2a_1j_n(k_2a_1)]'}{\mu_1\mu_2}\left\{j_n(k_1a_1)[k_1a_0h_n(k_1a_0)]' - h_n(k_1a_1)[k_1a_0j_n(k_1a_0)]'\right\} \end{aligned} \tag{2.18}$$

$$\begin{aligned} f_2 = {} & \frac{j_n(k_2a_1)}{\mu_0\mu_1}\left\{[k_1a_1j_n(k_1a_1)]'h_n(k_1a_0) - [k_1a_1h_n(k_1a_1)]'j_n(k_1a_0)\right\} \\ & + \frac{[k_2a_1j_n(k_2a_1)]'}{\mu_0\mu_2}[j_n(k_1a_0)h_n(k_1a_1) - j_n(k_1a_1)h_n(k_1a_0)] \end{aligned} \tag{2.19}$$

$$\begin{aligned} f_3 = {} & \frac{k_0^2[k_2a_1j_n(k_2a_1)]'}{\mu_0\mu_1k_2^2}\left\{j_n(k_1a_1)[k_1a_0h_n(k_1a_0)]' - h_n(k_1a_1)[k_1a_0j_n(k_1a_0)]'\right\} \\ & + \frac{k_2^2j_n(k_2a_1)}{\mu_0\mu_2k_1^2}\left\{[k_1a_0j_n(k_1a_0)]'[k_1a_1h_n(k_1a_1)]' - [k_1a_1j_n(k_1a_1)]'[k_1a_0h_n(k_1a_0)]'\right\} \end{aligned} \tag{2.20}$$

$$\begin{aligned} f_4 = {} & \frac{k_1^2[k_2a_1j_n(k_2a_1)]'}{\mu_1^2k_2^2}[j_n(k_1a_0)h_n(k_1a_1) - j_n(k_1a_1)h_n(k_1a_0)] \\ & + \frac{k_2^2j_n(k_2a_1)}{\mu_1\mu_2k_0^2}\left\{[k_1a_1j_n(k_1a_1)]'h_n(k_1a_0) - [k_1a_1h_n(k_1a_1)]'j_n(k_1a_0)\right\}. \end{aligned} \tag{2.21}$$

Note that the above formulas can be simplified if the inner sphere or core is a PEC that has an infinite conductivity of leading to $k_2 = \infty$. In this case, the surface impedance and admittance become

$$Z_n(k_0a_0) = ik_0a_0\frac{\mu_1}{\mu_0}\frac{j_n(k_1a_0)h_n(k_1a_1) - j_n(k_1a_1)h_n(k_1a_0)}{j_n(k_1a_1)[k_1a_0h_n(k_1a_0)]' - h_n(k_1a_1)[k_1a_0j_n(k_1a_0)]'} \tag{2.22}$$

$$Y_n(k_0a_0) = ik_0a_0\frac{\mu_0k_1^2}{\mu_1k_0^2}\frac{h_n(k_1a_0)[k_1a_1j_n(k_1a_1)]' - j_n(k_1a_0)[k_1a_1h_n(k_1a_1)]'}{[k_1a_0j_n(k_1a_0)]'[k_1a_1h_n(k_1a_1)]' - [k_1a_1j_n(k_1a_1)]'[k_1a_0h_n(k_1a_0)]'}. \tag{2.23}$$

Also, if $a_0 = a_1 = a$, the above case will be degenerated to the case of scattering by a homogeneous dielectric sphere and the surface impedance and admittance can be reduced to

$$Z_n(k_0a_0) = -i\frac{\mu_1k_0}{\mu_0k_1}\frac{k_1a_0j_n(k_1a_0)}{[k_1a_0j_n(k_1a_0)]'} \tag{2.24}$$

$$Y_n(k_0a_0) = \frac{\mu_0^2k_1^2}{\mu_1^2k_0^2}Z_n(k_0a_0). \tag{2.25}$$

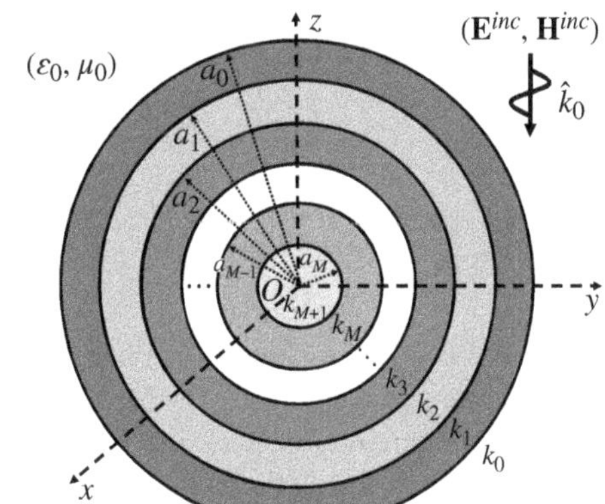

Figure 2.2 Scattering by an homogeneous dielectric spherical core coated with *M*-layer homogeneous dielectric coatings.

More generally, if the object is a homogeneous dielectric spherical core coated with M layers of homogeneous dielectric coatings, as shown in Figure 2.2, where the geometrical parameters and property of each layer's coating have been labelled, then the coefficients are in the same forms as those of Equation (2.15) and Equation (2.16), but the surface impedance and admittance are derived differently. We can use an iteration method to derive the surface impedance and admittance, beginning from the most inner interface with a radius of a_M, i.e. the boundary between the inner core and the first layer. If the surface impedance and admittance at the interface with a radius of a_l are labelled as $Z_n^{(l)}$ and $Y_n^{(l)}$, respectively, then the surface impedance and admittance at the most inner interface are $Z_n^{(M)}$ and $Y_n^{(M)}$, respectively, and they can be determined independently. Based on $Z_n^{(M)}$ and $Y_n^{(M)}$, we can determine $Z_n^{(M-1)}$ and $Y_n^{(M-1)}$ at the interface with a radius of a_{M-1} and the process continues until $Z_n^{(0)}$ and $Y_n^{(0)}$ at the most outer interface with a radius of a_0 are determined. With the found $Z_n^{(0)}$ and $Y_n^{(0)}$, we have

$$Z_n(k_0a_0) = -\frac{\mathrm{i}}{\eta_0}Z_n^{(0)}, \quad Y_n(k_0a_0) = -\mathrm{i}\eta_0 Y_n^{(0)}. \tag{2.26}$$

With the above strategy, the recursive formulas for $Z_n^{(l)}$ and $Y_n^{(l)}$ can be found as

$$\begin{aligned} Z_n^{(l)} &= \eta_{l+1}P_n(k_0m_{l+1}a_l)\left[1 - V_n^{(l)}\frac{1 - Z_n^{l+1}/\eta_{l+1}/P_n(k_0m_{l+1}a_{l+1})}{1 - Z_n^{l+1}/\eta_{l+1}/Q_n(k_0m_{l+1}a_{l+1})}\right] \\ &\times \left[1 - U_n^{(l)}\frac{Z_n^{(l+1)} - \eta_{l+1}P_n(k_0m_{l+1}a_{l+1})}{Z_n^{(l+1)} - \eta_{l+1}Q_n(k_0m_{l+1}a_{l+1})}\right]^{-1} \end{aligned} \tag{2.27}$$

$$\begin{aligned} Y_n^{(l)} &= \frac{P_n(k_0m_{l+1}a_l)}{\eta_{l+1}}\left[1 - V_n^{(l)}\frac{1 - \eta_{l+1}Y_n^{(l+1)}/P_n(k_0m_{l+1}a_{l+1})}{1 - \eta_{l+1}Y_n^{(l+1)}/Q_n(k_0m_{l+1}a_{l+1})}\right] \\ &\times \left[1 - U_n^{(l)}\frac{1 - P_n(k_0m_{l+1}a_{l+1})/\eta_{l+1}/Y_n^{l+1}}{1 - Q_n(k_0m_{l+1}a_{l+1})/\eta_{l+1}/Y_n^{l+1}}\right]^{-1} \end{aligned} \tag{2.28}$$

where l starts from $M-1$ and terminates at 0. Also, we define

$$V_n^{(l)} = \frac{\left[k_0m_{l+1}a_{l+1}j_n(k_0m_{l+1}a_{l+1})\right]\left[k_0m_{l+1}a_lh_n(k_0m_{l+1}a_l)\right]}{\left[k_0m_{l+1}a_lj_n(k_0m_{l+1}a_l)\right]\left[k_0m_{l+1}a_{l+1}h_n(k_0m_{l+1}a_{l+1})\right]} \tag{2.29}$$

$$U_n^{(l)} = \frac{\left[k_0 m_{l+1} a_{l+1} j_n(k_0 m_{l+1} a_{l+1})\right]' \left[k_0 m_{l+1} a_l h_n(k_0 m_{l+1} a_l)\right]'}{\left[k_0 m_{l+1} a_l j_n(k_0 m_{l+1} a_l)\right]' \left[k_0 m_{l+1} a_{l+1} h_n(k_0 m_{l+1} a_{l+1})\right]'} \tag{2.30}$$

$$P_n(x) = \frac{x j_n(x)}{\left[x j_n(x)\right]'} \tag{2.31}$$

$$Q_n(x) = \frac{x h_n(x)}{\left[x h_n(x)\right]'}. \tag{2.32}$$

Note that M can be an arbitrary positive integer and a continuously radially inhomogeneous sphere can be approximately discretized into an M-layer piecewise homogeneous sphere, so that we can use the above formulas to solve for its scattering characteristic. When M is large enough, the solution can be very accurate.

2.3 Numerical Methods

Numerical methods were developed in companion with the advent and rapid development of computer technology and they are thought of as "discrete" and approximate methods. Although the methods are approximate, they are very powerful and versatile and have become the primary topics of research in CEM. Theoretically, the methods can solve any kind of problem using CMS as long as the used computer's speed and storage allow, and many large and complicated problems that were not possible previously can now be easily solved through the methods. Although many numerical methods have been developed in CEM, they can mainly be grouped into three categories, i.e. the FDTD method, the FEM, and the MoM, which correspond to solving the original form of Maxwell's equations, vector wave equations, and integral equations, respectively. The FDTD method and FEM belong to differential equation solvers (DESs) while the MoM is one of integral equation solvers (IESs) because they solve differential equations and integral equations, respectively. The three mainstream methods include both advantages and disadvantages, and none of them shows an absolute superiority to the others, so they are all widely used at present. Also, there is a commonly-recognized comparison between the DESs and IESs. IESs usually require a smaller number of unknowns and have a better scaling property for computational costs. This is because IESs solve source distribution on boundaries or within objects first, instead of solving field distribution at spatial points directly. Therefore, there is no need to implement absorbing boundary conditions (ABCs) for open field domains and the solution domain is much smaller in IESs. However, IESs are more complicated to implement in general due to the need for the Green's function. The Green's function in wave physics is singular and much effort is required generally to evaluate matrix elements with singular kernels. Also, the system matrix is inherently dense and accelerators are needed for solving large problems with a comparative cost to that of DESs. In contrast, DESs are simpler to implement without any singularity problem and the system matrix is sparse. Moreover, DESs can easily account for the nonlinearity of the transient state in the time domain. Nevertheless, DESs usually entail worse cost-scaling properties since the field solutions are sought directly and may involve ABC implementation for open field domains. In addition, the presence of grid dispersion in DESs will further worsen their scaling properties for large-scale computations [1]. We now present a brief introduction to each of the three mainstream methods.

2.3.1 The Finite-Difference Time-Domain (FDTD) Method

The FDTD method for solving Maxwell's equations was first proposed by Yee in 1966 [12] and it was called Yee's method sometimes. The earlier work on the method can be attributed to Courant *et al.* in 1928 [13] and O'Brien *et al.* in 1950 [14], but their work was not for solving Maxwell's equations. In the context of CEM, the FDTD method is a numerical analysis technique used for modeling EM problems to find approximate solutions by an associated system of differential equations. Since it is a time-domain method, the FDTD solutions can cover a wide frequency range with a single simulation run, and treat nonlinear material properties in a natural way. The FDTD method belongs to the general class of grid-based differential numerical modeling methods or finite difference methods. Time-dependent Maxwell's equations, which are in a partial differential form, are discretized using central-difference approximations to the space and time partial derivatives. The resulting finite-difference equations are solved in either software or hardware in a leapfrog manner, i.e. the components of an electric field (E-field) vector in a volume of space are solved at a given instant in time, and then the components of a magnetic field (H-field) vector in the same spatial volume are solved at the next instant in time. The process is repeated over and over again until the desired transient or steady-state EM fields are fully determined [4].

When Maxwell's differential equations are examined, it can be seen that the change in the E-field in time (the time derivative) is dependent on the change in the H-field across space or the curl of the H-field. This results in the basic FDTD time-stepping relation, i.e. at any point in space the updated value of the E-field in time is dependent on the stored value of the E-field and the numerical curl of the local distribution of the H-field in space. The H-field is time-stepped in a similar manner, i.e. at any point in space, the updated value of the H-field in time is dependent on the stored value of the H-field and the numerical curl of the local distribution of the E-field in space. When three-dimensional (3D) problems are considered, calculating the numerical curl of fields can become complicated. Yee's seminal paper in 1966 proposed spatially staggering the components of the E-field and H-field vectors about rectangular unit cells of a Cartesian computational grid so that each component of E-field vector is located at the midpoint between a pair of components of the H-field vector, and vice versa. This scheme, known as Yee's lattice, has been proven to be very robust, and remains at the core of many current FDTD software frames [15].

Since the FDTD method requires that the entire computational domain be gridded, and the spatial discretization must be sufficiently fine to resolve both the smallest EM wavelength and the smallest geometrical feature in the model, very large computational domains may be needed sometimes. Also, the computational domain must be finite to permit its residence in the computer memory and this is achieved by inserting artificial boundaries into the simulation space. There are many effective ABCs to simulate an infinitely unbounded computational domain, but most modern FDTD implementations usually use a special absorbing "material", called a perfectly matched layer (PML), to implement absorbing boundaries [16].

2.3.2 The Finite Element Method (FEM)

The FEM is another robust numerical method for solving problems of engineering and mathematical physics. It is also referred to as finite element analysis (FEA) sometimes.

While it is difficult to quote a date for the invention of FEM, the method originated from the need to solve complex elasticity and structural analysis problems in civil and aeronautical engineering. Its development can be traced back to the work by Hrennikoff [17] and Courant [18] in the early 1940s. In the Union of Soviet Socialist Republics, the introduction of the practical application of the method was usually attributed to Leonard Oganesyan. In China, K. Feng proposed a systematic numerical method for solving partial differential equations in the computation of dam constructions in the later 1950s and early 1960s. The method was called the finite difference method based on the variation principle, which was another independent invention of the FEM. Although the approaches used by these pioneers are different, they share one essential characteristic, i.e. mesh discretization of a continuous domain into a set of discrete sub-domains, usually called elements. Hrennikoff's work discretizes the domain by using a lattice analogy, while Courant's approach divides the domain into finite triangular subregions to solve second-order elliptic PDEs that arise from the problem of torsion of a cylinder [19].

The typical problems of interest in the FEM include structural analysis, heat transfer, fluid flow, mass transport, and electromagnetic potential. The analytical solutions of these problems generally require the solutions to boundary value problems for PDEs, but the FEM formulation of the problems results in a system of algebraic equations. The FEM yields the approximate values of unknowns at chosen discrete points over the domain [20]. To solve the problem, it subdivides a large problem into smaller and simpler parts that are called finite elements. The simple equations that model these finite elements are then assembled into a larger system of equations that models the entire problem. The FEM then uses variational methods from the calculus of variations to approximate a solution by minimizing an associated error function. Note that there are many variant versions of the FEM, such as the applied element method (AEM), the generalized finite element method (GFEM), the mixed finite element method (MFEM), the hextended finite element method (EFEM), the smoothed finite element method (SFEM), the spectral element method (SEM), etc. and they have improved the performance of the traditional FEM in some aspects [21].

2.3.3 The Method of Moments (MoM)

The MoM, which is also known as the moment method, is a powerful numerical method for solving integral equations or integral-differential equations by transforming them into algebraic matrix equations. Although its prototype was proposed in the early stage of twentieth century, it was first introduced to electromagnetics by Mei and Van Bladel, Andreasen, Oshiro, Richmond, Harrington, etc. in the 1960s [22–28]. The method has become the most widely-used numerical method for solving EM integral equations and served as a foundation for many advanced numerical methods like the adaptive integral method and the fast multipole algorithm [25]. The essence of this method is to discretize the integral equations by using a basis function to represent an unknown function and then satisfy boundary conditions in an average sense by using a weighting function to test the equations. Since the method is closer to the Nyström method than the FDTD method and the FEM, we introduce this method with more details. To illustrate the method, we consider the following inhomogeneous integral equation

$$\mathcal{L}(f) = g \tag{2.33}$$

where $\mathcal{L}$ is a linear integral or integral-differential operator, g is a known function, and f is the unknown function to be solved. To solve the above equation, we expand the unknown function f in terms of a set of basis functions in the domain of $\mathcal{L}$, i.e.

$$f = \sum_{n=1}^{N} \alpha_n f_n \tag{2.34}$$

where α_n is the expansion coefficient, f_n is the expansion function or basis function, and N is the number of basis functions. Note that N should be large enough if an accurate solution is wanted. Substituting (2.34) into (2.33) and making use of the linearity of the $\mathcal{L}$ operator, we obtain

$$\sum_{n=1}^{N} \alpha_n \mathcal{L}(f_n) = g. \tag{2.35}$$

If we define an inner product $\langle f, g\rangle$, and select a set of weighting functions $[w_1, w_2, \cdots, w_N]$ in the range of the $\mathcal{L}$ operator to take the inner product of Equation (2.35), then we can get

$$\sum_{n=1}^{N} \alpha_n \langle w_m, \mathcal{L}(f_n)\rangle = \langle w_m, g\rangle \tag{2.36}$$

where $m = 1, 2, \cdots, N$. The above equation can be written into a matrix equation, i.e.

$$[Z_{mn}]\,[\alpha_n] = [b_m] \tag{2.37}$$

where

$$[Z_{mn}] = \begin{bmatrix} \langle w_1, \mathcal{L}(f_1)\rangle & \langle w_1, \mathcal{L}(f_2)\rangle & \cdots & \langle w_1, \mathcal{L}(f_N)\rangle \\ \langle w_2, \mathcal{L}(f_1)\rangle & \langle w_2, \mathcal{L}(f_2)\rangle & \cdots & \langle w_2, \mathcal{L}(f_N)\rangle \\ \vdots & \vdots & \vdots & \vdots \\ \langle w_N, \mathcal{L}(f_1)\rangle & \langle w_N, \mathcal{L}(f_2)\rangle & \cdots & \langle w_N, \mathcal{L}(f_N)\rangle \end{bmatrix} \tag{2.38}$$

$$[\alpha_n] = \begin{bmatrix} \alpha_1 \\ \alpha_2 \\ \vdots \\ \alpha_N \end{bmatrix} \tag{2.39}$$

$$[b_m] = \begin{bmatrix} \langle w_1, g\rangle \\ \langle w_2, g\rangle \\ \vdots \\ \langle w_N, g\rangle \end{bmatrix}. \tag{2.40}$$

If the matrix $[Z_{mn}]$ is non-singular, then the expansion coefficients can be obtained by

$$[\alpha_n] = [Z_{mn}]^{-1}\,[b_m]. \tag{2.41}$$

The selection of the weighting function can be flexible and is not required to be the same as the basis function. If they are the same, then the resultant method is named Galerkin's method. The accuracy of the MoM mainly depends on the selection of N and the basis function, and the complexity of its implementation mainly resides in the evaluation of matrix elements that may be singular due to the involvement of Green's function.

The most widely-used basis functions in the MoM for solving 3D problems are the Rao-Wilton-Glisson (RWG) basis function for surface integral equations (SIEs) and

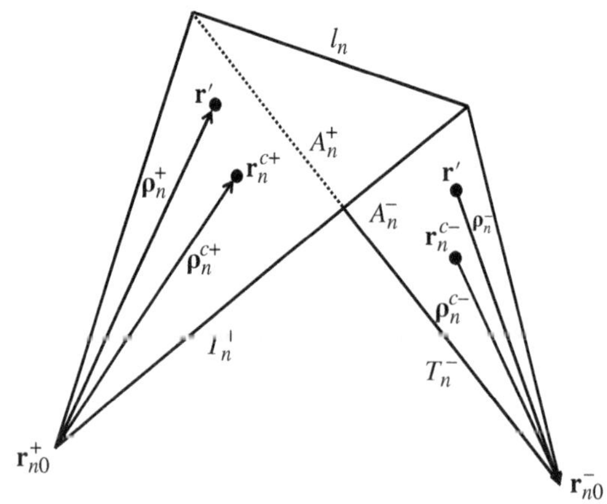

Figure 2.3 A pair of triangles that define the RWG basis function.

Schaubert-Wilton-Glisson (SWG) basis function for volume integral equations (VIEs). These two basis functions were proposed by Rao *et al.* in 1982 [26], and Schaubert *et al.* in 1984 [27], respectively. The RWG basis function is a vector basis function defined over a pair of triangles as shown in Figure 2.3 and can be written as

$$\boldsymbol{\Lambda}_n(\mathbf{r}') = \begin{cases} \frac{l_n}{2A_n^+}\boldsymbol{\rho}_n^+ = \frac{l_n}{2A_n^+}(\mathbf{r}' - \mathbf{r}_{n0}^+), & \mathbf{r}' \in T_n^+ \\ \frac{l_n}{2A_n^-}\boldsymbol{\rho}_n^- = \frac{l_n}{2A_n^-}(\mathbf{r}_{n0}^- - \mathbf{r}'), & \mathbf{r}' \in T_n^- \\ 0, \ \text{otherwise} \end{cases} \tag{2.42}$$

where $A_n^{\pm}$ is the area of the triangle pair $T_n^{\pm}$, l_n is the length of the common edge of $T_n^{\pm}$, $\mathbf{r}_{n0}^{\pm}$ is the free vertex of $T_n^{\pm}$, and $\boldsymbol{\rho}_n^{\pm}$ is the distance vector connecting an arbitrary point $\mathbf{r}'$ inside $T_n^{\pm}$ and the free vertex $\mathbf{r}_{n0}^{\pm}$. Also, $\boldsymbol{\rho}_n^{c\pm}$ is the distance vector connecting the centroid $\mathbf{r}_n^{c\pm}$ of $T_n^{\pm}$ and the free vertex $\mathbf{r}_{n0}^{\pm}$, which can be used to represent the averaged distance vector of $\boldsymbol{\rho}_n^{\pm}$ in an approximate calculation. When the RWG basis function is used to represent a surface current density in the SIEs, it has the following properties, i.e. (1) there is no normal component of the current density at the four non-common edges; (2) at the common edge l_n, the normal component of the current density is continuous and its magnitude is unitary; (3) the surface charge density is proportional to the divergence of the current density and is constant inside each of the two triangles, but the total charge is zero in the two triangles because the charge has the same magnitude with an opposite sign in the two triangles.

The SWG basis function can be viewed as an extension of the RWG basis function and is used to represent a volume current density in solving VIEs. It is also a vector basis function but is defined over a pair of tetrahedrons, as shown in Figure 2.4. The SWG basis function can be written as

$$\mathbf{f}_n(\mathbf{r}') = \begin{cases} \frac{a_n}{3V_n^+}\boldsymbol{\rho}_n^+ = \frac{a_n}{3V_n^+}(\mathbf{r}' - \mathbf{r}_{n0}^+), & \mathbf{r}' \in T_n^+ \\ \frac{a_n}{3V_n^-}\boldsymbol{\rho}_n^- = \frac{a_n}{3V_n^-}(\mathbf{r}_{n0}^- - \mathbf{r}'), & \mathbf{r}' \in T_n^- \\ 0, \ \text{otherwise} \end{cases} \tag{2.43}$$

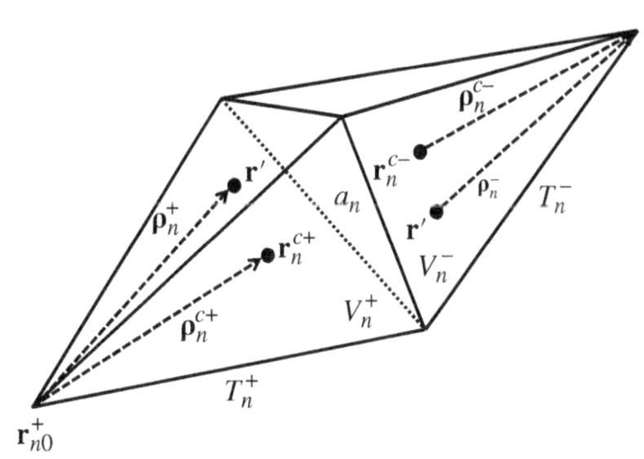

Figure 2.4 A pair of tetrahedrons that define the SWG basis function.

where $V_n^\pm$ is the volume of the tetrahedron pair $T_n^\pm$, a_n is the area of the common face of $T_n^\pm$, $\mathbf{r}_{n0}^\pm$ is the free vertex of $T_n^\pm$, and $\rho_n^\pm$ is the distance vector connecting an arbitrary point $\mathbf{r}'$ inside $T_n^\pm$ and the free vertex $\mathbf{r}_{n0}^\pm$. Also, $\rho_n^{c\pm}$ is the distance vector connecting the centroid $\mathbf{r}_n^{c\pm}$ of $T_n^\pm$ and the free vertex $\mathbf{r}_{n0}^\pm$, which can be used to represent the averaged distance vector of $\rho_n^\pm$ in an approximate calculation. When the SWG basis function is used to represent a flux density in the VIEs, it has the following properties, i.e. (1) there is no normal component of flux density at the three non-common faces; (2) at the common face a_n, the normal component of flux density is continuous and its magnitude is unitary; (3) the flux density is proportional to the divergence of the flux density and is constant inside each of the two tetrahedrons, but the total charge is zero in the two tetrahedrons because the charge has the same magnitude with an opposite sign in the two tetrahedrons.

2.4 Electromagnetic Integral Equations

Because the Nyström method is a numerical method of solving EM integral equations and this is the theme of this book, we need to take a look at these integral equations before addressing how to use the method to solve them. Although these equations will be presented in the chapters related to physical applications, it is necessary to summarize and clarify all these equations because some readers could very often be confused by them. One of distinctive features of EM is the diversity of the descriptive governing equations that can distinguish this area from other neighboring areas like acoustics and elastodynamics whose governing equations are much less diverse. This is both a curse and a blessing. The curse means that the diverse governing equations can easily confuse inexperienced researchers or students while the blessing means that the diverse governing equations can allow one to solve the same problems by different approaches with more observations from physics and mathematics. When people become more and more familiar with these equations, they will feel how magical and elegant they look, and how great those pioneers who proposed them were. The EM integral equations can be divided into three categories, i.e. SIEs, VIEs, and volume-surface integral equations (VSIEs) [1], and we will present each of them as follows.

2.4.1 Surface Integral Equations (SIEs)

The SIEs can be further divided into two categories and the first one is for impenetrable or conducting objects, while the second one is for penetrable or non-conducting objects. The impenetrable objects usually refer to perfectly electric conductors (PECs) or perfectly magnetic conductors (PMCs) but the PMCs are less addressed because they do not exist in reality and only exist by artificial construction. The penetrable objects can be all other objects except PECs and PMCs, but mostly refer to dielectric objects because dielectric materials so widely exist. The penetrable objects can be described by SIEs only if they are ideally homogeneous and isotropic. If they are inhomogeneous or anisotropic or nonlinear or dispersive or chiral, etc., then we have to use VIEs to describe them. With the above clarification, we now present the SIEs for the impenetrable and penetrable objects, respectively.

For Impenetrable Objects

The SIEs for impenetrable objects have three kinds of form, i.e. the electric field integral equation (EFIE), the magnetic field integral equation (MFIE), and the combined field integral equation (CFIE). If we take the EM scattering by a conducting (PEC) object as an example, as shown in Figure 2.5, the EFIE and MFIE can written as

$$\hat{n} \times \mathcal{L}\left[\mathbf{J}_S(\mathbf{r}')\right] = -\hat{n} \times \mathbf{E}^{\text{inc}}(\mathbf{r}), \quad \mathbf{r} \in S \tag{2.44}$$

$$\frac{1}{2}\mathbf{J}_S(\mathbf{r}) + \hat{n} \times \mathcal{K}\left[\mathbf{J}_S(\mathbf{r}')\right] = \hat{n} \times \mathbf{H}^{\text{inc}}(\mathbf{r}), \quad \mathbf{r} \in S \tag{2.45}$$

respectively, where $\mathcal{L}$ and $\mathcal{K}$ are the two operators defined by

$$\mathcal{L}\left[\mathbf{J}_S(\mathbf{r}')\right] = \mathrm{i}\omega\mu_0 \int_S \overline{\mathbf{G}}(\mathbf{r}, \mathbf{r}') \cdot \mathbf{J}_S(\mathbf{r}')\, dS' \tag{2.46}$$

$$\mathcal{K}\left[\mathbf{J}_S(\mathbf{r}')\right] = \fint_S \mathbf{J}_S(\mathbf{r}') \times \nabla g(\mathbf{r}, \mathbf{r}')\, dS'. \tag{2.47}$$

The CFIE, which is a weighted combination of EFIE and MFIE, can be written as

$$\begin{aligned} &\alpha\hat{n} \times \hat{n} \times \mathcal{L}\left[\mathbf{J}_S(\mathbf{r}')\right] + (1-\alpha)\eta_0 \left\{\frac{1}{2}\mathbf{J}_S(\mathbf{r}) + \hat{n} \times \mathcal{K}\left[\mathbf{J}_S(\mathbf{r}')\right]\right\} \\ &= \alpha\hat{n} \times \hat{n} \times \mathbf{E}^{\text{inc}}(\mathbf{r}) + (1-\alpha)\eta_0\hat{n} \times \mathbf{H}^{\text{inc}}(\mathbf{r}), \ \mathbf{r} \in S. \end{aligned} \tag{2.48}$$

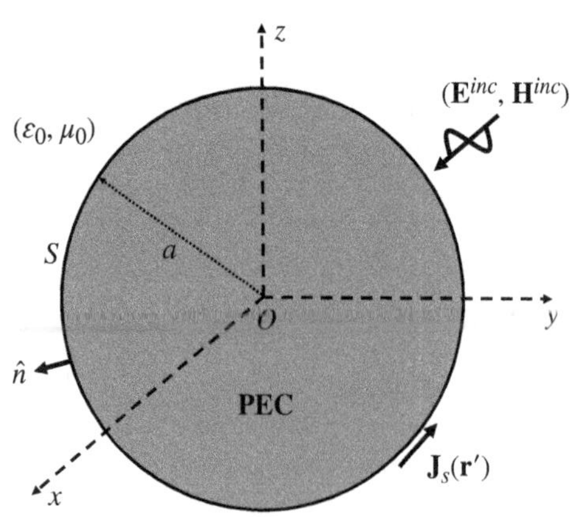

Figure 2.5 Scattering by an impenetrable PEC object.

In the above, S is the object surface, $\hat{n}$ is the unit normal vector on the surface, $\mathbf{J}_S(\mathbf{r}')$ is the induced surface current density, $\mathbf{E}^{\text{inc}}(\mathbf{r})$ and $\mathbf{H}^{\text{inc}}(\mathbf{r})$ are the incident electric field and magnetic field, respectively, and

$$\overline{\mathbf{G}}(\mathbf{r},\mathbf{r}') = \left(\overline{\mathbf{I}} + \frac{\nabla\nabla}{k_0^2}\right) g(\mathbf{r},\mathbf{r}') \tag{2.49}$$

is the dyadic Green's function in which $\overline{\mathbf{I}}$ is the identity dyad, $g(\mathbf{r},\mathbf{r}') = \mathrm{e}^{\mathrm{i}\kappa_0 R}/(4\pi R)$ is the scalar Green's function, and $R = |\mathbf{r} - \mathbf{r}'|$ is the distance between an observation point $\mathbf{r}$ and a source point $\mathbf{r}'$. Also, the integral in the $\mathcal{K}$ operator is defined in a Cauchy-principal-value (CPV) sense and α is named the combination parameter which is generally selected between 0.2 and 0.8 in the CFIE. The EFIE is known as a first kind of integral equation because the unknown function only appears inside the integral sign. The EFIE is more accurate than the MFIE if solved numerically in general, but it has a serious low-frequency breakdown problem. The MFIE is known as a second of the integral equation because the unknown function appears both inside and outside the integral sign and it is more well-conditioned than the EFIE, although it is less accurate in numerical solutions. Furthermore, the MFIE does not have an obvious low-frequency breakdown problem [1]. The CFIE is created by combining the EFIE and MFIE with a weight coefficient α and is for removing the internal resonance problem existing in both individual EFIE and MFIE.

For Penetrable Objects

For penetrable objects, they must be homogeneous and isotropic if SIEs are used to describe them. There are five types of SIEs in this case, i.e. they also include Poggio-Miller-Chang-Harrington-Wu-Tsai (PMCHWT) equations [28] and Müller equations [29] in addition to the EFIEs, MFIEs, and CFIEs which are in a different form from those of impenetrable objects. The EFIEs and MFIEs are fundamental and they can be derived by the equivalent electric and magnetic current method. Consider the scattering by a single 3D dielectric or magnetic body where there is only one interface. The original problem is illustrated in Figure 2.6(a). We designate the interior and exterior of the object as Region 1 and Region 2, respectively. Figure 2.6(b) and Figure 2.6(c) show the equivalent external problem and equivalent internal problem, respectively. In the equivalent external problem, the original electric and magnetic fields exist outside the object surface S and no fields exist inside S. To support such fields, there must exist equivalent electric and magnetic surface current densities $\mathbf{J}$ and $\mathbf{M}$ on the object surface or boundary S, which are unknowns to be solved. Since the tangential components of fields are continuous at the boundary, we can write the integral equations as follows

$$\hat{n} \times \mathbf{E}_2^{S^+}(\mathbf{J},\mathbf{M}) = -\hat{n} \times \mathbf{E}_{\text{inc}}^{S^+} \tag{2.50}$$

$$\hat{n} \times \mathbf{H}_2^{S^+}(\mathbf{J},\mathbf{M}) = -\hat{n} \times \mathbf{H}_{\text{inc}}^{S^+} \tag{2.51}$$

where $\hat{n}$ is the unit normal vector on the surface, and $\mathbf{E}$ and $\mathbf{H}$ are the electric and magnetic fields, respectively. The subscript "inc" represents an incident field and "2" denotes the region index of the fields. The superscript "S^+" denotes the interface where observation points are located and we use "+" and "−" on S to indicate the exterior side and interior side of S, respectively. In the equivalent internal problem described in

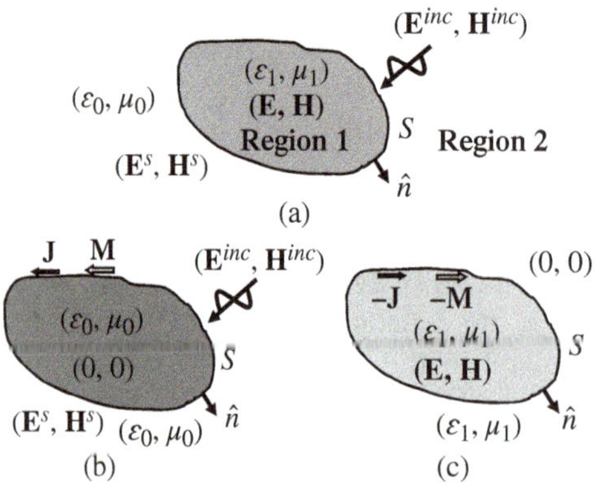

Figure 2.6 Scattering by a penetrable object that is homogeneous and istropic. (a) Original problem. (b) Equivalent external problem. (c) Equivalent internal problem.

Figure 2.6(c), the original fields exist inside the object surface S and none of fields exist outside S. There must exist an equivalent electric current density $-\mathbf{J}$ and magnetic current density $-\mathbf{M}$ on the surface S to produce the original fields inside the object. The boundary condition requires

$$\hat{n} \times \mathbf{E}_1^{S^-}(\mathbf{J}, \mathbf{M}) = 0 \tag{2.52}$$

$$\hat{n} \times \mathbf{H}_1^{S^-}(\mathbf{J}, \mathbf{M}) = 0 \tag{2.53}$$

which is another group of integral equations. The first two equations in (2.50) and (2.52) about the electric field are known as the EFIE or E-field formula, and the second two equations in (2.50) and (2.52) about the magnetic field are known as the MFIE or H-field formula. The electric field and magnetic field can be expressed in terms of the sources $\mathbf{J}$ and $\mathbf{M}$ by recalling the vector and scalar potentials, i.e.

$$\begin{aligned}\mathbf{E}(\mathbf{J}, 0) &= -\mathrm{i}\omega\mathbf{A}(\mathbf{J}) - \nabla\Phi(\mathbf{J}) \\ &= -\mathrm{i}\omega\mu \int_S \left[\bar{\mathbf{I}} + \frac{\nabla\nabla}{k^2}\right] g(\mathbf{r}, \mathbf{r}') \cdot \mathbf{J}(\mathbf{r}')\, \mathrm{d}S' = -\mathcal{L}(\mathbf{J})\end{aligned} \tag{2.54}$$

$$\begin{aligned}\mathbf{E}(0, \mathbf{M}) &= -\frac{1}{\epsilon}\nabla \times \mathbf{F}(\mathbf{M}) \\ &= \int_S \mathbf{M}(\mathbf{r}') \times \nabla g(\mathbf{r}, \mathbf{r}')\, \mathrm{d}S' = \mathcal{K}(\mathbf{M})\end{aligned} \tag{2.55}$$

$$\mathbf{H}(0, \mathbf{M}) = -\mathrm{i}\omega\mathbf{F}(\mathbf{M}) - \nabla\Psi(\mathbf{M}) = -\frac{\mathcal{L}(\mathbf{M})}{\eta^2} \tag{2.56}$$

$$\mathbf{H}(\mathbf{J}, 0) = \frac{1}{\mu}\nabla \times \mathbf{A}(\mathbf{J}) = -\mathcal{K}(\mathbf{J}) \tag{2.57}$$

where $\mathbf{A}$ and $\mathbf{F}$ are magnetic and electric vector potentials, and Φ and Ψ are electric and magnetic scalar potentials, respectively. Also, ϵ, μ, η, and k are the permittivity, permeability, wave impedance, and wavenumber of the related medium, respectively, $\bar{\mathbf{I}}$ is an identity dyad, and $g(\mathbf{r}, \mathbf{r}') = \mathrm{e}^{\mathrm{i}kR}/(4\pi R)$ is the scalar Green's function, where $R = |\mathbf{r} - \mathbf{r}'|$ is the distance between a field point $\mathbf{r}$ and a source point $\mathbf{r}'$. We have used $\mathcal{L}$ and $\mathcal{K}$ operators to represent the relationship between the fields and sources as in the integral equations of impenetrable objects. The above method and derivation can be extended to a more complicated or generalized case, i.e. a penetrable object with multilayered penetrable coatings, as we will see in Chapter 7 for applications.

If we add the internal-field equations to the external-field equations by a weighted-sum method [1], then we can obtain the PMCHWT equations and Müller equations, i.e.

$$\alpha_1 \hat{n} \times \mathbf{E}_2^{S^+}(\mathbf{J}, \mathbf{M}) + \alpha_2 \hat{n} \times \mathbf{E}_1^{S^-}(\mathbf{J}, \mathbf{M}) = -\alpha_1 \hat{n} \times \mathbf{E}_{\text{inc}}^{S^+} \tag{2.58}$$

$$\beta_1 \hat{n} \times \mathbf{H}_2^{S^+}(\mathbf{J}, \mathbf{M}) + \beta_2 \hat{n} \times \mathbf{H}_1^{S^-}(\mathbf{J}, \mathbf{M}) = -\beta_1 \hat{n} \times \mathbf{H}_{\text{inc}}^{S^+} \tag{2.59}$$

where α_1, α_2, β_1, and β_2 are the weight coefficients. The above equations are known as the PMCHWT equations when $\alpha_1 = \alpha_2 = \beta_1 = \beta_2 = 1$ while they are known as Müller formulas when $\alpha_1 = \beta_1 = 1$, $\alpha_2 = -\epsilon_1/\epsilon_2$, and $\beta_2 = \mu_1/\mu_2$. On the other hand, if we add the E-field equations to the H-field equations in the same regions with appropriate coefficients, then we can obtain the CFIE, i.e.

$$\alpha \hat{n} \times \hat{n} \times \mathbf{E}_2^{S^+}(\mathbf{J}, \mathbf{M}) + \eta_2(1-\alpha)\hat{n} \times \mathbf{H}_2^{S^+}(\mathbf{J}, \mathbf{M}) = -\alpha \hat{n} \times \hat{n} \times \mathbf{E}_{\text{inc}}^{S^+} - \eta_2(1-\alpha)\hat{n} \times \mathbf{H}_{\text{inc}}^{S^+} \tag{2.60}$$

$$\alpha \hat{n} \times \hat{n} \times \mathbf{E}_1^{S^-}(\mathbf{J}, \mathbf{M}) + \eta_1(1-\alpha)\hat{n} \times \mathbf{H}_1^{S^-}(\mathbf{J}, \mathbf{M}) = 0 \tag{2.61}$$

where α is a combination parameter as defined in the CFIE for impenetrable objects.

2.4.2 Volume Integral Equations (VIEs)

When penetrable objects are inhomogeneous or anisotropic, the SIEs cannot be used and the VIEs are indispensably employed to describe involved EM problems in the integral equation approach. The VIEs can be derived from the vector wave equation by introducing the dyadic Green's function. Consider the EM scattering by a 3D inhomogeneous anisotropic object embedded in free space with a permittivity ϵ_0 and permeability μ_0, see Figure 2.7, the VIEs can be written as

$$\begin{aligned}\mathbf{E}(\mathbf{r}) = \mathbf{E}^{\text{inc}}(\mathbf{r}) &+ \int_V \overline{\mathbf{G}}(\mathbf{r}, \mathbf{r}') \cdot \nabla' \times \mathbf{F}_\mu(\mathbf{r}')d\mathbf{r}' \\ &- k_0^2 \int_V \overline{\mathbf{G}}(\mathbf{r}, \mathbf{r}') \cdot \left[\overline{\mathbf{I}} - \overline{\epsilon}_r(\mathbf{r}')\right] \cdot \mathbf{E}(\mathbf{r}')d\mathbf{r}', \quad \mathbf{r} \in V\end{aligned} \tag{2.62}$$

$$\begin{aligned}\mathbf{H}(\mathbf{r}) = \mathbf{H}^{\text{inc}}(\mathbf{r}) &+ \int_V \overline{\mathbf{G}}(\mathbf{r}, \mathbf{r}') \cdot \nabla' \times \mathbf{F}_\epsilon(\mathbf{r}')d\mathbf{r}' \\ &- k_0^2 \int_V \overline{\mathbf{G}}(\mathbf{r}, \mathbf{r}') \cdot \left[\overline{\mathbf{I}} - \overline{\mu}_r(\mathbf{r}')\right] \cdot \mathbf{H}(\mathbf{r}')d\mathbf{r}', \quad \mathbf{r} \in V\end{aligned} \tag{2.63}$$

where $\mathbf{E}^{\text{inc}}(\mathbf{r})$ and $\mathbf{H}^{\text{inc}}(\mathbf{r})$ are the incident electric and magnetic fields, respectively, while $\mathbf{E}(\mathbf{r})$ and $\mathbf{H}(\mathbf{r})$ are the total electric and magnetic fields inside the object, respectively. Also, the integral kernel $\overline{\mathbf{G}}(\mathbf{r}, \mathbf{r}')$ is the 3D dyadic Green's function for the homogeneous background medium with a wavenumber $k_0 = \omega\sqrt{\mu_0\epsilon_0}$ and is defined as

$$\overline{\mathbf{G}}(\mathbf{r}, \mathbf{r}') = \left(\overline{\mathbf{I}} + \frac{\nabla\nabla}{k_0^2}\right) g(\mathbf{r}, \mathbf{r}') \tag{2.64}$$

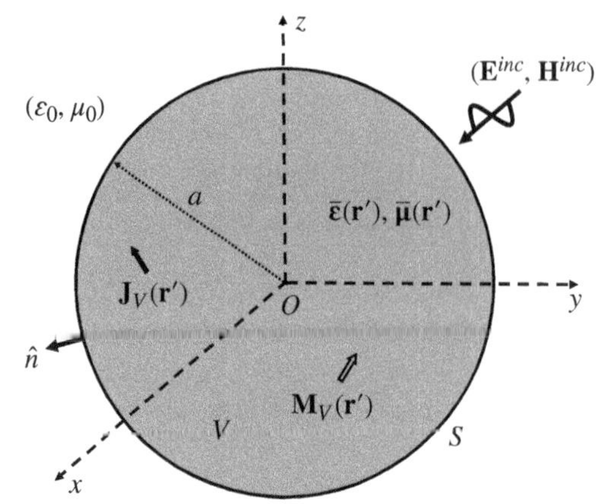

Figure 2.7 Scattering by an inhomogeneous anisotropic object embedded in free space. The volumetric electric current density $\mathbf{J}_V(\mathbf{r}')$ and magnetic current density $\mathbf{M}_V(\mathbf{r}')$ are induced inside the object, respectively.

where $g(\mathbf{r}, \mathbf{r}') = e^{ik_0R}/(4\pi R)$ is the 3D scalar Green's function in which $R = |\mathbf{r} - \mathbf{r}'|$ is the distance between an observation point $\mathbf{r}$ and a source point $\mathbf{r}'$, and $\bar{\mathbf{I}}$ is the identity dyad. In addition,

$$\mathbf{F}_\mu(\mathbf{r}') = \left[\bar{\mathbf{I}} - \bar{\mu}_r^{-1}(\mathbf{r}')\right] \cdot \nabla' \times \mathbf{E}(\mathbf{r}')$$
$$\mathbf{F}_\epsilon(\mathbf{r}') = \left[\bar{\mathbf{I}} - \bar{\epsilon}_r^{-1}(\mathbf{r}')\right] \cdot \nabla' \times \mathbf{H}(\mathbf{r}') \tag{2.65}$$

where $\bar{\mu}_r(\mathbf{r}') = \bar{\mu}(\mathbf{r}')/\mu_0$ is the relative permeability tensor and $\bar{\epsilon}_r(\mathbf{r}') = \bar{\epsilon}(\mathbf{r}')/\epsilon_0$ is the relative permittivity tensor. Note that the above equations are not computationally friendly and we can recast them into the following form [1]

$$\mathbf{E}(\mathbf{r}) = \mathbf{E}^{\text{inc}}(\mathbf{r}) + \int_V \nabla g(\mathbf{r}, \mathbf{r}') \times \bar{\mathbf{A}}_\mu(\mathbf{r}') \cdot \mathbf{M}_V(\mathbf{r}')d\mathbf{r}' + i\omega\mu_0 \int_V \bar{\mathbf{G}}(\mathbf{r}, \mathbf{r}') \cdot \mathbf{J}_V(\mathbf{r}')d\mathbf{r}', \quad \mathbf{r} \in V \tag{2.66}$$

$$\mathbf{H}(\mathbf{r}) = \mathbf{H}^{\text{inc}}(\mathbf{r}) - \int_V \nabla g(\mathbf{r}, \mathbf{r}') \times \bar{\mathbf{A}}_\epsilon(\mathbf{r}') \cdot \mathbf{J}_V(\mathbf{r}')d\mathbf{r}' + i\omega\epsilon_0 \int_V \bar{\mathbf{G}}(\mathbf{r}, \mathbf{r}') \cdot \mathbf{M}_V(\mathbf{r}')d\mathbf{r}', \quad \mathbf{r} \in V \tag{2.67}$$

where

$$\bar{\mathbf{A}}_\mu(\mathbf{r}') = \left[\bar{\mathbf{I}} - \bar{\mu}_r^{-1}(\mathbf{r}')\right] \cdot \bar{\mu}(\mathbf{r}') \cdot \left[\mu_0\bar{\mathbf{I}} - \bar{\mu}(\mathbf{r}')\right]^{-1} \tag{2.68}$$

$$\bar{\mathbf{A}}_\epsilon(\mathbf{r}') = \left[\bar{\mathbf{I}} - \bar{\epsilon}_r^{-1}(\mathbf{r}')\right] \cdot \bar{\epsilon}(\mathbf{r}') \cdot \left[\epsilon_0\bar{\mathbf{I}} - \bar{\epsilon}(\mathbf{r}')\right]^{-1} \tag{2.69}$$

are the two matrices related to the object's permeability tensor and permittivity tensor, respectively, and

$$\mathbf{J}_V(\mathbf{r}') = i\omega \left[\epsilon_0\bar{\mathbf{I}} - \bar{\epsilon}(\mathbf{r}')\right] \cdot \mathbf{E}(\mathbf{r}') \tag{2.70}$$

$$\mathbf{M}_V(\mathbf{r}') = i\omega \left[\mu_0\bar{\mathbf{I}} - \bar{\mu}(\mathbf{r}')\right] \cdot \mathbf{H}(\mathbf{r}') \tag{2.71}$$

are the induced volumetric electric and magnetic current densities inside the object, respectively. If the object is inhomogeneous but isotropic with a permittivity $\epsilon(\mathbf{r}')$ and permeability $\mu(\mathbf{r}')$, the corresponding VIEs can be written as [1]

$$\mathbf{E}(\mathbf{r}) = \mathbf{E}^{\text{inc}}(\mathbf{r}) + \mathrm{i}\omega\mu_0 \int_V \overline{\mathbf{G}}(\mathbf{r},\mathbf{r}') \cdot \mathbf{J}_V(\mathbf{r}')\mathrm{d}\mathbf{r}' - \nabla \times \int_V \overline{\mathbf{G}}(\mathbf{r},\mathbf{r}') \cdot \mathbf{M}_V(\mathbf{r}')\mathrm{d}\mathbf{r}', \ \mathbf{r} \in V \tag{2.72}$$

$$\mathbf{H}(\mathbf{r}) = \mathbf{H}^{\text{inc}}(\mathbf{r}) + \mathrm{i}\omega\epsilon_0 \int_V \overline{\mathbf{G}}(\mathbf{r},\mathbf{r}') \cdot \mathbf{M}_V(\mathbf{r}')\mathrm{d}\mathbf{r}' + \nabla \times \int_V \overline{\mathbf{G}}(\mathbf{r},\mathbf{r}') \cdot \mathbf{J}_V(\mathbf{r}')\mathrm{d}\mathbf{r}', \ \mathbf{r} \in V \tag{2.73}$$

where

$$\mathbf{J}_V(\mathbf{r}') = \mathrm{i}\omega \left[\epsilon_0 - \epsilon(\mathbf{r}')\right] \mathbf{E}(\mathbf{r}') \tag{2.74}$$
$$\mathbf{M}_V(\mathbf{r}') = \mathrm{i}\omega \left[\mu_0 - \mu(\mathbf{r}')\right] \mathbf{H}(\mathbf{r}') \tag{2.75}$$

are the induced volumetric electric and magnetic current densities inside the object, respectively. Furthermore, if we assume that the object has the same permeability as the background, which is usually true, then $\mathbf{M}_V(\mathbf{r}') = 0$ and the above two equations can be reduced to

$$\mathbf{E}(\mathbf{r}) = \mathbf{E}^{\text{inc}}(\mathbf{r}) + \mathrm{i}\omega\mu_0 \int_V \overline{\mathbf{G}}(\mathbf{r},\mathbf{r}') \cdot \mathbf{J}_V(\mathbf{r}')\mathrm{d}\mathbf{r}', \quad \mathbf{r} \in V \tag{2.76}$$

$$\mathbf{H}(\mathbf{r}) = \mathbf{H}^{\text{inc}}(\mathbf{r}) + \nabla \times \int_V \overline{\mathbf{G}}(\mathbf{r},\mathbf{r}') \cdot \mathbf{J}_V(\mathbf{r}')\mathrm{d}\mathbf{r}', \quad \mathbf{r} \in V \tag{2.77}$$

and we only need to solve one of the two equations to obtain the unknown current density $\mathbf{J}_V(\mathbf{r}')$. We can treat either $\mathbf{J}_V(\mathbf{r}')$, the total electric field $\mathbf{E}(\mathbf{r}')$ or electric flux density $\mathbf{D}(\mathbf{r}')$ inside the object as an unknown function to be solved in the above VIEs, but $\mathbf{D}(\mathbf{r}')$ is chosen in the MoM with the SWG basis function since the electric flux density is normally continuous across material interfaces.

2.4.3 Volume-Surface Integral Equations (VSIEs)

Real-world objects or structures usually include both impenetrable (conducting) and penetrable (non-conducting) materials, as shown in Figure 2.8. When the penetrable media are isotropic and homogeneous or even piecewise homogeneous, the SIEs are preferably used to describe these parts, but the VIEs can also be employed to describe these parts sometimes in mixed-material structures. The VIEs are the second kind of integral equations that can give rise to a better conditioning of system matrix compared with SIEs, so they may also be preferred in some applications like multiscale problems. The VIEs are of course indispensable if the penetrable media are inhomogeneous and anisotropic. When the impenetrable parts are governed by SIEs, while penetrable parts are described by VIEs, VSIEs will be formed to describe the EM feature of entire structures by the coupling of fields.

Consider the EM scattering by a composite object including both conducting media and penetrable media, which could be inhomogeneous and/or anisotropic, as shown in Figure 2.8. If the conducting part of the object is a PEC, we can use the EFIE to describe its EM feature when disregarding its coupling with the penetrable part, i.e.

$$\hat{n} \times \mathbf{E}^{\text{inc}}(\mathbf{r}) = -\hat{n} \times \mathrm{i}\omega\mu_0 \int_S \overline{\mathbf{G}}(\mathbf{r},\mathbf{r}') \cdot \mathbf{J}_S(\mathbf{r}')\,\mathrm{d}S', \quad \mathbf{r} \in S \tag{2.78}$$

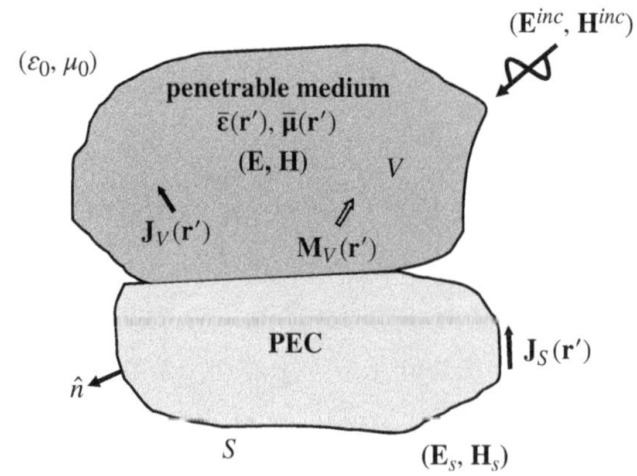

Figure 2.8 Scattering by an object with mixed PEC and penetrable media which could be inhomogeneous and/or anisotropic.

where all variables have been defined in Section 2.4.1. For the penetrable part of the object, which could be inhomogeneous and/or anisotropic, we can employ VIEs to describe its EM characteristic when disregarding its coupling with the conducting part

$$\mathbf{E}(\mathbf{r}) = \mathbf{E}^{\rm inc}(\mathbf{r}) - \int_V \nabla g(\mathbf{r},\mathbf{r}') \times \mathbf{M}_V(\mathbf{r}')\mathrm{d}\mathbf{r}' + \mathrm{i}\omega\mu_0 \int_V \overline{\mathbf{G}}(\mathbf{r},\mathbf{r}') \cdot \mathbf{J}_V(\mathbf{r}')\mathrm{d}\mathbf{r}', \quad \mathbf{r} \in V \tag{2.79}$$

$$\mathbf{H}(\mathbf{r}) = \mathbf{H}^{\rm inc}(\mathbf{r}) + \int_V \nabla g(\mathbf{r},\mathbf{r}') \times \mathbf{J}_V(\mathbf{r}')\mathrm{d}\mathbf{r}' + \mathrm{i}\omega\epsilon_0 \int_V \overline{\mathbf{G}}(\mathbf{r},\mathbf{r}') \cdot \mathbf{M}_V(\mathbf{r}')\mathrm{d}\mathbf{r}', \quad \mathbf{r} \in V \tag{2.80}$$

where all variables have been defined in Section 2.4.2. When considering the coupling of fields generated from the current density at the conducting surface and the volumetric current densities inside the penetrable media, we can form the following VSIEs

$$0 = \hat{n} \times \left[\mathbf{E}^{\rm inc}(\mathbf{r}) + \mathrm{i}\omega\mu_0 \int_S \overline{\mathbf{G}}(\mathbf{r},\mathbf{r}') \cdot \mathbf{J}_S(\mathbf{r}')\, \mathrm{d}S' - \int_V \nabla g(\mathbf{r},\mathbf{r}') \times \mathbf{M}_V(\mathbf{r}')\mathrm{d}\mathbf{r}' \right.$$
$$\left. + \mathrm{i}\omega\mu_0 \int_V \overline{\mathbf{G}}(\mathbf{r},\mathbf{r}') \cdot \mathbf{J}_V(\mathbf{r}')\mathrm{d}\mathbf{r}'\right], \quad \mathbf{r} \in S \tag{2.81}$$

$$\mathbf{E}(\mathbf{r}) = \mathbf{E}^{\rm inc}(\mathbf{r}) + \mathrm{i}\omega\mu_0 \int_S \overline{\mathbf{G}}(\mathbf{r},\mathbf{r}') \cdot \mathbf{J}_S(\mathbf{r}')\, \mathrm{d}S' - \int_V \nabla g(\mathbf{r},\mathbf{r}') \times \mathbf{M}_V(\mathbf{r}')\mathrm{d}\mathbf{r}'$$
$$+ \mathrm{i}\omega\mu_0 \int_V \overline{\mathbf{G}}(\mathbf{r},\mathbf{r}') \cdot \mathbf{J}_V(\mathbf{r}')\mathrm{d}\mathbf{r}', \quad \mathbf{r} \in V \tag{2.82}$$

$$\mathbf{H}(\mathbf{r}) = \mathbf{H}^{\rm inc}(\mathbf{r}) + \mathrm{i}\omega\mu_0 \int_S \overline{\mathbf{G}}(\mathbf{r},\mathbf{r}') \cdot \mathbf{J}_S(\mathbf{r}')\, \mathrm{d}S' + \int_V \nabla g(\mathbf{r},\mathbf{r}') \times \mathbf{J}_V(\mathbf{r}')d\mathbf{r}'$$
$$+ \mathrm{i}\omega\epsilon_0 \int_V \overline{\mathbf{G}}(\mathbf{r},\mathbf{r}') \cdot \mathbf{M}_V(\mathbf{r}')\mathrm{d}\mathbf{r}', \quad \mathbf{r} \in V. \tag{2.83}$$

If the penetrable media are non-magnetic or $\mathbf{M}_V(\mathbf{r}') = 0$, then the above equations can be simplified, i.e.

$$0 = \hat{n} \times \left[\mathbf{E}^{\text{inc}}(\mathbf{r}) + \mathrm{i}\omega\mu_0 \int_S \overline{\mathbf{G}}(\mathbf{r}, \mathbf{r}') \cdot \mathbf{J}_S(\mathbf{r}')\, \mathrm{d}S' + \mathrm{i}\omega\mu_0 \int_V \overline{\mathbf{G}}(\mathbf{r}, \mathbf{r}') \cdot \mathbf{J}_V(\mathbf{r}')\mathrm{d}\mathbf{r}' \right], \quad \mathbf{r} \in S \tag{2.84}$$

$$\mathbf{E}(\mathbf{r}) = \mathbf{E}^{\text{inc}}(\mathbf{r}) + \mathrm{i}\omega\mu_0 \int_S \overline{\mathbf{G}}(\mathbf{r}, \mathbf{r}') \cdot \mathbf{J}_S(\mathbf{r}')\, \mathrm{d}S' + \mathrm{i}\omega\mu_0 \int_V \overline{\mathbf{G}}(\mathbf{r}, \mathbf{r}') \cdot \mathbf{J}_V(\mathbf{r}')\mathrm{d}\mathbf{r}', \quad \mathbf{r} \in V. \tag{2.85}$$

The EFIE for the conducting part can of course be replaced by the MFIE and then we can form another group of VSIEs, but one seldom uses this kind of VSIEs.

2.5 Summary

Based on the macroscopic picture of the preceding chapter for EM, we have taken a special and fine look at the CEM in this chapter. The CEM has become an important branch of EM as CMS techniques have developed quickly and become more and more powerful. The CEM mainly studies the computational methods, particularly, numerical methods and relevant techniques for solving EM problems. The EM problems are governed by Maxwell's equations, which can be changed into different forms by mathematical manipulation, and different numerical methods and relevant techniques have been developed to solve them. The FDTD method, FEM, and MoM are three primary numerical methods that are used to solve the three different forms of Maxwell's equations and we have given a brief introduction for each of them. Although the available analytical methods are very few, we also introduce the Mie-series method, which is actually a semi-analytical method because its results are widely used as exact solutions for verifying numerical results. In fact, the numerical methods are just tools for transforming original continuous equations into discrete matrix equations and matrix equation solvers are needed to solve the matrix equations. For the Nyström method, which is also an important numerical method for solving EM integral equations, we will make a special introduction in next chapter because it is our theme in this book.

Bibliography

1 Chew, W.C., Tong, M.S., and Hu, B. (2008). *Integral Equation Methods for Electromagnetic and Elastic Waves*. San Rafael, CA: Morgan & Claypool.

2 Dubois, G. (2018). *Modeling and Simulation: Challenges and Best Practices for Industry*. Boca Raton: CRC Press.

3 Zeigler, B.P., Praehofer, H., and Kim, T. (2000). *Theory of Modeling and Simulation*, 2e. New York: Academic Press.

4 Taflove, A. and Hagness, S.C. (2005). *Computational Electrodynamics: The Finite-Difference Time-Domain Method*, 3e. Boston: Artech House.
5 Zienkiewicz, O.C. (1977). *The Finite Element Method*, 3e. New York: McGraw-Hill.
6 Harrington, R.F. (1968). *Field Computation by Moment Methods*. New York: Macmillan.
7 Mie, G. (1908). Beiträge zur optik trüber medien, speziell kolloidaler metallösungen. *Ann. Phys.* 25 (3): 377–445.
8 Balanis, C.A. (1989). *Advanced Engineering Electromagnetics*. New York: John Wiley & Sons.
9 Goodrich, R.F., Harrison, B.A., Kleinman, R.E., and Senior, T.B.A. (1961). *Studies in Radar Cross Sections XLVII-Diffraction and Scattering by Regular Bodies-I: The Sphere*. Ann Arbor: University of Michigan Radiation Laboratory.
10 Ruck, G.T., Barrick, D.E., Stuart, W.D., and Krichbaum, C.K. (1970). *Radar Cross Section Handbook*. New York: Plenum Press.
11 Wait, J.R. (1963). Electromagnetic scattering from radially inhomogeneous spheres. *Appl. Sci. Res.* 10 (7): 441.
12 Yee, K.S. (1966). Numerical solution of initial boundary value problems involving Maxwell's equations in isotropic media. *IEEE Trans. Antennas Propagat.* 14 (3): 302–307.
13 Courant, R., Friedrichs, K.O., and Lewy, H. (1928). Über die partiellen differenzengleichungen der mathematischen physik. *Mathematische Annalen (in German)*. 100 (1): 32–74.
14 O'Brien, G.G., Hyman, M.A., and Kaplan, S. (1950). A study of the numerical solution of partial differential equations. *J. Math. Phys.* 29 (1): 223–251.
15 Wikipedia, Finite-Difference Time-Domain Method. Internet: https://en.wikipedia.org/wiki/Finite-difference_time-domain_method.
16 Kunz, K.S. and Luebbers, R.J. (1993). *The Finite Difference Time Domain Method for Electromagnetics*. Boca Raton: CRC Press.
17 Hrennikoff, A. (1941). Solution of problems of elasticity by the framework method. *J. Appl. Mech.* 8 (4): 169–175.
18 Courant, R. (1943). Variational methods for the solution of problems of equilibrium and vibrations. *Bull. Am. Math. Soc.* 49: 1–23.
19 Wikipedia, Finite Element Method. Internet: https://en.wikipedia.org/wiki/Finite_element_method.
20 Reddy, J.N. (2006). *An Introduction to the Finite Element Method*, 3e. New York: McGraw-Hill.
21 Jin, J.-M. (2002). *The Finite Element Method in Electromagnetics*, 2e. New York: Wiley.
22 Mei, K.K. and Van Bladel, J. (1963). Scattering by perfectly conducting rectangular cylinders. *IEEE Trans. Antennas Propagat.* AP-11 (2): 185–192.
23 Andreasen, M.G. (1964). Scattering from parallel metallic cylinders with arbitrary cross section. *IEEE Trans. Antennas Propagat.* AP-12 (6): 746–754.
24 Richmond, J.H. (1965). Scattering by a dielectric cylinder of arbitrary cross section shape. *IEEE Trans. Antennas Propagat.* AP-14 (3): 334–341.
25 Chew, W.C., Jin, J.M., Michielssen, E., and Song, J.M. (2001). *Fast and Efficient Algorithms in Computational Electromagnetics*. Boston: Artech House.

26 Rao, S.M., Wilton, D.R., and Glisson, A.W. (1982). Electromagnetic scattering by surfaces of arbitrary shape. *IEEE Trans. Antennas Propagat.* 30 (3): 409–418.

27 Schaubert, D.H., Wilton, D.R., and Glisson, A.W. (1984). A tetrahedral modeling method for electromagnetic scattering by arbitrary shaped inhomogeneous dielectric bodies. *IEEE Trans. Antennas Propagat.* AP-32 (1): 77–85.

28 Poggio, A.J. and Miller, E.K. (1973). Integral equation solutions of three-dimensional scattering problems. In: *Computer Techniques for Electromagnetics* (ed. R. Mittra). Oxford: Pergamon Press.

29 Müller, C. (1969). *Foundations of the Mathematical Theory of Electromagnetic Waves.* Berlin: Springer-Verlag.

3

The Nyström Method

3.1 Introduction

As we have mentioned in the preceeding chapter, the Nyström method is a numerical method used to solve integral equations and was first proposed and improved by E. J. Nyström in 1928 [1, 2]. This method was introduced to the electromagnetics community in 1990 [3] and acquired significant developments later on, making it a good competitor or alternative for the method of moments (MoM) [4–9]. The method was originally designed for solving non-singular integral equations, but electromagnetic (EM) integral equations include singular kernels due to the existence of the Green's function, so one has to handle the singularity problem in solving EM integral equations if using this method.

The most distinct feature of the Nyström method is that it directly replaces the integral operators of integral equations with an algebraic summation under an appropriate quadrature rule when the operators are smooth. This is a process to transform or discretize continuous integral equations into discrete algebraic equations and the corresponding matrix entries can be generated by simply sampling integral kernels at quadrature points without involving numerical integrations. The primary merits of the method include a simple mechanism of implementation, ease of geometric discretization or lower requirement on the mesh quality that allows the use of non-conforming meshes, and removal of basis and testing functions. These merits could be very desirable in fast algorithms like the multilevel fast multipole algorithm (MLFMA) [10], because they can greatly simplify the implementation and reduce preprocessing costs.

The main difficulty of the Nyström method is the singularity treatment or local correction for singular or near-singular integrals appearing in self-interaction or near-interaction matrix elements since numerical quadrature rules cannot be applied directly. Unlike in the MoM, in which the well-designed basis and testing functions can help reduce the degree of singularity in the evaluation of singular or near-singular elements by using integration by parts, we have to handle the hypersingular integrals that come from the double gradient operator of the dyadic Green's function in the Nyström method. Fortunately, robust singularity treatment techniques or local correction schemes have been developed for all EM integral equations [11, 12]. Although the Nyström method may not be a first choice when the MoM can be easily implemented, it could be a good alternative to the MoM in some applications, especially when conforming meshes are hard to figure out due to a mixture of several very-different media.

The Nyström Method in Electromagnetics, First Edition. Mei Song Tong and Weng Cho Chew.

Note that there is a well-known Nyström method called "locally corrected Nyström method (LCNM)", which was first proposed by Canino *et al.* in 1998 [6] and then further developed by Gedney *et al.* [9]. The fundamental difference between the LCNM and the Nyström method presented in this book is that the LCNM relies on the redesign of the quadrature rule for singular integrands based on Strain's method [13] while the Nyström method in this book still relies on regular quadrature rules by regularizing singular integrands and trying to derive analytical formulas for subtracted singular integrals. The essence of local correction in the LCNM is to generate an adjustment scheme for the quadrature rule for singular integral kernels by computing the convolutions of the kernels with a suitable set of testing functions. Actually, this is a MoM-like procedure for determining the corresponding weights of the quadrature rule at different quadrature points. This adjustment is only required in the vicinity of the singularity and allows the quadrature rule to be as accurate near the singularity as it is far from it. Due to the above fundamental difference, we use the original and more common name "Nyström method" to refer to the method. Also, we usually use "singularity treatment" instead of "locally corrected" or "local correction" to refer to the techniques of handling singular integrals. Singularity treatment aim to regularize singular integrands while "local correction" aims to redesign quadrature rules for singular integrals.

3.2 Basic Principle

To solve an integral equation defined over an integral domain, which usually consists of the boundaries or volumes of an object or structure, we first need to discretize the integral domain and change the integral over the entire domain into a sum of integrals over small and regular domains. The discretization is needed because small domains allow a good approximate representation for unknown functions to be solved and quadrature rules are defined over small regular domains such as triangular, rectangular, circular, tetrahedral, cuboid domains, etc. After discretizing the domain or geometry, we need to appropriately represent or express the unknown functions because they are not defined, and integral operators cannot directly act on them. Usually, they are expressed in interpolation functions with unknown coefficients. Replacing the unknown functions with such an expression leads to the integration of the kernels with the interpolation functions for each matrix entry. The interpolation functions are chosen as shape functions of an element in an isoparametric scheme and the unknown coefficients are the system's unknowns to be solved. The Nyström method uses a different strategy to generate the matrix entries. If the integral kernels are smooth over an element such as a surface element, the Nyström method replaces the integral with a summation under a quadrature rule, i.e.

$$\int_{\Delta S} f(\mathbf{x}')\mathrm{d}S' = \sum_{j=1}^{P} w_j f(\mathbf{x}'_j) \tag{3.1}$$

where $f(\mathbf{x})$ is a general smooth function, P is the number of quadrature points and w_j is the jth weight over the surface element ΔS, which can be a triangle, square, circle, etc. We usually use the non-product quadrature rule instead of the product or repeated quadrature rule for a surface integral because the non-product rule has fewer quadrature points [3]. The typical non-product rules are the Gauss-Legendre rule with one, three or six quadrature points over a flat triangle [14]. The Nyström method wisely selects the values of the unknown function at quadrature points as the unknowns of the matrix equation and the matrix entries are just the directly-evaluated values of integral kernels times the weights of the quadrature rule. Consider a generalized three-dimensional (3D) surface integral equation (SIE)

$$\int_S F(\mathbf{x},\mathbf{x}')u(\mathbf{x}')dS' = -\phi(\mathbf{x}), \qquad \mathbf{x} \in S \tag{3.2}$$

where $F(\mathbf{x},\mathbf{x}')$ is the integral kernel related to the scalar Green's function, $u(\mathbf{x}')$ is the unknown function, say the component of a current density, and $\phi(\mathbf{x})$ represents an excitation, which is an incident field in wave scattering problems. After discretizing the surface S into N patches and applying the quadrature rule in (3.1), the SIE becomes

$$\sum_{i=1}^{N}\sum_{j=1}^{P_i} w_{ij}F(\mathbf{x},\mathbf{x}'_{ij})u(\mathbf{x}'_{ij}) = -\phi(\mathbf{x}), \qquad \mathbf{x} \in S \tag{3.3}$$

where P_i is the number of quadrature points at the ith element and w_{ij} represents the weight of the quadrature rule on the jth point of the ith element. Performing the collocation procedure on these quadrature points, i.e., choosing these quadrature points as observation or field points and matching or enforcing the above equation to be satisfied on these points, we can obtain the following matrix equation

$$\sum_{i=1}^{N}\sum_{j=1}^{P_i} w_{ij}g(\mathbf{x}_{mn},\mathbf{x}'_{ij})u(\mathbf{x}'_{ij}) = -\phi(\mathbf{x}_{mn}) \tag{3.4}$$

where $m = 1, 2, \cdots, N$ and $n = 1, 2, \cdots, P_m$. We can see that the whole process is very simple because it avoids the need for interpolations and numerical integrations. Since the integrands are usually smooth for far-interaction elements, the matrix entries corresponding to these elements can be created in such a way. This will greatly facilitate the implementation and save on matrix-filling time. For singular or near-singular elements, the integrands are singular or badly behaved and the quadrature rule in (3.1) cannot be applied directly. A special treatment called local correction is needed to generate those diagonal or near-diagonal entries. The local correction is actually a singularity manipulation technique and very similar to the procedure in the boundary element methods. The difference is that we interpolate the unknown function over a singular element based on the values of the unknown function at the quadrature points and we have developed analytical solutions for the resultant hypersingular and strongly-singular integrals under a Cauchy-principal-value (CPV) sense.

3.3 Singularity Treatment

The singularity problem appears because of the existence of the Green's function, which is inherently singular in EM integral equations. The use of the Green's function is both a blessing and a curse. The blessing means that it can bring many conveniences to solve EM problems, particularly the automatic satisfaction of radiation conditions leading to the removal of implementing absorbing boundary conditions and a great reduction in solution domains. The curse denotes that it gives rise to certain difficulties in solving the EM integral equations due to its singularity. The singularity treatment is a key in the Nyström method and different treatment techniques can actually characterize different Nyström schemes. We will specially address the physical techniques for the singularity treatment in Chapter 5, but we would like to present a brief introduction to the basic idea here.

There are three types of singularity in EM integral equations, i.e. $1/R$-type, $1/R^2$-type, and $1/R^3$-type, which come from the scalar Green's function $g(\mathbf{r},\mathbf{r}') = e^{ikR}/R$ itself, $\nabla g(\mathbf{r},\mathbf{r}')$, and $\nabla\nabla g(\mathbf{r},\mathbf{r}')$, respectively, where $R = |\mathbf{r} - \mathbf{r}'|$ is the distance between an observation point $\mathbf{r}$ and a source point $\mathbf{r}'$. On the other hand, there are two kinds of singularity treatment technique, i.e. singularity cancellation and singularity subtraction. The singularity cancellation, although preferable, can only be used to treat a $1/R$-type weak singularity while the singularity subtraction is more versatile and indispensable for handling a $1/R^2$-type strong singularity and a $1/R^3$-type hypersingularity for surface integrals. The basic idea of the singularity cancellation technique is to produce an extra zero point in a Jacobian by using an integral-variable change and use the extra zero point to cancel the zero point or singular point in the denominator of a singular integrand so that the integrand can be regularized. The most well-known method in such a technique is the Duffy's method [15], which has been widely used. Since one can only produce one extra zero point for surface integrals and two extra zero points for volume integrals at most by the integral-variable change, that is why the technique can only treat a $1/R$-type weak singularity for surface integrals and a $1/R^2$-type strong singularity for volume integrals.

The main idea of singularity subtraction is quite different from that of singularity cancellation. One needs to distinguish singular cores of a singular integrand by Taylor's series expansion and then subtract or extract the singular cores from the integrand so that the integrand can be regularized and become numerically integrable. The subtracted or extracted singular cores need to be handled specially and one has to derive analytical formulas for their integrals. The derivation of analytical formulas is very essential and could be very challenging, especially for hypersingular integrals. We have developed two different methods for accurately evaluating those singular integrals under a CPV sense in recent years. The first one is based on a subdivision of an integral domain and the second one is based on the transformation of a surface integral over a surface domain into a boundary integral along its boundary based on the Stokes' theorem or the Green–Gauss theorem. We will present the details of the techniques in Chapter 5.

3.4 Higher-Order Scheme

One of the distinct features in the Nyström method is the facilitation of using higher-order schemes to enhance numerical accuracy. Higher-order schemes are very desirable because they can use fewer unknowns to reach the designated accuracy or use the same number of unknowns to acquire higher accuracy. In fact, we usually have two choices to enhance the accuracy of numerical solutions: the first one is by reducing the size of meshes or increasing the density of meshes and the second one is by using higher-order quadrature rules. The former is called h-refinement while the latter is called p-refinement [6]. It can be proven that p-refinement can gives rise to a higher rate of convergence than h-refinement for numerical errors. Therefore, using higher-order schemes could be a good idea although it will bring some extra costs or inconveniences in implementation. Using fewer unknowns in higher-order schemes is realized by using coarse meshes for geometric discretization. If the geometry consists of flat faces, then using coarse flat meshes will not produce an extra error of geometric approximation. However, for curvilinear geometries, using coarse flat meshes will give rise to a large error of geometric approximation and curvilinear meshes should be employed. Using curvilinear meshes requires a geometric mapping to transform a curvilinear mesh into a flat mesh so that numerical integration can be performed. This is because quadrature rules are defined over flat elements rather than curvilinear elements and obviously this transformation will result in extra costs in implementation. We will illustrate the case in detail in Chapter 6.

3.5 Comparison to the Method of Moments

Both the Nyström method and the MoM belong to integral equation solvers or they are both numerical methods of transforming or discretizing integral equations into algebraic matrix equations. From a macroscopic point of view, the Nyström method can be seen as a special case or subset of the MoM, which uses a delta-like function as a basis function and testing function. However, from a microscopic angle of view, the two methods are quite different in mechanism and they have the following differences.

- The MoM usually uses a continuous basis function and testing function while the Nyström method uses a discontinuous or delta-like basis function and testing function. That is to say, the strategies of representing or sampling unknown functions are quite different.
- The MoM enforces the boundary condition to be satisfied over each element in an average sense while the Nyström method enforces the boundary condition to be satisfied over each element only at discrete points.
- The discrete points in the Nyström method are selected according to quadrature rules and this practice leads to a direct evaluation of integrands for generating far-interaction matrix elements in the Nyström method while one has to perform a double-fold integration for each matrix element in the MoM.

The above structural differences lead to the following different advantages and disadvantages.

- The MoM can be more stable while the Nyström method can be less stable due to the difference in the style of satisfying boundary conditions.
- The MoM is more tedious and double-fold integrations are required while the Nyström method is relatively simple due to the direct evaluation of integrands for generating far-interaction matrix elements and only a collocation or point-matching procedure is used in implementation.
- The MoM requires conforming meshes in geometric discretization while the Nyström method does not have this kind of requirement, leading to a big difference in preprocessing costs.

- The MoM uses specially-designed basis and testing functions while the Nyström method does not do so.
- It is easier to enhance numerical accuracy in the Nyström method than in the MoM because the Nyström method can use higher-order quadrature rules to improve the accuracy and it has been proven that p-refinement or using higher-order quadrature rules has a higher rate of convergence than h-refinement or using denser meshes in geometric discretization.
- The MoM only needs to treat weak and strong singularities due to the help of basis and testing functions while the Nyström method has to treat the hypersingularity, which cannot be reduced with the delta-like basis and testing functions.

In numerical implementation, the MoM may use a fewer number of unknowns while the Nyström method uses more unknowns for same geometric discretization, so more memory usage is required in the Nyström method in general. For SIEs, the number of unknowns is $N = 1.5 \times N_s$ (for one unknown function) in the MoM for a close structure while $N = 2.0 \times N_s$ in the Nyström method if a one-point quadrature rule is used, where N_s is the number of triangular patches. However, the numbers of unknowns of the two methods are very close to each other for VIEs. The number of unknowns is $N \approx 2.95 \times N_v$ (for one unknown function) in the MoM while $N = 3.0 \times N_v$ in the Nyström method if a one-point quadrature rule is used. Note that the faces of tetrahedrons which reside on the interfaces with the background or conductors also result in unknowns by assuming fictitious tetrahedrons inside the background or conductors in the MoM and that is why $N \approx 2.95 \times N_v$ because it is related to the number of tetrahedrons with one face at the material interfaces. Regarding the CPU time, the Nyström method usually uses less resource than the MoM for the same number of unknowns due to its simple method of generating most matrix elements. This feature also greatly facilitates the incorporation of Nyström method with fast algorithms like the MLFMA [10].

Furthermore, the Nyström method could also be much simpler in implementation for solving volume-surface integral equations (VSIEs), which include the interfaces between the penetrable materials and conductors [4]. From the equations, we can see that the material property only appears on the left-hand side (the material property is implied in the current densities on the right-hand side), so we can allow each tetrahedron to include inhomogeneous materials in the Nyström method, but that is not allowed in the MoM because the Schaubert-Wilton-Glisson (SWG) basis function [16] cannot be defined in such tetrahedrons. This merit can greatly facilitate the discretization of inhomogeneous materials in the Nyström method.

3.6 Comparison to the Point-Matching Method

In addition to the difference between the Nyström method and the MoM, many readers could also be concerned with the difference between the Nyström method and the point-matching method (PMM). Obviously, the Nyström method can be thought of as a kind of PMM, but it is different from the general PMM. The key characteristic in the Nystrom method is to use a quadrature rule to transform an integral into a summation. Due to the use of quadrature rules, the points in an element are not randomly selected while they are designated by the quadrature rules even if they are one-point quadrature rules. Also, the number of quadrature points should match the number of unknown coefficients in the polynomial representation of unknown functions in the singularity treatment, but there are not many good choices for higher-order quadrature rules in some elements and this can certainly reduce the flexibility of Nyström method. In the contrast, the general PMM can arbitrarily select the points without the constraint of quadrature rules, but the resultant matrix elements cannot be obtained through the direct evaluation of integrands and numerical integrations are needed.

The one-point quadrature rules are the simplest quadrature rules for a given element but they can still distinguish the Nyström method from the PMM because the unique point, which is usually the geometric center of the given element, is determined by the used quadrature rules in the Nyström method while it can be randomly selected in the PMM. In fact, there are different one-point quadrature rules and the point may not be the geometric center of element, especially for tetrahedral elements [14]. The geometric center of element may not necessarily be the best point to represent the whole element domain. The best point should be determined by the property of integrands over the element instead of the geometric characteristic of the element. Also, we have three different inner centers to be used for an element.

On the other hand, the PMM is more flexible to select the point and may be preferred in some situations. For example, in the scattering problem of two-dimensional (2D) open conducting object, the polarized current density close to an edge tends to be infinite and selecting a point nearby the edge in a triangle is better than the geometric center of the triangle because it can better represent the current density of the whole triangle. However, the PMM requires numerical integrations to calculate each matrix entry while the Nyström method does not need for far-interaction entries and this is the distinct advantage of Nyström method. We can actually combine the two methods together by individually applying their advantages to different regions when solving certain problems.

3.7 Summary

In this chapter, we systematically review the Nyström method, including its origin, basic principle, implementation, and comparison to the MoM. Although the method appeared very early, the EM community did not pay much attention to it until 1990, when Kot first used this method to solve EM integral equations. In principle, the method is quite simple, i.e. using an algebraic summation to replace an integral over a small regular domain under a quadrature rule so that the continuous integral equations can be discretized into algebraic matrix equations. However, as we have emphasized, the singularity treatment or local correction is the main difficulty in the implementation of this method. We introduce some robust singularity treatment techniques that we have developed in recent years but will be individually presented in Chapter 5. Also, we address higher-order Nyström schemes, which are less widely used but can be highly valuable and need to be further exploited. Finally, we make a comprehensive comparison between the Nyström method and the MoM and emphasize its advantages because many readers could naturally ask why we need to use this method when the MoM is so popular.

Bibliography

1 Nyström, E.J. (1928). Über die praktische auflösung von linearen integralgleichungen mit anwendungen auf randwertaufgaben der potentialtheorie. *Commentationes Physico-Mathematicae.* 4 (15): 1–52.
2 Nyström, E.J. (1930). Über die praktische auflösung von integralgleichungen mit anwendungen auf randwertaufgaben. *Acta Mathematica.* 1 (54): 185–204.
3 Kot, J.S. (1990). Solution of thin-wire integral equations by Nyström methods. *Microw. Opt. Technol. Lett.* 3: 393–396.
4 Chew, W.C., Tong, M.S., and Hu, B. (2008). *Integral Equation Methods for Electromagnetic and Elastic Waves.* San Rafael, CA: Morgan & Claypool.
5 Kot, J.S. (1992). Application of Nyström methods to electric- and magnetic-field integral equations. *Proceedings of 1992 URSI International Symposium on Electromagnetic Theory,* Sydney, Australia.
6 Canino, L.S., Ottusch, J.J., Stalzer, M.A. et al. (1998). Numerical solution of the Helmholtz equation in 2D and 3D using a high-order Nyström discretization. *J. Comput. Phys.* 146: 627–663.
7 Burghignoli, P., Di Nallo, C., Frezza, F., and Galli, A. (2003). A simple Nyström approach for the analysis of 3D arbitrarily shaped conducting and dielectric bodies. *Int. J. Numer. Model.* 16: 179–194.
8 Peterson, A.F. and Bibby, M.M. (2004). Higher-order numerical solutions of the MFIE for the linear dipole. *IEEE Trans. Antennas Propagat.* 52: 2684–2691.
9 Gedney, S.D. (2003). On deriving a locally corrected Nyström scheme from a quadrature sampled moment method. *IEEE Trans. Antennas Propagat.* 51: 2402–2412.
10 Chew, W.C., Jin, J.M., Michielssen, E., and Song, J.M. (2001). *Fast and Efficient Algorithms in Computational Electromagnetics.* Boston: Artech House.
11 Tong, M.S. and Chew, W.C. (2007). Super-hyper singularity treatment for solving 3D electric field integral equations. *Microw. Opt. Technol. Lett.* 49 (6): 1383–1388.
12 Tong, M.S., Qian, Z.G., and Chew, W.C. (2010). Nyström method solution of volume integral equations for electromagnetic scattering by 3D penetrable objects. *IEEE Trans. Antennas Propagat.* 58 (5): 1645–1652.
13 Strain, J. (1995). Locally-corrected multidimensional quadrature rules for singular functions. *SIAM J. Sci. Comput.* 16 (4): 992–1017.
14 Stroud, A.H. (1971). *Approximate Calculation of Multiple Integrals.* Englewood Cliffs, NJ: Prentice-Hall.
15 Duffy, M.G. (1982). Quadrature over a pyramid or cube of integrands with a singularity at a vertex. *SIAM J. Numer. Anal.* 19: 1260–1262.
16 Schaubert, D.H., Wilton, D.R., and Glisson, A.W. (1984). A tetrahedral modeling method for electromagnetic scattering by arbitrary shaped inhomogeneous dielectric bodies. *IEEE Trans. Antennas Propagat.* AP-32 (1): 77–85.

4

Numerical Quadrature Rules

4.1 Introduction

As we can see from the principle of the Nyström method in the preceding chapter, using the method strongly relies on numerical quadrature rules because the selection of discrete points in a patch or element is based on the used quadrature rules. This is also the reason why the matrix elements of far interactions can be so easily generated by only sampling the integrand and multiplying by the weight of the used quadrature rule [1]. This mechanism also distinguishes the method from other numerical methods. In this sense, the saying that some researchers believe, i.e. "the Nyström method is the same as the method of moments (MoM) if delta basis and testing functions are chosen" may not be true [2].

Because of the importance of quadrature rules, we specifically address them in this chapter. The quadrature rules have been extensively studied [3–12] and many good rules have been proposed or designed, especially the widely-used Gaussian quadrature rules [3–10]. The quadrature rules can be divided into two categories, i.e. product rules and non-product rules [13]. A product rule refers to a combination or product of one-fold or one-dimensional (1D) quadrature rules for a multiple-fold or multiple-dimensional integration that corresponds to a surface or volume integral. On the other hand, a non-product rule refers to a specially designed independent rule for a multiple-fold integration without using the one-fold quadrature rule. The product rules are readily available and have many choices in general, but they generally use many quadrature points or sampled points, although the integration accuracy can be very good since they could overly sample an integrand for a given quadrature accuracy. By contrast, the non-product rules can sample much fewer quadrature points for the same quadrature accuracy compared with the product rules, but the choice of rules is not as many as for the product rules. Since non-product rules are usually used in practice, we only address this kind of rules.

We begin from the definition and generation method for a quadrature rule, and then move to the specific quadrature rules defined for different meshes or cells. The meshes include segmental meshes for a line domain, triangular patch or quadrilateral patch for a surface domain, and tetrahedral element or cuboid or hexahedral element for a volumetric domain. There are other types of mesh, like hexagonal meshes for a surface domain and cylindrical meshes for a volumetric domain, but they are seldom used in practice. The quadrature rules defined over a triangular patch and tetrahedral element are most widely used and the rules defined over rectangular or cuboid meshes are less

The Nyström Method in Electromagnetics, First Edition. Mei Song Tong and Weng Cho Chew.

employed but could be needed in some cases. If the meshes are just 1D segments, as used for solving two-dimensional (2D) problems, then many quadrature rules can be used among which Gaussian quadrature rules could be mostly preferred. Note that all quadrature rules can only be applied for regular integrals whose integrands are bounded within their integral domains. A singularity treatment is required if their integrands are singular and this will be addressed in the next chapter.

4.2 Definition and Design

In numerical analysis, numerical integration is the calculation of integrals by certain numerical quadrature rules when their exact analytical expressions cannot be derived. A numerical quadrature rule is a kind of rule or method that approximately and numerically calculates the integration of a function, which certainly has some restrictions [13]. Usually, the rule can be stated as the weighted sum of sampled values of the function at some specific points or quadrature points within the domain of integration. For the 1D integrals defined over a segmental interval, the most-popularly used quadrature rule is Gaussian (or Gaussian-Legendre) quadrature rule, which was named after the great mathematician Carl Friedrich Gauss [3]. A Gaussian quadrature rule with n points is a quadrature rule that is constructed to yield an exact result of integration for a polynomial function in the order of $2n - 1$ or less by an appropriate choice of the points x_i and weights w_i for $i = 1, \cdots, n$ [3–6]. Mathematically, the rule can be written as

$$\int_{-1}^{1} f(x)\,\mathrm{d}x = \sum_{i=1}^{n} w_i f(x_i). \tag{4.1}$$

Note that the rule is usually defined over a standard interval $[-1, 1]$. For an arbitrary interval $[a, b]$, we need to make an interval transformation to use the rule, i.e.

$$\int_{a}^{b} f(x)\,\mathrm{d}x = \frac{b-a}{2}\int_{-1}^{1} f\left(\frac{b-a}{2}x + \frac{b+a}{2}\right)\,\mathrm{d}x. \tag{4.2}$$

The above quadrature rule can produce a good approximate result if an arbitrary function $f(x)$ can be well approximated by a $(2n - 1)$th-order polynomial function within the interval of integration, i.e.

$$\int_{-1}^{1} f(x)\,\mathrm{d}x \approx \sum_{i=1}^{n} w_i f(x_i) \tag{4.3}$$

or

$$\int_{a}^{b} f(x)\,\mathrm{d}x \approx \frac{b-a}{2}\sum_{i=1}^{n} w_i f\left(\frac{b-a}{2}x_i + \frac{b+a}{2}\right). \tag{4.4}$$

4.3 Quadrature Rules for a Segmental Mesh

A segmental mesh refers to the segmental discretization of a 1D or line integral domain. When the segmental mesh is fine enough, the numerical quadrature over each segmental mesh can be very accurate by using appropriate quadrature rules. Although there

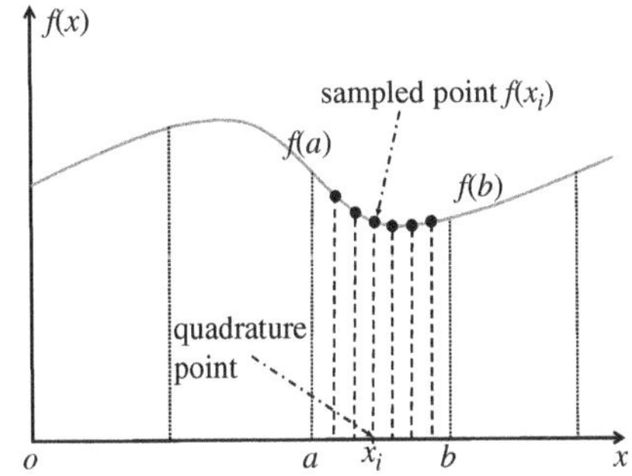

Figure 4.1 Illustration of quadrature rules defined over a segmental mesh.

are many types of quadrature rule, such as the trapezoid rule, the Simpson rule, and the Romberg rule, the Gaussian quadrature rule is most widely used due to its higher accuracy at the same number of sampled quadrature points. The commonly used Gaussian quadrature rules can be found in many mathematical handbooks, such as [14–16], and we illustrate one of these rules in Figure 4.1. We list some of commonly used rules in Table 4.1, which is edited from [14], because such a book may not be available for the readers in some countries. Note that the abscissas are actually the zeros of Legendre polynomials.

4.4 Quadrature Rules for a Surface Mesh

4.4.1 Quadrature Rules for a Triangular Patch

Triangular patches are the most widely used meshes for discretizing surface domains because they are most flexible in approximating arbitrary surface boundaries. There are many quadrature rules defined over a triangular patch and Table 4.2 shows some typical rules which are selected from [13] and could be useful in the Nyström method. The rules with one, three, six, and ten quadrature points as shown in Figure 4.2 correspond to zeroth-order, first-order, second-order, and third-order polynomial functions, respectively, for approximating unknown functions in the singularity treatment, but we cannot find any 10-point quadrature rules whose quadrature points are fully inside the patch. Note that all rules are defined over a standard right triangle, as shown in Figure 4.3, and the following mapping is needed to apply the rules to an arbitrary triangle $\Delta p_1 p_2 p_3$, as shown in the figure.

$$x = x_1 + (x_2 - x_1)u + (x_3 - x_1)v \tag{4.5}$$

$$y = y_1 + (y_2 - y_1)u + (y_3 - y_1)v \tag{4.6}$$

$$z = z_1 + (z_2 - z_1)u + (z_3 - z_1)v. \tag{4.7}$$

The most useful quadrature rules in Table 4.2 may be the three-point and six-point rules because they correspond to a linear polynomial and quadratic polynomial with three and six unknown coefficients, respectively, in the singularity treatment of the Nyström method. However, the quadrature rules with other numbers of quadrature points, such as the four-point rule and seven-point rule, can also be used. In this case, one of the points (usually the central point) can be unused in the linear polynomial or quadratic

Table 4.1 Gaussian quadrature rules with different points and orders.

Number of points	Abscissas $\pm x_i$	Weight factors w_i
$n = 2$	0.57735 02691 89626	1.00000 00000 00000
$n = 4$	0.33998 10435 84856	0.65214 51548 62546
	0.86113 63115 94053	0.34785 48451 37454
$n = 6$	0.23861 91860 83197	0.46791 39345 72691
	0.66120 93864 66265	0.36076 15730 48139
	0.93246 95142 03152	0.17132 44923 79170
$n = 8$	0.18343 46424 95650	0.36268 37833 78362
	0.52553 24099 16329	0.31370 66458 77887
	0.79666 64774 13627	0.22238 10344 53374
	0.96028 98564 97536	0.10122 85362 90376
$n = 10$	0.14887 43389 81631	0.29552 42247 14753
	0.43339 53941 29247	0.26926 67193 09996
	0.67940 95682 99024	0.21908 63625 15982
	0.86506 33666 88985	0.14945 13491 50581
	0.97390 65285 17172	0.06667 13443 08688
$n = 12$	0.12523 34085 11469	0.24914 70458 13403
	0.36783 14989 98180	0.23349 25365 38355
	0.58731 79542 86617	0.20316 74267 23066
	0.76990 26741 94305	0.16007 83285 43346
	0.90411 72563 70475	0.10693 93259 95318
	0.98156 06342 46719	0.04717 53363 86512
$n = 16$	0.09501 25098 37637 440185	0.18945 06104 55068 496285
	0.28160 35507 79258 913230	0.18260 34150 44923 588867
	0.45801 67776 57227 386342	0.16915 65193 95002 538189
	0.61787 62444 02643 748447	0.14959 59888 16576 732081
	0.75540 44083 55003 033895	0.12462 89712 55533 872052
	0.86563 12023 87831 743880	0.09515 85116 82492 784810
	0.94457 50230 73232 576078	0.06225 35239 38647 892863
	0.98940 09349 91649 932596	0.02715 24594 11754 094852
$n = 20$	0.07652 65211 33497 333755	0.15275 33871 30725 850698
	0.22778 58511 41645 078080	0.14917 29864 72603 746788
	0.37370 60887 15419 560673	0.14209 61093 18382 051329
	0.51086 70019 50827 098004	0.13168 86384 49176 626898
	0.63605 36807 26515 025453	0.11819 45319 61518 417312
	0.74633 19064 60150 792614	0.10193 01198 17240 435037
	0.83911 69718 22218 823395	0.08327 67415 76704 748725
	0.91223 44282 51325 905868	0.06267 20483 34109 063570
	0.96397 19272 77913 791268	0.04060 14298 00386 941331
	0.99312 85991 85094 924786	0.01761 40071 39152 118312

Table 4.2 Typical quadrature rules defined over a triangular patch.

Rules	Position	Weight	Degree
One-point	$\left(\frac{1}{3},\frac{1}{3}\right)$	$\frac{1}{2}$	1
Three-point	$\left(\frac{1}{6},\frac{1}{6}\right);\left(\frac{1}{6},\frac{2}{3}\right);\left(\frac{2}{3},\frac{1}{6}\right)$	$\frac{1}{6};\frac{1}{6};\frac{1}{6}$	2
Four-point	$\left(\frac{1}{3},\frac{1}{3}\right);\left(\frac{1}{5},\frac{1}{5}\right);\left(\frac{1}{5},\frac{3}{5}\right);\left(\frac{3}{5},\frac{1}{5}\right)$	$-\frac{9}{32};\frac{25}{96};\frac{25}{96};\frac{25}{96}$	3
Six-point	(0.10903 90090 72877, 0.23193 33685 53031)	$\frac{1}{12}$	3
	(0.23193 33685 53031, 0.10903 90090 72877)	$\frac{1}{12}$	
	(0.10903 90090 72877, 0.65902 76223 74092)	$\frac{1}{12}$	
	(0.65902 76223 74092, 0.10903 90090 72877)	$\frac{1}{12}$	
	(0.23193 33685 53031, 0.65902 76223 74092)	$\frac{1}{12}$	
	(0.65902 76223 74092, 0.23193 33685 53031)	$\frac{1}{12}$	
Six-point	(0.09157 62135 09771, 0.09157 62135 09771)	0.10995 17436 55322	4
	(0.09157 62135 09771, 0.81684 75729 80459)	0.10995 17436 55322	
	(0.81684 75729 80459, 0.09157 62135 09771)	0.10995 17436 55322	
	(0.44594 84909 15965, 0.44594 84909 15965)	0.22338 15896 78011	
	(0.44594 84909 15965, 0.09157 62135 09771)	0.22338 15896 78011	
	(0.09157 62135 09771, 0.44594 84909 15965)	0.22338 15896 78011	
Seven-point	$\left(\frac{1}{3},\frac{1}{3}\right)$	$\frac{9}{80}$	5
	$\left(\frac{6-\sqrt{15}}{21},\frac{6-\sqrt{15}}{21}\right)$	$\frac{155-\sqrt{15}}{2400}$	
	$\left(\frac{6-\sqrt{15}}{21},\frac{9+2\sqrt{15}}{21}\right)$	$\frac{155-\sqrt{15}}{2400}$	
	$\left(\frac{9+2\sqrt{15}}{21},\frac{6-\sqrt{15}}{21}\right)$	$\frac{155-\sqrt{15}}{2400}$	
	$\left(\frac{6+\sqrt{15}}{21},\frac{6+\sqrt{15}}{21}\right)$	$\frac{155+\sqrt{15}}{2400}$	
	$\left(\frac{6+\sqrt{15}}{21},\frac{9-2\sqrt{15}}{21}\right)$	$\frac{155+\sqrt{15}}{2400}$	
	$\left(\frac{9-2\sqrt{15}}{21},\frac{6+\sqrt{15}}{21}\right)$	$\frac{155+\sqrt{15}}{2400}$	

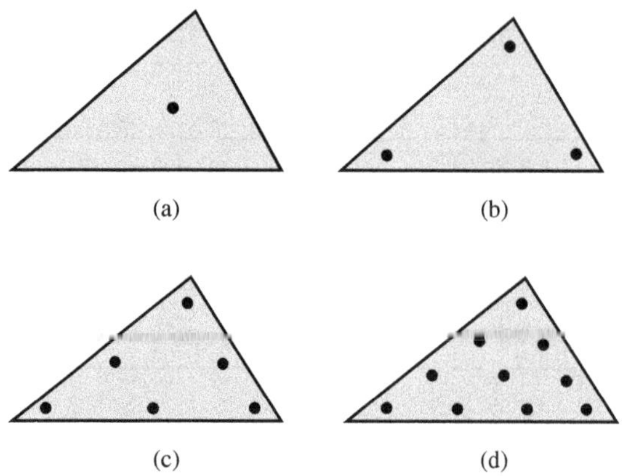

Figure 4.2 Distribution of quadrature points over a triangular patch for different quadrature rules. (a) One point. (b) Three points. (c) Six points. (d) Ten points.

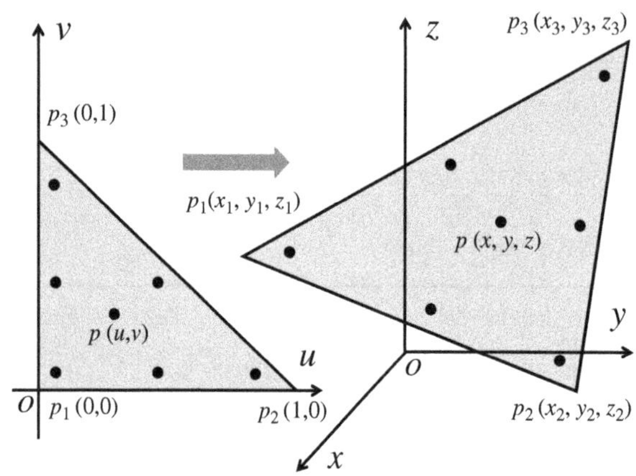

Figure 4.3 Geometric mapping for transforming a quadrature rule defined over a standard right triangle into that defined over an arbitrary triangle.

polynomial approximation of unknown functions in the singularity treatment of the Nyström method. Note that the quadrature points of chosen rules are all within the triangular patch because the unknown functions at some quadrature points may be undefined if those quadrature points are located at geometric boundaries or vertices. If the quadrature points are entirely or partially outside the triangular patch, those rules are also not a good choice because the distribution of quadrature points in the entire geometric domain will be irregular and unwanted. However, there are not many choices for desirable rules if we require the quadrature points to be inside the triangular patch and those three-point and six-point rules listed in Table 4.2 may be the only choice we can find.

4.4.2 Quadrature Rules for a Square Patch

In addition to the quadrature rules defined over a triangular patch, there are also quadrature rules defined over other surface meshes like a square patch, circular patch, and hexagonal patch. Since the quadrature rules defined over a square patch can also be useful when quadrilateral meshes are used to discretize a geometry, which may occur in many applications, we present some of these rules in Table 4.3, which are edited from [13]. Note that we cannot find the rules with six points but the seven-point rules can be

Table 4.3 Typical quadrature rules defined over a square patch.

Rules	Position	Weight	Degree
One-point	(0,0)	4	1
Three-point	$\left(\sqrt{\frac{2}{3}},0\right);\left(-\frac{1}{2}\sqrt{\frac{2}{3}},\frac{\sqrt{2}}{2}\right);$ $\left(-\frac{1}{2}\sqrt{\frac{2}{3}},-\frac{\sqrt{2}}{2}\right)$	$\frac{4}{3};\frac{4}{3};\frac{4}{3}$	2
Four-point	$\left(0,\sqrt{\frac{2}{3}}\right);\left(-\sqrt{\frac{2}{3}},0\right);$ $\left(0,-\sqrt{\frac{2}{3}}\right);\left(\sqrt{\frac{2}{3}},0\right)$	1; 1; 1; 1	3
Four-point	$\left(\frac{\sqrt{3}}{3},\frac{\sqrt{3}}{3}\right);\left(\frac{\sqrt{3}}{3},-\frac{\sqrt{3}}{3}\right);$ $\left(-\frac{\sqrt{3}}{3},\frac{\sqrt{3}}{3}\right);\left(-\frac{\sqrt{3}}{3},-\frac{\sqrt{3}}{3}\right)$	1; 1; 1; 1	3
Seven-point	$\left(\sqrt{\frac{3}{5}},\frac{1}{\sqrt{3}}\right)$	$\frac{5}{9}$	5
	$\left(\sqrt{\frac{3}{5}},-\frac{1}{\sqrt{3}}\right)$	$\frac{5}{9}$	
	$\left(-\sqrt{\frac{3}{5}},\frac{1}{\sqrt{3}}\right)$	$\frac{5}{9}$	
	$\left(-\sqrt{\frac{3}{5}},-\frac{1}{\sqrt{3}}\right)$	$\frac{5}{9}$	
	$\left(0,\sqrt{\frac{14}{15}}\right)$	$\frac{20}{63}$	
	$\left(0,-\sqrt{\frac{14}{15}}\right)$	$\frac{20}{63}$	
	(0,0)	$\frac{8}{7}$	

(*continued*)

Table 4.3 (Continued)

Rules	Position	Weight	Degree
Seven-point	$\left(\sqrt{\frac{7}{15}},\sqrt{\frac{7}{15}}\right)$	$\frac{25}{42}$	5
	$\left(-\sqrt{\frac{7}{15}},-\sqrt{\frac{7}{15}}\right)$	$\frac{25}{42}$	
	$\left(\sqrt{\frac{7+\sqrt{24}}{15}},-\sqrt{\frac{7-\sqrt{24}}{15}}\right)$	$\frac{5}{12}$	
	$\left(-\sqrt{\frac{7+\sqrt{24}}{15}},\sqrt{\frac{7-\sqrt{24}}{15}}\right)$	$\frac{5}{12}$	
	$\left(\sqrt{\frac{7-\sqrt{24}}{15}},-\sqrt{\frac{7+\sqrt{24}}{15}}\right)$	$\frac{5}{12}$	
	$\left(-\sqrt{\frac{7-\sqrt{24}}{15}},\sqrt{\frac{7+\sqrt{24}}{15}}\right)$	$\frac{5}{12}$	
	(0,0)	$\frac{8}{7}$	
Nine-point	$\left(\sqrt{\frac{3}{5}},\sqrt{\frac{3}{5}}\right);\left(\sqrt{\frac{3}{5}},-\sqrt{\frac{3}{5}}\right);$ $\left(-\sqrt{\frac{3}{5}},\sqrt{\frac{3}{5}}\right)$	$\frac{25}{81};\frac{25}{81};\frac{25}{81}$	5
	$\left(-\sqrt{\frac{3}{5}},-\sqrt{\frac{3}{5}}\right);\left(\sqrt{\frac{3}{5}},0\right);\left(-\sqrt{\frac{3}{5}},0\right)$	$\frac{25}{81};\frac{40}{81};\frac{40}{81}$	
	$\left(0,\sqrt{\frac{3}{5}}\right);\left(0,-\sqrt{\frac{3}{5}}\right);(0,0)$	$\frac{40}{81};\frac{40}{81};\frac{64}{81}$	
Twelve-point	(0.92582 00997 72551, 0)	0.24197 53086 41975	7
	(−0.92582 00997 72551, 0)	0.24197 53086 41975	
	(0, 0.92582 00997 72551)	0.24197 53086 41975	
	(0, −0.92582 00997 72551)	0.24197 53086 41975	
	(0.38055 44332 08316, 0.38055 44332 08316)	0.52059 29166 67394	

(continued)

Table 4.3 (Continued)

Rules	Position	Weight	Degree
	(0.38055 44332 08316, −0.38055 44332 08316)	0.52059 29166 67394	
	(−0.38055 44332 08316, 0.38055 44332 08316)	0.52059 29166 67394	
	(−0.38055 44332 08316, −0.38055 44332 08316)	0.52059 29166 67394	
	(0.80597 97829 18599, 0.80597 97829 18599)	0.23743 17746 90630	
	(0.80597 97829 18599, −0.80597 97829 18599)	0.23743 17746 90630	
	(−0.80597 97829 18599, 0.80597 97829 18599)	0.23743 17746 90630	
	(−0.80597 97829 18599, −0.80597 97829 18599)	0.23743 17746 90630	

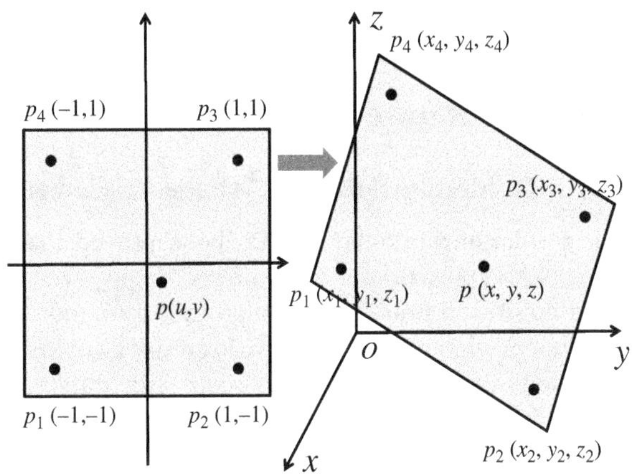

Figure 4.4 Geometric mapping for transforming a quadrature rule defined over a standard square into that defined over an arbitrary quadrilateral.

used for quadratic polynomial approximation of an unknown function in the singularity treatment of the Nyström method as we mentioned before. For the rules defined over a circular patch, hexagonal patch, or other patches, we do not address them because they are seldom used. When the quadrature rules are applied to an arbitrary quadrilateral, the geometric mapping is needed and illustrated in Figure 4.4. Assuming that the geometric mapping is governed by the following transformation [17]

$$x = a_x + b_x u + c_x v + d_x uv \tag{4.8}$$

$$y = a_y + b_y u + c_y v + d_y uv \tag{4.9}$$

$$z = a_z + b_z u + c_z v + d_z uv \tag{4.10}$$

we can find the unknown coefficients by matching the coordinates of four vertices of the standard square with those of arbitrary quadrilateral element and solving the resultant equations. With the coefficients, we can write the geometric mapping as

$$x = \sum_{i=1}^{4} N_i(u, v)x_i \tag{4.11}$$

$$y = \sum_{i=1}^{4} N_i(u, v)y_i \tag{4.12}$$

$$z = \sum_{i=1}^{4} N_i(u, v)z_i \tag{4.13}$$

where

$$N_i(u, v, w) = \frac{1}{4}(1 + u_i u)(1 + v_i v) \tag{4.14}$$

and (u_i, v_i) and (x_i, y_i, z_i) $(i = 1, 2, \cdots, 4)$ are the coordinates of four vertices of the standard square and arbitrary quadrilateral element in their own coordinate systems, respectively.

4.5 Quadrature Rules for a Volumetric Mesh

4.5.1 Quadrature Rules for a Tetrahedral Element

For a volumetric domain, it has been proved that tetrahedral elements are the best meshes to approximate an arbitrary shape [2] and they have been the most widely used meshes in practice, although they look more irregular than cuboid meshes. The tetrahedral elements are used to discretize a volume domain for solving volume integral equations and the quadrature rules with different quadrature points are shown in Figure 4.5. The commonly used quadrature rules are listed in Table 4.4 and they are also edited based on the rules of [13]. The most useful quadrature rule may be the four-point rule because it corresponds to a linear polynomial with four unknown coefficients in the singularity treatment of the Nyström method. Again, the quadrature points of chosen rules are all within the tetrahedral element. The quadrature rules are defined over a standard right tetrahedron, as shown in Figure 4.6. For an arbitrary tetrahedron, we

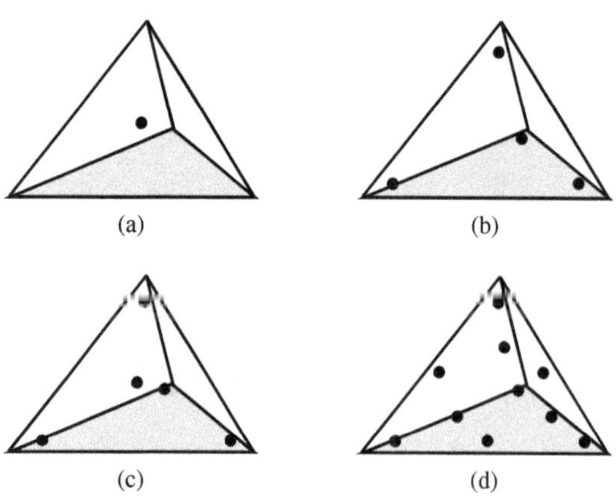

Figure 4.5 Distribution of quadrature points over a tetrahedral element for different quadrature rules. (a) One point. (b) Four points. (c) Five points. (d) Ten points.

Table 4.4 Typical quadrature rules defined over a tetrahedral element.

Rules	Position	Weight	Degree
One-point	$\left(\frac{1}{4}, \frac{1}{4}, \frac{1}{4}\right)$	$\frac{1}{6}$	1
Four-point	$\left(\frac{5-\sqrt{5}}{20}, \frac{5-\sqrt{5}}{20}, \frac{5-\sqrt{5}}{20}\right)$	$\frac{1}{24}$	2
	$\left(\frac{5-\sqrt{5}}{20}, \frac{5-\sqrt{5}}{20}, \frac{5+3\sqrt{5}}{20}\right)$	$\frac{1}{24}$	
	$\left(\frac{5-\sqrt{5}}{20}, \frac{5+3\sqrt{5}}{20}, \frac{5-\sqrt{5}}{20}\right)$	$\frac{1}{24}$	
	$\left(\frac{5+3\sqrt{5}}{20}, \frac{5-\sqrt{5}}{20}, \frac{5-\sqrt{5}}{20}\right)$	$\frac{1}{24}$	
Five-point	$\left(\frac{1}{4}, \frac{1}{4}, \frac{1}{4}\right)$	$-\frac{2}{15}$	3
	$\left(\frac{1}{6}, \frac{1}{6}, \frac{1}{6}\right)$	$\frac{3}{40}$	
	$\left(\frac{1}{6}, \frac{1}{6}, \frac{1}{2}\right)$	$\frac{3}{40}$	
	$\left(\frac{1}{6}, \frac{1}{2}, \frac{1}{6}\right)$	$\frac{3}{40}$	
	$\left(\frac{1}{2}, \frac{1}{6}, \frac{1}{6}\right)$	$\frac{3}{40}$	
Fifteen-point	$\left(\frac{1}{4}, \frac{1}{4}, \frac{1}{4}\right)$	$\frac{8}{405}$	5
	$\left(\frac{7-\sqrt{15}}{34}, \frac{7-\sqrt{15}}{34}, \frac{7-\sqrt{15}}{34}\right)$	$\frac{2665+14\sqrt{15}}{226800}$	
	$\left(\frac{7-\sqrt{15}}{34}, \frac{7-\sqrt{15}}{34}, \frac{13+3\sqrt{15}}{34}\right)$	$\frac{2665+14\sqrt{15}}{226800}$	
	$\left(\frac{7-\sqrt{15}}{34}, \frac{13+3\sqrt{15}}{34}, \frac{7-\sqrt{15}}{34}\right)$	$\frac{2665+14\sqrt{15}}{226800}$	
	$\left(\frac{13+3\sqrt{15}}{34}, \frac{7-\sqrt{15}}{34}, \frac{7-\sqrt{15}}{34}\right)$	$\frac{2665+14\sqrt{15}}{226800}$	
	$\left(\frac{7+\sqrt{15}}{34}, \frac{7+\sqrt{15}}{34}, \frac{7+\sqrt{15}}{34}\right)$	$\frac{2665-14\sqrt{15}}{226800}$	
	$\left(\frac{7+\sqrt{15}}{34}, \frac{7+\sqrt{15}}{34}, \frac{13-3\sqrt{15}}{34}\right)$	$\frac{2665-14\sqrt{15}}{226800}$	

(continued)

Table 4.4 (Continued)

Rules	Position	Weight	Degree
	$\left(\frac{7+\sqrt{15}}{34}0, \frac{13-3\sqrt{15}}{34}, \frac{7+\sqrt{15}}{34}\right)$	$\frac{2665-14\sqrt{15}}{226800}$	
	$\left(\frac{13-3\sqrt{15}}{34}, \frac{7+\sqrt{15}}{34}, \frac{7+\sqrt{15}}{34}\right)$	$\frac{2665-14\sqrt{15}}{226800}$	
	$\left(\frac{10-2\sqrt{15}}{40}, \frac{10-2\sqrt{15}}{40}, \frac{10+2\sqrt{15}}{40}\right)$	$\frac{5}{567}$	
	$\left(\frac{10-2\sqrt{15}}{40}, \frac{10+2\sqrt{15}}{40}, \frac{10-2\sqrt{15}}{40}\right)$	$\frac{5}{567}$	
	$\left(\frac{10+2\sqrt{15}}{40}, \frac{10-2\sqrt{15}}{40}, \frac{10-2\sqrt{15}}{40}\right)$	$\frac{5}{567}$	
	$\left(\frac{10+2\sqrt{15}}{40}, \frac{10-2\sqrt{15}}{40}, \frac{10+2\sqrt{15}}{40}\right)$	$\frac{5}{567}$	
	$\left(\frac{10+2\sqrt{15}}{40}, \frac{10+2\sqrt{15}}{40}, \frac{10-2\sqrt{15}}{40}\right)$	$\frac{5}{567}$	
	$\left(\frac{10-2\sqrt{15}}{40}, \frac{10+2\sqrt{15}}{40}, \frac{10+2\sqrt{15}}{40}\right)$	$\frac{5}{567}$	

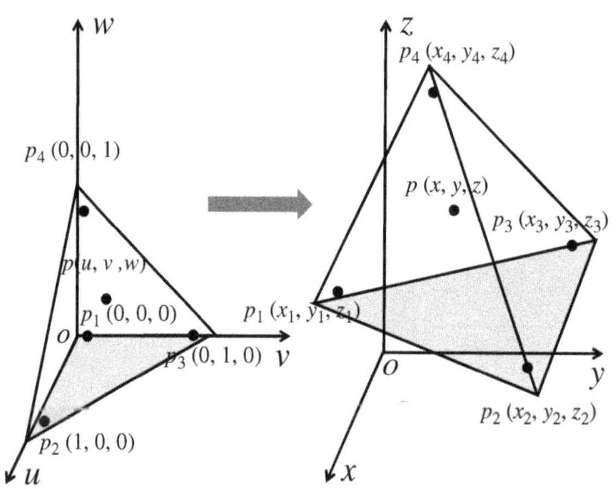

Figure 4.6 Geometric mapping for transforming a quadrature rule defined over a standard tetrahedron into that defined over an arbitrary tetrahedron.

Table 4.5 Typical quadrature rules defined over a cuboid element.

Rules	Position	Weight	Degree
One-point	(0,0,0)	8	1
Four-point	$\left(0, \sqrt{\frac{2}{3}}, -\frac{1}{\sqrt{3}}\right)$	2	2
	$\left(-\sqrt{\frac{2}{3}}, 0, \frac{1}{\sqrt{3}}\right)$	2	
	$\left(0, -\sqrt{\frac{2}{3}}, -\frac{1}{\sqrt{3}}\right)$	2	
	$\left(\sqrt{\frac{2}{3}}, 0, \frac{1}{\sqrt{3}}\right)$	2	
Six-point	$\left(\frac{1}{\sqrt{6}}, \frac{\sqrt{2}}{2}, -\frac{1}{\sqrt{3}}\right)$	$\frac{4}{3}$	3
	$\left(-\frac{1}{\sqrt{6}}, \frac{\sqrt{2}}{2}, \frac{1}{\sqrt{3}}\right)$	$\frac{4}{3}$	
	$\left(-\sqrt{\frac{2}{3}}, 0, -\frac{1}{\sqrt{3}}\right)$	$\frac{4}{3}$	
	$\left(-\frac{1}{\sqrt{6}}, -\frac{\sqrt{2}}{2}, -\frac{1}{\sqrt{3}}\right)$	$\frac{4}{3}$	
	$\left(\frac{1}{\sqrt{6}}, -\frac{\sqrt{2}}{2}, -\frac{1}{\sqrt{3}}\right)$	$\frac{4}{3}$	
	$\left(\sqrt{\frac{2}{3}}, 0, \frac{1}{\sqrt{3}}\right)$	$\frac{4}{3}$	
Eight-point	$\left(\frac{\sqrt{3}}{3}, \frac{\sqrt{3}}{3}, \frac{\sqrt{3}}{3}\right)$	1	3
	$\left(\frac{\sqrt{3}}{3}, \frac{\sqrt{3}}{3}, -\frac{\sqrt{3}}{3}\right)$	1	
	$\left(\frac{\sqrt{3}}{3}, -\frac{\sqrt{3}}{3}, \frac{\sqrt{3}}{3}\right)$	1	

(continued)

Table 4.5 (Continued)

Rules	Position	Weight	Degree
	$\left(\frac{\sqrt{3}}{3}, -\frac{\sqrt{3}}{3}, -\frac{\sqrt{3}}{3}\right)$	1	
	$\left(-\frac{\sqrt{3}}{3}, \frac{\sqrt{3}}{3}, \frac{\sqrt{3}}{3}\right)$	1	
	$\left(-\frac{\sqrt{3}}{3}, \frac{\sqrt{3}}{3}, -\frac{\sqrt{3}}{3}\right)$	1	
	$\left(-\frac{\sqrt{3}}{3}, -\frac{\sqrt{3}}{3}, \frac{\sqrt{3}}{3}\right)$	1	
	$\left(-\frac{\sqrt{3}}{3}, -\frac{\sqrt{3}}{3}, -\frac{\sqrt{3}}{3}\right)$	1	
Fourteen-point	$\left(\sqrt{\frac{19}{30}}, 0, 0\right)$	$\frac{320}{361}$	5
	$\left(0, \sqrt{\frac{19}{30}}, 0\right)$	$\frac{320}{361}$	
	$\left(0, 0, \sqrt{\frac{19}{30}}\right)$	$\frac{320}{361}$	
	$\left(-\sqrt{\frac{19}{30}}, 0, 0\right)$	$\frac{320}{361}$	
	$\left(0, -\sqrt{\frac{19}{30}}, 0\right)$	$\frac{320}{361}$	
	$\left(0, 0, -\sqrt{\frac{19}{30}}\right)$	$\frac{320}{361}$	
	$\left(\sqrt{\frac{19}{33}}, \sqrt{\frac{19}{33}}, \sqrt{\frac{19}{33}}\right)$	$\frac{121}{361}$	
	$\left(\sqrt{\frac{19}{33}}, \sqrt{\frac{19}{33}}, -\sqrt{\frac{19}{33}}\right)$	$\frac{121}{361}$	
	$\left(\sqrt{\frac{19}{33}}, -\sqrt{\frac{19}{33}}, \sqrt{\frac{19}{33}}\right)$	$\frac{121}{361}$	

(continued)

Table 4.5 (Continued)

Rules	Position	Weight	Degree
	$\left(\sqrt{\frac{19}{33}}, -\sqrt{\frac{19}{33}}, -\sqrt{\frac{19}{33}}\right)$	$\frac{121}{361}$	
	$\left(-\sqrt{\frac{19}{33}}, \sqrt{\frac{19}{33}}, \sqrt{\frac{19}{33}}\right)$	$\frac{121}{361}$	
	$\left(-\sqrt{\frac{19}{33}}, \sqrt{\frac{19}{33}}, -\sqrt{\frac{19}{33}}\right)$	$\frac{121}{361}$	
	$\left(-\sqrt{\frac{19}{33}}, -\sqrt{\frac{19}{33}}, \sqrt{\frac{19}{33}}\right)$	$\frac{121}{361}$	
	$\left(-\sqrt{\frac{19}{33}}, -\sqrt{\frac{19}{33}}, -\sqrt{\frac{19}{33}}\right)$	$\frac{121}{361}$	

need to use the following geometric mapping to transform the quadrature rules defined over a standard right tetrahedron into those defined over the arbitrary tetrahedron, where $p_1(x_1, y_1, z_1)$, $p_2(x_2, y_2, z_2)$, $p_3(x_3, y_3, z_3)$, and $p_4(x_4, y_4, z_4)$ are the four vertices of the arbitrary tetrahedron, see Figure 4.6.

$$x = x_1 + (x_2 - x_1)u + (x_3 - x_1)v + (x_4 - x_1)w \tag{4.15}$$

$$y = y_1 + (y_2 - y_1)u + (y_3 - y_1)v + (y_4 - y_1)w \tag{4.16}$$

$$z = z_1 + (z_2 - z_1)u + (z_3 - z_1)v + (z_4 - z_1)w. \tag{4.17}$$

4.5.2 Quadrature Rules for a Cuboid Element

Similar to surface meshes, cuboid elements could be useful in the discretization of volumetric domains in addition to tetrahedral elements and the quadrature rules defined over a cuboid element should be addressed. Again, we do not address quadrature rules defined over other elements like a spherical element and a cylindrical element because they are rarely used, although such rules are available. The quadrature rules defined over a cuboid element are listed in Table 4.5 and they are mainly edited from [13]. If the quadrature rules are applied to an arbitrary hexahedral element, the geometric mapping is needed and is shown in Figure 4.7. Assuming that the geometric mapping is governed by the following transformation [17]

$$x = a_x + b_x u + c_x v + d_x w + e_x uv + f_x vw + g_x wu + h_x uvw \tag{4.18}$$

$$y = a_y + b_y u + c_y v + d_y w + e_y uv + f_y vw + g_y wu + h_y uvw \tag{4.19}$$

$$z = a_z + b_z u + c_z v + d_z w + e_z uv + f_z vw + g_z wu + h_z uvw \tag{4.20}$$

we can find the unknown coefficients by matching the coordinates of eight vertices of the standard cuboid with those of arbitrary hexahedral element and by solving the resultant

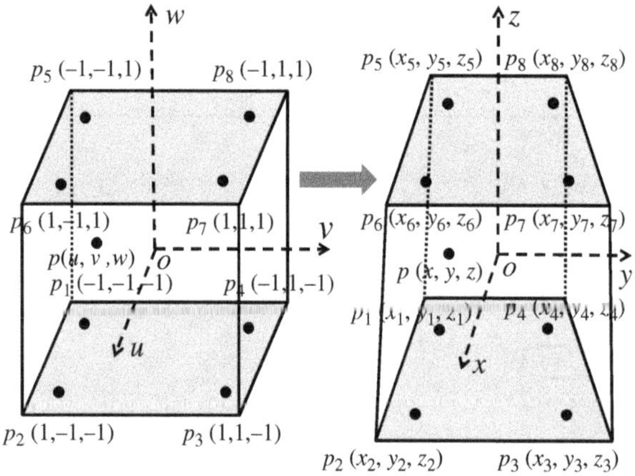

Figure 4.7 Geometric mapping for transforming a quadrature rule defined over a standard cuboid into that defined over an arbitrary hexahedral elements.

equations. With the coefficients, the geometric mapping can be expressed as

$$x = \sum_{i=1}^{8} N_i(u, v, w)x_i \tag{4.21}$$

$$y = \sum_{i=1}^{8} N_i(u, v, w)y_i \tag{4.22}$$

$$z = \sum_{i=1}^{8} N_i(u, v, w)z_i \tag{4.23}$$

where

$$N_i(u, v, w) = \frac{1}{8}(1 + u_i u)(1 + v_i v)(1 + w_i w) \tag{4.24}$$

and (u_i, v_i, w_i) and (x_i, y_i, z_i) $(i = 1, 2, \cdots, 8)$ are the coordinates of eight vertices of the standard cuboid and arbitrary hexahedral element in their own coordinate systems, respectively.

4.6 Summary

In this chapter, we address the quadrature rules that are very essential for the Nyström method because it uses the quadrature points as sampling points over meshes, and this is the key for its simple mechanism of generating matrix elements. We start with the definition of quadrature rules, and then present different types of quadrature rule in details. These rules are designed for segmental, surface, and volumetric meshes, respectively. For the segmental meshes, we mainly present the Gaussian quadrature rules in different orders with different points. For the surface and volumetric meshes, we primarily illustrate the widely used quadrature rules defined over triangular patches and tetrahedral elements, but we also consider the quadrature rules defined over square patches and cuboid elements, which are less employed but could be needed in certain cases. All quadrature rules are defined over a standard domain, i.e. $[-1, 1]$ for segmental meshes, a right triangle or square for surface meshes, and a right tetrahedron or cuboid for volumetric meshes. A geometric mapping is needed when the quadrature rules are applied to arbitrary meshes and we have provided the formulas for those geometric mappings.

Bibliography

1 Nyström, E.J. (1928). Über die praktische auflösung von linearen integralgleichungen mit anwendungen auf randwertaufgaben der potentialtheorie. *Commentationes Physico-Mathematicae.* 4 (15): 1–52.
2 Chew, W.C., Tong, M.S., and Hu, B. (2008). *Integral Equation Methods for Electromagnetic and Elastic Waves.* San Rafael, CA: Morgan & Claypool.
3 Gautschi, W. (1968). Construction of Gauss–Christoffel quadrature formulas. *Math. Comp.* 22 (102): 251–270.
4 Golub, G.H. and Welsch, J.H. (1969). Calculation of Gauss quadrature rules. *Math. Comp.* 23 (106): 221–230.
5 Gautschi, W. (1970). On the construction of Gaussian quadrature rules from modified moments. *Math. Comp.* 24 (110): 245–260.
6 Piessens, R. (1971). Gaussian quadrature formulas for the numerical integration of Bromwich's integral and the inversion of the Laplace transform. *J. Eng. Math.* 5 (1): 1–9.
7 Sagar, R.P. (1991). A Gaussian quadrature for the calculation of generalized Fermi–Dirac integrals. *Comput. Phys. Commun.* 66 (2–3): 271–275.
8 Yakimiw, E. (1996). Accurate computation of weights in classical Gauss–Christoffel quadrature rules. *J. Comput. Phys.* 129 (2): 406–430.
9 Laurie, D.P. (1999). Accurate recovery of recursion coefficients from Gaussian quadrature formulas. *J. Comput. Appl. Math.* 112 (1–2): 165–180.
10 Laurie, D.P. (2001). Computation of Gauss-type quadrature formulas. *J. Comput. Appl. Math.* 127 (1–2): 201–217.
11 Press, W.H., Teukolsky, S.A., Vetterling, W.T., and Flannery, B.P. (2007). *Numerical Recipes: The Art of Scientific Computing*, 3e. New York: Cambridge University Press.
12 Gil, A., Segura, J., and Temme, N.M. (2007). *Numerical Methods for Special Functions.* Philadelphia, PA: SIAM.
13 Stroud, A.H. (1971). *Approximate Calculation of Multiple Integrals.* Englewood Cliffs, NJ: Prentice-Hall.
14 Abramowitz, M. and Stegun, I.A. (eds.) (1964). *Handbook of Mathematical Functions with Formulas, Graphs, and Mathematical Tables.* Gaithersburg, MD: National Bureau of Standards.
15 Kahaner, D., Moler, C., and Nash, S. (1989). *Numerical Methods and Software.* Upper Saddle River, NJ: Prentice-Hall.
16 Quarteroni, A., Sacco, R., and Saleri, F. (2000). *Numerical Mathematics.* New York: Springer-Verlag.
17 Jin, J. (2002). *The Finite Element Method in Electromagnetics.* New York: John Wiley & Sons.

5

Singularity Treatment

5.1 Introduction

Although the Nyström method shows some advantages, as described in Chapter 3, it has an obvious and serious disadvantage, i.e. the singularity problem. The Nyström method was originally proposed for solving non-singular integral equations whose integral kernels are regular or non-singular [1], but it will meet the singularity problem when applied to solve wave problems. This is because the integral equations of wave problems include kernels composed of the scalar Green's function and/or certain partial differential operation on it. The Green's function represents the spatial field produced by a point source and is related to the inverse of distance R between the source point and a field point or observation point in space. However, the field point could coincide with the source point, resulting in a zero distance and an infinitely large value of the Green's function whose numerator will never be zero. The scalar Green's function itself includes a $1/R$ singularity, but the degree of singularity will certainly increase when partial differential operations are taken on it. The gradient operation on it gives a $1/R^2$ singularity as arising from the magnetic field integral equation (MFIE) while the double gradient operation on it results in a $1/R^3$ singularity as appearing in the electric field integral equation (EFIE). There could be a more singular integral kernel; for instance, there is a $1/R^4$ singularity that results from a triple gradient operation on the scalar Green's function in the elastic wave integral equations. Also, there exist many types of weaker singularities; for example, the kernel is of a logarithm singularity in 2D integral equations. We categorize a singularity as the weak singularity or an integral kernel as weakly-singular kernel if $\alpha \leq 1$ in $1/R^\alpha$, which is a measure of the degree of singularity. Similarly, we call a singularity as a strong singularity or an integral kernel as a strongly-singular kernel if $\alpha = 2$, and denominate a singularity as a hypersingularity or an integral kernel as a hypersingular kernel if $\alpha = 3$. Note that this definition is for surface integrals or two-fold integrals and the degree of singularity will decrease by 1 for volume integrals or three-fold integrals. Also, this category could be different from the convention of some researchers in the electromagnetics (EM) community but it is more reasonable and widely accepted in mechanical engineering. Some researchers in the EM community call a $1/R^2$ singularity for surface integrals as a hypersingularity [5] because they probably thought that a more singular kernel did not exist in EM integral equations when the appellation was formed.

In this chapter, we particularly address the singularity treatment techniques for the application of the Nyström method. The accurate and efficient evaluation of singular integrals involved in matrix elements is very essential because they have a major impact

The Nyström Method in Electromagnetics, First Edition. Mei Song Tong and Weng Cho Chew.

on final numerical solutions. Without an appropriate singularity treatment, the Nyström method will not work well for solving EM integral equations. We start with the regularization of singular integral kernels by singularity cancellation and singularity subtraction (or singularity extraction). Although the singularity cancellation like the Duffy's method [4] is more desirable, it can only be available for some weak singularities. The singularity subtraction is usually indispensable for strong singularities and hypersingularities. We will present several robust techniques we have developed for treating different singularities in solving surface integral equations (SIEs) and volume integral equations (VIEs) or their combination, volume-surface integral equations (VSIEs). The treatment is based on triangular patches in the SIEs and tetrahedral elements in the VIEs. Some special cases, such as the treatment of a logarithm singularity appearing in 2D integral equations, evaluation of singular Fourier expansion coefficients in body-of-revolution problems, and evaluation of high-conductivity singular integrals in conductive-medium problems, will be addressed in the relevant chapters about the applications of the Nyström method and they will not be presented in this chapter. In addition, we will also deal with the treatment techniques for near-singularity or near-singular integral kernels that are also important in the problems with very thin structures. Numerical experiments are presented to examine the effectiveness and accuracy of developed techniques and good performance can be seen.

5.2 Singularity Subtraction

When an integral kernel is singular, i.e. the value of kernel will become infinite at certain points within an integral domain, the quadrature rules cannot be applied directly to perform an integration because the quadrature rules are designed for regular integrands in general, except those specially designed for singular integrands. Designing special quadrature rules for singular integrands is one way to solve the singularity problem, but it is usually very tedious and also lacks versatility, so such a way is not very widely used in practice. A more widely used approach is the regularization of singular integrands by either singularity cancellation, which will be addressed in the next section, or singularity subtraction and then use regular quadrature rules to perform integrations. Singularity cancellation is more preferred in general, but it can only be applied to weakly-singular integrands. The singularity subtraction can be used for any singular integrands in theory, depending on whether the subtracted singular core can be integrated analytically or not, so it is more versatile and actually indispensable for strongly-singular and hypersingular integrands. We will illustrate the process of singularity subtraction for different integral kernels of EM integral equations.

5.2.1 Basic Principle

The basic principle of singularity subtraction can be illustrated by Figure 5.1, i.e. we subtract a singular core $f_0(R)$ from the original singular kernel $f_1(R)$ so that the resulting integrand $\Delta f(R)$ is regular, which is numerically integrable. The key point in the singularity subtraction is how we can find the singular core $f_0(R)$ because it should satisfy three requirements in general: (a) $f_0(R)$ should be in the same degree of singularity as $f_1(R)$ so that the $\Delta f(R)$ will be bounded at the singularity. If $f_0(R)$ has a lower degree of

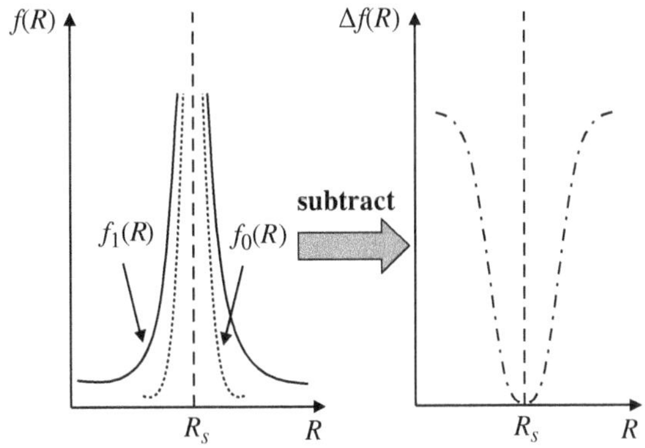

Figure 5.1 Illustration of singularity subtraction for a singular kernel, where $f_1(R)$ represents the original singular kernel, $f_0(R)$ denotes the subtracted similar singular core, and $\Delta f(R)$ is the resulting regular or weakened singular kernel.

singularity, then $\Delta f(R)$ is still singular but becomes weaker and a further treatment like using the singularity cancellation is required so that it can be numerically integrated; (b) $f_0(R)$ should be as similar as possible to $f_1(R)$ in the vicinity of singularity so that $\Delta f(R)$ will be regular near the singularity; (c) $f_0(R)$ should be integrable either analytically or numerically. For the integral equations of wave problems, since the singular kernel consists of the scalar Green's function and its partners resulting from certain differential operations on it, we can use a simple way to distinguish the singular cores. We first perform the differential operations on the scalar Green's function if needed, and then use the Taylor's series to expand the involved exponential function $e^{i\kappa R}$ in the expression so that the singular kernel becomes an infinite series of R. If one term in the final expression takes $1/R^\alpha$ form, where $\alpha > 0$, then it is a singular core that should be subtracted. When all such singular cores are subtracted, the resulting integrand is regular or bounded and can be numerically integrated. Note that the resulting integrand is usually discontinuous (non-smooth) at the singular point, although it has become bounded, so it is better to divide the integral domain (like a triangle) into several subdomains (like subtriangles) by connecting the singular point to each vertex of the integral domain and then perform a numerical integration for each subdomain. After such a singularity subtraction, the main work becomes how to evaluate the integration of the singular core $f_0(R)$, which can only be done by an analytical approach generally, i.e. how to derive an analytical or closed-form expression for such an integral. The singularity treatment sometimes refers to this part, i.e. the work after the singularity subtraction is done.

5.2.2 Subtraction for the Kernel of $\mathcal{L}$ Operator

Consider the EFIE of describing the EM interaction with conducting objects, which is an SIE and written in the form of $\mathcal{L}$ operator. Its integral kernel is the dyadic Green's function, i.e.

$$\overline{\mathbf{G}}(\mathbf{r},\mathbf{r}') = \left(\overline{\mathbf{I}} + \frac{\nabla\nabla}{k_0^2}\right) g(\mathbf{r},\mathbf{r}') \tag{5.1}$$

where $\overline{\mathbf{I}}$ is the identity dyad, $k_0 = \omega\sqrt{\epsilon_0\mu_0}$ is the wavenumber in the free space with a permittivity ϵ_0 and a permeability μ_0, and $g(\mathbf{r},\mathbf{r}') = e^{ik_0R}/(4\pi R)$ is the scalar Green's function in which $R = |\mathbf{r} - \mathbf{r}'|$ is the distance between an observation point $\mathbf{r}$ and a source point

$\mathbf{r}'$. The kernel includes two singular parts, i.e. the scalar Green's function itself and the expression resulting from the double gradient operation on it. If we expand the exponent function into a Taylor's series in the scalar Green's function, i.e.

$$g(\mathbf{r},\mathbf{r}') = \frac{e^{ikR}}{R} = \sum_{m=0}^{\infty} \frac{(i\kappa)^m R^{m-1}}{m!} \tag{5.2}$$

we can easily find that the singular core that should be subtracted is $1/R$, so the resulting integrand after the regularization is

$$g_R(\mathbf{r},\mathbf{r}') = \frac{e^{ikR}-1}{R} \tag{5.3}$$

which can be integrated numerically. The subtracted singular core should be added back to keep the original kernel unchanged and the evaluation of its integration is a key and will be addressed in the later section.

For the second part resulting from the double gradient operation of the scalar Green's function, which is named $\overline{\mathbf{G}}_0$, it has nine components if expanded, i.e.

$$\overline{\mathbf{G}}_0(\mathbf{r},\mathbf{r}') = \nabla\nabla g(\mathbf{r},\mathbf{r}') = \begin{bmatrix} G_{uu}\hat{u}\hat{u} & G_{uv}\hat{u}\hat{v} & G_{uw}\hat{u}\hat{w} \\ G_{vu}\hat{v}\hat{u} & G_{vv}\hat{v}\hat{v} & G_{vw}\hat{v}\hat{w} \\ G_{wu}\hat{w}\hat{u} & G_{wv}\hat{w}\hat{v} & G_{ww}\hat{w}\hat{w} \end{bmatrix} \tag{5.4}$$

but only six are independent due to its symmetry. Although the singularity subtraction or treatment can be performed in a global coordinate system (x, y, z), as shown in Figure 5.2, mostly it will be handled in a local coordinate system (u, v, w), as shown in Figure 5.3, so the components of $\overline{\mathbf{G}}_0$ are labelled in terms of these coordinates. The local coordinate system is established over a self triangular patch for the SIEs in which the origin is the projection of observation point on the triangle plane and the w axis is usually parallel to the normal vector of the plane. In the local coordinate system, the observation point is located at $(0, 0, w_0)$ while the source point is located at $(u, v, 0)$ on the triangle plane so that $R = \sqrt{u^2 + v^2 + w_0^2}$ and the components of $\overline{\mathbf{G}}_0$ can be found as

$$G_{uu} = \left[\frac{(3 - 3i\kappa R - \kappa^2 R^2)u^2}{R^5} + \frac{i\kappa R - 1}{R^3}\right] e^{i\kappa R} \tag{5.5}$$

$$G_{vv} = \left[\frac{(3 - 3i\kappa R - \kappa^2 R^2)v^2}{R^5} + \frac{i\kappa R - 1}{R^3}\right] e^{i\kappa R} \tag{5.6}$$

$$G_{ww} = \left[\frac{(3 - 3i\kappa R - \kappa^2 R^2)w^2}{R^5} + \frac{i\kappa R - 1}{R^3}\right] e^{i\kappa R} \tag{5.7}$$

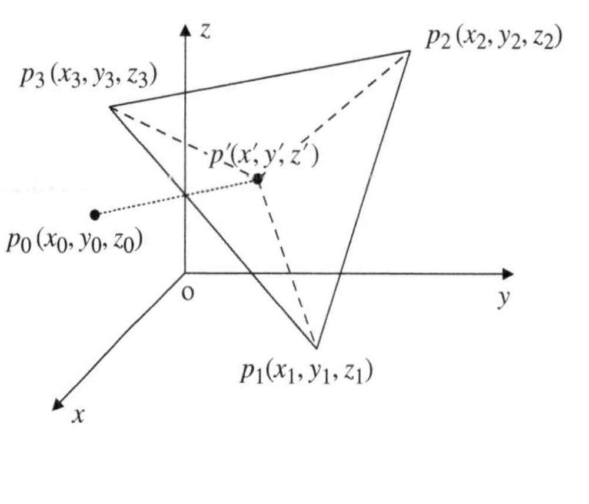

Figure 5.2 Geometry of a triangular patch in a global coordinate system (x, y, z).

Figure 5.3 Geometry of a triangular patch in a local coordinate system (u, v, w). The observation point p_0 is located at $(0, 0, w_0)$ and the source point is located at $(u, v, 0)$.

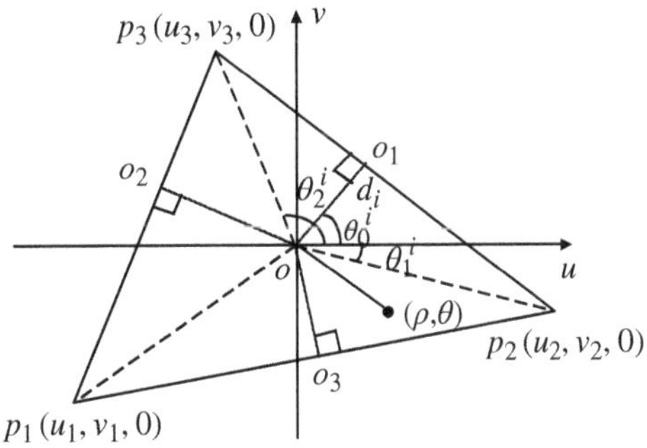

$$G_{uv} = \frac{(3 - 3\mathrm{i}\kappa R - \kappa^2 R^2)uv}{R^5} \mathrm{e}^{\mathrm{i}\kappa R} \tag{5.8}$$

$$G_{vw} = \frac{(3 - 3\mathrm{i}\kappa R - \kappa^2 R^2)vw_0}{R^5} \mathrm{e}^{\mathrm{i}\kappa R} \tag{5.9}$$

$$G_{wu} = \frac{(3 - 3\mathrm{i}\kappa R - \kappa^2 R^2)w_0 u}{R^5} \mathrm{e}^{\mathrm{i}\kappa R} \tag{5.10}$$

with $G_{vu} = G_{uv}$, $G_{wv} = G_{vw}$ and $G_{uw} = G_{wu}$.

Using the Taylor's series expansion of the exponential function in the above equations, we can easily distinguish the singular cores that should be subtracted. After the subtraction, the representative two components can be written as

$$\begin{aligned} G_{uu} = & \frac{3u^2}{R^5} + \frac{\kappa^2 u^2}{2R^3} + \frac{\kappa^4 u^2}{8R} + \frac{\mathrm{i}\kappa^5 u^2}{24} - \frac{\kappa^6 R u^2}{24} \\ & + \sum_{m=5}^{M} \frac{(\mathrm{i}\kappa)^m}{m!} \left(3R^{m-5} - 3\mathrm{i}\kappa R^{m-4} - \kappa^2 R^{m-3}\right) u^2 \\ & - \frac{1}{R^3} - \frac{\kappa^2}{2R} - \frac{\mathrm{i}\kappa^3}{2} + \sum_{m=3}^{M} \frac{(\mathrm{i}\kappa)^m}{m!} R^{m-3}(\mathrm{i}\kappa R - 1) \end{aligned} \tag{5.11}$$

$$\begin{aligned} G_{uv} = & \frac{3uv}{R^5} + \frac{\kappa^2 uv}{2R^3} + \frac{\kappa^4 uv}{8R} + \frac{\mathrm{i}\kappa^5 uv}{24} - \frac{\kappa^6 R uv}{24} \\ & + \sum_{m=5}^{M} \frac{(\mathrm{i}\kappa)^m}{m!} \left(3R^{m-5} - 3\mathrm{i}\kappa R^{m-4} - \kappa^2 R^{m-3}\right) uv \end{aligned} \tag{5.12}$$

and the other components can be written very similarly. From the above expressions, we can find that the singular integrals include the following eighteen forms

$$I_1 = \int_{\Delta S} \left(\frac{3u^2}{R^5} - \frac{1}{R^3}\right) \mathrm{d}S; \quad I_2 = \int_{\Delta S} \frac{uv}{R^5} \mathrm{d}S; \tag{5.13}$$

$$I_3 = \int_{\Delta S} \left(\frac{3v^2}{R^5} - \frac{1}{R^3}\right) \mathrm{d}S; \quad I_4 = \int_{\Delta S} \frac{w_0(-u)}{R^5} \mathrm{d}S; \tag{5.14}$$

$$I_5 = \int_{\Delta S} \left(\frac{3w_0^2}{R^5} - \frac{1}{R^3}\right) \mathrm{d}S; \quad I_6 = \int_{\Delta S} \frac{w_0(-v)}{R^5} \mathrm{d}S; \tag{5.15}$$

$$I_7 = \int_{\Delta S} \frac{u^2}{R^3} \mathrm{d}S; \quad I_8 = \int_{\Delta S} \frac{uv}{R^3} \mathrm{d}S; \tag{5.16}$$

$$I_9 = \int_{\Delta S} \frac{v^2}{R^3} \mathrm{d}S, \quad I_{10} = \int_{\Delta S} \frac{w_0(-u)}{R^3} \mathrm{d}S; \tag{5.17}$$

$$I_{11} = \int_{\Delta S} \frac{w_0^2}{R^3}\,\mathrm{d}S; \qquad I_{12} = \int_{\Delta S} \frac{w_0(-v)}{R^3}\,\mathrm{d}S; \tag{5.18}$$

$$I_{13} = \int_{\Delta S} \frac{u^2}{R}\,\mathrm{d}S; \qquad I_{14} = \int_{\Delta S} \frac{uv}{R}\,\mathrm{d}S; \tag{5.19}$$

$$I_{15} = \int_{\Delta S} \frac{v^2}{R}\,\mathrm{d}S; \qquad I_{16} = \int_{\Delta S} \frac{w_0(-u)}{R}\,\mathrm{d}S; \tag{5.20}$$

$$I_{17} = \int_{\Delta S} \frac{w_0^2}{R}\,\mathrm{d}S; \qquad I_{18} = \int_{\Delta S} \frac{w_0(-v)}{R}\,\mathrm{d}S. \tag{5.21}$$

All other terms are non-singular and the integrals including them can be easily evaluated numerically or analytically. In those eighteen singular integrals, the first five are hypersingular and need to be specially treated. Other integrals are weakly-singular and can be treated numerically or using a singularity cancellation technique like the Duffy's method to handle them, but we also derive their closed-form formulations which may be needed in some more accurate calculations.

Note that the current density will be represented by a polynomial function in the singularity treatment. In fact, the well-known Rao-Wilton-Glisson (RWG) basis function is also a kind of polynomial function in the method of moments (MoM). For instance, $J_u = a_0 + a_1 u + a_2 v$ for a linear approximation and $J_u = a_0 + a_1 u + a_2 v + a_3 u^2 + a_4 uv + a_5 v^2$ for a quadratic approximation. When the polynomial function is combined with the singular kernel, i.e. the dyadic Green's function, the strongest singularity will be from the combination of the constant term of the polynomial function with the singular kernel and all other terms of the polynomial function will weaken the degree of singularities, making the treatment for them become much easier. Therefore, those eighteen singular integrals resulting from the dyadic Green's function itself represent the worst case (the constant term of the polynomial function can be suppressed in the treatment) and other singular integrals resulting from the combination of other terms of the polynomial function with the dyadic Green's function can be much more easily treated, which we do not show here.

5.2.3 Subtraction for the Kernel of $\mathcal{K}$ Operator

The EM interaction with conducting objects can also be formulated by the MFIE, which is also an SIE, but written in the form of $\mathcal{K}$ operator. The kernel of $\mathcal{K}$ operator is the vector multiplication of the current density with the gradient of the scalar Green's function and the strongest singularity comes from the gradient of the scalar Green's function itself when the current density is represented by a polynomial function. Note that the singular integral in the MFIE is defined in a Cauchy-principal-value (CPV) sense. This means that the original integral over a self patch can be separated into two parts, i.e.

$$\begin{aligned} I &= \hat{n} \times \int_{\Delta S} \mathbf{J}(\mathbf{r}') \times \nabla g(\mathbf{r},\mathbf{r}')\,\mathrm{d}S' \\ &= \lim_{a\to 0}\left[\hat{n} \times \int_{\Delta S - S_0} \mathbf{J}(\mathbf{r}') \times \nabla g(\mathbf{r},\mathbf{r}')\,\mathrm{d}S' + \hat{n} \times \int_{S_0} \mathbf{J}(\mathbf{r}') \times \nabla g(\mathbf{r},\mathbf{r}')\,\mathrm{d}S'\right] \\ &= I_1 + I_2 \end{aligned} \tag{5.22}$$

where S_0 is a small circle centered at the projection of the observation point or the origin of a local coordinate system and a is its radius. We can prove that [2]

$$I_2 = \lim_{a\to 0}\left[\hat{n}\times\int_{S_0}\mathbf{J}(\mathbf{r}')\times\nabla g(\mathbf{r},\mathbf{r}')\,\mathrm{d}S'\right] = \frac{1}{2}\mathbf{J}(\mathbf{r}) \tag{5.23}$$

which is the residue of the singular integral and we no longer need to calculate it. We only need to calculate

$$I_1 = \lim_{a\to 0}\left[\hat{n}\times\int_{\Delta S-S_0}\mathbf{J}(\mathbf{r}')\times\nabla g(\mathbf{r},\mathbf{r}')\,\mathrm{d}S'\right] = \hat{n}\times\fint_{\Delta S}\mathbf{J}(\mathbf{r}')\times\nabla g(\mathbf{r},\mathbf{r}')\,\mathrm{d}S' \tag{5.24}$$

which is named an improper integral or CPV integral and it is also a kind of singularity treatment.

In a local coordinate system (u, v, w) established over the source triangular patch where the observation point is located at (u_0, v_0, w_0), the gradient of the scalar Green's function can be written as

$$\nabla g(\mathbf{r},\mathbf{r}') = g_u\hat{u} + g_v\hat{v} + g_w\hat{w} = \frac{\partial g}{\partial u_0}\hat{u} + \frac{\partial g}{\partial v_0}\hat{v} + \frac{\partial g}{\partial w_0}\hat{w} \tag{5.25}$$

where (g_u, g_v, g_w) are the three components of the gradient, and they can be expressed as

$$g_u = \frac{1}{R^3}(i\kappa R - 1)(u_0 - u)e^{i\kappa R} \tag{5.26}$$

$$g_v = \frac{1}{R^3}(i\kappa R - 1)(v_0 - v)e^{i\kappa R} \tag{5.27}$$

$$g_w = \frac{1}{R^3}(i\kappa R - 1)w_0 e^{i\kappa R} \tag{5.28}$$

after being expanded. If the observation point is located at $(0, 0, w_0)$ or the projection of the observation point over the triangular patch plane is chosen as the origin of the coordinate system, as we did previously, then $u_0 = 0$ and $v_0 = 0$ can be substituted in the above equations. If the exponent function is expanded into a Taylor's series as shown in Equation (5.12), then the three components can be written as

$$g_u = -\frac{u_0 - u}{R^3} + (i\kappa)^2\frac{u_0 - u}{2R} + (i\kappa)^3\frac{u_0 - u}{2} + (i\kappa R - 1)(u_0 - u)\sum_{m=3}^{\infty}\frac{(i\kappa)^m R^{m-3}}{m!} \tag{5.29}$$

$$g_v = -\frac{v_0 - v}{R^3} + (i\kappa)^2\frac{v_0 - v}{2R} + (i\kappa)^3\frac{v_0 - v}{2} + (i\kappa R - 1)(v_0 - v)\sum_{m=3}^{\infty}\frac{(i\kappa)^m R^{m-3}}{m!} \tag{5.30}$$

$$g_w = -\frac{w_0}{R^3} + (i\kappa)^2\frac{w_0}{2R} + (i\kappa)^3\frac{w_0}{2} + (i\kappa R - 1)w_0\sum_{m=3}^{\infty}\frac{(i\kappa)^m R^{m-3}}{m!}. \tag{5.31}$$

From the above equations, we can find the singular terms which should be subtracted, and the integrals with the subtracted singular terms take the following form

$$I_1 = \int_{\Delta S} \frac{u_0 - u}{R^3}\, \mathrm{d}S; \quad I_2 = \int_{\Delta S} \frac{v_0 - v}{R^3}\, \mathrm{d}S; \tag{5.32}$$

$$I_3 = \int_{\Delta S} \frac{w_0}{R^3}\, \mathrm{d}S; \quad I_4 = \int_{\Delta S} \frac{u_0 - u}{R}\, \mathrm{d}S; \tag{5.33}$$

$$I_5 = \int_{\Delta S} \frac{v_0 - v}{R}\, \mathrm{d}S; \quad I_6 = \int_{\Delta S} \frac{w_0}{R}\, \mathrm{d}S \tag{5.34}$$

where the first three integrals are strongly-singular integrals while the other three integrals are weakly-singualr integrals, which are different from those from the $\mathcal{L}$ operator.

5.2.4 Subtraction for the Kernels of VIEs

The VIEs for penetrable objects have been introduced in Chapter 2 and they actually include $\mathcal{L}$ operator and/or $\mathcal{K}$ operator as well, but the involved integrals are over a volume domain. The $\mathcal{L}$ operator includes an exactly same singular kernel, i.e. the dyadic Green's function, in the $\mathcal{L}$ operator of SIEs. However, the $\mathcal{K}$ operator is little different from that of SIEs because the involved integral is defined over a volume domain and no residue of singular integral exists at the singular point. The process of singularity subtraction is the same as that for the $\mathcal{L}$ operator and $\mathcal{K}$ operator of SIEs and the singular integrals of subtracted cores can be listed as follows

$$I_1 = \int_{\Delta V} \left(\frac{3u^2}{R^5} - \frac{1}{R^3} \right) \mathrm{d}V; \quad I_2 = \int_{\Delta V} \frac{uv}{R^5}\, \mathrm{d}V; \tag{5.35}$$

$$I_3 = \int_{\Delta V} \left(\frac{3v^2}{R^5} - \frac{1}{R^3} \right) \mathrm{d}V; \quad I_4 = \int_{\Delta V} \frac{w_0(-u)}{R^5}\, \mathrm{d}V; \tag{5.36}$$

$$I_5 = \int_{\Delta V} \left(\frac{3w_0^2}{R^5} - \frac{1}{R^3} \right) \mathrm{d}V; \quad I_6 = \int_{\Delta V} \frac{w_0(-v)}{R^5}\, \mathrm{d}V; \tag{5.37}$$

$$I_7 = \int_{\Delta V} \frac{u^2}{R^3}\, \mathrm{d}V; \quad I_8 = \int_{\Delta V} \frac{uv}{R^3}\, \mathrm{d}V; \tag{5.38}$$

$$I_9 = \int_{\Delta V} \frac{v^2}{R^3}\, \mathrm{d}V; \quad I_{10} = \int_{\Delta V} \frac{w_0(-u)}{R^3}\, \mathrm{d}V; \tag{5.39}$$

$$I_{11} = \int_{\Delta V} \frac{w_0^2}{R^3}\, \mathrm{d}V; \quad I_{12} = \int_{\Delta V} \frac{w_0(-v)}{R^3}\, \mathrm{d}V; \tag{5.40}$$

$$I_{13} = \int_{\Delta V} \frac{u^2}{R}\, \mathrm{d}V; \quad I_{14} = \int_{\Delta V} \frac{uv}{R}\, \mathrm{d}V; \tag{5.41}$$

$$I_{15} = \int_{\Delta V} \frac{v^2}{R}\, \mathrm{d}V; \quad I_{16} = \int_{\Delta V} \frac{w_0(-u)}{R}\, \mathrm{d}V; \tag{5.42}$$

$$I_{17} = \int_{\Delta V} \frac{w_0^2}{R}\, \mathrm{d}V; \quad I_{18} = \int_{\Delta V} \frac{w_0(-v)}{R}\, \mathrm{d}V. \tag{5.43}$$

We can see that all singular integrals are defined over a self volumetric element ΔV now and some singular integrals do not appear in the $\mathcal{L}$ operator of SIEs because they are vanishing there. Similarly, we can find the singular integrals for the $\mathcal{K}$ operator as follows

$$I_1 = \int_{\Delta V} \frac{u_0 - u}{R^3} \,\mathrm{d}V; \quad I_2 = \int_{\Delta V} \frac{v_0 - v}{R^3} \,\mathrm{d}V; \tag{5.44}$$

$$I_3 = \int_{\Delta V} \frac{w_0}{R^3} \,\mathrm{d}V; \quad I_4 = \int_{\Delta V} \frac{u_0 - u}{R} \,\mathrm{d}V; \tag{5.45}$$

$$I_5 = \int_{\Delta V} \frac{v_0 - v}{R} \,\mathrm{d}V; \quad I_6 = \int_{\Delta V} \frac{w_0}{R} \,\mathrm{d}V \tag{5.46}$$

where the first three integrals cannot be viewed as "strongly-singular" now because the integrals are three-fold integrals over a volumetric element.

5.3 Singularity Cancellation

In the MoM with a RWG basis function [3] for solving the EFIE, one only needs to handle the $1/R$ weak singularities. This is because one can move the gradient operator of the integral kernel, which is the dyadic Green's function, onto the basis function and testing function to lower the degree of singularity so as to facilitate the evaluation. The $1/R$ weak singularities can be handled by singularity cancellation techniques and the most well-known technique is the Duffy's method, which can be found in [4] for details and we do not repeat it here. The Duffy's method has been widely used in the MoM and it can also be used in the Nyström method for handling those weakly-singular integrals. The weak singularity has been widely studied and many efficient evaluation techniques have been developed [4–23] in addition to the Duffy's method. The basic principle of singularity cancellation is that one uses a variable change to produce an extra zero point in the Jacobian so that the singularity or zero point of denominator of integrands can be cancelled by the zero point of the Jacobian. The resultant integrals are then regular and can be accurately evaluated with numerical quadrature rules.

However, the Duffy's method and other similar methods require a two-fold numerical integration after regularizing the integral kernel and may be inconvenient in implementation. We present a different approach to evaluate the weakly-singular integrals by using a special polar coordinate system and it can automatically cancel the singularity without using a variable change or coordinate transform. Also, the approach can reduce the integral to a one-fold numerical integration when the constant term of the polynomial function in the local correction of Nyström method (the basis function in the MoM) is concerned or the observation point is inside the source triangle because the integral over the polar coordinate can be derived analytically. The one-fold numerical integration is for the angular coordinate and has a very simple integrand that can be easily evaluated. Compared with those existing methods, the approach could be simpler for implementation and more efficient in calculation. Numerical examples for EM scattering by various conducting objects are presented to demonstrate the approach and the accuracy of integrals is examined by comparing with available reference data.

5.3.1 Surface Integral Equation

To illustrate the approach for evaluating the weakly-singular integral, we consider the EM scattering problem for a three-dimensional (3D) conducting object embedded in the free space with a permittivity ϵ_0 and a permeability μ_0. The governing equation for the problem is the EFIE, which can be written as [2]

$$-\hat{n} \times \mathbf{E}^{\text{inc}}(\mathbf{r}) = \hat{n} \times \mathrm{i}\omega\mu_0 \int_S \overline{\mathbf{G}}(\mathbf{r}, \mathbf{r}') \cdot \mathbf{J}_S(\mathbf{r}')\, \mathrm{d}S', \quad \mathbf{r} \in S \tag{5.47}$$

where $\mathbf{E}^{\text{inc}}(\mathbf{r})$ is the incident electric field, $\mathbf{J}_S(\mathbf{r}')$ is the electric current induced on the conducting surface S, and $\hat{n}$ is the unit normal vector of the surface. Also, $\overline{\mathbf{G}}(\mathbf{r}, \mathbf{r}')$ is the dyadic Green's function defined by

$$\overline{\mathbf{G}}(\mathbf{r}, \mathbf{r}') = \left(\overline{\mathbf{I}} + \frac{\nabla\nabla}{k_0^2} \right) g(\mathbf{r}, \mathbf{r}') \tag{5.48}$$

where $\overline{\mathbf{I}}$ is the identity dyad, k_0 is the wavenumber in the free space, and $g(\mathbf{r}, \mathbf{r}') = \mathrm{e}^{ik_0R}/(4\pi R)$ is the 3D scalar Green's function in which $R = |\mathbf{r} - \mathbf{r}'|$ is the distance between an observation point $\mathbf{r}$ and a source point $\mathbf{r}'$.

To illustrate the evaluation of weakly-singular integrals by the singularity cancellation, we use the MoM to solve the above EFIE because the resultant weakly-singular integrals are more typical. Obviously, the technique can also be applied to the weakly-singular integrals from the Nyström method, which are either same as or similar to those from the MoM. In the MoM, the unknown current density $\mathbf{J}_S(\mathbf{r}')$ is expanded by the RWG basis function, i.e.

$$\mathbf{J}_S(\mathbf{r}') = \sum_{n=1}^{N} I_n \mathbf{\Lambda}_n(\mathbf{r}') \tag{5.49}$$

where $\mathbf{\Lambda}_n(\mathbf{r}')$ is the RWG basic function defined over the nth pair of triangular patches, I_n is the corresponding expansion coefficient, and N is the number of all pairs of triangular patches. If using the RWG basis function as a testing function to test the EFIE, then we can obtain the following matrix equation

$$\begin{aligned} -\left\langle \Lambda_m(\mathbf{r}), \mathbf{E}^{\text{inc}}(\mathbf{r}) \right\rangle &= \mathrm{i}\omega\mu_0 \sum_{n=1}^{N} I_n \left\langle \Lambda_m(\mathbf{r}), \overline{\mathbf{G}}(\mathbf{r}, \mathbf{r}'), \Lambda_n(\mathbf{r}') \right\rangle \\ &= \mathrm{i}\omega\mu_0 \sum_{n=1}^{N} I_n \left[\left\langle \Lambda_m(\mathbf{r}), g(\mathbf{r}, \mathbf{r}')\Lambda_n(\mathbf{r}') \right\rangle \right. \\ &\quad \left. + \left\langle \nabla \cdot \Lambda_m(\mathbf{r}), g(\mathbf{r}, \mathbf{r}'), \nabla' \cdot \Lambda_n(\mathbf{r}') \right\rangle \right] \end{aligned} \tag{5.50}$$

where we have moved the gradient operator in the dyadic Green's function onto the basis function and testing function, respectively. The integral kernel in the matrix equation is only in a $1/R$ weak singularity now and can be handled with the singularity cancellation technique introduced in the next subsection. Note that the same singularity appears in the integral equations for dielectric or composite objects, so the developed treatment technique can be applied to those scenarios as well.

5.3.2 Evaluation of the Weakly-Singular Integrals

From Equation (5.50), we can see that the singular integral takes two kinds of form resulting from the scalar potential and vector potential, respectively, i.e.

$$I_1 = \int_{\Delta S} \frac{e^{ik_0 R}}{R}\, dS, \quad I_2 = \int_{\Delta S} \Lambda_n(\mathbf{r}) \frac{e^{ik_0 R}}{R}\, dS \tag{5.51}$$

where ΔS is a triangular patch $\Delta p_1 p_2 p_3$ or integral domain, as shown in Figure 5.4. Note that the first integral including the scalar Green's function also appears in the Nyström method and the second integral including the product of basis function with the scalar Green's function is similar to that including the product of a polynomial function with the scalar Green's function in the Nyström method (the RWG basis function is also a polynomial function). Therefore, the singularity cancellation technique presented subsequently can fully be used in the Nyström method. We establish a local Cartesian coordinate system $\{u, v, w\}$ and polar coordinate system (ρ, θ) over the triangle plane, in which the projection of the observation point on the plane is chosen as the origin. Figure 5.4(a) and Figure 5.4(b) illustrate the situations when the origin is outside and inside the triangle, respectively. In such a coordinate system, the observation point is located at $(0, 0, w_0)$ while the source point lies on $(u, v, 0)$ or (ρ, θ) within the triangle, leading to $R = \sqrt{\rho^2 + w_0^2}$. Note that we have used the unprimed (u, v, w) or $\mathbf{r}$ to indicate a source point and (u_0, v_0, w_0) to denote an observation point as we did before. For the first integral I_1, there is no singularity in fact in the situation of Figure 5.4(a) since the origin is outside the triangle, but we can still derive a more friendly formulation to calculate it, i.e.

$$\begin{aligned} I_1 &= \sum_{j=1}^{2} \int_{\theta_j}^{\theta_{j+1}} d\theta \int_{\rho_1^j(\theta)}^{\rho_2^j(\theta)} \frac{e^{ik_0\sqrt{\rho^2+w_0^2}}}{\sqrt{\rho^2 + w_0^2}} \rho\, d\rho \\ &= \sum_{j=1}^{2} \int_{\theta_j}^{\theta_{j+1}} d\theta \int_{R_1^j(\theta)}^{R_2^j(\theta)} \frac{e^{ik_0 R}}{R} \sqrt{R^2 - w_0^2}\, d\left(\sqrt{R^2 - w_0^2}\right) \end{aligned}$$

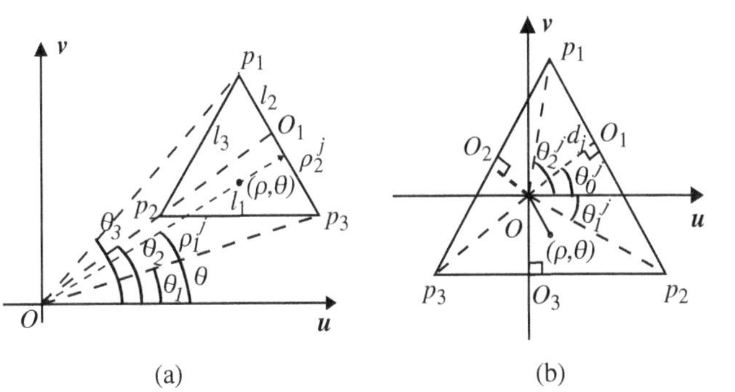

Figure 5.4 A local Cartesian coordinate system (u, v, w) and polar coordinate system (ρ, θ) are established over the plane of a triangular patch $\Delta p_1 p_2 p_3$. The projection of the observation point on the plane is chosen as the origin. (a) The origin is outside the triangle. (b) The origin is inside the triangle.

$$= \sum_{j=1}^{2} \int_{\theta_j}^{\theta_{j+1}} \mathrm{d}\theta \int_{R_1^j(\theta)}^{R_2^j(\theta)} \mathrm{e}^{\mathrm{i}k_0 R}\, \mathrm{d}R$$

$$= \sum_{j=1}^{2} \frac{1}{\mathrm{i}k_0} \int_{\theta_j}^{\theta_{j+1}} \left[\mathrm{e}^{\mathrm{i}k_0 R_2^j(\theta)} - \mathrm{e}^{\mathrm{i}k_0 R_1^j(\theta)}\right] \mathrm{d}\theta \tag{5.52}$$

where

$$R_1^j(\theta) = \sqrt{\rho_1^j(\theta) + w_0^2}, \quad R_2^j(\theta) = \sqrt{\rho_2^j(\theta) + w_0^2} \tag{5.53}$$

and $\rho_1^j(\theta)$ and $\rho_2^j(\theta)$ are the radial coordinates of the intersection points formed by the polar axis at the angle θ and two sides of the triangle, and they can be easily determined by the slopes and intercepts of the sides. If b_1 and b_2 denote the interceptions while k_1 and k_2 denote the slopes of the two sides, respectively, then they can be found as

$$\rho_\ell^j(\theta) = \frac{b_\ell}{\sin(\theta) - k_\ell \cos(\theta)}, \quad \ell = 1, 2. \tag{5.54}$$

From the formulas, we can see that the singularity, if it exists, moves to $R_1^j(\theta)$ or $R_2^j(\theta)$ when $\tan(\theta)$ is equal to one of the slopes of the sides. This situation occurs only when one of the sides coincides with the line connecting the point o and the point o_1. Fortunately, this singularity vanishes because the upper bound and lower bound of the integral with respect to θ are the same in this case and this result matches the reality, i.e. no singularity. On the other hand, there exists a singularity in the situation of Figure 5.4(b) because the origin is inside the triangle. We divide the triangular patch into three subtriangles by connecting the origin with three vertices of the triangle and each subtriangle is specified by the parameters d_j, θ_0^j, θ_1^j, and θ_2^j, as shown in the figure. In this case, the integral I_1 can be derived as

$$I_1 = \sum_{j=1}^{3} \int_{\theta_1^j}^{\theta_2^j} \mathrm{d}\theta \int_0^{d_j^\theta(\theta)} \frac{\mathrm{e}^{\mathrm{i}k_0\sqrt{\rho^2+w_0^2}}}{\sqrt{\rho^2 + w_0^2}} \rho\, \mathrm{d}\rho$$

$$= \sum_{j=1}^{3} \int_{\theta_1^j}^{\theta_2^j} \mathrm{d}\theta \int_{R_0}^{R_j(\theta)} \frac{\mathrm{e}^{\mathrm{i}k_0 R}}{R} \sqrt{R^2 - w_0^2}\, \mathrm{d}\left(\sqrt{R^2 - w_0^2}\right)$$

$$= \sum_{j=1}^{3} \int_{\theta_1^j}^{\theta_2^j} \mathrm{d}\theta \int_{R_0}^{R_j(\theta)} \mathrm{e}^{\mathrm{i}k_0 R}\, \mathrm{d}R$$

$$= \frac{1}{\mathrm{i}k_0} \sum_{j=1}^{3} \int_{\theta_1^j}^{\theta_2^j} \left[\mathrm{e}^{\mathrm{i}k_0 R_j(\theta)} - \mathrm{e}^{\mathrm{i}k_0 R_0}\right] \mathrm{d}\theta \tag{5.55}$$

where

$$R_0 = w_0, \quad R_j(\theta) = \sqrt{\left[d_j^\theta(\theta)\right]^2 + w_0^2} \tag{5.56}$$

with $d_j^\theta(\theta) = d_j / \cos(\theta - \theta_0^j)$. Obviously, there is no singularity in the above final formulation because $\theta - \theta_0^j$ cannot be 90° or $\cos(\theta - \theta_0^j)$ cannot be zero for a triangle and we can use a numerical quadrature rule to conveniently evaluate the one-fold integral with a very simple integrand. The above formulas are valid for all w_0, including $w_0 = 0$, which is the singular case.

For the second integral I_2, the involved RWG basis function is actually a first-order (linear) polynomial vector and its component could be written as $a_0 + a_1 u + a_2 v = a_0 + a_1 \rho \cos\theta + a_2 \rho \sin\theta$ in the local coordinate system (just like what we have used in the Nyström method), where a_0, a_1, a_2 are the known constants. When the polynomial function is combined with the scalar Green's function, the combination of the first term a_0 will result in the integral I_1, which we have addressed above. The combination of the second and third terms will yield the following integrals

$$
\begin{aligned}
I_{22} &= \sum_{j=1}^{2} \int_{\theta_j}^{\theta_{j+1}} \cos\theta \, \mathrm{d}\theta \int_{\rho_1^j(\theta)}^{\rho_2^j(\theta)} \rho \frac{\mathrm{e}^{\mathrm{i}k_0\sqrt{\rho^2+w_0^2}}}{\sqrt{\rho^2+w_0^2}} \rho \, \mathrm{d}\rho \\
&= \sum_{j=1}^{2} \int_{\theta_j}^{\theta_{j+1}} \cos\theta \, \mathrm{d}\theta \int_{R_1^j(\theta)}^{R_2^j(\theta)} \sqrt{R^2 - w_0^2} \mathrm{e}^{\mathrm{i}k_0 R} \, \mathrm{d}R
\end{aligned} \tag{5.57}
$$

$$
\begin{aligned}
I_{23} &= \sum_{j=1}^{2} \int_{\theta_j}^{\theta_{j+1}} \sin\theta \, \mathrm{d}\theta \int_{\rho_1^j(\theta)}^{\rho_2^j(\theta)} \rho \frac{\mathrm{e}^{\mathrm{i}k_0\sqrt{\rho^2+w_0^2}}}{\sqrt{\rho^2+w_0^2}} \rho \, \mathrm{d}\rho \\
&= \sum_{j=1}^{2} \int_{\theta_j}^{\theta_{j+1}} \sin\theta \, \mathrm{d}\theta \int_{R_1^j(\theta)}^{R_2^j(\theta)} \sqrt{R^2 - w_0^2} \mathrm{e}^{\mathrm{i}k_0 R} \, \mathrm{d}R
\end{aligned} \tag{5.58}
$$

respectively, for the case of Figure 5.4(a), and

$$
\begin{aligned}
I_{22} &= \sum_{j=1}^{3} \int_{\theta_1^j}^{\theta_2^j} \cos\theta \, \mathrm{d}\theta \int_{0}^{d_j^\theta(\theta)} \rho \frac{\mathrm{e}^{\mathrm{i}k_0\sqrt{\rho^2+w_0^2}}}{\sqrt{\rho^2+w_0^2}} \rho \, \mathrm{d}\rho \\
&= \sum_{j=1}^{3} \int_{\theta_1^j}^{\theta_2^j} \cos\theta \, \mathrm{d}\theta \int_{0}^{d_j^\theta(\theta)} \sqrt{R^2 - w_0^2} \mathrm{e}^{\mathrm{i}k_0 R} \, \mathrm{d}R
\end{aligned} \tag{5.59}
$$

$$
\begin{aligned}
I_{23} &= \sum_{j=1}^{3} \int_{\theta_1^j}^{\theta_2^j} \sin\theta \, \mathrm{d}\theta \int_{0}^{d_j^\theta(\theta)} \rho \frac{\mathrm{e}^{\mathrm{i}k_0\sqrt{\rho^2+w_0^2}}}{\sqrt{\rho^2+w_0^2}} \rho \, \mathrm{d}\rho \\
&= \sum_{j=1}^{3} \int_{\theta_1^j}^{\theta_2^j} \sin\theta \, \mathrm{d}\theta \int_{0}^{d_j^\theta(\theta)} \sqrt{R^2 - w_0^2} \mathrm{e}^{\mathrm{i}k_0 R} \, \mathrm{d}R
\end{aligned} \tag{5.60}
$$

respectively, for the case of Figure 5.4(b). For the non-singular case or $w_0 \neq 0$, we cannot derive a closed-form formula for the inner integrals in the above and they are evaluated with numerical quadrature rules. For the singular case, namely, $w_0 \to 0$ in the situation of Figure 5.4(b), we can derive a closed-form formula for the inner integrals in the above, i.e.

$$
I_{22}^s = \sum_{j=1}^{3} \lim_{w_0 \to 0} \int_{\theta_1^j}^{\theta_2^j} \cos\theta \, \mathrm{d}\theta \int_{0}^{d_j^\theta(\theta)} \sqrt{R^2 - w_0^2} \mathrm{e}^{\mathrm{i}k_0 R} \, \mathrm{d}R
$$

$$= \sum_{j=1}^{3} \int_{\theta_1^j}^{\theta_2^j} \cos\theta \left[\frac{e^{ik_0 d_j}}{ik_0} \left(d_j^\theta(\theta) - \frac{1}{ik_0} \right) - \frac{1}{k_0^2} \right] d\theta \tag{5.61}$$

$$I_{23}^s = \sum_{j=1}^{3} \lim_{w_0 \to 0} \int_{\theta_1^j}^{\theta_2^j} \sin\theta \, d\theta \int_0^{d_j^\theta(\theta)} \sqrt{R^2 - w_0^2} e^{ik_0 R} \, dR$$

$$= \sum_{j=1}^{3} \int_{\theta_1^j}^{\theta_2^j} \sin\theta \left[\frac{e^{ik_0 d_j}}{ik_0} \left(d_j^\theta(\theta) - \frac{1}{ik_0} \right) - \frac{1}{k_0^2} \right] d\theta. \tag{5.62}$$

Note that two special cases exist in the singular case, namely, the origin is occasionally located at an edge or corner of the triangle, but the formulas are still valid because the summation is just for two or one subtriangle in this situation.

5.3.3 Numerical Examples

To demonstrate the effectiveness of the approach for evaluating the weakly-singular integrals, we present several numerical examples for EM scattering by perfectly electric conducting (PEC) objects, but the technique can also be used in the problems with non-PEC objects. It is assumed that the incident wave is a plane wave with a frequency $f = 300$ MHz and is propagating along the $-z$ direction in free space. We calculate the bistatic radar cross section (RCS) of the objects observed along the principal cut ($\phi = 0°$ and $\theta = 0°$–$180°$) for the scatterers with both vertical polarization (VV) and horizontal polarization (HH). The first example is the scattering by a PEC sphere with a radius $a = 0.08\lambda$ (we do not show the geometry). The sphere is chosen because it has exact solutions or Mie-series solutions [24] that can be used to verify the numerical solutions. We discretize the spherical surface into 1988 triangular patches and Figure 5.6 plots the RCS solutions, which are very close to the analytical counterparts.

The subsequent several examples are the scattering by a PEC tetrahedron, diamond, thin circular disk, and triangular cylinder (prism), respectively, as shown in Figure 5.5.

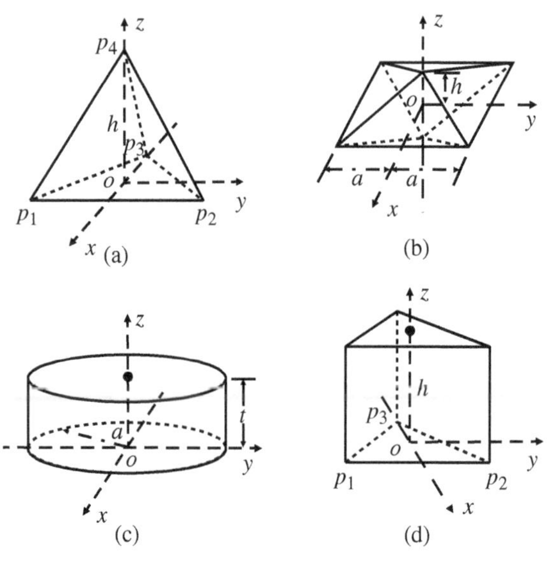

Figure 5.5 Geometries of conducting scatterers: (a) A tetrahedron; (b) A diamond; (c) A circular disk; (d) A triangular cylinder (prism).

The tetrahedron is defined by four vertices $p_1(0.25\lambda, -0.25\lambda, 0)$, $p_2(0.25\lambda, 0.25\lambda, 0)$, $p_3(-0.25\lambda, 0, 0)$, and $p_4(0, 0, 0.5\lambda)$ and we discretize its surface into 2620 triangular patches. The squared diamond is defined with $a = 0.4\lambda$ and $h = 0.02\lambda$ that is very thin near the edges and its surface is discretized into 1032 triangular patches. The thin circular disk is characterized with a radius of 1.0λ and a thickness of 0.05λ and its surface is meshed into 1372 triangular patches. The triangular cylinder has a base triangle defined by three vertices $p_1(0.25\lambda, -0.25\lambda, 0)$, $p_2(0.25\lambda, 0.25\lambda, 0)$, $p_3(-0.25\lambda, 0, 0)$, and has a height $h = 0.5\lambda$, and we mesh its surface into 1928 triangular patches. Figures 5.7–5.10 show the corresponding RCS solutions for the above four scatterers and they agree very well with the solutions based on the Duffy's method in the singularity treatment.

If the RCS solutions are not sensitive to the numerical errors of the integrals, we can individually check the accuracy of the above evaluation scheme. For the integral I_1, we choose the reference values of Table II (singular case) and Table III (near-singular case) shown in [16] to make a comparison. The source triangle is defined by three vertices $(0, 0, 0)$, $(1.0, 0, 0)$, and $(0, 1.0, 0)$ (the unit is meter and the wavelength is 10 meters). The observation point is located at $(0.1, 0.1, w_0)$ with $w_0 = 0$ for the singular case and $w_0 = 0.0001$ for the near-singular case. Figure 5.11 shows the convergence curve characterized by the number of significant digits (which is the same as that of the reference data) versus the number of sample points (for each subtriangle) in the singular and near-singular cases, respectively. Since the imaginary part of the integral is usually more accurate than the real part, we only show the results for the real part. The Gauss-Legendre quadrature rule is used in the numerical integration and we can see that the same accuracy is achieved with a fewer number of sampling points if compared with the Duffy's method. For the integral I_2, we use the reference values of Table III shown in [21] to verify. The source triangle is defined by three vertices $(0, 0, 0)$, $(2, 0, 0)$, and $(7.514 \times 10^{-4}, 0.1732, 0)$, and the observation

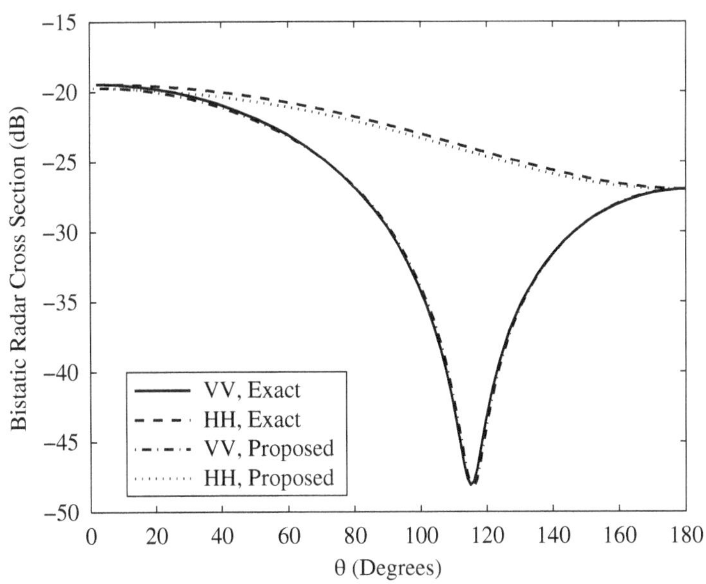

Figure 5.6 Bistatic RCS solutions for a PEC sphere with a radius $a = 0.08\lambda$.

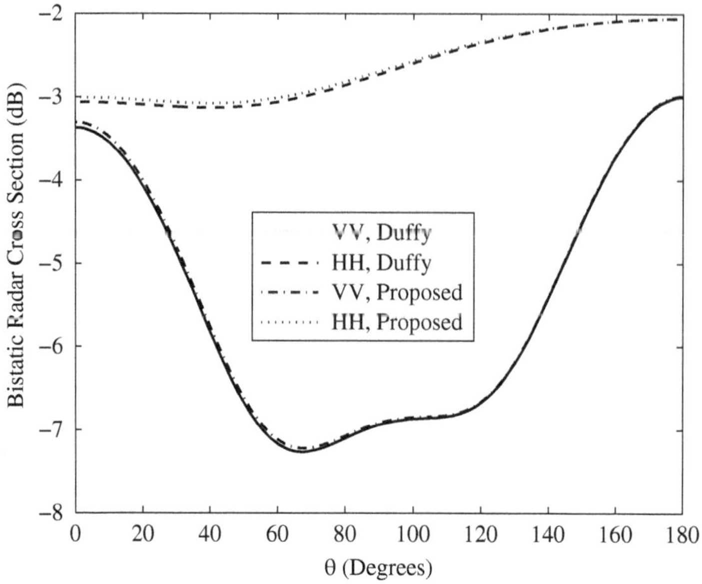

Figure 5.7 Bistatic RCS solutions for a PEC tetrahedron defined by four vertices $p_1(0.25\lambda, -0.25\lambda, 0)$, $p_2(0.25\lambda, 0.25\lambda, 0)$, $p_3(-0.25\lambda, 0, 0)$ and $p_4(0, 0, 0.5\lambda)$.

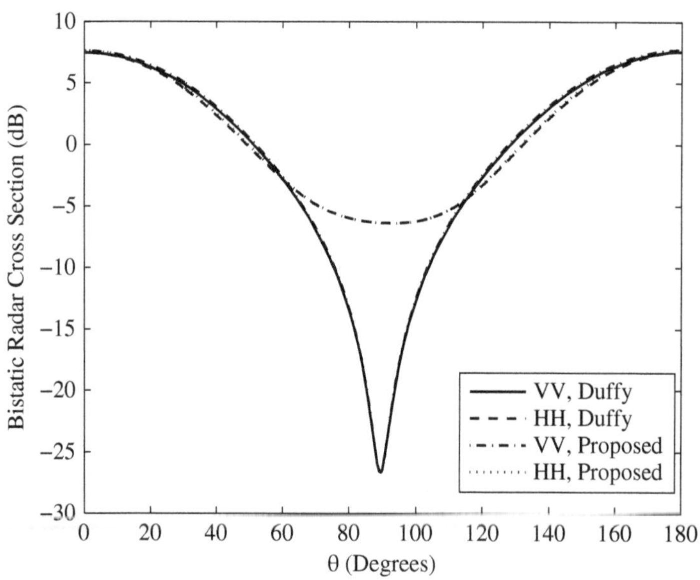

Figure 5.8 Bistatic RCS solutions for a PEC diamond defined with $a = 0.4\lambda$ and $h = 0.02\lambda$.

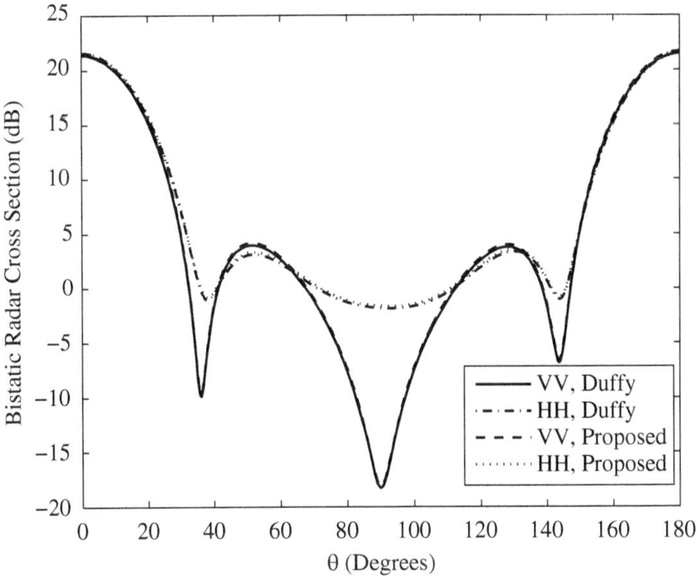

Figure 5.9 Bistatic RCS solutions for a PEC circular thin disk with $a = 1.0\lambda$ and $t = 0.05\lambda$.

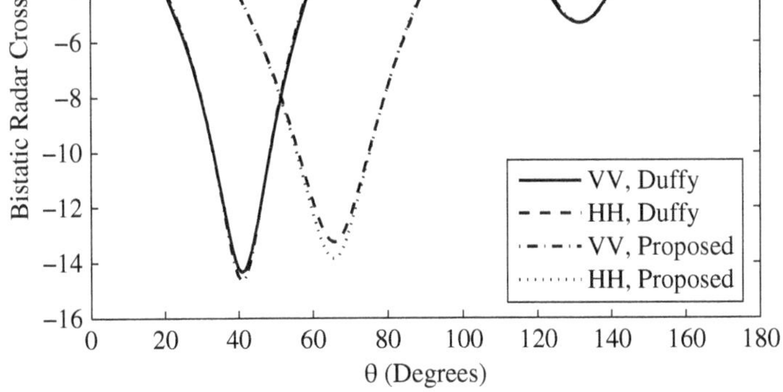

Figure 5.10 Bistatic RCS solutions for a PEC triangular cylinder (prism) that has a base triangle defined by the vertices $p_1(0.25\lambda, -0.25\lambda, 0)$, $p_2(0.25\lambda, 0.25\lambda, 0)$, and $p_3(-0.25\lambda, 0, 0)$, and has a height $h = 0.5\lambda$.

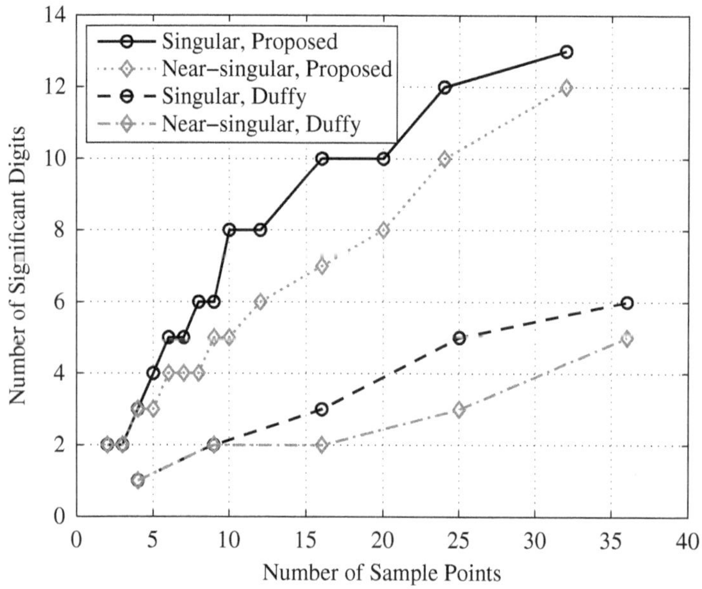

Figure 5.11 The convergence of the integral I_1 (real part) with a comparison to that from the Duffy's method.

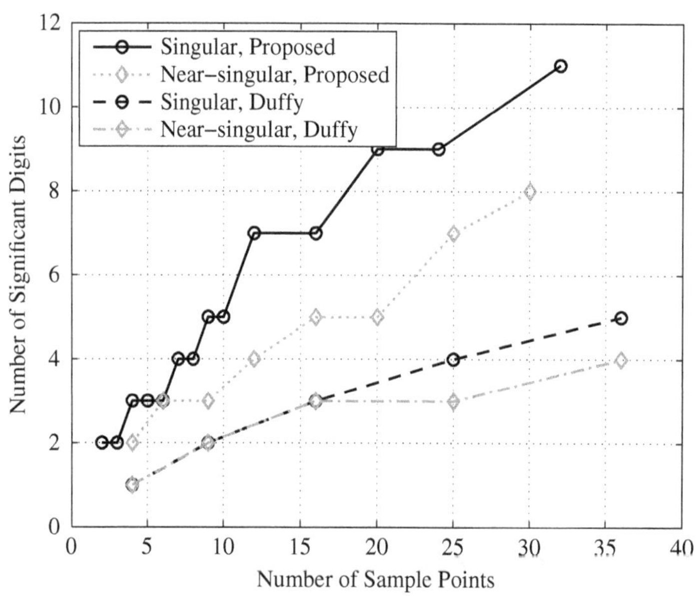

Figure 5.12 The convergence of the integral I_2 (real part) with a comparison to that from the Duffy's method.

point is located at $(7.514 \times 10^{-4}, 0.1732, w_0)$ with $w_0 = 0$ for the singular case and $w_0 = 0.01$ for the near-singular case. The linear source density is assumed to be $1 - 0.5u - 5.771502886836028v$ according to the relevant description in the paper. Figure 5.12 plots the convergence curve for the integral based on the Gauss-Legendre quadrature rule and again the same accuracy can be obtained with less number of sampling points when compared to the Duffy's method.

5.4 Evaluation of Hypersingular and Weakly-Singular Integrals over Triangular Patches

As we have mentioned in Chapter 3, the EFIE takes two forms with different integral kernels. In the first form, the kernel consists of the multiplication of the scalar Green's function by the current density, which represents the electric field produced by the current density or vector potential, and the product of the gradient of the scalar Green's function with the divergence of the current density, which represents the electric field produced by the charge density or scalar potential. In the second form, the kernel is composed of the dot product of the dyadic Green's function with the current density and the dyadic Green's function includes the double gradient of the scalar Green's function. The two forms are equivalent to each other based on the divergence identity and Gauss' divergence theorem with a vanishing contour integral due to the orthogonality between the current density's direction and normal direction of an object surface. The first form has a $1/R^2$ singularity in the kernel for a self interaction, where R is the distance between an observation point and a source point. This singularity, together with the $1/R$ singularity from the scalar Green's function itself in the kernel, can be efficiently handled by the singularity subtraction technique and Duffy's method [4–25]. Therefore, it is more widely used although the divergence of current density in the kernel requires at least linear approximation for the current density and there is no way to use a pulse-like basis function for the current density expansion if a simplicity is emphasized.

The second form removes the divergence of current density in the kernel and can localize the approximation of current density. This form not only allows a constant approximation for the current density over a discretized patch with a great simplicity but also provides a great ease in incorporation with fast algorithms, such as the fast multiple algorithm (FMA) [29]. However, this form includes a $1/R^3$ hypersingularity in the kernel generated by the double gradient of the scalar Green's function. Due to the higher degree of the singularity, there was no direct method to handle this kind of kernel in the past until we presented a robust technique in 2007 [19]. One usually changes the kernel back to the one of the first form for self interactions so that it can be handled and uses the kernel of the second form just for far interactions in the Nyström method [30–32]. This treatment is very inefficient because a boundary integral is involved for each self triangular patch and may not be convenient when working with the FMA.

It is obviously desirable to apply the second form without changing the kernel. The bottleneck is the efficient evaluation for the hypersingular and near-hypersingular integrals in generating self-interaction and near-interaction matrix elements. As we have shown in the singularity subtraction, there are many types of hypersingular integrals and they can only be defined or evaluated under a Cauchy-principal-value

(CPV) sense. This means that to evaluate a hypersingular integral, we first assume that the observation point is off the source triangular patch and then try to derive the corresponding analytical formula for the hypersingular integral. Once the analytical formula is available, we can then take a limit by letting the observation point approach the source triangular patch to obtain the formula of evaluating the hypersingular integral. Strictly speaking, all singular integrals should be performed in such a sense. Nevertheless, some weakly-singular integrals can be performed by assuming the observation point to be inside the source triangular patch first (i.e. taking the limit first) to simplify the derivation. The hypersingular integrals can only be performed under the CPV sense (taking the limit first is not allowed), otherwise they are not integrable. Based on the CPV sense, we can derive the closed-form formulas of evaluating all singular integrals even weakly-singular integrals and thus the bottleneck is removed. The analytical formulas for weakly-singular integrals are also important when accurate calculations for near-interaction elements are needed because numerical approaches may not be convenient or available sometimes.

In the numerical demonstration, we incorporate the evaluation formulas with the Nyström method to solve EM problems and flat triangular patches are used in the geometrical discretization. Numerical examples show that the evaluation scheme is easy to use and very efficient in implementation and computation.

5.4.1 Hypersingular Integrals

To derive the formulas of calculating the hypersingular integrals as shown in Equations (5.13)–(5.21), we can take a look at the local coordinate system as shown in Figure 5.2. In the figure, we divide the triangle patch into three subtriangles ΔT_i $(i = 1, 2, 3)$ by connecting the origin, or the projected point of the observation point on the triangle plane, to three vertices of the triangle patch. Each subtriangle is specified by four parameters d_i, θ_0^i, θ_1^i and θ_2^i, where d_i is the vertical distance $\overline{OO_i}$ between the origin and the ith side of the triangle patch, θ_0^i is the angle of $\overline{OO_i}$, and θ_1^i and θ_2^i are the angles of connecting lines from the origin to the first and second end point of the ith side, respectively. Under the polar coordinate system, the hypersingulr integral I_2 in Equation (5.13) can be written as

$$
\begin{aligned}
I_2 &= \int_{\Delta S} \frac{uv}{R^5}\, dS = \sum_{i=1}^{3} \frac{1}{2} \int_{\theta_1^i}^{\theta_2^i} d\theta \sin 2\theta \int_0^{d_\theta^i} \frac{\rho^3\, d\rho}{(\rho^2 + w_0^2)^{5/2}} \\
&= \sum_{i=1}^{3} \frac{1}{2} \int_{\theta_1^i}^{\theta_2^i} d\theta \sin 2\theta \left[\frac{-1}{\sqrt{d_i^2 + w_0^2}} + \frac{w_0^2}{3(d_i^2 + w_0^2)^{3/2}} + \frac{2}{3w_0^2} \right] \\
&= \sum_{i=1}^{3} \frac{1}{2} \left\{ \int_{\theta_{10}^i}^{\theta_{20}^i} d\phi \sin(2\phi + 2\theta_0^i) \left[\frac{-\cos\phi}{\sqrt{d_i^2 + w_0^2\cos^2\phi}} \right.\right. \\
&\quad \left.\left. + \frac{w_0^2\cos^3\phi}{3(d_i^2 + w_0^2\cos^2\phi)^{3/2}} \right] - \frac{2}{3w_0}(\cos 2\theta_2^i - \cos 2\theta_1^i) \right\}
\end{aligned}
$$

$$= \sum_{i=1}^{3} \frac{1}{2} \left[I_{2a}^i + I_{2b}^i - \frac{2}{3w_0} \left(\cos 2\theta_2^i - \cos 2\theta_1^i \right) \right] \tag{5.63}$$

where $d_\theta^i = \frac{d_i}{\cos(\theta-\theta_0^i)}$, $\theta_{10}^i = \theta_1^i - \theta_0^i$, $\theta_{20}^i = \theta_2^i - \theta_0^i$, and

$$\begin{aligned} I_{2a}^i &= \int_{\theta_{10}^i}^{\theta_{20}^i} \mathrm{d}\phi \sin 2\phi \cos 2\theta_0^i \left[\frac{-\cos\phi}{\sqrt{d_i^2 + w_0^2 \cos^2\phi}} + \frac{w_0^2 \cos^3\phi}{3(d_i^2 + w_0^2\cos^2\phi)^{3/2}} \right] \\ &= \frac{2}{3} \cos 2\theta_0^i \left(\frac{\cos^3\theta_{20}^i}{c_2} - \frac{\cos^3\theta_{10}^i}{c_1} \right) \end{aligned} \tag{5.64}$$

$$\begin{aligned} I_{2b}^i &= \int_{\theta_{10}^i}^{\theta_{20}^i} \mathrm{d}\phi \cos 2\phi \sin 2\theta_0^i \left[\frac{-\cos\phi}{\sqrt{d_i^2 + w_0^2 \cos^2\phi}} + \frac{w_0^2 \cos^3\phi}{3(d_i^2 + w_0^2\cos^2\phi)^{3/2}} \right] \\ &= \frac{2}{3} \sin 2\theta_0^i \left(\frac{\sin^3\theta_{20}^i}{c_2} - \frac{\sin^3\theta_{10}^i}{c_1} \right) - \frac{3d_i^2 + 2w_0^2}{3(d_i^2 + w_0^2)} \sin 2\theta_0^i \left(\frac{\sin\theta_{20}^i}{c_2} - \frac{\sin\theta_{10}^i}{c_1} \right). \end{aligned} \tag{5.65}$$

Note that we have set $c_1 = \sqrt{d_i^2 + w_0^2 \cos^2\theta_{10}^i}$ and $c_2 = \sqrt{d_i^2 + w_0^2 \cos^2\theta_{20}^i}$, and also applied $\sum_{i=1}^{3} (\cos 2\theta_2^i - \cos 2\theta_1^i) = 0$ in the above derivation. To facilitate the derivation for other hypersingular integrals, we introduce two intermediate integrals, whose expressions can be derived in a similar way, i.e.

$$\begin{aligned} I_t^i &= \int_{\theta_1^i}^{\theta_2^i} \mathrm{d}\theta \cos 2\theta \int_0^{d_\theta^i} \frac{\rho^3 \,\mathrm{d}\rho}{(\rho^2 + w_0^2)^{5/2}} \\ &= \cos 2\theta_0^i I_{2b}^i - \sin 2\theta_0^i I_{2a}^i \end{aligned} \tag{5.66}$$

$$\begin{aligned} I_s^i &= \int_{\theta_1^i}^{\theta_2^i} \mathrm{d}\theta \left[\frac{-\cos(\theta - \theta_0^i)}{\sqrt{d_i^2 + w_0^2 \cos^2(\theta - \theta_0^i)}} + \frac{w_0^2 \cos^3(\theta - \theta_0^i)}{3(d_i^2 + w_0^2 \cos^2(\theta - \theta_0^i))^{3/2}} \right] \\ &\quad + \frac{2}{3w_0}(\theta_2^i - \theta_1^i) \\ &= -\frac{2}{3w_0} \left[\arcsin(s_0 \sin\theta_{20}^i) - \arcsin(s_0 \sin\theta_{10}^i) \right] \\ &\quad - \frac{d_i^2}{3(d_i^2 + w_0^2)} \left(\frac{\sin\theta_{20}^i}{c_2} - \frac{\sin\theta_{10}^i}{c_1} \right) + \frac{2}{3w_0}(\theta_2^i - \theta_1^i) \end{aligned} \tag{5.67}$$

where $s_0 = \frac{w_0}{\sqrt{d_i^2+w_0^2}}$. Based on the intermediate integrals, we can find the first part of the hypersingular integral I_1 in Equation (5.13) as

$$\begin{aligned} I_{1a} &= \int_{\Delta S} \frac{3u^2}{R^5}\, \mathrm{d}S \\ &= \sum_{i=1}^{3} 3 \int_{\theta_1^i}^{\theta_2^i} \mathrm{d}\theta \cos^2\theta \int_0^{d_\theta^i} \frac{\rho^3\, \mathrm{d}\rho}{(\rho^2 + w_0^2)^{5/2}} \\ &= \sum_{i=1}^{3} \frac{3}{2}(I_t^i + I_s^i). \end{aligned} \tag{5.68}$$

Similarly, we can find the first part of the hypersingular integral I_3 in Equation (5.14) as

$$\begin{aligned} I_{3a} &= \int_{\Delta S} \frac{3v^2}{R^5}\, \mathrm{d}S \\ &= \sum_{i=1}^{3} 3 \int_{\theta_1^i}^{\theta_2^i} \mathrm{d}\theta \sin^2\theta \int_0^{d_\theta^i} \frac{\rho^3\, \mathrm{d}\rho}{(\rho^2 + w_0^2)^{5/2}} \\ &= \sum_{i=1}^{3} \frac{3}{2}(I_s^i - I_t^i). \end{aligned} \tag{5.69}$$

Note that the above and other followed derivations are also valid when the projection of the observation point on the triangle plane is outside the triangular patch, as shown in Figure 5.13. In this case, the integral over the triangular patch is still the sum of the integrals over three subtriangles with appropriate plus and minus signs. All expressions found above are convergent when taking a limit of $w_0 \to 0$ except the following term from I_s^i

$$A_0 = \frac{2}{3w_0} \sum_{i=1}^{3} (\theta_2^i - \theta_1^i). \tag{5.70}$$

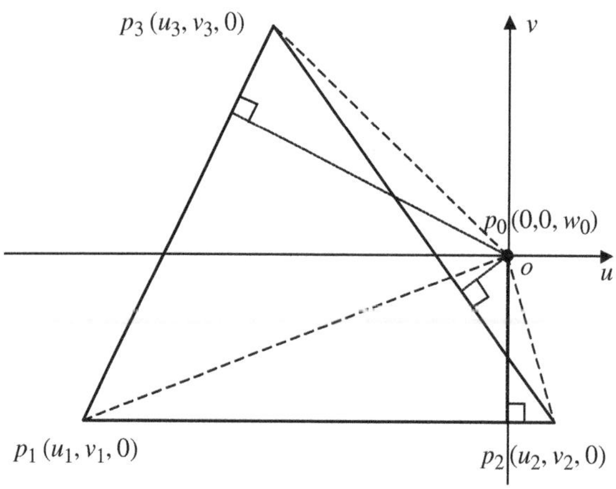

Figure 5.13 Subtriangles when the projection of observation point on the triangle plane is outside the triangular patch.

If the projection of the observation point is outside the triangular patch, we have $\sum_{i=1}^{3}(\theta_2^i - \theta_1^i) = 0$ and the term is vanishing. However, when the projection is inside the triangular patch, on the boundary of the triangular patch, or on the vertex of the triangular patch, $\sum_{i=1}^{3}(\theta_2^i - \theta_1^i) = 2\pi$, π, or the internal angle of that vertex, respectively. Due to the non-vanishing value of A_0 for these cases, the corresponding integrals I_{1a} and I_{3a} are divergent individually when taking the limit of $w_0 \to 0$. Fortunately, this divergent term A_0 can be cancelled by the same term generated from the hypersingular integral of $1/R^3$, i.e. the second part of I_1 and I_3

$$
\begin{aligned}
I_{1b} = I_{3b} &= \int_{\Delta S} \frac{1}{R^3} \mathrm{d}S = \sum_{i=1}^{3} \int_{\theta_1^i}^{\theta_2^i} \mathrm{d}\theta \int_0^{d_\theta^i} \frac{\rho\, \mathrm{d}\rho}{(\rho^2 + w_0^2)^{3/2}} \\
&= \frac{1}{w_0} \left[\arcsin(s_0 \sin\theta_{10}^i) - \arcsin(s_0 \sin\theta_{20}^i)\right] + \frac{1}{w_0} \sum_{i=1}^{3} (\theta_2^i - \theta_1^i).
\end{aligned} \tag{5.71}
$$

The above integral is also divergent individually when the observation point approaches the source triangular patch due to the divergent term $B_0 = \frac{1}{w_0} \sum_{i=1}^{3}(\theta_2^i - \theta_1^i)$. Since $\frac{3}{2}A_0 - B_0 = 0$, we have the following convergent hypersingular integrals

$$
\begin{aligned}
I_1 &= \int_{\Delta S} \left(\frac{3u^2}{R^5} - \frac{1}{R^3}\right) \mathrm{d}S \\
&= \sum_{i=1}^{3} \left[\cos 2\theta_0^i \left(\frac{\sin^3\theta_{20}^i}{c_2} - \frac{\sin^3\theta_{10}^i}{c_1}\right) - \sin 2\theta_0^i \left(\frac{\cos^3\theta_{20}^i}{c_2} - \frac{\cos^3\theta_{10}^i}{c_1}\right)\right. \\
&\quad \left. - \frac{(3d_i^2 + 2w_0^2)\cos 2\theta_0^i + d_i^2}{2(d_i^2 + w_0^2)} \left(\frac{\sin\theta_{20}^i}{c_2} - \frac{\sin\theta_{10}^i}{c_1}\right)\right]
\end{aligned} \tag{5.72}
$$

$$
\begin{aligned}
I_3 &= \int_{\Delta S} \left(\frac{3v^2}{R^5} - \frac{1}{R^3}\right) \mathrm{d}S \\
&= \sum_{i=1}^{3} \left[\sin 2\theta_0^i \left(\frac{\cos^3\theta_{20}^i}{c_2} - \frac{\cos^3\theta_{10}^i}{c_1}\right) - \cos 2\theta_0^i \left(\frac{\sin^3\theta_{20}^i}{c_2} - \frac{\sin^3\theta_{10}^i}{c_1}\right)\right. \\
&\quad \left. + \frac{(3d_i^2 + 2w_0^2)\cos 2\theta_0^i - d_i^2}{2(d_i^2 + w_0^2)} \left(\frac{\sin\theta_{20}^i}{c_2} - \frac{\sin\theta_{10}^i}{c_1}\right)\right].
\end{aligned} \tag{5.73}
$$

The limits of the integrals I_1, I_2 and I_3 can then be found easily from the above expressions

$$
\begin{aligned}
I_1^{(0)} &= \lim_{w_0 \to 0} \int_{\Delta S} \left(\frac{3u^2}{R^5} - \frac{1}{R^3}\right) \mathrm{d}S \\
&= \sum_{i=1}^{3} \left[\cos 2\theta_0^i s_3^i - \sin 2\theta_0^i c_3^i - 0.5(3\cos 2\theta_0^i + 1) s_1^i\right]
\end{aligned} \tag{5.74}
$$

$$
I_2^{(0)} = \lim_{w_0 \to 0} \int_{\Delta S} \frac{3uv}{R^5} \mathrm{d}S
$$

$$= \sum_{i=1}^{3} \left[\cos 2\theta_0^i c_3^i + \sin 2\theta_0^i s_3^i - 1.5 \sin 2\theta_0^i s_1^i\right] \tag{5.75}$$

$$I_3^{(0)} = \lim_{w_0 \to 0} \int_{\Delta S} \left(\frac{3v^2}{R^5} - \frac{1}{R^3}\right) \mathrm{d}S$$

$$= \sum_{i=1}^{3} \left[\sin 2\theta_0^i c_3^i - \cos 2\theta_0^i s_3^i + 0.5(3\cos 2\theta_0^i - 1)s_1^i\right] \tag{5.76}$$

where

$$s_3^i = \frac{1}{d_i}(\sin^3\theta_{20}^i - \sin^3\theta_{10}^i) \tag{5.77}$$

$$c_3^i = \frac{1}{d_i}(\cos^3\theta_{20}^i - \cos^3\theta_{10}^i) \tag{5.78}$$

$$s_1^i = \frac{1}{d_i}(\sin\theta_{20}^i - \sin\theta_{10}^i). \tag{5.79}$$

There are other three hypersingular integrals I_4, I_5, and I_6 in Equations (5.14) and (5.15), and I_4 and I_6 can be derived as follows

$$I_4 = \int_{\Delta S} \frac{w_0(-u)}{R^5} \mathrm{d}S$$

$$= \sum_{i=1}^{3}(-w_0) \int_{\theta_1^i}^{\theta_2^i} \mathrm{d}\theta \cos\theta \int_0^{d_\theta^i} \frac{\rho^2 \,\mathrm{d}\rho}{(\rho^2 + w_0^2)^{5/2}}$$

$$= -\frac{1}{3w_0} \sum_{i=1}^{3} \left[\frac{d_i^3 \cos\theta_0^i}{d_i^2 + w_0^2}\left(\frac{\sin\theta_{20}^i}{c_2} - \frac{\sin\theta_{10}^i}{c_1}\right) + d_i \sin\theta_0^i \left(\frac{\cos\theta_{20}^i}{c_2} - \frac{\cos\theta_{10}^i}{c_1}\right)\right] \tag{5.80}$$

$$I_6 = \int_{\Delta S} \frac{w_0(-v)}{R^5} \mathrm{d}S$$

$$= \sum_{i=1}^{3}(-w_0) \int_{\theta_1^i}^{\theta_2^i} \mathrm{d}\theta \sin\theta \int_0^{d_\theta^i} \frac{\rho^2 \,\mathrm{d}\rho}{(\rho^2 + w_0^2)^{5/2}}$$

$$= \frac{1}{3w_0} \sum_{i=1}^{3} \left[d_i \cos\theta_0^i \left(\frac{\cos\theta_{20}^i}{c_2} - \frac{\cos\theta_{10}^i}{c_1}\right) - \frac{d_i^3 \sin\theta_0^i}{d_i^2 + w_0^2}\left(\frac{\sin\theta_{20}^i}{c_2} - \frac{\sin\theta_{10}^i}{c_1}\right)\right]. \tag{5.81}$$

Note that I_4 and I_6 vanish when $w_0 \to 0$. This can be seen more clearly by separating each of the integrals into two parts, namely, over $\Delta S - S_0$ and over S_0, as shown in Figure 5.14. Taking I_4 as an example, we can find the limit of the integral as

$$\lim_{w_0 \to 0} I_4 = \lim_{w_0 \to 0} \left[(-w_0)\int_{\Delta S - S_0} \frac{u}{R^5} \mathrm{d}S - w_0 \int_0^{2\pi} \mathrm{d}\theta \cos\theta \int_0^a \frac{\rho^2 \,\mathrm{d}\rho}{(\rho^2 + w_0^2)^{5/2}}\right]$$

$$= \lim_{w_0 \to 0} \left[\frac{(-w_0)a^3}{3w_0^2(a^2 + w_0^2)^{3/2}}(\sin 2\pi - \sin 0)\right]$$

$$= 0. \tag{5.82}$$

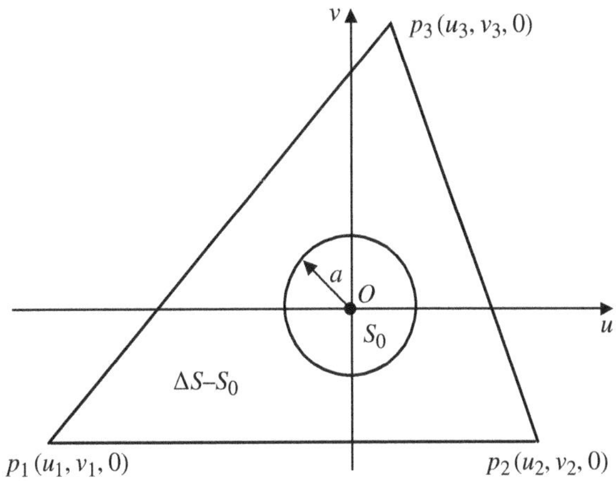

Figure 5.14 A triangular patch is divided into two parts, $\Delta S - S_0$ and S_0 for deriving the limits of the integrals I_4 and I_5.

In the above, the integral over $\Delta S - S_0$ is regular and vanishes when $w_0 \to 0$. For I_5, its first part can also be derived similarly

$$\begin{aligned} I_{5a} &= \int_{\Delta S} \frac{3w_0^2}{R^5}\, dS = \sum_{i=1}^{3} 3w_0^2 \int_{\theta_1^i}^{\theta_2^i} d\theta \int_0^{d_\theta^i} \frac{\rho\, d\rho}{(\rho^2 + w_0^2)^{5/2}} \\ &= \sum_{i=1}^{3} \left\{ \frac{-w_0^2}{(d_i^2 + w_0^2)} \left(\frac{\sin\theta_{20}^i}{c_2} - \frac{\sin\theta_{10}^i}{c_1} \right) - \frac{1}{w_0} \left[\arcsin(s_0 \sin\theta_{20}^i) - \arcsin(s_0 \sin\theta_{10}^i) \right] \right. \\ &\quad \left. + \frac{\sin\theta_{20}^i}{c_2} - \frac{\sin\theta_{10}^i}{c_1} \right\} + A_0. \end{aligned} \tag{5.83}$$

However, I_5 is not useful in the numerical solution because the surface current density does not have a w component in the local coordinate system. Again, the divergent term A_0 in I_{5a} will be cancelled by the same term from I_{1b} or I_{3b}, leading to the convergence of I_5 when $w_0 \to 0$.

5.4.2 Weakly-Singular Integrals

Except the first six hypersingular integrals, all other integrals are weakly-singular integrals in Equations (5.13)–(5.21). The first weakly-singular integral we want to derive is $I_{17} = w_0^2 I_0$, in which w_0^2 is a constant and we can only deal with the derivation of I_0. The integral I_0 has already been derived in [25] using a different strategy, but we present a simpler scheme here.

$$\begin{aligned} I_0 &= \int_{\Delta S} \frac{1}{R}\, dS = \sum_{i=1}^{3} \int_{\theta_1^i}^{\theta_2^i} d\theta \int_0^{d_\theta^i} \frac{\rho\, d\rho}{\sqrt{\rho^2 + w_0^2}} \\ &= \sum_{i=1}^{3} \int_{\theta_1^i}^{\theta_2^i} d\theta \sqrt{\frac{d_i^2}{\cos^2(\theta - \theta_0^i)} + w_0^2} \\ &= \sum_{i=1}^{3} \left[I_{0a}^i - w_0(\theta_2^i - \theta_1^i) \right] \end{aligned} \tag{5.84}$$

where

$$I_{0a}^{i} = \frac{1}{2}\left\{d_i(q_1 - q_2) - 2w_0\left[\arcsin(s_0 \sin\theta_{20}^{i}) - \arcsin(s_0 \sin\theta_{10}^{i})\right]\right.$$
$$\left. + s_2 - s_1 - s_4 + s_3\right\} \tag{5.85}$$

with

$$t_1 = 1 + \sin\theta_{10}^{i};\quad t_2 = 1 + \sin\theta_{20}^{i};\quad t_3 = 1 - \sin\theta_{10}^{i};\quad t_4 = 1 - \sin\theta_{20}^{i} \tag{5.86}$$

$$s_i = \sqrt{d_i^2 + (2t_i - t_i^2)w_0^2},\quad i = 1,2,3,4 \tag{5.87}$$

$$q_1 = \ln\left|\frac{t_1(d_i^2 + w_0^2 t_2 + d_i s_2)}{t_2(d_i^2 + w_0^2 t_1 + d_i s_1)}\right|;\quad q_2 = \ln\left|\frac{t_3(d_i^2 + w_0^2 t_4 + d_i s_4)}{t_4(d_i^2 + w_0^2 t_3 + d_i s_3)}\right|. \tag{5.88}$$

Also, $s_0 = \frac{w_0}{\sqrt{d_i^2 + w_0^2}}$ has been defined previously. To derive other weakly-singular integrals, we first derive several intermediate integrals.

$$I_{0b}^{i} = \int_{\theta_1^i}^{\theta_2^i} d\theta \cos 2\theta \sqrt{\frac{d_i^2}{\cos^2(\theta - \theta_0^i)} + w_0^2}$$
$$= p_1\left[\arcsin(s_0 \sin\theta_{20}^{i}) - \arcsin(s_0 \sin\theta_{10}^{i})\right] + p_2\left[\operatorname{arcsinh}(t_0 \cos\theta_{20}^{i}) - \right.$$
$$\left.\operatorname{arcsinh}(t_0 \cos\theta_{10}^{i})\right] + c_2 \sin(\theta_2^i + \theta_0^i) - c_1 \sin(\theta_1^i + \theta_0^i) - I_{0a}^{i} \cos 2\theta_0^i \tag{5.89}$$

where

$$p_1 = \frac{w_0 \cos 2\theta_0^i}{s_0^2};\quad p_2 = \frac{d_i^2 \sin 2\theta_0^i}{w_0};\quad t_0 = \frac{w_0}{d_i}. \tag{5.90}$$

$$I_{ua}^{i} = \int_{\theta_1^i}^{\theta_2^i} d\theta \frac{\left[d_i^2 + w_0^2 \cos^2(\theta - \theta_0^i)\right]^{3/2}}{\cos(\theta - \theta_0^i)}$$
$$= \frac{1}{6}(s_2^3 - s_1^3 - s_4^3 + s_3^3) - d_i^3(q_1 - q_2) + \frac{1}{2}(2d_i^2 - w_0^2)(s_2 - s_1 - s_4 + s_3)$$
$$+ \frac{1}{4}w_0^2(t_2 s_2 - t_1 s_1 - t_4 s_4 + t_3 s_3) - (3d_i^2 + w_0^2)w_0\left[\arcsin(s_0 \sin\theta_{20}^{i})\right.$$
$$\left. - \arcsin(s_0 \sin\theta_{10}^{i})\right] \tag{5.91}$$

$$I_{ub}^{i} = \int_{\theta_1^i}^{\theta_2^i} d\theta \sin(\theta - \theta_0^i) \frac{\left[d_i^2 + w_0^2 \cos^2(\theta - \theta_0^i)\right]^{3/2}}{\cos^2(\theta - \theta_0^i)}$$
$$= \frac{c_2^3}{\cos\theta_{20}^{i}} - \frac{c_1^3}{\cos\theta_{10}^{i}} - 1.5w_0^2(c_2 \cos\theta_{20}^{i} - c_1 \cos\theta_{10}^{i}) - 1.5w_0 d_i^2 \ln\left|\frac{w_0 \cos\theta_{20}^{i} + c_2}{w_0 \cos\theta_{10}^{i} + c_1}\right| \tag{5.92}$$

$$
\begin{aligned}
I^i_{uc} &= \int_{\theta^i_1}^{\theta^i_2} \mathrm{d}\theta \frac{\left[d_i^2 + w_0^2\cos^2(\theta-\theta^i_0)\right]^{3/2}}{\cos^3(\theta-\theta^i_0)} \\
&= \frac{1}{4d_i^2}\left(\frac{s_2^5}{t_2} - \frac{s_1^5}{t_1} + \frac{s_4^5}{t_4} - \frac{s_3^5}{t_3}\right) + \frac{w_0^2}{4d_i^2}\left[(2-t_2)s_2^3 - (2-t_1)s_1^3 + (2-t_4)s_4^3 - (2-t_3)s_3^3\right] \\
&+ \frac{1}{4}\Big\{1.5w_0^2\left[(3-t_2)s_2 - (3-t_1)s_1 + (3-t_4)s_4 - (3-t_3)s_3\right] - 3d_i w_0^2(q_2-q_1) \\
&-3w_0(d_i^2 - w_0^2)\left[\arcsin(s_0\sin\theta^i_{10}) - \arcsin(s_0\sin\theta^i_{10})\right]\Big\} + 2I^i_{ua}.
\end{aligned} \tag{5.93}
$$

Based on those intermediate integrals, we can find that

$$
\begin{aligned}
I^i_{u1} &= \int_{\theta^i_1}^{\theta^i_2} \mathrm{d}\theta\cos^2\theta \frac{\sqrt{d_i^2 + w_0^2\cos^2(\theta-\theta^i_0)}}{\cos(\theta-\theta^i_0)} \\
&= I^i_{0a} + I^i_{0b}
\end{aligned} \tag{5.94}
$$

$$
\begin{aligned}
I^i_{u2} &= \int_{\theta^i_1}^{\theta^i_2} \mathrm{d}\theta\cos^2\theta \frac{\left[d_i^2 + w_0^2\cos^2(\theta-\theta^i_0\right]^{3/2}}{\cos^3(\theta-\theta^i_0)} \\
&= 2\cos 2\theta^i_0 I^i_{ua} + 2\sin 2\theta^i_0 I^i_{ub} + (1+\cos 2\theta^i_0)I^i_{uc}
\end{aligned} \tag{5.95}
$$

and finally we can obtain the integral I_{13} in Equation (5.19) as

$$
\begin{aligned}
I_{13} &= \int_{\Delta S} \frac{u^2}{R}\mathrm{d}S = \sum_{i=1}^{3}\int_{\theta^i_1}^{\theta^i_2}\mathrm{d}\theta\int_0^{d^i_\theta}\frac{\rho^3\,\mathrm{d}\rho}{\sqrt{\rho^2+w_0^2}} \\
&= \sum_{i=1}^{3}\left(\frac{1}{3}I^i_{u2} - w_0^2 I^i_{u1}\right).
\end{aligned} \tag{5.96}
$$

Similarly, we can find the integral I_7 in Equation (5.16) as

$$
\begin{aligned}
I_7 &= \int_{\Delta S} \frac{u^2}{R^3}\,\mathrm{d}S = \sum_{i=1}^{3}\int_{\theta^i_1}^{\theta^i_2}\mathrm{d}\theta\int_0^{d^i_\theta}\frac{\rho^3\,\mathrm{d}\rho}{(\rho^2+w_0^2)^{3/2}} \\
&= \sum_{i=1}^{3}(I^i_{u1} + I^i_{u3})
\end{aligned} \tag{5.97}
$$

where

$$
\begin{aligned}
I^i_{u3} &= \int_{\theta^i_1}^{\theta^i_2} \mathrm{d}\theta\cos^2\theta \frac{w_0^2\cos(\theta-\theta^i_0)}{\sqrt{d_i^2 + w_0^2\cos^2(\theta-\theta^i_0)}} \\
&= I^i_x + I^i_y
\end{aligned} \tag{5.98}
$$

$$I_y^i = \int_{\theta_1^i}^{\theta_2^i} \mathrm{d}\theta \frac{w_0^2 \cos(\theta - \theta_0^i)}{\sqrt{d_i^2 + w_0^2\cos^2(\theta - \theta_0^i)}} = w_0^2 \left[\arcsin(s_0 \sin\theta_{20}^i) - \arcsin(s_0 \sin\theta_{10}^i)\right] \tag{5.99}$$

$$\begin{aligned} I_x^i &= \int_{\theta_1^i}^{\theta_2^i} \mathrm{d}\theta \cos 2\theta \frac{w_0^2 \cos(\theta - \theta_0^i)}{\sqrt{d_i^2 + w_0^2\cos^2(\theta - \theta_0^i)}} \\ &= \cos 2\theta_0^i I_y^i + \cos 2\theta_0^i \left\{ c_2 \sin\theta_{20}^i - c_1 \sin\theta_{10}^i - \frac{w_0}{s_0^2} \left[\arcsin(s_0 \sin\theta_{20}^i) - \arcsin(s_0 \sin\theta_{10}^i)\right] \right\} \\ &\quad + \sin 2\theta_0^i \left\{ c_2 \cos\theta_{20}^i - c_1 \cos\theta_{10}^i - \frac{d_i^2}{w_0} \left[\operatorname{arcsinh}(t_0 \sin\theta_{20}^i) - \operatorname{arcsinh}(t_0 \sin\theta_{10}^i)\right] \right\}. \end{aligned} \tag{5.100}$$

Obviously, other weakly-singular integrals can be derived in a similar way, but we omit them here for simplicity.

5.4.3 Non-Singular Integrals

The singularity subtraction will produce some non-singular integrals which take the following forms

$$I_0^N = \int_{\Delta S} R^m \,\mathrm{d}S; \qquad I_1^N = \int_{\Delta S} u^2 R^m \,\mathrm{d}S; \tag{5.101}$$

$$I_2^N = \int_{\Delta S} uvR^m \,\mathrm{d}S; \qquad I_3^N = \int_{\Delta S} v^2 R^m \,\mathrm{d}S; \tag{5.102}$$

$$I_4^N = \int_{\Delta S} w_0(-u)R^m \,\mathrm{d}S; \quad I_5^N = \int_{\Delta S} w_0(-v)R^m \,\mathrm{d}S \tag{5.103}$$

where $m > 0$. Usually, one does not specially treat such non-singular or regular integrals because they can be integrated numerically. However, we still need to see these integrals magnificently if a high accuracy is desired. When the projection of observation point is outside the source triangular patch, the integrands of all integrals are continuous and smooth, so the numerical integration with available quadrature rules (like Gauss-Legendre quadrature rules) can give rise to very accurate results. However, when the projection of observation point is inside the triangular patch and m is odd, the integrands may not be smooth although they are continuous since the derivative of R may be singular when $w_0 = 0$. In this case, the analytical formulas are more accurate than numerical approaches because the n-point Gaussian quadrature rule requires the existence of the $(2n - 1)$th derivative of the integrand. Since there is no singularity in the above integrals, we can take the limit of $w_0 \to 0$ first in the integrals to simplify the derivation. When doing so, we have (for an odd m)

$$\begin{aligned} I_0^N &= \int_{\Delta S} \rho^m \,\mathrm{d}S = \sum_{i=1}^{3} \int_{\theta_1^i}^{\theta_2^i} \mathrm{d}\theta \int_0^{d_\theta^i} \rho^{m+1} \,\mathrm{d}\rho \\ &= \sum_{i=1}^{3} \left[\frac{d_i^{m+2}}{m+2} \int_{\theta_{10}^i}^{\theta_{20}^i} \frac{\mathrm{d}\phi}{\cos^{m+2}\phi} \right] \\ &= \sum_{i=1}^{3} \left[\frac{d_i^{m+2}}{m+2} I_{m+2}^i \right] \end{aligned} \tag{5.104}$$

where I^i_{m+2} can be found from the following formula

$$
\begin{aligned}
I_{2\ell} &= \int \frac{\mathrm{d}\phi}{\cos^{2\ell}\phi} \\
&= \frac{\sin\phi}{2\ell-1}\left[\sec^{2\ell-1}\phi + \sum_{k=1}^{\ell-1}\frac{2^k q_0}{\sec\phi}\prod_{p=1}^{k}\frac{r}{(2r-1)}\right] \\
I_{2\ell+1} &= \int \frac{\mathrm{d}\phi}{\cos^{2\ell+1}\phi} \\
&= \frac{\sin\phi}{2\ell}\left[\sec^{2\ell}\phi + \sum_{k=1}^{\ell-1}\frac{q_0}{2^k}\prod_{p=1}^{k}\frac{(2r+1)}{r}\right] + \frac{(2\ell-1)!!}{2^\ell \ell!}\ln\left[\tan\left(\frac{\pi}{4}+\frac{\phi}{2}\right)\right]
\end{aligned} \tag{5.105}
$$

with $r = \ell - p$ and $q_0 = \sec^{2\ell-2k}\phi$. Similarly, we can derive

$$
\begin{aligned}
I_1^N &= \int_{\Delta S} u^2\rho^m \,\mathrm{d}S \\
&= \sum_{i=1}^{3}\int_{\theta_1^i}^{\theta_2^i}\mathrm{d}\theta\cos^2\theta\int_0^{d_\theta^i}\rho^{m+3}\,\mathrm{d}\rho \\
&= \sum_{i=1}^{3}\left\{\frac{d_i^{m+4}}{2(m+4)}\left[\cos 2\theta_0^i I^i_{m+2} - \frac{2\sin 2\theta_0^i}{m+2}f^i_{m+2} + (1-\cos 2\theta_0^i)I^i_{m+4}\right]\right\}
\end{aligned} \tag{5.106}
$$

$$
\begin{aligned}
I_2^N &= \int_{\Delta S} uv\rho^m \,\mathrm{d}S \\
&= \sum_{i=1}^{3}\int_{\theta_1^i}^{\theta_2^i}\mathrm{d}\theta\cos\theta\sin\theta\int_0^{d_\theta^i}\rho^{m+3}\,\mathrm{d}\rho \\
&= \sum_{i=1}^{3}\left\{\frac{d_i^{m+4}}{2(m+4)}\left[\sin 2\theta_0^i(2I^i_{m+2} - I^i_{m+4}) + \frac{2\cos 2\theta_0^i}{m+2}f^i_{m+2}\right]\right\}
\end{aligned} \tag{5.107}
$$

$$
\begin{aligned}
I_4^N &= \int_{\Delta S} (-w_0 u)\rho^m \,\mathrm{d}S \\
&= \sum_{i=1}^{3}(-w_0)\int_{\theta_1^i}^{\theta_2^i}\mathrm{d}\theta\cos\theta\int_0^{d_\theta^i}\rho^{m+2}\,\mathrm{d}\rho \\
&= \sum_{i=1}^{3}\left\{\frac{-w_0 d_i^{m+3}}{2(m+3)}\left[\cos\theta_0^i I^i_{m+2} - \frac{\sin\theta_0^i}{m+2}f^i_{m+2}\right]\right\}
\end{aligned}
$$

with

$$
f^i_{m+2} = \frac{1}{\cos^{m+2}\theta^i_{20}} - \frac{1}{\cos^{m+2}\theta^i_{10}}. \tag{5.108}
$$

We omit I_3^N and I_5^N because they are similar to I_1^N and I_4^N, respectively. Note that $I_4^N = I_5^N = 0$ when $w_0 \to 0$ in fact.

5.4.4 Numerical Examples

We present several numerical examples to demonstrate the effectiveness of the developed approach. We consider the EM scattering by different conducting scatterers as shown in Figure 5.15 and the incident wave is a plane wave which has a frequency of $f = 300$ MHz and propagates along the $-z$ direction. In the first example, the scatterer is a conducting sphere as shown in Figure 5.15(a) and it has a radius of $a = 0.1\lambda$, where λ denotes the free-space wavelength. Figure 5.16 plots the induced current density distribution, which is characterized by J_θ and J_ϕ components along the principal cut ($\phi = 0°$ and $\theta = 0° - 180°$) for vertical and horizontal polarizations, respectively. They are very close to the exact solution obtained from the Mie series formula. Note that the locations of the displayed current densities are at the centroids of triangular patches closest to the principal cut and may not be uniform.

In the following examples, the scatterers are the conducting cube, conducting thin disk, and conducting thin diamond, which are selected from [35, 36], and they are sketched in Figure 5.15(b), (c), (d), respectively. We calculate the bistatic radar cross section (RCS) along the principal cut for these scatterers in both vertical (VV) and horizonal (HH) polarizations. Figure 5.17 shows the solutions of the cube with a side length of $s = 0.75\lambda$ and Figure 5.18 plots the solutions of the thin disk with a radius of $a = 1.0\lambda$ and a thickness of $t = 0.05\lambda$. Also, Figure 5.19 illustrates the solutions of the thin diamond with a half side length of $a = 0.4\lambda$ and a half central height of $h = 0.02\lambda$. We compare these solutions with those from the MoM with the RWG basis and testing functions and they are in a good agreement.

5.5 Different Scheme for Evaluating Strongly-Singular and Hypersingular Integrals Over Triangular Patches

The SIEs include both $\mathcal{L}$ and $\mathcal{K}$ operators if the objects are penetrable or include one of them if impenetrable. As we have seen, the kernel of the $\mathcal{L}$ operator is the dyadic Green's function, which includes a double gradient operation on the scalar Green's function and

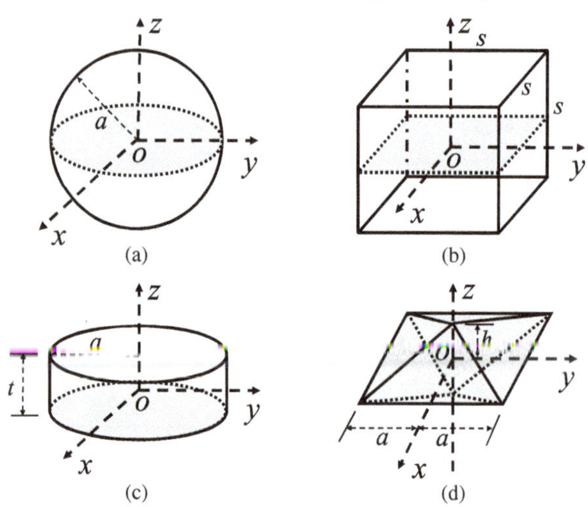

Figure 5.15 Geometries of scatterers. (a) A conducting sphere. (b) A conducting cube. (c) A conducting thin disk. (d) A conducting thin diamond.

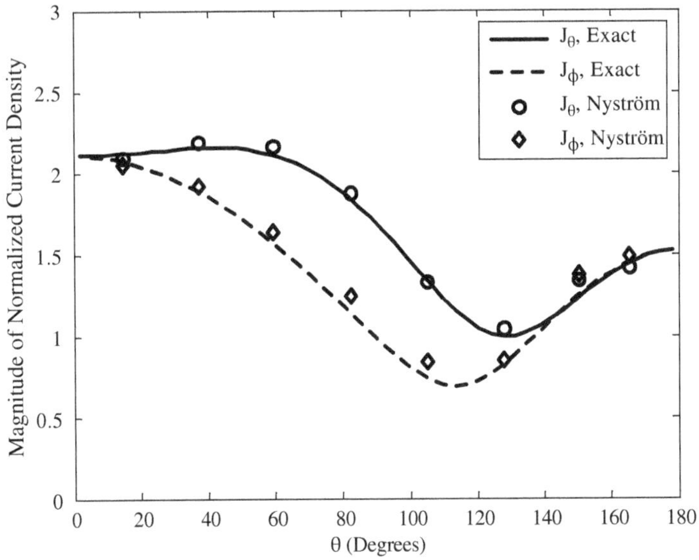

Figure 5.16 Current density distribution along the principal cut for EM scattering by a conducting sphere in vertical and horizonal polarizations, $\theta_i = 0^\circ$ and $\phi_i = 0^\circ$.

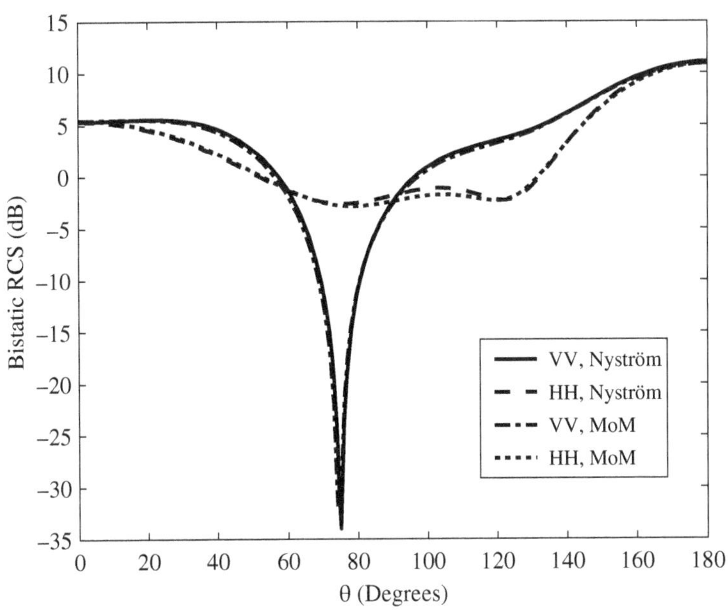

Figure 5.17 Bistatic RCS solutions for EM scattering by a conducting cube with a side length of 0.75λ in vertical and horizontal polarizations, $\theta_i = 0^\circ$ and $\phi_i = 0^\circ$.

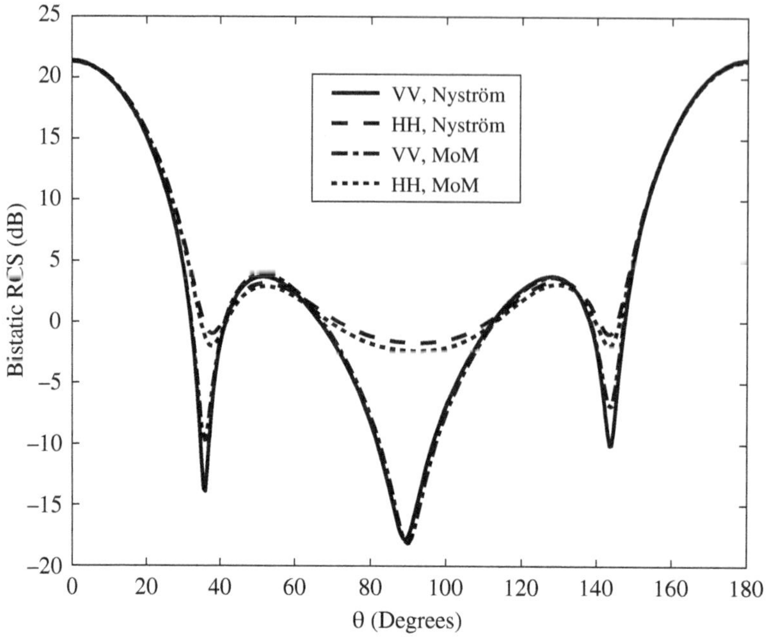

Figure 5.18 Bistatic RCS solutions for EM scattering by a conducting thin disk with a radius of 1.0λ and a thickness of 0.05λ in vertical and horizontal polarizations, $\theta_i = 0°$ and $\phi_i = 0°$.

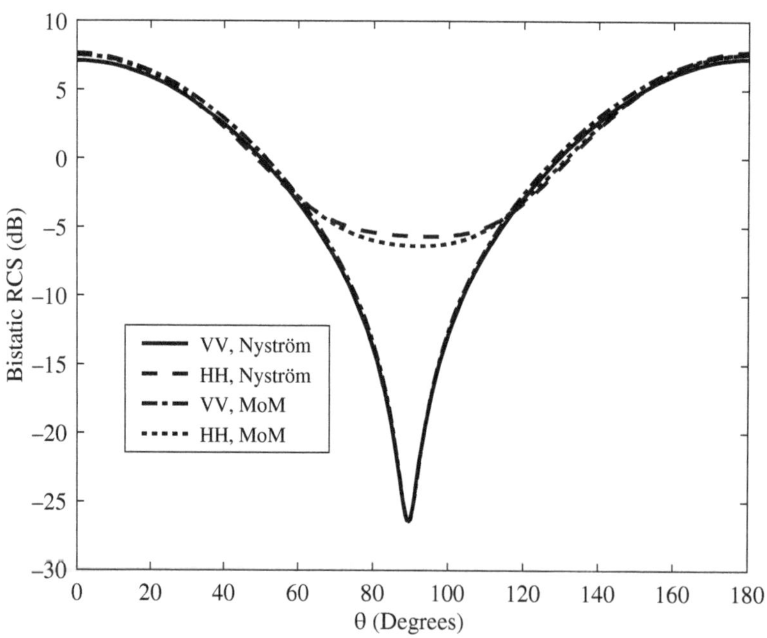

Figure 5.19 Bistatic RCS solutions for EM scattering by a conducting thin diamond with a half side length of 0.4λ and a half central height of 0.02λ in vertical and polarizations, $\theta_i = 0°$ and $\phi_i = 0°$.

results in $1/R^3$ hypersingular integrals (HSIs) in evaluating matrix entries, where R is the distance between a source point and an observation point or field point. However, the HSIs could be reduced to $1/R$ weakly-singular integrals (WSIs) in the MoM solution if a divergence conforming basis function such as the RWG basis function [3] is used as an expansion and testing function. Without the help of a RWG-like basis function, we must carefully handle the HSIs and this happens in the implementation of the Nyström method or boundary element method (BEM) because they do not use any particularly-designed basis or testing function. In the $\mathcal{K}$ operator, the kernel is a single gradient operator on the scalar Green's function, yielding $1/R^2$ strongly-singular integrals (SSIs) in the matrix elements. Note that we have adopted the convention of mechanical engineering [40] and mathematics [41] to categorize the degrees of singularity in the above. This convention could be different from the convention of EM community in which the $1/R^2$ singular integrals are called HSIs sometimes [20], and also different from the definition in [42] where HSIs are named Cauchy singular integrals and SSIs are called WSIs. The SSIs always exist in the $\mathcal{K}$ operator, even in the MoM implementation. They could also exist in the $\mathcal{L}$ operator in the MoM when the RWG-like basis function represents both the electric and magnetic current densities in penetrable structures and the $\hat{n}\times$ RWG basis function has to be used as a testing function for producing a well-conditioned impedance matrix [43]. The accurate and efficient evaluation for the HSIs and SSIs is essential in solving SIEs because they have a significant impact on the numerical solutions and we must treat them wisely.

There have been many publications dealing with the singular integrals for EM integral equations [5–23], but they have mainly focused on the WSIs and SSIs. For the HSIs, we first presented a systematic evaluation scheme as shown in the last section [19]. Although the formulas in those publications are effective, there is certain inconvenience in their implementations due to the introduction of a polar coordinate system and extra coordinate transformation. Later on, we developed a different or novel scheme for evaluating those HSIs and SSIs and it will be introduced in this section. The new scheme is based on the Stokes' theorem [49], which changes a surface integral over a surface patch into a line integral along its boundary. We assume that the surface patches are the flat triangular patches that are most widely used in geometrical discretization. To derive the HSIs and SSIs, we also assume that the observation point is initially off the source patch, so the corresponding integrals are non-singular or regular. We then derive analytical solutions for the line integrals after applying the Stokes' theorem and take a limit of the derived formulas by letting the observation point approach the source patch to obtain the solutions of the HSIs and SSIs. This is also known as the CPV-like approach for handling the HSIs and SSIs. Compared with other approaches, the new approach is simpler and the resultant formulas are more convenient in implementation since no polar coordinates and extra coordinate transformation are involved. Numerical experiments are presented to demonstrate the validity of the new approach by comparing its results with those from our previous approach as presented in the last section. We also show several numerical examples to illustrate the applications of the new formulas to solve EM problems.

5.5.1 Strongly-Singular and Hypersingular Integrals

Although the strongly-singular and hypersingular integrals have been presented in the previous sections, they are expressed in a local coordinate system whose origin is the projection of observation point on the source patch plane. In fact, those integrals can be written in a global coordinate system which gives a bigger picture and their formulas can be more conveniently used in evaluating near-singular integrals which will be addressed later. In a global coordinate system $\{x, y, z\}$, the double gradient of scalar Green's function takes the following form

$$\nabla\nabla g(\mathbf{r}, \mathbf{r}') = g_{xx}\hat{x}\hat{x} + g_{yx}\hat{x}\hat{y} + g_{zx}\hat{x}\hat{z} + g_{xy}\hat{y}\hat{x} + g_{yy}\hat{y}\hat{y} + g_{zy}\hat{y}\hat{z} + g_{xz}\hat{z}\hat{x} + g_{yz}\hat{z}\hat{y} + g_{zz}\hat{z}\hat{z} \tag{5.109}$$

where

$$g_{st} = \frac{\partial^2 g}{\partial s \partial t}, \quad s = x, y, z; \quad t = x, y, z \tag{5.110}$$

and only six components are independent due to its symmetry. The singularity subtraction for the above integrand can be illustrated by using the first component g_{xx} as follows

$$\begin{aligned} I_{xx} &= \int_{\Delta S} g_{xx}\, \mathrm{d}S' = \left[Q(x-x')^2 + \frac{\mathrm{i}kR-1}{R^3}\right] \mathrm{e}^{\mathrm{i}kR}\, \mathrm{d}S' \\ &= \int_{\Delta S} \left\{ \left[Q(x-x')^2 + \frac{\mathrm{i}kR-1}{R^3}\right] \mathrm{e}^{\mathrm{i}kR} - \frac{k^2(x-x')^2}{2R^3} - \frac{k^4(x-x')^2}{8R} + \frac{k^2}{2R} \right. \\ &\quad \left. - \left[\frac{3(x-x')^2}{R^5} - \frac{1}{R^3}\right] \right\} \mathrm{d}S' + \int_{\Delta S} \left[\frac{k^2(x-x')^2}{2R^3} + \frac{k^4(x-x')^2}{8R} - \frac{k^2}{2R}\right] \mathrm{d}S' \\ &\quad + \int_{\Delta S} \left[\frac{3(x-x')^2}{R^5} - \frac{1}{R^3}\right] \mathrm{d}S' \end{aligned} \tag{5.111}$$

where (x, y, z) is an observation point and (x', y', z') is a source point within a source patch ΔS. Also, $R = \sqrt{(x-x')^2 + (y-y')^2 + (z-z')^2}$ is the distance between the observation point and source point and $Q = (3 - 3\mathrm{i}kR - k^2R^2)/R^5$. Obviously, the third integral in the above

$$I_1 = \int_{\Delta S} \left[\frac{3(x-x')^2}{R^5} - \frac{1}{R^3}\right] \mathrm{d}S' \tag{5.112}$$

is a HSI which we want to evaluate. From other components in (5.109), we can find other HSIs by the singularity subtraction as follows

$$I_2 = \int_{\Delta S} \frac{(x-x')(y-y')}{R^5}\, \mathrm{d}S'; \quad I_3 = \int_{\Delta S} \left[\frac{3(y-y')^2}{R^5} - \frac{1}{R^3}\right] \mathrm{d}S'; \tag{5.113}$$

$$I_4 = \int_{\Delta S} \frac{(z-z')(x-x')}{R^5}\, \mathrm{d}S'; \quad I_5 = \int_{\Delta S} \frac{(y-y')(z-z')}{R^5}\, \mathrm{d}S'; \tag{5.114}$$

$$I_6 = \int_{\Delta S} \left[\frac{3(z-z')^2}{R^5} - \frac{1}{R^3}\right] \mathrm{d}S'. \tag{5.115}$$

For the kernel $\nabla g(\mathbf{r}, \mathbf{r}')$ in the $\mathcal{K}$ operator

$$\begin{aligned} \nabla g(\mathbf{r}, \mathbf{r}') &= g_x\hat{x} + g_y\hat{y} + g_z\hat{z} \\ &= \frac{\partial g}{\partial x}\hat{x} + \frac{\partial g}{\partial y}\hat{y} + \frac{\partial g}{\partial z}\hat{z} \end{aligned} \tag{5.116}$$

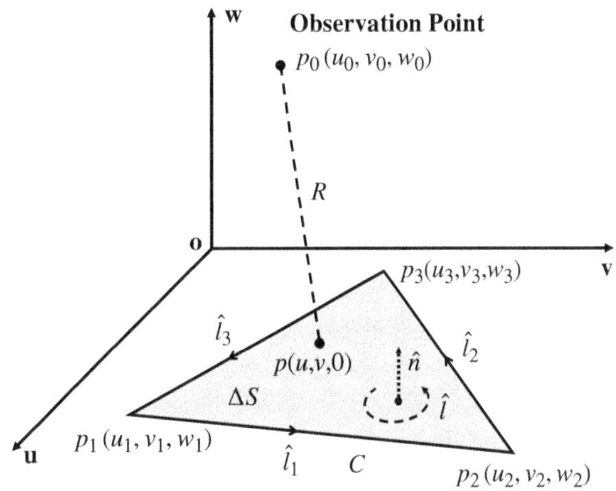

Figure 5.20 A local coordinate system (u, v, w) whose uv plane coincides with the flat triangular patch and the w axis is parallel to the unit normal vector of the patch.

we can find the SSIs by the singularity subtraction in a similar way, i.e.

$$I_7 = \int_{\Delta S} \frac{(x-x')}{R^3} \, dS'; \quad I_8 = \int_{\Delta S} \frac{(y-y')}{R^3} \, dS'; \tag{5.117}$$

$$I_9 = \int_{\Delta S} \frac{(z-z')}{R^3} \, dS'. \tag{5.118}$$

To derive the formulas of evaluating the above HSIs and SSIs, we still need to introduce a local coordinate system $\{u, v, w\}$ as sketched in Figure 5.20, but its origin is not the projection of observation point on the source patch plane now. Nevertheless, the flat source triangular patch plane is chosen as the uv plane and the w axis is parallel to the unit normal vector of the patch, which points to the side of observation point. The coordinate system is chosen so because we want to apply the Stokes' theorem to derive the formulas of evaluating the HSIs and SSIs. In such a local coordinate system, the observation point is located at (u_0, v_0, w_0), the source point is located at $(u, v, 0)$ within the triangular patch, and $R = \sqrt{(u_0 - u)^2 + (v_0 - v)^2 + w_0^2}$. The expressions of the above HSIs and SSIs should also be changed by replacing $x - x'$ with $u_0 - u$, $y - y'$ with $v_0 - v$, $z - z'$ with w_0, and dS' with dS. Once the formulas in the local coordinate system are found, their expressions in the global coordinate system can be found conveniently by a coordinate transformation.

5.5.2 Stokes' Theorem

The existing techniques of evaluating singular integrals are mostly based on the multiple integrals for surface integrations [33]. We present a different approach here to derive the HSIs and SSIs based on the Stokes' theorem, which changes a surface integral over a patch into a line integral along its boundary. This change can dramatically simplify the derivation of analytical solutions for the singular integrals. The Stokes' theorem can be written as [49]

$$\int_{\Delta S} (\nabla \times \mathbf{A}) \cdot \hat{n} \, dS = \oint_C \mathbf{A} \cdot \, d\mathbf{l} \tag{5.119}$$

where $\mathbf{A} = A_u\hat{u} + A_v\hat{v} + A_w\hat{w}$ is a vector defined over a surface patch ΔS, $\hat{n}$ is the unit normal vector on ΔS, C is the boundary of ΔS, and $\hat{l}$ is the unit tangential vector along the boundary C, see Figure 5.20. For a flat patch, $\hat{n} = n_u\hat{u} + n_v\hat{v} + n_w\hat{w}$ is a constant vector and $\mathbf{dl} = \mathrm{d}u\hat{u} + \mathrm{d}v\hat{v} + \mathrm{d}w\hat{w}$, so we have

$$\int_{\Delta S}\left[\left(\frac{\partial A_w}{\partial v} - \frac{\partial A_v}{\partial w}\right)n_u + \left(\frac{\partial A_u}{\partial w} - \frac{\partial A_w}{\partial u}\right)n_v + \left(\frac{\partial A_v}{\partial u} - \frac{\partial A_u}{\partial v}\right)n_w\right]\mathrm{d}S$$
$$= \int_{\Delta S}\left[\left(\frac{\partial A_u}{\partial w}n_v - \frac{\partial A_u}{\partial v}n_w\right) + \left(\frac{\partial A_v}{\partial u}n_w - \frac{\partial A_v}{\partial w}n_u\right) + \left(\frac{\partial A_w}{\partial v}n_u - \frac{\partial A_w}{\partial u}n_v\right)\right]\mathrm{d}S$$
$$= \oint_C (A_u\mathrm{d}u + A_v\mathrm{d}v + A_w\mathrm{d}w). \tag{5.120}$$

From the above equation, we can prove that [49]

$$\int_{\Delta S}\left(\frac{\partial A_u}{\partial w}n_v - \frac{\partial A_u}{\partial v}n_w\right)\mathrm{d}S = \oint_C A_u\mathrm{d}u \tag{5.121}$$
$$\int_{\Delta S}\left(\frac{\partial A_v}{\partial u}n_w - \frac{\partial A_v}{\partial w}n_u\right)\mathrm{d}S = \oint_C A_v\mathrm{d}v \tag{5.122}$$
$$\int_{\Delta S}\left(\frac{\partial A_w}{\partial v}n_u - \frac{\partial A_w}{\partial u}n_v\right)\mathrm{d}S = \oint_C A_w\mathrm{d}w. \tag{5.123}$$

If we choose the patch plane as the uv plane and $\hat{w} = \hat{n}$, then we have $n_u = 0$, $n_v = 0$, and $n_w = 1$, and the above equations can be simplified as

$$\int_{\Delta S}\frac{\partial A_u}{\partial v}\mathrm{d}S = -\oint_C A_u\mathrm{d}u \tag{5.124}$$
$$\int_{\Delta S}\frac{\partial A_v}{\partial u}\mathrm{d}S = \oint_C A_v\mathrm{d}v \tag{5.125}$$
$$\oint_C A_w dw = 0. \tag{5.126}$$

The first two equations in the above can be used to evaluate the HSIs and SSIs.

5.5.3 Derivation of New Formulas for HSIs and SSIs

Before deriving the formulas for evaluating the HSIs and SSIs, we have to derive the corresponding formulas for those integrals when the observation point p_0 is not inside the source triangular patch $\Delta p_1p_2p_3$, see Figure 5.20. The HSIs and SSIs in the local coordinate system take the following forms and we add a prime to distinguish them from those in the previous sections

$$I_1' = \int_{\Delta S}\frac{(u_0-u)^2}{R^5}\mathrm{d}S;\quad I_2' = \int_{\Delta S}\frac{(v_0-v)^2}{R^5}\mathrm{d}S;\quad I_3' = \int_{\Delta S}\frac{(u_0-u)(v_0-v)}{R^5}\mathrm{d}S; \tag{5.127}$$

$$I_4' = \int_{\Delta S}\frac{(u_0-u)}{R^5}\mathrm{d}S;\quad I_5' = \int_{\Delta S}\frac{(v_0-v)}{R^5}\mathrm{d}S;\quad I_6' = \int_{\Delta S}\frac{1}{R^5}\mathrm{d}S; \tag{5.128}$$

$$I_7' = \int_{\Delta S}\frac{(u_0-u)}{R^3}\mathrm{d}S;\quad I_8' = \int_{\Delta S}\frac{(v_0-v)}{R^3}\mathrm{d}S;\quad I_9' = \int_{\Delta S}\frac{1}{R^3}\mathrm{d}S. \tag{5.129}$$

From the above integrals, the integrals in Equations (5.110)–(5.113) and (5.115)–(5.116) can be obtained by a suitable combination. The above integrals can be derived by using the two transformation equations in (5.122) and (5.123). With the help of integral tables in finding the original functions A_u or A_v [37], we can obtain

$$\begin{aligned} I_1' &= \oint_C \frac{(u_0 - u)^3}{3\left[(v_0 - v)^2 + w_0^2\right] R^3}\, dv \\ &= \sum_{i=1}^{3} \left\{ \frac{1}{w_0} \left[\tan^{-1}(t_1^i) - \tan^{-1}(t_2^i)\right] + \frac{2}{3w_0} \left[-\tan^{-1}(t_3^i) + \tan^{-1}(t_4^i)\right] \right. \\ &\quad \left. + \frac{1}{3} \left[s_1^i - \frac{s_2^i}{l_i}(v_{i+1} - v_i)^2 \right] \right\} \end{aligned} \tag{5.130}$$

$$\begin{aligned} I_2' &= -\oint_C \frac{(v_0 - v)^3}{3\left[(u_0 - u)^2 + w_0^2\right] R^3}\, du \\ &= \sum_{i=1}^{3} \left\{ \frac{2}{3w_0} \left[-\tan^{-1}(t_1^i) + \tan^{-1}(t_2^i)\right] + \frac{1}{w_0} \left[\tan^{-1}(t_3^i) - \tan^{-1}(t_4^i)\right] \right. \\ &\quad \left. - \frac{1}{3} \left[s_1^i + \frac{s_2^i}{l_i}(u_{i+1} - u_i)^2 \right] \right\} \end{aligned} \tag{5.131}$$

$$\begin{aligned} I_3' &= \oint_C \frac{v_0 - v}{3R^3}\, dv \\ &= \frac{1}{3} \sum_{i=1}^{3} \left\{ \frac{v_{i+1} - v_i}{l_i^2 r_i^2 - b_i^2} \left[(v_0 - v_i) \left(\frac{l_i^2 + b_i}{r_{i+1}} - \frac{b_i}{r_i} \right) + (v_{i+1} - v_i) \left(\frac{b_i + r_i^2}{r_{i+1}} - r_i \right) \right] \right\} \end{aligned} \tag{5.132}$$

$$\begin{aligned} I_4' &= \oint_C \frac{1}{3R^3}\, dv \\ &= \frac{1}{3} \sum_{i=1}^{3} \left\{ \frac{v_{i+1} - v_i}{l_i^2 r_i^2 - b_i^2} \left(\frac{l_i^2 + b_i}{r_{i+1}} - \frac{b_i}{r_i} \right) \right\} \end{aligned} \tag{5.133}$$

$$\begin{aligned} I_5' &= -\oint_C \frac{1}{3R^3}\, du \\ &= -\frac{1}{3} \sum_{i=1}^{3} \left\{ \frac{u_{i+1} - u_i}{l_i^2 r_i^2 - b_i^2} \left(\frac{l_i^2 + b_i}{r_{i+1}} - \frac{b_i}{r_i} \right) \right\} \end{aligned} \tag{5.134}$$

$$\begin{aligned} I_6' &= -\oint_C \frac{(u_0 - u)}{R\left[(v_0 - v)^2 + w_0^2\right]^2} \left[1 - \frac{(u_0 - u)^2}{3R^2} \right] dv \\ &= \sum_{i=1}^{3} \left\{ \frac{1}{6w_0^3} \left[\tan^{-1}(t_1^i) - \tan^{-1}(t_2^i) + \tan^{-1}(t_3^i) - \tan^{-1}(t_4^i)\right] + \frac{l_i s_2^i}{3w_0^2} \right\} \end{aligned} \tag{5.135}$$

$$\begin{aligned} I_7' &= \oint_C \frac{1}{R}\, dv \\ &= \sum_{i=1}^{3} \left\{ \frac{v_{i+1} - v_i}{l_i} \ln \left| \frac{l_i r_{i+1} + l_i^2 + b_i}{l_i r_i + b_i} \right| \right\} \end{aligned} \tag{5.136}$$

$$I_8' = -\oint_C \frac{1}{R}\, du$$

$$
= \sum_{i=1}^{3} \left\{ \frac{u_{i+1} - u_i}{l_i} \ln \left| \frac{l_i r_i + b_i}{l_i r_{i+1} + l_i^2 + b_i} \right| \right\} \tag{5.137}
$$

$$
I_9' = -\oint_C \frac{(u_0 - u)}{\left[(v_0 - v)^2 + w_0^2\right] R} \, dv
$$

$$
= \sum_{i=1}^{3} \left\{ \frac{1}{2w_0} \left[\tan^{-1}(t_1^i) - \tan^{-1}(t_2^i) + \tan^{-1}(t_3^i) - \tan^{-1}(t_4^i)\right] \right\} \tag{5.138}
$$

where

$$
l_i = \sqrt{(u_{i+1} - u_i)^2 + (v_{i+1} - v_i)^2} \tag{5.139}
$$

$$
r_i = \sqrt{(u_0 - u_i)^2 + (v_0 - v_i)^2 + w_0^2} \tag{5.140}
$$

$$
a_i = (u_i - u_0)(v_{i+1} - v_i) - (v_i - v_0)(u_{i+1} - u_i) \tag{5.141}
$$

$$
b_i = (u_{i+1} - u_i)(u_i - u_0) + (v_{i+1} - v_i)(v_i - v_0) \tag{5.142}
$$

$$
c_i = (u_{i+1} - u_i)(u_{i+1} - u_0) + (v_{i+1} - v_i)(v_{i+1} - v_0) \tag{5.143}
$$

$$
r_{i+1} = \sqrt{l_i^2 + 2b_i + r_i^2} = \sqrt{(u_0 - u_{i+1})^2 + (v_0 - v_{i+1})^2 + w_0^2} \tag{5.144}
$$

$$
t_1^i = -\frac{(u_i - u_0)a_i + (v_{i+1} - v_i)w_0^2}{(u_{i+1} - u_i)r_i w_0} \tag{5.145}
$$

$$
t_2^i = -\frac{(u_{i+1} - u_0)a_i + (v_{i+1} - v_i)w_0^2}{(u_{i+1} - u_i)r_{i+1} w_0} \tag{5.146}
$$

$$
t_3^i = -\frac{(v_i - v_0)a_i - (u_{i+1} - u_i)w_0^2}{(v_{i+1} - v_i)r_i w_0} \tag{5.147}
$$

$$
t_4^i = -\frac{(v_{i+1} - v_0)a_i - (u_{i+1} - u_i)w_0^2}{(v_{i+1} - v_i)r_{i+1} w_0} \tag{5.148}
$$

$$
s_1^i = \frac{(u_{i+1} - u_i)(v_{i+1} - v_i)(r_i - r_{i+1})}{r_i r_{i+1} l_i^2} \tag{5.149}
$$

$$
s_2^i = \frac{a_i(c_i r_i - b_i r_{i+1})}{r_i r_{i+1}(a_i^2 + l_i^2 w_0^2)} \tag{5.150}
$$

and it is assumed that $i + 1 = \mathrm{mod}(i, 3) + 1$. Note that we can derive the formulas for more integrals, such as for the WSIs as shown in Equations (5.13)–(5.21) by using this approach.

The derived formulas are effective no matter how close the observation point is to the source triangular patch and they can be used to accurately evaluate the near-interaction matrix elements. When the observation point is inside the source patch, i.e. for the HSIs and SSIs, we only need to take a limit of those formulas by letting $w_0 \to 0$. By doing so, we can find the solutions for the integrals in Equations (5.110)–(5.113) and (5.115)–(5.116) as

$$
I_1^L = \lim_{w_0 \to 0} \int_{\Delta S} \left[\frac{3(u_0 - u)^2}{R^5} - \frac{1}{R^3} \right] dS
$$

$$= \sum_{i=1}^{3} \left[h_1^i - \frac{h_2^i}{l_i}(v_{i+1} - v_i)^2 \right] \tag{5.151}$$

$$I_2^L = \lim_{w_0 \to 0} \int_{\Delta S} \frac{(u_0 - u)(v_0 - v)}{R^5} \, \mathrm{d}S$$

$$= \frac{1}{3} \sum_{i=1}^{3} \left\{ \frac{v_{i+1} - v_i}{l_i^2 d_i^2 - b_i^2} \left[(v_0 - v_i) \left(\frac{l_i^2 + b_i}{d_{i+1}} - \frac{b_i}{d_i} \right) + (v_{i+1} - v_i) \left(\frac{b_i + d_i^2}{d_{i+1}} - d_i \right) \right] \right\} \tag{5.152}$$

$$I_3^L = \lim_{w_0 \to 0} \int_{\Delta S} \left[\frac{3(v_0 - v)^2}{R^5} - \frac{1}{R^3} \right] \mathrm{d}S$$

$$= -\sum_{i=1}^{3} \left[h_1^i + \frac{h_2^i}{l_i}(u_{i+1} - u_i)^2 \right] \tag{5.153}$$

$$I_4^L = \lim_{w_0 \to 0} \int_{\Delta S} \frac{w_0(u_0 - u)}{R^5} \, \mathrm{d}S$$

$$= 0 \tag{5.154}$$

$$I_5^L = \lim_{w_0 \to 0} \int_{\Delta S} \frac{w_0(v_0 - v)}{R^5} \, \mathrm{d}S$$

$$= 0 \tag{5.155}$$

$$I_6^L = \lim_{w_0 \to 0} \int_{\Delta S} \left[\frac{3w_0^2}{R^5} - \frac{1}{R^3} \right] \mathrm{d}S$$

$$= \sum_{i=1}^{3} (l_i h_2^i) \tag{5.156}$$

$$I_7^L = \lim_{w_0 \to 0} \int_{\Delta S} \frac{(u_0 - u)}{R^3} \, \mathrm{d}S$$

$$= \sum_{i=1}^{3} \left\{ \frac{v_{i+1} - v_i}{l_i} \ln \left| \frac{l_i d_{i+1} + l_i^2 + b_i}{l_i d_i + b_i} \right| \right\} \tag{5.157}$$

$$I_8^L = \lim_{w_0 \to 0} \int_{\Delta S} \frac{(v_0 - v)}{R^3} \, \mathrm{d}S$$

$$= \sum_{i=1}^{3} \left\{ \frac{u_{i+1} - u_i}{l_i} \ln \left| \frac{l_i d_i + b_i}{l_i d_{i+1} + l_i^2 + b_i} \right| \right\} \tag{5.158}$$

$$I_9^L = \lim_{w_0 \to 0} \int_{\Delta S} \frac{w_0}{R^3} \, \mathrm{d}S$$

$$= 0 \tag{5.159}$$

where

$$d_i = \sqrt{(u_0 - u_i)^2 + (v_0 - v_i)^2} \tag{5.160}$$

$$h_1^i = \frac{(u_{i+1} - u_i)(v_{i+1} - v_i)(d_i - d_{i+1})}{d_i d_{i+1} l_i^2} \tag{5.161}$$

$$h_2^i = \frac{(c_i r_i - b_i r_{i+1})}{a_i d_i d_{i+1}} \tag{5.162}$$

and the superscript L denotes the limit.

5.5.4 Numerical Tests

To examine the effectiveness and accuracy of the above formulas, we perform some numerical tests by comparing the results from the new formulas with the results from the numerical integrations and the results from our previous or old formulas as shown in the last section [19]. The source triangular patch in the tests is defined by randomly choosing three vertices $P_1(0.7, 0.1, 0)$, $P_2(0.4, 0.3, 0)$ and $P_3(0.1, 0.2, 0)$, and the observation point is located at $P_0(0.4, 0.2, w_0)$, where w_0 can be changed and the projection of observation point is inside the source triangular patch. The numerical integrations are performed with a 24-point product Gaussian quadrature rule. Tables 5.1, 5.2 and 5.3 show the results for the integrals $-(3I_1' - I_9')$, $-I_2'$ and I_8', respectively, when the observation point is approaching the source triangular patch (the minus sign is added in the first two integrals so that their values are mostly positive in the tables and the width of tables can be reduced). It can be seen that the results of the new formulas are almost the same (first 14 digits) as those of our previous formulas no matter how close the observation point is to the source patch. However, the results by numerical integrations are only the same as those of the analytical formulas when the observation points are far away from the source patch as expected and there is a big difference between them when the observation point is close to the source patch, indicating that the results of numerical integrations are incorrect and they cannot be used.

5.5.5 Numerical Examples

We use the derived formulas to evaluate the HSIs or SSIs in solving the EFIEs via the Nyström method for EM scattering by PEC, dielectric, or composite objects. Both $\mathcal{L}$ and $\mathcal{K}$ operators appear in the EFIEs for the scattering by dielectric or composite objects, so all those formulas will be used in the solutions. The geometry of objects is chosen to be spherical so that the available analytical solutions or Mie series solutions can be used for comparison. It is assumed that the incident wave has a frequency $f = 300$ MHz and

Table 5.1 Comparison between the results of new formula and those of old formula and numerical integration for $-(3I_1' - I_9')$.

w_0	Numerical	New formula	Old formula
1.0	0.04207506901143	0.04207506901143	0.04207506901143
0.5	0.28270434706987	0.28270434706987	0.28270434706987
0.2	2.13621444759592	2.13621444759785	2.13621444759784
0.1	5.76804426575391	5.76805036402819	5.76805036402820
0.05	9.78504726946795	9.82558791932322	9.82558791932321
0.01	−213.34976469790	12.9739082926520	12.9739082926520
0.001	−746.80626185120	13.1603576935886	13.1603576935886
0	−758.35676592578	13.1622776601684	13.1622776601684

Table 5.2 Comparison between the results of new formula and those of old formula and numerical integration for $-I'_2$.

w_0	Numerical	New formula	Old formula
1.0	0.00010219825220	0.00010219825220	0.00010219825220
0.5	0.00254755788576	0.00254755788576	0.00254755788576
0.2	0.08688236050388	0.08688236050398	0.08688236050398
0.1	0.54969556889504	0.54969625809616	0.54969625809616
0.05	1.60691429859838	1.60941914276084	1.60941914276083
0.01	3.25565772338300	3.05001385461441	3.05001385461442
0.001	5.64047815442650	3.16111509468861	3.16111509468861
0	5.71007197111685	3.16227766016838	3.16227766016839

Table 5.3 Comparison between the results of new formula and those of old formula and numerical integration for I'_8.

w_0	Numerical	New formula	Old formula
1.0	0.00001876773386	0.00001876773386	0.00001876773386
0.5	0.00049038289149	0.00049038289149	0.00049038289149
0.2	0.01910021007892	0.01910021007891	0.01910021007891
0.1	0.13090134214621	0.13090132308192	0.13090132308192
0.05	0.39011701056815	0.39002750128922	0.39002750128922
0.01	0.86082762832819	0.71607179698974	0.71607179698974
0.001	1.08910451047110	0.73936584680719	0.73936584680719
0	1.09305561678139	0.73960807782012	0.73960807782012

is propagating along $-z$ direction in free space in all examples. We then calculate the bistatic RCS observed along the principal cut ($\phi = 0°$ and $\theta = 0°–180°$) for the scatterers with both vertical (VV) and horizontal (HH) polarizations. The solutions are compared with the analytical Mie-series solutions which can be found in [24]. The first example is the scattering by a PEC sphere with a radius $a = 0.25\lambda$ and we discretize its surface into 2146 triangular patches. The bistatic RCS solutions are shown in Figure 5.21 and they are close to the corresponding analytical solutions.

The second example is the scattering by a dielectric sphere with a radius $a = 0.25\lambda$ and a relative permittivity $\varepsilon_r = 4.0$ (the relative permeability $\mu_r = 1.0$ is assumed). We use the same geometrical discretization as for the PEC sphere and the bistatic RCS solutions are plotted in Figure 5.22. It can be seen that the solutions are also close to the corresponding analytical solutions. The third example illustrates the scattering by a composite sphere, i.e. a PEC sphere fully coated with one-layer dielectric material. The radii of the inner and outer interfaces are $a_1 = 0.2\lambda$ and $a_2 = 0.25\lambda$, respectively, and the relative permittivity of the coating is $\varepsilon_r = 4.0$. The discretization for the outer interface is the same as for the PEC sphere and the inner interface is meshed into 1782

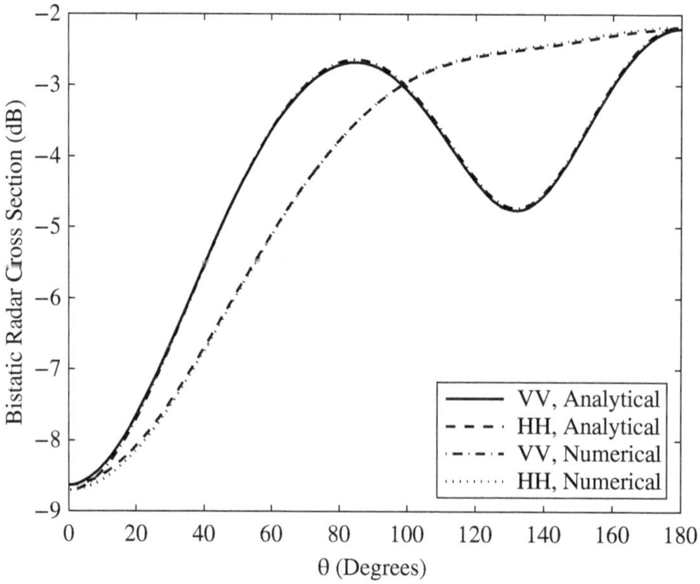

Figure 5.21 Bistatic RCS solutions for a PEC sphere with a radius $a = 0.25\lambda$.

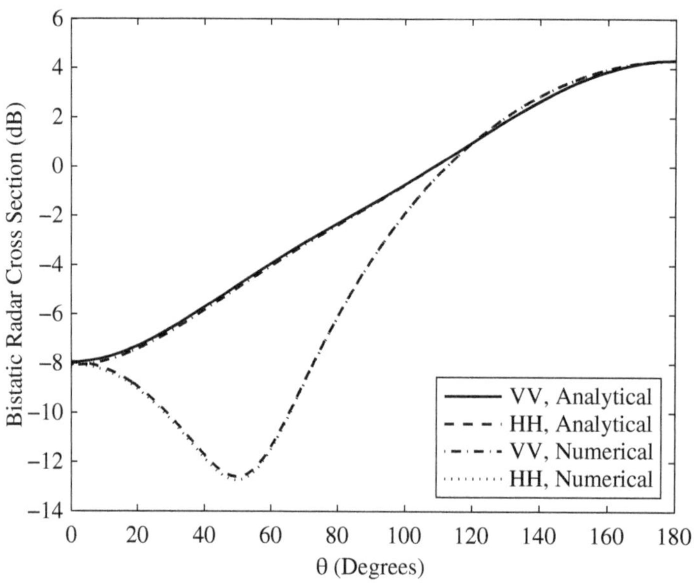

Figure 5.22 Bistatic RCS solutions for a dielectric sphere with a radius $a = 0.25\lambda$ and a relative permittivity $\varepsilon_r = 4.0$.

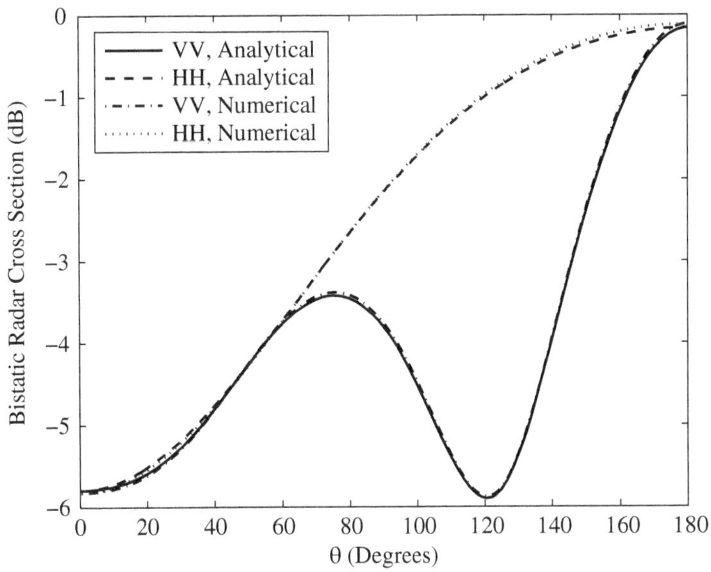

Figure 5.23 Bistatic RCS solutions for a PEC sphere with a one-layer full dielectric coating. The inner radius is $a_1 = 0.2\lambda$, the outer radius is $a_2 = 0.25\lambda$, and the relative permittivity of coating material is $\varepsilon_r = 4.0$.

triangular patches. The bistatic RCS solutions are sketched in Figure 5.23 and again the solutions agree well with the corresponding analytical counterparts.

5.6 Evaluation of Singular Integrals Over Volume Domains

To solve volume integral equations (VIEs), we have to efficiently and accurately evaluate singular integrals over volume domains which are usually tetrahedral elements. The singular integrals over tetrahedral elements were already shown in the previous section and they are similar to those over triangular patches in the form. Since the integrals are over a volume domain, they can produce two zero points in the Jacobian when a variable change is taken, so the $1/R^2$ strongly-singular integrals can also be handled by a singularity cancellation method like the Duffy's method. For $1/R^3$ hypersingular integrals, their degree of singularity is equivalent to that of $1/R^2$ strongly-singular integrals over a surface domain, but it is still very challenging to handle them because those techniques for calculating surface singular integrals may not be directly applicable to volumetric singular integrals. We propose a simple scheme to handle them by making use of the formulations of evaluating surface singular integrals. We first divide a volumetric singular integral into a line integral along a chosen height and a surface integral over a triangular bottom which is changeable along the height in the tetrahedral element. The line integral is always regular and can be evaluated numerically. The singularity resides in the surface integral and the observation point could be far away from, close to, and overlapping with the triangular bottom when it is moving along the height. Regardless of which situation, we have derived closed-form formulations for them in the treatment for surface singular integrals and they can be used as integrands in the line integral.

If the volumetric domain is a cuboid element or hexagonal column, which could be used in some situations, the treatment is quite similar to that for the tetrahedral element. We only need to change the triangular bottom into a rectangular or hexagonal bottom which will be divided into subtriangles in handling surface singular integrals and the resulting treatment may be simpler than that for the tetrahedral element. In certain rare cases, one could use cylindrical cells to discretize an object and we also consider the volumetric singular integrals over a cylindrical domain. Due to the axial symmetry of the domain, we can derive purely closed-form formulations for such singular integrals and this will greatly facilitate the treatment for them.

5.6.1 Representation of Volume Current Density

As for handling surface singular integrals, we need to appropriately represent the volume current densities in the treatment of volumetric singularities. In our strategy, the unknown current densities are represented by a polynomial interpolation function based on the quadrature points in the Nyström method. For the VIEs, the unknown current densities are in three dimensions (3D) and the interpolation function for the zeroth-order, first-order and second-order polynomials in a local coordinate system (u, v, w) can be written as

$$f(u, v, w) = a \qquad \text{(zeroth-order)} \tag{5.163}$$

$$f(u, v, w) = a + bu + cv + dw \qquad \text{(first-order)} \tag{5.164}$$

$$\begin{aligned} f(u, v, w) &= a + bu + cv + dw + euv \\ &\quad + fvw + gwu + hu^2 + iv^2 + jw^2 \quad \text{(second-order).} \end{aligned} \tag{5.165}$$

The above polynomials require the one-point, four-point and ten-point quadrature rules, respectively, to uniquely determine the coefficients because there are $(n + 1)(n + 2)(n + 3)/6$ coefficients or terms in the nth order polynomial [59]. If the quadrature rules with other numbers of quadrature points are used, then the polynomial coefficients have to be determined approximately by the least square method [60]. As an example, we consider the coefficients of the first-order polynomial

$$f(u, v, w) = a + bu + cv + dw = \mathbf{u}^t \cdot \mathbf{a} \tag{5.166}$$

where $\mathbf{u}^t = [1, u, v, w]$, $\mathbf{a}^t = [a, b, c, d]$ and the superscript t represents the transpose. By matching the polynomial function of representing unknown current densities at the four quadrature points of a quadrature rule, we have

$$f_i = \mathbf{u}_i^t \cdot \mathbf{a}, \quad i = 1, \cdots, 4 \tag{5.167}$$

where $f_i = f(u_i, v_i, w_i)$ is the polynomial function's value at the quadrature point (u_i, v_i, w_i) and $\mathbf{u}_i^t = [1, u_i, v_i, w_i]$, $i = 1, \cdots, 4$. The above equations can be written into a matrix form

$$\mathbf{f} = \overline{\mathbf{u}} \cdot \mathbf{a} \tag{5.168}$$

and the solution for the polynomial coefficients is found as

$$\mathbf{a} = \overline{\mathbf{u}}^{-1} \cdot \mathbf{f} \tag{5.169}$$

where $\mathbf{f}^t = [f_1, f_2, f_3, f_4]$ and $\overline{\mathbf{u}}^t = [\mathbf{u}_1, \mathbf{u}_2, \mathbf{u}_3, \mathbf{u}_4]$. Using the found coefficients, the polynomial function of representing unknown current densities can be expressed as

$$\begin{aligned} f(u, v, w) &= \mathbf{u}^t \cdot \overline{\mathbf{u}}^{-1} \cdot \mathbf{f} \\ &= \mathbf{T}^t \cdot \mathbf{f} \\ &= \sum_{j=1}^{4} T_j(u, v, w) f_j \end{aligned} \tag{5.170}$$

where $\mathbf{T}^t = \left[T_1(u, v, w), T_2(u, v, w), T_3(u, v, w), T_4(u, v, w)\right]$ is the interpolation function. With the above interpolation representation of the unknown current densities, the real unknown coefficients in the matrix equation become the values of unknown current densities at the quadrature points, which should be solved. The above interpolation representation for the unknown current densities can be easily extended to higher-order polynomials and the higher-order accuracy can be obtained.

5.6.2 Evaluation of Singular Integrals

Over a Tetrahedral Element

The primary integral kernel in the VIEs is the same as that in the SIEs, i.e. the dyadic Green's function, which has nine components if expanded but only six of them are independent due to its symmetry. After using the polynomial interpolation representation of the unknown current densities, the strongest-singular or hypersingular integrals come from the combination of the dyadic Green's function with the constant term in the polynomial. The combination of the kernel with other terms in the polynomial will weaken the degree of singularities, leading to an easier treatment for resultant singular integrals. After a singularity subtraction process is performed, all resultant singular integrals have been presented in Section 5.2.4. Some of them are strongly-singular or weakly-singular and we do not address them here because they can be easily handled. We only focus on the treatment for hypersingular integrals which can be listed again as

$$I_1 = \int_{\Delta V} \left(\frac{3u^2}{R^5} - \frac{1}{R^3} \right) \mathrm{d}V; \quad I_2 = \int_{\Delta V} \frac{uv}{R^5} \mathrm{d}V; \tag{5.171}$$

$$I_3 = \int_{\Delta V} \left(\frac{3v^2}{R^5} - \frac{1}{R^3} \right) \mathrm{d}V; \quad I_4 = \int_{\Delta V} \frac{-u(w_0 - w)}{R^5} \mathrm{d}V; \tag{5.172}$$

$$I_5 = \int_{\Delta V} \frac{-v(w_0 - w)}{R^5} \mathrm{d}V; \quad I_6 = \int_{\Delta V} \left[\frac{3(w_0 - w)^2}{R^5} - \frac{1}{R^3} \right] \mathrm{d}V. \tag{5.173}$$

Note that the sixth integral I_6 is not needed for solving SIEs but it is needed for solving VIEs because the unknown function is a 3D vector in a volume. To evaluate those hypersingular integrals, we first establish a local coordinate system as shown in Figure 5.24: one of the four facets of a tetrahedron is chosen as the base triangle whose three vertices are labelled as P_1, P_2 and P_3, respectively, and the left vertex (or free vertex) of the tetrahedron is labelled as P_4. Choosing the projection of the observation point P_0 on the base triangle plane as the origin of the local coordinate system and choosing the unit normal vector of the base triangle plane pointing to the P_4 side as the $+w$ axis. Choosing the unit vector parallel to the line from P_1 to P_2 as the $+u$ axis. The $+v$ axis is determined by

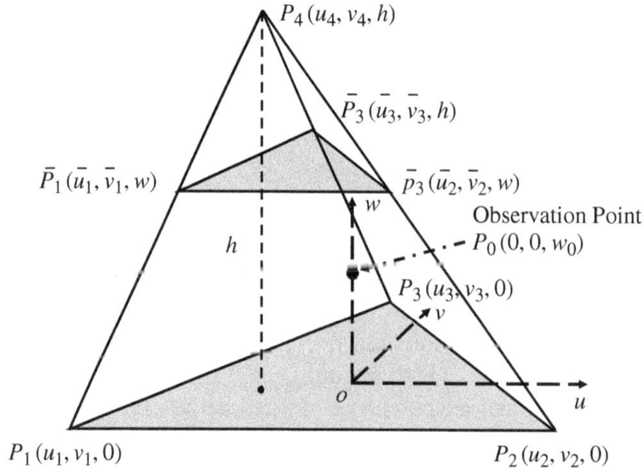

Figure 5.24 Derivation of hypersingular integrals in a local coordinate system $\{u, v, w\}$ whose origin is the projection of the observation point on the base triangle plane and $+w$ axis is parallel to the inward normal vector of the base triangle.

the $+w$ axis and the $+u$ axis. In such a local coordinate system, the above hypersingular integrals over a tetrahedron, for instance, the first integral in (5.169), can be written as

$$I_1 = \int_0^h \mathrm{d}w \int_{\triangle S} \left(\frac{3u^2}{R^5} - \frac{1}{R^3} \right) \mathrm{d}S. \tag{5.174}$$

The outer integral with respect to w in I_1 is regular and is evaluated by a numerical quadrature rule (such as Gaussian quadrature rule). The inner surface integral could be regular, near-singular or singular, depending on the spatial relation between the observation point P_0 and the contributing triangle $\triangle S = \triangle \overline{P}_1\overline{P}_2\overline{P}_3$, which is dependent on w. However, no matter how close the observation point P_0 is to the contributing triangle $\triangle S$, the surface integral is always integrable under the CPV sense, as we have derived in the previous sections. Thus, the above hypersingular integrals over a tetrahedron can be evaluated conveniently by using those analytical formulas we derived for surface integrals.

Over a Cylindrical Element

When the integral domain is a cylindrical element, we can establish a local coordinate system $\{u, v, w\}$ as shown in Figure 5.25, where the observation point is located at the origin or the center of the cylindrical element with a radius a and a height $2h$ while the source point is located at (u, v, w) inside the cylindrical element. With such a coordinate system, the hypersingular volume integrals take the following form after a singularity subtraction is performed

$$I_1 = \int_{V_0} \left(\frac{3u^2}{R^5} - \frac{1}{R^3} \right) \mathrm{d}V; \quad I_2 = \int_{V_0} \left(\frac{3v^2}{R^5} - \frac{1}{R^3} \right) \mathrm{d}V; \tag{5.175}$$

$$I_3 = \int_{V_0} \left(\frac{3w^2}{R^5} - \frac{1}{R^3} \right) \mathrm{d}V; \quad I_4 = \int_{V_0} \frac{uv}{R^5} \mathrm{d}V; \tag{5.176}$$

$$I_5 = \int_{V_0} \frac{wu}{R^5} \mathrm{d}V; \qquad I_6 = \int_{V_0} \frac{wv}{R^5} \mathrm{d}V. \tag{5.177}$$

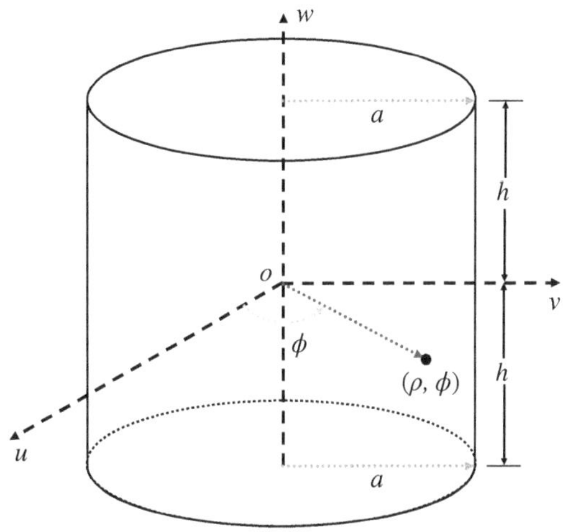

Figure 5.25 A cylindrical element for evaluating hypersingular integrals.

The evaluation of those hypersingular integrals usually requires a numerical integration for outer line integrals and this may not be very convenient in applications. In fact, we can derive new formulas that are purely in a closed-form for those integrals by making use of the axis symmetry of integral domain. In the cylindrical element, we also establish a polar coordinate system (ρ, ϕ) over its cross section, thus we have $\rho^2 = u^2 + v^2$ and $R^2 = \rho^2 + w^2$. The differential of cylindrical volume domain is $\rho \mathrm{d}\rho \mathrm{d}\phi \mathrm{d}w$ but the integration over ϕ can be easily obtained. The above hypersingular integrals and some useful weakly-singular or strongly-singular integrals can be evaluated by using the following integrals either individually or jointly and we have derived closed-form expressions for them [61].

$$\int_{V_0} \frac{\mathrm{d}V}{4\pi R^3} = \int_0^{2\pi} \mathrm{d}\phi \int_0^h \int_0^a \frac{\rho \mathrm{d}\rho \mathrm{d}w}{(\rho^2 + w^2)^{3/2}} = -\ln\left[\frac{1}{w}\left(w + \sqrt{w^2 + a^2}\right)\right]\Bigg|_0^h \tag{5.178}$$

$$\int_{V_0} \frac{u^2 \mathrm{d}V}{4\pi R^3} = \int_0^{2\pi} \mathrm{d}\phi \cos^2\phi \int_0^h \int_0^a \frac{\rho^3 \mathrm{d}\rho \mathrm{d}w}{2(\rho^2 + w^2)^{3/2}} = \frac{h}{2}\left[\sqrt{a^2 + h^2} - h\right] \tag{5.179}$$

$$\int_{V_0} \frac{w^2 \mathrm{d}V}{4\pi R^3} = \int_0^{2\pi} \mathrm{d}\phi \int_0^h \int_0^a \frac{\rho w^2 \mathrm{d}\rho \mathrm{d}w}{2(\rho^2 + w^2)^{3/2}} = \frac{1}{2}\left\{h^2 - h\sqrt{a^2 + h^2} + a^2 \ln\left[\frac{h}{a} + \sqrt{1 + \frac{h^2}{a^2}}\right]\right\} \tag{5.180}$$

$$\int_{V_0} \frac{u^2 \mathrm{d}V}{4\pi R^4} = \int_0^{2\pi} \mathrm{d}\phi \cos^2\phi \int_0^h \int_0^a \frac{\rho^3 \mathrm{d}\rho \mathrm{d}w}{2(\rho^2 + w^2)^2} = \frac{1}{4}\left[h \ln\left(1 + \frac{a^2}{h^2}\right) + a \tan^{-1}\left(\frac{h}{a}\right)\right] \tag{5.181}$$

$$\int_{V_0} \frac{w^2 \mathrm{d}V}{4\pi R^4} = \int_0^{2\pi} \mathrm{d}\phi \int_0^h \int_0^a \frac{\rho w^2 \mathrm{d}\rho \mathrm{d}w}{(\rho^2 + w^2)^2}$$
$$= \frac{a}{2}\tan^{-1}\left(\frac{h}{a}\right) \tag{5.182}$$

$$\int_{V_0} \frac{u^2 \mathrm{d}V}{4\pi R^5} = \int_0^{2\pi} \mathrm{d}\phi \cos^2\phi \int_0^h \int_0^a \frac{\rho^3 \mathrm{d}\rho \mathrm{d}w}{2(\rho^2 + w^2)^{5/2}}$$
$$= -\frac{1}{6}\left\{ 2\ln\left[1 + \sqrt{1 + \frac{a^2}{w^2}}\right] + \frac{w}{\sqrt{a^2 + w^2}} \right\}\Bigg|_0^h$$
$$= \int_{V_0} \frac{v^2 \mathrm{d}V}{4\pi R^5} \tag{5.183}$$

$$\int_{V_0} \frac{uv \mathrm{d}V}{4\pi R^5} = \int_0^{2\pi} \mathrm{d}\phi \cos\phi \sin\phi \int_0^h \int_0^a \frac{\rho^3 \, \mathrm{d}\rho \, \mathrm{d}w}{4\pi(\rho^2 + w^2)^{5/2}} = 0 \tag{5.184}$$

$$\int_{V_0} \frac{w^2 \mathrm{d}V}{4\pi R^5} = \int_0^{2\pi} \mathrm{d}\phi \int_0^h \int_0^a \frac{\rho w^2 \mathrm{d}\rho \mathrm{d}w}{(\rho^2 + w^2)^{5/2}}$$
$$= \frac{1}{3}\left\{ \frac{w}{\sqrt{a^2 + w^2}} - \ln\left[1 + \sqrt{1 + \frac{a^2}{w^2}}\right] \right\}\Bigg|_0^h \tag{5.185}$$

$$\int_{V_0} \frac{w(-u)\mathrm{d}V}{4\pi R^5} = \int_0^{2\pi} \mathrm{d}\phi \cos\phi \int_0^h \int_0^a \frac{\rho^2 w \, \mathrm{d}\rho \, \mathrm{d}w}{4\pi(\rho^2 + w^2)^{5/2}} = 0 \tag{5.186}$$

$$\int_{V_0} \frac{w(-v)\mathrm{d}V}{4\pi R^5} = \int_0^{2\pi} \mathrm{d}\phi \sin\phi \int_0^h \int_0^a \frac{\rho^2 w \, \mathrm{d}\rho \, \mathrm{d}w}{4\pi(\rho^2 + w^2)^{5/2}} = 0. \tag{5.187}$$

Note that some hypersingular integrals in the above are individually divergent when $w \to 0$ (so we do not expand them by substituting the upper limit and lower limit of their integrals), but their appropriate combination appearing in the singularity subtraction will cancel the divergent terms, leading to convergent expressions, i.e.

$$\int_{V_0} \frac{3u^2 - R^2}{4\pi R^5} \mathrm{d}V = \int_{V_0} \frac{3v^2 - R^2}{4\pi R^5} \mathrm{d}V = -\frac{h}{2\sqrt{a^2 + h^2}} \tag{5.188}$$

$$\int_{V_0} \frac{3w^2 - R^2}{4\pi R^5} \mathrm{d}V = \frac{h}{\sqrt{a^2 + h^2}}. \tag{5.189}$$

We have also reduced the integral interval over w due to its symmetry in the above derivation.

5.6.3 Numerical Examples

To demonstrate the proposed approach, we use the Nyström method to solve the VIEs for EM scattering by different penetrable objects. In all examples, the incident wave is a plane wave propagating along the $-z$ direction in free space and has a frequency $f = 300$ MHz. We calculate the bistatic RCS for the objects with both vertical (VV) and horizontal (HH) polarizations. The first example is the scattering by a dielectric sphere with a radius $a = 0.15\lambda$ and a relative permittivity $\epsilon_r = 4.0$ (the relative permeability $\mu_r = 1.0$ is assumed except stated otherwise). We discretize the sphere into 4376 tetrahedral elements and a one-point quadrature rule is used. Since the volume current density has three components in space, the number of unknowns is then three times the number of tetrahedrons. Figure 5.26 shows the bistatic RCS solutions observed along the principal cut ($\phi = 0°$ and $\theta = 0°–180°$) and they are compared to the analytical Mie-series solutions. It can be seen that the solutions from the two approaches are in excellent agreement (NM represents the Nyström method in all figures).

The second example is the scattering by a dielectric cylinder with a radius of cross section $r = 0.1\lambda$, a height $h = 0.5\lambda$, and a relative permittivity $\epsilon_r = 4.0$. We discretize the cylinder into 5732 tetrahedral elements and a one-point quadrature rule is also used. The bistatic RCS solutions are shown in Figure 5.27 and they are close to the solutions obtained by solving the same VIE using the conventional MoM with the Schaubert-Wilton-Glisson (SWG) basis function.

To compare the convergence of the Nyström method with the one of the MoM, we calculate the bistatic RCSs for a dielectric cube with a side length $s = 0.2\lambda$ and a relative permittivity $\epsilon_r = 4.0$ in the third example. Since there is no curvilinear surface on the object, thus there is no geometrical approximation in the discretization and the numerical errors are only related to or controlled by the mesh size and used quadrature rules. We first discretize the cube into 1657 tetrahedral elements and a one-point quadrature rule is used, leading to 4971 unknowns in the Nyström method. We then discretize the cube into 2282 tetrahedral elements in the conventional MoM with the SWG basis function, leading to 4978 unknowns (the number of unknowns is roughly 2.18 times the number of tetrahedrons in the conventional MoM), which is very close to the number of unknowns in the Nyström method. We also use the solutions obtained at

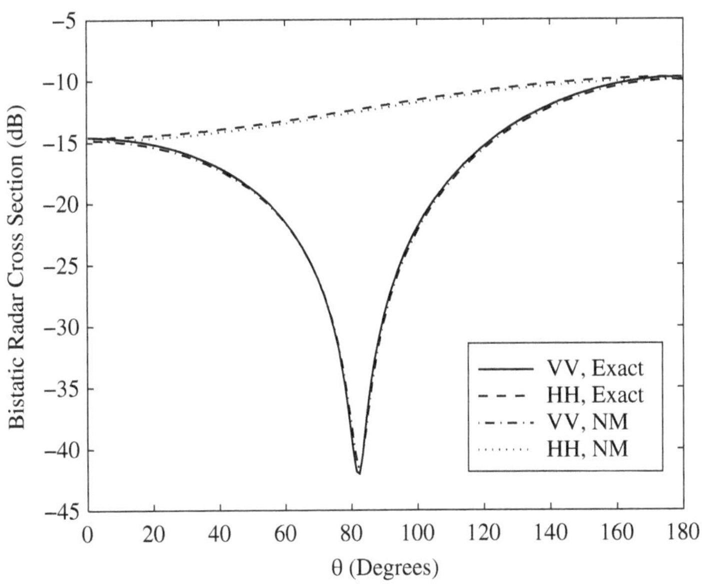

Figure 5.26 Bistatic RCS solutions for a dielectric sphere with a radius $a = 0.15\lambda$ and a relative permittivity $\epsilon_r = 4.0$.

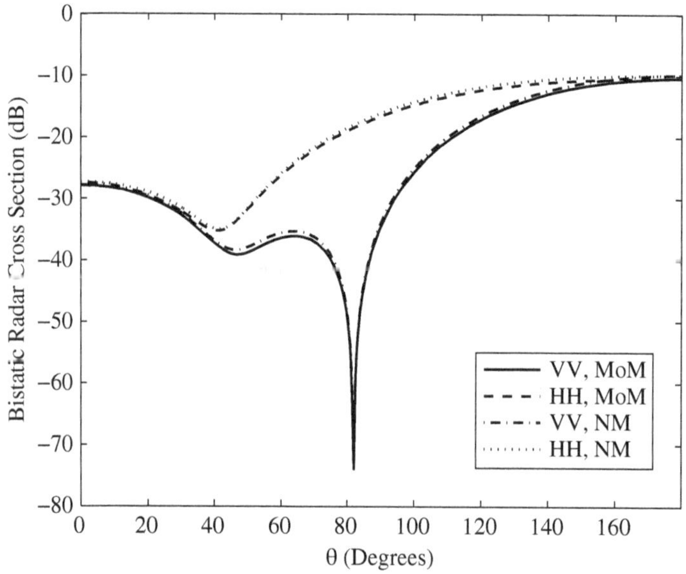

Figure 5.27 Bistatic RCS solutions for a dielectric cylinder with a radius of cross section $r = 0.1\lambda$, a height $h = 0.5\lambda$, and a relative permittivity $\epsilon_r = 4.0$.

a very fine level of geometrical discretization (12586 tetrahedral elements in the MoM) as the reference solutions [32] and calculate the root-mean-square (RMS) errors for each case. It is found that the RMS error of the Nyström method is 4.31×10^{-2}, which is larger than that of the MoM, 5.47×10^{-3}. However, if we use a four-point quadrature rule and use 415 tetrahedral elements, leading to 4980 unknowns, which is similar to the number of unknowns in the MoM, the RMS error of the Nyström method becomes 1.68×10^{-3}, which is smaller than that of the MoM. This indicates that the higher-order Nyström scheme could be better in convergence than the conventional MoM. Despite this, the higher-order Nyström scheme may not be widely used because it relies on the use of coarse meshes that could be difficult to minimize the geometrical approximation for curvilinear objects in implementation. The bistatic RCS solutions of the cube obtained from both the Nyström method and the MoM are illustrated in Figure 5.28.

To demonstrate the new formulas for singular integrals over cylindrical elements, we consider the EM scattering by a homogeneous dielectric sphere with a radius r and a relative permittivity ϵ_r. The problem is solved by a Nyström method incorporating with a meshless scheme as described in [62] and the derived formulas are used to evaluate the hypersingular integrals over a cylindrical element. The incident wave is the same as the previous cases and we also calculate the bistatic RCS in both VV and HH polarizations. Figure 5.29 shows the solutions when $r = 0.4\lambda$ and $\epsilon_r = 4.0$ and it can be seen that the solutions agree with the exact Mie-series solutions very well.

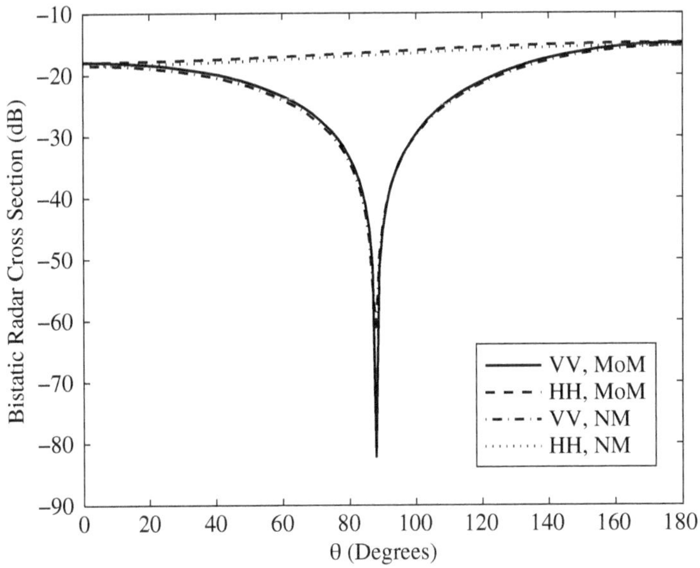

Figure 5.28 Bistatic RCS solutions for a dielectric cube with a side length $s = 0.2\lambda$ and a relative permittivity $\epsilon_r = 4.0$.

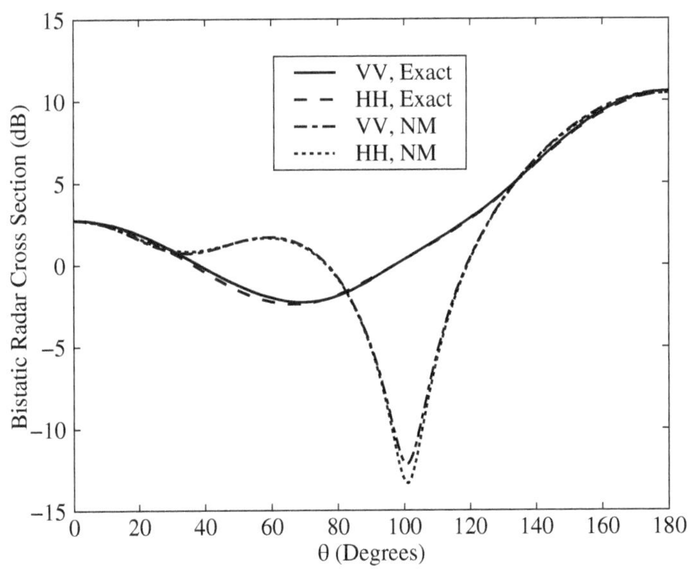

Figure 5.29 Bistatic RCS solutions for EM scattering by a dielectric sphere with a radius $r = 0.4\lambda$ and a relative permittivity $\epsilon_r = 4.0$.

5.7 Evaluation of Near-Singular Integrals

In transforming an EM integral equation into an algebraic matrix equation by using a numerical method, one has to evaluate the matrix elements, which may represent the far interactions (FIs), near interactions (NIs) or self interactions (SIs) between a field point and a source patch. The FIs include regular or non-singular integrals, NIs include near-singular integrals, and SIs include singular integrals, respectively. The near-singular integrals are not individually treated in general when the degree of singularities is not very strong and they are mostly treated as non-singular integrals which can be evaluated by numerical quadrature rules. However, when the degree of singularities is strong as in the strongly-singular integrals (SSIs) and hypersingular integrals (HSIs), and a high accuracy is pursued, special care is needed not only for the SI elements but also for NI elements. Both SI and NI elements need to be regularized by singularity subtraction before applying numerical integrations and then we have to derive analytical formulas for evaluating those subtracted singular integrals. Generally, we may fall into a dilemma for the NI elements with highly singular kernels because we do not have a clear definition for the NIs and we do not know whether we can use numerical integrations to evaluate NI elements or not.

In this section, we address these questions by investigating the effect of the evaluation of those NI elements on the numerical solutions of EM integral equations and we take the EM scattering by thin conducting objects as an example to study because there are many NI elements in the system matrix resulting from such objects. We first define the NI between a field point and a source patch by using a nearness factor (NF). If the geometrical surfaces are discretized into triangular patches, the NF is defined as the ratio of the area sum of three side triangles to the area of the bottom triangle when the field point and the source patch form a tetrahedron and the source patch is chosen as the bottom triangle. Based on the analytical formulas of calculating the SSIs and HSIs we have derived, we can show the accuracy of numerical integrations for the near-singular integrals of NI elements by numerical tests. It is found that there exists a transition area or turning point on the NF before and after which the convergence behaviors of those integrations are significantly different. We define the NF value at the 1.0% relative error as the turning point NF_0 and define $NF \leq NF_0$ as the NI range. By extensive numerical tests, it is found that the NI ranges determined by turning points are roughly unchanged with the shape and size of triangular patches and the location of observation points for given SSIs and HSIs. We find that $NF_0 = 1.4$ could be a conservative bound above which the accuracy of numerical integrations for those integrals can surely be below a 1.0% relative error. Within the NI ranges, the numerical integrations of those integrals could be extremely inaccurate and analytical formulas should be used in the evaluation of NI elements. The quadrature rule used in the numerical integrations of the tests is the 10-point product Gaussian quadrature rule that is applied to each fold in a two-fold integration [33]. To see the overall influence of NI elements in the numerical solutions, we solve the combined field integral equation (CFIE) for scattering by thin conducting objects because it is preferred for close thin structures [67]. We then compare the solutions with NI treatments (using analytical formulas) to the solutions without NI treatments (using numerical integrations) and the obvious differences between them can be observed. This can demonstrate the significance of NI treatments and the results may be desirable in pursuing more accurate numerical solutions for thin-structure problems.

5.7.1 Integral Equations and Near-Singular Integrals

Although near-singular integrals exist in all EM integral equations, we take the CFIE as an example and consider the EM scattering by conducting thin objects to demonstrate the importance of taking care of the NI elements. The CFIE includes both EFIE and MFIE, so the near-singular integrals of both $\mathcal{L}$ and $\mathcal{K}$ operators can be examined. Also, the thin objects include much more NI matrix elements, so the near-singular integrals could play a more important role. Considering the EM scattering by PEC objects, we can write the EFIE as [29]

$$\hat{n}\times\mathcal{L}(\mathbf{J}) = -\hat{n}\times\mathbf{E}^{\mathrm{inc}}(\mathbf{r}), \quad \mathbf{r}\in S \tag{5.190}$$

and the corresponding MFIE as

$$\frac{1}{2}\mathbf{J}(\mathbf{r}) + \hat{n}\times\mathcal{K}(\mathbf{J}) = \hat{n}\times\mathbf{H}^{\mathrm{inc}}(\mathbf{r}), \quad \mathbf{r}\in S \tag{5.191}$$

where $\mathbf{J}(\mathbf{r})$ denotes the unknown surface current density and the integral operators $\mathcal{L}$ and $\mathcal{K}$ are defined as

$$\begin{aligned}\mathcal{L}(\mathbf{J}) &= \mathrm{i}\kappa\eta\int_S\left[\bar{\mathbf{I}}+\frac{\nabla\nabla}{\kappa^2}\right]g(\mathbf{r},\mathbf{r}')\cdot\mathbf{J}(\mathbf{r}')\,\mathrm{d}S'\\ &= \mathrm{i}\kappa\eta\int_S\bar{\mathbf{G}}(\mathbf{r},\mathbf{r}')\cdot\mathbf{J}(\mathbf{r}')\,\mathrm{d}S'\end{aligned} \tag{5.192}$$

$$\mathcal{K}(\mathbf{J}) = \fint_S \mathbf{J}(\mathbf{r}')\times\nabla g(\mathbf{r},\mathbf{r}')\,\mathrm{d}S'. \tag{5.193}$$

In the above equations, S is the object surface, κ is the wave number, η is the wave impedance of free space, $\hat{n}$ is the outward unit vector at the object surface and $(\mathbf{E}^{\mathrm{inc}}, \mathbf{H}^{\mathrm{inc}})$ are the incident electric and magnetic fields. Also, $g(\mathbf{r},\mathbf{r}') = \mathrm{e}^{\mathrm{i}\kappa R}/(4\pi R)$ is the 3D scalar Green's function, where $R = |\mathbf{r}-\mathbf{r}'|$ is the distance between a field point $\mathbf{r}$ and a source point $\mathbf{r}'$, and $\bar{\mathbf{G}}(\mathbf{r},\mathbf{r}')$ is the dyadic Green's function, where $\bar{\mathbf{I}}$ denotes the identity dyad. The above integral equations have been presented in Chapter 2 and we repeat them here for a reading convenience. The CFIE is a linear combination of the above EFIE and MFIE and can be written as [29]

$$\begin{aligned}&\alpha\hat{n}\times\hat{n}\times\mathcal{L}(\mathbf{J}) + (1-\alpha)\eta\left[\frac{1}{2}\mathbf{J}(\mathbf{r}) + \hat{n}\times\mathcal{K}(\mathbf{J})\right]\\ &= \alpha\hat{n}\times\hat{n}\times\mathbf{E}^{\mathrm{inc}}(\mathbf{r}) + (1-\alpha)\eta\hat{n}\times\mathbf{H}^{\mathrm{inc}}(\mathbf{r}), \quad \mathbf{r}\in S\end{aligned} \tag{5.194}$$

where α is the combination parameter and is generally selected between 0.2 and 0.8.

The expressions of near-singular integrals are the same as those of singular integrals as shown in Subsections 5.2.2 and 5.2.3, just when the observation point is not inside a source patch but very close to it. We will use a special index named "nearness factor (NF)" to measure how close an observation point is to a source triangular patch in the next subsection.

As we have mentioned, the formulas derived for evaluating those singular integrals can also be used for evaluating near-singular integrals, in which the limit of observation point approaching the source patch is not taken. In addition, only the NIs in strongly-singular and hypersingular integrals need to be handled in general, and the NIs in weakly-singular integrals can be treated as non-singular integrals which can be accurately evaluated by numerical quadrature rules.

For the $1/R^2$ strongly-singular integral kernels generated from the gradient of the Green's function in $\mathcal{K}$ operator, they are not treated in Subsection 5.4 which mainly focuses on the evaluation of hypersingular and weakly singular integrals for the $\mathcal{L}$ operator, but we can derive the solutions for them by using a similar way. They take the following forms

$$
\begin{aligned}
I_a &= \int_{\Delta S} \frac{-u}{R^3}\, \mathrm{d}S \\
&= \sum_{i=1}^{3} \left[\int_{\theta_{10}^i}^{\theta_{20}^i} \frac{d_i \cos(\phi + \theta_0^i)}{\sqrt{d_i^2 + w_0^2 \cos^2\phi}}\, \mathrm{d}\phi - \int_{\theta_{10}^i}^{\theta_{20}^i} \cos(\phi + \theta_0^i) \ln\left(d_i + \sqrt{d_i^2 + w_0^2\cos^2\phi} \right) \mathrm{d}\phi \right. \\
&\quad \left. + \int_{\theta_{10}^i}^{\theta_{20}^i} \cos(\phi + \theta_0^i) \ln(\cos\phi)\, \mathrm{d}\phi \right] \\
&= I_a^{(1)} + I_a^{(2)} + I_a^{(3)}
\end{aligned} \tag{5.195}
$$

$$
\begin{aligned}
I_b &= \int_{\Delta S} \frac{-v}{R^3}\, \mathrm{d}S \\
&= \sum_{i=1}^{3} \left[\int_{\theta_{10}^i}^{\theta_{20}^i} \frac{d_i \sin(\phi + \theta_0^i)}{\sqrt{d_i^2 + w_0^2 \cos^2\phi}}\, \mathrm{d}\phi - \int_{\theta_{10}^i}^{\theta_{20}^i} \sin(\phi + \theta_0^i) \ln\left(d_i + \sqrt{d_i^2 + w_0^2\cos^2\phi} \right) \mathrm{d}\phi \right. \\
&\quad \left. + \int_{\theta_{10}^i}^{\theta_{20}^i} \sin(\phi + \theta_0^i) \ln(\cos\phi)\, \mathrm{d}\phi \right] \\
&= I_b^{(1)} + I_b^{(2)} + I_b^{(3)}
\end{aligned} \tag{5.196}
$$

$$
\begin{aligned}
I_c &= \int_{\Delta S} \frac{w_0}{R^3}\, \mathrm{d}S \\
&= \sum_{i=1}^{3} \left[\arcsin(s_0 \sin\theta_{10}^i) - \arcsin(s_0 \sin\theta_{20}^i) + (\theta_2^i - \theta_1^i) \right]
\end{aligned} \tag{5.197}
$$

where $s_0 = w_0/\sqrt{d_i^2 + w_0^2}$. In I_a and I_b, the second integrals $I_a^{(2)}$ and $I_b^{(2)}$ are not analytically integrable, but they are regular and high-order accuracy can be obtained by using higher-order numerical integrations. The first and third integrals in I_a and I_b can be found in a closed-form. For example,

$$
\begin{aligned}
I_a^{(1)} &= \frac{d_i}{w_0} \left\{ \cos\theta_0^i \left[\arcsin(s_0 \sin\theta_{20}^i) - \arcsin(s_0 \sin\theta_{10}^i) \right] \right. \\
&\quad \left. + \sin\theta_0^i \left[\sinh^{-1}(s_1 \cos\theta_{20}^i) - \sinh^{-1}(s_1 \cos\theta_{10}^i) \right] \right\}
\end{aligned} \tag{5.198}
$$

$$
I_a^{(3)} = 0.5 \cos\theta_0^i \left[(1 + \sin\theta_{20}^i) \ln(1 + \sin\theta_{20}^i) - (1 - \sin\theta_{20}^i) \ln(1 - \sin\theta_{20}^i) \right.
$$

$$-(1 + \sin\theta^i_{10})\ln(1 + \sin\theta^i_{10}) + (1 - \sin\theta^i_{10})\ln(1 - \sin\theta^i_{10})]$$
$$-2\left(\sin\theta^i_{20} - \sin\theta^i_{10}\right) \tag{5.199}$$

where $s_1 = w_0/d_i$. These formulas can be used as exact solutions for strongly near-singular integrals although they cannot be so for strongly-singular integrals because there are not closed-form expressions for $I_a^{(2)}$ and $I_b^{(2)}$. The accuracy of the formulas has been verified by comparing with that of the formulas for the same integrals in Subsection 5.5.3.

5.7.2 Evaluation

In evaluating the matrix elements, the NI between a field point and a source patch can widely exist in thin structures such as thin plates, sharp corners, or thin wires. Also, if any two patches share a common edge in arbitrary geometries, the quadrature points as field points in one patch can be very close to the other patch or source patch when numerical integrations are performed. Since there are a large number of NI elements in those thin structures, the accuracy of their evaluation may significantly impact the numerical solutions and special treatments are needed.

In order to measure how a field point is close to a source patch, we may define a nearness factor (NF). If we connect the field point with the three vertices of the source triangular patch to form a tetrahedron, see Figure 5.30(a), the NF is defined as the ratio of the area sum of the three side triangles of the tetrahedron to the area of the bottom triangle which is the source triangle, i.e.

$$\text{NF} = \frac{S_1 + S_2 + S_3}{S_0}. \tag{5.200}$$

Obviously, when the field point is inside the source triangular patch as shown in Figure 5.30(b), NF = 1, and this is the singular case. When the field point is off the source triangular patch, NF > 1, and moreover, the larger NF is, the farther the field point is away from the source triangular patch. Therefore, NF reflects the spatial relation

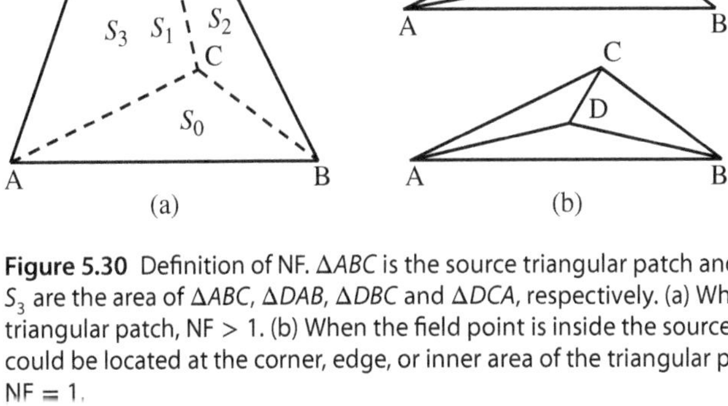

Figure 5.30 Definition of NF. ΔABC is the source triangular patch and D is the field point. S_0, S_1, S_2 and S_3 are the area of ΔABC, ΔDAB, ΔDBC and ΔDCA, respectively. (a) When the field point is off the source triangular patch, NF > 1. (b) When the field point is inside the source triangular patch (the field point could be located at the corner, edge, or inner area of the triangular patch as shown in the figure), NF = 1.

between the field point and its contributing source triangular patch. We can define the NI range with such a NF for the hypersinglar and strongly-singular integrals within which numerical integrations are not suitable and analytical formulas should be used.

We use the formulas in Subsections 5.4.1 and 5.7.1 as the exact solutions for the hypersingular and strongly-singular integrals to check the accuracy of numerical integrations and find the NI ranges for them by numerical tests. The numerical integrations are performed with a 10-point product Gaussian quadrature rule [33]. The source triangular patch is randomly defined by three vertices $P_1(0.1, 0.1, 0)$, $P_2(0.6, 0.2, 0)$, and $P_3(0.3, 0.5, 0)$, and the field point is located at $P_0(0.3, 0.3, w_0)$, where w_0 can be changed. Tables 5.4, 5.5 and 5.6 show the exact values and numerical integration values of the hypersingular integrals I_1, I_2, and I_4, respectively, when the field point is approaching the source triangular patch. It can be seen that the numerical integrations are not accurate when $w_0 \to 0$. In fact, the numerical integrations of those hypersingular integrals tend to be divergent within a certain range of NF and the errors of the numerical integrations are not controllable within that range. To find the range, we plot the dependence relationship of relative errors on NF for those integrals over arbitrarily chosen four different triangular patches

Table 5.4 Comparison of numerical integration values with exact values for I_1.

No.	w_0	Exact	Numerical	NF
1	1.0	−0.08922223917026	−0.08922223917026	7.5541
2	0.5	−0.60446704732328	−0.60446704733170	3.8783
3	0.2	−4.48709772694627	−4.48706311513697	1.8074
4	0.1	−10.72520436059866	−10.67990660788759	1.2546
5	0.05	−15.62877534534030	−14.87967889372427	1.0700
6	0.02	−17.91310842228915	40.58944875509436	1.0116
7	0.01	−18.29742488380399	307.2505588546849	1.0029
8	0	−18.42942165897242	688.8205359879435	1.0000

Table 5.5 Comparison of numerical integration values with exact values for I_2.

No.	w_0	Exact	Numerical	NF
1	1.0	−0.00007305038914	−0.00007305038914	7.5541
2	0.5	−0.00191131066771	−0.00191131066695	3.8783
3	0.2	−0.07730819947236	−0.07731164032032	1.8074
4	0.1	−0.46969844744423	−0.47381421579674	1.2546
5	0.05	−1.06311324955844	−1.01794659745086	1.0700
6	0.02	−1.43258451131781	14.04058811268205	1.0116
7	0.01	−1.50059539682674	66.60085458974463	1.0029
8	0	−1.52434084144863	134.2293939199608	1.0000

Table 5.6 Comparison of numerical integration values with exact values for I_4.

No.	w_0	Exact	Numerical	NF
1	1.0	−0.00006804204916	−0.00006804204916	7.5541
2	0.5	−0.00291975884530	−0.00291975884387	3.8783
3	0.2	−0.12221988036152	−0.12221652585852	1.8074
4	0.1	−0.46801574792989	−0.49439875956351	1.2546
5	0.05	−0.56740894120208	−3.72282451546325	1.0700
6	0.02	−0.31238031200878	−60.12021431411318	1.0116
7	0.01	−0.16411614121412	−111.6588917572541	1.0029
8	0.001	−0.01668594459804	−21.04418362217304	1.0000

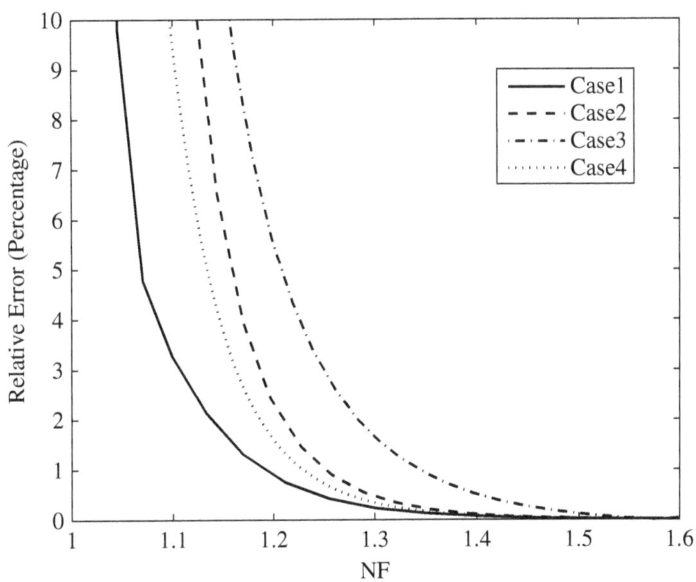

Figure 5.31 Dependence of relative error on NF for I_1.

as shown in Figures 5.31, 5.32 and 5.33, respectively. The four triangular patches with observation points are defined as $P_1(0.1, 0.1, 0)$, $P_2(0.6, 0.2, 0)$, $P_3(0.2, 0.5, 0)$, $P_0(0.3, 0.3, w_0)$ (Case 1), $P_1(0, 0, 0)$, $P_2(0.8, 0.5, 0)$, $P_3(0.2, 0.7, 0)$, $P_0(0.3, 0.3, w_0)$ (Case 2), $P_1(0, 0, 0)$, $P_2(0.8, 0.1, 0)$, $P_3(0.1, 0.8, 0)$, $P_0(0.4, 0.4, w_0)$ (Case 3), and $P_1(-0.5, -0.8, 0)$, $P_2(1.6, -0.2, 0)$, $P_3(0.2, 2.5, 0)$, $P_0(0.1, 0.7, w_0)$ (Case 4), respectively. From these figures, we can see that there exists a transition area or turning point on NF before and after which the convergence behaviors of those integrals are dramatically different. When NF is greater than the turning point, the errors change very slightly in terms of NF and when NF is smaller than the turning point, the errors change rapidly in terms of NF. We can choose the NF value at the relative error of 1.0% as the turning point NF_0. The figures show that the turning point is roughly $\text{NF}_0 = 1.25$ for I_1, and $\text{NF}_0 = 1.35$ for I_2 and I_4.

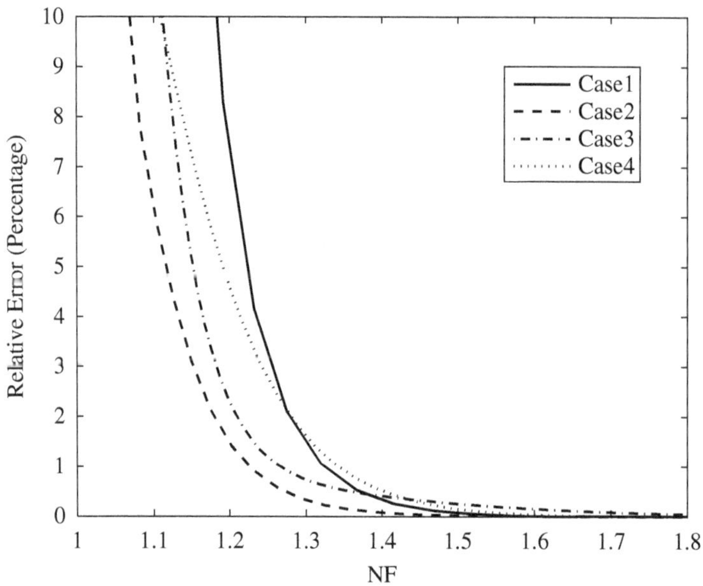

Figure 5.32 Dependence of relative error on NF for I_2.

We define the range with NF $\leq$ NF$_0$ as the NI range because the numerical errors within this range tend to be uncontrollable and the closed-form formulas should be resorted to evaluate those integrals.

For the $1/R^2$ strongly-singular integrals as shown in Subsection 5.7.1, Tables 5.7 and 5.8 illustrate the comparison between the exact values and numerical integration values with a 10-point product Gaussian quadrature rule for I_a and I_c, respectively. Since there is no closed-form expression for $I_a^{(2)}$, we use a 24-point product Gaussian quadrature rule to calculate $I_a^{(2)}$ and the resultant values can be treated as exact values for this part. It is clear again that the numerical integrations are not accurate for those integrals when $w_0 \to 0$. We also plot the dependence relationship of relative errors on NF for the two integrals over the chosen four different triangular patches, as shown in Figures 5.34 and

Table 5.7 Comparison of numerical integration values with exact values for I_a.

No.	w_0	Exact	Numerical	NF
1	1.0	−0.00004261563002	−0.00004261563803	7.5541
2	0.5	−0.00102113320885	−0.00102113323147	3.8783
3	0.2	−0.02708541659837	−0.02708517065008	1.8074
4	0.1	−0.10718757276342	−0.10801005094243	1.2546
5	0.05	−0.18971257730353	−0.28765019440544	1.0700
6	0.02	−0.23167402751778	−2.08637121331199	1.0116
7	0.01	−0.23887879424923	−4.73966715166153	1.0029
8	0	−0.24136123758909	−7.05598924054385	1.0000

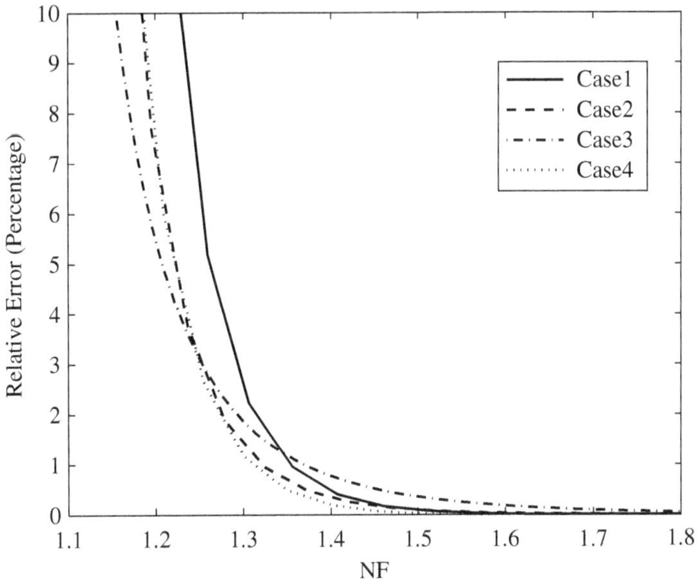

Figure 5.33 Dependence of relative error on NF for I_4.

Table 5.8 Comparison of numerical integration values with exact values for I_c.

No.	w_0	Exact	Numerical	NF
1	1.0	0.09226432468145	0.09226432468145	7.5541
2	0.5	0.34078627816568	0.34078627816597	3.8783
3	0.2	1.44972230240104	1.44972181191605	1.8074
4	0.1	2.96238734854490	2.96171449127705	1.2546
5	0.05	4.37964855170492	4.36761381860790	1.0700
6	0.02	5.48371259742470	5.31253242115307	1.0116
7	0.01	5.88042749489848	4.44878943069956	1.0029
8	0.001	6.24280789858813	0.58817022766023	1.0000

5.35, respectively. It can be seen that the turning points at the relative error of 1.0% for I_a and I_c are roughly $\text{NF}_0 = 1.25$ and $\text{NF}_0 = 1.16$, respectively. We define $\text{NF} \leq 1.25$ and $\text{NF} \leq 1.16$ as the NI range for these two integrals, respectively, and analytical formulas should be used within this range.

Note that the above determination for NI ranges based on the turning points only depends on the type of integrals according to our extensive numerical tests. The NI ranges are roughly unchanged with the shape and size of triangular patches, and the location of observation points. If we choose $\text{NF}_0 = 1.4$ conservatively, then the accuracy of the numerical integrations for those hypersingular and strongly-singular integrals can be surely controlled within the relative error of 1.0% when $\text{NF} \geq \text{NF}_0$. Therefore, the NI ranges defined with the NF can be used as a guideline for determining when we should

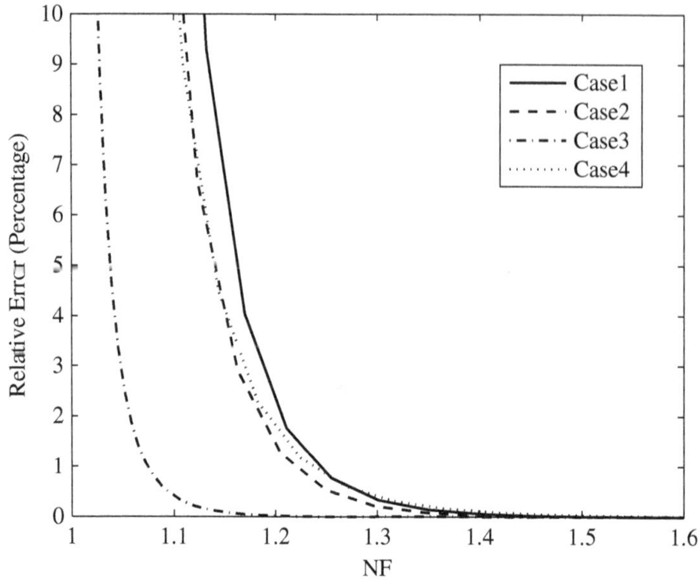

Figure 5.34 Dependence of relative error on NF for I_a

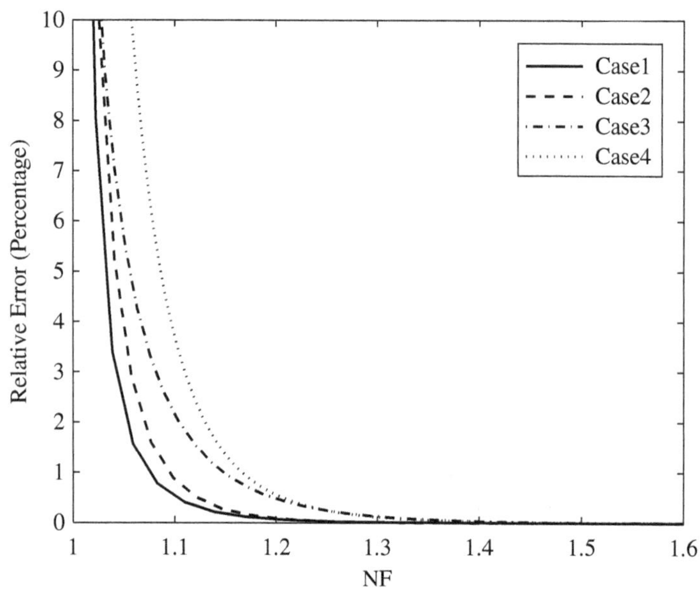

Figure 5.35 Dependence of relative error on NF for I_c.

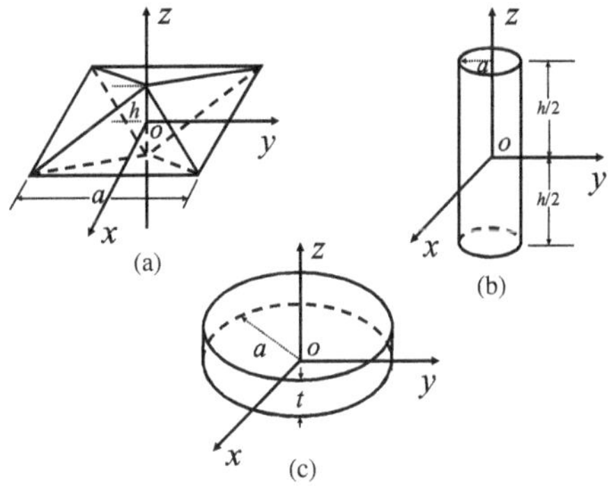

Figure 5.36 Geometries of thin objects. (a) Diamond. (b) Wire. (c) Disk.

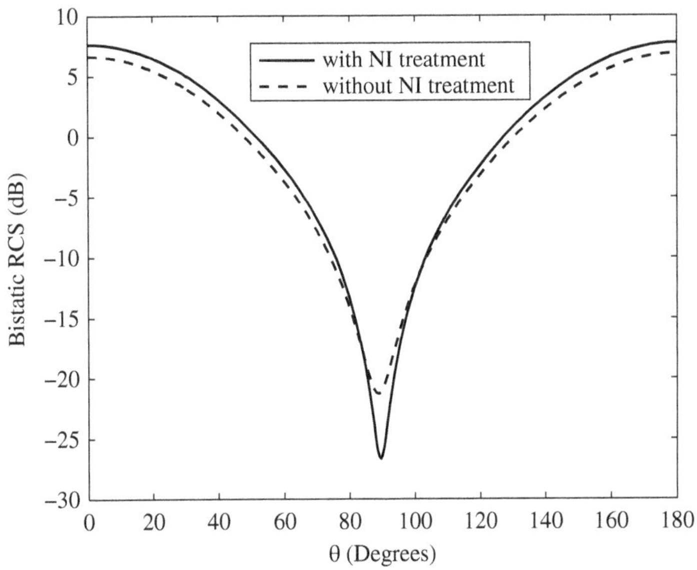

Figure 5.37 Bistatic RCS solutions of a thin diamond scatterer in vertical polarization.

not use numerical integrations and should resort to analytical formulas to evaluate those NI elements in numerical implementation.

5.7.3 Numerical Examples

Although the overall impact of the accuracy of those hypersingular and strongly-singular integrals in a NI range on the final numerical solutions of solved problems is not predictable and depends on case by case, we can still use the following numerical examples for EM scattering by thin structures to illustrate such an impact. By comparing the solutions with a NI treatment, i.e. using the analytical formulas to calculate the hypersingular and strongly-singular integrals, with the solutions without a NI treatment, we can know how large the errors of numerical solutions could be when there is no NI treatment for NI elements.

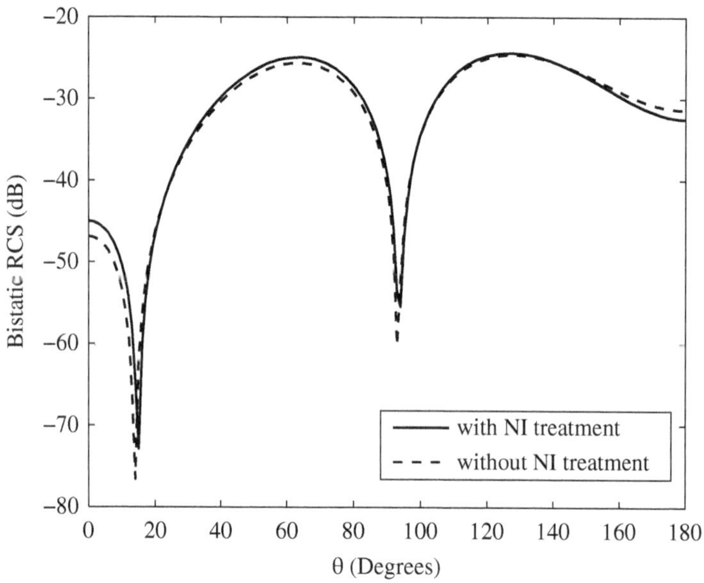

Figure 5.38 Bistatic RCS solutions of a thin wire scatterer in horizontal polarization.

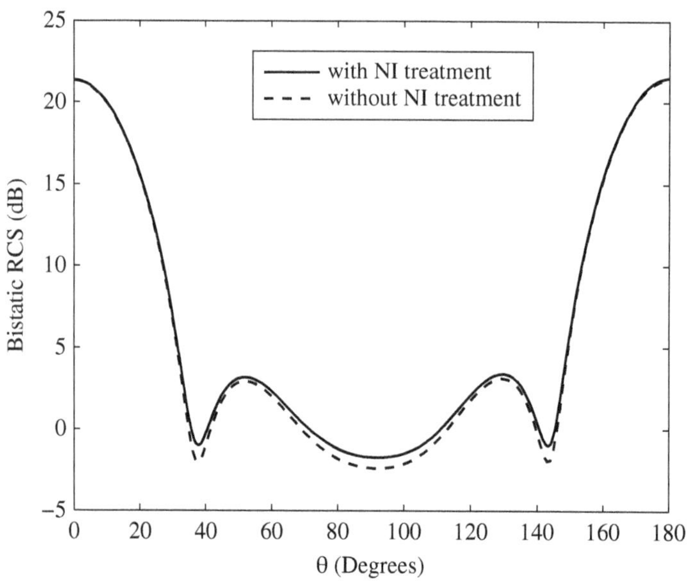

Figure 5.39 Bistatic RCS solutions of a thin disk scatterer in horizontal polarization.

We consider the EM scattering by three thin objects, i.e. diamond, wire, and disk as shown in Figure 5.36. The geometrical dimensions are $a = 0.4\lambda$ and $h = 0.02\lambda$ for the diamond, $h = 1.0\lambda$ and $a = 0.05\lambda$ for the wire, and $a = 1.0\lambda$ and $t = 0.05\lambda$ for the disk, respectively, where λ is the wavelength in free space. We use the Nyström method to solve the CFIE and calculate the bistatic RCS in vertical or horizontal polarization for these objects. The solutions with and without NI treatments are plotted in Figures 5.37, 5.38, and 5.39, respectively. The obvious difference between the two types of solution can clearly be seen, indicating that the NI treatment is definitely needed for accurate solutions.

5.8 Summary

In this chapter, we present some robust singularity treatment techniques that are needed to solve EM integral equations by the Nyström method. Certainly, these techniques can also be applied to other numerical methods. We start with the introduction of the singularity subtraction method, which is indispensable for strongly-singular and hypersingular integrals. By using the Tayor series expansion of the scalar Green's function, we can easily distinguish the singular cores from singular integrands and then subtract them to regularize the original singular integrals. We then address the singularity cancellation method, which is preferable but can only be applied to weakly-singular integrals. Although many techniques have been developed for evaluating the weakly-singular integrals, we provide an alternative approach to calculate them. The approach can automatically cancel the singularity without using a variable change or coordinate transformation and reduce the integrals to a one-fold numerical integration with a very simple integrand. Compared with the Duffy's method, which requires a two-fold numerical integration, the approach could be more convenient in implementation and more efficient in calculation as illustrated by the numerical examples.

The singularity treatment sometimes refers to the evaluation of integrals with the subtracted singular cores after a singularity subtraction has been performed and we particularly address the methods in different sections for different cases. We first present a robust technique to efficiently handle the hypersingularity in the $\mathcal{L}$ operator of SIEs, which was originally thought of as impossible and the work could be the first attempt in EM. The approach is based on the derivation of closed-form formulas for the hypersingular integrals under a CPV sense, but requires a subdivision for triangular patches and may not be very convenient in implementation, especially for those very slim patches. In addition, the approach can also derive closed-form formulations for the weakly-singular integrals. We then present a more robust and convenient approach for evaluating the strongly-singular and hypersingular integrals. The approach is based on the Stokes' theorem, which changes a surface integral over a surface patch into a line integral along its boundary and the line integral can be evaluated in a closed form more conveniently. The approach appears simpler and more friendly in implementation because the coordinate system transformation and subdivision of patches widely used in other approaches are not required. The derived formulations are particularly suitable for evaluating the singular integrals over very slim patches or low-quality meshes.

For the singularity treatment of the VIEs, the strategy and their subtracted singular integrals are similar to those for the SIEs because their kernels are similar in form, just defined over a volumetric domain which is usually a tetrahedral domain. The evaluation of singular volume integrals is divided into a line integral along the height of a tetrahedral element and a surface integral over the triangular cross section moving along the height. The line integral is always regular and can be evaluated numerically and the singularity resides in the surface integral. When the triangular cross section is moving along the height, the observation point could be far away from, close to, or inside it, but we have derived closed-form formulations for evaluating the surface integral in the singularity treatment of SIEs no matter which situation occurs and those formulations can be directly used for the numerical integration of line integrals. In addition, if the volumetric domain is a cylindrical domain, we also derive full closed-form formulas for those singular integrals by making use of the axis symmetry of the domain.

We finally consider the treatment for near-singular integrals, which are seldom addressed. Although the treatment of near singularities in NI elements may not be necessary for weakly-singular integrals because their influence is limited, it is needed for hypersingular and strongly-singular integrals whose influence was originally unclear. We study the accuracy of numerical integrations in evaluating the NI elements with hypersingular and strongly-singular kernels by numerical tests and illustrate their impact on the final numerical solutions of solved problems. We first propose the concept of "nearness factor (NF)" to measure the degree of closeness between a field point and a contributing source patch. With this factor, we find that there is a transition area or turning point on NF before and after which the convergence behaviors of the numerical integrations for those hypersingular and strongly-singular integrals are quite different. Also, it is found that the NI ranges are roughly unchanged with the shape and size of triangular patches and the location of observation points. In particular, the numerical integrations could be extremely inaccurate in the NI ranges and the analytical formulations should be resorted to evaluate the NI elements. Those findings can be used to clarify the influence of near singularities and guide the treatment for them. All the above conclusions have been verified through appropriate numerical examples or numerical tests.

Bibliography

1 Nyström, E.J. (1928). Über die praktische auflösung von linearen integralgleichungen mit anwendungen auf randwertaufgaben der potentialtheorie. *Commentationes Physico-Mathematicae.* 4 (15): 1–52.

2 Chew, W.C., Tong, M.S., and Hu, B. (2008). *Integral Equation Methods for Electromagnetic and Elastic Waves.* San Rafael, CA: Morgan & Claypool.

3 Rao, S.M., Wilton, D.R., and Glisson, A.W. (1982). Electromagnetic scattering by surfaces of arbitrary shape. *IEEE Trans. Antennas Propagat.* AP-30 (3): 409–418.

4 Duffy, M.G. (1982). Quadrature over a pyramid or cube of integrands with a singularity at a vertex. *SIAM J. Numer. Anal.* 19: 1260–1262.

5 Wilton, D.R., Rao, S.M., Glisson, A.W. et al. (1984). Potential integrals for uniform and linear source distributions on polygonal and polyhedral domains. *IEEE Trans. Antennas Propagat.* 32: 276–281.

6 Graglia, R.D. (1987). Static and dynamic potential integrals for linearly varying source distributions in two- and three-dimensional problems. *IEEE Trans. Antennas Propagat.* 35: 662–669.

7 Graglia, R.D. (1993). On the numerical integration of the linear shape functions times the 3-D Green's function or its gradient on a plane triangle. *IEEE Trans. Antennas Propagat.* 41: 1448–1455.

8 Caorsi, S., Moreno, D., and Sidoti, F. (1993). Theoretical and numerical treatment of surface integrals involving the free-space Green's function. *IEEE Trans. Antennas Propagat.* 41 (9): 1296–1301.

9 Eibert, T.F. and Hansen, V. (1995). On the calculation of potential integrals for linear source distributions on triangular domains. *IEEE Trans. Antennas Propagat.* 43: 1499–1502.

10 Arcioni, P., Bressan, M., and Perregrini, L. (1997). On the evaluation of the double surface integrals arising in the application of the boundary integral method to 3-D problems. *IEEE Trans. Microw. Theory Tech.* 45: 436–439.

11 Bluck, M.J., Pocock, M.D., and Walker, S.P. (1997). An accurate method for the calculation of singular integrals arising in time-domain integral equation analysis of electromagnetic scattering. *IEEE Trans. Antennas Propagat.* 45 (12): 1793–1798.

12 Rossi, L. and Cullen, P.J. (1999). On the fully numerical evaluation of the linear-shape function times the 3-D Green's function on a plane triangle. *IEEE Trans. Microw. Theory Tech.* 47: 398–402.

13 Herschlein, A., Hagen, J.V., and Wiesbeck, W. (2002). Methods for the evaluation of regular, weakly-singular and strongly-singular surface reaction integrals arising in method of moments. *ACES J.* 17 (1): 63–73.

14 Järvenpää, S., Taskinen, M., and Ylä-Oijala, P. (2003). Singularity extraction technique for integral equation methods with higher order basis functions on plane triangles and tetrahedra. *Int. J. Numer. Meth. Eng.* 58: 1149–1165.

15 Taylor, D.J. (2003). Accurate and efficient numerical integration of weakly-singular integrals in Galerkin EFIE solutions. *IEEE Trans. Antennas Propagat.* 51: 1630–1637.

16 Khayat, M.A. and Wilton, D.R. (2005). Numerical evaluation of singular and near-singular potential integrals. *IEEE Trans. Antennas Propagat.* 53: 3180–3190.

17 Järvenpää, S., Taskinen, M., and Ylä-Oijala, P. (2006). Singularity subtraction technique for high-order polynomial vector basis functions on planar triangles. *IEEE Trans. Antennas Propagat.* 54 (1): 42–49.

18 Tang, W.-H. and Gedney, S.D. (2006). An efficient evaluation of near singular surface integrals via the Khayat–Wilton transformation. *Microw. Opt. Technol. Lett.* 48 (8): 1583–1586.

19 Tong, M.S. and Chew, W.C. (2007). Super-hyper singularity treatment for solving 3D electric field integral equations. *Microw. Opt. Technol. Lett.* 49: 1383–1388.

20 Fink, P.W., Wilton, D.R., and Khayat, M.A. (2008). Simple and efficient numerical evaluation of near-hypersingular integrals. *IEEE Antennas Wireless Propagat. Lett.* 7: 469–472.

21 Khayat, M.A., Wilton, D.R., and Fink, P.W. (2008). An improved transformation and optimized sampling scheme for the numerical evaluation of singular and near-singular potentials. *IEEE Antennas Wireless Propagat. Lett.* 7: 377–380.

22 Graglia, R.D. and Lombardi, G. (2008). Machine precision evaluation of singular and nearly singular potential integrals by use of Gauss quadrature formulas for rational functions. *IEEE Trans. Antennas Propagat.* 56: 981–998.

23 Polimeridis, A.G. and Yioultsis, T.V. (2008). On the direct evaluation of weakly-singular integrals in Galerkin mixed potential integral equation formulations. *IEEE Trans. Antennas Propagat.* 56 (9): 3011–3019.

24 Ruck, G.T., Barrick, D.E., Stuart, W.D., and Krichbaum, C.K. (1970). *Radar Cross Section Handbook*. New York: Plenum Press.

25 Wilton, D.R., Rao, S.M., Glisson, A.W. et al. (1984). Potential integrals for uniform and linear source distributions on polygonal and polyhedral domains. *IEEE Trans. Antennas Propagat. AP-32*: 276–281.

26 Cowper, G.R. (2003). Singularity extraction technique for integral equation methods with higher order basis functions on plane triangles and tetrahedra. *Int. J. Numer. Meth. Eng.* 58: 1149–1165.

27 Burghignoli, P., Di Nallo, C., Frezza, F., and Galli, A. (2003). A simple Nyström approach for the analysis of 3D arbitrarily shaped conducting and dielectric bodies. *Int. J. Numer. Model.* 16: 179–194.

28 Ylä-Oijala, P. and Taskinen, M. (2003). Calculation of CFIE impedance matrix elements with RWG and n×RWG functions. *IEEE Trans. Antennas Propagat.* 51: 1837–1846.

29 Chew, W.C., Jin, J.M., Michielssen, E., and Song, J.M. (2001). *Fast and Efficient Algorithms in Computational Electromagnetics.* Boston: Artech House.

30 Canino, L.S., Ottusch, J.J., Stalzer, M.A. et al. (1998). Numerical solution of the Helmholtz equation in 2D and 3D using a high-order Nyström discretization. *J. Comput. Phys.* 146: 627–663.

31 Gedney, S.D. (2003). On deriving a locally corrected Nyström scheme from a quadrature sampled moment method. *IEEE Trans. Antennas Propagat.* 51: 2402–2412.

32 Gedney, S.D. (2000). Application of the high-order Nyström scheme to the integral equation solution of electromagnetic interaction problems. *IEEE International Symposium on Electromagnetic Compatibility*, Washington, DC, vol. 1, pp. 289–294.

33 Stroud, A.H. (1971). *Approximate Calculation of Multiple Integrals.* Englewood Cliffs: Prentice-Hall.

34 Glisson, A.W. and Wilton, D.R. (1980). Simple and efficient numerical methods for problems of electromagnetic radiation and scattering from surfaces. *IEEE Trans. Antennas Propagat.* AP-28: 593–603.

35 Penno, R.P., Thiele, G.A., and Pasala, K.M. (1989). Scattering from a perfectly conducting cube. *Proc. IEEE.* 77: 815–823.

36 Lu, C.C, Chew, W.C., and Song, J.M. (1996). A study of disparate grid sizes for an irregular-shape scatterer on EFIE, MFIE, and CFIE. *IEEE Antennas and Propagation Society International Symposium*, Baltimore, vol. 3, pp. 1746–1749.

37 Dwight, H.B. (1961). *Tables of Integrals and Other Mathematical Data*, 4e. New York: Macmillan.

38 Poggio, A.J. and Miller, E.K. (1973). Integral equation solutions of three-dimensional scattering problems. In: *Computer Techniques for Electromagnetics* (ed. R. Mittra). Oxford: Pergamon Press, Chapter 4.

39 Müller, C. (1969). *Foundations of the Mathematical Theory of Electromagnetic Waves.* Berlin: Springer-Verlag.

40 Gray, L.J., Salvadori, A., Phan, A.-V., and Mantic, A. (2006). Direct evaluation of hypersingular Galerkin surface integrals, *II. Electron J. Boundary Elem.* 4 (3): 105–130.

41 Krommer, A.R. and Ueberhuber, C.W. (1998). *Computational Integration.* Philadelphia: SIAM.

42 Hsiao, G.C., Kleinman, R.E., and Wang, D.-Q. (1999). Applications of boundary integral equation methods in 3D electromagnetic scattering. *J. Comput. Appl. Math.* 104 (2): 89–110.

43 Sheng, X.Q., Jin, J.M., Song, J.M. et al. (1998). Solution of combined-field integral equation using multilevel fast multipole algorithm for scattering by homogeneous bodies. *IEEE Trans. Antennas Propagat.* 46 (11): 1718–1726.

44 Schwab, C. and Wendland, W.L. (1992). On numerical cubatures of singular surface integrals in boundary element methods. *Numer. Math.* 62: 343–369.

45 Klees, R. (1996). Numerical calculation of weakly-singular surface integrals. *J. Geod.* 70 (11): 781–797.

46 Cai, W., Yu, Y., and Yuan, X.C. (2002). Singularity treatment and high-order RWG basis functions for integral equations of electromagnetic scattering. *Int. J. Numer. Meth. Eng.* 53: 31–47.

47 Johnston, B.M. and Johnston, P.R. (2003). A comparison of transformation methods for evaluation two-dimensional weakly-singular integrals. *Int. J. Numer. Meth. Eng.* 56: 589–607.

48 Gray, L.J., Glaeser, J.M., and Kaplan, T. (2004). Direct evaluation of hypersingular Galerkin surface integrals. *SIAM J. Sci. Comput.* 25 (5): 1534–1556.

49 Kellogg, O.D. (1944). *Foundations of Potential Theory*. New York: Frederick Ungar.

50 Nie, X.C., Yuan, N., Li, L.W. et al. (2006). A fast combined field volume integral equation solution to EM scattering by 3-D dielectric objects of arbitrary permittivity and permeability. *IEEE Trans. Antennas Propagat.* 54 (3): 961–969.

51 Sun, L.E. and Chew, W.C. (2009). A novel formulation of the volume integral equation for electromagnetic scattering. *Wave. Random Complex.* 19 (1): 162–180.

52 Schaubert, D.H., Wilton, D.R., and Glisson, A.W. (1984). A tetrahedral modeling method for electromagnetic scattering by arbitrary shaped inhomogeneous dielectric bodies. *IEEE Trans. Antennas Propagat.* AP-32 (1): 77–85.

53 Botha, M.M. (2006). Solving the volume integral equations of electromagnetic scattering. *J. Comput. Phys.* 218 (1): 141–158.

54 Kot, J.S. (1990). Solution of thin-wire integral equations by Nyström methods. *Microw. Opt. Technol. Lett.* 3: 393–396.

55 Kot, J.S. (1992). Application of Nyström methods to electric- and magnetic-field integral equations. *Proceedings of the 1992 URSI International Symposium on Electromagnetic Theory*, Sydney, Australia (August 1992).

56 Peterson, A.F. and Bibby, M.M. (2004). Higher-order numerical solutions of the MFIE for the linear dipole. *IEEE Trans. Antennas Propagat.* 52: 2684–2691.

57 Liu, G. and Gedney, S.D. (2000). High-order Nyström solution of the volume-EFIE for TM-wave scattering. *Microw. Opt. Technol. Lett.* 25: 8–11.

58 Liu, G. and Gedney, S.D. (2001). High-order Nyström solution of the volume-EFIE for TE-wave scattering. *Electromagnetics.* 21: 1–14.

59 Silvester, P. (1972). Tetrahedronal polynomial finite elements for the Helmholtz equation. *Int. J. Numer. Meth. Eng.* 4 (3): 405–413.

60 Björck, Å. (1996). *Numerical Methods for Least Squares Problems*. Philadelphia, PA: SIAM.

61 Wang, P.C., Zhou, Z.G., Hong, J.X. et al. (2014). Evaluation of hypersingular volume integrals over a cylindrical domain in meshless methods. *IEEE Antennas and Propagation Society International Symposium*, Memphis, Tennessee (July 2014).

62 Tong, M.S. (2012). Meshfree solutions of volume integral equations for electromagnetic scattering by anisotropic objects. *IEEE Trans. Antennas Propagat.* 60 (9): 4249–4258.

63 Harrington, R.F. (1968). *Field Computations by Moment Methods*. New York: Macmillan.

64 Miller, E.K., Medgyesi-Mitschang, L., and Newman, E.H. (eds.) (1992). *Computational Electromagnetics: Frequency-Domain Method of Moments*. New York: IEEE Press.

65 Gibson, W.C. (2008). *The Method of Moments in Electromagnetics*. Boca Raton: Chapman & Hall/CRC.

66 Peterson, A.F. and Bibby, M.M. (2003). Three discretizations of the MFIE for the linear dipole. *Proc. IEEE Antennas Propagat.* 3: 6–9.

67 Song, J.M., Lu, C.C., and Chew, W.C. (1997). Multilevel fast multipole algorithm for electromagnetic scattering by large complex objects. *IEEE Trans. Antennas Propagat.* 45: 1488–1493.

6

Application to Conducting Media

6.1 Introduction

The previous chapters are mainly dedicated to the complete delineation of the Nyström method, from its origin, history, basic principles, quadrature rules, and singularity treatment. With this necessary information and components, we can now apply this method to solve real electromagnetics (EM) problems. We begin with conducting objects, which are the simplest materials, and then move to penetrable and composite objects, which are the complicated materials. This chapter only considers conducting objects, and penetrable and composite objects will be addressed in the next chapter.

For conducting objects, we first consider two-dimensional (2D) problems which are approximate in geometry. The problems include large concave structures and open structures with incorporation of edge conditions, and we illustrate how to solve them by using the Nyström method. We then extend to solve body-of-revolution (BOR) problems which can be reduced from a three-dimensional (3D) form to a 2D form by applying a Fourier expansion. For regular 3D problems, we use the electric field integral equation (EFIE), magnetic field integral equation (MFIE), and combined field integral equation (CFIE), respectively, to describe the problems, and illustrate the solutions of those equations by the Nyström method. In particular, we consider the very-thin or super-thin conducting structures using the MFIE and CFIE because one could meet several unfavorable factors in solving for such structures [1]. Although we have designated Chapter 5 to specially address the singularity treatment, it may be better to present some singularity treatment techniques in incorporation with their applications and we will do so in this chapter. Finally, we present appropriate numerical examples to demonstrate the method and provide necessary numerical analysis and comparison.

6.2 Solution for 2D Structures

The Nyström method has been used to solve EM problems since 1990 [2]. As an initial demonstration for the method, we consider the simplest 2D cases, i.e. the objects are assumed to be infinitely long and the 3D governing integral equations can be reduced to 2D versions by a mathematical manipulation. Also, we only need to discretize the contours of cross sections of objects because the unknown current density is assumed to be unchanged in the axial direction and we only need to determine its distribution along the contours of cross sections of objects. Both transverse magnetic (TM) and transverse

The Nyström Method in Electromagnetics, First Edition. Mei Song Tong and Weng Cho Chew.

electric (TE) cases are considered and the singularity treatment techniques developed in Chapter 5 are applied in the demonstration. In particular, an electrically large object with a concave shape is considered in numerical examples so that the performance of the method can be fully examined because the solution of such an object is very oscillatory and an extra attention should be paid to the numerical implementation.

Compared with the method of moments (MoM), the Nyström method may use more unknowns for the same geometrical discretization, as mentioned before. To reduce the number of unknowns, we can use a higher-order scheme with a coarse geometrical discretization if the accuracy is not sacrificed. It has been proved that higher-order quadrature rules or p-refinement can give a higher rate of convergence than geometrical refinement or h-refinement [3, 4]. In addition to the generalized approach, we can also make use of some *a priori* information to further reduce the number of unknowns. For example, we can incorporate the edge condition (EC) to control the edge behavior of current density for 2D open structures. Since the EC correctly reflects the change of current density near geometrical edges, incorporating the EC can greatly coarsen the discretization near the edges without loss of accuracy.

The behavior of EM fields near an edge was first studied by Bouwkamp [5] and the so-called EC was named by his study. The EC states that the EM energy density must be integrable over any finite domain, even if this domain contains a singularity of an EM field [6]. This condition requires that the tangential component of current density be on the order of $\rho^{-1/2}$, where ρ is the distance from the edge, while the normal component be on the order of $\rho^{1/2}$. The EC was taken into account in the MoM solutions very early [7, 8], but it had not been incorporated with the Nyström method until our work was done. Also, we apply the EC to higher-order schemes so that its merits can be further exploited.

To investigate the function of EC in the Nyström method, we need to greatly coarsen the discretization near edges. This will require us to use more terms in the approximation of the singular kernel, which is the zeroth-order Hankel function in the TM case, and zeroth- and second-order Hankel functions in the TE case. To this end, we derive closed-form formulas for the integration of the singular kernel with arbitrary-term approximation multiplied by polynomials with or without the EC. Although there are simpler series-approximation formulas for singular kernels [10], we use their original series definition to emphasize the approximation accuracy. The polynomials are used to represent the current density over a self or near segment in the local correction. The higher-order polynomials will generate a higher-order rate of convergence when other numerical errors are negligible. It is demonstrated by our numerical experiments that the numerical solutions will converge much faster when using the EC than without using the EC. In addition, the use of the EC is more significant in TM polarization than in TE polarization due to the singular behavior of TM current density near edges. The numerical errors for TM current density may be uncontrollable without using the EC.

6.2.1 General 2D Structures

Consider TM^z wave scattering by a 2D conducting scatterer as shown in Figure 6.1. The problem can be represented by the following EFIE [9]

$$E_z^i(\rho) = \frac{\kappa\eta}{4}\int_C J_z(\rho')H_0^{(2)}(\kappa|\rho-\rho'|)\mathrm{d}\ell' \qquad \rho \text{ and } \rho' \text{ on } C \tag{6.1}$$

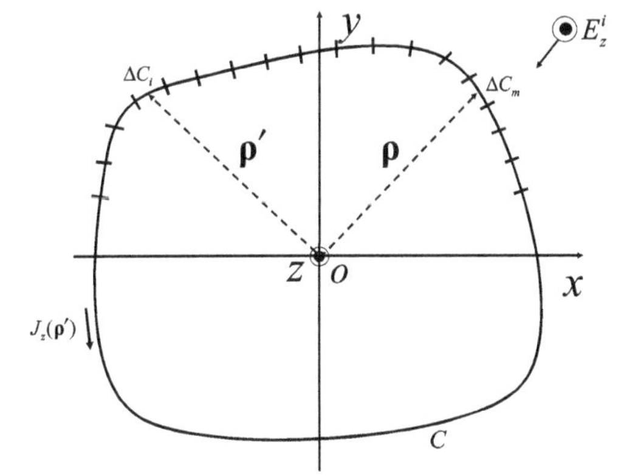

Figure 6.1 TM^z wave scattering by a 2D conducting scatterer.

where E_z^i denotes the incident electric field, κ is the wavenumber of free space, η is the wave impedance of free space, J_z denotes the unknown current density on the contour C of the scatterer and $H_0^{(2)}$ denotes the second kind of Hankel function of order 0. Also, ρ and ρ' are the position vectors of observation point and source point at the contour, respectively. To solve the equation with the unknown current density J_z by using the Nyström method, we first discretize the contour of the scatterer into N segments with a step size of h, and then the EFIE becomes

$$E_z^i(\rho) = \sum_{i=1}^{N} \int_{\Delta C} g(\rho, \rho') J_z(\rho') d\ell' \quad \rho \text{ and } \rho' \text{ on } C \tag{6.2}$$

where $g(\rho, \rho') = \frac{\kappa\eta}{4} H_0^{(2)}(\kappa|\rho - \rho'|)$. In each segment, the integration can be replaced by the summation using an appropriate quadrature rule, say Gauss–Legendre quadrature rule. We then obtain

$$E_z^i(\rho) = \sum_{i=1}^{N} \sum_{j=1}^{P} g(\rho, \rho'_{ij}) J_z(\rho'_{ij}) w_{ij} \tag{6.3}$$

where P is the number of quadrature points on each segment and w_{ij} is the weight of the quadrature rule on the jth point of the ith segment. Choosing the quadrature points as the observation points as well and performing the point-matching procedure on these quadrature points results in the following algebraic matrix equation

$$E_z^i(\rho_{mn}) = \sum_{i=1}^{N} \sum_{j=1}^{P} g(\rho_{mn}, \rho'_{ij}) J_z(\rho'_{ij}) w_{ij} \tag{6.4}$$

where $m = 1, 2, \cdots, N$ and $n = 1, 2, \cdots, P$. This is an $NP \times NP$ matrix equation with NP unknown values of the current density on NP quadrature points. The resulting matrix equation is also solved using iteration methods when the matrix size is very large. The problem in this process is that the diagonal elements in the matrix cannot be determined because the observation points coincide the source points with a logarithm singularity resulting from the Green's function. This problem can be solved by using a local correction based on the singularity treatment in [11]. Due to the complexity of the local correction with a high order, We adopt a simpler but efficient procedure proposed in [12] to evaluate the diagonal elements. We rewrite (6.4) by extracting the self term and obtain

$$E_z^i(\rho_{mn}) = \sum_{\substack{i=1\\ i\neq m}}^{N} \sum_{j=1}^{P} g(\rho_{mn}, \rho'_{ij}) J_z(\rho'_{ij}) w_{ij} + A \tag{6.5}$$

where

$$A = \sum_{j=1}^{P} g(\rho_{mn}, \rho'_{mj}) J_z(\rho'_{mj}) w_{mj} \tag{6.6}$$

is the self term. The self term can be further rewritten as

$$\begin{aligned} A &= \sum_{\substack{j=1\\ j\neq n}}^{P} g(\rho_{mn}, \rho'_{mj}) J_z(\rho'_{mj}) w_{mj} + g(\rho_{mn}, \rho'_{mn}) J_z(\rho'_{mn}) w_{mn} \\ &= \sum_{j=1}^{P} \left[J_z(\rho'_{mj}) - J_z(\rho'_{mn}) \right] g(\rho_{mn}, \rho'_{mj}) w_{mj} + J_z(\rho'_{mn}) B \end{aligned} \tag{6.7}$$

where

$$B = \int_{\Delta C_m} g(\rho_{mn}, \rho') \mathrm{d}\ell'. \tag{6.8}$$

The integral (6.8) can be evaluated in a closed form by using the small argument approximation of the Hankel function [9], namely

$$B = \Delta C_m \left[1 - j\frac{2}{\pi} \ln\left(\frac{1.781\kappa\Delta C_m}{4e} \right) \right] \tag{6.9}$$

where $e = 2.718$. This integral can also be efficiently evaluated using the generalized quadrature rule proposed in [11].

6.2.2 2D Open Structures with Edge Conditions

Consider the EM scattering by a 2D conducting strip sketched in Figure 6.2. The EFIE to describe this problem can be written as [9]

$$\frac{\kappa\eta}{4} \int_0^w J_z(x') H_0^{(1)}(\kappa\rho)\, \mathrm{d}x' = \mathrm{e}^{\mathrm{i}\kappa x \cos\phi_i} \tag{6.10}$$

$$\frac{\kappa\eta}{8} \int_0^w J_x(x') \left[H_0^{(1)}(\kappa\rho) + H_2^{(1)}(\kappa\rho) \cos 2\psi \right]\, \mathrm{d}x' = \sin\phi_i \mathrm{e}^{\mathrm{i}\kappa x \cos\phi_i} \tag{6.11}$$

for TM polarization and TE polarization, respectively. In the above, κ is the wavenumber in the free space, η is the wave impedance in the free space, J_z and J_x are the induced current densities along the z and x directions corresponding to the TM polarization and TE polarization, respectively. Also, $H_0^{(1)}$ and $H_2^{(1)}$ are the first kind of Hankel function of order 0 and order 2, respectively, and $\rho = |x - x'|$ is the distance between a source point at x' and an observation point at x on the strip. The angles ϕ_i and ψ are defined in

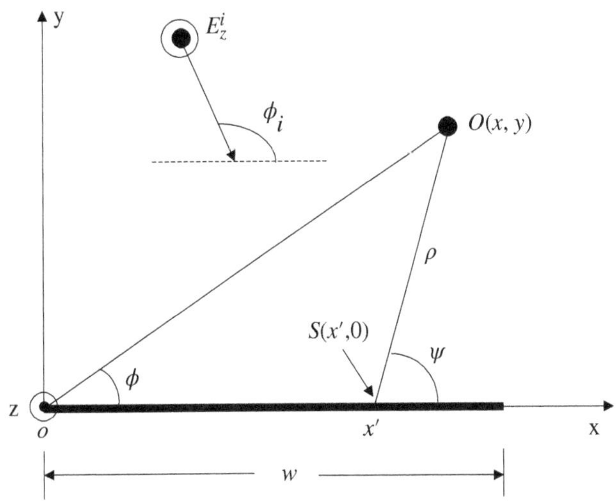

Figure 6.2 Geometry of scattering by a 2D conducting strip with a finite width.

Figure 6.2. To solve the two EFIEs, we discretize the strip into N segments along the x direction and the EFIEs can be uniformly written as

$$\sum_{i=1}^{N} \int_{x_i}^{x_{i+1}} J(x')g(x, x')\, \mathrm{d}x' = V(x) \tag{6.12}$$

where $J(x')$ is either $J_z(x')$ (TM) or $J_x(x')$ (TE), $V(x)$ stands for the corresponding right-hand side, and $g(x, x')$ denotes the integral kernels of EFIEs. The Nyström method states that the integration over a segment can be replaced with a summation defined by a quadrature rule, i.e.

$$I = \int_{x_i}^{x_{i+1}} J(x')g(x, x')\, \mathrm{d}x' = \sum_{j=1}^{P} w_{ij} g(x, x'_{ij}) J_{ij} \tag{6.13}$$

where P is the number of quadrature points, w_{ij} is the jth weight of a quadrature rule within the ith segment and $J_{ij} = J(x'_{ij})$ is the current density at the jth quadrature point within the ith segment. J_{ij} $(i = 1, \cdots, N;\ j = 1, \cdots, P)$ are also the unknowns we want to solve for in the matrix equation. We choose the Gauss–Legendre quadrature rule for the integration in (6.13) for a higher-order accuracy. However, the quadrature rule can only be applied when the integrand is smooth or the source point is far away from the observation point. If the source point coincides or approaches the observation point, the kernel is singular or near-singular and a local correction is needed. We have developed a simple higher-order local correction scheme for the scattering by 3D close structures [14]. This scheme can also be used for 2D open structures by incorporating the EC. On the non-end self or near segments, the current density is approximated by using a polynomial, i.e.

$$J^N(x') = \sum_{k=0}^{q} a_k x'^k \tag{6.14}$$

where q is the order of the polynomial and a_k $(k = 0, 1, \cdots, q)$ are the coefficients. The superscript N implies the non-end segment and no subscript in J denotes that the approximation is valid for both TM and TE cases. For the self and near segments at two ends, the EC is taken into account and the current density is approximated as

$$J_z^L(x') = \frac{1}{\sqrt{x'}} \sum_{k=0}^{q} a_k x'^k \tag{6.15}$$

$$J_z^R(x') = \frac{1}{\sqrt{w - x'}} \sum_{k=0}^{q} a_k x'^k \tag{6.16}$$

$$J_x^L(x') = \sqrt{x'} \sum_{k=0}^{q} a_k x'^k \tag{6.17}$$

$$J_x^R(x') = \sqrt{w - x'} \sum_{k=0}^{q} a_k x'^k \tag{6.18}$$

where the subscripts z and x in J correspond to TM and TE polarizations, and the superscripts L and R correspond to the left end and right end, respectively. Since the EC in (6.15)–(6.18) allows the approximation of the current density to be much closer to the real current density, we may dramatically coarsen the discretization and thus reduce the number of unknowns. The polynomial coefficients can be expressed into current densities at the quadrature points by matching those current densities at those quadrature points. This requires $q = P - 1$, i.e., the order of polynomial is one less than the number of quadrature points, to determine the coefficients uniquely. For $q = 1$ (linear approximation), the two-point quadrature rule is used and after determining those polynomial coefficients, the current density can be written as

$$J^N(x') = \frac{(x_2' - x')J_1 + (x' - x_1')J_2}{x_2' - x_1'} \tag{6.19}$$

$$J_z^L(x') = \frac{\sqrt{x_1'}(x_2' - x')J_1 + \sqrt{x_2'}(x' - x_1')J_2}{\sqrt{x'}(x_2' - x_1')} \tag{6.20}$$

$$J_z^R(x') = \frac{\sqrt{w - x_1'}(x_2' - x')J_1 + \sqrt{w - x_2'}(x' - x_1')J_2}{\sqrt{w - x'}(x_2' - x_1')} \tag{6.21}$$

$$J_x^L(x') = \frac{\sqrt{x'}}{x_2' - x_1'} \left[\frac{x_2' - x'}{\sqrt{x_1'}} J_1 + \frac{x' - x_1'}{\sqrt{x_2'}} J_2 \right] \tag{6.22}$$

$$J_x^R(x') = \frac{\sqrt{w - x'}}{x_2' - x_1'} \left[\frac{x_2' - x'}{\sqrt{w - x_1'}} J_1 + \frac{x' - x_1'}{\sqrt{w - x_2'}} J_2 \right] \tag{6.23}$$

where x_1' and x_2' are the coordinates of the two quadrature points on each segment and J_1 and J_2 are the current densities on those quadrature points. The current density expressions using a quadratic approximation ($q = 2$) and three-point quadrature rule can be derived in a similar way. Substituting the above current density approximation into the self and near segments in the discretized EFIE (6.12), we obtain

$$I^N = \sum_{k=0}^{q} J_{ik} \cdot \int_{x_i}^{x_{i+1}} a_k x'^k g(x, x')\, \mathrm{d}x' \tag{6.24}$$

$$I_z^L = \sum_{k=0}^{q} J_{ik} \cdot \int_{x_i}^{x_{i+1}} a_k x'^{k-0.5} g(x, x') \, \mathrm{d}x' \tag{6.25}$$

$$I_z^R = \sum_{k=0}^{q} J_{ik} \cdot \int_{x_i}^{x_{i+1}} \frac{a_k x'^k}{\sqrt{w - x'}} g(x, x') \, \mathrm{d}x' \tag{6.26}$$

$$I_x^L = \sum_{k=0}^{q} J_{ik} \cdot \int_{x_i}^{x_{i+1}} a_k x'^{k+0.5} g(x, x') \, \mathrm{d}x' \tag{6.27}$$

$$I_x^R = \sum_{k=0}^{q} J_{ik} \cdot \int_{x_i}^{x_{i+1}} a_k x'^k \sqrt{w - x'} g(x, x') \, \mathrm{d}x'. \tag{6.28}$$

In the above equations, we have assumed that the observation points are chosen arbitrarily on the strip. To create the matrix equation to solve for the unknown current densities on quadrature points, a point-matching process is needed by choosing the quadrature points as the observation points, i.e. $x = x'_{mn}$ $(m = 1, \cdots, N;\ n = 1, \cdots, P)$. Once the observation points are chosen, the far, near and self segments can be defined accordingly. In the determination of matrix elements, the accurate evaluation of the integrations on the right-hand sides of (6.24)–(6.28) is very critical. These integrations correspond to the diagonal and near-diagonal elements of the impedance matrix and their accuracy is tightly associated with the accuracy of solution. The other elements of the matrix, which reflect the far interactions between source points and observation points, are determined using (6.13) directly if the source points are not inside end segments. If the source points fall into end segments, numerical integrations are needed because the current densities are expressed as the polynomials with the EC. In this case, the Gauss–Jacobi quadrature rule designed for the following integral is used [15]

$$I = \int_{-1}^{1} (1 - x)^{\alpha} (1 + x)^{\beta} f(x) \, \mathrm{d}x = \sum_{j=1}^{P} w_j f(x_j) \tag{6.29}$$

where we take $\alpha = -0.5$ (TM) or 0.5 (TE) and $\beta = 0$.

6.2.3 Evaluation of Singular and Near-Singular Integrations

Those integrals in (6.24)–(6.28) are singular or near-singular. For the TM case, the kernel is the first kind of Hankel function of order 0, which includes an integrable logarithmic singularity. The Lin-Log quadrature rule [16] has been designed to integrate such kind of integrands numerically. Nevertheless, analytical formulas are more desirable due to their accuracy, simplicity and flexibility in implementation. We derive such kind of closed-form expressions for those integrals based on the series approximation of the Hankel function. For TE case, the kernel possesses a hypersingularity in addition to the logarithmic singularity due to the appearance of the first kind of Hankel function of order 2. In this case, the integral associated with the hypersingular term is defined in a Cauchy-principal-value sense and the integration is performed by assuming that the observation point is initially located at a point (x_0, y_0) with $y_0 \neq 0$, and taking the $y_0 \to 0$ limit after finding the analytical expression for the integration [13]. Since we will use very coarse discretization to reach a higher-order accuracy, the commonly used small-argument approximation for the Hankel functions by taking the first series term

[13] may not be accurate for us. For this reason, we derive the analytical formulas for the integration of Hankel functions with arbitrary-term approximation multiplied by a polynomial with or without the EC.

TM case

In the TM case, we need to perform the following three integrals found in (6.24)–(6.28)

$$I^N = \int_{x_i}^{x_{i+1}} x'^k H_0^{(1)}(\kappa|x_0 - x'|)\, dx' \tag{6.30}$$

$$I^L = \int_{x_i}^{x_{i+1}} x'^{k-0.5} H_0^{(1)}(\kappa|x_0 - x'|)\, dx' \tag{6.31}$$

$$I^R = \int_{x_i}^{x_{i+1}} \frac{x'^k}{\sqrt{w - x'}} H_0^{(1)}(\kappa|x_0 - x'|)\, dx' \tag{6.32}$$

where x_0 is the observation point falling into the integral interval (for self segments) or neighboring integral intervals (for near segments). The first kind of Hankel function of order 0 can be expressed in a series form [10]

$$H_0^{(1)}(z) = \frac{i2}{\pi} \ln\left(\frac{z}{2}\right) \sum_{n=0}^{\infty} \frac{\left(-\frac{1}{4}z^2\right)^n}{(n!)^2} + \sum_{n=0}^{\infty} \left[1 - \frac{i2}{\pi}\psi(n+1)\right] \frac{\left(-\frac{1}{4}z^2\right)^n}{(n!)^2} \tag{6.33}$$

and the singularity comes from the $\ln\left(\frac{z}{2}\right)$ term. Note that the calculation of the Hankel function by the series expansion is numerically inefficient when the argument is very large [17]. However, the argument is usually smaller than 15 because the above series only applies to the self and near interactions between a source point and an observation point and the series converges very fast. For example, the numerical error of the first kind of Hankel functions of both order 0 and order 2 will be below 10^{-10} when the number of terms in the series reaches 20 if $z = 15$. Substituting the series expansion for the Hankel function to (6.30)–(6.32), the singular parts of the integrals take the following form for the nth term

$$\begin{aligned} I_s^N = &\int_{x_i}^{x_0} x'^k (x_0 - x')^{2n} \ln(x_0 - x')\, dx' \\ &+ \int_{x_0}^{x_{i+1}} x'^k (x' - x_0)^{2n} \ln(x' - x_0)\, dx' \end{aligned} \tag{6.34}$$

$$\begin{aligned} I_s^L = &\int_{x_i}^{x_0} x'^{k-0.5} (x_0 - x')^{2n} \ln(x_0 - x')\, dx' \\ &+ \int_{x_0}^{x_{i+1}} x'^{k-0.5} (x' - x_0)^{2n} \ln(x' - x_0)\, dx' \end{aligned} \tag{6.35}$$

$$\begin{aligned} I_s^R = &\int_{x_i}^{x_0} (w - x')^{-0.5} x'^k (x_0 - x')^{2n} \ln(x_0 - x')\, dx' \\ &+ \int_{x_0}^{x_{i+1}} (w - x')^{-0.5} x'^k (x' - x_0)^{2n} \ln(x' - x_0)\, dx' \end{aligned} \tag{6.36}$$

where the subscript s means the singular part. With the aid of (610.9) and (621.9) in [18], we can derive the following closed-form expressions for these integrals

$$I_s^N = \sum_{p=0}^{k} C_k^p x_0^p \left[(-1)^p \frac{\delta_1^\ell}{\ell} \left(\ln \delta_1 - \frac{1}{\ell} \right) + \frac{\delta_2^\ell}{\ell} \left(\ln \delta_2 - \frac{1}{\ell} \right) \right] \tag{6.37}$$

$$I_s^L = \sum_{p=0}^{2n} C_{2n}^p x_0^{2n-p} (-1)^p \frac{1}{u} \left[x_{i+1}^u \ln \delta_2 - x_i^u \ln \delta_1 + I_1^{(\alpha)} \right] \tag{6.38}$$

where

$$I_1^{(\alpha)} = \frac{1}{u} \left(x_i^u - x_{i+1}^u \right) + x_0 I_1^{(\alpha-1)} \tag{6.39}$$

$$I_1^{(0)} = 2 \left(\sqrt{x_i} - \sqrt{x_{i+1}} \right) + \sqrt{x_0} \left(\ln \left| \frac{\sqrt{x_i} - \sqrt{x_0}}{\sqrt{x_i} + \sqrt{x_0}} \right| - \ln \left| \frac{\sqrt{x_{i+1}} - \sqrt{x_0}}{\sqrt{x_{i+1}} + \sqrt{x_0}} \right| \right) \tag{6.40}$$

and

$$I_s^R = \sum_{p=0}^{2n} \sum_{r=0}^{k} C_{2n}^p C_k^r t_0^{4n-2p} (-1)^\beta w^{k-r} \frac{2}{v} \left[t_1^v \ln \delta_1 - t_2^v \ln \delta_2 + I_2^{(\beta)} \right] \tag{6.41}$$

where

$$I_2^{(\beta)} = \frac{2}{v} \left(t_2^v - t_1^v \right) + t_0^2 I_2^{(\beta-1)} \tag{6.42}$$

$$I_2^{(0)} = 2 \left(t_2 - t_1 \right) + t_0 \left(\ln \left| \frac{t_2 - t_0}{t_2 + t_0} \right| - \ln \left| \frac{t_1 - t_0}{t_1 + t_0} \right| \right). \tag{6.43}$$

In the above, $\ell = 2n + k + 1 - p$, $\delta_1 = x_0 - x_i$, $\delta_2 = x_{i+1} - x_0$, $u = p + k + 0.5$, $\alpha = p + k$, $v = 2(p + r) + 1$, $\beta = p + r$, $t_0 = \sqrt{w - x_0}$, $t_1 = \sqrt{w - x_i}$, $t_2 = \sqrt{w - x_{i+1}}$, and C_k^p, C_{2n}^p and C_k^r are the binomial expansion coefficients. Also, $I_1^{(\alpha)}$ and $I_2^{(\beta)}$ are calculated recursively, in which $I_1^{(0)}$ and $I_2^{(0)}$ are their initial points, respectively. The non-singular part of the Hankel function is a polynomial and the integration with this part in (6.30)–(6.32) can be easily derived. For the integrations over near segments in (6.30)–(6.32), there is no singularity in the kernel and the corresponding formulas can be derived in a similar way.

TE case

In the TE case, the singular and near-singular integrations consist of two parts. The first part comes from the first kind of Hankel function of order 0 as shown in (6.33) and the singular integrals can be expressed as

$$I_s^N = \int_{x_i}^{x_0} x'^k (x_0 - x')^{2n} \ln(x_0 - x') \, dx' + \int_{x_0}^{x_{i+1}} x'^k (x' - x_0)^{2n} \ln(x' - x_0) \, dx' \tag{6.44}$$

$$I_s^L = \int_{x_i}^{x_0} x'^{k+0.5} (x_0 - x')^{2n} \ln(x_0 - x') \, dx' + \int_{x_0}^{x_{i+1}} x'^{k+0.5} (x' - x_0)^{2n} \ln(x' - x_0) \, dx' \tag{6.45}$$

$$I_s^R = \int_{x_i}^{x_0} \sqrt{w - x'} x'^k (x_0 - x')^{2n} \ln(x_0 - x') \, dx' + \int_{x_0}^{x_{i+1}} \sqrt{w - x'} x'^k (x' - x_0)^{2n} \ln(x' - x_0) \, dx'. \tag{6.46}$$

The closed-form expressions for these integrals are the same as those in (6.37)–(6.43) except for the following two formulas which are the initial points of the two recursive formulas

$$I_1^{(0)} = \frac{2}{3}\left(x_i^{1.5} - x_{i+1}^{1.5}\right) + 2x_0\left(\sqrt{x_i} - \sqrt{x_{i+1}}\right) + x_0^{1.5}\left(\ln\left|\frac{\sqrt{x_i} - \sqrt{x_0}}{\sqrt{x_i} + \sqrt{x_0}}\right| - \ln\left|\frac{\sqrt{x_{i+1}} - \sqrt{x_0}}{\sqrt{x_{i+1}} + \sqrt{x_0}}\right|\right) \tag{6.47}$$

$$I_2^{(0)} = \frac{2}{3}\left(t_2^{1.5} - t_1^{1.5}\right) + 2t_0^2(t_2 - t_1) + t_0^{1.5}\left(\ln\left|\frac{t_2 - t_0}{t_2 + t_0}\right| - \ln\left|\frac{t_1 - t_0}{t_1 + t_0}\right|\right) \tag{6.48}$$

where $u = p + k + 1.5$ and $v = 2(p + r) + 3$.

The second part comes from the first kind of Hankel function of order 2 and the integrals take the following forms

$$I_s^N = \int_{x_i}^{x_{i+1}} x'^k H_2^{(1)}(\kappa|x_0 - x'|) \cos 2\psi \, dx' \tag{6.49}$$

$$I_s^L = \int_{x_i}^{x_{i+1}} x'^{k+0.5} H_2^{(1)}(\kappa|x_0 - x'|) \cos 2\psi \, dx' \tag{6.50}$$

$$I_s^R = \int_{x_i}^{x_{i+1}} x'^k \sqrt{w - x'} H_2^{(1)}(\kappa|x_0 - x'|) \cos 2\psi \, dx'. \tag{6.51}$$

By checking the series expression of the first kind of Hankel function of order 2 [9]

$$H_2^{(1)}(z) = -\frac{i}{\pi} - \frac{i4}{\pi z^2} + \frac{i2}{\pi}\ln\left(\frac{z}{2}\right)\sum_{k=0}^{\infty}\frac{z^2\left(-\frac{1}{4}z^2\right)^k}{4k!(k+2)!} + \sum_{k=0}^{\infty}\left\{1 - \frac{i}{\pi}\left[\psi(k+1) + \psi(k+3)\right]\right\}\frac{z^2\left(-\frac{1}{4}z^2\right)^k}{4k!(k+2)!} \tag{6.52}$$

we find that the singular integrations come from the terms including $\frac{1}{z^2}$ and $\ln\left(\frac{z}{2}\right)$. For the integration associated with $\frac{1}{z^2}$, it is defined in a Cauchy-principal-value sense. We assume that the observation point is initially located at the point (x_0, y_0) with $y_0 \neq 0$, which results in a non-singular integration and the original integration is the $y_0 \to 0$ limit of the non-singular integration. This can be written as

$$I_s^N = \lim_{y_0 \to 0} \int_{x_i}^{x_{i+1}} \frac{x'^k}{(x_0 - x')^2 + y_0^2} \cos 2\psi \, dx' = \lim_{y_0 \to 0} \text{Re}\left\{\int_{x_i}^{x_{i+1}} \frac{x'^k}{(z_0 - x')^2} \, dx'\right\} \tag{6.53}$$

$$I_s^L = \lim_{y_0 \to 0} \int_{x_i}^{x_{i+1}} \frac{x'^{k+0.5}}{(x_0 - x')^2 + y_0^2} \cos 2\psi \, dx' = \lim_{y_0 \to 0} \text{Re}\left\{\int_{x_i}^{x_{i+1}} \frac{x'^{k+0.5}}{(z_0 - x')^2} \, dx'\right\} \tag{6.54}$$

$$I_s^R = \lim_{y_0 \to 0} \int_{x_i}^{x_{i+1}} \frac{x'^k \sqrt{w - x'}}{(x_0 - x')^2 + y_0^2} \cos 2\psi \, dx' = \lim_{y_0 \to 0} \text{Re}\left\{\int_{x_i}^{x_{i+1}} \frac{x'^k \sqrt{w - x'}}{(z_0 - x')^2} \, dx'\right\} \tag{6.55}$$

where we have used

$$\cos 2\psi = \frac{(x_0 - x')^2 - y_0^2}{(x_0 - x')^2 + y_0^2} \tag{6.56}$$

and introduced a complex number $z_0 = x_0 + \mathrm{i}y_0$ corresponding to the position of observation point on the complex plane. The integrals in the above are all analytically integrable with the aid of (194.1) and (194.2) in [18]. For $k = 0, 1, 2$, we can derive (the subscript of I denotes the value of k)

$$I_0^N = -\left(\frac{1}{\delta_1} + \frac{1}{\delta_2}\right) \tag{6.57}$$

$$I_1^N = \ln\left|\frac{\delta_2}{\delta_1}\right| - x_0\left(\frac{1}{\delta_1} + \frac{1}{\delta_2}\right) \tag{6.58}$$

$$I_2^N = \delta_1 + \delta_2 + 2x_0 \ln\left|\frac{\delta_2}{\delta_1}\right| - x_0^2\left(\frac{1}{\delta_1} + \frac{1}{\delta_2}\right) \tag{6.59}$$

$$I_0^L = \frac{r_3}{2\sqrt{x_0}} - \left(\frac{r_2}{\delta_2} + \frac{r_1}{\delta_1}\right) \tag{6.60}$$

$$I_1^L = -\frac{1}{x_0}\left(\frac{r_2^5}{\delta_2} + \frac{r_1^5}{\delta_1}\right) + \frac{r_2^3 - r_1^3}{x_0} + 3(r_2 - r_1) + \frac{3r_3\sqrt{x_0}}{2} \tag{6.61}$$

$$I_2^L = -\frac{1}{x_0}\left(\frac{r_2^7}{\delta_2} + \frac{r_1^7}{\delta_1}\right) + \frac{r_2^5 - r_1^5}{x_0} + \frac{5}{3}(r_2^3 - r_1^3) + 5x_0(r_2 - r_1) + \frac{5x_0\sqrt{x_0}}{2}r_3 \tag{6.62}$$

$$I_0^R = -\left(\frac{s_2}{\delta_2} + \frac{s_1}{\delta_1}\right) - \frac{s_3}{2s_0} \tag{6.63}$$

$$I_1^R = 2(s_2 - s_1) + s_0 s_3 + x_0 I_0^R \tag{6.64}$$

$$I_2^R = -\frac{2}{3}\left(s_2^3 - s_1^3\right) + 2x_0 I_1^R + x_0^2 I_0^R \tag{6.65}$$

where

$$r_1 = \sqrt{x_0 - \delta_1} \tag{6.66}$$

$$r_2 = \sqrt{x_0 + \delta_2} \tag{6.67}$$

$$r_3 = \ln\left|\frac{r_2 - \sqrt{x_0}}{r_2 + \sqrt{x_0}}\right| - \ln\left|\frac{r_1 - \sqrt{x_0}}{r_1 + \sqrt{x_0}}\right| \tag{6.68}$$

$$s_0 = \sqrt{w - x_0} \tag{6.69}$$

$$s_1 = \sqrt{w - x_0 + \delta_1} \tag{6.70}$$

$$s_2 = \sqrt{w - x_0 - \delta_2} \tag{6.71}$$

$$s_3 = \ln\left|\frac{s_2 - s_0}{s_2 + s_0}\right| - \ln\left|\frac{s_1 - s_0}{s_1 + s_0}\right|. \tag{6.72}$$

For the integrations associated with the $\ln\left(\frac{z}{2}\right)$ term, the observation points are chosen on the strip directly and $\cos 2\psi = 1$, so they can be written as

$$I_s^N = \int_{x_i}^{x_{i+1}} x'^k (x_0 - x')^{2n+2} \ln|x_0 - x'|\,\mathrm{d}x' \tag{6.73}$$

$$I_s^L = \int_{x_i}^{x_{i+1}} x'^{k+0.5} (x_0 - x')^{2n+2} \ln|x_0 - x'|\,\mathrm{d}x' \tag{6.74}$$

$$I_s^R = \int_{x_i}^{x_{i+1}} \sqrt{w - x'x'}^k (x_0 - x')^{2n+2} \ln |x_0 - x'| \, dx' \tag{6.75}$$

and the corresponding formulas are obtained by replacing $2n$ with $2n+2$ in (6.44)–(6.46). For integrations over near segments, there is no singularity and $\cos 2\psi = 1$, and all formulas can be derived in a similar way.

6.2.4 Numerical Examples

To illustrate the Nyström method for solving 2D problems, we first consider the scattering by a 2D conducting L-shaped scatterer, as shown in Figure 6.3. We select $\ell = 250\lambda$ here to enlarge the size of the scatterer with a circumference of 1000λ, where λ is the wavelength in free space. Figure 6.4 shows the corresponding numerical solution for the normalized current density on the contour of the scatterer in the TM case. The contour of the scatterer is discretized into 5000 segments with 4 quadrature points in each segment and the solution is close to that of the MoM with the same number of unknowns, as shown in Figure 6.5.

To check the function of the EC in the Nyström method, we take $w = 2\lambda$ and 5λ, respectively, and choose $\phi_i = 90°$ in Figure 6.2 to solve for the current density on the strip surface for both TM and TE polarizations. Figures 6.6–6.9 show the corresponding solutions with and without the EC based on a quadratic polynomial approximation ($q = 2$) for the current densities in the local correction. The solutions when $N = 500$ for $w = 2\lambda$ and $N = 1000$ for $w = 5\lambda$ are accurate enough and can be treated as exact solutions for comparison purpose (the root-mean-square or RMS error will be below 10^{-8} when the mesh size decreases further). From these figures, we can see that the solutions with the EC are much more accurate than without the EC for the same discretization. Also,

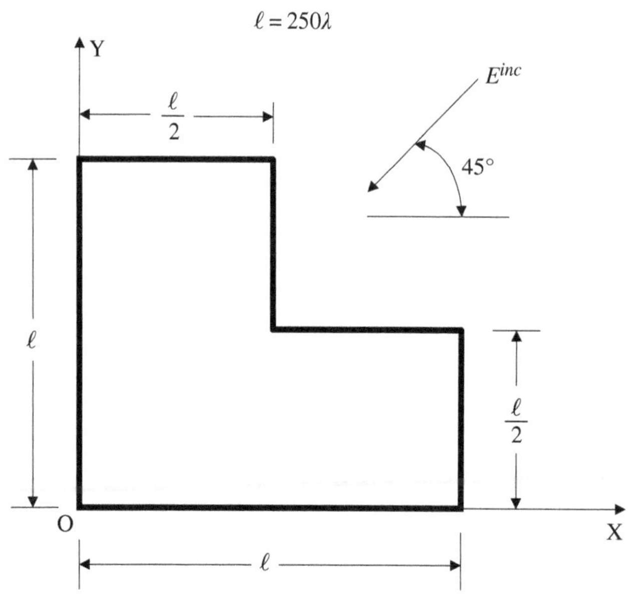

Figure 6.3 Geometry of a 2D conducting L-shape cylinder.

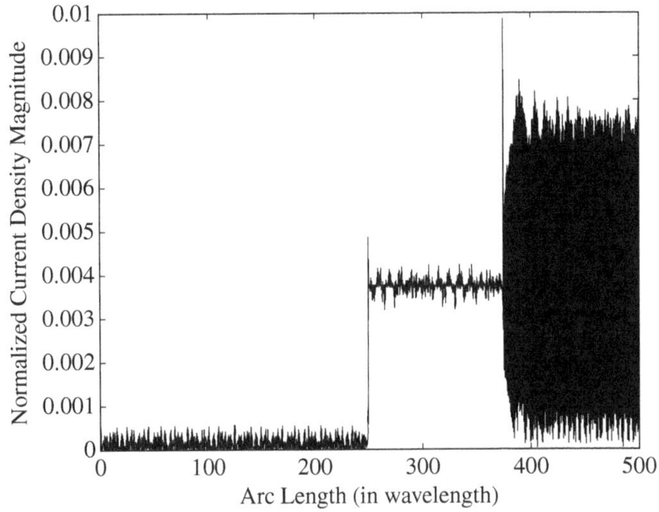

Figure 6.4 Nyström solution with 5000 × 4 unknowns for the current density on the L-shape cylinder.

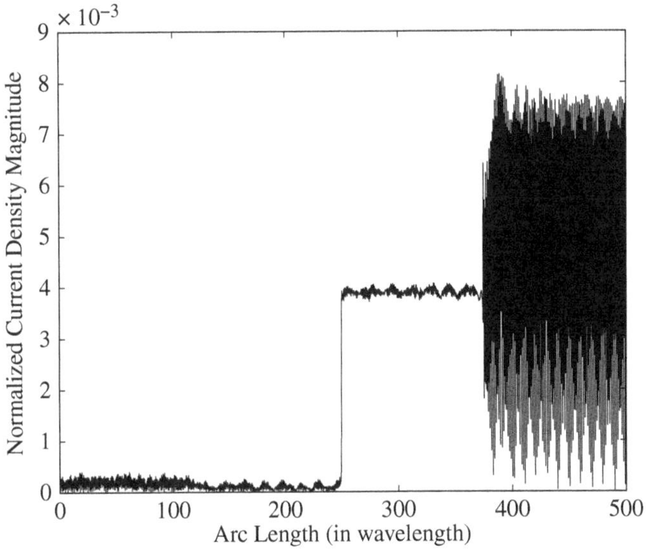

Figure 6.5 MoM solution with 20000 unknowns for the current density on the L-shape cylinder.

if we compare Figure 6.8 with Figure 2 in [13] for the TE case with $w = 2\lambda$, the solutions with $N = 4$ are much better than those MoM solutions with $N = 20$. This fact indicates that the higher-order Nyström method can use much fewer unknowns than the MoM to achieve a similar accuracy.

Figures 6.10 and 6.11 illustrate the current density distribution obtained by interpolation using $N = 10$ and quadratic approximation. These solutions almost coincide with the exact solutions, indicating that the highly accurate current density value at an arbitrary point can be achieved by interpolation even using very coarse discretization. The

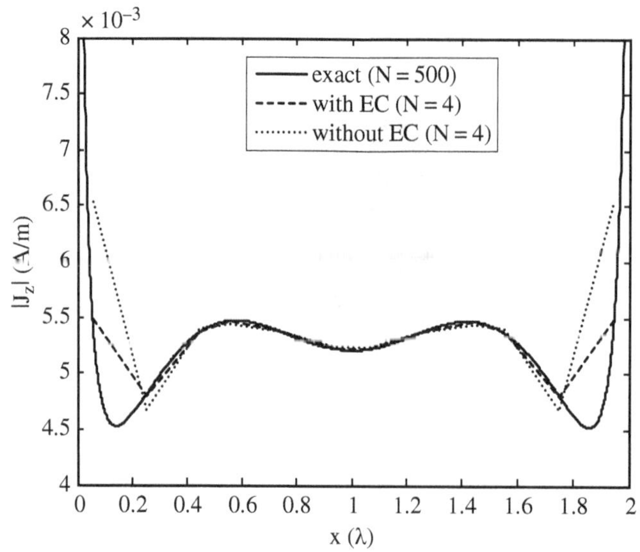

Figure 6.6 Current density distribution on the strip surface for TM polarization, $w = 2\lambda$.

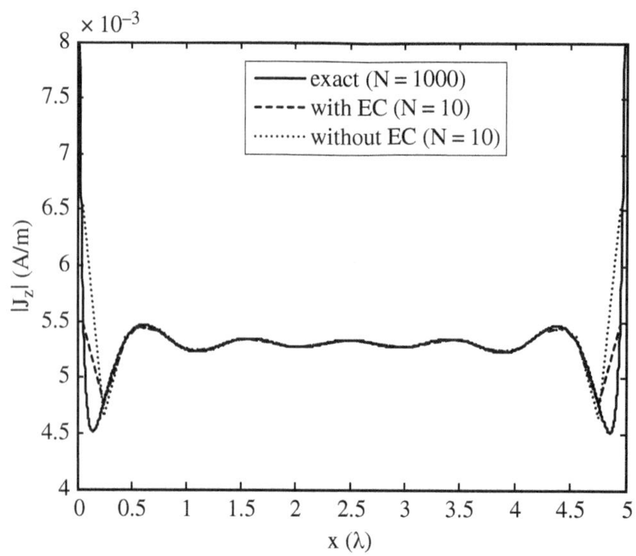

Figure 6.7 Current density distribution on the strip surface for TM polarization, $w = 5\lambda$.

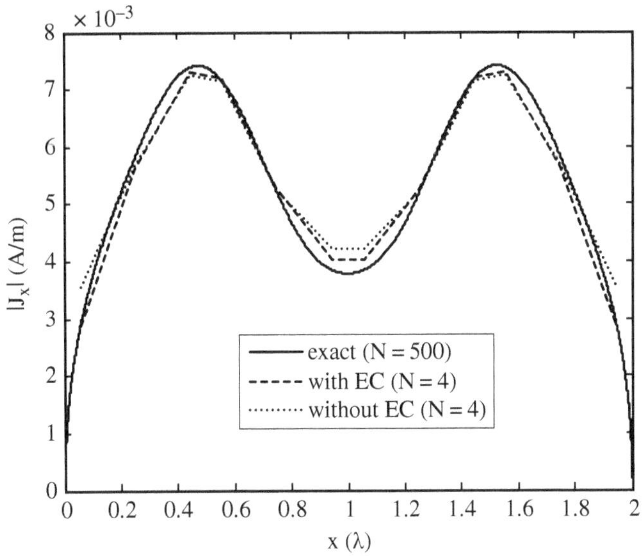

Figure 6.8 Current density distribution on the strip surface for TE polarization, $w = 2\lambda$.

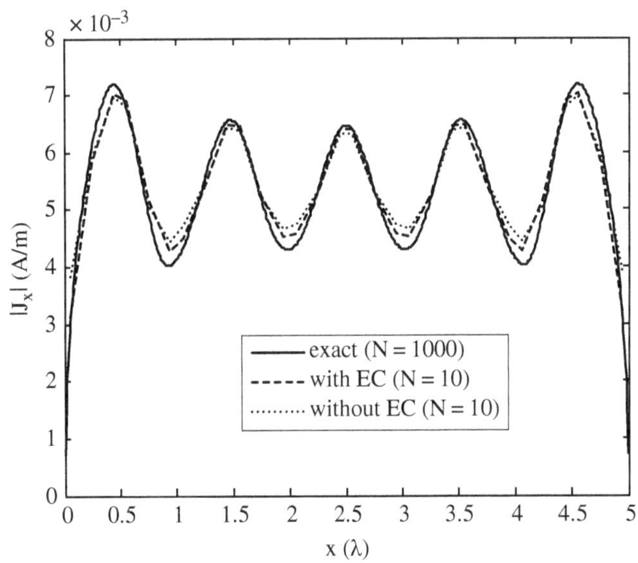

Figure 6.9 Current density distribution on the strip surface for TE polarization, $w = 5\lambda$.

interpolation formulas are the polynomials (6.14) and (6.15)–(6.18) in the local correction scheme.

Figures 6.12–6.15 present the specific numerical errors for those solutions in terms of the RMS error. In the comparison, the exact current density values on non-quadrature points in the exact solutions are obtained by the interpolation based on (6.14) and (6.15)–(6.18). These figures demonstrate that the solutions with an EC converge

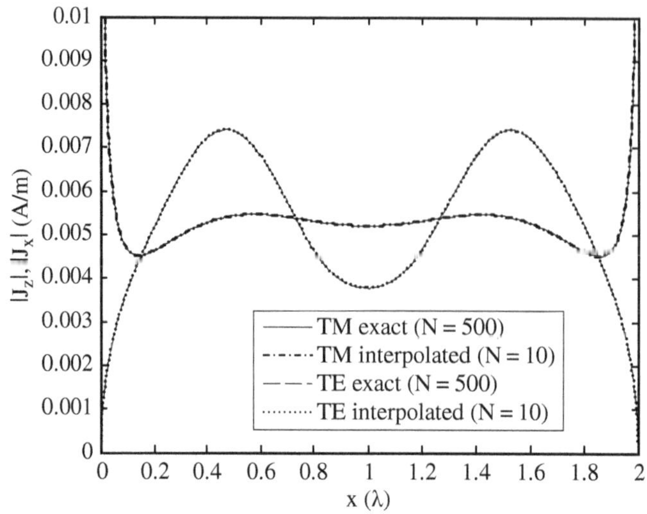

Figure 6.10 Interpolated current density distribution using very coarse meshes, $w = 2\lambda$.

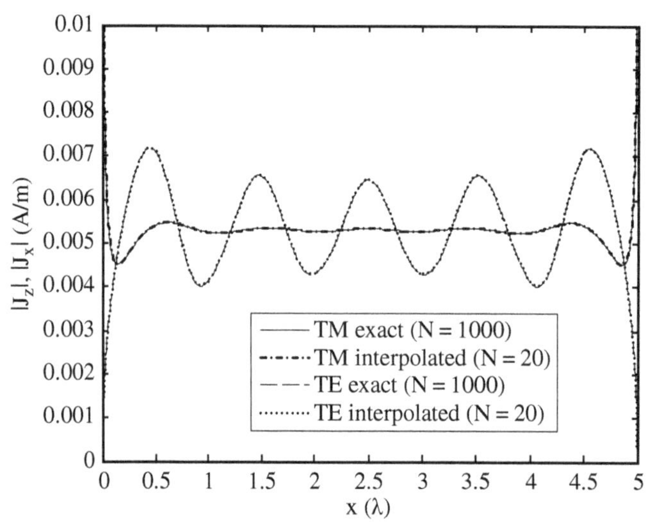

Figure 6.11 Interpolated current density distribution using very coarse meshes, $w = 5\lambda$.

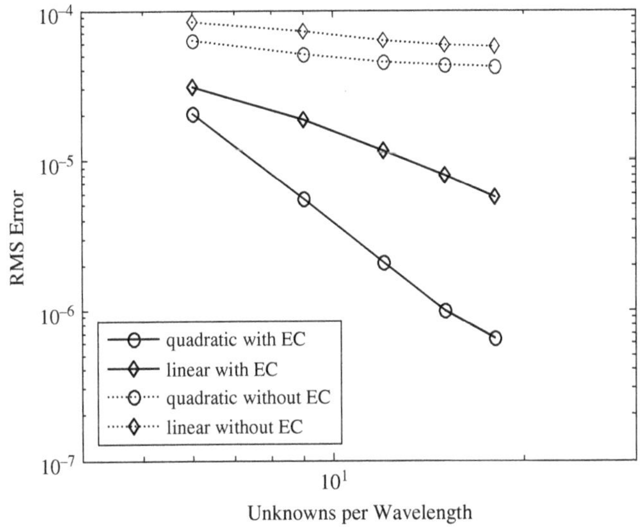

Figure 6.12 RMS error of the current density for TM polarization, $w = 2\lambda$.

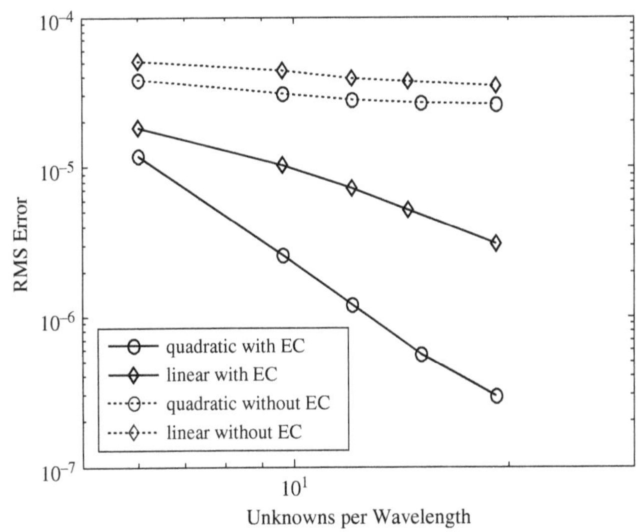

Figure 6.13 RMS error of the current density for TM polarization, $w = 5\lambda$.

much faster than without an EC, especially for the TM case. There is also an obvious p-refinement with the use of an EC, i.e. a higher-order quadrature rule or higher-order polynomial approximation for the current densities in the local correction yields a higher-order convergence rate. Note that we remove two end points in the calculation of errors for the TM without the EC. This is because the errors on these two points, if included, will dominate the total errors that increase slowly as the mesh size decreases.

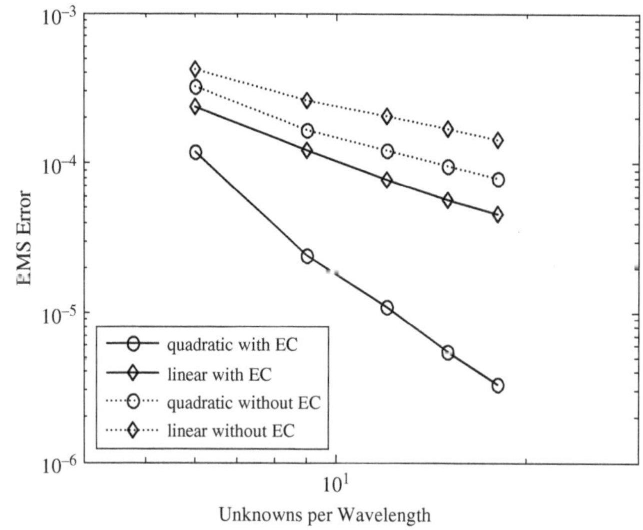

Figure 6.14 RMS error of the current density for TE polarization, $w = 2\lambda$.

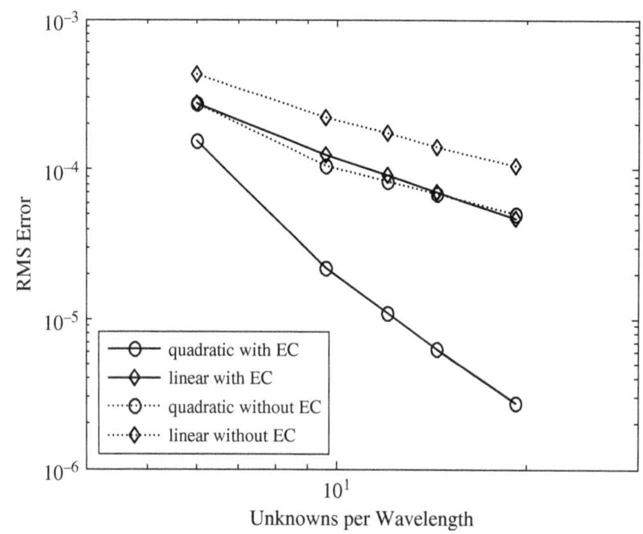

Figure 6.15 RMS error of the current density for TE polarization, $w = 5\lambda$.

Figure 6.16 depicts the numerical error curves for this case. This phenomenon reflects that the use of an EC is more essential in the TM case than in the TE case and the numerical error may be uncontrollable without using an EC in the TM case. The reason for this is that the TM current density is singular near edges and it is very easy to cause a large numerical error near edges without an EC restriction. By contrast, the EC is not as essential in the TE case as in the TM case because of the regular current density

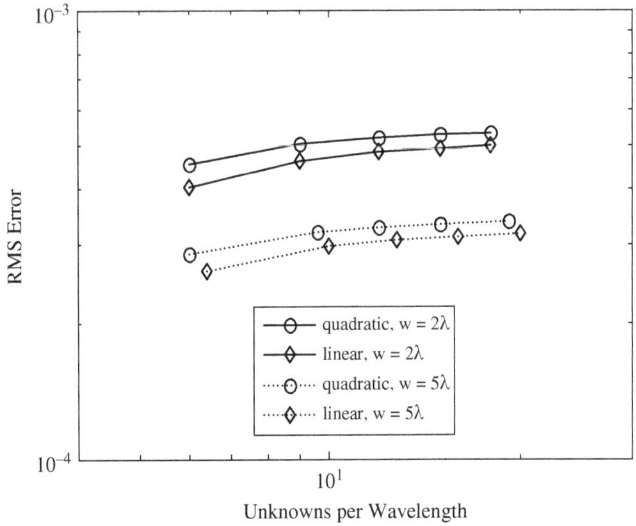

Figure 6.16 RMS error of the current density for TM polarization without the EC. Two end points are counted.

behavior near edges. The solutions can still converge fast without using an EC in the TE case, but apparently the use of an EC can accelerate the convergence.

6.3 Solution for Body-of-Revolution (BOR) Structures

EM scattering by a conducting BOR object was first studied by Andreasen [19], followed by many other researchers, based on the MoM [20–31]. Due to the geometrical symmetry, one can make use of Fourier expansions in the direction of revolution for the quantities in the governing equations of the problem and reduce the 3D equations to 2D equations. This reduction will dramatically lower computational costs and allow one to solve very large BOR problems with moderate expenses. The BOR problems can also be solved using different methods, such as the finite element method (FEM), the finite difference time domain (FDTD) method or some hybrid methods, and many papers have addressed the solution techniques for various BOR structures [32–41].

We will use the Nyström method to solve the 2D BOR integral equations. However, as in the MoM approach, one will encounter the problem on how to evaluate the singular Fourier expansion coefficients in the solution. Since the integral kernels related to the 3D Green's function are singular for the self interactions between source points and field points, the involved Fourier expansion coefficients are also singular and a special treatment technique is needed. The efficient evaluation for the singular Fourier expansion coefficients plays a critical role in the solution process because it is related to the generation of diagonal entries in the system matrix, which have a significant influence on the solution. Unfortunately, there had not been an explicit description for the indispensable treatment technique in those earlier literatures [19–25] until Glisson's dissertation appeared [26]. Glisson developed a systematic treatment scheme for various kernels in the MoM approach in his dissertation and the scheme has been demonstrated to be effective in applications [27, 28]. Abdelmageed [29] also presented an evaluation method for those Fourier expansion coefficients, but the method was only for the regular Fourier expansion coefficients for which the numerical quadrature rules can be used conveniently.

There are two kinds of singular Fourier expansion coefficients. One is related to the Green's function itself, called the modal Green's function (MGF) [29] and the other is generated by the gradient of the Green's function, called the MGF's derivatives. The MGF has a weak $1/R$ singularity and the MGF's derivatives include a strong $1/R^2$ singularity. To handle these singular integrals, we need to regularize the kernels by subtracting the singularities. This can be done by constructing a singular integrand that coincides with the kernel at the singularity and similar to it in the vicinity of the singularity. The constructed singular integrand that is subtracted from the kernel should be tractable either analytically or numerically and the regularized kernel should be as smooth as possible so that ordinary numerical quadrature rules can be applied to perform integrations.

We propose a new evaluation scheme for the singular Fourier expansion coefficients which can be used in both the Nyström method and MoM. The scheme differs from Glisson's scheme in several aspects. First, the scheme incorporates the Fourier expansion coefficients with the line integral in the 2D integral equation and performs the integrals over generating arc segments first with analytical expressions. Second, the scheme gets rid of the elliptical integral that is handled approximately in Glisson's scheme. Third, when the outer integral over the revolution angle is performed, the singular parts are further distinguished from the regular parts in the kernels and evaluated exactly in a closed form. Fourth, we construct a different singular integrand in subtracting the kernel of MGF's derivatives so that the kernel is better regularized. Fifth, we use a first-order polynomial function to expand the unknown current density and incorporate the constant term and first-order term of the polynomial function with the evaluation, respectively, yielding different evaluation formulations. Due to the above improvements, the scheme is simpler in implementation and more accurate in calculation. Numerical examples have demonstrated the effectiveness and robustness of the scheme.

6.3.1 2D Integral Equations

EM scattering by arbitrarily shaped conducting objects can be described by 3D integral equations which can be the EFIE, MFIE, or CFIE. Since the treatment for the singular kernels in those equations are quite similar, we exemplarily consider the EFIE for scattering by a conducting BOR object. The EFIE is given by

$$\hat{t}_0 \cdot \mathrm{i}\kappa\eta \int_S \left[g(\mathbf{r},\mathbf{r}')\mathbf{J}(\mathbf{r}') + \frac{1}{\kappa^2}\nabla g(\mathbf{r},\mathbf{r}')\nabla' \cdot \mathbf{J}(\mathbf{r}') \right] \mathrm{d}S' = \hat{t}_0 \cdot \mathbf{E}^{\mathrm{inc}} \tag{6.76}$$

where κ and η are the wavenumber and wave impedance of free space, respectively, and $g(\mathbf{r},\mathbf{r}') = \frac{e^{\mathrm{i}\kappa R}}{4\pi R}$ is the 3D scalar Green's function in which $R = |\mathbf{r} - \mathbf{r}'|$ is the distance between a field or observation point and a source point. Also, $\mathbf{J}(\mathbf{r}')$ is the unknown surface current density induced on the surface of conducting object S, $\mathbf{E}^{\mathrm{inc}}$ is the incident electric field or excitation and $\hat{t}_0$ is the unit tangential vector at an observation point on S. If the object is a BOR as shown in Figure 6.17, we can expand the integral kernel and unknown function into Fourier series over the revolution direction. Since the current density is a surface vector, we can decompose it into two tangential components along $\hat{t}$

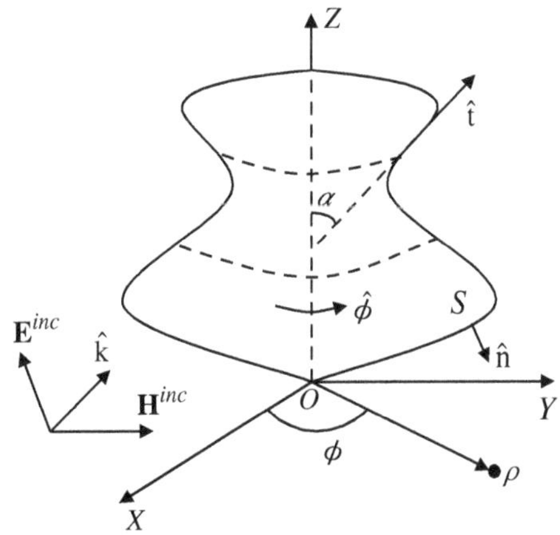

Figure 6.17 Geometry of a conducting body of revolution (BOR) where $\hat{t}$ and $\hat{\phi}$ are the two unit tangential vectors at the surface along the generating arc and revolution direction, respectively.

and $\hat{\phi}$, which are the unit tangential vectors at the generating arc and in the revolution direction of the BOR, respectively, i.e.

$$\mathbf{J}(\mathbf{r}') = J_t(\mathbf{r}')\hat{t} + J_\phi(\mathbf{r}')\hat{\phi} \tag{6.77}$$

and its divergence is given by [26]

$$\nabla' \cdot \mathbf{J}(\mathbf{r}') = \frac{1}{\rho'}\left[\frac{\partial(\rho' J_t)}{\partial t'} + \frac{\partial J_\phi}{\partial \phi'}\right]. \tag{6.78}$$

Multiplying (6.76) by $\hat{t}$ and $\hat{\phi}$ with a dot product, we can form the following two scalar equations

$$\mathrm{i}\kappa\eta \int_S \left\{ g(\mathbf{r},\mathbf{r}')J_t(\mathbf{r}') + \frac{1}{\kappa^2}\left[\hat{t}\cdot\nabla g(\mathbf{r},\mathbf{r}')\right]\nabla'\cdot\mathbf{J}(\mathbf{r}') \right\} \mathrm{d}S' = E_t^{\mathrm{inc}} \tag{6.79}$$

$$\mathrm{i}\kappa\eta \int_S \left\{ g(\mathbf{r},\mathbf{r}')J_\phi(\mathbf{r}') + \frac{1}{\kappa^2}\left[\hat{\phi}\cdot\nabla g(\mathbf{r},\mathbf{r}')\right]\nabla'\cdot\mathbf{J}(\mathbf{r}') \right\} \mathrm{d}S' = E_\phi^{\mathrm{inc}}. \tag{6.80}$$

We then expand all quantities in the above equations into Fourier series, i.e.

$$g(\mathbf{r},\mathbf{r}') = \sum_{n=-\infty}^{\infty} g_n(\rho, z, \rho', z')\mathrm{e}^{\mathrm{i}n(\phi-\phi')} \tag{6.81}$$

$$\hat{t}\cdot\nabla g(\mathbf{r},\mathbf{r}') = \sum_{n=-\infty}^{\infty} G_n^t(\rho, z, \rho', z')\mathrm{e}^{\mathrm{i}n(\phi-\phi')} \tag{6.82}$$

$$\hat{\phi}\cdot\nabla g(\mathbf{r},\mathbf{r}') = \sum_{n=-\infty}^{\infty} G_n^\phi(\rho, z, \rho', z')\mathrm{e}^{\mathrm{i}n(\phi-\phi')} \tag{6.83}$$

$$E_t^{\mathrm{inc}}(\mathbf{r}) = \sum_{n=-\infty}^{\infty} E_{tn}^{\mathrm{inc}}(\rho, z)\mathrm{e}^{\mathrm{i}n\phi} \tag{6.84}$$

$$E_\phi^{\mathrm{inc}}(\mathbf{r}) = \sum_{n=-\infty}^{\infty} E_{\phi n}^{\mathrm{inc}}(\rho, z)\mathrm{e}^{\mathrm{i}n\phi} \tag{6.85}$$

$$J_t(\mathbf{r}') = \sum_{n'=-\infty}^{\infty} J_{n'}^t(\rho', z')e^{in'\phi'} \tag{6.86}$$

$$J_\phi(\mathbf{r}') = \sum_{n'=-\infty}^{\infty} J_{n'}^\phi(\rho', z')e^{in'\phi'} \tag{6.87}$$

where

$$\begin{aligned} g_n(\rho, z, \rho', z') &= \frac{1}{2\pi}\int_0^{2\pi} d\beta g(\mathbf{r}, \mathbf{r}')e^{-in\beta} \\ &= \frac{1}{2\pi}\int_0^{2\pi} d\beta \frac{e^{i\kappa R}}{4\pi R}e^{-in\beta} \end{aligned} \tag{6.88}$$

$$\begin{aligned} G_n^t(\rho, z, \rho', z') &= \frac{1}{2\pi}\int_0^{2\pi} d\beta \left[\hat{t}\cdot\nabla g(\mathbf{r}, \mathbf{r}')\right] e^{-in\beta} \\ &= \frac{1}{2\pi}\int_0^{2\pi} d\beta v_1 v_2 e^{-in\beta} \end{aligned} \tag{6.89}$$

$$\begin{aligned} G_n^\phi(\rho, z, \rho', z') &= \frac{1}{2\pi}\int_0^{2\pi} d\beta \left[\hat{\phi}\cdot\nabla g(\mathbf{r}, \mathbf{r}')\right] e^{-in\beta} \\ &= \frac{1}{2\pi}\int_0^{2\pi} d\beta v_1 \rho' \sin\beta e^{-in\beta} \end{aligned} \tag{6.90}$$

$$E_{tn}^{\mathrm{inc}}(\rho, z) = \frac{1}{2\pi}\int_0^{2\pi} d\phi E_t^{\mathrm{inc}}(\mathbf{r})e^{-in\phi} \tag{6.91}$$

$$E_{\phi n}^{\mathrm{inc}}(\rho, z) = \frac{1}{2\pi}\int_0^{2\pi} d\phi E_\phi^{\mathrm{inc}}(\mathbf{r})e^{-in\phi}. \tag{6.92}$$

In the above, $J_n^t(\rho', z')$ and $J_n^\phi(\rho', z')$ (only $n' = n$ terms exist) are the unknown Fourier expansion coefficients for the current densities to be solved, and we have defined $\beta = \phi - \phi'$, $R = \sqrt{\rho^2 + \rho'^2 - 2\rho\rho'\cos\beta + (z - z')^2}$, $v_1 = (i\kappa R - 1)e^{i\kappa R}/(4\pi R^3)$ and $v_2 = (\rho - \rho'\cos\beta)\sin\alpha + (z - z')\cos\alpha$. Also, g_n is known as the MGF with the modal number n, and G_n^t and G_n^ϕ are the two tangential components of the MGF's gradient. Here, we use (ρ, ϕ, z) to denote a field or observation point and (ρ', ϕ', z') to denote a source point in a cylindrical coordinate system. Substituting those Fourier expansions into (6.79) and (6.80) and making use of the orthogonality of Fourier basis, we obtain

$$2\pi\int_C d\ell' \left(\rho' g_n J_n^t + \frac{Q_n}{\kappa^2}G_n^t\right) = \frac{E_{tn}^{\mathrm{inc}}}{i\kappa\eta} \tag{6.93}$$

$$2\pi\int_C d\ell' \left(\rho' g_n J_n^\phi + \frac{Q_n}{\kappa^2}G_n^\phi\right) = \frac{E_{\phi n}^{\mathrm{inc}}}{i\kappa\eta} \tag{6.94}$$

where C is the generating arc of the BOR as shown in Figure 6.18. Also,

$$Q_n = \sin\alpha J_n^t + \rho'\frac{\partial J_n^t}{\partial t'} + inJ_n^\phi \tag{6.95}$$

where α is the angle between the tangential direction of the generating arc and the $+z$ axis, as defined in Figure 6.17. The above equations are 2D integral equations from which the unknown J_n^t and J_n^ϕ can be solved by using the Nyström method or MoM.

Figure 6.18 Discretization of a conducting BOR's generating arc where (ρ_1, z_1) and (ρ_2, z_2) are the two end points of a segment and (ρ, z) is an observation point on the segment.

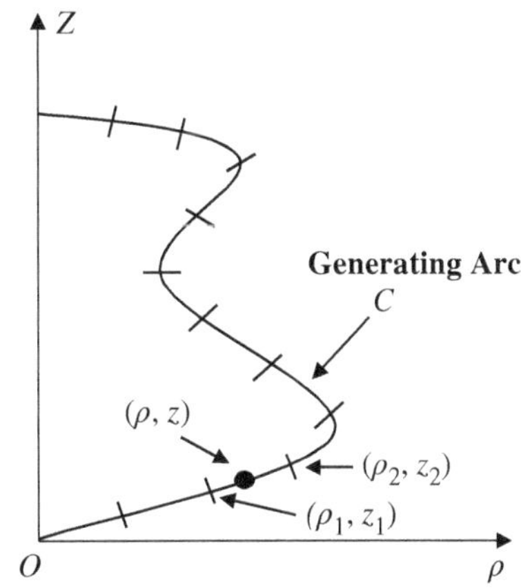

6.3.2 Evaluation of Singular Fourier Expansion Coefficients

Since the Fourier expansion coefficients with regular kernels can be evaluated by numerical integration conveniently, we only consider the evaluation for singular Fourier expansion coefficients. They are the MGF and its gradients as shown in (6.88)–(6.90), and we treat them separately as follows.

For the MGF

We assume that the generating arc of a BOR is discretized into straight line segments, as shown in Figure 6.18. Since the singularity appears when $\rho = \rho'$, $z = z'$ and $\phi = \phi'$ or $\beta = 0$, we incorporate the evaluation of the MGF and its gradient with the outer integral over the self segment. For the MGF, the integral will be, with the consideration of a constant term in the polynomial function of expanding current densities,

$$\begin{aligned} I_g^0 &= \int_{\rho_1}^{\rho_2} \rho' \, \mathrm{d}\rho' \int_0^{\pi} \mathrm{d}\beta \frac{\mathrm{e}^{\mathrm{i}\kappa R}}{R} \cos(n\beta) \\ &= \int_{\rho_1}^{\rho_2} \rho' \, \mathrm{d}\rho' \int_0^{\pi} \mathrm{d}\beta \left[\frac{\mathrm{e}^{\mathrm{i}\kappa R}}{R} \cos(n\beta) - \frac{1}{R} \right] + \int_{\rho_1}^{\rho_2} \rho' \, \mathrm{d}\rho' \int_0^{\pi} \frac{\mathrm{d}\beta}{R}. \end{aligned} \tag{6.96}$$

In the above, we have used the even property of the integrand on β and $\mathrm{d}\ell' = J_a \, \mathrm{d}\rho'$ where J_a is the Jacobian. Since J_a is a constant for the straight line approximation of an arc segment, we have suppressed it in all expressions. The first term in the above integral is regular now because the singular kernel has been subtracted with a similar singular integrand and the numerical integration can be applied. The second term is the subtracted similar singular integrand and we need to handle it specially. As shown in Section 5.2.1, the construction of similar singular integrand $f_0(R)$ should satisfy three requirements and we may construct $f_0(R)$ for the kernel in the above equation by expanding $\mathrm{e}^{\mathrm{i}\kappa R}$ into a Taylor-series form and extract the singular term, which is $\cos(n\beta)/R$, as the similar singular integrand. Such a $f_0(R)$ is more similar to $f_1(R)$ in the neighborhood of the singularity than $f_0(R) = 1/R$, which was chosen in [26] and [29] by letting $\beta = 0$. The reason for selecting $f_0(R) = 1/R$ instead of $f_0(R) = \cos(n\beta)/R$ in [26] and [29] is probably due to the inability of integrating $f_0(R) = \cos(n\beta)/R$, but we have derived the solution for the integration of $f_0(R) = \cos(n\beta)/R$. However, since the singularity is weak in this case, the two choices for $f_0(R)$ above do not show an obvious difference in the results, and we also choose the latter one to facilitate the treatment. Switching the order of integrations in the second term, we have

$$\begin{aligned} I_g^{0s} &= \int_0^{\pi} d\beta \int_{\rho_1}^{\rho_2} \frac{\rho'\, d\rho'}{R} \\ &= \int_0^{\pi} d\beta \int_{\rho_1 - b_0}^{\rho_2 - b_0} \frac{(x' + b_0)\, dx'}{\sqrt{x'^2 + a_0^2}} \\ &- \int_0^{\pi} d\beta (q_2 - q_1) + \int_0^{\pi} d\beta b_0 \ln(f_2) - \int_0^{\pi} d\beta b_0 \ln(f_1) \\ &= I_1^{0s} + I_2^{0s} + I_3^{0s} \end{aligned} \tag{6.97}$$

where I_k^{0s} $(k = 1, 2, 3)$ is the kth integral in the preceding equation and

$$f_j = \rho_j + q_j - b_0; \quad q_j = \sqrt{(\rho_j - b_0)^2 + a^2}; \tag{6.98}$$

$$b = \frac{z_2 - z_1}{\rho_2 - \rho_1} = \frac{z - z'}{\rho - \rho'}; \quad b_0 = \frac{\rho(\cos\beta + b^2)}{h}; \tag{6.99}$$

$$a = \frac{\rho}{h}\sqrt{h^2 - (\cos\beta + b^2)^2}; \quad g = \sqrt{1 + b^2}; \quad h = g^2 \tag{6.100}$$

where $j = 1, 2$. The integrals in (6.97) include both regular and singular parts, but I_1^{0s} is aways regular. If $\rho_2 < \rho < \rho_1$, then I_2^{0s} is singular and I_3^{0s} is regular, while I_2^{0s} is regular and I_3^{0s} is singular if $\rho_2 > \rho > \rho_1$. The regular parts are evaluated numerically and the singular parts are performed analytically. The singular parts in I_2^{0s} or I_3^{0s} can be extracted further

$$I_2^{0s} = \int_0^{\pi} d\beta b_0 \ln a_0^2 - \int_0^{\pi} d\beta (q_2 - \rho_2 + b_0) \tag{6.101}$$

$$I_3^{0s} = \int_0^{\pi} d\beta b_0 \ln a_0^2 - \int_0^{\pi} d\beta (q_1 - \rho_1 + b_0). \tag{6.102}$$

The second term in the above equations is regular and evaluated numerically, but the first term is singular and can be evaluated analytically with the aid of [42]

$$\begin{aligned} I_0^{0s} &= \int_0^{\pi} d\beta b_0 \ln a_0^2 \\ &= \frac{2\rho\pi}{h} \left\{ b^2 \ln\left[\frac{\rho(g + |b|)}{2h}\right] - \frac{|b|}{g + |b|} \right\}. \end{aligned} \tag{6.103}$$

If the MGF incorporates the first-order term in the polynomial function of expanding current densities, the subtracted singular integral is

$$\begin{aligned} I_g^{1s} &= \int_0^{\pi} d\beta \int_{\rho_1}^{\rho_2} \frac{\rho'^2\, d\rho'}{R} \\ &= \int_0^{\pi} d\beta \int_{\rho_1 - b_0}^{\rho_2 - b_0} \frac{(x' + b_0)^2\, dx'}{\sqrt{x'^2 + a_0^2}} \\ &= \int_0^{\pi} d\beta \int_{\rho_1 - b_0}^{\rho_2 - b_0} \frac{x'^2\, dx'}{\sqrt{x'^2 + a_0^2}} + \int_0^{\pi} d\beta \int_{\rho_1 - b_0}^{\rho_2 - b_0} \frac{b_0(2x' + b_0)\, dx'}{\sqrt{x'^2 + a_0^2}} \\ &= I_1^{1s} + I_2^{1s}. \end{aligned} \tag{6.104}$$

The second integral I_2^{1s} in the above is very similar to I_g^{0s} in (6.97) and is handled in a similar way. The first integral I_1^{1s} can be written as

$$I_1^{1s} = 0.5\int_0^{\pi} \mathrm{d}\beta[(\rho_2 - b_0)q_2 - (\rho_1 - b_0)q_1] + 0.5\int_0^{\pi} \mathrm{d}\beta a_0 \ln(f_2) - 0.5\int_0^{\pi} \mathrm{d}\beta a_0 \ln(f_1)$$
$$= 0.5(I_a^{1s} + I_b^{1s} + I_c^{1s}). \tag{6.105}$$

Similar to the last three integrals in (6.97), I_b^{1s} is singular and I_c^{1s} is regular if $\rho_2 < \rho < \rho_1$, while I_b^{1s} is regular and I_c^{1s} is singular if $\rho_2 > \rho > \rho_1$, and I_a^{1s} is always regular. The singular parts in I_b^{1s} or I_c^{1s} can be extracted by

$$I_b^{1s} = \int_0^{\pi} \mathrm{d}\beta a_0^2 \ln a_0^2 - \int_0^{\pi} \mathrm{d}\beta(q_2 - \rho_2 + b_0) \tag{6.106}$$

$$I_c^{1s} = \int_0^{\pi} \mathrm{d}\beta a_0^2 \ln a_0^2 - \int_0^{\pi} \mathrm{d}\beta(q_1 - \rho_1 + b_0) \tag{6.107}$$

where the second integral is regular and can be evaluated numerically while the first integral is singular and can be evaluated analytically with the aid of [42]

$$I_0^{1s} = \int_0^{\pi} \mathrm{d}\beta a_0^2 \ln a_0^2$$
$$= \frac{\pi\rho^2}{h^2}\left\{(1+4b^2)\ln\left[\frac{\rho(g+|c|)}{2h}\right] + 4|b|b^2(g-|b|) + \frac{1}{4}(h^2+b^2-2g|b|)^2 + \frac{1}{4}\right\}. \tag{6.108}$$

For the MGF's Derivatives

When incorporating the outer integral over an arc segment with the constant term in the polynomial function of expanding current densities, the t component of the MGF's derivatives can be written as

$$I_{Gt}^{0} = \int_{\rho_1}^{\rho_2} \mathrm{d}\rho' \int_0^{\pi} \mathrm{d}\beta v_1 v_2 \cos(n\beta)$$
$$= \int_{\rho_1}^{\rho_2} \mathrm{d}\rho' \int_0^{\pi} \mathrm{d}\beta \left[v_1 v_2 \cos(n\beta) + \frac{d_0(\rho-\rho')}{R_0^3}\right] - d_0 \int_0^{\pi} \mathrm{d}\beta \int_{\rho_1}^{\rho_2} \frac{(\rho-\rho')\mathrm{d}\rho'}{R_0^3} \tag{6.109}$$

where $R_0 = \sqrt{h(\rho-\rho')^2 + \rho\rho'\pi^2}$ and $d_0 = \sin\alpha + b\cos\alpha$. The first integral in the above is regular now and can be evaluated numerically. The second integral is singular and will be handled specially. The above singularity subtraction is based on the fact that $1 - \cos\beta = 2\sin^2(\beta/2) \approx \beta^2$ when $\beta \to 0$. The subtracted similar singular integrand $f_0(R)$ in this case is also different from the one in Glisson's method [27] in which $\rho \approx \rho'$ is used in R_0 to enable the derivation of closed-form solutions. Since our subtracted singular integrand $f_0(R)$ is more similar to the kernel $f_1(R)$ nearby the singularity, the regularized kernel is numerically integrable and the required quadrature rule can be of lower order in numerical integration. Because the singularity is strong in this case, the difference in the subtracted similar singular integrand is important. The singular integral in the above can be found in a closed form as [42]

$$I_{Gt}^{0s} = \int_0^{\pi} \mathrm{d}\beta \int_{\rho_1}^{\rho_2} \mathrm{d}\rho' \frac{\rho-\rho'}{R_0^3}$$
$$= \frac{1}{\rho g} \ln\left|\frac{(2R_2 + \pi\rho^2 + \pi\rho\rho_2)(\rho-\rho_1)}{(2R_1 + \pi\rho^2 + \pi\rho\rho_1)(\rho-\rho_2)}\right| \tag{6.110}$$

where $R_j = \sqrt{h(\rho - \rho_j)^2 + \rho\rho_j\pi^2}$, $j = 1, 2$. Similarly, we can derive the subtracted singular integral when G_n^t incorporates the first-order term in the polynomial function of expanding current densities [42], i.e.

$$I_{Gt}^{1s} = \int_0^\pi \mathrm{d}\beta \int_{\rho_1}^{\rho_2} \mathrm{d}\rho' \frac{\rho'(\rho - \rho')}{R_0^3}$$
$$= \rho I_{Gt}^{0s} + \frac{\pi}{g^3} \ln \left| \frac{2gR_2 + 2h(\rho - \rho_2) - \rho\pi^2}{2gR_1 + 2h(\rho - \rho_1) - \rho\pi^2} \right|. \tag{6.111}$$

For the ϕ component of the MGF's derivatives incorporating the constant term in the polynomial function of expanding current densities, we have

$$I_{G\phi}^0 = \int_{\rho_1}^{\rho_2} \mathrm{d}\rho' \int_0^\pi \mathrm{d}\beta v_1 \rho' \sin(n\beta)$$
$$= i \int_{\rho_1}^{\rho_2} \mathrm{d}\rho' \int_0^\pi \mathrm{d}\beta \left[v_1\rho' \sin(n\beta) + \frac{n\rho'\beta^2}{R_0^3} \right] + in \int_0^\pi \mathrm{d}\beta \int_{\rho_1}^{\rho_2} \frac{\rho'\beta^2 \,\mathrm{d}\rho'}{R_0^3}. \tag{6.112}$$

The extracted singular integral is

$$I_{G\phi}^{0s} = \int_0^\pi \mathrm{d}\beta \int_{\rho_1}^{\rho_2} \mathrm{d}\rho' \frac{\rho'\beta^2}{R_0^3}$$
$$= \int_{\rho_1}^{\rho_2} \mathrm{d}\rho' \left[\frac{1}{\rho\sqrt{\rho\rho'}} \ln\left(\sqrt{\rho\rho'}\pi + R_0\right) - \frac{\pi}{\rho R_0} \right] - \frac{1}{\rho\sqrt{\rho}} \int_{\rho_1}^{\rho_2} \frac{\mathrm{d}\rho'}{\sqrt{\rho'}} \ln\left(g|\rho - \rho'|\right) \tag{6.113}$$

where the first integral is numerically integrable and evaluated numerically. The second integral is singular and evaluated analytically as

$$I_{G\phi 2}^{0s} = \int_{\rho_1}^{\rho_2} \frac{\mathrm{d}\rho'}{\sqrt{\rho'}} \ln\left(g|\rho - \rho'|\right)$$
$$= 2s_0 \ln(g) + 2\left\{ s_1 \ln|\rho - \rho_1)| + s_2 \ln|\rho_2 - \rho| - 2s_0 + 2\sqrt{\rho}\left[\ln(s_4) - \ln(s_3)\right] \right\} \tag{6.114}$$

where

$$s_0 = \rho_2^{\frac{1}{2}} - \rho_1^{\frac{1}{2}};\ \ s_1 = \rho^{\frac{1}{2}} - \rho_1^{\frac{1}{2}};\ \ s_2 = \rho_2^{\frac{1}{2}} - \rho^{\frac{1}{2}};\ \ s_3 = \rho^{\frac{1}{2}} + \rho_1^{\frac{1}{2}};\ \ s_4 = \rho_2^{\frac{1}{2}} + \rho^{\frac{1}{2}}. \tag{6.115}$$

If the ϕ component of the MGF's derivatives incorporates the first-order term in the polynomial function of expanding current densities, the extracted singular integral becomes

$$I_{G\phi}^{1s} = \int_0^\pi \mathrm{d}\beta \int_{\rho_1}^{\rho_2} \mathrm{d}\rho' \frac{\rho'^2\beta^2}{R_0^3}$$
$$= \int_{\rho_1}^{\rho_2} \mathrm{d}\rho' \left[\frac{\sqrt{\rho'}}{\rho\sqrt{\rho}} \ln\left(\sqrt{\rho\rho'}\pi + R_0\right) - \frac{\pi\rho'}{\rho R_0} \right] - \frac{1}{\rho\sqrt{\rho}} \int_{\rho_1}^{\rho_2} \mathrm{d}\rho' \sqrt{\rho'} \ln\left(g|\rho - \rho'|\right) \tag{6.116}$$

The first integral is regular again and evaluated numerically. The second integral is singular and evaluated analytically as

$$I_{G\phi 2}^{1s} = \int_{\rho_1}^{\rho_2} \mathrm{d}\rho' \sqrt{\rho'} \ln\left(g|\rho - \rho'|\right)$$
$$= \frac{2}{3}\left[t_0 \ln(g) + t_1 \ln|s_1| + t_2 \ln|s_2| - t_3 \ln|s_3| + t_4 \ln|s_4| - \frac{2}{3}t_0 - 2\rho s_0 \right] \tag{6.117}$$

where

$$t_0 = \rho_2^{\frac{3}{2}} - \rho_1^{\frac{3}{2}};\ \ t_1 = \rho^{\frac{3}{2}} - \rho_1^{\frac{3}{2}};\ \ t_2 = \rho_2^{\frac{3}{2}} - \rho^{\frac{3}{2}};\ \ t_3 = \rho^{\frac{3}{2}} + \rho_1^{\frac{3}{2}};\ \ t_4 = \rho_2^{\frac{3}{2}} + \rho^{\frac{3}{2}}. \quad (6.118)$$

6.3.3 Numerical Examples

The Nyström method with the new evaluation scheme of singular Fourier expansion coefficients is used to solve for the EM scattering by three different conducting objects, as shown in Figure 6.19. The incident wave is a horizontally or vertically polarized plane wave propagating along the $(\theta^{\text{inc}}, \phi^{\text{inc}})$ direction. The sphere has a radius of $a = 0.5\lambda$ while the cone has a base radius of $r = 0.5\lambda$ and a height of $h = 2\lambda$. The step-radius cylinder consists of four segments with different radii and thicknesses. The radii are $r_1 = 0.6\lambda$, $r_2 = 0.4\lambda$, $r_3 = 0.2\lambda$, and $r_4 = 0.8\lambda$, respectively, and the thicknesses are $h_1 = 0.3\lambda$, $h_2 = 0.5\lambda$, $h_3 = 0.8\lambda$, and $h_4 = 0.4\lambda$, respectively. Figure 6.20 shows the normalized magnitude of the tangential components J_t (vertical polarization) and J_ϕ (horizontal polarization) of the surface current density along the principal cut ($\phi = 0°$ and $\theta = 0° - 180°$) when $(\theta^{\text{inc}}, \phi^{\text{inc}}) = (30°, 30°)$ and using 80 segments in the discretization of generating arc. We can see that the BOR solutions agree well with the exact solutions derived from the Mie series. Figures 6.21 and 6.22 plot the bistatic radar cross section (RCS) solutions for the EM scattering by the conducting cone and step-radius cylinder, respectively. The propagating direction of the incident wave is $(30°, 30°)$ for the cone and $(60°, 60°)$ for the step-radius cylinder, and the observation for the scattered wave is also taken along the principal cut. The BOR solutions with 100 segments for the cone and 200 segments for the step-radius cylinder in the discretization of generating arc are compared to the MoM solutions obtained by treating the objects as full 3D scatterers. The cone and step-radius cylinder are discretized with 3080 and 4146 triangle patches, respectively, in the MoM

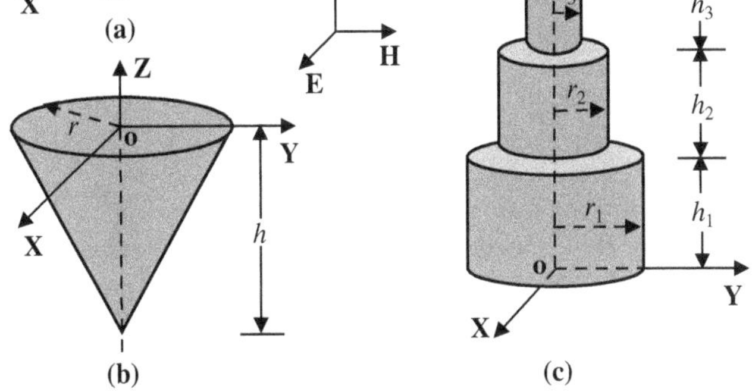

Figure 6.19 Geometries of three conducting BOR scatterers: (a) sphere; (b) cone; (c) step-radius cylinder.

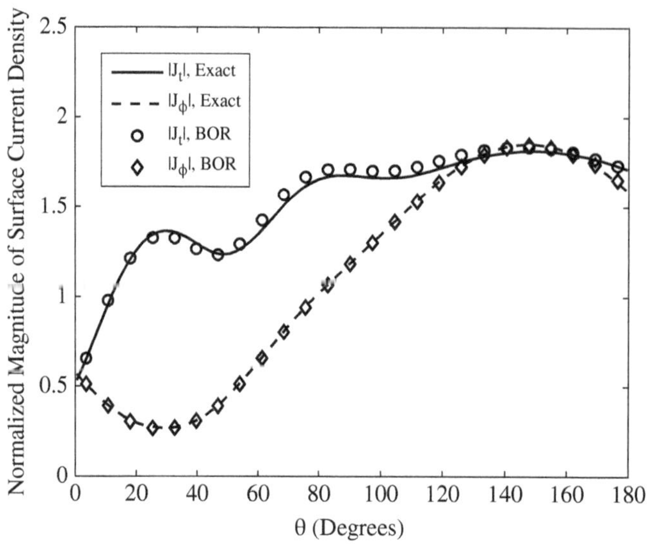

Figure 6.20 Current density distribution along the principal cut at the surface of a conducting sphere for EM scattering by the sphere.

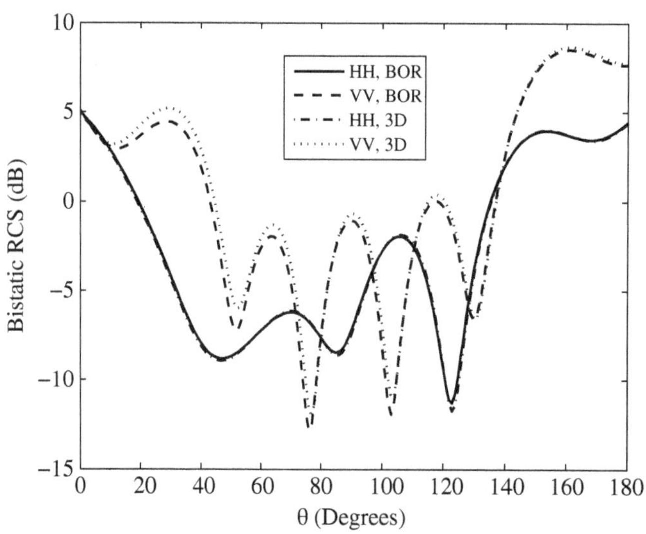

Figure 6.21 Bistatic RCS solutions for EM scattering by a conducting cone in horizontal (HH) and vertical (VV) polarizations.

solutions. It can be seen that the solutions based on the two approaches are in good agreement with each other, except for the case of the cone in vertical polarization. The reason for this case is probably that the geometry has a sharp circular corner, which is recognized as a challenging geometry sometimes. Due to the difficulty of generating good meshes around the corner, the solution from the full 3D formulations may not be accurate enough and we believe that the BOR solution is more accurate. Since

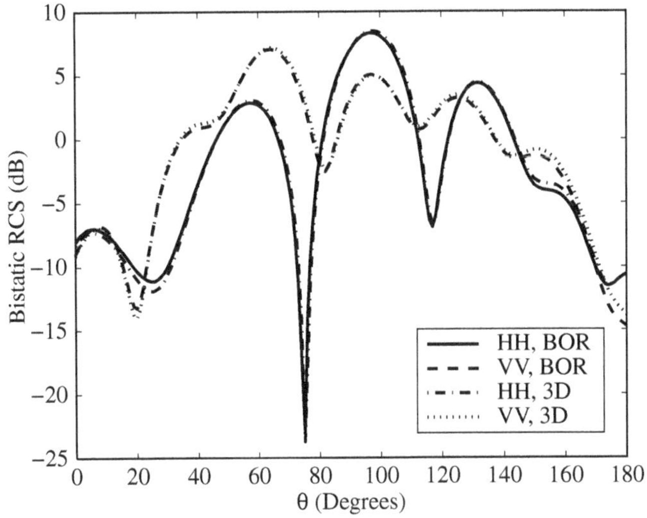

Figure 6.22 Bistatic RCS solutions for EM scattering by a conducting step-radius cylinder in horizontal (HH) and vertical (VV) polarizations.

the geometrical singularity or discontinuity is in the θ direction, it may have a larger influence on the solution of the vertical polarization.

6.4 Solutions of the Electric Field Integral Equation

The surface integral equations (SIEs) for governing EM problems with conducting objects are mostly solved using the MoM with appropriate basis functions. The Nyström method is an alternative approach to solve the SIEs with a simpler implementation and adaptability to defective meshes. The Nyström method employs the point-based discretization scheme and simply evaluates the integrands at quadrature points to fill the system matrix. This only requires $\mathcal{O}(P)$ cost where P is the number of quadrature points. However, this elegance in the Nyström method is not fully exploited in electromagnetics because the canonical Nyström method was designed for regular integral kernels whereas EM problems suffer from singular kernels related to the Green's function. The higher-order accuracy in a Nyström scheme can only be achieved by efficiently treating the singular terms in the numerical integrations for filling the impedance matrix. Fortunately, some higher-order Nyström schemes have been proposed for EM applications in the past years [3, 4]. In [3], very systematic treatments for various singular kernels were designed, known as local correction. The essence of the local correction is to generate an adjustment scheme for the quadrature rule for singular integral kernels by computing convolutions of the kernels with a suitable set of testing functions. This is a MoM-like procedure for determining the corresponding weights of the quadrature rule at each singular point. This adjustment is only required in the vicinity of the singularity and allows the quadrature rule to be as accurate near the singularity as it is far from it. Similarly, the local correction of the Nyström scheme in [4] is also based on a MoM-like procedure to find the weights for a specialized local quadrature rule. However, the MoM-like procedure there is different from that in [3] and is derived by a more traditional approach.

We develop a different Nyström scheme with a higher-order accuracy for EM scattering by arbitrary conducting surfaces. Unlike the treatments in [3, 4], the local correction in our Nyström scheme is based on the Lagrange interpolation for unknown functions without using a local MoM-like procedure. The resulting singular kernel is evaluated using a singularity extraction and Duffy's transformation [43]. We also apply a super-parametric geometrical representation for curvilinear patches to minimize the geometrical approximation [46]. In solving the resultant matrix equation, we use the lower-upper triangular decomposition (LUD) method for small problems and the iterative method for large problems. This procedure will not introduce uncontrollable errors when the impedance matrix is well-conditioned within an appropriate frequency range and mesh size. Compared to those higher-order Nyström schemes based on a MoM-like procedure for the local correction, the scheme is easier to implement and more efficient for controlling numerical errors. The SIEs include the electric field integral equation (EFIE), magnetic field integral equation (MFIE), and combined field integral equation (CFIE), but we only focus on the EFIE in this section. The solutions of MFIE and CFIE will be addressed in next sections. We present some numerical examples to demonstrate the scheme and good results have been obtained.

6.4.1 Higher-order Nyström method

Higher-order methods are desirable because they can use fewer unknowns to reach the designated accuracy. To obtain a higher-order accuracy in the Nyström method, the implementation should be strongly error-controllable. In general, the errors in a numerical method come from the geometrical approximation of objects, matrix representation or discretization of integral equations and solutions of matrix solvers. To accurately solve the EFIE with a conducting object, we first minimize the error of geometrical approximation by selecting a super-parametric geometrical mapping for curvilinear patches. We then derive an accurate matrix representation for the original EFIE by using a Lagrange-interpolator-based local correction procedure for the singular integral kernels. The matrix solver usually contributes a negligible error when the system matrix is well-conditioned. This can be guaranteed by carefully limiting the frequency range and discretization size.

Geometrical Mapping

Since we consider to solve the problems with arbitrary conducting surfaces, we assume that the objects possess curvilinear surfaces and flat surfaces are just a special case in the generalized surfaces. In our scheme, the curvilinear surface of an object is discretized into small curvilinear triangular patches and the geometrical mapping is conducted to transform a curvilinear triangular patch into a standard right-angle flat triangle over which quadrature rules are defined. This will greatly facilitate the manipulation of singular kernels and the implementation of numerical integrations. The geometrical mapping can be a lower-cost and lower-order one with a larger error, or a higher-cost and higher-order one with smaller error. To obtain an optimal h-convergence rate, we have to approximate the geometry by using the elements which are one order higher than the approximation degree for unknown functions and such elements are known

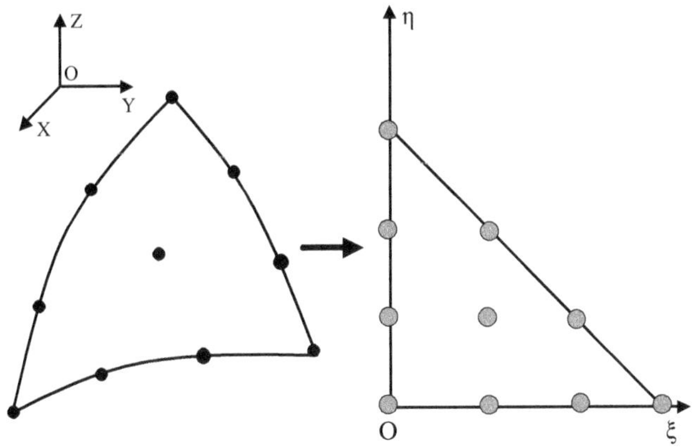

Figure 6.23 A canonical geometrical mapping for a curvilinear triangular patch.

as super-parametric elements [46]. A typical geometrical mapping for a curvilinear triangular patch with 10 sampling points is shown in Figure 6.23. The mapping can be expressed as

$$\mathbf{r} = \sum_{j=1}^{N_g} \phi_j(\xi, \eta, \zeta)\mathbf{r}_j, \quad \xi + \eta + \zeta = 1 \tag{6.119}$$

where N_g is the number of sampling points and ϕ_j are the shape functions. The shape functions have the same form as the interpolation functions for representing an unknown function as shown in the next subsection, provided that we let $L_2^i = \xi$, $L_3^i = \eta$ and $L_1^i = 1 - \xi - \eta$ [47]. With the parametric representation of the coordinates, the integration can be performed over a standard right-angle triangle with the following differential surface element

$$\mathrm{d}S' = |\mathcal{J}'|\mathrm{d}\xi'\mathrm{d}\eta' \tag{6.120}$$

where $|\mathcal{J}'| = \sqrt{g_{11}g_{22} - g_{12}g_{21}}$ is the Jacobian with

$$g_{11} = \frac{\partial \mathbf{r}'}{\partial \xi'} \cdot \frac{\partial \mathbf{r}'}{\partial \xi'} \tag{6.121}$$

$$g_{22} = \frac{\partial \mathbf{r}'}{\partial \eta'} \cdot \frac{\partial \mathbf{r}'}{\partial \eta'} \tag{6.122}$$

$$g_{12} = g_{21} = \frac{\partial \mathbf{r}'}{\partial \xi'} \cdot \frac{\partial \mathbf{r}'}{\partial \eta'}. \tag{6.123}$$

Manipulation of Singular Kernels

The EFIE is the most common integral equation for governing 3D EM problems with conducting objects. For EM scattering by a conducting object, the EFIE can be written as

$$\hat{t} \cdot i\omega\mu \int_S \left[g(\mathbf{r}, \mathbf{r}')\mathbf{J}(\mathbf{r}') + \nabla g(\mathbf{r}, \mathbf{r}')\nabla' \cdot \mathbf{J}(\mathbf{r}')\right] \mathrm{d}S' = \hat{t} \cdot \mathbf{E}^{\mathrm{inc}}(\mathbf{r}), \quad \mathbf{r} \in S \tag{6.124}$$

where $\mathbf{J}(\mathbf{r}')$ is the induced surface current density on the object surface, $\hat{t}$ is a unit tangential vector at an observation point, $\mathbf{E}^{\text{inc}}(\mathbf{r})$ is the incident electric field, and

$$g(\mathbf{r}, \mathbf{r}') = \frac{e^{ikR}}{4\pi R} \tag{6.125}$$

is the scalar Green's function in which $R = |\mathbf{r} - \mathbf{r}'|$ is the distance between an observation point $\mathbf{r}$ and a source point $\mathbf{r}'$. Also, $k = \omega\sqrt{\epsilon\mu}$ is the wavenumber in the surrounding medium with a permittivity ϵ and a permeability μ, and ω is the angular frequency of incident wave. The canonical kernel in the equation can be written as

$$g_0(\mathbf{r}, \mathbf{r}') = \frac{f(\mathbf{r}, \mathbf{r}')}{|\mathbf{r} - \mathbf{r}'|^{\alpha}} \tag{6.126}$$

which is related to the scalar Green's function. The above kernel is singular when $\mathbf{r} = \mathbf{r}'$ and α denotes the degree of singularity which is 1 or 2 in the EFIE. Also, $f(\mathbf{r}, \mathbf{r}')$ is a smooth function which could be related to a geometrical mapping. Since the Duffy's transformation can only remove the first-order singularity for a double-fold integral, we must transfer the second-order or strong singularity to a first-order singularity before applying the Duffy's transformation to solve the EFIE. This can be completed by utilizing a singularity extraction technique. In the EFIE, the extraction for the strongly-singular term yields

$$\begin{aligned} &\int_{\Delta S_i} \hat{t} \cdot \nabla g(\mathbf{r}, \mathbf{r}') \nabla' \cdot \mathbf{J}(\mathbf{r}') \mathrm{d}S' \\ &= \int_{\Delta S_i} \hat{t} \cdot \nabla g(\mathbf{r}, \mathbf{r}') \left[\nabla' \cdot \mathbf{J}(\mathbf{r}') - \nabla \cdot \mathbf{J}(\mathbf{r})\right] \mathrm{d}S' - \nabla \cdot \mathbf{J}(\mathbf{r}) \int_{\partial S_i} g(\mathbf{r}, \mathbf{r}') \hat{t} \cdot \hat{l}' \mathrm{d}l' \end{aligned} \tag{6.127}$$

where the first integral on the right-hand side has a $1/R$ singularity now and can be evaluated by Duffy's method. The second integral on the right-hand side comes from the Gauss' theorem and ∂S_i denotes the boundary of ΔS_i where $\hat{l}'$ is the unit tangential vector along it. The contour integral along ∂S_i is regular when the quadrature points are located inside a triangular patch to avoid meeting geometrical discontinuities.

After the geometrical mapping for the preprocessed kernel is incorporated, we can transform the integration over each triangular patch into that over a standard planar right-angle triangle in the coordinate system (ξ, η). The kernel in each component can then be written as

$$g_0(\mathbf{r}, \mathbf{r}') = \frac{f(\xi, \eta, \xi', \eta')}{R^{\alpha}}. \tag{6.128}$$

In the coordinate system (ξ, η), the component of unknown current density in the ith triangle is $J_i(\mathbf{r}') = J_i(\xi', \eta')$ and can be expressed as the interpolation based on its values at quadrature points, namely

$$J_i(\xi', \eta') = \sum_{j=1}^{P_i} N_j^i(\xi', \eta') J_{ij} \tag{6.129}$$

where $N_j^i(\xi', \eta')$ is the interpolation function and J_{ij} is the value of the component of unknown current density at the jth quadrature point of the ith triangle that we want to seek. The interpolation function is an unbiased combination of Lagrange interpolation polynomials [48]

$$N_j^i(\xi', \eta') = P_I^n(L_1^i)P_J^n(L_2^i)P_K^n(L_3^i) \tag{6.130}$$

where $P_I^n(L_1^i)$, $P_J^n(L_2^i)$ and $P_K^n(L_3^i)$ are the Lagrange interpolation polynomials and (L_1^i, L_2^i, L_3^i) denote the area coordinates at the point (ξ', η'). $P_I^n(L_1^i)$ is defined as

$$P_I^n(L_1^i) = \frac{1}{I!}\prod_{k=0}^{I-1}(nL_1^i - k) \tag{6.131}$$

with $P_0^n = 1$, and $P_J^n(L_2^i)$ and $P_K^n(L_3^i)$ can be defined in a similar way. Here, $n = I + J + K$ is the order of the interpolation and (I, J, K) is a node- numbering scheme. With the help of (6.128) and (6.129), the canonical integral over a triangle can be written as

$$\int_{\Delta S_i} g_0(\mathbf{r}, \mathbf{r}')J(\mathbf{r}')\mathrm{d}S' = \sum_{j=1}^{P_i} J_{ij}\int_{\Delta T_i}\frac{f(\xi, \eta, \xi', \eta')N_j^i(\xi', \eta')}{R^\alpha}|\mathcal{J}|\mathrm{d}\xi'\mathrm{d}\eta' \tag{6.132}$$

where ΔT_i represents the standard right-angle triangle over which quadrature rules are defined. The above integral is accurately evaluated by means of triangle subdivision and Duffy's transformation. The resulting matrix equation can then represent the original EFIE accurately.

Matrix Solvers

One of the widely used matrix solvers is the LUD with a direct solution for small problems. Iterative methods like the conjugate gradient (CG) method and the bi-conjugate gradient (BiCG) method are preferred for electrically large problems. If the impedance matrix is well-conditioned, the error in this process is controllable. After minimizing the error of geometrical approximation and carefully treating the singular kernels, the resulting matrix equation can accurately represent the original problem. The resulting impedance matrix is generally well-conditioned if the operating frequency is not too low and not close to a resonant frequency, which will cause a breakdown. In solving the matrix equation, the CG method is used when the number of unknowns is large. The memory usage can be reduced to $\mathcal{O}(\mathcal{N})$ when only the diagonal elements are stored, where $\mathcal{N}$ is the number of unknowns. The total CPU time is also reduced although the matrix-filling time is increased because the non-diagonal terms are calculated as needed in each iteration.

6.4.2 Numerical Examples

We present two numerical examples with sharp-corner objects to demonstrate the higher-order convergence of the Nyström scheme. The first example is the scattering by a conducting ogive scatterer whose geometry is shown in Figure 6.24(a). The ogive has a central radius $a = 1$ inch and a half height $d = 5$ inches. The monostatic RCS solutions in horizontal (HH) and vertical (VV) polarizations at the frequency $f = 1.18$ GHz are plotted in Figure 6.25. The results are validated using the benchmark solutions in [49]. If we use the solutions from a very fine discretization of the geometry and the highest-order quadrature rule as an accurate solution, we can obtain the curves

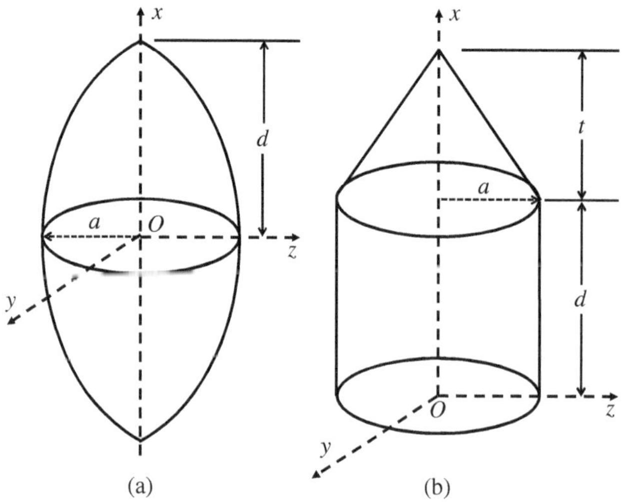

Figure 6.24 Geometries of two scatterers. (a) A conducting ogive scatterer. (b) A conducting pencil-like target.

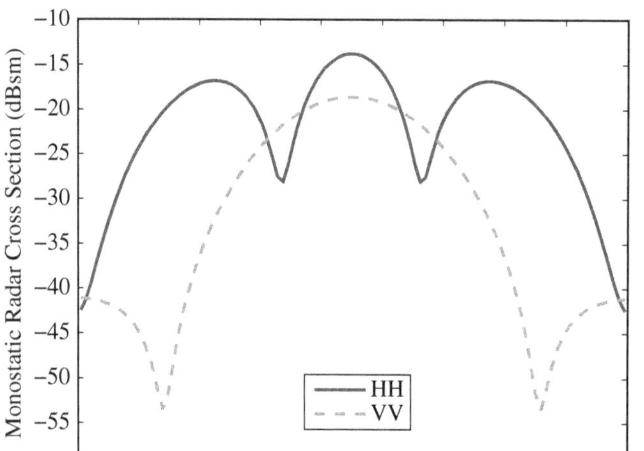

Figure 6.25 Monostatic RCS solutions of an ogive scatterer in HH and VV polarizations, respectively.

of the root-mean-square (RMS) error versus the density of unknowns, as shown in Figure 6.26. The very fine discretization has 360 triangular patches, resulting in 2160 unknowns and a density of unknowns of 71.6 unknowns per wavelength. The rule1, rule2 and rule3 in the legend denote the used quadrature rules of the second-degree 3 points, third-degree 6 points and fourth-degree 6 points, respectively [50]. The higher-order convergence can be clearly seen when the geometrical discretization is refined and this is called h-refinement in [3]. The second example is the scattering by a conducting pencil-like target whose geometry is shown in Figure 6.24(b). The

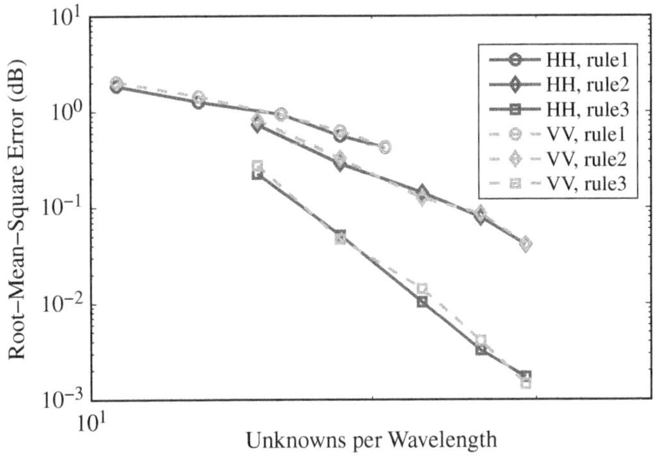

Figure 6.26 RMS Error versus density of unknowns for the monostatic RCS solution of an ogive scatterer.

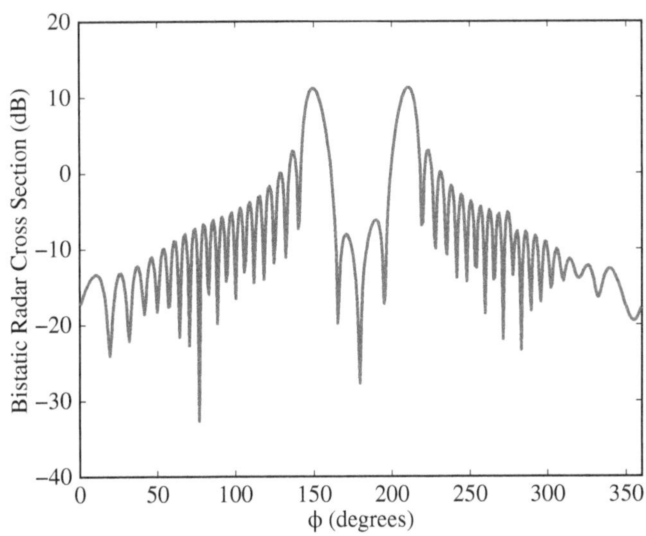

Figure 6.27 Bistatic RCS solution for a pencil-like target in HH polarization.

target has a radius $a = 0.1$ m and a height $d = 3$ m for the cylindrical part, and a height $t = 0.173$ m for the conical part. Figure 6.27 depicts the bistatic RCS solution in HH polarization (the solution for VV polarization is omitted) when the incident wave has a frequency $f = 1.0$ GHz and an incident angle $(\theta^i, \phi^i) = (30^\circ, 0^\circ)$. The solutions have been validated by using the well-known fast Illinois solver code (FISC). In a similar way, we use the solution based on a highly refined discretization of geometry and the highest-order quadrature rule as accurate solutions and calculate the corresponding RMS errors whose curves are shown in Figure 6.28. The very fine discretization for this case has 2600 triangular patches, resulting in 15600 unknowns and a density of

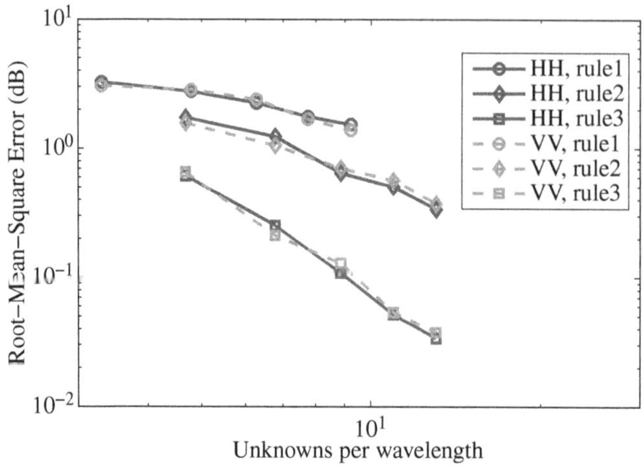

Figure 6.28 RMS error versus density of unknowns for the bistatic RCS solutions of a pencil-like target.

unknowns of 29.6 unknowns per wavelength. The h-refinement is also achieved before reaching a low frequency breakdown in this case.

6.5 Solutions of the Magnetic Field Integral Equation

In this section, we focus on the Nyström method solution for the magnetic field integral equation (MFIE) and particularly consider to solve the problems with very thin conducting objects. Accurate analysis for EM interaction with very thin conducting objects has a wide application and could be required in, for instance, the scattering by some antennas or shielding structures of electronic devices designed for preventing EM interference [1]. In the integral equation approach for modeling the problem, surface integral equations (SIEs) are preferred to describe involved EM features when the objects can be approximately treated as perfectly electric conductors (PECs) [51]. In the numerical solution for the problem, however, accurate solutions may not be easily obtained if the objects are very thin because the polarized component of induced current density tends to be singular when approaching geometric edges. Also, many slim meshes with very low quality will be produced on the side faces of thin objects in geometric discretization. When the SIEs are transformed into matrix equations by a numerical method, the evaluation of singular and near-singular integrals in matrix elements is very essential for accurate solutions and those low-quality meshes will give an extra difficulty for numerical implementation.

Traditionally, one uses the EFIE to describe very thin objects by treating them as zero-thickness objects and taking a vector sum of the current density on the two opposite sides of objects as a new unknown current density imposed at one side to be solved. This will ignore the current density on side walls and the solution may not be very accurate in some applications. Also, the EFIE can result in a worse conditioning of the system matrix after discretization because it is the first kind of integral equation and may suffer from a serious low-frequency breakdown problem [52]. Since the thin structures are of multiscale property (the thickness is much smaller than other dimensions), which could produce a low-frequency effect, the conditioning of the system matrix will greatly deteriorate in the EFIE, which is vulnerable to it.

To overcome the drawbacks in the traditional method, we need to develop a robust method that can treat an arbitrarily thin object as an exact 3D or closed object. We have developed a Nyström scheme to solve the EFIE [53] and CFIE [54], respectively, for thin conducting objects, but the objects were not very thin. Here, we use the MFIE individually to formulate the problem and consider super-thin structures. The MFIE is less accurate than the EFIE in general because its operator is non-smooth while the operator of the EFIE is smooth [55], but the accuracy can be greatly improved by using a curl-conforming basis function in the MoM solution [56]. The most remarkable advantage of the MFIE is that it can result in a better conditioning of the system matrix since it is the second kind of integral equation. Also, the MFIE does not have an obvious low-frequency breakdown or low-frequency effect due to its magnetostatic feature [57] and this may be very desirable for super-thin structures. The aforementioned geometric approximation by viewing a very thin object as a 2D open structure can only be valid for the EFIE and it cannot be applied to the MFIE since the directions of unit normal vectors on the top and bottom sides are opposite and the current density on the two sides cannot be summed up in the MFIE. This also requires a strict 3D treatment for super-thin objects.

The MFIE is solved by the Nyström method which had not been used for solving the problems with super-thin structures until the work in [53]. Since the evaluation of singular or near-singular integrals is very essential to obtain accurate numerical solutions and the widely used Duffy's method [43] may not be suitable for calculating strongly singular or near-singular integrals appearing in the MFIE, we particularly develop a robust treatment technique for them based on the Green's lemma, which changes a surface integral over a surface patch into a line integral along its boundary. Also, the near-singular integrals will play a vital role because there are many near-interaction matrix elements in the super-thin structures. Unlike existent singularity treatment techniques [43, 59–72], which usually need to further subdivide triangular patches, the formulations of the developed approach are expressed in terms of the coordinates of vertices of triangular patches without a subdivision. Such formulations are very friendly in numerical implementation, especially for those slim triangular meshes appearing in the side faces of super-thin structures. Some numerical examples for EM scattering by super-thin conducting objects are presented to illustrate the scheme and good results have been observed. In particular, we study the asymptotic behavior when the thickness of objects becomes smaller and smaller and a threshold is suggested at which the zero-thickness or open-structure approximation can be taken without an obvious error compared to the corresponding closed structure.

6.5.1 Integral Equations

Consider a super-thin conducting object that is assumed to be a PEC and is embedded in the free space with a permittivity ϵ_0 and permeability μ_0. The EM scattering by such an object can be formulated by the MFIE, which results from the corresponding boundary condition [1]

$$\hat{n} \times \left[\mathbf{H}^{\mathrm{inc}}(\mathbf{r}) + \mathbf{H}^{\mathrm{sca}}(\mathbf{r}) \right] = \mathbf{J}_s(\mathbf{r}), \quad \mathbf{r} \in S \tag{6.133}$$

or

$$\frac{1}{2}\mathbf{J}_s(\mathbf{r}) - \hat{n} \times \fint_S \nabla g(\mathbf{r}, \mathbf{r}') \times \mathbf{J}_s(\mathbf{r}')\, \mathrm{d}S' = \hat{n} \times \mathbf{H}^{\mathrm{inc}}(\mathbf{r}), \quad \mathbf{r} \in S \tag{6.134}$$

where $\mathbf{H}^{\mathrm{inc}}(\mathbf{r})$ is the incident magnetic field, $\mathbf{H}^{\mathrm{sca}}(\mathbf{r})$ is the scattered magnetic field, $\mathbf{J}_s(\mathbf{r})$ is the induced current density at the surface of conducting object S, and $\hat{n}$ is the unit normal vector at S. Also,

$$g(\mathbf{r}, \mathbf{r}') = \frac{\mathrm{e}^{\mathrm{i}\kappa_0 R}}{4\pi R} \tag{6.135}$$

is the scalar Green's function in which $\kappa_0 = \omega\sqrt{\mu_0\epsilon_0}$ is the free-space wavenumber, $R = |\mathbf{r} - \mathbf{r}'|$ is the distance between an observation point $\mathbf{r}$ and a source point $\mathbf{r}'$, and ω is the angular frequency of the incident wave. Note that the integral in the MFIE is defined under a Cauchy-principal-value (CPV) sense and the first term in the left-hand side has included the residue contribution of the singular integral. Since the unknown current density appears both inside and outside the integral, it is known as the second kind of integral equation, which can usually lead to a better conditioning of the system matrix after discretization.

6.5.2 Singularity and Near-Singularity Treatment

Although we have specially addressed the singularity and near-singularity treatment techniques in Chapter 5, some supplementary techniques will be presented in incorporation with their applications and this section presents one of them. In the above matrix equation, we need to accurately evaluate matrix elements that include different integrals. For far-interaction elements, the integrals are regular and we can use numerical quadrature rules to evaluate them. For self-interaction elements or when $m = n$, the integrals are singular and we need to carefully handle them. Since we work on super-thin objects, the near-interaction elements including near-singular integrals are popular and play an

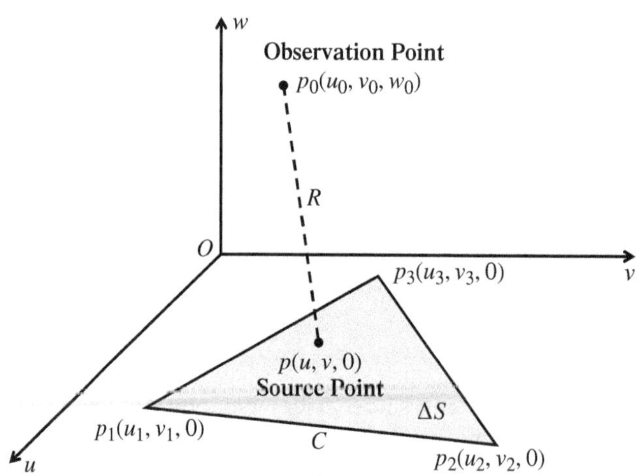

Figure 6.29 A surface integral over a triangular patch can be changed into a line integral along its boundary based on the Green's lemma.

essential role for accurate solutions, but they are less addressed [71, 72]. We develop a robust treatment technique for evaluating those singular or near-singular integrals by using the Green's lemma to derive two identities that can be used to transform a surface integral over a triangular patch into a line integral along its boundary, as shown in Figure 6.29. The derived identities are actually the same as those derived based on Stokes' theorem in Section 5.5 in Chapter 5 and we just want to provide another approach to do the derivation. Furthermore, the Green's lemma is also the same as the 2D version of Green-Gauss theorem and the identities look simpler than those from Stokes' theorem. As shown in Section 5.5 in Chapter 5, the resulting formulations for evaluating singular and near-singular integrals do not require a subdivision of triangles with a calculation of inner angles and they are more friendly and convenient in implementation, especially for those very slim triangular patches in thin structures.

The Green's lemma can be written as [73]

$$\int_{\Delta S} \left[\frac{\partial f_1(u,v)}{\partial u} - \frac{\partial f_2(u,v)}{\partial v} \right] \mathrm{d}S = \int_C \left[f_2(u,v)\, du + f_1(u,v)\, \mathrm{d}v \right] \tag{6.136}$$

where ΔS is a surface domain and C is its boundary, respectively, as shown in Figure 6.29. The functions $f_1(u,v)$ and $f_2(u,v)$ are defined in a local coordinate system (u,v,w) and they must be continuous. If we let one of these two functions be zero, then we can obtain the following two integral identities

$$\int_{\Delta S} \frac{\partial f_1(u,v)}{\partial u} \mathrm{d}S = \oint_C f_1(u,v)\, \mathrm{d}v \tag{6.137}$$

$$\int_{\Delta S} \frac{\partial f_2(u,v)}{\partial v} \mathrm{d}S = -\oint_C f_2(u,v)\, \mathrm{d}u \tag{6.138}$$

which can be used to transform a surface integral over a surface domain into a line integral along its boundary for a continuous function. The strategy has been used for getting rid of domain discretization in a meshless method [74], but we use it for singularity or near-singularity treatment here.

In the MFIE, there are many types of singular or near-singular integrals produced from the singularity subtraction for the integrand that includes the gradient of the scalar Green's function. The weakly- singular or near-singular integrals can be easily handled by many traditional methods [43, 59–72] and we do not address them here. The strongly singular or near-singular integrals result from the combination of the gradient of the scalar Green's function with the constant term of the polynomial function of representing the unknown current density and they take the following forms

$$I_1 = \int_{\Delta S} \frac{(u_0 - u)}{R^3} \mathrm{d}S \tag{6.139}$$

$$I_2 = \int_{\Delta S} \frac{(v_0 - v)}{R^3} \mathrm{d}S \tag{6.140}$$

$$I_3 = w_0 \int_{\Delta S} \frac{1}{R^3} \mathrm{d}S \tag{6.141}$$

where (u_0, v_0, w_0) represents an observation point and $R = \sqrt{(u_0 - u)^2 + (v_0 - v)^2 + w_0^2}$ is the distance between the observation point and a source point $(u, v, 0)$ within the triangular patch ΔS. If we apply Equation (6.137) or Equation (6.138) to the above integrals by selecting the original function of the corresponding integrand as $f_1(u, v)$ or $f_2(u, v)$, those integrals can be derived analytically with the help of integral formulas in [42] and they can be expressed as

$$\begin{aligned} I_1 &= \oint_C \frac{1}{R} \, dv \\ &= \sum_{i=1}^{3} \left\{ \frac{v_{i+1} - v_i}{d_i} \ln \left| \frac{d_i r_{i+1} + d_i^2 + b_i}{d_i r_i + b_i} \right| \right\} \end{aligned} \tag{6.142}$$

$$\begin{aligned} I_2 &= - \oint_C \frac{1}{R} \, du \\ &= \sum_{i=1}^{3} \left\{ \frac{u_{i+1} - u_i}{d_i} \ln \left| \frac{d_i r_i + b_i}{d_i r_{i+1} + d_i^2 + b_i} \right| \right\} \end{aligned} \tag{6.143}$$

$$\begin{aligned} I_3 &= - w_0 \oint_C \frac{(u_0 - u)}{\left[(v_0 - v)^2 + w_0^2\right] R} \, dv \\ &= \sum_{i=1}^{3} \left\{ \frac{1}{2} \left[\tan^{-1}(s_1^i) - \tan^{-1}(s_2^i) + \tan^{-1}(s_3^i) - \tan^{-1}(s_4^i) \right] \right\} \end{aligned} \tag{6.144}$$

where

$$d_i = \sqrt{(u_{i+1} - u_i)^2 + (v_{i+1} - v_i)^2} \tag{6.145}$$

$$r_i = \sqrt{(u_0 - u_i)^2 + (v_0 - v_i)^2 + w_0^2} \tag{6.146}$$

$$a_i = (u_i - u_0)(v_{i+1} - v_i) - (v_i - v_0)(u_{i+1} - u_i) \tag{6.147}$$

$$b_i = (u_{i+1} - u_i)(u_i - u_0) + (v_{i+1} - v_i)(v_i - v_0) \tag{6.148}$$

$$r_{i+1} = \sqrt{d_i^2 + 2b_i + r_i^2} = \sqrt{(u_0 - u_{i+1})^2 + (v_0 - v_{i+1})^2 + w_0^2} \tag{6.149}$$

$$s_1^i = -\frac{(u_i - u_0)a_i + (v_{i+1} - v_i)w_0^2}{(u_{i+1} - u_i)r_i w_0} \tag{6.150}$$

$$s_2^i = -\frac{(u_{i+1} - u_0)a_i + (v_{i+1} - v_i)w_0^2}{(u_{i+1} - u_i)r_{i+1} w_0} \tag{6.151}$$

$$s_3^i = -\frac{(v_i - v_0)a_i - (u_{i+1} - u_i)w_0^2}{(v_{i+1} - v_i)r_i w_0} \tag{6.152}$$

$$s_4^i = -\frac{(v_{i+1} - v_0)a_i - (u_{i+1} - u_i)w_0^2}{(v_{i+1} - v_i)r_{i+1} w_0} \tag{6.153}$$

and it is assumed that $i+1 = \text{mod}(i, 3)+1$. The above formulations can be used no matter how close the observation point is to the source triangular patch, i.e., they can also be used to calculate the regular integrals of far-interaction elements (but the numerical integration may be more convenient in this case). To obtain the formulations of singular integrals, we only need to take a limit of $w_0 \to 0$ in the above formulations and the resultant formulations are exactly the same as those in Subsection 5.5.3. It can be seen that the above formulations can be more easily implemented in the calculation of singular and near-singular integrals since they only relate to the coordinates of vertices of each intact triangle. Such formulations are particularly suitable for slim or low-quality triangular meshes.

6.5.3 Numerical Examples

We present several numerical examples to demonstrate the solutions of the MFIE for EM scattering by super-thin conducting objects (PECs). The incident wave is a plane wave propagating along the $-z$ direction and it has a frequency of $f = 300$ MHz. The bistatic radar cross section (RCS), which is observed along the principal cut defined by $\phi = 0°$ and $\theta = 0° - 180°$, is calculated for the scatterers and both vertical (VV) and horizontal (HH) polarizations are considered.

The first example is the scattering by a thin hexagonal plate as shown in Figure 6.30(a) and its geometry is characterized by the side length of each edge $a = 0.8\lambda$ and the thickness $t = 0.006\lambda$, where λ is the wavelength in free space. We discretize the surface into 3288 triangular patches. The subsequent two examples are the scattering by an annulus plate and a dumbbell-like plate as shown in Figure 6.30(b) and Figure 6.30(c), respectively. The annulus plate is defined by an external radius $r_1 = 2.0\lambda$, an inner radius $r_2 = 1.5\lambda$, and a thickness $t = 0.02\lambda$, and its surface is discretized into 1456 triangular

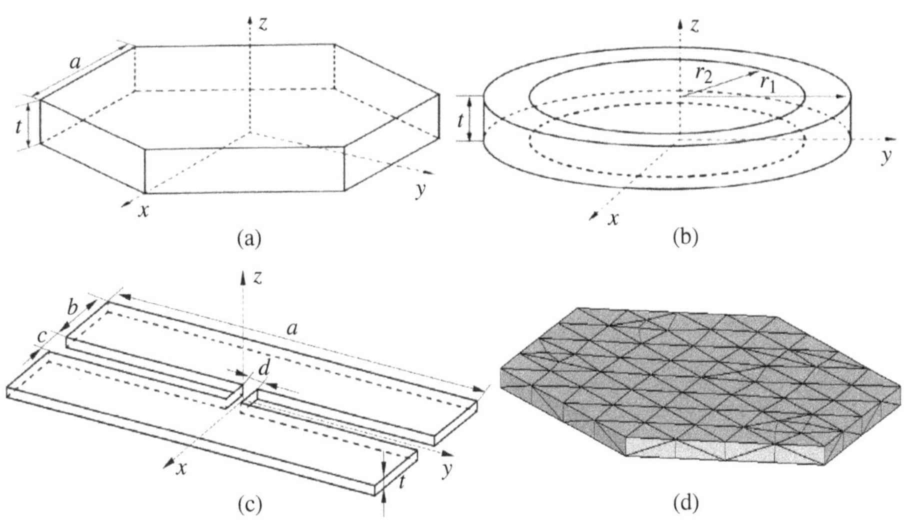

Figure 6.30 Geometries of three super-thin conducting scatterers and a sample of meshed object. (a) A hexagonal plate. (b) An annulus plate. (c) A dumbbell-like plate. (d) a sample of meshed hexagonal plate.

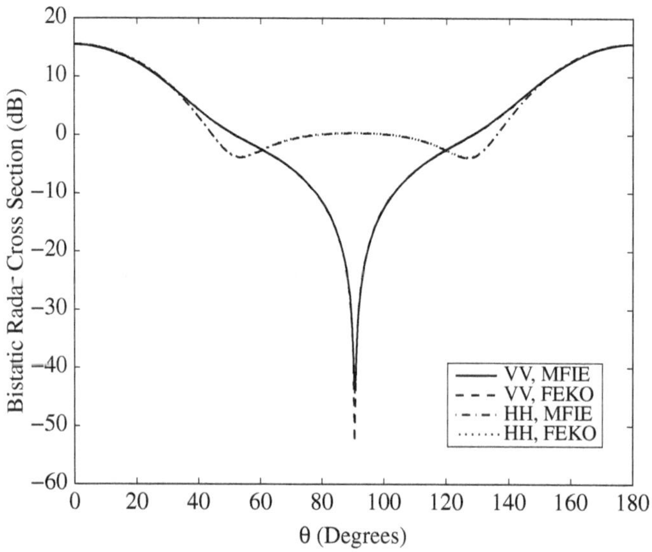

Figure 6.31 Bistatic RCS solutions for a super-thin conducting hexagonal plate with a side length $a = 0.8\lambda$ and a thickness $t = 0.006\lambda$.

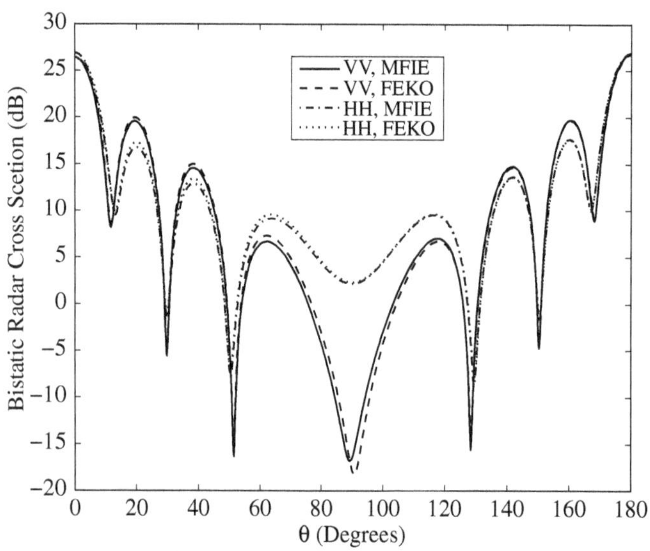

Figure 6.32 Bistatic RCS solutions for a super-thin conducting annulus plate with an external radius $r_1 = 2\lambda$, an inner radius $r_2 = 1.5\lambda$, and a thickness $t = 0.02\lambda$.

patches. The dumbbell-like plate is defined by $a = 0.2\lambda$, $b = 0.1\lambda$, $c = 0.04\lambda$, $d = 0.02\lambda$, and a thickness $t = 0.001\lambda$, and we mesh its surface into 2108 triangular patches. As an example, we present the meshes of the super-thin hexagonal plate in Figure 6.30(d) and it can be seen that the triangular meshes at the side faces are very slim (the thickness is not drawn in scale). Usually, the quality of triangular meshes is measured

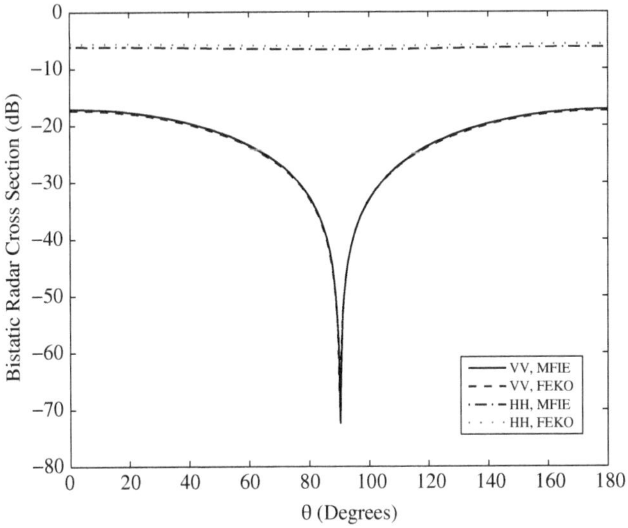

Figure 6.33 Bistatic RCS solutions for a super-thin conducting dumbbell-like plate with $a = 0.2\lambda$, $b = 0.1\lambda$, $c = 0.04\lambda$, $d = 0.02\lambda$, and $t = 0.001\lambda$.

by the quality factor q, which will be addressed in the next section. The minimum values of q in the above three examples are 0.2837, 0.3415, and 0.2178, respectively, indicating that the quality of their meshes are quite poor. Figures 6.31–6.33 show the corresponding RCS solutions for the three scatterers and they are in good agreement with the corresponding solutions from the well-known commercial software FEKO.

It is well known that the MFIE is less accurate than the EFIE, but we cannot prove this for super-thin structures since no exact solution can be used as a standard to make

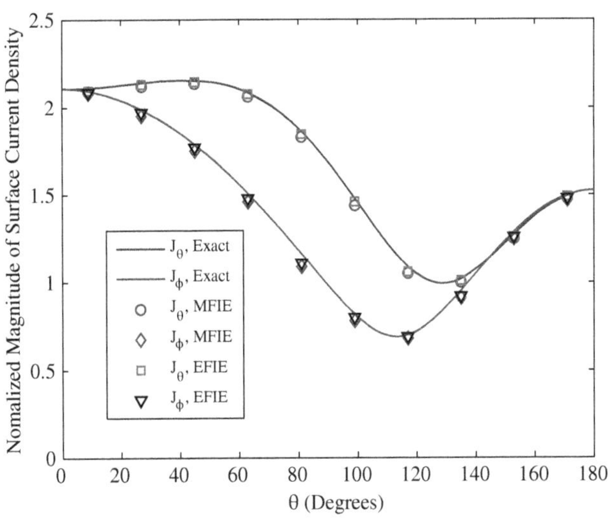

Figure 6.34 Comparison of current density solutions from the MFIE and EFIE for a conducting sphere with a radius $a = 0.2\lambda$. The observed points are sampled equally along the principal cut.

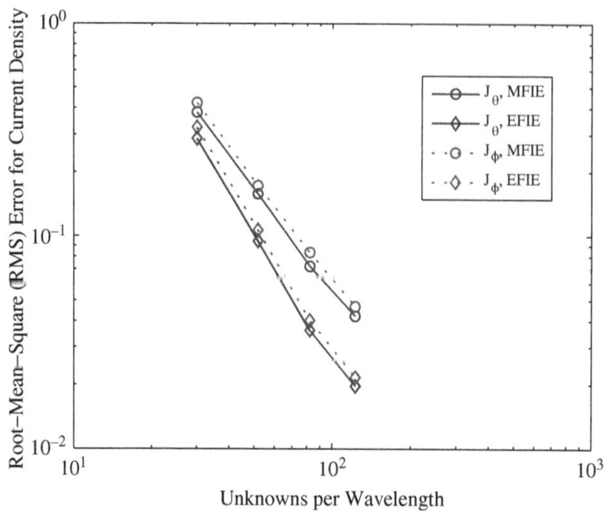

Figure 6.35 Comparison of error convergence of current density solutions from the MFIE and EFIE for a conducting sphere with a radius $a = 0.2\lambda$.

Table 6.1 Comparison of condition numbers (CNs) of impedance matrices for three numerical examples.

Cases		Example 1	Example 2	Example 3
CNs	EFIE	6.5312×10^7	5.2736×10^5	1.3623×10^8
	MFIE	4.7257×10^4	3.1685×10^3	7.0291×10^4

a reasonable comparison. We use the scattering by a conducting sphere as an example to make such a comparison because the example has an exact Mie-series solution to compare with. The sphere has a radius $a = 0.2\lambda$ and we calculate the distribution of current density along the principal cut at the surface of sphere. Figure 6.34 shows the comparison of current density between the two formulations and the MFIE results are less accurate than the EFIE results, as expected. The accuracy comparison between the two approaches is further performed by checking their internal convergence based on the refinement of meshes for the example. Figure 6.35 shows the convergence of RMS errors of current density as the mesh size reduces or mesh density increases, where the number of triangular meshes is 452, 1346, 3386, and 7556, respectively, and it can be seen that the convergence rate (slope) of the MFIE approach is smaller than that of the EFIE approach. Since the sphere has a curvilinear surface, the geometric approximation will produce some errors when flat triangular meshes are used, especially, when the surface is discretized with coarse meshes. As the mesh density increases, the errors will rapidly decrease, as shown in the figure. To demonstrate the merit of the MFIE, i.e. it has a better conditioning of impedance matrix than the EFIE, we present a comparison of the condition numbers resulting from the two formulations for the first three examples.

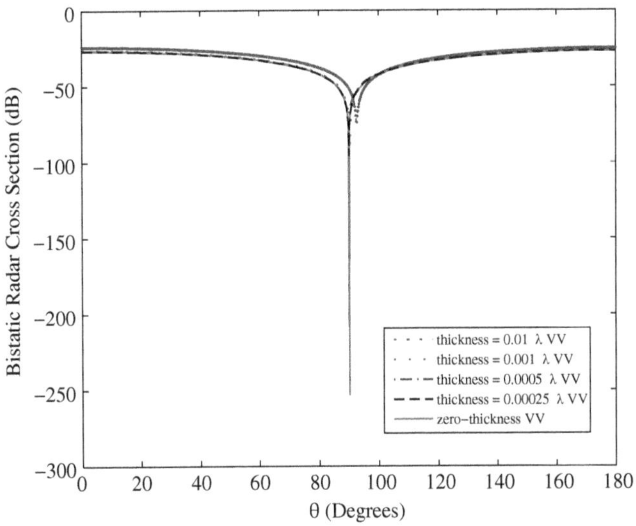

Figure 6.36 Bistatic RCS solutions for a square plate with different thicknesses (vertical polarization). The side length of plate is $l = 0.15\lambda$ and its thickness changes from $t = 0.01\lambda$ to $t = 0.00025\lambda$.

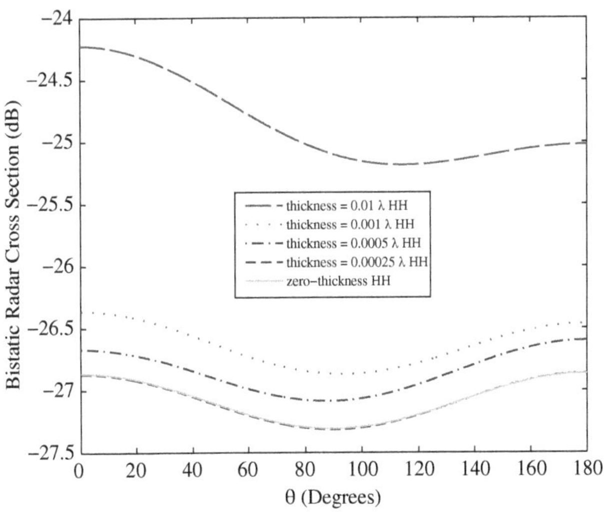

Figure 6.37 Bistatic RCS solutions for a square plate with different thicknesses (horizontal polarization). The side length of plate is $l = 0.15\lambda$ and its thickness changes from $t = 0.01\lambda$ to $t = 0.00025\lambda$.

As shown in Table 6.1, it is clear that the condition numbers of the MFIE approach are much smaller than those of the EFIE approach.

As we have pointed out before, if an object could be treated as a zero-thickness one, then the current density at two opposite sides can be added together to result in an open-structure problem in the EFIE and the solving process can be greatly simplified. To find out what thickness can make the zero-thickness approximation be appropriate in the MFIE, we present a fourth example to study it. The fourth example is scattering by a square plate (no figure is plotted) with a side length $l = 0.15\lambda$ and its thickness changes from $t = 0.01\lambda$ to $t = 0.00025\lambda$. The RCS solutions in the vertical (VV) and horizontal (HH) polarizations for the square plate with different thicknesses are shown in Figures 6.36 and 6.37, respectively, and we can see that as the thickness of the square plate reduces, the solution curve approaches the zero-thickness one. When the thickness goes to $t = 0.00025\lambda$, its solution curve exactly coincides with the zero-thickness one in both VV and HH. From our extensive numerical experiments, we believe that the thickness $t = 0.001\lambda$ is a threshold below which the zero-thickness approximation can be used without causing an obvious numerical error.

6.6 Solutions of the Combined Field Integral Equation

We illustrate the solution of the CFIE by also considering the very thin structures because one could encounter some unfavorable factors in solving for such structures. The CFIE is a linear combination of the EFIE and MFIE and was proposed to remove the internal resonance problem in the EFIE and MFIE, but it may also be more suitable for solving the problems with thin structures. For thin objects, the individual EFIE may not be able to give accurate solutions if they are strictly treated as 3D objects [76]. As a limiting case of thick conducting objects, the individual EFIE will degenerate to the dependent equations about the top and bottom surfaces when the thickness reduces [77]. This degeneration will dramatically deteriorate the conditioning of impedance matrix and can lead to the failure of the EFIE in the limit. In contrast, the individual MFIE will not degenerate to the dependent equations about the top and bottom surfaces when the thickness unlimitedly reduces because the directions of unit normal vectors at the two opposite surfaces are opposite. The EFIE can only work for thin objects by taking the vector sum of the current density on the opposite sides of the objects as a new unknown current density and treat the thin objects as open structures. The MFIE cannot follow the same way as the EFIE to work for thin objects due to the opposite unit normal directions on the two opposite sides and must treat the objects as closed objects. Due to including the MFIE, the CFIE can always apply to thin objects no matter how thin the objects are. This is because the opposite directions of unit normal vectors on the two opposite sides of a thin object will make the CFIE reduce to a sum and difference of the EFIE and MFIE on the two opposite sides. Since the sum and difference are linearly independent, the CFIE is always valid for thin objects just like the MFIE and may improve the conditioning of impedance matrix although it requires more computational effort. Also, as shown in the last section, the thin objects generally require much more evaluations of near-singular integrals in filling the impedance matrix because the near interactions between observation points and source patches prevail. The near-interaction elements are very important to the accuracy of numerical solutions and they have to be carefully treated, just like self-interaction elements. In addition, the thin objects will produce many high-aspect-ratio triangular patches in the geometric discretization and these low-quality patches could result in a difficulty for the accurate evaluation of singular and near-singular matrix elements, although such an evaluation could be avoided sometimes [78].

Aiming to mitigate these unfavorable factors, we develop an efficient Nyström method for solving thin-structure problems based on the CFIE. As mentioned before, the singularity treatment is a key in the method and the numerical errors come from three sources, namely, geometric approximation, matrix representation approximation, and solution approximation of matrix equations. Accurate solutions require each type of error to be localized and controllable and any of them could dominate the total error if it is inappropriately handled. Compared with the other two kinds of approximation, which are usually more straightforward, the matrix representation approximation for the original integral equation is less explicit and the appropriate treatment for singular and near-singular matrix elements is very essential to achieve a good matrix representation. Although many singularity treatment techniques have been developed in recent years [43–88], they mainly deal with weak or strong singularities. We developed an efficient hypersingularity treatment scheme in our previous work [61] and it has been used in many scenarios. However, the formulations in the technique require a subdivision of triangular patches and they may not be very suitable for the high-aspect-ratio patches that are common in thin objects. In this section, we develop some different formulations for evaluating singular and hypersingular integrals without subdividing triangles, and they are more friendly and appropriate for those low-quality patches. The formulations are derived by transforming a surface integral into a boundary integral based on the Stokes theorem [89] and they only need the information of three vertices of a triangular patch, which is much simpler to handle. We present several numerical examples for EM scattering by thin conducting objects to demonstrate the method, and good results can be observed.

6.6.1 Integral Equations

Owing to the geometric characteristic of thin objects, we use the CFIE as a governing equation to describe the EM interaction with them. Consider the EM scattering by a conducting object embedded in the free space with a wavenumber k_0, the EFIE and MFIE can be written as [1]

$$\hat{n} \times \mathcal{L}\left[\mathbf{J}(\mathbf{r}')\right] = \hat{n} \times \mathbf{E}^{\text{inc}}(\mathbf{r}), \quad \mathbf{r} \in S \tag{6.154}$$

$$\frac{1}{2}\mathbf{J}(\mathbf{r}) + \hat{n} \times \mathcal{K}\left[\mathbf{J}(\mathbf{r}')\right] = \hat{n} \times \mathbf{H}^{\text{inc}}(\mathbf{r}), \quad \mathbf{r} \in S \tag{6.155}$$

where S is the object surface, $\hat{n}$ is the unit normal vector on the surface, $\mathbf{J}(\mathbf{r}')$ is the induced surface current density, $\mathbf{E}^{\text{inc}}(\mathbf{r})$ and $\mathbf{H}^{\text{inc}}(\mathbf{r})$ are the incident electric field and magnetic field, respectively, and the integral operators $\mathcal{L}$ and $\mathcal{K}$ are defined as

$$\mathcal{L}\left[\mathbf{J}(\mathbf{r}')\right] = \mathrm{i}k_0\eta_0 \int_S \overline{\mathbf{G}}(\mathbf{r}, \mathbf{r}') \cdot \mathbf{J}(\mathbf{r}')\, \mathrm{d}S' \tag{6.156}$$

$$\mathcal{K}\left[\mathbf{J}(\mathbf{r}')\right] = \fint_S \mathbf{J}(\mathbf{r}') \times \nabla g(\mathbf{r}, \mathbf{r}')\, \mathrm{d}S' \tag{6.157}$$

respectively. In the above, η_0 is the wave impedance of the free space and

$$\overline{\mathbf{G}}(\mathbf{r}, \mathbf{r}') = \left(\overline{\mathbf{I}} + \frac{\nabla\nabla}{k_0^2}\right) g(\mathbf{r}, \mathbf{r}') \tag{6.158}$$

is the dyadic Green's function in which $\overline{\mathbf{I}}$ is the identity dyad, $g(\mathbf{r}, \mathbf{r}') = \mathrm{e}^{\mathrm{i}kR}/(4\pi R)$ is the scalar Green's function, and $R = |\mathbf{r} - \mathbf{r}'|$ is the distance between an observation point

$\mathbf{r}$ and a source point $\mathbf{r}'$. Combining the EFIE and MFIE with an appropriate weight, we have the following CFIE [1]

$$\alpha\hat{n}\times\hat{n}\times\mathcal{L}\left[\mathbf{J}(\mathbf{r}')\right]+(1-\alpha)\eta_0\left\{\frac{1}{2}\mathbf{J}(\mathbf{r})+\hat{n}\times\mathcal{K}\left[\mathbf{J}(\mathbf{r}')\right]\right\}$$
$$=\alpha\hat{n}\times\hat{n}\times\mathbf{E}^{\text{inc}}(\mathbf{r})+(1-\alpha)\eta_0\hat{n}\times\mathbf{H}^{\text{inc}}(\mathbf{r}),\ \ \mathbf{r}\in S \tag{6.159}$$

where α is the combination parameter and is generally selected between 0.2 and 0.8. The CFIE can not only improve the conditioning of impedance matrix but also remove the internal resonance though requiring higher computational costs. Also, since $\hat{n}$ is oriented on the two opposite sides for a thin object, there will be an opposite operation in $\hat{n}\times\mathcal{K}\left[\mathbf{J}(\mathbf{r}')\right]$ and $\hat{n}\times\mathbf{H}^{\text{inc}}(\mathbf{r})$ in the above equation on the two sides [1].

6.6.2 Quality of Triangular Patches

In thin objects, the thickness is usually much smaller than other geometric dimensions and this feature could yield a difficulty for generating good triangular meshes that are close to equilateral triangles. Good quality of meshes can optimize the performance of numerical solvers and how to generate good meshes has been extensively studied in the FEM [75]. In the integral equation method, although the shape of triangular meshes may not be very significant in the MoM with the Rao-Wilton-Glisson (RWG) basis function [90] because the singularity is weak and many mature techniques can fully handle it [43–88], it is important in the point-matching method, as in the Nyström method, because there exists a hypersingularity and the mesh quality could affect the accuracy of evaluating the hypersingular integrals. The quality of triangular meshes can be measured with the aspect ratio, which can be defined in several ways [91] and the simplest way is to use the ratio of the longest side to the minimal height in a triangle. Alternatively, we can use the more reasonable quality factor proposed in [75] to measure the quality of meshes, as we have mentioned in the last section. The quality factor q is

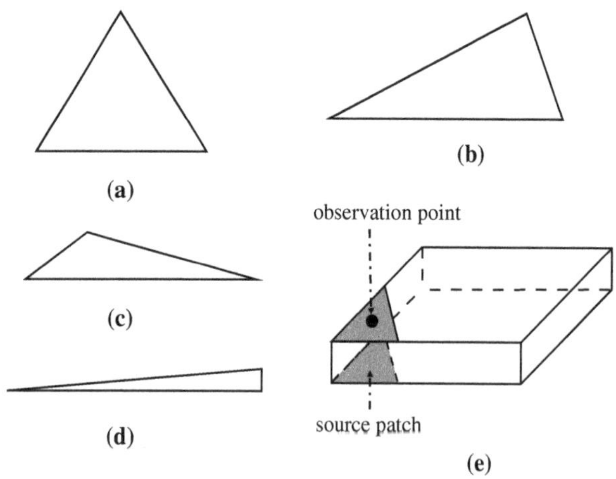

Figure 6.38 Quality of triangular meshes which is measured by the quality factor q and near interaction between non-neighboring patches. (a) $q = 1.0$. (b) $q = 0.7623$. (c) $q = 0.2585$. (d) $q = 0.0862$. (e) Near interaction between an observation point on the top and a source patch on the bottom.

defined as the ratio of the diameter of the inscribed circle to the radius of the circumscribed circle for a given triangle, or equivalently

$$q = \frac{8}{abc}(s-a)(s-b)(s-c) \tag{6.160}$$

where a, b, and c are the lengths of the three sides of the triangle, respectively, and $s = (a+b+c)/2$. The equilateral triangle has a maximum $q = 1.0$ while $q = 0$ corresponds to a colinear triangle. High aspect ratio or low quality factor means that the triangular mesh is poor or is far away from an equilateral triangle, as shown in Figure 6.38. Because of the small thickness in thin objects, high-aspect-ratio or low-quality-factor meshes can be easily created in geometric discretization. In the evaluation of singular or hypersingular integrals, the formulations usually require a subdivision of triangular patches and involve many calculations related to the inner angles of triangles. The high aspect ratio or low quality factor of meshes could reduce the accuracy of evaluating hypersingular or singular integrals and it is desirable to find alternative formulations suitable for those slim meshes.

6.6.3 Nyström Discretization

The basic principle of the Nyström method has been presented in Chapter 3 and its implementation for solving the EFIE and MFIE has also been addressed in the previous sections. Since the CFIE is just a linear combination of the EFIE and MFIE, the Nyström discretization for the CFIE is also the linear combination of the Nyström discretizations for the EFIE and MFIE. However, we consider a higher-order scheme of discretization with a geometric mapping for the EFIE previously, see Section 6.4. Now we present a different discretization scheme from that in Section 6.4, in which flat triangular patches are used but higher-order quadrature rules can still be employed if needed. This scheme is more like that for the MFIE, so they can be combined together to form the scheme of discretizating the CFIE.

As we have seen, the EFIE can actually be written in two kinds of form with different kernels: one is the sum of the scalar Green's function times the current density and the gradient of the scalar Green's function times the divergence of current density as shown in Section 6.4, and another is the dyadic Green's function including the double gradient of the scalar Green's function, which is a tensor, times the current density. These two kinds of form are equivalent to each other in fact and they can be mutually transformed by applying the divergence identity and Gauss' divergence theorem with a vanishing contour integral. In practice, one usually uses the first kind of form since it only includes the $1/R$ weak singularity from the scalar Green's function and the $1/R^2$ strong singularity from the gradient of the Green's function in the self-interaction matrix elements. The $1/R^2$ strong singularity can be reduced to the $1/R$ weak singularity either by applying the Gauss' divergence theorem to make a transformation in the MoM with RWG basis and testing functions, or by using the singularity subtraction to lower the degree of singularity as shown in Section 6.4. The $1/R$ weak singularity can be easily handled by the Duffy's method as we have done previously.

The second kind of form, which has been shown in Equation (6.154) with an $\mathcal{L}$ operator, possesses a $1/R^3$ hypersingularity from the double gradient of the scalar Green's function in its self-interaction matrix elements and there was no direct way to deal with it previously until we developed some techniques. Before our techniques appeared, the second kind of form was also solved sometimes, but the kernel with the double gradient of the scalar Green's function was transformed back to that of the first kind of form in order to treating the singularities, because of the inability in treating the hypersingularity at that time. Although the second kind of form has a hypersingularity which was thought of as prohibitive to treat before, it looks more friendly because the unknown current density is free of the divergence or partial-differential operation and it can be represented by any kind of function, including a constant. After the hypersingularity can be efficiently handled, the second kind of form has been preferred in many applications, especially when incorporated with the fast algorithms like multilevel fast multipole algorithm (MLFMA).

The second kind of form as shown in Equation (6.154) can be rewritten in a more explicit and implementable form

$$\hat{t} \cdot i\kappa_0\eta_0 \int_S \overline{\mathbf{G}}(\mathbf{r}, \mathbf{r}') \cdot \mathbf{J}(\mathbf{r}')\, dS' = \hat{t} \cdot \mathbf{E}^{\text{inc}}(\mathbf{r}), \quad \mathbf{r} \in S \tag{6.161}$$

where $\hat{t}$ is a tangential vector at the observation point $\mathbf{r}$. The Nyström discretization for the above EFIE has been shown in Sections 5.4 and 5.5 in Chapter 5 when we address the treatment of involved hypersingularities and we do not repeat it here. However, we use the formulations developed in Section 5.5 in Chapter 5 to evaluate the strongly singular or near-singular and hypersingular or near-hypersingular integrals for self-interaction or near-interaction matrix elements. This is because we solve the problems with thin structures now and those formulations are more friendly and convenient to implement as we have emphasized.

6.6.4 Numerical Examples

We present three numerical examples to demonstrate the Nyström solutions of the CFIE for EM scattering by thin conducting objects. It is assumed that the objects are PECs and the incident wave with a frequency $f = 300$ MHz is a plane wave propagating along the $-z$ direction in free space. We calculate the bistatic RCS observed along the principal cut ($\phi = 0°$ and $\theta = 0° - 180°$) for the scatterers with a vertical polarization (VV) or horizontal polarization (HH). The first example is the scattering by a thin conducting rectangular diamond as shown in Figure 6.39(a) and its geometry is characterized by the side length $a = b = 0.8\lambda$ and half central height $h = 0.02\lambda$, where λ is the wavelength in free space. Figure 6.40 plots the bistatic RCS solutions of the scatterer based on the Nyström method and the MoM with the number of triangular patches $N = 4326$, and they are close to each other. In the MoM, the CFIE is also solved and the RWG basis function is used to expand the unknown current density. The degree of singularity can be reduced in the MoM and the well-known Duffy's method is employed to evaluate the involved singular integrals.

The second example is the scattering by a thin conducting circular disk, as shown in Figure 6.39(b) and the geometry is defined by the radius $a = 1.0\lambda$ and the thickness $t = 0.05\lambda$. We use the MoM solution with very fine meshes ($N_0 = 18572$) as an exact solution and compare the Nyström solutions with three different numbers of meshes, which are $N_1 = 1246$, $N_2 = 5368$, and $N_3 = 8190$, respectively, to it. The solutions are

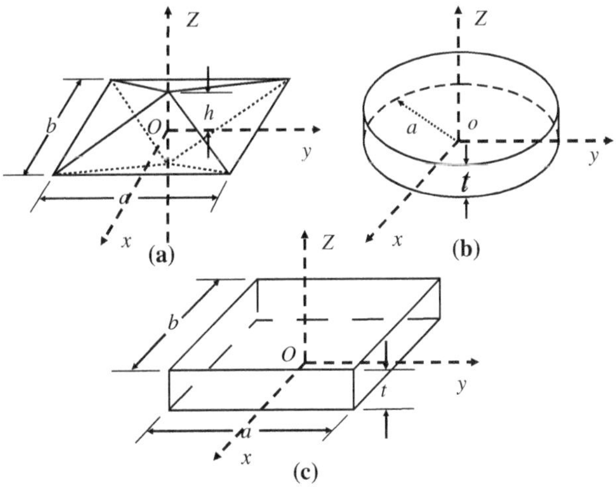

Figure 6.39 Geometries of scatterers. (a) A thin rectangular diamond with a length a, a width b, and a half central height h. (b) A thin circular disk with a radius a and a thickness t. (c) A thin rectangular plate with a length a, a width b, and a thickness t.

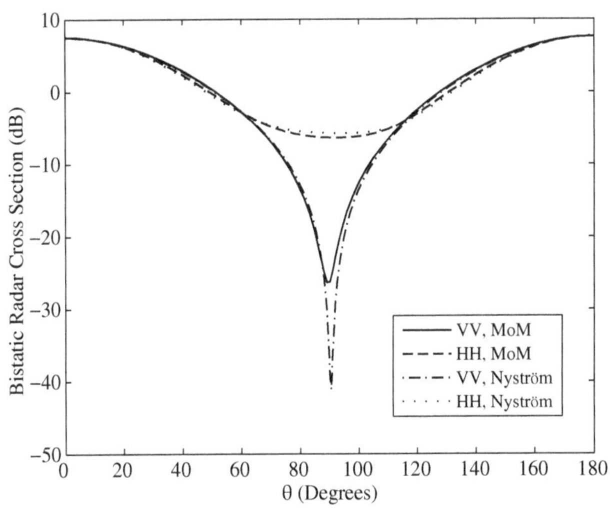

Figure 6.40 Bistatic RCS solutions for a thin conducting rectangular diamond with a side length $a = b = 0.8\lambda$ and half central height $h = 0.02\lambda$. The Nyström solutions are close to the MoM solutions with the same geometric discretization.

sketched in Figure 6.41 and it is found that the Nyström solutions converge well to the MoM solutions as the number of meshes increases.

In the third example, we study the scattering by a very thin conducting rectangular plate as shown in Figure 6.39(c) and the scatterer has a side length $a = b = 0.15\lambda$ and a thickness $t = 0.001\lambda$. Due to the small thickness, one usually treats the scatterer as an open structure and uses the EFIE to solve the problem. We now treat the scatterer as a 3D object without a geometric approximation and use the Nyström method to solve the

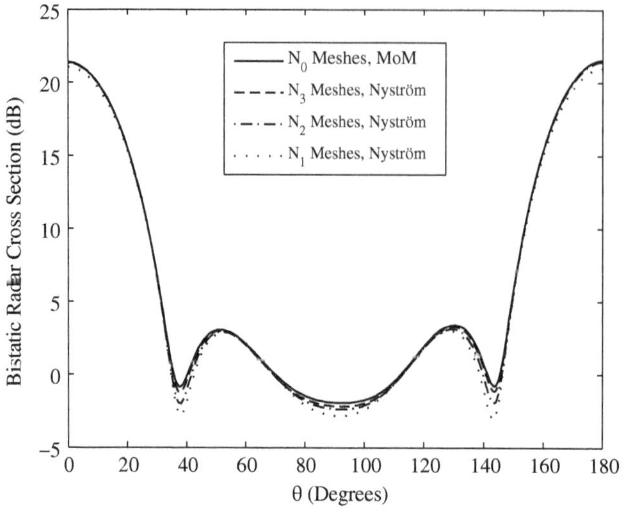

Figure 6.41 Bistatic RCS solutions (horizontal polarization) for a thin conducting circular disk with a radius $a = 1.0\lambda$ and thickness $t = 0.05\lambda$. The Nyström solutions converge to the MoM solution with very fine meshes when the number of meshes increases.

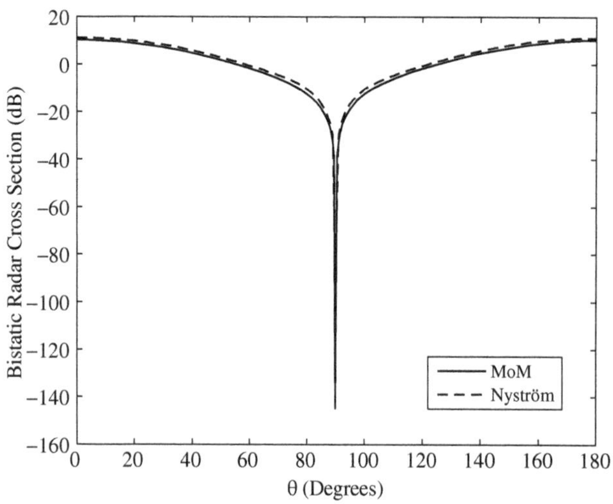

Figure 6.42 Bistatic RCS solutions (vertical polarization) for a thin conducting rectangular plate with a side length $a = b = 0.15\lambda$ and a thickness $t = 0.001\lambda$.

CFIE for the problem. Figure 6.42 depicts the bistatic RCS solutions and it can be seen that the Nyström solution and the MoM solution agree well with each other.

To demonstrate the usefulness of the proposed formulas for evaluating the singular integrals over a low-quality triangular patch, we perform a numerical test as shown in Table 6.1. The triangular patch is defined by three vertices $P_1(0.1, 0, 0)$, $P_2(0.14, 0.03, 0)$ and $P_3(0, 0.6, 0)$, leading to $q = 0.1312$, and the observation point is located at $P_0(0.08, 0.21, w_0)$, where w_0 can be changed and whose projection is inside

Table 6.2 Accuracy comparison between numerical integration, new formula and old formula for $-(3I_1 - I_3)$

w_0	Numerical	New formula	Old formula
1.0	0.01308718554113	0.01308718554113	0.01308718554113
0.5	0.09663099654479	0.09663099654479	0.09663099654479
0.2	1.07378959209505	1.07378959209509	1.07378959209512
0.1	5.14490060208385	5.14490061452641	5.14490065173382
0.05	20.6107923951961	20.6100537847676	20.6105126653252
0.01	172.578120266687	176.009802873133	177.821843693247
0.001	−7355.32664002998	252.835047148456	256.384179023236
0	−12809.5082089069	253.954032048001	257.721803562472

the patch. The numerical integrations are performed with a 24-point product Gaussian quadrature rule and Table 6.2 shows the results for the integral $-(3I_1 - I_3)$ (The minus sign is added so that most of values are positive in the table) when the observation point is approaching the patch. The new formula in the table refers to those developed in this section while the old formula refers to those based on the subdivision of source triangular patch as shown in [61]. It can be seen that the results of the numerical integration, new formula, and old formula are identical when the observation point is far away from the patch, as expected. However, the numerical integration will produce different results from the analytical formulas when the observation point is close to the patch and its results are actually incorrect. Among the new formula and old formula, the results are also little different when the observation point is close to the patch and we believe that the new formula is more accurate because it avoids the subdivision of patch and simplifies the calculation.

6.7 Summary

In this chapter, we start to consider the application of the Nyström method to solve EM problems and first aim at the problems with conducting objects. The 2D structures with concave shapes of cross section and large sizes are first considered and good solutions can be obtained by the method even though there are strong oscillations in them. We then develop the EC-based Nyström method for 2D open structures by incorporating the EC in formulations. The EC allows us to greatly coarsen the discretization because it represents the behavior of unknown current density nearby edges, which could be dramatically changeable. To enhance the accuracy to a higher order for very coarse meshes, we derive the closed-form formulations for evaluating singular integrals including the polynomial function of representing unknown current densities with or without the EC. Numerical results show that the application of EC will substantially enhance the convergence rate and a higher-order accuracy can be achieved with the higher-order polynomial approximation of unknown current densities in the local correction. We also employ the Nyström method to solve for BOR structures and the problem is reduced from a 3D form to a 2D form by using Fourier series expansion. The key part in the solving process is the accurate evaluation of singular Fourier expansion coefficients and we propose a different scheme for it by incorporating the Fourier expansion coefficients with the integration over segments of discretizing the generating arc of BOR and performing the segment integration first with closed-form expressions. This allows us to avoid the approximate calculation of the elliptical integrals in other methods. For the kernels of the MGF's derivative, we regularize them by using more similar singular integrands to subtract, resulting in more friendly subtracted kernels which are analytically integrable.

Regarding regular 3D conducting objects, we first develop a higher-order Nyström scheme to solve the EFIE. After performing a super-parametric geometric mapping for curvilinear patches and Lagrange interpolation for unknown functions, we can efficiently handle the singular integral kernels by using the singularity extraction and Duffy's method. Since this kind of local correction avoids solving a local MoM-like matrix equation, the scheme is easier to implement and the resulting error control is more efficient. We then particularly consider super-thin or very-thin conducting structures by using the MFIE and CFIE to formulate, respectively. Due to the geometric feature of thin objects, the polarized component of current density will change rapidly and tend to be infinite when approaching an edge at the top or bottom surface. Also, many low-quality or high-aspect-ratio triangular patches exist in geometric discretization and matrix elements include many near-singular integrals that require a careful treatment. Those unfavorable factors lead to a difficulty to obtain accurate numerical solutions. Although the MFIE is less accurate than the EFIE in general, it can result in a well-conditioned impedance matrix without an obvious low-frequency effect and this is very important to super-thin structures. We develop robust formulations for evaluating those near-singular integrals based on the Green's lemma or the Stokes' theorem, which transforms a surface integral over a triangular patch into a line integral along its boundary. The formulations can be easily used and are particularly preferred for those slim triangular patches because they are only related to the coordinates of vertices of original triangular patches. Finally, the thin-structure problems are further solved by using the CFIE which was proposed for removing the internal resonance. Since the individual EFIE will degenerate when the thickness reduces and the individual MFIE is usually less accurate, the CFIE could be preferred if the thin structure is treated as a closed structure. Numerical examples for different cases are presented to illustrate the method and good results have been observed.

Bibliography

1 Chew, W.C., Tong, M.S., and Hu, B. (2008). *Integral Equation Methods for Electromagnetic and Elastic Waves*. San Rafael, CA: Morgan & Claypool.

2 Kot, J.S. (1990). Solution of thin-wire integral equations by Nyström methods. *Microw. Opt. Technol. Lett.* 3: 393–396.

3 Canino, L.S., Ottusch, J.J., Stalzer, M.A. et al. (1998). Numerical solution of the Helmholtz equation in 2D and 3D using a high-order Nyström discretization. *J. Comput. Phys.* 146: 627–663.

4 Gedney, S.D. (2003). On deriving a locally corrected Nyström scheme from a quadrature sampled moment method. *IEEE Trans. Antennas Propagat.* 51: 2402–2412.

5 Bouwkamp, C. (1946). A note on singularities occurring at sharp edges in electromagnetic diffraction theory. *Physica.* 12: 467.

6 Meixner, J. (1972). The behavior of electromagnetic fields at edges. *IEEE Trans. Antennas Propagat.* AP-20: 442–446.

7 Faraji-Dana, R. and Chow, Y. (1990). Edge condition of the field and a.c. resistance of a rectangular strip conductor. *IEEE Proc. H.* 137 (2): 133–140.

8 Lavretsky, E.L. (1994). Taking into account the edge condition in the problem of scattering from the circular aperture in circular-to-rectangular and rectangular-to-rectangular waveguide junctions. *IEEE Proc. Microw. Antennas Propagat.* 141 (1): 45–51.

9 Abramowitz, M. and Stegun, I.A. (1964). *Handbook of Mathematical Functions with Formulas, Graphs and Mathematical Tables.* New York: Dover.

10 Balanis, C.A. (1989). *Advanced Engineering Electromagnetics,* 2e. New York: Wiley.

11 Kolm, P. and Rokhlin, V. (2001). Numerical quadratures for singular and hypersingular integrals. *Comput. Math. Appl.* 41 (3): 327–352.

12 Burghignoli, P., Di Nallo, C., Frezza, F., and Galli, A. (2003). A simple Nyström approach for the analysis of 3D arbitrarily shaped conducting and dielectric bodies. *Int. J. Numer. Model.* 16 (2): 179–194.

13 Liu, K. and Balanis, C.A. (1991). Simplified formulations for two-dimensional TE-polarization field computations. *IEEE Trans. Antennas Propagat.* 39: 259–262.

14 Tong, M.S. and Chew, W.C. (2005). A higher-order Nyström scheme for electromagnetic scattering by arbitrarily shaped surfaces. *IEEE Antennas Wireless Propagat. Lett.* 4: 277–280.

15 Stroud, A.H. and Secrest, D. (1966). *Gaussian Quadrature Formulas.* Englewood Cliffs: Prentice-Hall.

16 Ma, J.-H., Rokhlin, V., and Wandzura, S.M. (1996). Generalized Gaussian quadrature rules for systems of arbitrary functions. *SIAM J. Numer. Anal.* 33: 971–996.

17 Press, W.H., Flannery, B.P., Teukolsky, S.A., and Vetterling, W.T. (1987). *Numerical Recipes, the Art of Scientific Computing.* Cambridge: Cambridge University Press.

18 Dwight, H.B. (1961). *Tables of Integrals and Other Mathematical Data,* 4e. New York: Macmillan.

19 Andreasen, M.G. (1965). Scattering from bodies of revolution. *IEEE Trans. Antennas Propagat.* AP-13: 303–310.

20 Mautz, J.R. and Harrington, R.F. (1969). Radiation and scattering from bodies of revolution. *Appl. Sci. Res.* 20: 405–434.

21 Wu, T.K. and Tsai, L.L. (1977). Scattering from arbitrarily-shaped lossy dielectric bodies of revolution. *Radio Sci.* 12 (5): 709–718.

22 Mautz, J.R. and Harrington, R.F. (1978). H-field, E-field, and combined-field solutions for conducting bodies of revolution. *Arch. Elektron. Uebertragungstech.* 32 (4): 157–164.

23 Mautz, J.R. and Harrington, R.F. (1979). Electromagnetic scattering from a homogeneous material body of revolution. *Arch. Elektron. Uebertragungstech.* 33 (2): 71–80.

24 Morgan, M.A. and Mei, K.K. (1979). Finite-element computation of scattering by inhomogeneous penetrable bodies of revolution. *IEEE Trans. Antennas Propagat.* 27 (2): 202–213.

25 Mautz, J.R. and Harrington, R.F. (1982). An improved E-field solution for a conducting body of revolution. *Arch. Elektron. Uebertragungstech.* 36 (5): 198–206.

26 Glisson, A.W. (1978). On the development of numerical techniques for treating arbitrarily-shaped surface. Ph.D. Dissertation, University of Mississippi.

27 Glisson, A.W. and Wilton, D.R. (1980). Simple and efficient numerical techniques for treating bodies of revolution. *IEEE Trans. Antennas Propagat.* AP-28: 593–603.

28 Govind, S., Wilton, D.R., and Glisson, A.W. (1984). Scattering from inhomogeneous penetrable bodies of revolution. *IEEE Trans. Antennas Propagat.* 32 (11): 1163–1173.

29 Abdelmageed, A.K. (2000). Efficient evaluation of modal Green's functions arising in EM scattering by bodies of revolution. *Prog. Electromagn. Res.* 27: 337–356.

30 Datthanasombat, S. and Prata, A. Jr., (1999). A compact moment method for determining the electromagnetic scattering of bodies of revolution. *IEEE Antennas Propagat. Soc. Int. Symp.* 4: 2558–2561.

31 Kucharski, A.A. (2000). A method of moments solution for electromagnetic scattering by inhomogeneous dielectric bodies of revolution. *IEEE Trans. Antennas Propagat.* 48 (8): 1202–1210.

32 Khebir, A., D'Angelo, J., and Joseph, J. (1993). A new finite element formulation for RF scattering by complex bodies of revolution. *IEEE Trans. Antennas Propagat.* 41 (5): 534–541.

33 Greenwood, A.D. and Jin, J.M. (1999). A novel efficient algorithm for scattering from a complex BOR using mixed finite elements and cylindrical PML. *IEEE Trans. Antennas Propagat.* 47 (4): 620–629.

34 Dunn, E.A., Byun, J.K., Branch, E.D., and Jin, J.M. (2006). Numerical simulation of BOR scattering and radiation using a higher order FEM. *IEEE Trans. Antennas Propagat.* 54 (3): 945–952.

35 Mittra, R. and Gordon, R.K. (1989). Radar scattering from bodies of revolution using an efficient partial differential equation algorithm. *IEEE Trans. Antennas Propagat.* 37 (5): 538–545.

36 Sun, E.-Y. and Rusch, W.V.T. (1994). EFIE time-marching scattering from bodies of revolution and its applications. *IEEE Trans. Antennas Propagat.* 42 (3): 412–417.

37 Yang, B. and Hesthaven, J.S. (1999). A pseudospectral method for time-domain computation of electromagnetic scattering by bodies of revolution. *IEEE Trans. Antennas Propagat.* 47 (1): 132–141.

38 van der Vorst, M.J.M. and de Maagt, P.J.I. (2002). Efficient body of revolution finite-difference time-domain modeling of integrated lens antennas. *IEEE Microw. Wireless Compon. Lett.* 12 (7): 258–260.

39 Medgyesi-Mitschang, L. and Wang, D.-S. (1983). Hybrid solutions for scattering from perfectly conducting bodies of revolution. *IEEE Trans. Antennas Propagat.* 31 (4): 570–583.

40 Gedney, S.D. and Mittra, R. (1990). The use of the FFT for the efficient solution of the problem of electromagnetic scattering by a body of revolution. *IEEE Trans. Antennas Propagat.* 38 (3): 313–322.

41 Altman, Z. and Mittra, R. (1996). Combining an extrapolation technique with the method of moments for solving large scattering problems involving bodies of revolution. *IEEE Trans. Antennas Propagat.* 44 (4): 548–553.

42 Gradshteyn, I.S. and Ryzhik, I.M. (1965). *Table of Integrals, Series, and Products.* New York: Academic Press.

43 Duffy, M.G. (1982). Quadrature over a pyramid or cube of integrands with a singularity at a vertex. *SIAM J. Numer. Anal.* 19: 1260–1262.

44 Donepudi, K.C., Jin, J.M., Velamparambil, S. et al. (2001). A higher order parallelized multilevel fast multipole algorithm for 3-D scattering. *IEEE Trans. Antennas Propagat.* 49: 1069–1078.

45 Kang, G., Song, J.M., Chew, W.C. et al. (2001). A novel grid-robust higher order vector basis function for the method of moments. *IEEE Trans. Antennas Propagat.* 49: 908–915.

46 Nedelec, J.C. (1976). Curved finite element methods for the solution of singular integral equations on surfaces in R^3. *Comput. Meth. Appl. Mech. Eng.* 8: 61–80.

47 Jin, J.M. (2002). *The Finite Element Method in Electromagnetics*, 2e. New York: Wiley.

48 Silvester, P.P. and Ferrari, R.L. (1990). *Finite Elements for Electrical Engineers*, 2e. Cambridge: Cambridge University Press.

49 Volakis, J.L. (1993). Benchmark Radar targets for the validation of computational electromagnetics program. *IEEE Antennas Propagat. Mag.* 35: 84–89.

50 Stroud, A.H. (1971). *Approximate Calculation of Multiple Integrals.* Englewood Cliffs: Prentice-Hall.

51 Chiang, I.-T. and Chew, W.C. (2006). A coupled PEC-TDS surface integral equation approach for electromagnetic scattering and radiation from composite metallic and thin dielectric objects. *IEEE Trans. Antennas Propagat.* 54 (11): 3511–3516.

52 Chew, W.C., Davis, C.P., Warnick, K.F. et al. (2008). *EFIE and MFIE, why the difference?* IEEE Antennas and Propagation Society International Symposium, San Diego, USA (July 2008).

53 Yan, S.C. and Tong, M.S. (2016). Accurate solution of electromagnetic scattering by very thin conducting objects. *Progress in Electromagnetic Research Symposium*, Shanghai, China (August 2016), pp. 2486–2490.

54 Tong, M.S., Zhang, J., Chen, X.Z. et al. (2013). On the Nyström solutions for electromagnetic scattering by thin conducting objects. *IEEE Trans. Antennas Propagat.* 61 (11): 5813–5817.

55 Hsiao, G.C. and Kleinman, R.E. (1997). Mathematical foundations for error estimation in numerical solutions of integral equations in electromagnetics. *IEEE Trans. Antennas Propagat.* 45 (3): 316–328.

56 Cools, K., Andriulli, F.P., Olyslager, F., and Michielssen, E. (2009). Improving the MFIE's accuracy by using a mixed discretization. *IEEE Antennas and Propagation Society International Symposium*, Charleston, USA (June 2009).

57 Zhang, Y., Cui, T.J., Chew, W.C., and Zhao, J.S. (2003). Magnetic field integral equation at very low frequencies. *IEEE Trans. Antennas Propagat.* 51 (8): 1864–1871.

58 Johnston, B.M. and Johnston, P.R. (2003). A comparison of transformation methods for evaluation two-dimensional weakly singular integrals. *Int. J. Numer. Meth. Eng.* 56: 589–607.

59 Taylor, D.J. (2003). Accurate and efficient numerical integration of weakly singular integrals in Galerkin EFIE solutions. *IEEE Trans. Antennas Propagat.* 51: 1630–1637.

60 Polimeridis, A.G. and Yioultsis, T.V. (2008). On the direct evaluation of weakly singular integrals in Galerkin mixed potential integral equation formulations. *IEEE Trans. Antennas Propagat.* 56 (9): 3011–3019.

61 Tong, M.S. and Chew, W.C. (2007). Super-hyper singularity treatment for solving 3D electric field integral equations. *Microw. Opt. Technol. Lett.* 49: 1383–1388.

62 Tong, M.S. and Chew, W.C. (2010). A novel approach for evaluating hypersingular and strongly singular surface integrals in electromagnetics. *IEEE Trans. Antennas Propagat.* 58 (11): 3593–3601.

63 Caorsi, S., Moreno, D., and Sidoti, F. (1993). Theoretical and numerical treatment of surface integrals involving the free-space Green's function. *IEEE Trans. Antennas Propagat.* 41 (9): 1296–1301.

64 Klees, R. (1996). Numerical calculation of weakly singular surface integrals. *J. Geod.* 70 (11): 781–797.

65 Arcioni, P., Bressan, M., and Perregrini, L. (1997). On the evaluation of the double surface integrals arising in the application of the boundary integral method to 3-D problems. *IEEE Trans. Microw. Theory Tech.* 45: 436–439.

66 Rossi, L. and Cullen, P.J. (1999). On the fully numerical evaluation of the linear-shape function times the 3-D Green's function on a plane triangle. *IEEE Trans. Microw. Theory Tech.* 47: 398–402.

67 Herschlein, A., Hagen, J.V., and Wiesbeck, W. (2002). Methods for the evaluation of regular, weakly singular and strongly singular surface reaction integrals arising in method of moments. *ACES J.* 17 (1): 63–73.

68 Cai, W., Yu, Y., and Yuan, X.C. (2002). Singularity treatment and high-order RWG basis functions for integral equations of electromagnetic scattering. *Int. J. Numer. Meth. Eng.* 53: 31–47.

69 Järvenpää, S., Taskinen, M., and Ylä-Oijala, P. (2003). Singularity extraction technique for integral equation methods with higher order basis functions on plane triangles and tetrahedra. *Int. J. Numer. Meth. Eng.* 58: 1149–1165.

70 Khayat, M.A. and Wilton, D.R. (2005). Numerical evaluation of singular and near-singular potential integrals. *IEEE Trans. Antennas Propagat.* 53: 3180–3190.

71 Tang, W.-H. and Gedney, S.D. (2006). An efficient evaluation of near singular surface integrals via the Khayat-Wilton transformation. *Microw. Opt. Technol. Lett.* 48 (8): 1583–1586.

72 Graglia, R.D. and Lombardi, G. (2008). Machine precision evaluation of singular and nearly singular potential integrals by use of Gauss quadrature formulas for rational functions. *IEEE Trans. Antennas Propagat.* 56: 981–998.

73 Wylie, C.R. and Barrett, L.C. (1995). Advanced Engineering Mathematics, *6e*. New York, USA: McGraw-Hill.

74 Tong, M.S. and Chew, W.C. (2012). A novel meshless scheme for solving surface integral equations with flat integral domains. *IEEE Trans. Antennas Propagat.* 60 (7): 3285–3293.

75 Lindholm, D.A. (1983). Automatic triangular mesh generation on surfaces of polyhedra. *IEEE Trans. Magn.* MAG-19: 2539–2542.

76 Wilton, D.R. (1992). Review of current status and trends in the use of integral equations in computational electromagnetics. *Electromagnetics.* 12: 287–341.

77 Mittra, R., Rahmat-Samii, Y., Jamnejad, D.V., and Davis, W.A. (1973). A new look at the thin plate scattering problem. *Radio Sci.* 8: 869–875.

78 Vipiana, F., Polemi, A., Maci, S., and Vecchi, G. (2008). A mesh-adapted closed-form regular kernel for 3D singular integral equations. *IEEE Trans. Antennas Propagat.* 56 (6): 1687–1698.

79 Nyström, E.J. (1928). Über die praktische auflösung von linearen integralgleichungen mit anwendungen auf randwertaufgaben der potentialtheorie. *Commentationes Physico-Mathematicae.* 4 (15): 1–52.

80 Kot, J.S. (2010). A higher-order Nyström method for dielectric scattering. *International Conference on Electromagnetics in Advanced Applications*, Sydney, Australia (September 2010), pp. 374–377.

81 Gedney, S.D. (2000). Application of the high-order Nyström scheme to the integral equation solution of electromagnetic interaction problems. *IEEE Int. Symp. Electromagn. Compat.* 1: 289–294.

82 Peterson, A.F. and Bibby, M.M. (2004). Higher-order numerical solutions of the MFIE for the linear dipole. *IEEE Trans. Antennas Propagat.* 52: 2684–2691.

83 Wilton, D.R., Rao, S.M., Glisson, A.W. et al. (1984). Potential integrals for uniform and linear source distributions on polygonal and polyhedral domains. *IEEE Trans. Antennas Propagat.* 32: 276–281.

84 Graglia, R.D. (1987). Static and dynamic potential integrals for linearly varying source distributions in two- and three-dimensional problems. *IEEE Trans. Antennas Propagat.* 35: 662–669.

85 Schwab, C. and Wendland, W.L. (1992). On numerical cubatures of singular surface integrals in boundary element methods. *Numer. Math.* 62: 343–369.

86 Graglia, R.D. (1993). On the numerical integration of the linear shape functions times the 3-D Green's function or its Gradient on a plane triangle. *IEEE Trans. Antennas Propagat.* 41: 1448–1455.

87 Khayat, M.A., Wilton, D.R., and Fink, P.W. (2008). An improved transformation and optimized sampling scheme for the numerical evaluation of singular and near-singular potentials. *IEEE Antennas Wireless Propagat. Lett.* 7: 377–380.

88 Vipiana, F. and Wilton, D.R. (2013). Numerical evaluation via singularity cancellation schemes of near-singular integrals involving the gradient of Helmholtz-type potentials. *IEEE Trans. Antennas Propagat.* 61 (3): 1255–1265.

89 Kellogg, O.D. (1944). *Foundations of Potential Theory.* New York: Frederick Ungar.

90 Rao, S.M., Wilton, D.R., and Glisson, A.W. (1982). Electromagnetic scattering by surfaces of arbitrary shape. *IEEE Trans. Antennas Propagat.* 30: 409–418.

91 Asano, T. (2007). Aspect-ratio voronoi diagram and its complexity bounds. *Inf. Process. Lett.* 105: 26–31.

7

Application to Penetrable Media

7.1 Introduction

Scattered fields by objects are the fundamental quantity in studying the interaction between electromagnetic (EM) waves and objects, such as EM compatibility (EMC) analysis. Accurate solutions for the EM scattering by arbitrarily-shaped three-dimensional (3D) composite objects have aroused much interest due to their relevance to practical situations and needs [1–13]. In fact, there are seldom pure perfectly-electric-conducting (PEC) objects. Practical objects are always coated either fully or partially by dielectric materials. Therefore, it is very desirable to accurately analyze EM scattering by such complex-material objects.

In the integral equation methods for solving EM scattering by composite objects, the dominant solver is the method of moments (MoM). However, since surface integral equations (SIEs) with dielectric materials include both equivalent electric and magnetic current densities as unknowns, this method will encounter the problem of how to represent the magnetic current density if the electric current density is expanded using the robust Rao-Wilton-Glisson (RWG) basis function [14]. One may utilize the RWG basis function again to expand the magnetic current density [15], but the resultant system matrix may be ill-conditioned and the solution may not be stable. This is because the equivalent magnetic current density is approximately orthogonal to the electric current density (they come from $\hat{n} \times \mathbf{E}$ and $\hat{n} \times \mathbf{H}$, respectively) and it is not suitable to use the same RWG basis function to represent both. One may also use the $\hat{n}\times$ RWG basis function to represent the magnetic current density so that the rough orthogonality between the electric and magnetic current densities can be reflected. However, such a representation is also deficient because the $\hat{n}\times$ RWG basis function is not continuous across the boundaries of triangle meshes and a fictitious line charge will be created on the boundaries. The dual basis function proposed by Chen and Wilton in 1990 [16] has overcome the drawbacks of the above basis functions and can more accurately capture the characteristics of the current densities, but this basis function is not convenient in its implementation and requires much higher costs to obtain solutions. In addition, the MoM with a RWG basis function requires good mesh discretization for geometries and some deficient meshes are not tolerated.

Although EM scattering by complex objects can be solved by the volume integral equations (VIEs) with the SIE for PEC parts or forming volume-surface integral equations (VSIEs), we will use the SIEs to solve the problem since SIEs only discretize the surface of dielectrics instead of their volumes. This will reduce the number of unknowns dramatically. The SIEs for dielectrics can be formed by Huygens' equivalence principle and the unknowns are the equivalent electric current density and magnetic current density on the boundaries or interfaces of the dielectrics. There are five types of SIEs, i.e. the electric field integral equations (EFIEs) or E-field formulations, the magnetic field integral equations (MFIEs) or H-field formulations, and the combined field integral equations (CFIEs) or combined-field formulations. The fourth type of equations is the Poggio-Miller-Chang-Harrington-Wu-Tsai (PMCHWT) formulations and the fifth type of equations is the Müller formulations. The last two types of SIEs are formed by adding internal fields to external fields with a suitable weight [3] and are more widely used in practice because they are more stable and free of internal resonance problems. The Nyström method has also been used to solve these two SIEs [13], but we will solve the first three SIEs that are never or less frequently solved by the Nyström method. We find that the Nyström method can also provide a stable and accurate solution for EFIEs and MFIEs if not close to the resonant frequency. The EFIEs and MFIEs are simpler and require less costs in implementation compared with the PMCHWT and Müller formulations. Numerical examples are used to demonstrate the robustness of the Nyström method in solving those equations for EM scattering by composite-material bodies.

7.2 Surface Integral Equations for Homogeneous and Isotropic Media

For homogeneous and isotropic media, the SIEs are preferred to govern the EM feature because we only need to discretize material interfaces in seeking numerical solutions. The SIEs include both equivalent electric and magnetic current densities as unknowns since penetrable media exist. In the MoM solution for the SIEs, one may encounter the problem of how to represent the magnetic current density using an appropriate basis function if the electric current density is represented by the RWG basis function. Compared with the MoM, the Nyström method is simpler to implement and, most importantly, it can get rid of the problems in the MoM. We employ this method to solve the E-field, H-field, and combined-field SIEs with efficient local correction schemes. Numerical examples show that the Nyström method can give stable and efficient solutions for both near and far fields, when away from the resonant frequencies in E-field and H-field formulations, even for relatively complicated structures.

7.2.1 Surface Integral Equations

The integral equations for scattering by penetrable objects can be derived by the equivalent electric and magnetic current method. Consider first the scattering by a single 3D dielectric or magnetic body where there is only one interface. The original problem is illustrated in Figure 7.1(a). We designate the interior and exterior of the object as Region 1 and Region 2, respectively. Figure 7.1(b) and Figure 7.1(c) show the equivalent external problem and equivalent internal problem, respectively. In the equivalent external problem, the original electric and magnetic fields exist outside the object surface S and no fields exist inside S. To support such fields, there must exist equivalent electric and

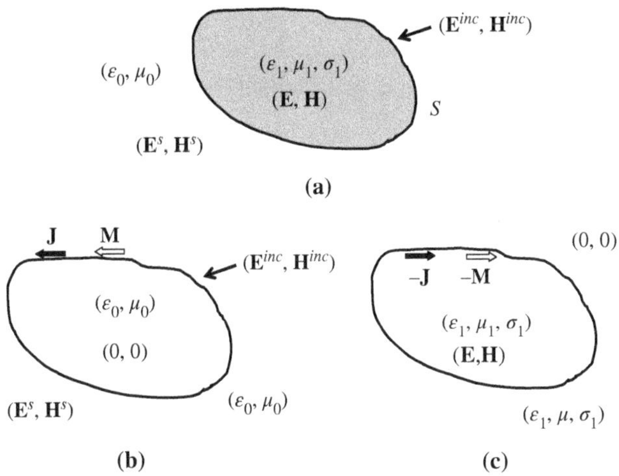

Figure 7.1 Scattering by a single penetrable body. (a) Original problem. (b) Equivalent external problem. (c) Equivalent internal problem.

magnetic surface current densities **J** and **M** on the boundary S, which are unknowns to be solved. Since the tangential components of fields are continuous at the boundary, we can write the integral equations as follows

$$\hat{n} \times \mathbf{E}_2^{S^+}(\mathbf{J}, \mathbf{M}) = -\hat{n} \times \mathbf{E}_{\text{inc}}^{S^+} \tag{7.1}$$

$$\hat{n} \times \mathbf{H}_2^{S^+}(\mathbf{J}, \mathbf{M}) = -\hat{n} \times \mathbf{H}_{\text{inc}}^{S^+} \tag{7.2}$$

where $\hat{n}$ is the unit normal vector on the surface, and **E** and **H** are the electric field and magnetic field, respectively. The subscript "inc" represents an incident field and "2" denotes the region index of the fields. The superscript "S^+" denotes the interface where observation points are located and we use "+" and "−" on S to indicate the exterior side and interior side of the interface, respectively. In the equivalent internal problem described in Figure 7.1(c), the original fields exist inside the object surface S and none of fields exist outside S. There must exist an equivalent electric current density −**J** and magnetic current density −**M** on the surface S to produce the original fields inside the object. The boundary condition requires

$$\hat{n} \times \mathbf{E}_1^{S^-}(\mathbf{J}, \mathbf{M}) = 0 \tag{7.3}$$

$$\hat{n} \times \mathbf{H}_1^{S^-}(\mathbf{J}, \mathbf{M}) = 0 \tag{7.4}$$

which is another group of integral equations. The two equations about the electric field in (7.1) and (7.3) are known as the EFIEs or E-field formulations, and the two equations about the magnetic field in (7.2) and (7.4) are known as the MFIEs or H-field formulations. The electric field and magnetic field can be expressed in terms of the sources, **J** and **M** by recalling the vector and scalar potentials, i.e.

$$\begin{aligned}\mathbf{E}(\mathbf{J}, 0) &= -i\omega\mathbf{A}(\mathbf{J}) - \nabla\Phi(\mathbf{J}) \\ &= -i\omega\mu \int_S \left[\bar{\mathbf{I}} + \frac{\nabla\nabla}{k^2}\right] g(\mathbf{r}, \mathbf{r}') \cdot \mathbf{J}(\mathbf{r}')\, dS' \\ &= -ik\eta\mathcal{L}(\mathbf{J})\end{aligned} \tag{7.5}$$

$$\begin{aligned}\mathbf{E}(0,\mathbf{M}) &= -\frac{1}{\epsilon}\nabla\times\mathbf{F}(\mathbf{M})\\ &= \int_S \mathbf{M}(\mathbf{r}')\times\nabla g(\mathbf{r},\mathbf{r}')\,\mathrm{d}S'\\ &= \mathcal{K}(\mathbf{M})\end{aligned} \tag{7.6}$$

$$\begin{aligned}\mathbf{H}(0,\mathbf{M}) &= -\mathrm{i}\omega\mathbf{F}(\mathbf{M}) - \nabla\Psi(\mathbf{M})\\ &= -\frac{\mathrm{i}k}{\eta}\mathcal{L}(\mathbf{M})\end{aligned} \tag{7.7}$$

$$\mathbf{H}(\mathbf{J},0) = \frac{1}{\mu}\nabla\times\mathbf{A}(\mathbf{J}) = -\mathcal{K}(\mathbf{J}) \tag{7.8}$$

where $\mathbf{A}$ and $\mathbf{F}$ are the magnetic and electric vector potentials, and Φ and Ψ are the electric and magnetic scalar potentials, respectively. Also, ϵ, μ, η, and k are the permittivity, permeability, wave impedance, and wave number of the related medium, respectively, $\bar{\mathbf{I}}$ is an identity dyad, and $g(\mathbf{r},\mathbf{r}') = \mathrm{e}^{\mathrm{i}kR}/(4\pi R)$ is the 3D scalar Green's function, where $R = |\mathbf{r}-\mathbf{r}'|$ is the distance between a field point $\mathbf{r}$ and a source point $\mathbf{r}'$. We have used $\mathcal{L}$ and $\mathcal{K}$ operators to represent the relationship between the fields and sources in the above.

The above derivation of integral equations can be extended to the generalized case, i.e. a dielectric body with L-layer full dielectric coatings, see Figure 7.2. Now we have $L+1$ interfaces where the equivalent electric and magnetic current densities $(\mathbf{J}_\ell, \mathbf{M}_\ell)$ $(\ell = 1, 2, \cdots, L+1)$ reside. By decomposing the original problem into the equivalent internal and external problems layer by layer and applying the boundary conditions at each interface, we can derive the following EFIEs and MFIEs:

$$\hat{n}\times\mathbf{U}_1^{S_1^-}(\mathbf{J}_1,\mathbf{M}_1) = 0 \tag{7.9}$$

$$\vdots$$

$$\hat{n}\times[\mathbf{U}_\ell^{S_{\ell-1}^+}(\mathbf{J}_{\ell-1},\mathbf{M}_{\ell-1}) - \mathbf{U}_\ell^{S_{\ell-1}^+}(\mathbf{J}_\ell,\mathbf{M}_\ell)] = 0 \tag{7.10}$$

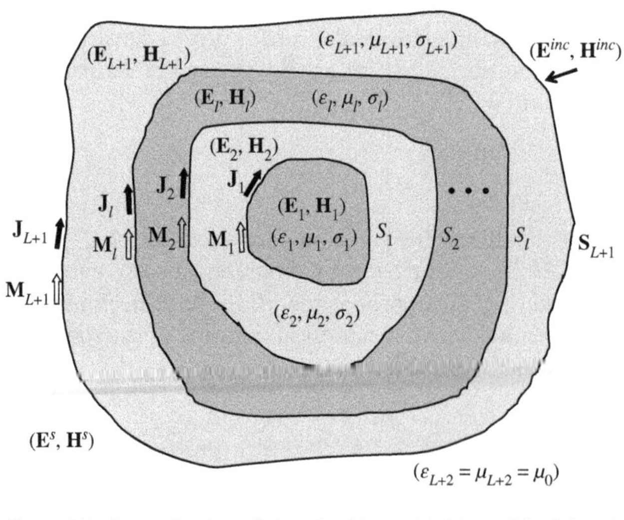

Figure 7.2 Scattering by a dielectric object with L-layer full dielectric coatings.

$$\hat{n} \times [\mathbf{U}_{\ell}^{S_{\ell}^-}(\mathbf{J}_{\ell-1}, \mathbf{M}_{\ell-1}) - \mathbf{U}_{\ell}^{S_{\ell}^-}(\mathbf{J}_{\ell}, \mathbf{M}_{\ell})] = 0 \tag{7.11}$$

$$\vdots$$

$$\hat{n} \times \mathbf{U}_{L+2}^{S_{L+1}^+}(\mathbf{J}_{L+1}, \mathbf{M}_{L+1}) = -\hat{n} \times \mathbf{U}_{\text{inc}}^{S_{L+1}^+} \tag{7.12}$$

$$\ell = 2, 3, \cdots, L+1 \tag{7.13}$$

where $\mathbf{U} = \mathbf{E}$ is for the EFIEs and $\mathbf{U} = \mathbf{H}$ is for the MFIEs, and this is implied later on.

If the scatterer is a fully coated conductor, then the original problem can also be decomposed into an equivalent external problem and an equivalent internal problem, as shown in Figure 7.3. By applying the boundary conditions, the corresponding EFIEs and MFIEs can be written as

$$\hat{n} \times [\mathbf{U}_1^{S_1^+}(\mathbf{J}_1, 0) - \mathbf{U}_1^{S_1^+}(\mathbf{J}_2, \mathbf{M}_2)] = 0, \tag{7.14}$$

$$\hat{n} \times [\mathbf{U}_1^{S_2^-}(\mathbf{J}_1, 0) - \mathbf{U}_1^{S_2^-}(\mathbf{J}_2, \mathbf{M}_2)] = 0, \tag{7.15}$$

$$\hat{n} \times \mathbf{U}_2^{S_2^+}(\mathbf{J}_2, \mathbf{M}_2) = -\hat{n} \times \mathbf{U}_{\text{inc}}^{S_2^+}. \tag{7.16}$$

When the scatterer is a partially coated body, as shown in Figure 7.4(a), which is a more general case, the decomposition is illustrated in Figures 7.4(b) and 7.4(c), respectively. In this case, the original field in free space is generated not only by the equivalent current densities $\mathbf{J}_2$ and $\mathbf{M}_2$ on the coating surface S_2 but also by the conduction current density $\mathbf{J}_1$ at the conducting surface S_1. The original field inside the dielectric coating is still due to the equivalent current densities $\mathbf{J}_2$ and $\mathbf{M}_2$ on the coating surface. The corresponding EFIEs and MFIEs can be formulated by the boundary conditions as

$$\hat{n} \times [\mathbf{U}_1^{S_1^+}(\mathbf{J}_1, 0) - \mathbf{U}_1^{S_1^+}(\mathbf{J}_2, \mathbf{M}_2)] = -\hat{n} \times \mathbf{U}_{\text{inc}}^{S_1^+} \tag{7.17}$$

$$\hat{n} \times [\mathbf{U}_1^{S_2^+}(\mathbf{J}_1, 0) - \hat{n} \times \mathbf{U}_1^{S_2^+}(\mathbf{J}_2, \mathbf{M}_2)] = -\hat{n} \times \mathbf{U}_{\text{inc}}^{S_2^+} \tag{7.18}$$

$$\hat{n} \times \mathbf{U}_2^{S_2^-}(\mathbf{J}_2, \mathbf{M}_2) = 0 \tag{7.19}$$

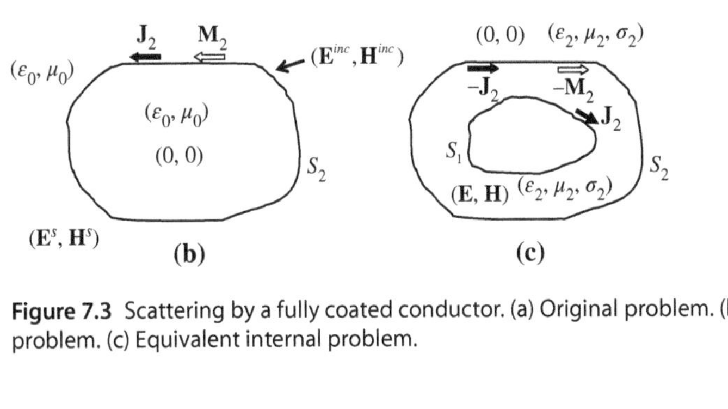

Figure 7.3 Scattering by a fully coated conductor. (a) Original problem. (b) Equivalent external problem. (c) Equivalent internal problem.

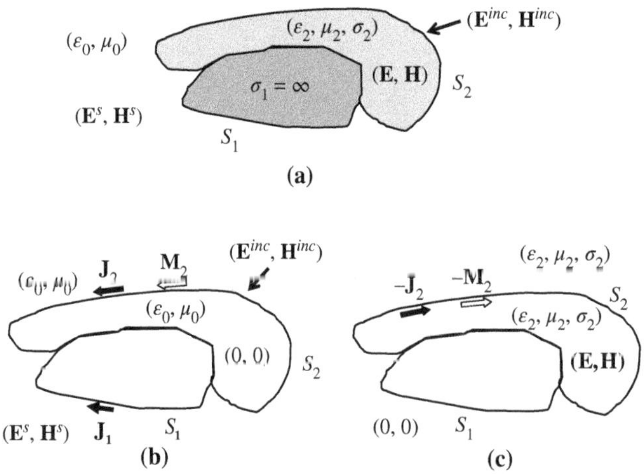

Figure 7.4 Scattering by a partially coated conductor. (a) Original problem. (b) Equivalent external problem. (c) Equivalent internal problem.

If we add the internal field equations to the external field equations by a weighted-sum method [3], then we can obtain the PMCHWT equations and Müller formulations. For example, for the partially coated scatterer, the following equations

$$\alpha_1 \hat{n} \times \left[\mathbf{E}_2^{S_2^+}(\mathbf{J}_1, 0) + \mathbf{E}_2^{S_2^+}(\mathbf{J}_2, \mathbf{M}_2) \right] + \alpha_2 \hat{n} \times \mathbf{E}_1^{S_2^-}(\mathbf{J}_2, \mathbf{M}_2) = -\alpha_1 \hat{n} \times \mathbf{E}_{\text{inc}}^{S_2^+} \tag{7.20}$$

$$\beta_1 \hat{n} \times \left[\mathbf{H}_2^{S_2^+}(\mathbf{J}_1, 0) + \mathbf{H}_2^{S_2^+}(\mathbf{J}_2, \mathbf{M}_2) \right] + \beta_2 \hat{n} \times \mathbf{H}_1^{S_2^-}(\mathbf{J}_2, \mathbf{M}_2) = -\beta_1 \hat{n} \times \mathbf{H}_{\text{inc}}^{S_2^+} \tag{7.21}$$

are the PMCHWT equations when $\alpha_1 = \alpha_2 = \beta_1 = \beta_2 = 1$ and Müller formulations when $\alpha_1 = \beta_1 = 1$, $\alpha_2 = -\epsilon_1/\epsilon_2$, and $\beta_2 = \mu_1/\mu_2$, respectively. The third equation in the PMCHWT and Müller formulations can be chosen as one of the equations in the EFIEs or MFIEs. If we add the E-field equations to the H-field equations in the same regions with appropriate coefficients, then the CFIEs are formed. The following are the CFIEs for scattering by a partially coated scatterer

$$\begin{aligned} &\alpha \hat{n} \times \hat{n} \times \left[\mathbf{E}_2^{S_1^+}(\mathbf{J}_1, 0) + \mathbf{E}_2^{S_1^+}(\mathbf{J}_2, \mathbf{M}_2) \right] + \eta_2 (1-\alpha) \hat{n} \times \left[\mathbf{H}_2^{S_1^+}(\mathbf{J}_1, 0) + \mathbf{H}_2^{S_1^+}(\mathbf{J}_2, \mathbf{M}_2) \right] \\ &= -\alpha \hat{n} \times \hat{n} \times \mathbf{E}_{\text{inc}}^{S_1^+} - \eta_2 (1-\alpha) \hat{n} \times \mathbf{H}_{\text{inc}}^{S_1^+} \end{aligned} \tag{7.22}$$

$$\begin{aligned} &\alpha \hat{n} \times \hat{n} \times \left[\mathbf{E}_2^{S_2^+}(\mathbf{J}_1, 0) + \mathbf{E}_2^{S_2^+}(\mathbf{J}_2, \mathbf{M}_2) \right] + \eta_2 (1-\alpha) \hat{n} \times \left[\mathbf{H}_2^{S_2^+}(\mathbf{J}_1, 0) + \mathbf{H}_2^{S_2^+}(\mathbf{J}_2, \mathbf{M}_2) \right] \\ &= -\alpha \hat{n} \times \hat{n} \times \mathbf{E}_{\text{inc}}^{S_2^+} - \eta_2 (1-\alpha) \hat{n} \times \mathbf{H}_{\text{inc}}^{S_2^+} \end{aligned} \tag{7.23}$$

$$\alpha \hat{n} \times \hat{n} \times \mathbf{E}_1^{S_2^-}(\mathbf{J}_2, \mathbf{M}_2) + \eta_1 (1-\alpha) \hat{n} \times \mathbf{H}_1^{S_2^-}(\mathbf{J}_2, \mathbf{M}_2) = 0 \tag{7.24}$$

where α is a constant about 0.5. We will solve the above EFIEs, MFIEs, and CFIEs for scattering by penetrable or composite bodies using the Nyström method. However, we have to notice the enforcement of the current continuity condition in the junction areas of different materials in the numerical implementation. This can be realized by using the same geometrical meshes and only defining one group of unknowns for the two sides of the junctions.

7.2.2 Nyström Discretization

As shown in Chapter 3, the Nyström method transforms an integral equation into a matrix equation by replacing the integral operator with a summation under an appropriate quadrature rule when the operator is smooth. Consider a surface integration under a quadrature rule

$$\int_{\Delta S} f(\mathbf{r}')\mathrm{d}S' = \sum_{j=1}^{P} w_j f(\mathbf{r}_j'), \tag{7.25}$$

where $f(\mathbf{r}')$ is a general smooth function, P is the number of quadrature points, and w_j is the jth weight over the surface triangle patch ΔS. We usually choose the non-product quadrature rule instead of product or repeated quadrature rule for a surface integral because the non-product rule has fewer quadrature points. The typical non-product rules are the Gauss–Legendre rules with one, three, or six quadrature points over a flat triangle, as shown in Figure 7.5 [17]. By selecting the quadrature points as the field points and performing a collocation procedure, the matrix entries are simply the direct evaluation of integral kernels times the weights of the quadrature rule at those points. Since the integrands are usually smooth for far-interaction elements, the matrix entries corresponding to these elements can be generated in such a way. This will greatly facilitate the implementation and save on matrix-filling time. For singular or near-singular elements, the integrands are singular or badly behaved and the quadrature rule in (7.25) cannot be applied directly. A special singularity treatment or local correction is needed to generate these self or near-interaction entries.

Usually the EM integral equations are solved using the MoM whereas the Nyström method is less used. This is probably due to the robustness of the MoM with the use of the divergence-conforming RWG basis function, especially for solving the EFIE or

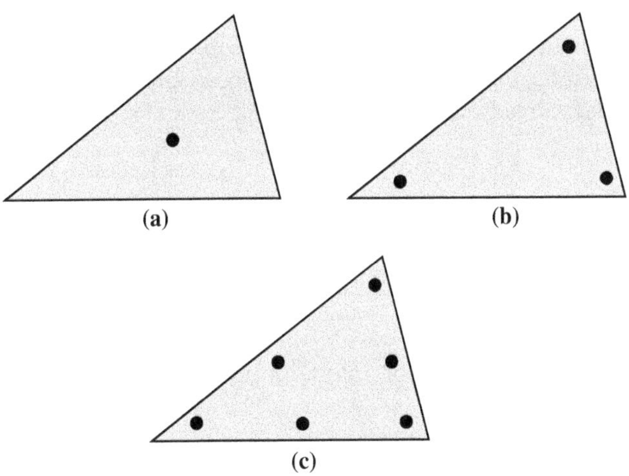

Figure 7.5 Typical non-product quadrature rules used in the Nyström method and corresponding to the zeroth, first, and second order polynomial interpolation functions. (a) One-point rule. (b) Three-point rule. (c) Six-point rule.

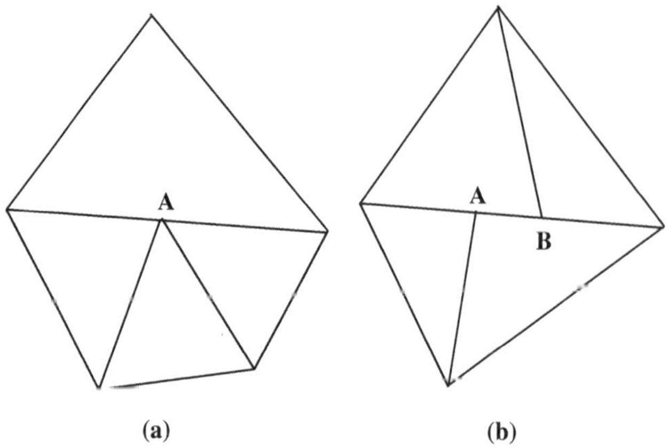

Figure 7.6 Defective meshes in the MoM. (a) One vertex A of a triangle is located on a side of another triangle. (b) Two vertices A and B of two triangles are located on a side of another triangle.

MFIE with a single $\mathcal{L}$ or $\mathcal{K}$ operator or PEC objects. However, when complex materials exist, the MoM may suffer from the difficulty of basis function choice for representing the magnetic current density if the RWG basis function is used to represent the electric current density. The EFIE and MFIE in this case are sensitive to the employed basis functions and testing schemes [15], and some choices like RWG, $\hat{n}\times$ RWG, or dual basis for representing the magnetic current density may have instability, fictitious charge, or high cost problems, respectively. Compared with the MoM, the Nyström method is simpler to implement and, most importantly, it does not have the basis function selection problem. In addition, the Nyström method has fewer requirements on geometrical meshes. The defective meshes as shown in Figure 7.6 are not allowed in the MoM since they cannot form a triangle pair with an identical common edge, but they can be used in the Nyström method. Due to the requirement on meshes, an extra mesh examination is usually required in the implementation of the MoM. The only problem in the Nyström method is the requirement of careful local correction for self or near-interaction terms, but several efficient schemes have already been developed to take care of it.

7.2.3 Numerical Examples

Several numerical examples are presented here to demonstrate the Nyström method in solving EM scattering by composite bodies. The numerical implementation can be summarized as follows:

- Given an object, discretize its boundaries or interfaces into triangular meshes (patches) and choose the integral equation.
- Create a local coordinate system for each triangular patch, choose a quadrature rule, and define unknown surface current densities on quadrature points.
- Transform the integral equation into a matrix equation by replacing the integration over each triangular patch with a summation under the quadrature rule for far-interaction elements. The unknown coefficients are the current density values on quadrature points.

- For self and near-interaction elements, perform a local correction scheme to produce the corresponding matrix elements.
- Solve the matrix equation to get the current densities defined on quadrature points in the local coordinate system. With the obtained current densities, the scattered and total fields at arbitrary points in space can be calculated.

We consider three scatterers as sketched in Figures 7.7(a)–(c) and they are a fully coated PEC sphere, a dielectric (or magnetic) sphere with two-layer full dielectric (or magnetic) coatings, and a partially coated PEC sphere, respectively. The incident wave is a plane wave propagating along the $+z$ direction with a wavelength of 1.0 m. The first example is the scattering by the fully coated PEC sphere. We choose $a_1 = 0.1\lambda$ and $a_2 = 0.15\lambda$, where λ is the wavelength in free space. The complex material for the coating is characterized by the relative permittivity $\epsilon_r = 3.0$ and relative permeability $\mu_r = 1.0$. Figures 7.8 and 7.9 illustrate the near-zone scattered electric field by the scatterer in the vertical polarization (VV) and horizontal polarization (HH), respectively. The observation is made along the principal cut ($\phi = 0°$ and $\theta = 0° - 180°$) at the $r = 1.0\lambda$ surface (referred to as the "near zone"). We use the one-point and three-point quadrature rules to solve the EFIE in the same discretization and the solutions are compared to the analytical solutions [24]. It can be seen that the solutions using a higher-order quadrature rule are more accurate than when using a lower-order quadrature rule. However, the costs will be higher if the higher-order quadrature rules are used. Hence, we will use a one-point quadrature rule, which requires similar costs to those in the MoM if the geometrical discretization is the same.

The second example considers the scattering by the dielectric (or magnetic) sphere with two-layer full dielectric (or magnetic) coatings, as shown in Figure 7.7(b). We choose $a_1 = 0.1\lambda$, $a_2 = 0.15\lambda$, and $a_3 = 0.2\lambda$ for the geometry. Figures 7.10, 7.11, and 7.12 show the near-zone scattered electric fields by solving the EFIE, MFIE, and CFIE, respectively, with different material properties. We choose $\epsilon_{r1} = 2.0$, $\epsilon_{r2} = 4.0$, $\epsilon_{r3} = 3.0$, and $\mu_{r1} = \mu_{r2} = \mu_{r3} = 1.0$ for the first case (lossless dielectrics), $\mu_{r1} = 4.0$, $\mu_{r2} = 3.0$, $\mu_{r3} = 2.0$, and $\epsilon_{r1} = \epsilon_{r2} = \epsilon_{r3} = 1.0$ for the second case (lossless magnetic material), and $\epsilon_{r1} = 2.0$, $\epsilon_{r2} = 4.0$, $\epsilon_{r3} = 3.0$, $\tan\delta = 0.5$ (for each layer), and $\mu_{r1} = \mu_{r2} = \mu_{r3} = 1.0$ for the third case (lossy dielectrics), respectively. The solutions are close to the analytical counterparts [24].

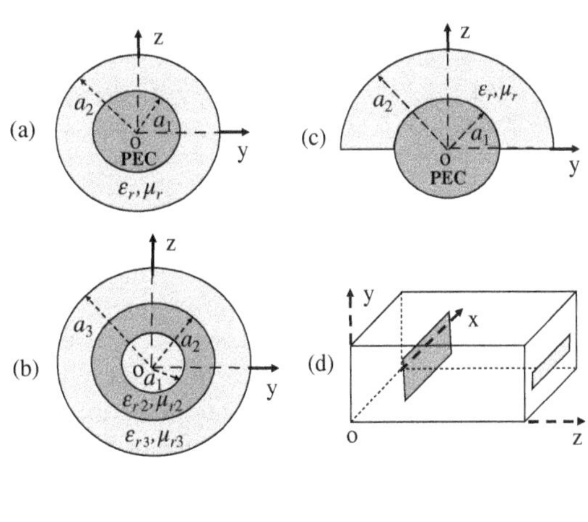

Figure 7.7 Geometries of four objects. (a) Fully coated PEC sphere. (b) Fully coated dielectric sphere. (c) Partially coated PEC sphere. (d) Loaded rectangular enclosure with an aperture.

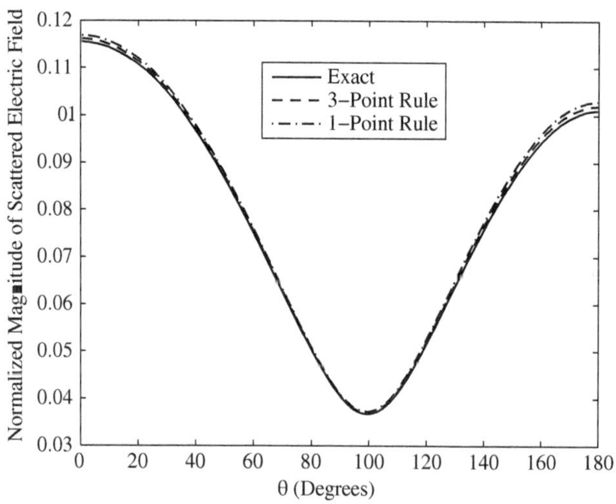

Figure 7.8 Normalized magnitude of scattered electric field by a fully coated PEC sphere in vertical polarization. $a_1 = 0.1\lambda$, $a_2 = 0.15\lambda$, $\epsilon_r = 3.0$, and $\mu_r = 1.0$. The observation is made along the principal cut at $r = 1.0\lambda$.

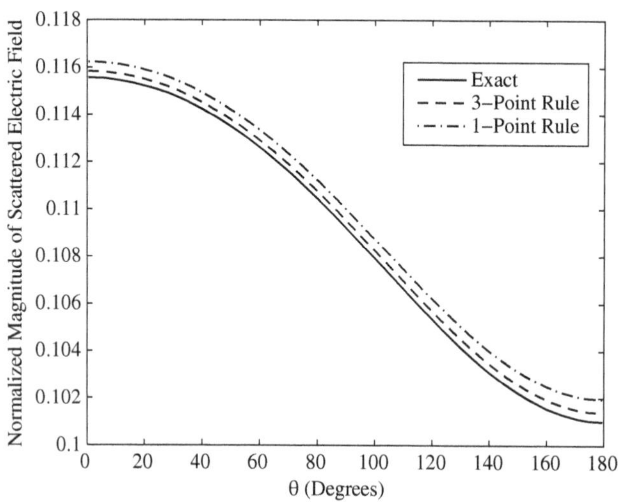

Figure 7.9 Normalized magnitude of scattered electric field by a fully coated PEC sphere in horizontal polarization. $a_1 = 0.1\lambda$, $a_2 = 0.15\lambda$, $\epsilon_r = 3.0$, and $\mu_r = 1.0$. The observation is made along the principal cut at $r = 1.0\lambda$.

The third example is the scattering by a partially coated PEC sphere, as shown in Figure 7.7(c). The geometry is defined with either $a_1 = 0.1\lambda$ and $a_2 = 0.11\lambda$, or $a_1 = 0.1\lambda$ and $a_2 = 0.13\lambda$. The dielectric coating is actually a half spherical shell and its property is defined by $\epsilon_r = 3.0$ and $\mu_r = 1.0$. We solve the EFIE and CFIE for this problem and the far-zone feature of scattered fields represented by the bistatic radar cross section (RCS) (normalized with the cross section area) is presented in Figures 7.13 and 7.14, respectively. The solutions are in good agreement with the results in [16].

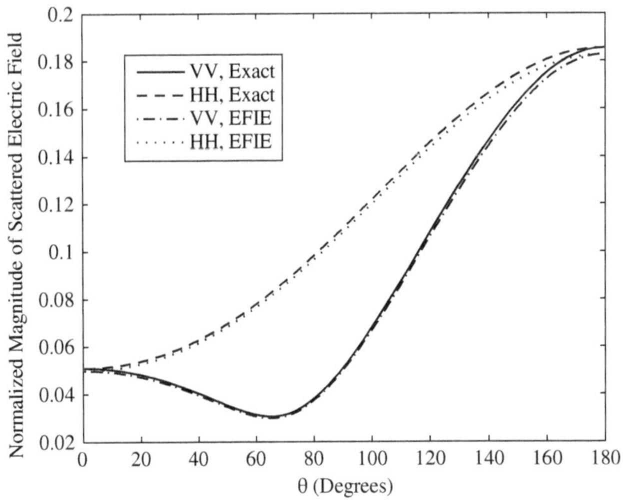

Figure 7.10 Normalized magnitude of scattered electric field by a dielectric sphere with two-layer full dielectric coatings. $a_1 = 0.1\lambda$, $a_2 = 0.15\lambda$, $a_3 = 0.2\lambda$, $\epsilon_{r1} = 2.0$, $\epsilon_{r2} = 4.0$, $\epsilon_{r3} = 3.0$, and $\mu_{r1} = \mu_{r2} = \mu_{r3} = 1.0$. The observation is made along the principal cut at $r = 1.0\lambda$.

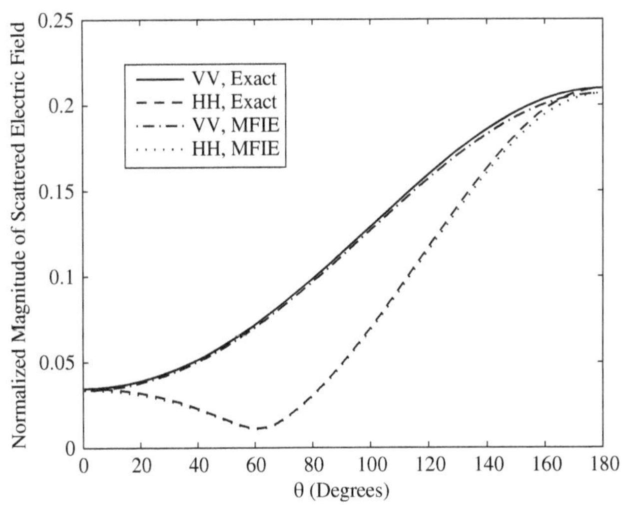

Figure 7.11 Normalized magnitude of scattered electric field by a magnetic sphere with two-layer full magnetic coatings. $a_1 = 0.1\lambda$, $a_2 = 0.15\lambda$, $a_3 = 0.2\lambda$, $\mu_{r1} = 4.0$, $\mu_{r2} = 3.0$, $\mu_{r3} = 2.0$, and $\epsilon_{r1} = \epsilon_{r2} = \epsilon_{r3} = 1.0$. The observation is made along the principal cut at $r = 1.0\lambda$.

To demonstrate the robustness of the Nyström method, we make a simple comparison between the Nyström method and the MoM. Figure 7.15 shows the solutions of the near-zone scattered electric field from these two approaches for the fully coated PEC sphere in the first example with $a_1 = 0.2\lambda$ and $a_2 = 0.3\lambda$. In the MoM, the electric current density is expanded with the RWG basis function and the magnetic current density is represented by the dual basis function. The geometry is discretized into

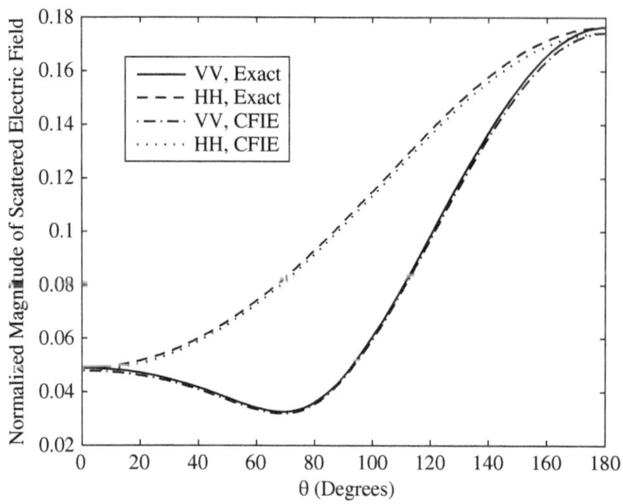

Figure 7.12 Normalized magnitude of scattered electric field by a lossy dielectric sphere with two-layer full lossy dielectric coatings. $a_1 = 0.1\lambda$, $a_2 = 0.15\lambda$, $a_3 = 0.2\lambda$, $\epsilon_{r1} = 2.0$, $\epsilon_{r2} = 4.0$, $\epsilon_{r3} = 3.0$, $\tan\delta = 0.5$, and $\mu_{r1} = \mu_{r2} = \mu_{r3} = 1.0$. The observation is made along the principal cut at $r = 1.0\lambda$.

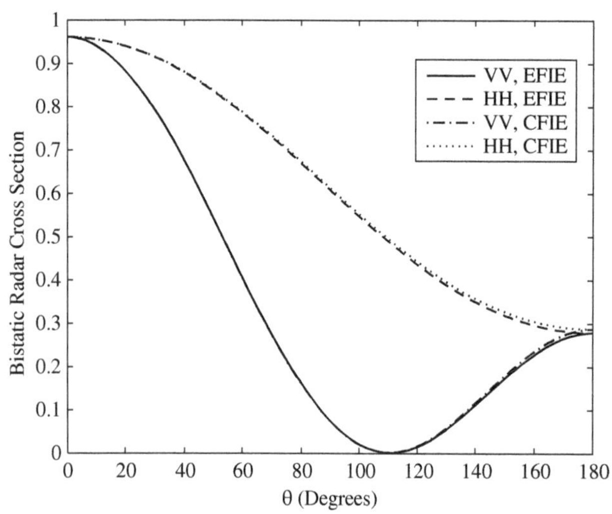

Figure 7.13 Bistatic RCS solutions for a partially coated PEC sphere with $a_1 = 0.1\lambda$, $a_2 = 0.11\lambda$, $\epsilon_r = 3.0$ and $\mu_r = 1.0$.

$N_1 = 1158$ patches on the inner interface and $N_2 = 2582$ patches on the outer interface. The used CPU time and memory on a Dell Precision 690 machine are $T = 658$ seconds and $M = 271$ MB for the Nyström method, and $T = 821$ seconds and $M = 246$ MB for the MoM, respectively. The costs of the two approaches are similar, but the Nyström method is simpler in the implementation.

The Nyström method can of course be applied to analyze other problems like electromagnetic compatibility (EMC) problem. For example, the electric shielding effectiveness

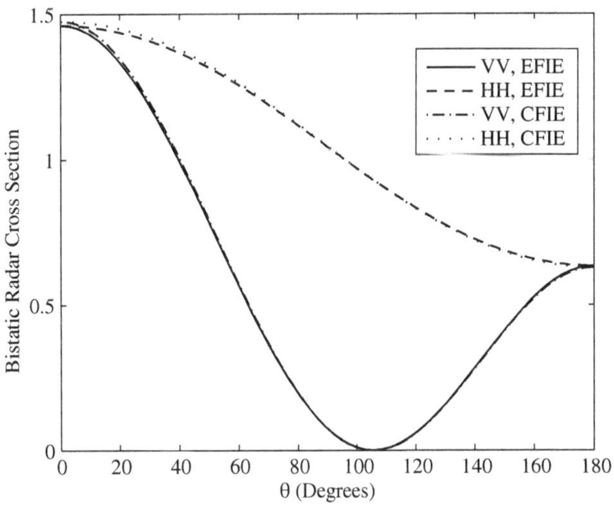

Figure 7.14 Bistatic RCS solutions for a partially coated PEC sphere with $a_1 = 0.1\lambda$, $a_2 = 0.13\lambda$, $\epsilon_r = 3.0$ and $\mu_r = 1.0$.

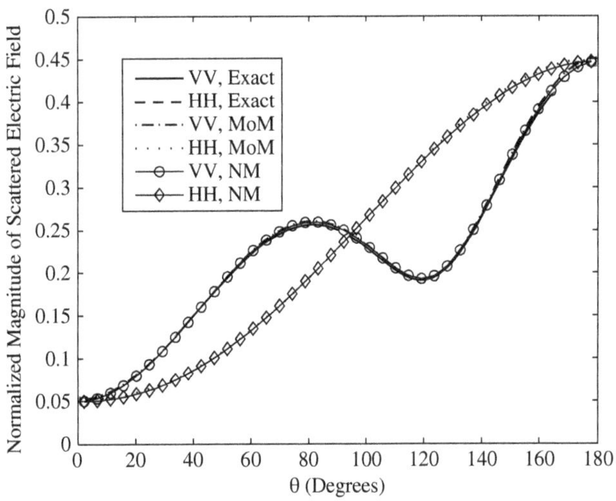

Figure 7.15 A comparison of the Nyström method (NM) solutions with MoM solutions for the scattered electric field by a fully coated PEC sphere. $a_1 = 0.2\lambda$, $a_2 = 0.3\lambda$, $\epsilon_r = 3.0$, and $\mu_r = 1.0$. The observation is made along the principal cut at $r = 1.0\lambda$.

SE_E of a loaded rectangular enclosure with an aperture, as sketched in Figure 7.7(d), which is chosen from Figure 8 in [25], can be solved by this method. The enclosure has a dimension $\ell_x = 30$ cm, $\ell_y = 12$ cm, and $\ell_z = 30$ cm in the x, y, and z directions, respectively, and the rectangular aperture at the center of one side wall has a dimension $a_x = 20$ cm in the x direction and $a_y = 3$ cm in the y direction. Also, a rectangular PEC plate with a size of 15 cm × 9 cm in the xy plane is loaded inside the enclosure. The center of the plate is located at (12.5, 5.5, 10) cm and the observation point is at the center

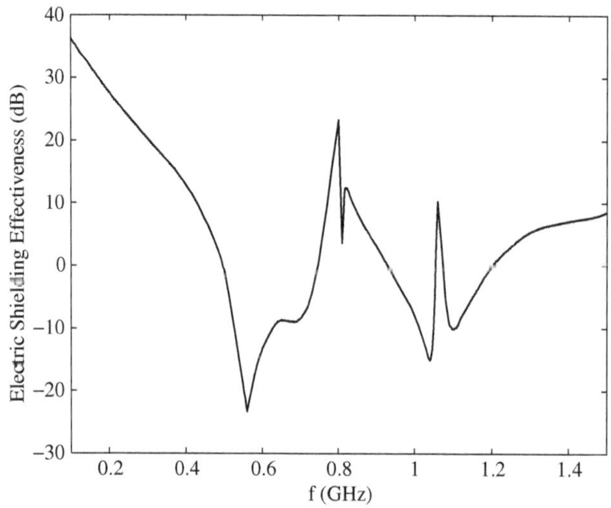

Figure 7.16 The electric shielding effectiveness SE_E of a rectangular PEC enclosure with a rectangular aperture at the center of one side wall.

of the enclosure. The incident plane wave is polarized horizontally and is incident along the $-z$ direction. Figure 7.16 shows the SE_E solution by solving the EFIE and we can see that it is close to that in Figure 8 in [25]. Note that we also calculate 401 frequency points as in [25] and spend about 32 hours to obtain the solution on our workstation. The used CPU time is comparable to the one of the commercial software FEKO as provided in [25] because FEKO is also an integral equation solver using the MoM technique and we also treat the enclosure and its load as an arbitrary object in the geometrical discretization.

7.3 Volume Integral Equations for Homogeneous and Isotropic Media†

The integral equation method is an efficient numerical method for solving EM problems [3]. For the structures with penetrable media, the governing integral equations can be either SIEs or VIEs, but the SIEs have to assume that the media are homogeneous and isotropic. The SIEs include the EFIEs, MFIEs, CFIEs, PMCHWT formulation [1] and Müller formulation [2]. SIEs are preferred whenever available and have been widely solved because they only discretize the boundaries or interfaces of media and require much fewer unknowns. Nevertheless, VIEs cannot be replaced by SIEs and are indispensable for solving problems with inhomogeneous and/or anisotropic media. Conventionally, VIEs refer to the electric field VIE (EFVIE) when the objects are dielectric and not magnetic, or the magnetic field VIE (MFVIE) when the objects are magnetic and not dielectric because only a single electric or magnetic current density exists in the objects. If the objects are both dielectric and magnetic, the VIEs refer to the combination of EFVIE and MFVIE or the combined field VIEs (CFVIEs) [26] since both current

† Part of contents in this section were reused from "Chapter 4: Volume Integral Equation Solvers for Electromagnetic Scattering by Penetrable Objects" authored by Mei Song Tong, in the book "Electromagnetic Scattering: A Remote Sensing Perspective" Edited by Yang Du, World Scientific, Singapore, 2017.

densities exist and two equations are needed for solving two unknown current densities. However, Sun and Chew have solved the single EFVIE when both current densities exist by using a curl-conforming basis function [27]. This new approach can dramatically reduce the number of unknowns compared with the conventional approach.

The VIEs have been extensively solved by the MoM [26–30]. Although the MoM is thought of as very robust in general, the method strongly relies on the well-designed basis and testing functions and also requires numerical integrations in the testing procedure. For example, the divergence-conforming or Schaubert-Wilton-Glisson (SWG) basis function is used in [26, 29, 30] while the curl-conforming basis function is required in [27]. These basis functions are usually defined in element pairs and they have a higher requirement on geometrical discretization. An extra mesh inspection is usually required to guarantee that no defective meshes are involved in the implementation of the MoM.

The Nyström method as an alternative for the MoM has shown much potential and has been widely used to solve SIEs [31–36], but it is less used for solving VIEs, especially for 3D EM problems. In [37] and [38], the authors employed the method to solve the two-dimensional (2D) VIE for transverse magnetic (TM) and transverse electric (TE) cases by their local correction scheme. In [39], the author mentioned the implementation of the Nyström method for solving a 3D VIE, but there was no numerical validation. Therefore, it is necessary to demonstrate this method for solving 3D VIEs.

The main difficulty in the Nyström method is the local correction for self or near-interaction elements since the numerical quadrature rules cannot be applied directly. Since the local correction is vital in the Nyström method implementation, its quality has become an indicator of characterizing the method. In [35] and [36], the authors developed a very systematic local correction scheme for various singular kernels in the SIEs. The scheme is based on generating an adjustment scheme for the quadrature rule for singular integral kernels by computing convolutions of the kernels with a suitable set of testing functions. We also developed a different local correction scheme for the SIEs [19] in which the unknown functions are first interpolated and then the resultant singular integrals are derived analytically after the treatment of singularity subtraction. This scheme may be simpler and more efficient in implementation because it does not require a MoM-like procedure to redesign the quadrature rule for singular kernels as performed in [35] and [36]. We now extend this scheme to the local correction for the VIEs.

Unlike in the MoM in which the well-designed basis and testing functions can be resorted to reduce the degree of singularity in evaluating the singular elements using integration by parts, we have to derive the hypersingular integrals over a volumetric element, which is usually a tetrahedron, in the local correction. The hypersingular integrals are created by the double gradient of the scalar Green's function in the kernel and need to be carefully handled in the local correction. As in the SIEs, we may first apply a singularity subtraction process to treat the kernels and then derive the closed-form solution for the integrals with the subtracted hypersingular cores under a Cauchy-principal-value (CPV) sense. The derivation of integrals over a tetrahedron for the subtracted hypersingular cores can be divided into a regular line integral along the height direction of the tetrahedron and a singular or near singular surface integral over the base areas of the tetrahedron, which depends on the height. The surface integrals can then be derived using the strategy in the SIEs. Numerical examples for EM scattering by 3D penetrable objects are used to demonstrate the effectiveness of the scheme for solving the 3D VIEs and its convergence property is also studied based on the non-curvilinear object.

7.3.1 Volume Integral Equations

The VIE for EM scattering by a penetrable object embedded in a homogeneous background with permittivity ϵ_b and permeability μ_b can be written as [28]

$$\mathbf{E}(\mathbf{r}) = \mathbf{E}^{\text{inc}}(\mathbf{r}) + \mathrm{i}\omega\mu_b \int_V \overline{\mathbf{G}}(\mathbf{r}, \mathbf{r}'; k_b) \cdot \mathbf{J}_V(\mathbf{r}')\mathrm{d}\mathbf{r}', \quad \mathbf{r} \in V \tag{7.26}$$

where $\mathbf{E}^{\text{inc}}(\mathbf{r})$ is the incident electric field, $\mathbf{E}(\mathbf{r})$ is the total electric field, and $\mathbf{J}_V(\mathbf{r}') = \mathrm{i}\omega\left[\epsilon_b - \epsilon(\mathbf{r}')\right]\mathbf{E}(\mathbf{r}')$ is the equivalent volumetric electric current density inside the object. Here, we assume that the object has the same permeability as the background and its permittivity is $\epsilon(\mathbf{r}')$. The integral kernel $\overline{\mathbf{G}}$ is the 3D dyadic Green's function for the homogeneous background medium with a wavenumber k_b and is defined as

$$\overline{\mathbf{G}}(\mathbf{r}, \mathbf{r}'; k_b) = \left(\overline{\mathbf{I}} + \frac{\nabla\nabla}{k_b^2}\right) g(\mathbf{r}, \mathbf{r}'; k_b) \tag{7.27}$$

where $g(\mathbf{r}, \mathbf{r}'; k_b) = \mathrm{e}^{\mathrm{i}k_bR}/(4\pi R)$ is the 3D scalar Green's function and $R = |\mathbf{r} - \mathbf{r}'|$ is the distance between an observation point $\mathbf{r}$ and a source point $\mathbf{r}'$. We can treat either $\mathbf{J}_V$, the total electric field $\mathbf{E}$ or electric flux density $\mathbf{D}$ inside the object as an unknown function in the above VIE.

If the object also has a different permeability $\mu(\mathbf{r}')$ from the background medium, then a volume magnetic current density $\mathbf{M}_V$ exists in addition to $\mathbf{J}_V$ inside the object. In this case, the following EFVIE and MFVIE

$$\begin{aligned}\mathbf{E}(\mathbf{r}) = {} & \mathbf{E}^{\text{inc}}(\mathbf{r}) + \mathrm{i}\omega\mu_b \int_V \overline{\mathbf{G}}(\mathbf{r}, \mathbf{r}'; k_b) \cdot \mathbf{J}_V(\mathbf{r}')\mathrm{d}\mathbf{r}' \\ & - \nabla \times \int_V \overline{\mathbf{G}}(\mathbf{r}, \mathbf{r}'; k_b) \cdot \mathbf{M}_V(\mathbf{r}')\mathrm{d}\mathbf{r}', \quad \mathbf{r} \in V\end{aligned} \tag{7.28}$$

$$\begin{aligned}\mathbf{H}(\mathbf{r}) = {} & \mathbf{H}^{\text{inc}}(\mathbf{r}) + \mathrm{i}\omega\epsilon_b \int_V \overline{\mathbf{G}}(\mathbf{r}, \mathbf{r}'; k_b) \cdot \mathbf{M}_V(\mathbf{r}')\mathrm{d}\mathbf{r}' \\ & + \nabla \times \int_V \overline{\mathbf{G}}(\mathbf{r}, \mathbf{r}'; k_b) \cdot \mathbf{J}_V(\mathbf{r}')\mathrm{d}\mathbf{r}', \quad \mathbf{r} \in V\end{aligned} \tag{7.29}$$

are solved simultaneously for the two unknown current density functions. In the above, $\mathbf{H}^{\text{inc}}(\mathbf{r})$ represents the incident magnetic field and $\mathbf{H}(\mathbf{r})$ represents the total magnetic field, respectively. The magnetic current density is related to the total magnetic field inside the object by

$$\mathbf{M}_V(\mathbf{r}') = \mathrm{i}\omega\left[\mu_b - \mu(\mathbf{r}')\right]\mathbf{H}(\mathbf{r}') \tag{7.30}$$

and the last term in the above equations can be simplified as

$$\nabla \times \int_V \overline{\mathbf{G}}(\mathbf{r}, \mathbf{r}'; k_b) \cdot \mathbf{A}_V(\mathbf{r}')\mathrm{d}\mathbf{r}' = \int_V \nabla g(\mathbf{r}, \mathbf{r}'; k_b) \times \mathbf{A}_V(\mathbf{r}')\mathrm{d}\mathbf{r}' \tag{7.31}$$

where $\mathbf{A}$ is either $\mathbf{J}$ or $\mathbf{M}$.

7.3.2 Nyström Discretization

Consider a volume integration under a quadrature rule

$$\int_{\Delta V} f(\mathbf{r}')\mathrm{d}V' = \sum_{j=1}^{P} w_j f(\mathbf{r}_j') \tag{7.32}$$

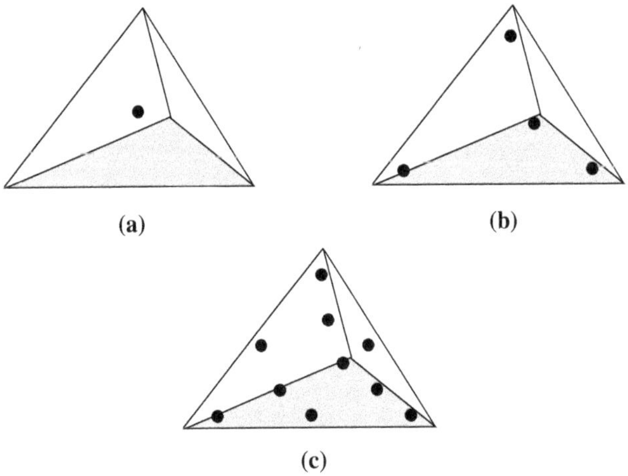

Figure 7.17 Quadrature rules for a tetrahedron. (a) One point. (b) Four points. (c) Ten points. They correspond to constant, linear and quadratic polynomial interpolation functions in the local correction scheme, respectively.

where $f(\mathbf{r}')$ is a general smooth function, P is the number of quadrature points and w_j is the jth weight of the quadrature rule over a volume element ΔV. The non-product quadrature rules are usually used in the Nyström method because they use fewer quadrature points to achieve a higher accuracy of integration. The typical non-product rules for volume integrals over a flat tetrahedron are the one-point, four-point and ten-point quadrature rules, as shown in Figure 7.17. Note that although the one-point and four-point quadrature rules are common, the ten-point quadrature rule is very rare and we only find it in [17] as shown in Table 7.1, where $t = 0.5$, $r = 0.1438564719$, $s = 0.5684305843$, $A = 0.02148995341V$, $B = 0.2177650699V$, and $V = 1/6$. The above three types of quadrature rules correspond to the zeroth-order (constant), first-order (linear), and second-order (quadratic) polynomial interpolation functions, respectively, in the local correction scheme in the Nyström method. Other quadrature rules with different number of points could be used but a least-square approximation should be employed in the polynomial interpolation in the local correction. After dividing the integral domain into small elements, which are usually tetrahedrons in the VIEs, we can perform a collocation procedure by selecting the quadrature points as the observation points in each element to form the matrix equations. If the integral kernels are regular, the matrix entries are simply the direct evaluation of integral kernels times the weights of the quadrature rule at those points. For a typical 3D VIE

$$\int_V F(\mathbf{r}, \mathbf{r}')u(\mathbf{r}')\mathrm{d}V' = -\phi(\mathbf{r}) \qquad \mathbf{r} \in V \tag{7.33}$$

where $F(\mathbf{r}, \mathbf{r}')$ is the integral kernel related to the scalar Green's function in wave problems, $u(\mathbf{r}')$ is the unknown function to be solved, and $\phi(\mathbf{r})$ represents an excitation or incident wave in wave scattering problems. The collocation procedure yields the following matrix equation

$$\sum_{i=1}^{N} \sum_{j=1}^{P_i} w_{ij} F(\mathbf{r}_{mn}, \mathbf{r}'_{ij}) u(\mathbf{r}'_{ij}) = -\phi(\mathbf{r}_{mn}) \tag{7.34}$$

Table 7.1 A ten-point quadrature rule for tetrahedral element.

Rule	Sampling points	Weights
Ten-point	(0,0,t)	A
(degree 3)	(0,t,0)	A
	(t,0,0)	A
	(0,t,t)	A
	(t,0,t)	A
	(t,t,0)	A
	(r,r,r)	B
	(r,r,s)	B
	(r,s,r)	B
	(s,r,r)	B

where P_i is the number of quadrature points in the ith element, w_{ij} represents the weight of the quadrature rule on the jth point of the ith element, $m = 1, 2, \cdots, N$, $n = 1, 2, \cdots, P_m$, and N is the number of elements after discretizing the integral domain. The above process is very simple because it does not involve any numerical integrations. The majority of matrix entries can be created in such a way because they correspond to the far-interaction between an observation point and a source element and the integral kernels are smooth. However, the integral kernels are singular or near singular in the matrix entries representing the self or near-interaction between an observation point and a source element. Thus the above simple process cannot be performed for those entries since the quadrature rule in (7.32) cannot be applied directly. In this case, a local correction is needed to generate these matrix entries.

It is obvious that the Nyström method is less popular in solving EM problems compared with the MoM. Nevertheless, it could still be preferred in some situations due to its advantages. First, this method is simpler overall in implementation, even if the local correction is involved. This feature may be particularly desirable in fast algorithms because it can quickly set up the patterns of algorithms. Second, this method has a lower requirement on the quality of meshes. The defective meshes as illustrated in Figure 7.18 are not allowed in the MoM but they can be used in the Nyström method. The defective meshes can be encountered very frequently in the discretization of complicated structures, such as vehicles and aircrafts, with many details. A careful inspection for the mesh quality is usually needed in the MoM implementation and remeshing is required if defective meshes are found. Note that remeshing (manual remeshing is needed sometimes) is not a minor thing and reducing the remeshing costs has become the motivation for developing meshless methods. By contrast, the Nyström method usually does not require remeshing since those defective meshes in the MoM can be used. Third, the Nyström method could be more convenient than the MoM in solving inhomogeneous problems. For example, in the reconstruction of objects by inverse scattering, the unknown objects are always assumed to be inhomogeneous and each tetrahedral element has a different property from its neighboring elements. Thus there will be surface charge distribution on the common face of each tetrahedron pair in the MoM and this requires extra effort in the implementation.

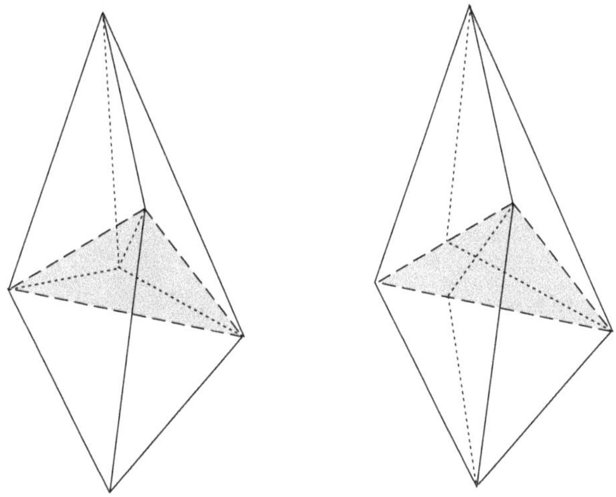

Figure 7.18 Typical defective tetrahedral elements in the MoM. The shadowed triangles usually belong to the junction area between two materials. (a) The lower tetrahedron cannot form a tetrahedral pair with the subtetrahedrons of the upper tetrahedron. (b) The subtetrahedrons on both sides cannot form a pair with each other.

However, the Nyström method usually employs more unknowns than the MoM for the same geometrical discretization because there is no help from the basis function in representing unknowns. To reduce the number of unknowns, one can use a higher-order Nyström scheme in which very coarse meshes with higher-order quadrature rules are used. If the geometry is non-curvilinear or the geometrical approximation for curvilinear objects is minimized by using higher-order geometrical mapping, it is found that using coarse meshes with higher-order quadrature rules (i.e. using the p-refinement) could obtain a better convergence rate than using fine meshes with lower-order quadrature rules (i.e. using h-refinement) [35]. However, for arbitrarily curvilinear objects, it is usually very difficult to minimize the geometrical approximation of curvilinear tetrahedral elements in order to use the coarse meshes. This is because a higher-order geometrical mapping with many sampling points is needed since the meshes are coarse. Also, taking sampling points from a curvilinear surface is equivalent to manually remeshing the curvilinear elements and it is also very difficult for an arbitrary curvilinear surface like a vehicle's surface. Therefore, based on the practical applicability, we only consider the flat-facet tetrahedral elements currently.

7.3.3 Local Correction Scheme

As in the SIEs, the main difficulty in the implementation of the Nyström method is the local correction for self or near-interaction elements. Some efficient local correction schemes have been proposed for solving SIEs [19, 35, 36]. In the scheme we developed [19], the unknown function is first represented with a polynomial interpolation function based on the quadrature points and then the resulting $1/R$ weakly singular integrals are evaluated by Duffy's method [20] and the $1/R^2$ strongly singular or $1/R^3$ hypersingular integrals generated from the singularity subtraction are derived analytically. For the local correction of VIEs, the unknown function is in three dimensions and the interpolation functions for the zeroth-order, first-order and second-order polynomials in a local coordinate system (u, v, w) can be written as

$$f(u,v,w) = a \qquad \text{(zeroth-order)} \tag{7.35}$$

$$f(u,v,w) = a + bu + cv + dw \qquad \text{(first-order)} \tag{7.36}$$

$$\begin{aligned} f(u,v,w) &= a + bu + cv + dw + euv + fvw + gwu \\ &\quad + hu^2 + iv^2 + jw^2 \qquad \text{(second-order).} \end{aligned} \tag{7.37}$$

The above polynomials require the one-point, four-point and ten-point quadrature rules, respectively, to uniquely determine the coefficients because there are $(n+1)(n+2)(n+3)/6$ coefficients or terms in the nth order polynomial [40]. If using the quadrature rules with other numbers of quadrature points, then the polynomial coefficients have to be determined approximately by the least square method [41]. As an example, we consider the coefficients of the first-order polynomial

$$f(u,v,w) = a + bu + cv + dw = \mathbf{u}^t \cdot \mathbf{a} \tag{7.38}$$

where $\mathbf{u}^t = [1, u, v, w]$, $\mathbf{a}^t = [a, b, c, d]$ and the superscript t represents a transpose. By matching the unknown function's values at the four quadrature points, we have

$$f_i = \mathbf{u}_i^t \cdot \mathbf{a}, \quad i = 1, \cdots, 4 \tag{7.39}$$

where $f_i = f(u_i, v_i, w_i)$ is the unknown function's value at the quadrature point (u_i, v_i, w_i), $i = 1, \cdots, 4$, and $\mathbf{u}_i^t = [1, u_i, v_i, w_i]$. The above equations can be written into a matrix form

$$\mathbf{f} = \overline{\mathbf{u}} \cdot \mathbf{a} \tag{7.40}$$

and the solution for the polynomial coefficients is found as

$$\mathbf{a} = \overline{\mathbf{u}}^{-1} \cdot \mathbf{f} \tag{7.41}$$

where $\mathbf{f}^t = [f_1, f_2, f_3, f_4]$ and $\overline{\mathbf{u}}^t = [\mathbf{u}_1, \mathbf{u}_2, \mathbf{u}_3, \mathbf{u}_4]$. Using the found coefficients, the unknown function can be expressed as

$$\begin{aligned} f(u,v,w) &= \mathbf{u}^t \cdot \overline{\mathbf{u}}^{-1} \cdot \mathbf{f} \\ &= \mathbf{T}^t \cdot \mathbf{f} \\ &= \sum_{j=1}^{4} T_j(u,v,w) f_j \end{aligned} \tag{7.42}$$

where $\mathbf{T}^t = \left[T_1(u,v,w), T_2(u,v,w), T_3(u,v,w), T_4(u,v,w)\right]$ is the interpolation function. With the above interpolation representation of the unknown function, the real unknown coefficients in the matrix equation become the unknown function's values at the quadrature points. The above interpolation representation for the unknown function can be easily extended to higher-order polynomials and higher-order accuracy can be obtained.

The primary integral kernel in the VIEs is the same as in the SIEs, i.e. the dyadic Green's function which has nine components if expanded but only six of them are independent due to its symmetry. After using the polynomial interpolation representation of the unknown function, the strongest singular or hypersingular integrals come from the combination of the dyadic Green's function with the constant term in the polynomial. The combination of the kernel with other terms weakens the degree of singularity and can be handled easily, so we do not address them here. For the hypersingular integrals, whose kernel is, for example, the second component G_{vv} of the dyadic Green's function (the constant term from the polynomial is omitted), we have to first perform a singularity subtraction process, i.e.

$$\begin{aligned}
I &= \int_{\Delta V} \left[\frac{(3 - 3i\kappa R - \kappa^2 R^2)v^2}{R^5} + \frac{i\kappa R - 1}{R^3}\right] e^{i\kappa R}\, dV \\
&= \int_{\Delta V} \left\{\left[\frac{(3 - 3i\kappa R - \kappa^2 R^2)v^2}{R^5} + \frac{i\kappa R - 1}{R^3}\right] e^{i\kappa R}\right. \\
&\quad \left. - \frac{\kappa^2 v^2}{2R^3} - \frac{\kappa^4 v^2}{8R} + \frac{\kappa^2}{2R} - \left(\frac{3v^2}{R^5} - \frac{1}{R^3}\right)\right\} dV \\
&\quad + \int_{\Delta V} \left[\frac{\kappa^2 v^2}{2R^3} + \frac{\kappa^4 v^2}{8R} - \frac{\kappa^2}{2R}\right] dV \\
&\quad + \int_{\Delta V} \left(\frac{3v^2}{R^5} - \frac{1}{R^3}\right) dV.
\end{aligned} \tag{7.43}$$

The first integral above is regular or bounded and can be evaluated using numerical quadrature rules. The second integral only includes $1/R$ or a more weakly singular core and can be evaluated using the Duffy method. We do not address these two integrals because they can be handled easily and their accuracy can be controlled by using higher-order quadrature rules. We only treat the third integral which includes a hypersingular core. By referring to [19], all the hypersingular cores from the above singularity subtraction process can be listed as follows:

$$I_1 = \int_{\Delta V} \left(\frac{3u^2}{R^5} - \frac{1}{R^3}\right) dV; \quad I_2 = \int_{\Delta V} \frac{uv}{R^5}\, dV \tag{7.44}$$

$$I_3 = \int_{\Delta V} \left(\frac{3v^2}{R^5} - \frac{1}{R^3}\right) dV; \quad I_4 = \int_{\Delta V} \frac{w_0(-u)}{R^5}\, dV \tag{7.45}$$

$$I_5 = \int_{\Delta V} \frac{w_0(-v)}{R^5}\, dV; \quad I_6 = \int_{\Delta V} \frac{3w_0^2}{R^5}\, dV \tag{7.46}$$

where we have followed the convention in [19], i.e. the integrals are defined in a local coordinate system (u, v, w), as shown in Figure 7.19, and unprimed variables are related to the source point while the variables with a subscript 0 are pertinent to the observation point. Note that the sixth integral I_6 is not needed for solving SIEs but it is needed for solving VIEs because the unknown function is a 3D vector in a volume. The local coordinate system is created in such a way: one of the four facets of a tetrahedron is chosen as the base triangle whose three vertices are labelled as P_1, P_2 and P_3, respectively, and the remaining vertex (or free vertex) of the tetrahedron is labelled as P_4. Choose the projection of the observation point P_0 on the base triangle plane as the origin of the local coordinate system and choose the unit normal vector of the base triangle plane pointing to P_4 side as the $+w$ axis. Also, choose the unit vector parallel to the line from P_1 to P_2 as the $+u$ axis. The $+v$ axis is determined by the $+w$ axis and $+u$ axis. In such a local coordinate system, the above hypersingular integrals over a tetrahedron, for instance,

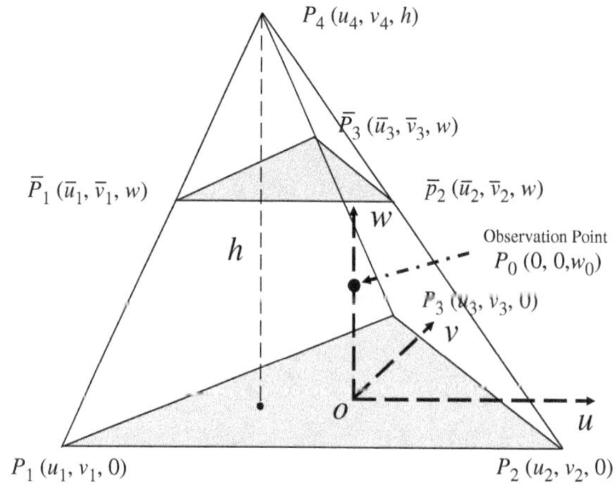

Figure 7.19 Derivation of hypersingular integrals in a local coordinate system (u, v, w) whose origin is the projection of the observation point on the base triangle plane and $+w$ coincides with the normal vector of the base triangle.

the first integral in (7.46), can be written as

$$I_1 = \int_0^h \mathrm{d}w \int_{\Delta S} \left(\frac{3u^2}{R^5} - \frac{1}{R^3} \right) \mathrm{d}S. \tag{7.47}$$

The outer integral with respect to w in I_1 is regular and is evaluated by a numerical quadrature rule (like Gauss-Legendre quadrature rule). The inner surface integral could be regular, near-singular or singular, depending on the spatial relation between the observation point P_0 and the contributing triangle $\Delta S = \Delta\overline{P}_1\overline{P}_2\overline{P}_3$, which is dependent on w. However, no matter how close the observation point P_0 is to the contributing triangle ΔS, the surface integral is always integrable under the CPV sense, as we derived in [19]. Thus, the solutions of above hypersingular integrals over a tetrahedron can be obtained conveniently.

7.3.4 Numerical Examples

To demonstrate the approach, we use the Nyström method to solve the VIEs for EM scattering by different penetrable objects. In all examples, the incident wave is a plane wave propagating along the $-z$ direction in free space and has a frequency $f = 300$ MHz. We calculate the bistatic RCS for the objects with both vertically and horizontally polarized incident wave. The first example is the scattering by a dielectric sphere with a radius $a = 0.15\lambda$ and a relative permittivity $\epsilon_r = 4.0$ (the relative permeability is assumed to be 1.0 except when stated otherwise). We discretize the sphere into 4376 tetrahedral elements and a one-point quadrature rule is used. Since the volume current density has three components in space, the number of unknowns is then three times the number of tetrahedrons. Figure 7.20 shows the bistatic RCS solutions with a comparison to the analytical solutions and they are in excellent agreement (NM represents the Nyström method in all figures). To verify that the Nyström method can work with the defective meshes, which are not allowed in the MoM, we can perform a numerical experiment using this example. The discretization was done by the commercial software IDEAS and there are no defective meshes originally. Now we create some defective meshes by subdividing the first and last tetrahedral elements of the original discretization in this

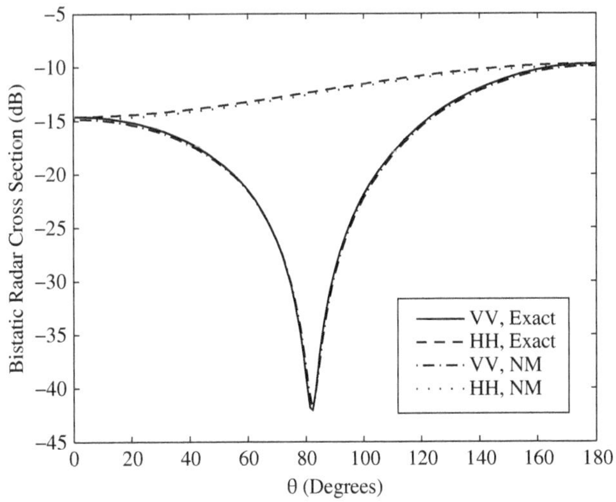

Figure 7.20 Bistatic RCS solutions for a dielectric sphere with $a = 0.15\lambda$ and $\epsilon_r = 4.0$.

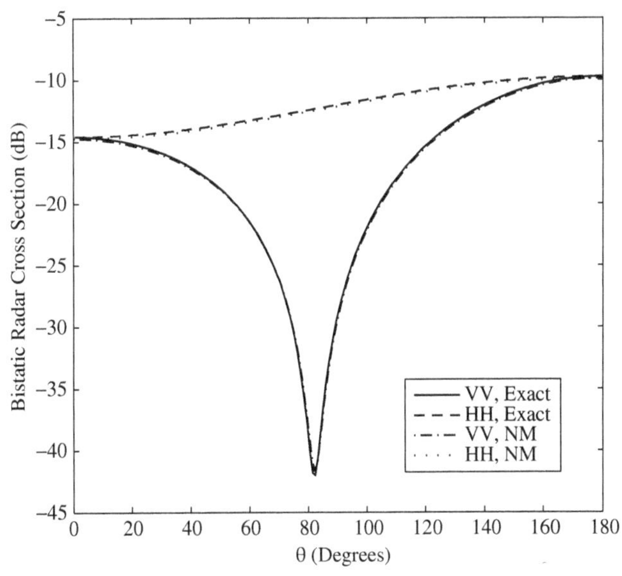

Figure 7.21 Bistatic RCS solutions for a dielectric sphere with $a = 0.15\lambda$ and $\epsilon_r = 4.0$. The geometrical discretization includes defective meshes.

way, namely, choosing one facet, which does not coincide with the sphere surface, on the two tetrahedral elements, finding the centroid for each facet, and connecting the four vertices with the centroid in each tetrahedron. Thus each of subtetrahedrons in the two tetrahedrons cannot form a tetrahedral element pair with the neighboring tetrahedron sharing the chosen facet. The bistatic RCS solutions with such defective meshes are shown in Figure 7.21 and we can see that the new solutions are almost the same as the original solutions shown in Figure 7.20.

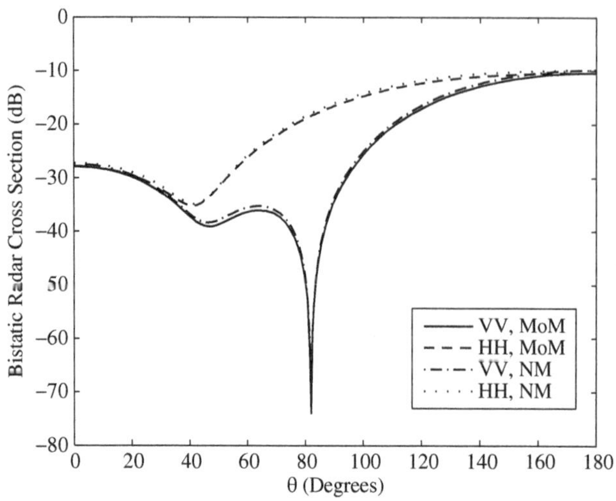

Figure 7.22 Bistatic RCS solutions for a dielectric cylinder with $r = 0.1\lambda$, $h = 0.5\lambda$, and $\epsilon_r = 4.0$.

The second example is the scattering by a dielectric cylinder with a radius $r = 0.1\lambda$ at its cross section, a height $h = 0.5\lambda$, and a relative permittivity $\epsilon_r = 4.0$. We discretize the cylinder into 5732 tetrahedral elements and a one-point quadrature rule is also used. The bistatic RCS solutions are shown in Figure 7.22 and they are close to the solutions obtained by solving the same VIE using the conventional MoM with the SWG basis function.

To compare the convergence of the Nyström method with the one of the MoM, we calculate the bistatic RCSs for a dielectric cube with a side length $s = 0.2\lambda$ and a relative permittivity $\epsilon_r = 4.0$ in the third example. Since there is no curvilinear surface on the object, thus there is no geometrical approximation in the discretization, the numerical errors are only related to or controlled by the mesh size and used quadrature rules. We first discretize the cube into 1657 tetrahedral elements and a one-point quadrature rule is used, leading to 4971 unknowns in the Nyström method. We then discretize the cube into 2282 tetrahedral elements in the conventional MoM with the SWG basis function, leading to 4978 unknowns (the number of unknowns is roughly 2.18 times the number of tetrahedrons in the conventional MoM), which are very close to the number of unknowns in the Nyström method. We also use the solutions obtained at a very fine level of geometrical discretization (12586 tetrahedral elements in the MoM) as the reference solutions [39] and calculate the root-mean-square (RMS) errors for each case. It is found that the RMS error of the Nyström method is 4.31×10^{-2}, which is larger than that of the MoM, 5.47×10^{-3}. However, if we use a four-point quadrature rule and use 415 tetrahedral elements, leading to 4980 unknowns, which is similar to the number of unknowns in the MoM, the RMS error of the Nyström method becomes 1.68×10^{-3}, which is smaller than that of the MoM. This indicates that the higher-order Nyström scheme could be better in convergence than the conventional MoM. Despite this, the higher-order Nyström scheme may not be widely used because it relies on the use of coarse meshes, which could be very difficult for curvilinear objects in implementation. The bistatic RCS solutions of the cube obtained from both the Nyström method and the MoM are illustrated in Figure 7.23.

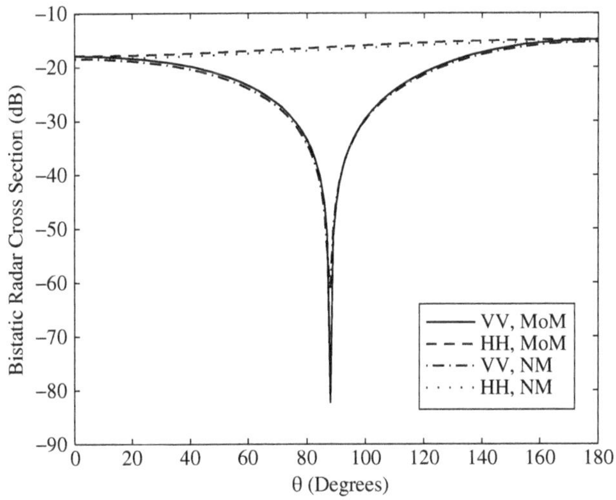

Figure 7.23 Bistatic RCS solutions for a dielectric cube with $s = 0.2\lambda$ and $\epsilon_r = 4.0$.

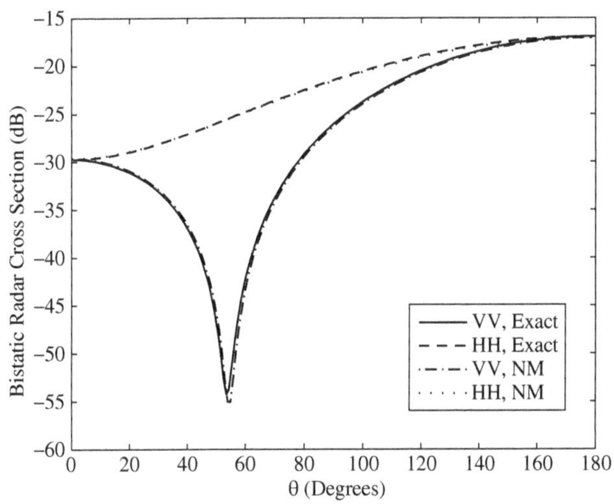

Figure 7.24 Bistatic RCS solutions for a sphere with $a = 0.1\lambda$, $\epsilon_r = 4.0$, and $\mu_r = 2.0$.

When the scatterer also has a different permeability from the background, the volume magnetic current density will exist inside the scatterer and the coupled EFVIE and MFVIE should be solved simultaneously in this case. Figure 7.24 illustrates the bistatic RCS solutions in such a case for a sphere with a radius $a = 0.1\lambda$, a relative permittivity $\epsilon_r = 4.0$, and a relative permeability $\mu_r = 2.0$. The sphere is discretized into 3875 tetrahedral elements and a one-point quadrature rule is used. The results are in good agreement with the corresponding analytical solutions.

To further study the convergence of the Nyström method solutions, we use different mesh densities of geometrical discretization and different quadrature rules to calculate the bistatic RCSs for the dielectric cube in the third example. Figure 7.25 depicts

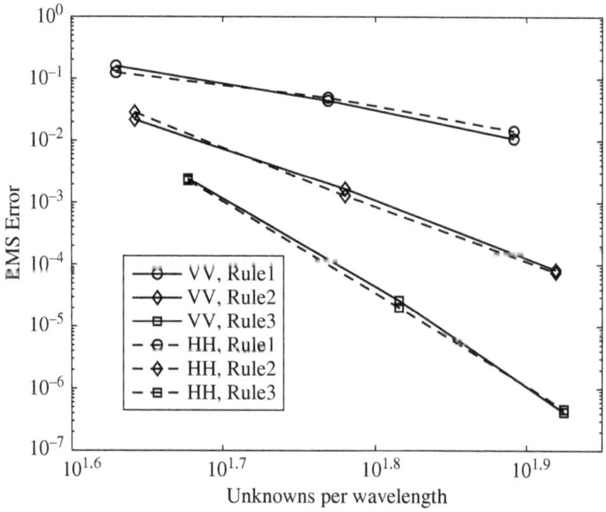

Figure 7.25 RMS error of the bistatic RCS solutions for a dielectric cube with $s = 0.2\lambda$ and $\epsilon_r = 4.0$, which is in a h-refinement. Rule1, Rule2 and Rule3 in the legend correspond to the one-point, four-point and ten-point quadrature rules, respectively.

the convergence of the RMS errors when the quadrature rules are fixed and the mesh density is increased. This is the h-refinement [35] behavior and it can be seen that the higher-order quadrature rules give rise to higher-order convergence rates. However, the roughly straight lines in the log-log coordinate system indicate that they only have a geometrical convergence rate. On the other hand, Figure 7.26 illustrates the convergence of the RMS errors when the mesh size is fixed and the order of quadrature rules

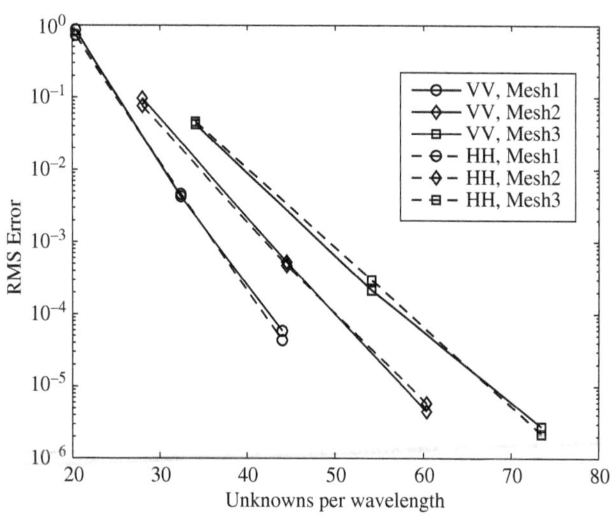

Figure 7.26 RMS error of the bistatic RCS solutions for a dielectric cube with $s = 0.2\lambda$ and $\epsilon_r = 4.0$, which is in a p-refinement. Mesh1, Mesh2 and Mesh3 in the legend correspond to 68, 176 and 317 tetrahedral elements, respectively.

is increased. This is the p-refinement behavior [35] and the roughly straight lines in the semi-log plot indicate that they are in an exponential convergence rate. It is found that the convergence rate is reduced when the mesh size decreases. This is because the integrands within each element vary more slowly (tend to be constant) when the mesh size is reduced and the higher-order quadrature rules do not significantly improve the integral accuracy anymore.

7.4 Volume Integral Equations for Inhomogeneous or/and Anisotropic Media

The integral equation approach has been widely employed to solve EM problems [3] due to its unique merits compared with other numerical approaches like the finite-difference time-domain method [42] and the finite element method (FEM) [43]. For composite objects with inhomogeneous or anisotropic penetrable media that could be encountered in many applications, the problems are described with VIEs or VSIEs if conductors exist simultaneously. Although SIEs can also be applied to penetrable materials in addition to the conducting surfaces and may be preferred when available, they require a piecewise homogeneity in the materials and may not be suitable for arbitrarily inhomogeneous media. SIEs include the EFIE, MFIE, or CFIE for conducting surfaces. For the interfaces of piecewise homogeneous penetrable materials, there are also the PMCHWT formulation [1] and Müller formulation [2] in addition to those three SIEs for the conducting surfaces. If the penetrable materials are arbitrarily inhomogeneous or include many material interfaces, we have to resort to the VIEs which include the EFVIE and MFVIE to describe the problems [3]. Coupled with the SIEs for conducting surfaces, we can form the VSIEs for the objects including both conductors and penetrable materials. If the penetrable materials are also anisotropic, the resultant VSIEs become complicated and seeking their accurate solutions efficiently could be challenging for some objects [44–46].

Traditionally, the above VIEs or VSIEs are solved with the MoM in which the RWG basis function is used to expand the surface current density on the conducting surface with a triangular tessellation [14] while the SWG basis function is applied to represent the volume current densities in the penetrable materials with a tetrahedral discretization [29]. Though the MoM with these well-designed basis functions is robust, it is found that its implementation could be inconvenient for some complex objects [47]. This is because the RWG basis function is defined over a triangle pair and the SWG basis function is defined over a tetrahedron pair, and both basis functions require a higher mesh quality or conformal meshes in geometric discretization. The MoM is sensitive to defective (non-conformal) meshes that cannot form an effective triangle pair or tetrahedron pair, and they cannot be tolerated [48]. The defective meshes can be encountered frequently near the geometric junctions in the complex objects including many different parts and a tedious remeshing process should be performed when this happens. Also, the SWG basis function does not allow the inclusion of material interfaces in tetrahedral elements because of the violation of boundary conditions, prohibiting it from being used in arbitrarily inhomogeneous problems. If we have to clearly locate the material interfaces in geometric discretization, the discretization is tedious and solving VIEs could be meaningless (SIEs are preferred). In addition, we need to take care of the surface charge

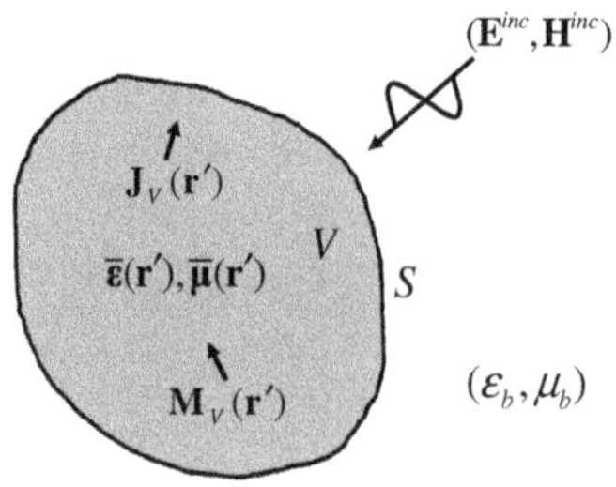

Figure 7.27 Scattering by a 3D inhomogeneous anisotropic object embedded in a homogeneous background with a constant permittivity ϵ_b and constant permeability μ_b. The equivalent volumetric electric current density $\mathbf{J}_V(\mathbf{r}')$ and magnetic current density $\mathbf{M}_V(\mathbf{r}')$ exist inside the object.

density residing on a common face separating two dissimilar media in the SWG basis function for the VIEs [29]. Treating the surface charge density requires an extra care and it could be very inconvenient in composite objects with many material interfaces [49].

We develop an efficient Nyström scheme as an alternative to the MoM for solving those VIEs or VSIEs. Although the Nyström method has been widely used to solve EM problems in recent years [31–36], it is mostly used to solve SIEs for simple homogeneous isotropic objects and seldom applied to the EM analysis involving both conductors and inhomogeneous or anisotropic penetrable media. The Nyström method has some distinct merits as described before and these merits could be especially desirable in fast algorithms like multilevel fast multipole algorithm (MLFMA) [28], because they can greatly simplify the implementation. However, the Nyström method requires an efficient local correction scheme for evaluating self-interaction (singular) or near-interaction (near singular) matrix elements. Fortunately, we have developed a robust local correction scheme for both SIEs and VIEs [18, 50] and it can be used for more complicated cases by a revision. Though the Nyström method may not be preferred when the MoM can be easily implemented, it could be a good alternative to the MoM for complicated objects such as those including both conducting and inhomogeneous or/and anisotropic penetrable materials. Numerical examples for EM scattering by such objects are presented to illustrate the scheme and its merits can be observed.

7.4.1 Volume Integral Equations

Although the VSIEs are needed when conductors co-exist, we only focus on the VIEs which are used to describe inhomogeneous or/and anisotropic problems and can be derived from the vector wave equation by introducing the dyadic Green's function. Consider the EM scattering by a 3D inhomogeneous anisotropic object embedded in a homogeneous background with a constant permittivity ϵ_b and constant permeability μ_b, see Figure 7.27; the VIEs can be written as

$$\begin{aligned}\mathbf{E}(\mathbf{r}) = \mathbf{E}^{\rm inc}(\mathbf{r}) &+ \int_V \overline{\mathbf{G}}(\mathbf{r}, \mathbf{r}'; k_b) \cdot \nabla' \times \mathbf{F}_\mu(\mathbf{r}')d\mathbf{r}' \\ &- k_b^2 \int_V \overline{\mathbf{G}}(\mathbf{r}, \mathbf{r}'; k_b) \cdot \left[\overline{\mathbf{I}} - \overline{\epsilon}_r(\mathbf{r}')\right] \cdot \mathbf{E}(\mathbf{r}')d\mathbf{r}', \quad \mathbf{r} \in V \end{aligned} \tag{7.48}$$

$$\begin{aligned}\mathbf{H}(\mathbf{r}) = \mathbf{H}^{\rm inc}(\mathbf{r}) &+ \int_V \overline{\mathbf{G}}(\mathbf{r}, \mathbf{r}'; k_b) \cdot \nabla' \times \mathbf{F}_\epsilon(\mathbf{r}')d\mathbf{r}' \\ &- k_b^2 \int_V \overline{\mathbf{G}}(\mathbf{r}, \mathbf{r}'; k_b) \cdot \left[\overline{\mathbf{I}} - \overline{\mu}_r(\mathbf{r}')\right] \cdot \mathbf{H}(\mathbf{r}')d\mathbf{r}', \quad \mathbf{r} \in V \end{aligned} \tag{7.49}$$

where $\mathbf{E}^{\text{inc}}(\mathbf{r})$ and $\mathbf{H}^{\text{inc}}(\mathbf{r})$ are the incident electric and magnetic fields, respectively, while $\mathbf{E}(\mathbf{r})$ and $\mathbf{H}(\mathbf{r})$ are the total electric and magnetic fields inside the object, respectively. Also, the integral kernel $\overline{\mathbf{G}}(\mathbf{r},\mathbf{r}';k_b)$ is the 3D dyadic Green's function for the homogeneous background medium with a wavenumber $k_b = \omega\sqrt{\mu_b\epsilon_b}$ and is defined as

$$\overline{\mathbf{G}}(\mathbf{r},\mathbf{r}';k_b) = \left(\overline{\mathbf{I}} + \frac{\nabla\nabla}{k_b^2}\right) g(\mathbf{r},\mathbf{r}';k_b) \tag{7.50}$$

where $g(\mathbf{r},\mathbf{r}';k_b) = \mathrm{e}^{\mathrm{i}k_bR}/(4\pi R)$ is the 3D scalar Green's function in which $R = |\mathbf{r} - \mathbf{r}'|$ is the distance between an observation point $\mathbf{r}$ and a source point $\mathbf{r}'$, and $\overline{\mathbf{I}}$ is the identity dyad. In addition,

$$\mathbf{F}_\mu(\mathbf{r}') = \left[\overline{\mathbf{I}} - \overline{\boldsymbol{\mu}}_r^{-1}(\mathbf{r}')\right] \cdot \nabla' \times \mathbf{E}(\mathbf{r}') \tag{7.51}$$

$$\mathbf{F}_\epsilon(\mathbf{r}') = \left[\overline{\mathbf{I}} - \overline{\boldsymbol{\epsilon}}_r^{-1}(\mathbf{r}')\right] \cdot \nabla' \times \mathbf{H}(\mathbf{r}') \tag{7.52}$$

where $\overline{\boldsymbol{\mu}}_r(\mathbf{r}') = \overline{\boldsymbol{\mu}}(\mathbf{r}')/\mu_b$ is the relative permeability tensor, and $\overline{\boldsymbol{\epsilon}}_r(\mathbf{r}') = \overline{\boldsymbol{\epsilon}}(\mathbf{r}')/\epsilon_b$ is the relative permittivity tensor. Note that the above equations are not computationally friendly and we can recast them into the following form [3]

$$\begin{aligned}\mathbf{E}(\mathbf{r}) = \mathbf{E}^{\text{inc}}(\mathbf{r}) &+ \int_V \nabla g(\mathbf{r},\mathbf{r}';k_b) \times \overline{\mathbf{A}}_\mu(\mathbf{r}') \cdot \mathbf{M}_V(\mathbf{r}')\mathrm{d}\mathbf{r}' \\ &+ \mathrm{i}\omega\mu_b \int_V \overline{\mathbf{G}}(\mathbf{r},\mathbf{r}';k_b) \cdot \mathbf{J}_V(\mathbf{r}')\mathrm{d}\mathbf{r}', \quad \mathbf{r} \in V\end{aligned} \tag{7.53}$$

$$\begin{aligned}\mathbf{H}(\mathbf{r}) = \mathbf{H}^{\text{inc}}(\mathbf{r}) &- \int_V \nabla g(\mathbf{r},\mathbf{r}';k_b) \times \overline{\mathbf{A}}_\epsilon(\mathbf{r}') \cdot \mathbf{J}_V(\mathbf{r}')\mathrm{d}\mathbf{r}' \\ &+ \mathrm{i}\omega\epsilon_b \int_V \overline{\mathbf{G}}(\mathbf{r},\mathbf{r}';k_b) \cdot \mathbf{M}_V(\mathbf{r}')\mathrm{d}\mathbf{r}', \quad \mathbf{r} \in V\end{aligned} \tag{7.54}$$

where

$$\overline{\mathbf{A}}_\mu(\mathbf{r}') = \left[\overline{\mathbf{I}} - \overline{\boldsymbol{\mu}}_r^{-1}(\mathbf{r}')\right] \cdot \overline{\boldsymbol{\mu}}(\mathbf{r}') \cdot \left[\mu_b\overline{\mathbf{I}} - \overline{\boldsymbol{\mu}}(\mathbf{r}')\right]^{-1} \tag{7.55}$$

$$\overline{\mathbf{A}}_\epsilon(\mathbf{r}') = \left[\overline{\mathbf{I}} - \overline{\boldsymbol{\epsilon}}_r^{-1}(\mathbf{r}')\right] \cdot \overline{\boldsymbol{\epsilon}}(\mathbf{r}') \cdot \left[\epsilon_b\overline{\mathbf{I}} - \overline{\boldsymbol{\epsilon}}(\mathbf{r}')\right]^{-1} \tag{7.56}$$

are the two matrices related to the object's permeability tensor and permittivity tensor, respectively, and

$$\mathbf{J}_V(\mathbf{r}') = \mathrm{i}\omega\left[\epsilon_b\overline{\mathbf{I}} - \overline{\boldsymbol{\epsilon}}(\mathbf{r}')\right] \cdot \mathbf{E}(\mathbf{r}') \tag{7.57}$$

$$\mathbf{M}_V(\mathbf{r}') = \mathrm{i}\omega\left[\mu_b\overline{\mathbf{I}} - \overline{\boldsymbol{\mu}}(\mathbf{r}')\right] \cdot \mathbf{H}(\mathbf{r}') \tag{7.58}$$

are the equivalent volumetric electric current density and magnetic current density inside the object, respectively.

If the object is inhomogeneous but isotropic with a permittivity $\epsilon(\mathbf{r}')$ and a permeability $\mu(\mathbf{r}')$, the corresponding VIEs can be written as [3]

$$\begin{aligned}\mathbf{E}(\mathbf{r}) = \mathbf{E}^{\text{inc}}(\mathbf{r}) &+ \mathrm{i}\omega\mu_b \int_V \overline{\mathbf{G}}(\mathbf{r},\mathbf{r}';k_b) \cdot \mathbf{J}_V(\mathbf{r}')\mathrm{d}\mathbf{r}' \\ &- \nabla \times \int_V \overline{\mathbf{G}}(\mathbf{r},\mathbf{r}';k_b) \cdot \mathbf{M}_V(\mathbf{r}')\mathrm{d}\mathbf{r}', \quad \mathbf{r} \in V\end{aligned} \tag{7.59}$$

$$\mathbf{H}(\mathbf{r}) = \mathbf{H}^{\text{inc}}(\mathbf{r}) + \mathrm{i}\omega\epsilon_b \int_V \overline{\mathbf{G}}(\mathbf{r}, \mathbf{r}'; k_b) \cdot \mathbf{M}_V(\mathbf{r}')\mathrm{d}\mathbf{r}'$$
$$+ \nabla \times \int_V \overline{\mathbf{G}}(\mathbf{r}, \mathbf{r}'; k_b) \cdot \mathbf{J}_V(\mathbf{r}')\mathrm{d}\mathbf{r}', \ \mathbf{r} \in V \tag{7.60}$$

where

$$\mathbf{J}_V(\mathbf{r}') = \mathrm{i}\omega \left[\epsilon_b - \epsilon(\mathbf{r}')\right] \mathbf{E}(\mathbf{r}') \tag{7.61}$$
$$\mathbf{M}_V(\mathbf{r}') = \mathrm{i}\omega \left[\mu_b - \mu(\mathbf{r}')\right] \mathbf{H}(\mathbf{r}') \tag{7.62}$$

are the equivalent volumetric electric and magnetic current densities inside the object, respectively. Furthermore, if we assume that the object has the same permeability as the background, which is usually true, then $\mathbf{M}_V(\mathbf{r}') = 0$ and the above two equations can be reduced to

$$\mathbf{E}(\mathbf{r}) = \mathbf{E}^{\text{inc}}(\mathbf{r}) + \mathrm{i}\omega\mu_b \int_V \overline{\mathbf{G}}(\mathbf{r}, \mathbf{r}'; k_b) \cdot \mathbf{J}_V(\mathbf{r}')\mathrm{d}\mathbf{r}', \quad \mathbf{r} \in V \tag{7.63}$$

$$\mathbf{H}(\mathbf{r}) = \mathbf{H}^{\text{inc}}(\mathbf{r}) + \nabla \times \int_V \overline{\mathbf{G}}(\mathbf{r}, \mathbf{r}'; k_b) \cdot \mathbf{J}_V(\mathbf{r}')\mathrm{d}\mathbf{r}', \quad \mathbf{r} \in V. \tag{7.64}$$

We only need to solve one of the two equations to obtain the unknown current density $\mathbf{J}_V(\mathbf{r}')$ and usually the first equation is solved because it is simpler than the second equation. We can treat either $\mathbf{J}_V(\mathbf{r}')$, the total electric field $\mathbf{E}(\mathbf{r}')$ or electric flux density $\mathbf{D}(\mathbf{r}')$ inside the object as an unknown function to be solved in the above VIEs, but $\mathbf{D}(\mathbf{r}')$ is chosen in the MoM with the SWG basis function since the flux density is normally continuous across material interfaces.

7.4.2 Inconvenience of the Method of Moments

The above VIEs or VSIEs are usually solved with the MoM in which the unknown electric current density on the conducting surface is expanded with the RWG basis function while the unknown volumetric electric and magnetic current densities inside the penetrable materials are represented with the SWG basis function. Since the used basis functions are defined over a triangle pair with a common edge or tetrahedron pair with a common face, they require a higher mesh quality or conformal meshes in geometric discretization. However, it is possible to apply a discontinuous element-wise basis function defined over non-conformal meshes or its higher-order version to represent unknown functions in the MoM [52] and using the SWG function is not the sole choice. Nevertheless, the SWG-based MoM is popular for solving the VIEs or VSIEs and non-conformal or defective meshes are not allowed in this scenario. The non-conformal meshes could be easily produced in complex objects and removing them is very tedious in general. Also, since discontinuous functions should not be expanded with continuous basis functions, for example, the volume current densities and the electric or magnetic field with a tangential continuity should not be expanded with the SWG function which is normally continuous across the common face, only the flux densities with a normal continuity can be chosen as unknown functions and expanded with the SWG function. By doing so, i.e.

$$\mathbf{D}(\mathbf{r}') = \sum_{n=1}^{N} D_n \mathbf{f}_n(\mathbf{r}') \tag{7.65}$$

where $\mathbf{D}(\mathbf{r}')$ is the unknown flux density, $\mathbf{f}_n(\mathbf{r}')$ is the SWG basis function defined over the nth tetrahedron pair, N is the total number of common faces or tetrahedron pairs, and D_n is the nth expansion coefficient, then we can find a charge density [29]

$$\rho(\mathbf{r}') = -\sum_{n=1}^{N} D_n \kappa(\mathbf{r}') \nabla' \cdot \mathbf{f}_n(\mathbf{r}') - \sum_{n=1}^{N} D_n \mathbf{f}_n(\mathbf{r}') \cdot \nabla' \kappa(\mathbf{r}') \tag{7.66}$$

where $\kappa(\mathbf{r}') = \left[\epsilon(\mathbf{r}') - \epsilon_0\right] / \epsilon(\mathbf{r}')$ is the contrast ratio of permittivity (the material is assumed to be non-magnetic, i.e. its permeability is the same as the background). The charge density includes two terms and the first term represents the induced volume charge density inside tetrahedrons while the second term denotes the induced surface charge density on the faces separating dissimilar media. For piecewise homogeneous objects, $\kappa(\mathbf{r}')$ is discontinuous across material interfaces and the surface charge density can only be expressed by a generalized function, which is usually represented with the difference of contrast ratios. Moreover, we have to assume a constant property of material or constant $\kappa(\mathbf{r}')$ in each tetrahedron and this will cause an extra error for inhomogeneous or anisotropic materials. In addition, if tetrahedrons stride across the material interfaces, which could happen in thin-layered objects, it is hard to account for the surface charge density because there could be an irregular inhomogeneity in the interior of tetrahedrons. Such an irregular inhomogeneity inside tetrahedrons could be more complicated in solving inverse scattering problems for reconstructing unknown penetrable objects. Although the above problem could be avoided with an appropriate rearrangement of the basis functions, so-called solenoidal basis functions as proposed in [53], the implementation is very tedious.

7.4.3 Nyström Discretization

To solve the VIEs as shown in Equations (7.53) and (7.54), or their simplified version, we first use Equations (7.57) and (7.58) to replace the total electric field and magnetic field with the electric current density and magnetic current density, respectively. We then choose the two current densities as unknown functions to be solved. If we discretize the penetrable material into N_p tetrahedral elements, then we have $6N_p$ unknowns under a one-point quadrature rule and will produce the same number of equations after performing a point-matching procedure over the quadrature points. Equations (7.53) and (7.54) can be changed into a matrix equation by taking each component of the fields and collocating over those quadrature points. For those far-interaction entries that are the major elements of the system matrix, the above procedure is quite simple because we only need to evaluate the integrands to generate the matrix entries. For those self-interaction or near-interaction entries, we can resort to the previous local correction scheme to generate diagonal or near diagonal matrix elements.

Compared to the traditional MoM, the Nyström method could be much simpler in implementation. Note that the faces of tetrahedrons that touch the background or conductors also result in unknowns by assuming fictitious tetrahedrons inside the background or conductors. From Equations (7.53) and (7.54) we can see that the material property only appears on the left-hand side (the material property is implied in the current densities on the right-hand side), so we can allow the inhomogeneity of materials in each tetrahedron in the Nyström method, but that is not allowed in the MoM because the SWG basis function cannot be defined. This merit can greatly facilitate the discretization of inhomogeneous materials in the Nyström method.

7.4.4 Numerical Examples

We present three numerical examples to demonstrate the Nyström scheme for solving the VIEs or VSIEs describing EM scattering by composite objects with inhomogeneous or anisotropic materials. The geometries of scatterers are illustrated in Figure 7.28 and they are: (a) a dielectric sphere with two-layer dielectric coatings; (b) an inhomogeneous penetrable box (cuboid) consisting of two different anisotropic dielectric materials each of which occupies a half of the box; and (c) a PEC cylinder with a PEC hemisphere on its two ends that is fully coated with an anisotropic dielectric material. It is assumed that the incident wave is a plane wave with a frequency $f = 300$ MHz and is propagating along the $-z$ direction in free space. We calculate the scattered near electric field or bistatic RCS observed along the principal cut ($\phi = 0°$ and $\theta = 0° - 180°$) for the scatterers with a vertical polarization (VV) or horizontal polarization (HH). In the first example, the scatterer is shown in Figure 7.28(a) and the problem has an exact Mie-series solution that can be used to verify the numerical solution. The problem can actually be solved with pure SIEs more conveniently, but we use the VIEs to solve it and compare the results with the exact solutions so that the performance of the Nyström scheme can be evaluated. The radii of interfaces are $a_1 = 0.2\lambda$, $a_2 = 0.25\lambda$, and $a_3 = 0.3\lambda$ (λ is the wavelength in free space), respectively. The relative permittivity of each dielectric layer is $\epsilon_{r1} = 3.0$, $\epsilon_{r2} = 7.0$, and $\epsilon_{r3} = 5.0$, respectively (the relative permeability $\mu_r = 1.0$ is assumed except when stated otherwise). We discretize the whole object tetrahedrally without considering the inner interfaces of different materials, resulting in 5286 tetrahedral elements. Figure 7.29 shows the solutions of scattered near electric field at the $r = 2.0\lambda$ surface

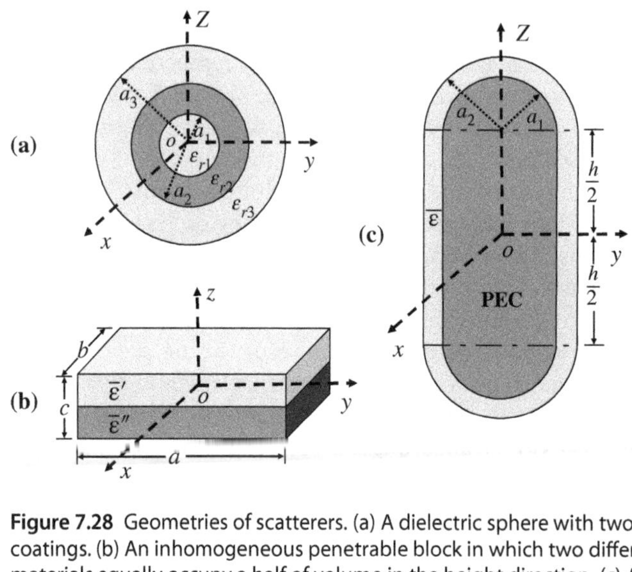

Figure 7.28 Geometries of scatterers. (a) A dielectric sphere with two-layer concentric dielectric coatings. (b) An inhomogeneous penetrable block in which two different anisotropic dielectric materials equally occupy a half of volume in the height direction. (c) A PEC cylinder with a PEC hemisphere on its two ends which is fully coated with an anisotropic dielectric material.

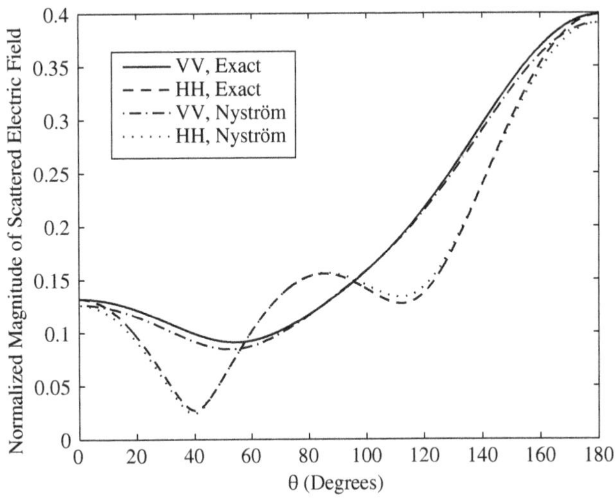

Figure 7.29 Solution of scattered near electric field at $r = 2.0\lambda$ for the scatterer defined in the first example.

for the object when a one-point quadrature rule is used and they are close to the corresponding exact solutions.

The second example considers the scattering by the object as shown in Figure 7.28(b). The cuboid is sized with $a = 2.0\lambda$ in length, $b = 1.0\lambda$ in width, and $c = 0.1\lambda$ in height. In the height direction, the cuboid is equally divided into two parts and the upper half is made of an anisotropic dielectric material characterized by the relative permittivity tensor $\overline{\epsilon}'$, while its lower half is made of another anisotropic dielectric material featured by the relative permittivity tensor $\overline{\epsilon}''$. The tensor $\overline{\epsilon}'$ has non-zero elements $\epsilon_{11} = \epsilon_{22} = 2.2 - i0.2$, $\epsilon_{33} = 3.0 - i0.1$, and $\epsilon_{21} = -\epsilon_{12} = i$, while the non-zero elements are $\epsilon_{11} = \epsilon_{22} = 3.0 - i0.1$, $\epsilon_{33} = 2.2 - i0.1$, and $\epsilon_{21} = -\epsilon_{12} = i$ in the tensor $\overline{\epsilon}''$. We discretize the cuboid as a whole into 9672 tetrahedral elements without considering the material interface. Figure 7.30 depicts the bistatic RCS solution of the object in the vertical polarization by the Nyström method and it is close to the one from the traditional MoM. Note that we have to use the material interface as a boundary in the geometric discretization in the MoM, resulting in 11230 tetrahedral elements.

In the third example, the scatterer includes both conductor and anisotropic material, as shown in Figure 7.28(c), and the VSIEs are required to describe the problem. The VSIEs can be formed by coupling the SIE of conductor parts and the VIES of penetrable parts, see Section 7.6. The geometry of the scatterer is characterized with $a_1 = 0.3\lambda$, which is the inner radius, $a_2 = 0.35\lambda$, which is the outer radius, and $h = 0.6\lambda$, which is the height of the cylinder. The coating is an anisotropic dielectric material characterized by the relative permittivity tensor $\overline{\epsilon}$ in which the non-zero elements are $\epsilon_{11} = \epsilon_{22} = 5.0$, $\epsilon_{33} = 7.0$, and $\epsilon_{12} = -\epsilon_{21} = i$, respectively. We discretize the inner conducting surface into 5976 triangular patches and the anisotropic dielectric coating into 8162 tetrahedral elements. Figure 7.31 plots the bistatic RCS solutions of the object in the vertical polarization by both the Nyström method and the traditional MoM with the same meshes and they are in good agreement.

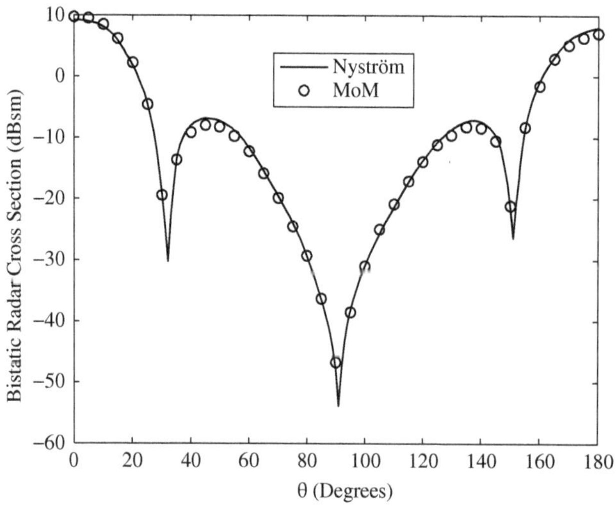

Figure 7.30 Bistatic RCS solutions (vertical polarization) for the scatterer defined in the second example.

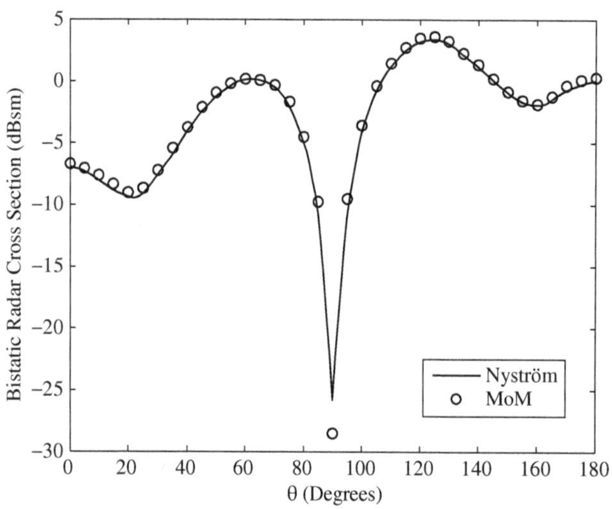

Figure 7.31 Bistatic RCS solutions (vertical polarization) for the scatterer defined in the third example.

To compare the Nyström method and the traditional MoM for computational costs, we summarize the CPU time (Second) and memory usage (GB) for those three examples in Table 7.2. Since the MoM does not allow the inhomogeneity of material in tetrahedrons, we rediscretize the first scatterer without striding across material interfaces, resulting in 5232 tetrahedrons. The second example uses 11230 elements so that the two methods are compared under the same geometric discretization. Note that the magnetic current density is zero in the equations since we assume non-magnetic materials in the examples. The matrix equations are solved on a T7500 Dell Precision workstation (3.06-GHz CPU and 64-GB RAM) with the generalized minimal residual

Table 7.2 A comparison of computational costs between the Nyström method and traditional MoM.

	Costs	Example 1	Example 2	Example 3
Nyström	CPU time (Second)	347	779	1437
	Memory usage (GB)	0.82	1.95	2.41
MoM	CPU time (Second)	365	861	1576
	Memory usage (GB)	0.86	2.17	2.35

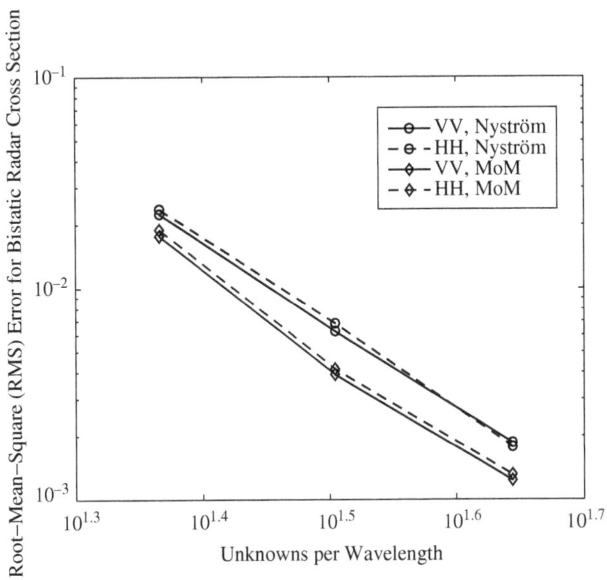

Figure 7.32 A comparison of convergence for the bistatic RCS solutions between the Nyström method and the traditional MoM.

(GMRES) iterative method. It can be seen that the Nyström method consumes less computational resource in the first two examples and similar resource in the third example compared to the MoM, but the Nyström method is more flexible in implementation due to its inherent features. Figure 7.32 is a comparison of convergence for the two methods based on the first scatterer. We discretize the scatterer with different mesh densities and calculate the numerical errors by comparing to the exact solutions in the RMS sense. It can seen that the convergence rate of the two methods is quite similar but the Nyström method has a little larger numerical error and this is because we use a low-order quadrature rule in it.

7.5 Volume Integral Equations for Conductive Media

Conductors are fundamental parts in electronic devices and systems and revealing their EM features is very essential to optimally design relevant products. Although most conductors have a very high conductivity and they are frequently treated as PECs in analysis, such approximations may not be acceptable in many applications. Accurate solution of EM problems with conductive media requires considering the loss or finite conductivity of the media in modeling and simulation, especially when the frequency is low or the conductivity is small since the skin depth is large [3]. While the problems can be solved with the robust FEM or other differential equation methods (DEMs) [56], we use the integral equation method (IEM) to solve them here because the IEM has a smaller solution domain, a better scaling property for computational costs, and no need for implementing an absorbing boundary condition compared with the DEMs [57]. In the IEM, one usually relies on SIEs with an approximate impedance boundary condition (IBC) [58]. The IBC has been widely studied [59–64] because it can dramatically simplify the solved problems, but it also requires that the skin depth be small for conductive media [65]. For the large skin depth caused by low frequency or small conductivity, one can treat the conductive media as dielectric-like objects and use a two-region scheme to formulate the SIEs [66, 67].

However, the regular SIEs, including the robust PMCHWT equations, which can sustain a high contrast of media, cannot be directly used to solve high-conductivity problems because the related Green's function is of rapidly oscillating nature, caused by the large magnitude of the wavenumber, and wisely recombining the SIEs is necessary [66]. The SIEs are usually solved by the MoM [68] with the RWG basis function at high frequencies [14] or loop-star/tree basis functions at low frequencies [69], but the solutions may not be valid for a wide range of frequencies and conductivities because the integral kernels are related to those parameters and the conditioning of SIEs is susceptible to them.

In this section, we treat the conductive media as penetrable objects and use VIEs to exactly formulate the problem. The VIEs that allow an inhomogeneity of materials belong to the second kind of integral equations, so their matrix equations are well conditioned in general [3]. Also, the VIEs only have one unknown function (electric current density) to be solved for non-magnetic objects while two unknown functions (both electric and magnetic current densities) appear in the SIEs and two basis functions for representing them with a wisely chosen testing scheme are needed in the MoM solutions [15]. To solve the VIEs, we do not use the traditional MoM with the SWG basis function [29] but propose a Nyström scheme that does not use any basis and testing functions and allows a geometric discretization of non-conforming meshes, resulting in much convenience in implementation.

Furthermore, the scheme can choose current densities instead of flux densities as the unknowns to be solved so that the integral kernels of VIEs are free of the material parameters of media because the material parameters have been implied in the current densities. The integrals of matrix elements can thus be evaluated in an identical and regular way no matter how the material property varies and the numerical solutions can endure a wide change in frequency and conductivity. If the skin depth is small or the current densities do not penetrate over entire media, we can only discretize the skin domain with a non-zero current density so that the number of unknowns can be reduced. The skin depth can be calculated approximately in terms of the frequency and conductivity and the current source region can be determined by drawing an inner surface whose distance from the outer interface of media is the skin depth. We do not need to mesh the region with a zero current density inside the media, leading to a surface-like discretization.

The scheme requires an accurate evaluation of hypersingular integrals resulting from the double gradient operation in the dyadic Green's function. We have developed a robust technique to handle those integrals in the Nyström method [19, 70] and it can also be employed here. The technique first subtracts the singular core of an integrand to regularize the integrand and then derives a closed form expression for the added-back integral with the subtracted core under a CPV sense [3]. In addition to the hypersingular integrals, weakly and strongly singular integrals also exist, but they can be easily handled and many techniques have been developed for treating them [20–23,54,77]. For electrically large conductive media, we also incorporate the MLFMA [28], which will be addressed detailedly in Chapter 8, to accelerate the solution process. Several numerical examples are presented to illustrate the approach and good results can be observed.

7.5.1 Volume Integral Equations

When the conductive media are treated as penetrable objects, we can use the VIEs of the integral equation method to govern their interaction with EM waves. Consider the EM scattering by a 3D penetrable object in the free space with a permittivity ϵ_0 and a permeability μ_0; the VIEs can be written as [3]

$$\mathbf{E}(\mathbf{r}) = \mathbf{E}^{\text{inc}}(\mathbf{r}) + \mathrm{i}\omega\mu_0 \int_V \overline{\mathbf{G}}(\mathbf{r}, \mathbf{r}') \cdot \mathbf{J}_V(\mathbf{r}')\mathrm{d}V' - \nabla \times \int_V \overline{\mathbf{G}}(\mathbf{r}, \mathbf{r}') \cdot \mathbf{M}_V(\mathbf{r}')\mathrm{d}V', \quad \mathbf{r} \in V \tag{7.67}$$

$$\mathbf{H}(\mathbf{r}) = \mathbf{H}^{\text{inc}}(\mathbf{r}) + \mathrm{i}\omega\epsilon_0 \int_V \overline{\mathbf{G}}(\mathbf{r}, \mathbf{r}') \cdot \mathbf{M}_V(\mathbf{r}')\mathrm{d}V' + \nabla \times \int_V \overline{\mathbf{G}}(\mathbf{r}, \mathbf{r}') \cdot \mathbf{J}_V(\mathbf{r}')\mathrm{d}V', \quad \mathbf{r} \in V \tag{7.68}$$

where $\mathbf{E}^{\text{inc}}(\mathbf{r})$ and $\mathbf{H}^{\text{inc}}(\mathbf{r})$ are the incident electric field and magnetic field, respectively, while $\mathbf{E}(\mathbf{r})$ and $\mathbf{H}(\mathbf{r})$ are the total electric field and magnetic field inside the object, respectively. The integral kernel is the dyadic Green's function given by

$$\overline{\mathbf{G}}(\mathbf{r}, \mathbf{r}') = \left(\overline{\mathbf{I}} + \frac{\nabla\nabla}{k_0^2} \right) g(\mathbf{r}, \mathbf{r}') \tag{7.69}$$

where $\overline{\mathbf{I}}$ is the identity dyad, $k_0 = \omega\sqrt{\epsilon_0\mu_0}$ is the free-space wavenumber (ω is the angular frequency of incident wave), and $g(\mathbf{r}, \mathbf{r}') = \mathrm{e}^{\mathrm{i}k_0 R}/(4\pi R)$ is the scalar Green's function in which $R = |\mathbf{r} - \mathbf{r}'|$ is the distance between an observation point $\mathbf{r}$ and a source point $\mathbf{r}'$. The unknown functions to be solved are the volumetric electric current density and magnetic current density inside the object, which are related to the total electric field and magnetic field by

$$\mathbf{J}_V(\mathbf{r}') = \mathrm{i}\omega \left[\epsilon_0 - \epsilon(\mathbf{r}') \right] \mathbf{E}(\mathbf{r}') \tag{7.70}$$

$$\mathbf{M}_V(\mathbf{r}') = \mathrm{i}\omega \left[\mu_0 - \mu(\mathbf{r}') \right] \mathbf{H}(\mathbf{r}') \tag{7.71}$$

where $\epsilon(\mathbf{r}')$ and $\mu(\mathbf{r}')$ are the permittivity and permeability of the object, respectively. The above VIEs can be applicable to any penetrable objects including conductive media although the property of resulting matrix equations could be quite different due to the significant difference in material parameters. When we choose the current densities as the unknown functions to be solved, the above VIEs can be changed into

$$\frac{1}{\mathrm{i}\omega\left[\epsilon_0 - \epsilon(\mathbf{r}')\right]}\mathbf{J}_V(\mathbf{r}) - \mathrm{i}\omega\mu_0 \int_V \overline{\mathbf{G}}(\mathbf{r},\mathbf{r}')\cdot\mathbf{J}_V(\mathbf{r}')\mathrm{d}V' + \int_V \nabla g(\mathbf{r},\mathbf{r}')\times\mathbf{M}_V(\mathbf{r}')\mathrm{d}V'$$
$$= \mathbf{E}^{\mathrm{inc}}(\mathbf{r}), \quad \mathbf{r}\in V \tag{7.72}$$

$$\frac{1}{\mathrm{i}\omega\left[\mu_0 - \mu(\mathbf{r}')\right]}\mathbf{M}_V(\mathbf{r}) - \mathrm{i}\omega\epsilon_0 \int_V \overline{\mathbf{G}}(\mathbf{r},\mathbf{r}')\cdot\mathbf{M}_V(\mathbf{r}')\mathrm{d}V' - \int_V \nabla g(\mathbf{r},\mathbf{r}')\times\mathbf{J}_V(\mathbf{r}')\mathrm{d}V'$$
$$= \mathbf{H}^{\mathrm{inc}}(\mathbf{r}), \quad \mathbf{r}\in V \tag{7.73}$$

where $\epsilon(\mathbf{r}') = \epsilon_r'\epsilon_0 + i\frac{\sigma}{\omega} = \epsilon_c$ in which ϵ_r' is the real part of relative permittivity and σ is the conductivity while $\mu(\mathbf{r}')$ can be assumed to be a constant μ_c for conductive media. Note that we have simplified the second integral of each equation in the above by using a vector identity. The two integrals correspond to the $\mathcal{L}$ operator and $\mathcal{K}$ operator, respectively, as in the SIEs [3]. If the conductive media are non-magnetic or $\mu(\mathbf{r}') = \mu_0$, then the magnetic current density vanishes and the VIEs can be reduced to

$$\frac{1}{\mathrm{i}\omega\left[\epsilon_0 - \epsilon(\mathbf{r}')\right]}\mathbf{J}_V(\mathbf{r}) - \mathrm{i}\omega\mu_0 \int_V \overline{\mathbf{G}}(\mathbf{r},\mathbf{r}')\cdot\mathbf{J}_V(\mathbf{r}')\mathrm{d}\mathbf{r}' = \mathbf{E}^{\mathrm{inc}}(\mathbf{r}), \quad \mathbf{r}\in V \tag{7.74}$$

which can be solved with much convenience.

7.5.2 Nyström Discretization

The above VIEs for conductive media can be solved with the Nyström method. The Nyström method is one of point-matching methods but its points are designated by a quadrature rule. We do not have many choices on quadrature rules for a tetrahedron or other volume elements, so the Nyström method is not very flexible for VIEs and the generalized point-matching method without the constraint of quadrature rules may be more flexible. Due to this reason, we implement the Nyström method more like a point-matching method in this section. Since the current densities are chosen as the unknown functions to be solved, the integrands of the VIEs do not include the material parameters of media and the numerical implementation can be greatly facilitated. Although the method is only applied to conductive media in this section, it can also be used for generalized media as described in Equations (7.67) and (7.68). The VIEs are first expressed into a scalar form since we do not use any vector basis function to represent the unknown current densities. After discretizing the conductors into N small tetrahedral elements, we can assume that the current densities in each tetrahedron is constant and the VIEs in Equations (7.72) and (7.73) can be changed into the following matrix equations

$$\sum_{n=1}^{N}\int_{\Delta V_n}\mathrm{d}V'\left\{\mathrm{i}\omega\mu_0\overline{\mathbf{G}}(\mathbf{r}_m^{\mathrm{c}},\mathbf{r}')\cdot\begin{bmatrix} J_V^x(\mathbf{r'}_n^{\mathrm{c}}) \\ J_V^y(\mathbf{r'}_n^{\mathrm{c}}) \\ J_V^z(\mathbf{r'}_n^{\mathrm{c}}) \end{bmatrix} - \begin{bmatrix} g_y(\mathbf{r}_m^{\mathrm{c}},\mathbf{r}')M_V^z(\mathbf{r'}_n^{\mathrm{c}}) - g_z(\mathbf{r}_m^{\mathrm{c}},\mathbf{r}')M_V^y(\mathbf{r'}_n^{\mathrm{c}}) \\ g_z(\mathbf{r}_m^{\mathrm{c}},\mathbf{r}')M_V^x(\mathbf{r'}_n^{\mathrm{c}}) - g_x(\mathbf{r}_m^{\mathrm{c}},\mathbf{r}')M_V^z(\mathbf{r'}_n^{\mathrm{c}}) \\ g_x(\mathbf{r}_m^{\mathrm{c}},\mathbf{r}')M_V^y(\mathbf{r'}_n^{\mathrm{c}}) - g_y(\mathbf{r}_m^{\mathrm{c}},\mathbf{r}')M_V^x(\mathbf{r'}_n^{\mathrm{c}}) \end{bmatrix}\right\}$$

$$-\frac{1}{i\omega(\epsilon_0-\epsilon_c)}\begin{bmatrix} J_V^x(\mathbf{r}_m^c)\\ J_V^y(\mathbf{r}_m^c)\\ J_V^z(\mathbf{r}_m^c)\end{bmatrix} = -\begin{bmatrix} E_x^{\text{inc}}(\mathbf{r}_m^c)\\ E_y^{\text{inc}}(\mathbf{r}_m^c)\\ E_z^{\text{inc}}(\mathbf{r}_m^c)\end{bmatrix}, \quad m=1,2,\cdots,N \tag{7.75}$$

$$\sum_{n=1}^{N}\int_{\Delta V_n} dV'\left\{ i\omega\epsilon_0\overline{\mathbf{G}}(\mathbf{r}_m^c,\mathbf{r}')\cdot\begin{bmatrix} M_V^x(\mathbf{r}'^c_n)\\ M_V^y(\mathbf{r}'^c_n)\\ M_V^z(\mathbf{r}'^c_n)\end{bmatrix} + \begin{bmatrix} g_y(\mathbf{r}_m^c,\mathbf{r}')J_V^z(\mathbf{r}'^c_n)-g_z(\mathbf{r}_m^c,\mathbf{r}')J_V^y(\mathbf{r}'^c_n)\\ g_z(\mathbf{r}_m^c,\mathbf{r}')J_V^x(\mathbf{r}'^c_n)-g_x(\mathbf{r}_m^c,\mathbf{r}')J_V^z(\mathbf{r}'^c_n)\\ g_x(\mathbf{r}_m^c,\mathbf{r}')J_V^y(\mathbf{r}'^c_n)-g_y(\mathbf{r}_m^c,\mathbf{r}')J_V^x(\mathbf{r}'^c_n)\end{bmatrix}\right\}$$

$$-\frac{1}{i\omega(\mu_0-\mu_c)}\begin{bmatrix} M_V^x(\mathbf{r}_m^c)\\ M_V^y(\mathbf{r}_m^c)\\ M_V^z(\mathbf{r}_m^c)\end{bmatrix} = -\begin{bmatrix} H_x^{\text{inc}}(\mathbf{r}_m^c)\\ H_y^{\text{inc}}(\mathbf{r}_m^c)\\ H_z^{\text{inc}}(\mathbf{r}_m^c)\end{bmatrix}, \quad m=1,2,\cdots,N \tag{7.76}$$

where $\mathbf{r}_m^c$ represents the center of the mth tetrahedron (observation element) and $\mathbf{r}'^c_n$ denotes the center of the nth tetrahedron (source element). Also, $g_x(\mathbf{r},\mathbf{r}')$, $g_y(\mathbf{r},\mathbf{r}')$, and $g_z(\mathbf{r},\mathbf{r}')$ are the three components of the gradient of the scalar Green's function. Note that the multiple points with a higher-order approximation for current densities in each tetrahedron can also be chosen in the scheme but the implementing cost will be much higher. The matrix elements in the above can be obtained by integrating the components of the dyadic Green's function or the gradient of the scalar Green's function over each small tetrahedral element ΔV_n $(n=1,2,\cdots,N)$ for different observation points $\mathbf{r}_m^c$ $(m=1,2,\cdots,N)$. The dyadic Green's function has nine components, i.e.

$$\overline{\mathbf{G}}(\mathbf{r}_m^c,\mathbf{r}') = \begin{bmatrix} G_{xx}(\mathbf{r}_m^c,\mathbf{r}') & G_{yx}(\mathbf{r}_m^c,\mathbf{r}') & G_{zx}(\mathbf{r}_m^c,\mathbf{r}')\\ G_{xy}(\mathbf{r}_m^c,\mathbf{r}') & G_{yy}(\mathbf{r}_m^c,\mathbf{r}') & G_{zy}(\mathbf{r}_m^c,\mathbf{r}')\\ G_{xz}(\mathbf{r}_m^c,\mathbf{r}') & G_{yz}(\mathbf{r}_m^c,\mathbf{r}') & G_{zz}(\mathbf{r}_m^c,\mathbf{r}')\end{bmatrix} \tag{7.77}$$

but only six are independent due to its symmetry. If the media are non-magnetic or $\mathbf{M}_V(\mathbf{r}')=0$, then the above matrix equation can be reduced to

$$i\omega\mu_0\sum_{n=1}^{N}\left\{\left[\int_{\Delta V_n} dV'\overline{\mathbf{G}}(\mathbf{r}_m^c,\mathbf{r}')\right]\cdot\begin{bmatrix} J_V^x(\mathbf{r}'^c_n)\\ J_V^y(\mathbf{r}'^c_n)\\ J_V^z(\mathbf{r}'^c_n)\end{bmatrix}\right\} - \frac{1}{i\omega(\epsilon_0-\epsilon_c)}\begin{bmatrix} J_V^x(\mathbf{r}_m^c)\\ J_V^y(\mathbf{r}_m^c)\\ J_V^z(\mathbf{r}_m^c)\end{bmatrix} = -\begin{bmatrix} E_x^{\text{inc}}(\mathbf{r}_m^c)\\ E_y^{\text{inc}}(\mathbf{r}_m^c)\\ E_z^{\text{inc}}(\mathbf{r}_m^c)\end{bmatrix},$$

$$m=1,2,\cdots,N. \tag{7.78}$$

The integral kernels are regular when $m \neq n$ in the above matrix elements and we can use a numerical quadrature rule like the Gauss–Legendre quadrature rule to evaluate them. When $m=n$, however, the integral kernels are hypersingular or strongly singular and we need to specially treat them. Since the integral kernels do not include the material parameters of media, we can use the singularity treatment technique developed for regular dielectric media to evaluate the relevant integrals [19].

7.5.3 Numerical Examples

To verify the approach, we study several numerical examples related to EM interaction with different conductive media whose geometries are sketched in Figure 7.33 and non-conforming meshes are used in geometric discretization. In the first example, we consider the EM scattering by a conductive sphere with a radius $a=1.0$ m and a conductivity $\sigma=32.0$ S m^{-1}, as shown in Figure 7.33(a). The relative permeability $\mu_r=1.0$

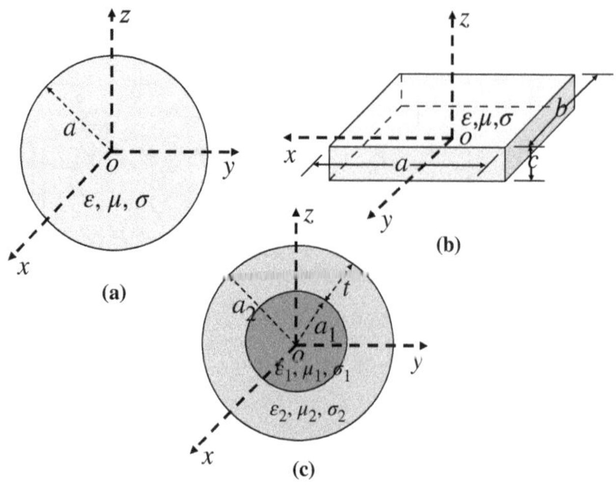

Figure 7.33 Geometries of lossy conductor structures. (a) A conductive sphere. (b) A rectangular conductive plate. (c) A conductive sphere coated with a dielectric layer.

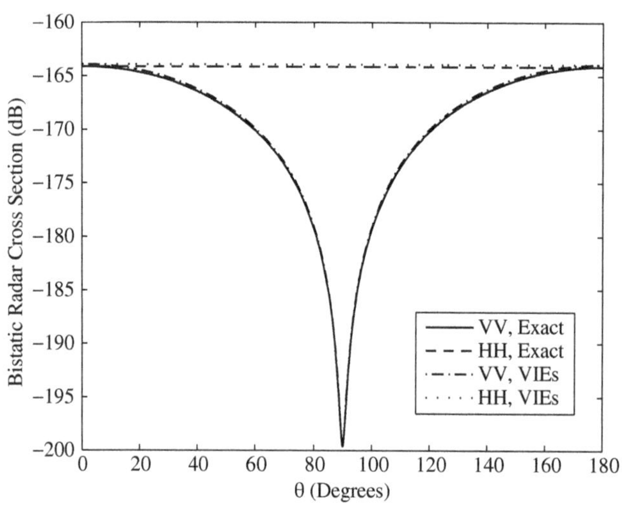

Figure 7.34 Bistatic RCS solutions of a conductive sphere with a radius $a = 1.0$ m and a conductivity $\sigma = 32.0$ S m^{-1} when the frequency is $f = 2.0$ KHz.

is assumed (non-magnetic) and the real part of relative permittivity is chosen as $\epsilon_r' = 3.0$. The incident wave is a x-polarized plane wave propagating along the $+z$ direction and it has a frequency $f = 2.0$ KHz. We calculate the bistatic RCS observed along the principal cut ($\phi = 0^\circ$ and $\theta = 0^\circ$–180°) for the scatterer in both vertical polarization (VV) and horizontal polarization (HH) and the results as shown in Figure 7.34 agree with the exact Mie-series solutions [79] very well. The skin depth is about $\delta = 1.9894$ m in this example and we discretize the entire object into 10718 tetrahedral elements. If we choose $f = 200.0$ KHz and $\sigma = 8.0$ S m^{-1}, then the skin depth is about $\delta = 0.3979$ m and we can only discretize the skin area with a non-zero current density. The number of

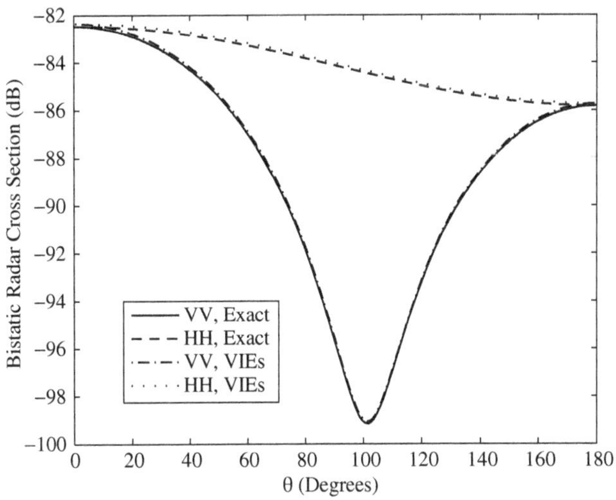

Figure 7.35 Bistatic RCS solutions of a conductive sphere with a radius $a = 1.0$ m and a conductivity $\sigma = 8.0$ S m^{-1} when the frequency is $f = 200.0$ KHz. We only discretize the skin area with a non-zero current density.

tetrahedral elements is 7562 in such a discretization. Figure 7.35 shows the bistatic RCS solutions for such a try and they can also be in good agreement with the exact solutions. Note that double-precision variables are used in programming due to the small electrical size of the object and the skin depth is approximately calculated by $\delta = \sqrt{\frac{2}{\omega\mu\sigma}}$ [67].

In the second example, we also consider the EM scattering by a conductive sphere but it is embedded in a dielectric medium with a relative permittivity $\epsilon_r = 3.0$. The sphere has a radius $a = 5.0$ mm, a relative permeability $\mu_r = 20.0$, and a conductivity $\sigma = 2.0 \times 10^6$ S m^{-1}. Also, the incident wave is changed into the one excited by a z-polarized magnetic dipole with a unit moment and a frequency $f = 1.0$ KHz. The dipole is located at the $+z$ axis and is 50.0 mm away from the center of the sphere. The object is discretized into 5835 tetrahedrons and we calculate the scattered electric field in a near zone, namely, the observation is taken along a line defined by $x = [-10, 10]$ mm, $y = 0$, and $z = 10$ mm. Figure 7.36 depicts the result when $x = [0, 10]$ mm since the result for $x = [-10, 0]$ mm is fully symmetric to the one for $x = [0, 10]$ mm and we can see that the result also agrees well with the exact Mie-series solution.

In the third example, we analyze the dissipated power of a conductive rectangular plate as shown in Figure 7.33(b). The plate has a length $a = 0.5$ mm, a width $b = 0.5$ mm, and a height $c = 0.02$ mm, and is excited by a plane wave propagating along the $-x$ direction. The incident wave has a frequency $f = 0.1$ GHz and is polarized in the y direction or $E_y^{\text{inc}} = 1.0$ V m^{-1}. The plate is meshed into 12526 tetrahedral elements and we calculate the time-average dissipated power versus a wide range of conductivities as sketched in Figure 7.37. The dissipated power is computed by

$$P_{\text{d}} = \int_V \sigma |\mathbf{E}|^2 \, \mathrm{d}V = \int_V \frac{1}{\sigma} |\mathbf{J}_V|^2 \, \mathrm{d}V \approx \frac{1}{\sigma} \sum_{n=1}^{N_t} \Delta V_n |\mathbf{J}_{V_n}|^2 \tag{7.79}$$

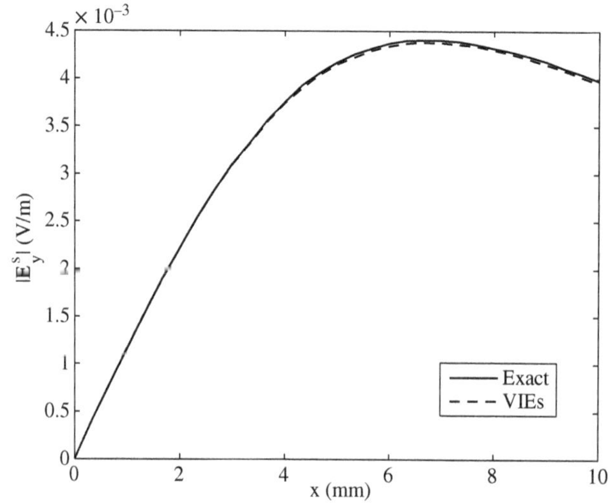

Figure 7.36 Scattered electric field (*y*-component) by a conductive sphere with a radius $a = 5.0$ mm and conductivity $\sigma = 2.0 \times 10^6$ S/m. The observation is taken along a line defined by $x = [-10.0, 10.0]$ mm, $y = 0$, and $z = 10.0$ mm, but we only show the result within $x = [0, 10.0]$ mm due to its symmetry. The incident field is excited by a *z*-polarized magnetic dipole with a unit moment and a frequency $f = 1.0$ KHz. The dipole is located at $z = 50.0$ mm on the *z* axis.

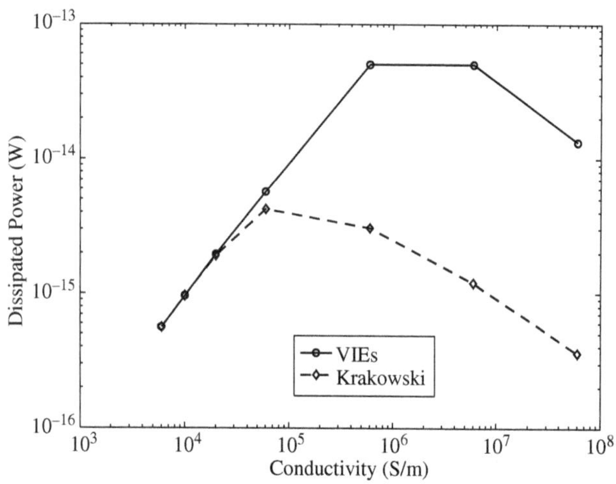

Figure 7.37 The dissipated power of a rectangular conductive plate versus its conductivity at the frequency $f = 0.1$ GHz. The plate has a length $a = 0.5$ mm, a width $b = 0.5$ mm, and a height $c = 0.02$ mm. The incident wave is a plane wave propagating along the $-x$ direction and is polarized in the *y* direction with a unit magnitude.

and the result is compared with the one based on Krakowski's formulation [80] that has an assumption that the skin depth is much larger than the thickness of the plate. It can be seen that the solutions from the two different approaches are consistent with each other when the conductivity is below 2.0×10^4 S m^{-1} at which the skin depth is roughly 17.8 times the thickness of the plate. As the conductivity increases, the skin depth decreases and the Krakowski's formulation becomes invalid, leading to very different results between the two approaches.

The structures in the above are electrically small and we can incorporate the MLFMA to accelerate the solution process for electrically large problems. As the fourth example, we consider the EM scattering by a conductive sphere coated with a dielectric layer as shown in Figure 7.33(c). The sphere has a radius $a_1 = 5.0$ m and a conductivity $\sigma = 1.0$ S m^{-1} while the concentric dielectric layer has a thickness $t = 3.0$ m and a relative permittivity $\epsilon_r = 3.0$. The incident wave is the same as in the first example except that the frequency is $f = 300$ MHz now, leading to $a_1 = 5.0\lambda$ and $t = 3.0\lambda$, where λ is the wavelength in free space. The conductive sphere and its dielectric layer are discretized into 123576 and 269538 tetrahedrons, respectively, resulting in 1179342 unknowns and we use the MLFMA with the generalized minimal residual method (GMRES) as an iterative solver to accelerate the solution. Figure 7.38 illustrates the bistatic RCS solution in the vertical polarization (VV) and the solution is very close to the exact Mie-series solution.

To examine the convergence of the scheme, we present the numerical error versus the number of unknowns per wavelength for the last example, which has an exact solution to compare with. The numerical error is defined by the relative RMS error, i.e.

$$e_{\text{RMS}} = \frac{\sqrt{\frac{1}{N_0}\sum_{n=1}^{N_0}\left[\gamma_n^{\text{numer}} - \gamma_n^{\text{exact}}\right]^2}}{\sqrt{\frac{1}{N_0}\sum_{n=1}^{N_0}\left[\gamma_n^{\text{exact}}\right]^2}} \tag{7.80}$$

where γ denotes the bistatic RCS and N_0 is the number of observation points in the RCS. Figure 7.39 displays the convergence curve and it can be seen that the numerical error

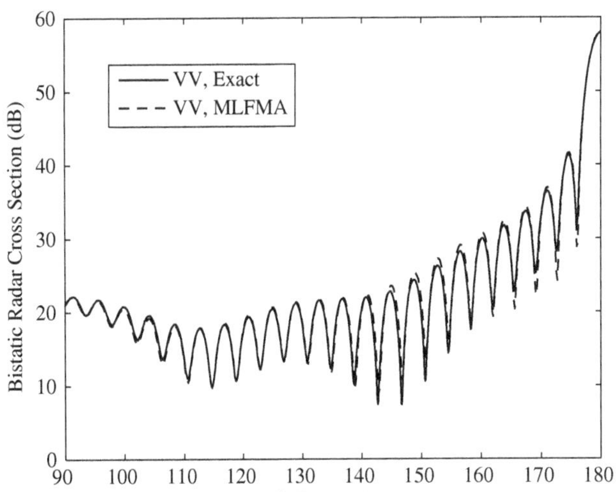

Figure 7.38 Bistatic RCS solutions (VV) of a conductive sphere coated with a dielectric layer. The sphere has a radius $a = 5.0$ m and a conductivity $\sigma = 1.0$ S m^{-1}. The dielectric layer has a thickness $t = 3.0$ m and the frequency is $f = 300.0$ MHz.

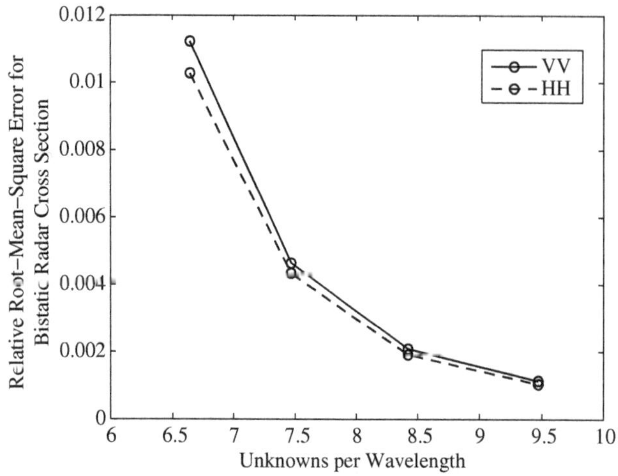

Figure 7.39 Convergence of radar cross section solutions based on the relative RMS error for the last example.

can be reduced exponentially by increasing the density of unknowns or reducing the size of meshes.

7.6 Volume-Surface Integral Equations for Mixed Media

Efficient and accurate EM analysis plays a central role for understanding the electrical performance of interconnect and packaging structures in microelectronic or nanoelectronic devices [81]. However, performing an EM analysis for such structures is a non-trivial task due to the inherent structural characteristics. The structures are usually very small compared with the wavelength within a certain frequency range, so extra care is required for numerical accuracy and double-precision variables should be used in programming. Also, the structures include multiscale features, namely, some parts are much smaller than others in dimensions [82, 83]. Moreover, the EM analysis usually requires covering a wide range of frequencies with significant low-frequency components and it will suffer from low-frequency effects [84]. These factors remarkably distinguish the packaging problems from other EM problems and seriously deteriorate the conditioning of the system matrix in the numerical procedure. Consequently, it is essential to choose appropriate governing equations, a robust numerical method, and a wise implementation scheme so that the best numerical solutions can be achieved. In practice, one has two choices for describing the involved EM nature of the structures, namely, the differential equation approaches and the integral equation approaches. Although the former have been widely applied to the EM modeling and simulation of interconnect and packaging structures [85–91], the latter are preferred in many scenarios due to their distinctive merits and have also been extensively employed in the EM analysis of those structures [3, 67, 82, 84]. The integral equation approaches include SIEs for both impenetrable and penetrable materials and VIEs for penetrable materials as governing equations, but SIEs are usually preferred whenever available because they can result in a fewer number of unknowns in the geometric discretization [55].

However, the SIEs for penetrable materials require a homogeneity of materials and their scope of applications could be confined. The SIEs include the EFIE, MFIE, CFIE, PMCHWT formulation [1] and Müller formulation [2]. All these equations include both electric current density and magnetic current density as unknown functions on material interfaces or boundaries and one has to represent both current densities with appropriate basis functions in the MoM [68] solution. Although the well-known RWG basis function [14] is a natural choice to represent the electric current density, how one should represent the magnetic current density is less obvious. One could employ the RWG basis function again or $\hat{n}\times$RWG basis function, where $\hat{n}$ is a unit normal vector on material interfaces, to represent the magnetic current density, but both choices present some problems [92]. One could also use the dual basis function proposed by Chen and Wilton in 1990 to remedy the problems [16], but its implementation is quite sophisticated. The good representation of magnetic current density is very important because it is related to the conditioning of the resultant system matrix, which seriously affects the iterative-method-based fast algorithms like the MLFMA [28]. By contrast, VIEs are more generalized and allow the involvement of inhomogeneous materials. Moreover, they are more insensitive for low-frequency effects because they belong to the second kind of integral equations whose system matrix is diagonally dominant and well conditioned in general. Also, only one unknown function, i.e. the electric flux density, needs to be represented if the materials are non-magnetic and the well-known SWG basis function [29] is enough to expand it in the MoM. In addition, using the VIEs might facilitate geometric discretization for the structures with many material interfaces because the SIEs need to individually discretize two sides with conforming meshes for each interface.

Owing to the above reasons, we use the VIEs to replace the SIEs for the penetrable media and they are coupled with the SIEs for the conductors to form the VSIEs for describing the EM features of interconnect and packaging structures. The VSIEs are usually solved with the MoM in which the RWG basis function is used to expand the surface current density on the conductors with a triangular tessellation while the SWG basis function is applied to represent the flux density in the penetrable media with a tetrahedral discretization. VSIEs have been extensively studied and applied to solve EM problems [93–96], but they are less used to analyze interconnect and packaging problems that include multiscale features and low-frequency effects. Since the traditional MoM implementation of the VIEs has to assume a homogeneous material in each tetrahedral element and also requires taking care of the surface charge density on the common faces of paired tetrahedrons, we propose a mixed scheme in which the traditional MoM is kept to discretize the SIE for conductor parts while the Nyström method replaces the MoM to discretize the VIEs for penetrable parts. We can avoid the inconvenience caused by the charge density in the VIEs in this way though requiring a good treatment for the hypersingularity in the dyadic Green's function. With the techniques of singularity treatment [19, 70], the proposed scheme is feasible and can show certain merits of implementation. Numerical examples for inhomogeneous interconnect and packaging structures are presented to illustrate the approach and good results have been observed.

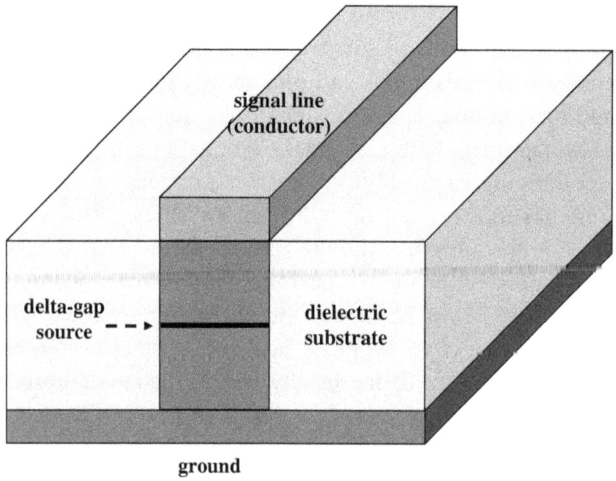

Figure 7.40 A typical interconnect and packaging structure. The signal lines (transmission lines) and ground are conductors while the substrates are dielectric.

7.6.1 Volume-Surface Integral Equations

The interconnect and packaging structures include both conducting signal lines (transmission lines) and ground and dielectric substrates as sketched in Figure 7.40. We assume that the signal lines and ground are PECs and the involved EM features can be described by the following EFIE

$$-\hat{n} \times \mathbf{E}^{\mathrm{ex}}(\mathbf{r}) = \hat{n} \times \mathrm{i}\omega\mu_0 \int_S \overline{\mathbf{G}}(\mathbf{r}, \mathbf{r}') \cdot \mathbf{J}_S(\mathbf{r}')\, \mathrm{d}S', \quad \mathbf{r} \in S \tag{7.81}$$

where $\mathbf{J}_S(\mathbf{r}')$ is the electric current density induced on the conductor surface S whose unit normal vector is $\hat{n}$ and $\mathbf{E}^{\mathrm{ex}}(\mathbf{r})$ represents a delta-gap excitation at an appropriate position on the conductors. Also, $\overline{\mathbf{G}}(\mathbf{r}, \mathbf{r}')$ is the dyadic Green's function defined by

$$\overline{\mathbf{G}}(\mathbf{r}, \mathbf{r}') = \left(\overline{\mathbf{I}} + \frac{\nabla\nabla}{k_0^2}\right) g(\mathbf{r}, \mathbf{r}') \tag{7.82}$$

where $\overline{\mathbf{I}}$ is the identity dyad, $g(\mathbf{r}, \mathbf{r}') = \mathrm{e}^{\mathrm{i}k_0 R}/(4\pi R)$ is the 3D scalar Green's function in which $R = |\mathbf{r} - \mathbf{r}'|$ is the distance between an observation point $\mathbf{r}$ and a source point $\mathbf{r}'$, and k_0 is the wavenumber of the free space with a permittivity ϵ_0 and a permeability μ_0. For the substrate, which could consist of inhomogeneous materials with a permittivity $\epsilon(\mathbf{r}')$ and a permeability $\mu(\mathbf{r}')$, we can use the VIEs to describe its EM characteristics, i.e.

$$\begin{aligned} \mathbf{E}(\mathbf{r}) = \mathbf{E}^{\mathrm{ex}}(\mathbf{r}) &+ \mathrm{i}\omega\mu_0 \int_V \overline{\mathbf{G}}(\mathbf{r}, \mathbf{r}') \cdot \mathbf{J}_V(\mathbf{r}')\mathrm{d}\mathbf{r}' \\ &- \nabla \times \int_V \overline{\mathbf{G}}(\mathbf{r}, \mathbf{r}') \cdot \mathbf{M}_V(\mathbf{r}')\mathrm{d}\mathbf{r}', \quad \mathbf{r} \in V \end{aligned} \tag{7.83}$$

$$\begin{aligned} \mathbf{H}(\mathbf{r}) = \mathbf{H}^{\mathrm{ex}}(\mathbf{r}) &+ \mathrm{i}\omega\epsilon_0 \int_V \overline{\mathbf{G}}(\mathbf{r}, \mathbf{r}') \cdot \mathbf{M}_V(\mathbf{r}')\mathrm{d}\mathbf{r}' \\ &+ \nabla \times \int_V \overline{\mathbf{G}}(\mathbf{r}, \mathbf{r}') \cdot \mathbf{J}_V(\mathbf{r}')\mathrm{d}\mathbf{r}', \quad \mathbf{r} \in V \end{aligned} \tag{7.84}$$

where $\mathbf{E}^{ex}(\mathbf{r}) = \mathbf{H}^{ex}(\mathbf{r}) = 0$ in the substrate in general, and

$$\mathbf{J}_V(\mathbf{r}') = i\omega \left[\epsilon_0 - \epsilon(\mathbf{r}')\right] \mathbf{E}(\mathbf{r}') \tag{7.85}$$

$$\mathbf{M}_V(\mathbf{r}') = i\omega \left[\mu_0 - \mu(\mathbf{r}')\right] \mathbf{H}(\mathbf{r}') \tag{7.86}$$

are the induced volumetric electric and magnetic current densities inside the substrate, respectively. If the substrate is non-magnetic or has the same permeability as the background, which is usually true, then $\mathbf{M}_V(\mathbf{r}') = 0$ and only Equation (7.83) is needed, which can be reduced to

$$\mathbf{E}(\mathbf{r}) = i\omega\mu_0 \int_V \overline{\mathbf{G}}(\mathbf{r}, \mathbf{r}') \cdot \mathbf{J}_V(\mathbf{r}')d\mathbf{r}', \quad \mathbf{r} \in V. \tag{7.87}$$

When considering the coupling of fields produced by the surface current density on the conductors and the volume current density inside the substrate, we can form the following VSIEs

$$0 = \hat{n} \times \left[\mathbf{E}^{ex}(\mathbf{r}) + i\omega\mu_0 \int_S \overline{\mathbf{G}}(\mathbf{r}, \mathbf{r}') \cdot \mathbf{J}_S(\mathbf{r}')\, dS' + i\omega\mu_0 \int_V \overline{\mathbf{G}}(\mathbf{r}, \mathbf{r}') \cdot \mathbf{J}_V(\mathbf{r}')d\mathbf{r}'\right], \quad \mathbf{r} \in S \tag{7.88}$$

$$\mathbf{E}(\mathbf{r}) = i\omega\mu_0 \int_S \overline{\mathbf{G}}(\mathbf{r}, \mathbf{r}') \cdot \mathbf{J}_S(\mathbf{r}')\, dS' + i\omega\mu_0 \int_V \overline{\mathbf{G}}(\mathbf{r}, \mathbf{r}') \cdot \mathbf{J}_V(\mathbf{r}')d\mathbf{r}', \quad \mathbf{r} \in V \tag{7.89}$$

from which the unknown current densities can be solved and equivalent circuit parameters of the structure can subsequently be extracted.

7.6.2 Nyström-Based Mixed Scheme for Solving the VSIEs

We assume that the dielectric substrate in the structure is non-magnetic (which is usually true) so that the simplified VSIEs can be used. We use a mixed scheme combining the MoM with the Nyström method to solve the VSIEs. The surface of conducting interconnects and ground is discretized into N_c triangular patches while the dielectric substrate is discretized into N_d tetrahedrons. The surface current density on the conductor surface is expanded through the RWG basis function while the volumetric current density inside each tetrahedron in the dielectric substrate is represented by the one at the center of that tetrahedron, i.e.

$$\mathbf{J}_S(\mathbf{r}') = \sum_{n=1}^{N_t} J_{Sn}\mathbf{\Lambda}_n(\mathbf{r}') \tag{7.90}$$

$$\mathbf{J}_V(\mathbf{r}') = \sum_{n=1}^{N_d} \mathbf{J}_{Vn}\delta_n(\mathbf{r}') \tag{7.91}$$

where $\mathbf{\Lambda}_n(\mathbf{r}')$ is the RWG basis function defined over the nth pair of triangular patches $\Delta T_n^{\pm}$ and J_{Sn} is the corresponding expansion coefficient. Also, N_t is the number of the pairs of triangular patches on the conductor surface, which is 1.5 times the number of patches N_c when the thickness of conductors is considered. To facilitate the description of the mixed scheme for the dielectric part, we introduce a delta-like basis function defined by

$$\delta_n(\mathbf{r}') = \begin{cases} 1, & \text{if } \mathbf{r}' \in \Delta V_n \\ 0, & \text{otherwise} \end{cases} \tag{7.92}$$

where $\mathbf{J}_{Vn} = \mathbf{J}_V(\mathbf{r}'_n)$ acts as an expansion coefficient, which is actually the volumetric current density at the center of the nth tetrahedron ΔV_n. We use the MoM to discretize the conducting part and use the Nyström method to discretize the dielectric part. Substituting the expansions of Equations (7.90) and (7.91) to Equation (7.88) and using the RWG basis function as a testing function to test Equation (7.88), we obtain

$$\begin{aligned}&\sum_{n=1}^{N_t} J_{Sn} \int_{\Delta T_m^{\pm}} \mathrm{d}S\boldsymbol{\Lambda}_m(\mathbf{r}) \cdot \int_{\Delta T_n^{\pm}} \overline{\mathbf{G}}(\mathbf{r},\mathbf{r}') \cdot \boldsymbol{\Lambda}_n(\mathbf{r}')\,\mathrm{d}S' + \\ &\sum_{n=1}^{N_d} \int_{\Delta T_m^{\pm}} \mathrm{d}S\boldsymbol{\Lambda}_m(\mathbf{r}) \cdot \int_{\Delta V_n} \overline{\mathbf{G}}(\mathbf{r},\mathbf{r}')\,\mathrm{d}V' \cdot \mathbf{J}_{Vn} = -\frac{1}{\mathrm{i}\omega\mu_0} \int_{\Delta T_m^{\pm}} \boldsymbol{\Lambda}_m(\mathbf{r}) \cdot \mathbf{E}^{\mathrm{ex}}(\mathbf{r})\,\mathrm{d}S, \\ &\quad m = 1, 2, \cdots, N_t. \end{aligned} \tag{7.93}$$

Substituting the expansions of Equations (7.90) and (7.91) to Equation (7.89) and performing a point-matching procedure on those centers of tetrahedrons in the dielectric substrate, or equivalently, using the delta-like basis function as a testing function to test Equation (7.89), we have

$$\begin{aligned}t_0\mathbf{J}_{Vm} &= \sum_{n=1}^{N_t} J_{Sn} \int_{\Delta T_n^{\pm}} \overline{\mathbf{G}}(\mathbf{r}_m,\mathbf{r}') \cdot \boldsymbol{\Lambda}_n(\mathbf{r}')\,\mathrm{d}S' + \sum_{n=1}^{N_d} \int_{\Delta V_n} \overline{\mathbf{G}}(\mathbf{r}_m,\mathbf{r}')\,\mathrm{d}V' \cdot \mathbf{J}_{Vn}, \\ m &= 1, 2, \cdots, N_d \end{aligned} \tag{7.94}$$

where $t_0 = \left[k_0^2(\epsilon_r - 1)\right]^{-1}$, which is obtained by using Equation (7.86) in Equation (7.89) and contains the permittivity information for different dielectric materials. Equations (7.93) and (7.94) in the above can also be written into a scalar form which can be easily changed into a matrix form

$$\sum_{n=1}^{N_t} a_{mn} J_{Sn} + \sum_{n=1}^{N_d} \left[b_{mn}^x J_{Vn}^x + b_{mn}^y J_{Vn}^y + b_{mn}^z J_{Vn}^z\right] = A_m, \quad m = 1, 2, \cdots, N_t \tag{7.95}$$

$$\sum_{n=1}^{N_t} c_{mn}^x J_{Sn} + \sum_{n=1}^{N_d} \left[(d_{mn}^{xx} - t_0\delta_{mn}) J_{Vn}^x + d_{mn}^{xy} J_{Vn}^y + d_{mn}^{xz} J_{Vn}^z\right] = 0, \quad m = 1, 2, \cdots, N_d \tag{7.96}$$

$$\sum_{n=1}^{N_t} c_{mn}^y J_{Sn} + \sum_{n=1}^{N_d} \left[d_{mn}^{yx} J_{Vn}^x + (d_{mn}^{yy} - t_0\delta_{mn}) J_{Vn}^y + d_{mn}^{yz} J_{Vn}^z\right] = 0, \quad m = 1, 2, \cdots, N_d \tag{7.97}$$

$$\sum_{n=1}^{N_t} c_{mn}^z J_{Sn} + \sum_{n=1}^{N_d} \left[d_{mn}^{zx} J_{Vn}^x + d_{mn}^{zy} J_{Vn}^y + (d_{mn}^{zz} - t_0\delta_{mn}) J_{Vn}^z\right] = 0, \quad m = 1, 2, \cdots, N_d \tag{7.98}$$

where

$$a_{mn} = \int_{\Delta T_m^{\pm}} \mathrm{d}S\boldsymbol{\Lambda}_m(\mathbf{r}) \cdot \int_{\Delta T_n^{\pm}} \overline{\mathbf{G}}(\mathbf{r},\mathbf{r}') \cdot \boldsymbol{\Lambda}_n(\mathbf{r}')\,\mathrm{d}S' \tag{7.99}$$

$$\mathbf{b}_{mn} = \int_{\Delta T_m^{\pm}} \mathrm{d}S\boldsymbol{\Lambda}_m(\mathbf{r}) \cdot \int_{\Delta V_n} \overline{\mathbf{G}}(\mathbf{r},\mathbf{r}')\,\mathrm{d}V' \tag{7.100}$$

$$\mathbf{c}_{mn} = \int_{\Delta T_n^{\pm}} \overline{\mathbf{G}}(\mathbf{r}_m, \mathbf{r}') \cdot \mathbf{\Lambda}_n(\mathbf{r}')\, \mathrm{d}S' \tag{7.101}$$

$$\overline{\mathbf{d}}_{mn} = \int_{\Delta V_n} \overline{\mathbf{G}}(\mathbf{r}_m, \mathbf{r}')\, \mathrm{d}V' \tag{7.102}$$

$$A_m = -\frac{1}{\mathrm{i}\omega\mu_0} \int_{\Delta T_m^{\pm}} \mathbf{\Lambda}_m(\mathbf{r}) \cdot \mathbf{E}^{\mathrm{ex}}(\mathbf{r})\, \mathrm{d}S \tag{7.103}$$

$$\delta_{mn} = \begin{cases} 1, & \text{if } m = n \\ 0, & \text{if } m \neq n \end{cases} \tag{7.104}$$

and $\mathbf{r}_m$ is the position of the center of the mth tetrahedron. In the above derivation, Equation (7.95) is obtained by rewriting Equation (7.93) which is generated by applying a standard MoM (the RWG function is used as both the basis function and testing function) to Equation (7.88). Equations (7.96–7.98) are obtained by writing Equation (7.94) into a scalar form and moving the left-hand side to the right-hand side with a combination of J_{Vm}. Equation (7.94) is obtained by using the delta function defined in Equation (7.92) as a testing function to test Equation (7.89) and the testing is equivalent to a point-matching or collocation procedure. We have used the superscripts x, y, z to denote the individual components of volume current density along different coordinate axes in the above scalar form and the total number of unknowns is $N_t + 3N_d$ which is close to that of a traditional MoM. In the existing approach or traditional MoM approach, the electric flux density instead of volumetric electric current density is chosen as an unknown function in the dielectric medium and it is expanded by using the SWG basis function, which is also employed as a testing function to test Equation (7.89). The resulting number of unknowns is roughly $N_t + 2.95N_d$, depending on the number of half-paired tetrahedrons at the surface of dielectric medium. This difference will result in a quite different form in Equations (7.93) and (7.94). Note that there are several singularities in the matrix elements resulting from the dyadic Green's function when $m = n$. For the weak and strong singularities, many mature techniques have been developed to handle them [20–23,54,77]. For the hypersingularity, we can use the developed technique in Chapter 5 to treat it for tetrahedrons.

7.6.3 Numerical Examples

To illustrate the proposed approach, we present several numerical examples for the EM analysis of typical interconnect and packaging structures that might include inhomogeneous materials. The geometries of the structures are shown in Figure 7.41 and we consider three cases, i.e. (a) one conducting signal line with a bridge at the top of a one-layer dielectric substrate, (b) one cross-shaped conducting signal line at the top of a one-layer dielectric substrate, and (c) two straight conducting signal lines at the top of a two-layer dielectric substrate. It is assumed that the signal lines and ground are made of PECs and the dielectric substrates are also lossless but could be inhomogeneous. The structures are strictly in 3D and both the interconnects and ground have a non-negligible thickness.

The dimensions of the geometries are characterized as follows (all in millimeters). In the first case (a), the length, width, and height of the substrate are $l = 10.0$, $w = 5.0$, and $h = 0.3$, respectively, and the thickness of the ground is $d = 0.1$. The signal line has a width $s = 0.2$, a thickness $t = 0.05$, and a bridge with a length $a = 1.0$ and a height

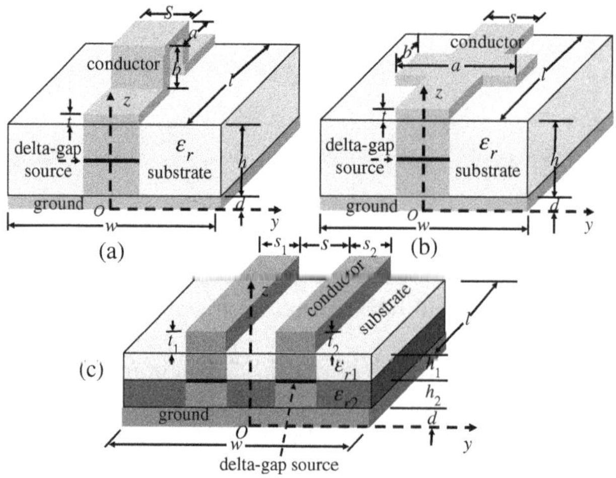

Figure 7.41 Geometries of typical interconnect and packaging structures. (a) One conducting signal line with a bridge at the top of a one-layer dielectric substrate. (b) One cross-shaped conducting signal line at the top of a one-layer dielectric substrate. (c) Two straight conducting signal lines at the top of a two-layer dielectric substrate.

$b = 0.2$. In the second case (b), the profile of the substrate and ground as well as the width and thickness of the signal line are the same as in the first case (a). The extension of the signal line on the two sides in the y direction has a length $a = 1.0$ and a width $b = 0.2$. In the third case (c), the width and thickness of the two straight signal lines are $s_1 = s_2 = 0.2$ and $t_1 = t_2 = 0.05$, respectively, and the spacing between the two lines is $s = 0.2$. The profile of the substrate and ground is the same as that in the previous cases, but there are two dielectric layers with an equal height $h_1 = h_2 = 0.3$ in the substrate now. The relative permittivity of the substrate is $\varepsilon_r = 3.0$ in the first and second cases while it is $\varepsilon_{r1} = 3.0$ for the upper layer and $\varepsilon_{r2} = 5.0$ for the lower layer in the third case (the relative permeability $\mu_r = 1.0$ is assumed in all cases). All the signal lines are located symmetrically at the top of the substrate in the y direction and the bridge or extension of the signal line is centered at the signal line in the x direction. The signal propagating in the lines is excited with a delta-gap source located at the center of the side wall of the substrate (in the y–z plane). We discretize the surfaces of conductors into triangular patches and volumes of substrates into tetrahedral elements for the three structures, resulting in 5278, 5642, and 7165 unknowns, respectively. Figures 7.42–7.44 plot the solutions of S parameters for those structures based on the proposed approach and they are in good agreement with the corresponding solutions obtained from the traditional MoM of SIEs. Note that the proposed approach can still give rise to good solutions even at the relatively low frequencies between $f = 0.01$ GHz and $f = 0.1$ GHz while the traditional MoM of SIEs may not produce correct solutions, as shown in Figure 7.45 if the dual basis function is not used [92]. Since the low-frequency problem is approximately scale-invariant, obtaining the accurate solution at the frequency roughly $f = 0.1$ GHz is a key [84].

The SIEs are invalid if the substrate includes inhomogeneous materials and the VSIEs are indispensable. We consider this case to demonstrate the robustness of the proposed approach now. We assume that the relative permittivity of the dielectric substrate obeys

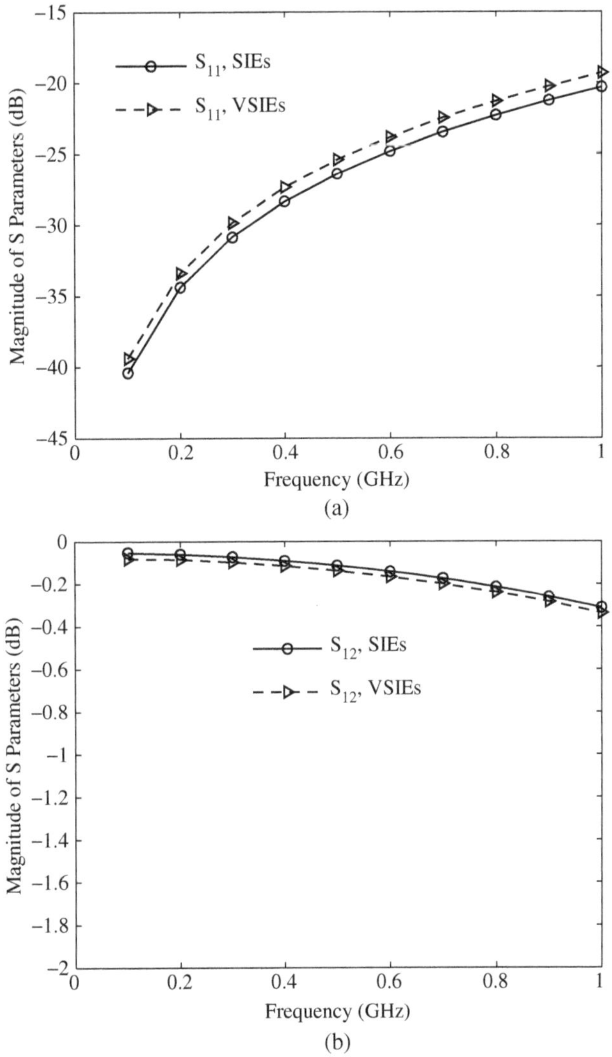

Figure 7.42 S parameters for the interconnect and packaging structure including one signal line with a bridge and a one-layer dielectric substrate. The geometry is defined with $l = 10.0$, $w = 5.0$, $d = 0.1$, $s = 0.2$, $t = 0.05$, $a = 1.0$, $b = 0.2$, and $h = 0.3$, all in millimeters, and the substrate has a relative permittivity $\varepsilon_r = 3.0$. (a) S_{11}. (b) S_{12}.

a basin-shape function defined by

$$\varepsilon_r(y) = \cos(2.0\pi|y|/w) + 3.0 \tag{7.105}$$

in the y direction for the first and second structures. In the third structure, the upper layer has the same relative permittivity as the above, namely, $\varepsilon_{r1}(y) = \cos(2.0\pi|y|/w) + 3.0$, but the lower layer possesses a relative permittivity defined by

$$\varepsilon_{r2}(y) = \cos(2.0\pi|y|/w) + 5.0. \tag{7.106}$$

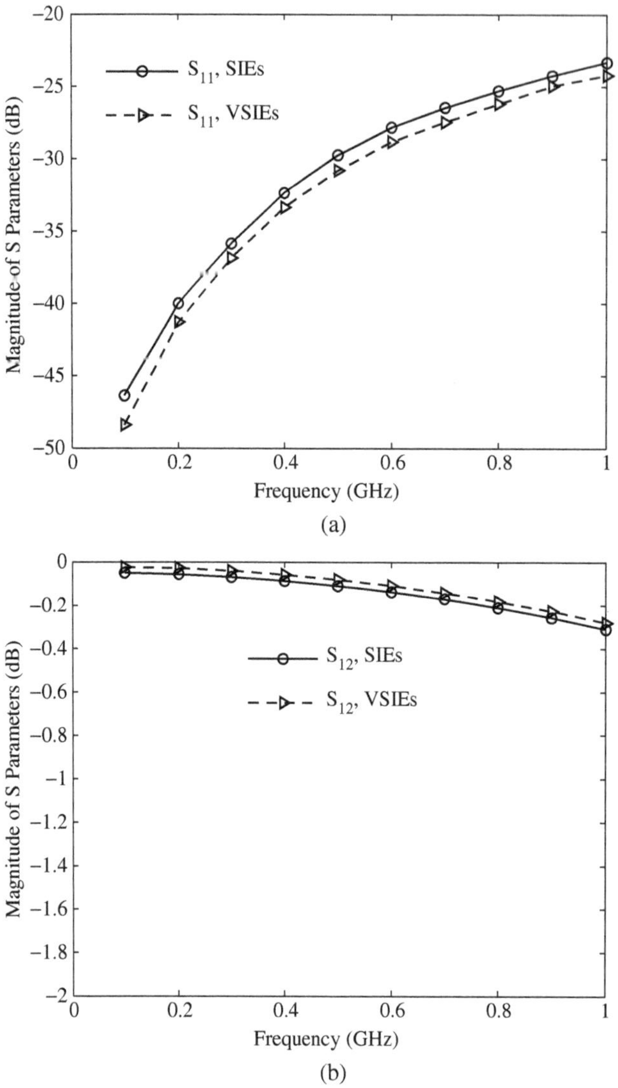

Figure 7.43 S parameters for the interconnect and packaging structure including one cross-shaped signal line and a one-layer dielectric substrate. The geometry is defined with $l = 10.0$, $w = 5.0$, $d = 0.1$, $s = 0.2$, $t = 0.05$, $a = 1.0$, $b = 0.2$, and $h = 0.3$, all in millimeters, and the substrate has a relative permittivity $\varepsilon_r = 3.0$. (a) S_{11}. (b) S_{12}.

Figures 7.46–7.48 display the solutions of the corresponding S parameters for those structures with the inhomogeneous materials in the substrate and these results are verified with an FEM solver. Usually, it is not necessary to compare the integral equation solver with a differential equation solver like the FEM solver because their advantages and disadvantages have already been commonly recognized [3]. Nevertheless, we

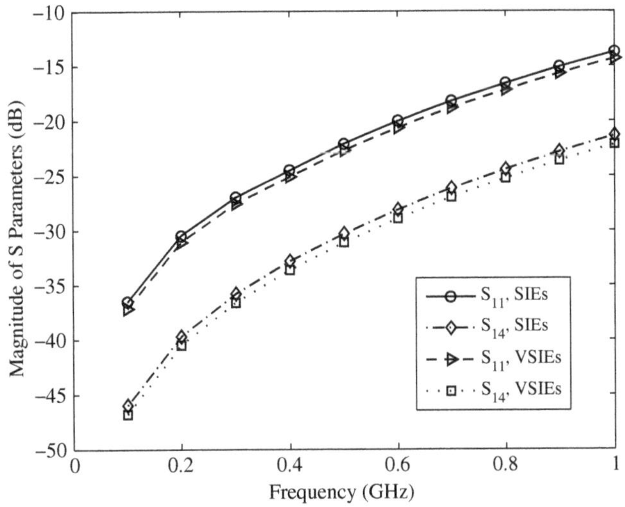

Figure 7.44 S parameters for the interconnect and packaging structure including two straight signal lines and a two-layer dielectric substrate. The geometry is defined with $l = 10.0$, $w = 5.0$, $d = 0.1$, $s = s_1 = s_2 = 0.2$, $t_1 = t_2 = 0.05$, and $h_1 = h_2 = 0.3$, all in millimeters. The upper and lower layers of the substrate have a relative permittivity $\varepsilon_{r1} = 3.0$ and $\varepsilon_{r2} = 5.0$, respectively.

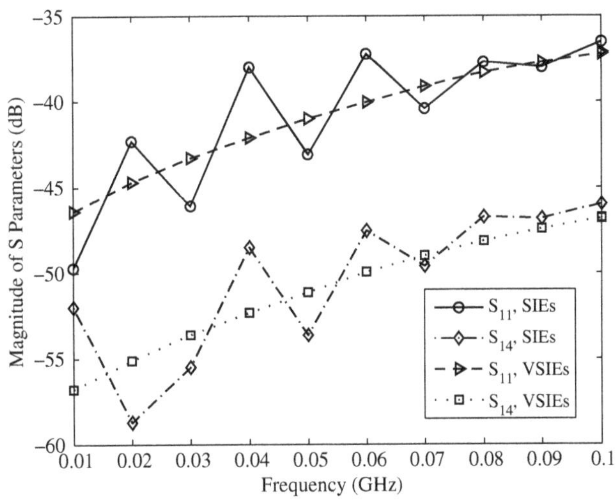

Figure 7.45 S parameters at low frequencies for the interconnect and packaging structure including two straight signal lines and a two-layer dielectric substrate. The geometry is defined with $l = 10.0$, $w = 5.0$, $d = 0.1$, $s = s_1 = s_2 = 0.2$, $t_1 = t_2 = 0.05$, and $h_1 = h_2 = 0.3$, all in millimeters. The upper and lower layers of the substrate have a relative permittivity $\varepsilon_{r1} = 3.0$ and $\varepsilon_{r2} = 5.0$, respectively.

provide a comparison between these approaches for their computational costs, which are shown in Table 7.3, and it can be seen that the proposed approach requires less CPU time but more memory usage.

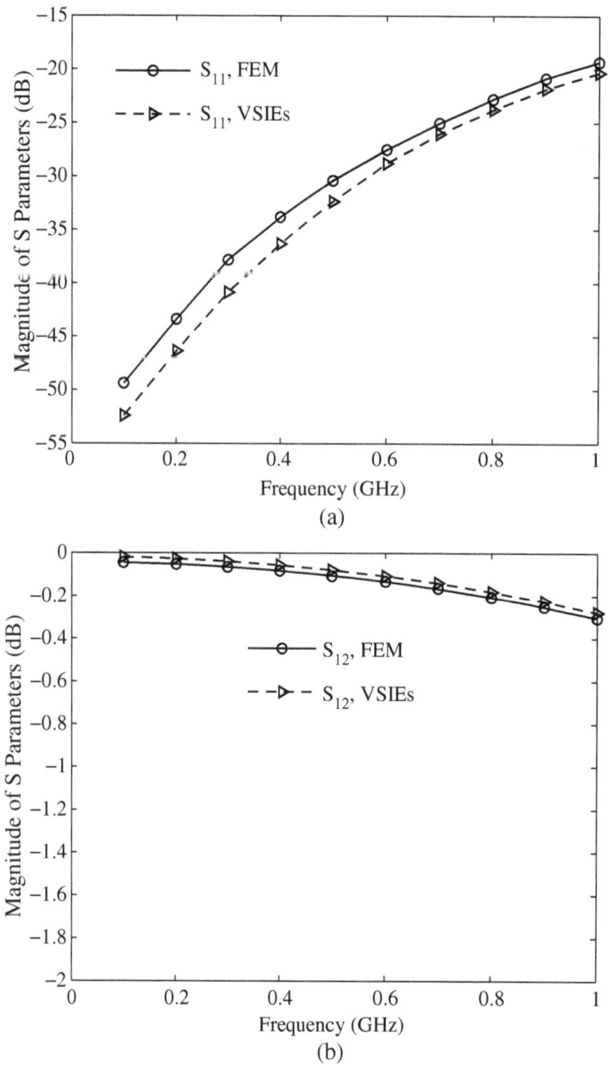

Figure 7.46 *S* parameters for the interconnect and packaging structure including one signal line with a bridge and a one-layer dielectric substrate. The geometry is defined with $l = 10.0$, $w = 5.0$, $d = 0.1$, $s = 0.2$, $t = 0.05$, $a = 1.0$, $b = 0.2$, and $h = 0.3$, all in millimeters, and the substrate has an inhomogeneous relative permittivity $\epsilon_r(y) = \cos(2.0\pi|y|/w) + 3.0$ in the y direction. (a) S_{11}. (b) S_{12}.

7.7 Summary

In this chapter, we apply the Nyström method or Nyström-based mixed scheme to solve EM problems with penetrable media. When the media are homogeneous and isotropic, we first use SIEs, which are in the forms of EFIEs, MFIEs, and CFIEs, to describe the problems. The Nyström method is employed to solve these equations and both the near-zone and far-zone fields are calculated with a comparison to the exact solutions whenever available. These SIEs are widely solved with the MoM but less frequently

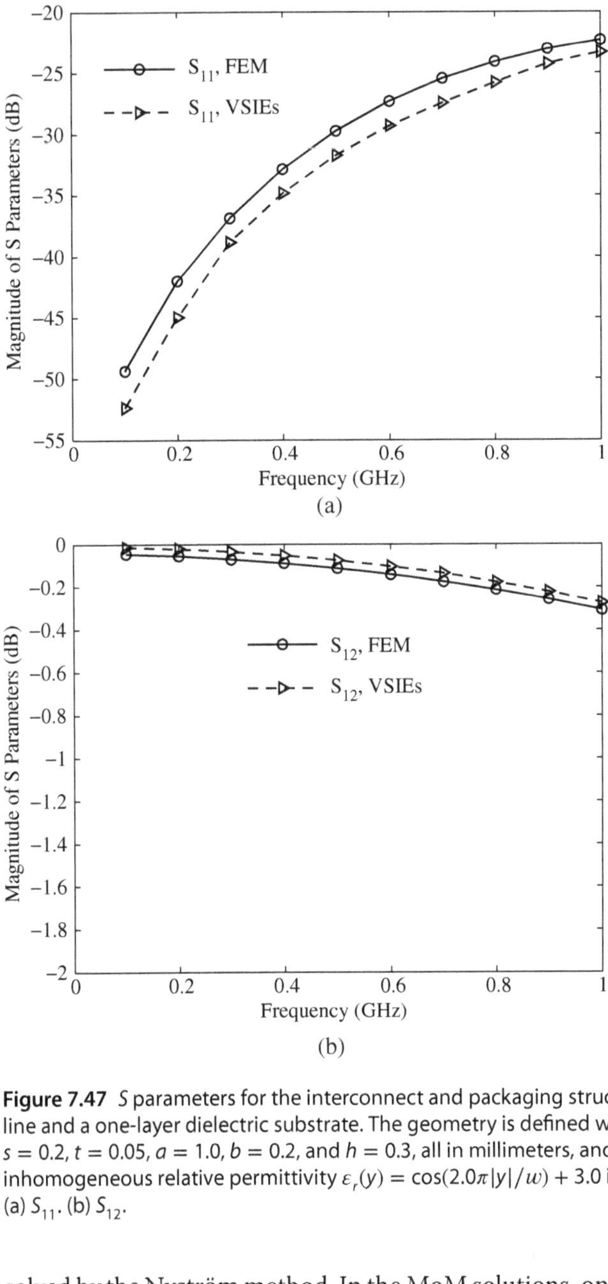

Figure 7.47 S parameters for the interconnect and packaging structure with one cross-shaped signal line and a one-layer dielectric substrate. The geometry is defined with $l = 10.0$, $w = 5.0$, $d = 0.1$, $s = 0.2$, $t = 0.05$, $a = 1.0$, $b = 0.2$, and $h = 0.3$, all in millimeters, and the substrate has an inhomogeneous relative permittivity $\varepsilon_r(y) = \cos(2.0\pi|y|/w) + 3.0$ in the y direction. (a) S_{11}. (b) S_{12}.

solved by the Nyström method. In the MoM solutions, one may suffer from the difficulty of choosing an appropriate basis function to represent the magnetic current density if the primary electric current density is represented using the RWG basis function. The EFIEs and MFIEs are sensitive to the choice of basis functions and testing schemes as shown in [15] and the existent choices for the magnetic current density, like RWG, $\hat{n}\times$ RWG, and dual basis functions may not be ideal. The Nyström method does not have these problems and the solutions are also accurate and stable when the internal

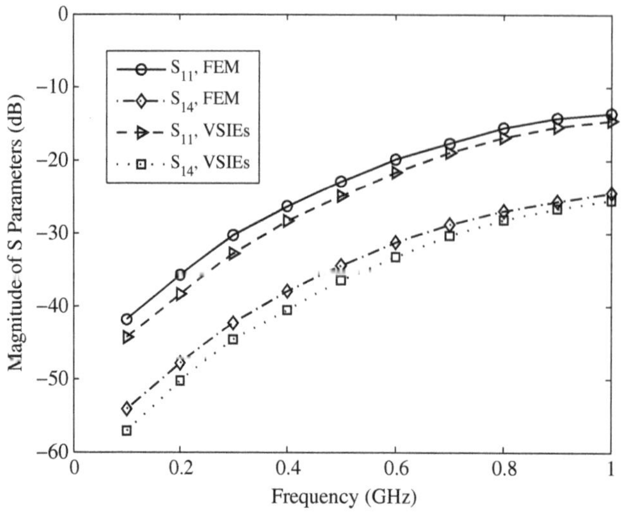

Figure 7.48 S parameters for the interconnect and packaging structure with two straight signal lines and a two-layer dielectric substrate. The geometry is defined with $l = 10.0$, $w = 5.0$, $d = 0.1$, $s = s_1 = s_2 = 0.2$, $t_1 = t_2 = 0.05$, and $h_1 = h_2 = 0.3$, all in millimeters. The upper and lower layer of the substrate have an inhomogeneous relative permittivity defined by $\varepsilon_{r1}(y) = \cos(2.0\pi|y|/w) + 3.0$ and $\varepsilon_{r2}(y) = \cos(2.0\pi|y|/w) + 5.0$, respectively, in the y direction.

Table 7.3 A comparison of computational costs between the VSIEs approach and the FEM approach for three interconnect structures.

	Costs	Structure (a)	Structure (b)	Structure (c)
VSIEs	CPU time (Second)	187	196	267
	Memory usage (MB)	552	587	791
FEM	CPU time (Second)	225	238	372
	Memory usage (MB)	435	468	612

resonances are not involved in the EFIEs and MFIEs. The CFIEs can be employed as in the MoM if the internal resonances cannot be avoided.

The Nyström method is mainly used for solving SIEs with relatively simple structures and it is rarely applied to solve VIEs with penetrable media. We fully exploit this method for solving VIEs by developing an efficient local correction scheme for singular or near singular integrals over tetrahedral elements. The scheme also uses the strategy in the SIEs, i.e. interpolating the unknown functions and deriving the resultant hypersingular integrals analytically after applying the singularity subtraction. The derivation of hypersingular integrals divides the volume integration into a regular line integral along the height direction of the tetrahedron and a singular or near singular surface integral over the base areas of the tetrahedron, which are height-dependent. The surface integrals can then be handled as in the SIEs. This derivation can minimize the overhead for evaluating self and near-interaction matrix elements and make the scheme simple and compact. In addition to those merits in the SIEs, the Nyström method also allows the inhomogeneity of material in tetrahedrons and there is no surface charge problem for inhomogeneous media in the VIEs. We demonstrate the method for solving VIES by considering the homogeneous and isotropic, inhomogeneous and/or anisotropic, and conductive media, respectively. Although the homogeneous and isotropic media can be formulated by the SIEs preferably, the VIEs are indispensable for inhomogeneous and/or anisotropic media. For conductive media, one usually uses the SIEs of conductors with the approximation of impedance boundary conditions to describe the problem, but this is only valid when the skin depth is small and the VIEs are required when skin depth is large. We also consider the mixed media with both penetrable dielectrics and impenetrable conductors and the VSIEs are solved with the Nyström-based mixed scheme for analyzing interconnect and packaging problems. The VIEs are usually well-conditioned due to the feature of the second kind of integral equations and only have one unknown function to be solved for non-magnetic media. Also, the Nyström method can choose the current densities instead of flux densities as unknowns to be solved and the integrands of VIEs are free of the material parameters of the media, resulting in much convenience in evaluating matrix elements and incorporating fast algorithms for these complex media. These merits make the VSIEs and Nyström-based mixed scheme be very suitable for the analysis of interconnect and packaging problems which have multiscale features and low-frequency effects. We present different numerical examples to demonstrate the Nyström method or Nyström-based mixed scheme and their effectiveness and robustness have been observed. Moreover, we demonstrate the convergence of the Nyström method in both h-refinement and p-refinement and compare it to the convergence of conventional MoM with the SWG basis function in some examples.

Bibliography

1 Poggio, A.J. and Miller, E.K. (1973). Integral equation solutions of three-dimensional scattering problems. In: *Computer Techniques for Electromagnetics* (ed. R. Mittra). Oxford: Pergamon Press, Chapter 4.

2 Müller, C. (1969). *Foundations of the Mathematical Theory of Electromagnetic Waves*. Berlin: Springer-Verlag.

3 Chew, W.C., Tong, M.S., and Hu, B. (2008). *Integral Equation Methods for Electromagnetic and Elastic Waves*. San Rafael, CA: Morgan & Claypool.

4 Miller, E.K., Medgyesi-Mitschang, L., and Newman, E.H. (eds.) (1992). *Computational Electromagnetics: Frequency-Domain Method of Moments*. New York: IEEE Press.

5 Kolundžija, B.M. and Djordjević, A.R. (2002). *Electromagnetic Modeling of Composite Metallic and Dielectric Structures*. Boston: Artech House.

6 Mautz, J.R. and Harrington, R.F. (1979). Electromagnetic scattering from a homogeneous material body of revolution. *Arch. Electron. Übertragungstechn.* 33 (2): 71–80.

7 Medgyesi-Mitschang, L.N., Putnam, J.M., and Gedera, M.B. (1994). Generalized method of moments for three-dimensional penetrable scatterers. *J. Opt. Soc. Am. A.* 11: 1383–1398.

8 Song, J.M., Lu, C.C., and Chew, W.C. (1997). Multilevel fast multipole algorithm for electromagnetic scattering by large complex objects. *IEEE Trans. Antennas Propagat.* 45: 1488–1493.

9 Kolundžija, B.M. (1999). Electromagnetic modeling of composite metallic and dielectric structures. *IEEE Trans. Microw. Theory Tech.* 47 (7): 1021–1032.

10 Ylä-Oijala, P. and Taskinen, M. (2005). Application of combined field integral equation for electromagnetic scattering by dielectric and composite objects. *IEEE Trans. Antennas Propagat.* 53 (3): 1168–1173.

11 Ylä-Oijala, P., Taskinen, M., and Järvenpää, S. (2005). Surface integral equation formulations for solving electromagnetic scattering problems with iterative methods. *Radio Sci.* 40: RS6002.

12 Ylä-Oijala, P., Taskinen, M., and Sarvas, J. (2005). Surface integral equation method for general composite metallic and dielectric structures with junctions. *Prog. Electromagn. Res.* 52: 81–108.

13 Zhu, A., Gedney, S.D., and Visher, J.L. (2005). A study of combined field formulations for material scattering for a locally corrected Nyström discretization. *IEEE Trans. Antennas Propagat.* 53 (12): 4111–4120.

14 Rao, S.M., Wilton, D.R., and Glisson, A.W. (1982). Electromagnetic scattering by surfaces of arbitrary shape. *IEEE Trans. Antennas Propagat.* AP-30: 409–418.

15 Sheng, X.Q., Jin, J.M., Song, J.M. et al. (1998). Solution of combined-field integral equation using multilevel fast multipole algorithm for scattering by homogeneous bodies. *IEEE Trans. Antennas Propagat.* 46: 1718–1726.

16 Chen, Q. L. (1990). Electromagnetic modeling of three-dimensional piecewise homogeneous material bodies of arbitrary composition and geometry. Ph.D. Dissertation, University of Houston.

17 Stroud, A.H. (1971). *Approximate Calculation of Multiple Integrals.* Englewood Cliffs: Prentice-Hall.

18 Tong, M.S. and Chew, W.C. (2005). A higher-order Nyström scheme for electromagnetic scattering by arbitrarily shaped surfaces. *IEEE Antennas Wireless Propagat. Lett.* 4: 277–280.

19 Tong, M.S. and Chew, W.C. (2007). Super-hyper singularity treatment for solving 3D electric field integral equations. *Microw. Opt. Technol. Lett.* 49: 1383–1388.

20 Duffy, M.G. (1982). Quadrature over a pyramid or cube of integrands with a singularity at a vertex. *SIAM J. Numer. Anal.* 19: 1260–1262.

21 Graglia, R.D. (1987). Static and dynamic potential integrals for linearly varying source distributions in two- and three-dimensional problems. *IEEE Trans. Antennas Propagat.* 35: 662–669.

22 Graglia, R.D. (1993). On the numerical integration of the linear shape functions times the 3-D Green's function or its Gradient on a plane triangle. *IEEE Trans. Antennas Propagat.* 41: 1448–1455.

23 Järvenpää, S., Taskinen, M., and Ylä-Oijala, P. (2003). Singularity extraction technique for integral equation methods with higher order basis functions on plane triangles and tetrahedra. *Int. J. Numer. Meth. Eng.* 58: 1149–1165.

24 Ruck, G.T., Barrick, D.E., Stuart, W.D., and Krichbaum, C.K. (1970). *Radar Cross Section Handbook.* New York: Plenum Press.

25 Araneo, R. and Lovat, G. (2009). Fast MoM analysis of the shielding effectiveness of rectangular enclosures with apertures, metal plates, and conducting objects. *IEEE Trans. Electromagn. Compat.* 51: 274–283.

26 Nie, X.C., Yuan, N., Li, L.W. et al. (2006). A fast combined field volume integral equation solution to EM scattering by 3-D dielectric objects of arbitrary permittivity and permeability. *IEEE Trans. Antennas Propagat.* 54 (3): 961–969.

27 Sun, L.E. and Chew, W.C. (2009). A novel formulation of the volume integral equation for electromagnetic scattering. *Wave. Random Complex.* 19 (1): 162–180.

28 Chew, W.C., Jin, J.M., Michielssen, E., and Song, J.M. (2001). *Fast and Efficient Algorithms in Computational Electromagnetics.* Boston, MA: Artech House.

29 Schaubert, D.H., Wilton, D.R., and Glisson, A.W. (1984). A tetrahedral modeling method for electromagnetic scattering by arbitrary shaped inhomogeneous dielectric bodies. *IEEE Trans. Antennas Propagat.* AP-32 (1): 77–85.

30 Botha, M.M. (2006). Solving the volume integral equations of electromagnetic scattering. *J. Comput. Phys.* 218 (1): 141–158.

31 Kot, J.S. (1990). Solution of thin-wire integral equations by Nyström methods. *Microw. Opt. Technol. Lett.* 3: 393–396.

32 Burghignoli, P., Di Nallo, C., Frezza, F., and Galli, A. (2003). A simple Nyström approach for the analysis of 3D arbitrarily shaped conducting and dielectric bodies. *Int. J. Numer. Model.* 16: 179–194.

33 Peterson, A.F. and Bibby, M.M. (2003). Three discretizations of the MFIE for the linear dipole. *Proc. IEEE Antennas Propagat.* 3: 6–9.

34 Peterson, A.F. and Bibby, M.M. (2004). Higher-order numerical solutions of the MFIE for the linear dipole. *IEEE Trans. Antennas Propagat.* 52: 2684–2691.

35 Canino, L.S., Ottusch, J.J., Stalzer, M.A. et al. (1998). Numerical solution of the Helmholtz equation in 2D and 3D using a high-order Nyström discretization. *J. Comput. Phys.* 146: 627–663.

36 Gedney, S.D. (2003). On deriving a locally corrected Nyström scheme from a quadrature sampled moment method. *IEEE Trans. Antennas Propagat.* 51: 2402–2412.

37 Liu, G. and Gedney, S.D. (2000). High-order Nyström solution of the volume-EFIE for TM-wave scattering. *Microw. Opt. Technol. Lett.* 25: 8–11.

38 Liu, G. and Gedney, S.D. (2001). High-order Nyström solution of the volume-EFIE for TE-wave scattering. *Electromagnetics.* 21: 1–14.

39 Gedney, S.D. (2000). Application of the high-order Nyström scheme to the integral equation solution of electromagnetic interaction problems. *IEEE Int. Symp. Electromagn. Compat.* 1: 289–294.

40 Silvester, P. (1972). Tetrahedral polynomial finite elements for the Helmholtz equation. *Int. J. Numer. Meth. Eng.* 4 (3): 405–413.

41 Björck, Å. (1996). *Numerical Methods for Least Squares Problems.* Philadelphia, PA: SIAM.

42 Taflove, A. and Hagness, S.C. (2005). *Computational Electrodynamics: The Finite-Difference Time-Domain Method,* 3e. Boston: Artech House.

43 Jin, J.-M. (2002). *The Finite Element Method in Electromagnetics,* 2e. New York: Wiley.

44 Nadobny, J., Wust, P., Seebass, M. et al. (1996). A volume-surface integral equation method for solving Maxwell's equations in electrically inhomogeneous media using tetrahedral grids. *IEEE Trans. Microw. Theory Tech.* 44 (4): 543–554.

45 Nie, X.-C., Yuan, N., Li, L.-W. et al. (2005). A fast volume-surface integral equation solver for scattering from composite conducting-dielectric objects. *IEEE Trans. Antennas Propagat.* 53 (2): 818–824.

46 Usner, B.C., Sertel, K., Carr, M.A., and Volakis, J.L. (2006). Generalized volume-surface integral equation for modeling inhomogeneities within high contrast composite structures. *IEEE Trans. Antennas Propagat.* 54 (1): 68–75.

47 He, M., Chen, Q., Yuan, Q. et al. (2009). A simple strip model in the volume-surface integral equation for analysis of arbitrary probe-fed conformal microstrip antennas. *IEEE Antennas Wireless Propagat. Lett.* 8: 530–533.

48 Wust, P., Nadobny, J., Seebass, M. et al. (1993). 3-D computation of E fields by the volume-surface integral equation (VSIE) method in comparison with the finite-integration theory (FIT) method. *IEEE Trans. Biomed. Eng.* 40 (8): 745–759.

49 Kim, O.S., Meincke, P., Breinbjerg, O., and Jøgensen, E. (2007). Solution of volume-surface integral equations using higher-order hierarchical Legendre basis functions. *Radio Sci.* 42: RS4023.

50 Tong, M.S., Qian, Z.G., and Chew, W.C. (2010). Nyström method solution of volume integral equations for electromagnetic scattering by 3D penetrable objects. *IEEE Trans. Antennas Propagat.* 58 (5): 1645–1652.

51 Kobidze, G. and Shanker, B. (2004). Integral equation based analysis of scattering from 3-D inhomogeneous anisotropic bodies. *IEEE Trans. Antennas Propagat.* 52 (10): 2650–2658.

52 Ozdemir, N.A. and Lee, J.-F. (2006). A nonconformal volume integral equation for electromagnetic scattering from anisotropic materials. *IEEE International Symposium on Antennas and Propagation*, Albuquerque, NM (July 2006), pp. 2889–2892.

53 Botha, M.M. (2007). Fully hierarchical divergence-conforming basis functions on tetrahedral cells, with applications. *Int. J. Numer. Meth. Eng.* 71 (2): 127–148.

54 Khayat, M.A. and Wilton, D.R. (2005). Numerical evaluation of singular and near-singular potential integrals. *IEEE Trans. Antennas Propagat.* 53: 3180–3190.

55 Chew, W.C. (1995). *Waves and Fields in Inhomogeneous Media*. New York: IEEE Press.

56 Volakis, J.L., Chatterjee, A., and Kempel, L.C. (1998). *Finite Element Method Electromagnetics: Antennas, Microwave Circuits, and Scattering Applications*. New York: John Wiley & Sons.

57 Chew, W.C., Chiang, I.T., Davis, C.P. et al. (2006). Integral equation solvers for real world applications–some challenge problems. *IEEE Antennas and Propagation Society International Symposium*, Albuquerque, NM (July 2006), vol. 1, pp. 91–94.

58 Glisson, A.W. (1992). Electromagnetic scattering by arbitrarily shaped surfaces with impedance boundary conditions. *Radio Sci.* 27 (6): 935–943.

59 Hoppe, D.J. and Rahmat-Samii, Y. (1995). *Impedance Boundary Conditions in Electromagnetics*. Bristol, PA: Taylor & Francis.

60 Senior, T.B.A. and Volakis, J.L. (1995). *Approximate Boundary Conditions in Electromagnetics*. London: IEEE Press.

61 Ammari, H. and Nédélec, J.C. (1999). Generalized impedance boundary conditions for the Maxwell's equations as singular perturbation problems. *Commun. Partial Diff. Eq.* 24: 821–850.

62 Marceaux, O. and Stupfel, B. (2000). High-order impedance boundary conditions for multilayer coated 3D objects. *IEEE Trans. Antennas Propagat.* 48 (3): 429–436.

63 Haddar, H., Joly, P., and Nguyen, H.-M. (2008). Generalized impedance boundary conditions for scattering problems from strongly absorbing obstacles: the case of Maxwell's equations. *Math. Models Meth. Appl. Sci.* 18 (10): 1787–1827.

64 Stupfel, B. and Poget, D. (2011). Sufficient uniqueness conditions for the solution of the time harmonic Maxwell's equations associated with surface impedance boundary conditions. *J. Comp. Phys.* 230 (12): 4571–4587.

65 Coperich, K.M., Ruehli, A.E., and Cangellaris, A. (2000). Enhanced skin effect for partial-element equivalent-circuit (PEEC) models. *IEEE Trans. Microw. Theory Tech.* 48 (9): 1435–1442.

66 Chu, Y.H. and Chew, W.C. (2005). A robust surface-integral-equation formulation for conductive media. *Microw. Opt. Technol. Lett.* 46 (2): 109–114.

67 Qian, Z.G. and Chew, W.C. (2007). Generalized impedance boundary condition for conductor modeling in surface integral equation. *IEEE Trans. Antennas Propagat.* 55 (11): 2354–2364.

68 Harrington, R.F. (1993). *Field Computation by Moment Methods*. Piscataway, NJ: IEEE Press.

69 Andriulli, F.P. (2012). Loop-star and loop-tree decompositions: analysis and efficient algorithms. *IEEE Trans. Antennas Propagat.* 60 (5): 2347–2356.

70 Tong, M.S. and Chew, W.C. (2010). A novel approach for evaluating hypersingular and strongly singular surface integrals in electromagnetics. *IEEE Trans. Antennas Propagat.* 58 (11): 3593–3601.

71 Arcioni, P., Bressan, M., and Perregrini, L. (1997). On the evaluation of the double surface integrals arising in the application of the boundary integral method to 3-D problems. *IEEE Trans. Microw. Theory Tech.* 45: 436–439.

72 Bluck, M.J., Pocock, M.D., and Walker, S.P. (1997). An accurate method for the calculation of singular integrals arising in time-domain integral equation analysis of electromagnetic scattering. *IEEE Trans. Antennas Propagat.* 45 (12): 1793–1798.

73 Rossi, L. and Cullen, P.J. (1999). On the fully numerical evaluation of the linear-shape function times the 3-D Green's function on a plane triangle. *IEEE Trans. Microw. Theory Tech.* 47: 398–402.

74 Herschlein, A., Hagen, J.V., and Wiesbeck, W. (2002). Methods for the evaluation of regular, weakly singular and strongly singular surface reaction integrals arising in method of moments. *ACES J.* 17 (1): 63–73.

75 Cai, W., Yu, Y., and Yuan, X.C. (2002). Singularity treatment and high-order RWG basis functions for integral equations of electromagnetic scattering. *Int. J. Numer. Meth. Eng.* 53: 31–47.

76 Järvenpää, S., Taskinen, M., and Ylä-Oijala, P. (2003). Singularity extraction technique for integral equation methods with higher order basis functions on plane triangles and tetrahedra. *Int. J. Numer. Meth. Eng.* 58: 1149–1165.

77 Polimeridis, A.G. and Yioultsis, T.V. (2008). On the direct evaluation of weakly singular integrals in Galerkin mixed potential integral equation formulations. *IEEE Trans. Antennas Propagat.* 56 (9): 3011–3019.

78 Board, J. and Schulten, K. (2000). The fast multipole algorithm. *IEEE Comput. Sci. Eng.* 2 (1): 76–79.

79 Ruck, G.T., Barrick, D.E., Stuart, W.D., and Krichbaum, C.K. (1970). *Radar Cross Section Handbook*. New York: Plenum Press.

80 Krakowski, M. (1982). Eddy-current losses in thin circular and rectangular plates. *Archiv für Elekrotechnik.* 64: 307–311.
81 Ulrich, R.K. and Brown, W.D. (2006). *Advanced Electronic Packaging,* 2e. New York: IEEE Press.
82 Li, M.K. and Chew, W.C. (2008). Multiscale simulation of complex structures using equivalence principle algorithm with high-order field point sampling scheme. *IEEE Trans. Antennas Propagat.* 56 (8): 2389–2397.
83 Qian, Z.-G. and Chew, W.C. (2009). Fast full-wave surface integral equation solver for multiscale structure modeling. *IEEE Trans. Antennas Propagat.* 57 (11): 3594–3601.
84 Jiang, L., Rubin, B.J., Liu, Y. et al. (2007). Electromagnetic simulation for inhomogeneous interconnect and packaging structures. *Proceedings of 16th IEEE Topical Meeting on Electrical Performance of Electronic Packaging,* Atlanta, Georgia (October 2007), pp. 357–360.
85 Li, E.-P., Wei, X.-C., Cangellaris, A.C. et al. (2010). Progress review of electromagnetic compatibility analysis technologies for packages, printed circuit boards and novel interconnects. *IEEE Trans. Electromagn. Compat.* 52 (2): 248–265.
86 Okhmatovski, V.I., Morsey, J.D., and Cangellaris, A.C. (2004). Loop-tree implementation of the adaptive integral method (AIM) for numerically-stable, broadband, fast electromagnetic modeling. *IEEE Trans. Antennas Propagat.* 52 (8): 2130–2140.
87 Okhmatovski, V.I., Morsey, J.D., and Cangellaris, A.C. (2004). Enhancement of the numerical stability of the adaptive integral method at low frequencies through a loop-charge formulation of the method-of-moments approximation. *IEEE Trans. Microw. Theory Tech.* 52 (3): 962–970.
88 Wu, H. and Cangellaris, A.C. (2008). A finite-element domain-decomposition methodology for electromagnetic modeling of multilayer of high-speed interconnects. *IEEE Trans. Adv. Packag.* 31 (2): 339–350.
89 Kollia, V. and Cangellaris, A.C. (2010). A domain decomposition approach for efficient electromagnetic analysis of the power distribution network of packaged electronic systems. *IEEE Trans. Electromagn. Compat.* 52 (2): 320–331.
90 Wang, C., Mao, J., Selli, G. et al. (2006). An efficient approach for power delivery network design with closed-form expressions for parasitic interconnect inductances. *IEEE Trans. Adv. Packag.* 29 (2): 320–334.
91 Choi, M.J. and Cangellaris, A.C. (2002). A quasi three-dimensional distributed electromagnetic model for complex power distribution networks. *IEEE Trans. Adv. Packag.* 25 (1): 28–34.
92 Tong, M.S. (2011). Efficient electromagnetic analysis for interconnect and packaging structures using dual basis function in the method of moments. *IEEE Trans. Compon. Packag. Manufact. Tech.* 1 (7): 1089–1097.
93 Vaupel, T. and Hansen, V. (1999). Electrodynamic analysis of combined microstrip and coplanar/slotline structure with 3-D components based on a surface/volume integral equation approach. *IEEE Trans. Microw. Theory Tech.* 47 (9): 1788–1800.
94 Lu, C.C. and Chew, W.C. (2000). A coupled surface-volume integral equation approach for the calculation of electromagnetic scattering from composite metallic and material targets. *IEEE Trans. Antennas Propagat.* 48 (12): 1866–1868.

95 Ewe, W.-B., Li, L.-W., and Leong, M.-S. (2004). Fast solution of mixed dielectric/conducting scattering problem using volume-surface adaptive integral method. *IEEE Trans. Antennas Propagat.* 52 (11): 3071–3077.

96 Usner, B.C., Sertel, K., and Volakis, J.L. (2007). Doubly periodic volume-surface integral equation formulation for modelling metamaterials. *IET Microw. Antennas Propagat.* 1 (1): 150–157.

8

Incorporation with Multilevel Fast Multipole Algorithm

8.1 Introduction

Although the integral equation method (IEM) has many advantages, it also has an obvious disadvantage, i.e. it results in dense impedance matrices [1], and one has to use fast algorithms to accelerate the solution process for large or complicated problems [2]. The traditional method of moments (MoM) requires a cost of $\mathcal{O}(N^2)$ in both memory usage and CPU time for iterative solutions, where N is the number of unknowns, and this is prohibitive for general computational resources when N is large. In recent decades, however, several fast algorithms have been developed for solving large electromagnetics (EM) problems, such as the conjugate gradient fast Fourier transform method [3], the fast multipole method (FMM) [4], the adaptive integral method [5], the precorrected fast Fourier transform method [6], etc. The FMM was first proposed by Rokhlin for rapidly calculating Coulombic potentials in large-scale particle systems in 1985 [7] and was thought of as one of the most significant and influential algorithms developed in the 20th century for science and engineering [8]. Later on, it was generalized by Greengard [9] to seek fast numerical solutions for 2D and 3D Laplace equations in particle simulations. This method was mainly applied to particle simulations in astrophysics and molecular dynamics at the beginning [10, 11] and was extended to an acoustic wave problem that follows from the Helmholtz equation [12]. In EM, the method was also studied extensively, and was extended to a multilevel scheme for solving large EM problems, known as the multilevel fast multipole algorithm (MLFMA), by Lu and Chew in 1994 [13]. The MLFMA has been rapidly developed and has allowed solving extremely large problems with parallel computing techniques in recent years [14–17].

The most distinctive feature or elegance of MLFMA resides in the fast evaluation for matrix-vector multiply in solving dense matrix equations by iterative methods. Also, the MLFMA does not store matrix elements explicitly except for the near-interaction elements, leading to a matrix-free scheme with dramatic reduction on memory requirement. The matrix-vector multiply is the bottleneck of limiting the speed in the iterative solution of dense matrix equations and requires $\mathcal{O}(N^2)$ computational costs. The MLFMA decomposes the matrix entries representing far-interactions between a source point and a field point into radiation pattern, translator and receiving pattern based on the addition theorem for the Green's function and performs the matrix-vector multiply through these three stages. The radiation pattern, translator and receiving pattern can be constructed in advance and stored with much less memory. With the

The Nyström Method in Electromagnetics, First Edition. Mei Song Tong and Weng Cho Chew.

use of the multilevel tree structure, the diagonalization of the translation operator, and the interpolation or anterpolation for wave expansions at different levels, the MLFMA can reduce the original $\mathcal{O}(N^2)$ complexity, both in CPU time and in memory usage, to $\mathcal{O}(N \log N)$ complexity. This is a significant reduction in computational costs when N is very large [2], making it possible to solve very large problems on common computers.

However, MLFMA implementation is usually based on the MoM [18] discretization of integral equations. With the robust Rao-Wilton-Glisson (RWG) basis function [19] or Schaubert-Wilton-Glisson (SWG) basis function [20], the MoM has been the most widely used tool in transforming surface integral equations (SIEs) or volume integral equations (VIEs) into matrix equations. Nevertheless, the MoM may not be the best choice in some applications and the Nyström method could be a good alternative to the MoM, as demonstrated in the preceding chapters. We use the Nyström method to discretize integral equations and incorporate it with the MLFMA to solve for EM problems with electrically large objects. The Nyström method has been increasingly used to solve EM integral equations due to its unique features and with the help of efficient local correction schemes developed in recent years [21–29]. When combined with the MLFMA, its well-known merits appear to be more significant because the simple mechanism of implementation results in a fast setup, the flexibility on meshes removes the need of remeshing large and complex structures, and the removal of basis and testing functions avoid the embarrassment of selecting two basis functions to represent two current densities at the boundaries of penetrable media in the SIEs. Furthermore, the Nyström method can select volume current densities instead of volume flux densities as unknowns to be solved in the VIEs, making integral kernels free of material parameters. This is particularly desirable when incorporating with the MLFMA because the MLFMA structure can be kept unchanged for different materials.

This chapter starts with a brief introduction to the MLFMA and then presents the details of its incorporation with the Nyström method discretization for different integral equations with different property objects. Particular attention is drawn to the difference between the MoM-based MLFMA and Nyström-based MLFMA. Typical numerical examples of solving electrically large problems for each case are provided to demonstrate the effectiveness and robustness of the Nyström-based MLFMA and good results have been observed.

8.2 Multilevel Fast Multipole Algorithm

To accelerate the solution process of a dense matrix equation, the MLFMA performs the matrix-vector multiply, which is a bottleneck in the iterative solution of dense matrix equations, through three steps, i.e. aggregation, translation and disaggregation for the far-interactions between the grouped source points and field points. Consider the following typical matrix equation resulting from EM integral equations

$$\sum_{i=1}^{N} A_{ji} a_i = b_j, \quad j = 1, \cdots, N \tag{8.1}$$

where N is the total number of unknowns after discretizing the integral equations, b_j is the known wave excitation, a_i is the unknown coefficient to be solved, and A_{ji} is the matrix element. For EM integral equations, the matrix element A_{ji} is related to the scalar Green's function, which can be expanded as, after applying the addition theorem [2],

$$\frac{e^{ikr_{ji}}}{r_{ji}} = \int d^2\hat{k} e^{i\mathbf{k}\cdot(\mathbf{r}_{jm}-\mathbf{r}_{im'})} \alpha_{mm'}(\mathbf{k}, \mathbf{r}_{mm'}) \tag{8.2}$$

where

$$\alpha_{mm'}(\mathbf{k}, \mathbf{r}_{mm'}) = \frac{ik}{4\pi} \sum_{l=0}^{L} i^l (2l+1) h_l^{(1)}(kr_{mm'}) P_l(\hat{r}_{mm'} \cdot \hat{k}) \tag{8.3}$$

is the translator in which $h_l^{(1)}$ is the spherical Hankel function of the first kind of order l, P_l is the Legendre polynomial of the first kind of order l, and L is the number of series terms after truncation. Also, $\mathbf{r}_m$ and $\mathbf{r}_{m'}$ represent the centers of the mth and m'th groups in which the field point $\mathbf{r}_j$ and source point $\mathbf{r}_i$ reside, respectively, $\mathbf{r}_{jm}$ denotes the distance vector from the field point to its group center, $\mathbf{r}_{im'}$ denotes the distance vector from the source point to its group center, and $\mathbf{r}_{mm'}$ denotes the distance vector from the field group center to the source group center. Usually the integral kernels in EM integral equations include the gradient operation on the scalar Green's function. This gradient operation ∇ in space domain is replaced with $i\mathbf{k}$ in the $\hat{k}$-domain.

In the context of Nyström discretization for SIEs, the number of unknowns is $N = N_t P$, where N_t is the number of triangular patches and P is the number of quadrature points on each patch. The observation point (or field point) and source point are both chosen as the quadrature points, so i or $j = P(n-1) + p$ represents the pth quadrature point on the nth patch and this is understood throughout. Based on (8.2), the field (one component) on the jth field point, which is contributed from the current density (one component) over the nth source patch, can be calculated through

$$F_{jn} = \int d^2\hat{k} V_{fmj}(\hat{k}) \cdot \alpha_{mm'}(\mathbf{k}, \mathbf{r}_{mm'}) V_{sm'n}(\hat{k}) \tag{8.4}$$

where

$$V_{fmj}(\hat{k}) = \frac{1}{4\pi} e^{i\mathbf{k}\cdot\mathbf{r}_{jm}} \tag{8.5}$$

is the receiving pattern at the jth field point and

$$\begin{aligned} V_{sm'n}(\hat{k}) &= \int_{\Delta S_n} dS' e^{-i\mathbf{k}\cdot\mathbf{r}_{sm'}} f_c(\hat{k}) J_s \\ &= \sum_{p=1}^{P} e^{-i\mathbf{k}\cdot\mathbf{r}_{im'}} f_c(\hat{k}) w_i J_i \\ &= \sum_{p=1}^{P} V_{sm'i}(\hat{k}) J_i. \end{aligned} \tag{8.6}$$

is the radiation pattern related to the ith source point. In the above, the integration is with respect to the position variable $\mathbf{r}_s$ of a source point within the nth patch and J_s is a current component on that point. This integration is replaced with a summation under a quadrature rule, and w_i and J_i are the weight and current value on the ith quadrature point, respectively. Also, $f_c(\hat{k})$ is a component of $\mathbf{f}(\hat{k})$ that denotes the $\hat{k}$-domain representation of the gradient operation acting on the scalar Green's function. The above

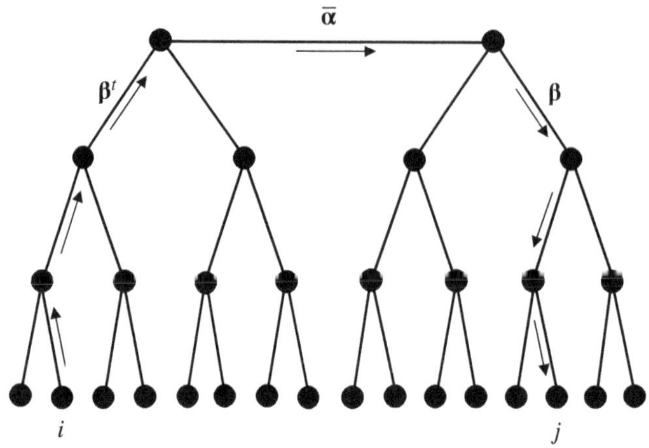

Figure 8.1 Inverted tree structure for calculating the matrix-vector multiply of far-interaction terms in the MLFMA.

illustrates the relation between the field and current source for one component and the matrix element can be written as

$$A_{ji} = \int \mathrm{d}^2\hat{k} V_{fmj}(\hat{k}) \cdot \alpha_{mm'}(\mathbf{k}, \mathbf{r}_{mm'}) V_{sm'i}(\hat{k}) \tag{8.7}$$

while $a_i = J_i$ in (8.1).

The implementation of MLFMA can be illustrated in principle by using an inverted tree structure as shown in Figure 8.1. For a multilevel scheme, the outgoing wave expansions in the radiation pattern are calculated at the finest level and then the expansions for higher levels are obtained by interpolation. Similarly, the anterpolation or transpose interpolation is used to get the incoming wave expansion at a lower level from that at its parent level in the receiving pattern. The interpolation or anterpolation is recognized as one of the key techniques for reducing costs in the MLFMA.

8.3 Surface Integral Equations for Conducting Objects

As we have discussed in the previous chapters, the SIEs for conducting media include the electric field integral equation (EFIE), the magnetic field integral equation (MFIE), and the combined field integral equation (CFIE), and they include either a $\mathcal{L}$ operator or $\mathcal{K}$ operator, or both, as integral kernels. In Nyström discretization, there are no particularly-designed basis and testing functions and the relationship between the operators and unknown current density is algebraic. The operators are calculated as a matrix or vector and stored individually, making the incorporation with the MLFMA much easier. Since the MLFMA only works for far-interaction matrix elements, the near terms including singular terms are the same as those before the acceleration of MLFMA and are calculated and stored in advance.

8.3.1 Integral Equations

Although the integral equations have been presented in the previous chapters, we rewrite them in an operator form here for convenience. Consider the EM scattering by a conducting object embedded in the free space with permittivity ϵ_0 and permeability μ_0. The object has a boundary or surface S and an outward unit normal vector $\hat{n}$. The incident wave with an electric field $\mathbf{E}^{\mathrm{inc}}$ and magnetic field $\mathbf{H}^{\mathrm{inc}}$ is impinging upon the object and there is an unknown surface current density $\mathbf{J}$ induced on the surface of object. The corresponding EFIE and MFIE can be written as

$$\hat{n} \times \mathcal{L}\left[\mathbf{J}(\mathbf{r}')\right] = \hat{n} \times \mathbf{E}^{\mathrm{inc}}(\mathbf{r}), \quad \mathbf{r} \in S \tag{8.8}$$

$$\frac{1}{2}\mathbf{J}(\mathbf{r}) + \hat{n} \times \mathcal{K}\left[\mathbf{J}(\mathbf{r}')\right] = \hat{n} \times \mathbf{H}^{\mathrm{inc}}(\mathbf{r}), \quad \mathbf{r} \in S \tag{8.9}$$

respectively, where the integral operators $\mathcal{L}$ and $\mathcal{K}$ are defined as

$$\mathcal{L}\left[\mathbf{J}(\mathbf{r}')\right] = -\mathrm{i}k_0\eta_0 \int_S \overline{\mathbf{G}}(\mathbf{r}, \mathbf{r}') \cdot \mathbf{J}(\mathbf{r}')\, \mathrm{d}S' \tag{8.10}$$

$$\mathcal{K}\left[\mathbf{J}(\mathbf{r}')\right] = -\fint_S \nabla g(\mathbf{r}, \mathbf{r}') \times \mathbf{J}(\mathbf{r}')\, \mathrm{d}S'. \tag{8.11}$$

In the above, $k_0 = \omega\sqrt{\epsilon_0\mu_0}$ is the wavenumber and $\eta_0 = \sqrt{\mu_0/\epsilon_0}$ is the wave impedance of free space, and ω is the frequency of the incident wave. Also,

$$\overline{\mathbf{G}}(\mathbf{r}, \mathbf{r}') = \left(\overline{\mathbf{I}} + \frac{\nabla\nabla}{k_0^2}\right) g(\mathbf{r}, \mathbf{r}') \tag{8.12}$$

is the dyadic Green's function in which $\overline{\mathbf{I}}$ is the identity dyad, $g(\mathbf{r}, \mathbf{r}') = \mathrm{e}^{\mathrm{i}k_0R}/(4\pi R)$ is the scalar Green's function, and $R = |\mathbf{r} - \mathbf{r}'|$ is the distance between an observation point $\mathbf{r}$ and a source point $\mathbf{r}'$. Combining the EFIE and MFIE with an appropriate weight, we have the following CFIE [1]

$$\begin{aligned} &\alpha\hat{n} \times \hat{n} \times \mathcal{L}\left[\mathbf{J}(\mathbf{r}')\right] + (1-\alpha)\eta_0 \left\{\frac{1}{2}\mathbf{J}(\mathbf{r}) + \hat{n} \times \mathcal{K}\left[\mathbf{J}(\mathbf{r}')\right]\right\} \\ &= \alpha\hat{n} \times \hat{n} \times \mathbf{E}^{\mathrm{inc}}(\mathbf{r}) + (1-\alpha)\eta_0\hat{n} \times \mathbf{H}^{\mathrm{inc}}(\mathbf{r}), \ \mathbf{r} \in S \end{aligned} \tag{8.13}$$

where α is named the combination parameter and is generally selected between 0.2 and 0.8.

8.3.2 Nyström Discretization and MLFMA Acceleration

When the surface of the object is discretized into N triangular patches and there are Q quadrature points in each patch, the EFIE can be discretized into

$$\begin{aligned} &\sum_{n=1}^{N} \mathrm{i}\omega\mu_0 \int_{\Delta S_n} \mathrm{d}S' \left[\hat{t}_{mp}^{(l)} \cdot \overline{\mathbf{G}}(\mathbf{r}_{mp}, \mathbf{r}'_{nq}) \cdot \mathbf{J}(\mathbf{r}')\right] = -\hat{t}_{mp}^{(l)} \cdot \mathbf{E}^{\mathrm{inc}}(\mathbf{r}_{mp}) \\ &m = 1, 2, \cdots, N; \quad p = 1, 2, \cdots, Q \end{aligned} \tag{8.14}$$

while the MFIE can be discretized into

$$\sum_{n=1}^{N} \int_{\Delta S_n} \mathrm{d}S' \left[\hat{t}_{mp}^{(l)} \cdot \hat{n}_{mp} \times \nabla g(\mathbf{r}_{mp}, \mathbf{r}') \times \mathbf{J}(\mathbf{r}') \right] - \frac{1}{2} \hat{t}_{mp}^{(l)} \cdot \mathbf{J}(\mathbf{r}) = -\hat{t}_{mp}^{(l)} \cdot \hat{n}_{mp} \times \mathbf{H}^{\mathrm{inc}}(\mathbf{r}_{mp})$$
$$m = 1, 2, \cdots, N; \quad p = 1, 2, \cdots, Q \tag{8.15}$$

where $\hat{n}_{mp}$ is a unit normal vector at the pth quadrature point inside the mth observation triangle while $\hat{t}_{mp}^{(l)}$ is a unit tangential vector at that point. We need to take $l = 1, 2$, meaning that two orthogonal tangential vectors are taken at each quadrature point in the observation triangle. Note that the current density is a surface vector and there are only two independent components (J^u, J^v) in a local coordinate system $\{u, v, w\}$ established over a source triangle. The three components of current in the global coordinate system $\{x, y, z\}$ can be obtained by the following transformation

$$\begin{bmatrix} J_x \\ J_y \\ J_z \end{bmatrix} = \overline{\boldsymbol{\beta}}_n \cdot \begin{bmatrix} J^u \\ J^v \\ 0 \end{bmatrix} = \begin{bmatrix} \beta_x(n,1) & \beta_x(n,2) \\ \beta_y(n,1) & \beta_y(n,2) \\ \beta_z(n,1) & \beta_z(n,2) \end{bmatrix} \begin{bmatrix} J^u \\ J^v \end{bmatrix} \tag{8.16}$$

where

$$\overline{\boldsymbol{\beta}}_n = \begin{bmatrix} \beta_x(n,1) & \beta_x(n,2) & \beta_x(n,3) \\ \beta_y(n,1) & \beta_y(n,2) & \beta_y(n,3) \\ \beta_z(n,1) & \beta_z(n,2) & \beta_z(n,3) \end{bmatrix} \tag{8.17}$$

is a transformation matrix between the local coordinate system and the global coordinate system for the nth patch. Here, $\left[\beta_x(n,k), \beta_y(n,k), \beta_z(n,k)\right]$ $(k = 1, 2, 3)$ are the direction cosines of describing the kth axis of the local coordinate system defined over the nth triangular patch in the global coordinate system. The transformation is necessary because the MLFMA frame is usually created in a global coordinate system.

When the Nyström scheme is applied to the integral over the nth source triangle ΔS_n, the above EFIE can be written as

$$\sum_{n=1}^{N} \sum_{q=1}^{Q} \mathrm{i}\omega\mu_0 w_{nq} \left[\hat{t}_{mp}^{(l)} \cdot \overline{\mathbf{G}}(\mathbf{r}_{mp}, \mathbf{r}'_{nq}) \cdot \mathbf{J}(\mathbf{r}'_{nq}) \right] = -\hat{t}_{mp}^{(l)} \cdot \mathbf{E}^{\mathrm{inc}}(\mathbf{r}_{mp})$$
$$m = 1, 2, \cdots, N; \quad p = 1, 2, \cdots, Q \tag{8.18}$$

and the MFIE can be expressed as

$$\sum_{n=1}^{N} \sum_{q=1}^{Q} w_{nq} \left[\hat{t}_{mp}^{(l)} \cdot \hat{n}_{mp} \times \nabla g(\mathbf{r}_{mp}, \mathbf{r}'_{nq}) \times \mathbf{J}(\mathbf{r}'_{nq}) \right] - \frac{1}{2} \hat{t}_{mp}^{(l)} \cdot \mathbf{J}(\mathbf{r}_{mp}) = -\hat{t}_{mp}^{(l)} \cdot \hat{n}_{mp} \times \mathbf{H}^{\mathrm{inc}}(\mathbf{r}_{mp})$$
$$m = 1, 2, \cdots, N; \quad p = 1, 2, \cdots, Q \tag{8.19}$$

where w_{nq} is the weight of a quadrature rule at the qth quadrature point in the nth triangle and

$$\begin{bmatrix} J_x(\mathbf{r}'_{nq}) \\ J_y(\mathbf{r}'_{nq}) \\ J_z(\mathbf{r}'_{nq}) \end{bmatrix} = \overline{\boldsymbol{\beta}}_n \cdot \begin{bmatrix} J_{nq}^u \\ J_{nq}^v \\ 0 \end{bmatrix}$$

$$= \begin{bmatrix} \beta_x(n,1) & \beta_x(n,2) \\ \beta_y(n,1) & \beta_y(n,2) \\ \beta_z(n,1) & \beta_z(n,2) \end{bmatrix} \begin{bmatrix} J^u_{nq} \\ J^v_{nq} \end{bmatrix}. \tag{8.20}$$

In the above, J^u_{nq} and J^v_{nq} at the qth quadrature point in the nth source triangle are the two independent components of unknown current density to be solved and two groups of EFIE or MFIE will be formed when $l = 1, 2$ is taken.

In the MLFMA acceleration, the matrix element related to $\mathcal{L}$ operator can be found

$$A^{(l)}_{ji} = \int \mathrm{d}^2\hat{k}\mathbf{V}^{(l)}_{fmj}(\hat{k}) \cdot \alpha_{mm'}(\mathbf{k}, \mathbf{r}_{mm'})\mathbf{V}_{sm'i}(\hat{k}) \tag{8.21}$$

where

$$\mathbf{V}^{(l)}_{fmj}(\hat{k}) = \frac{1}{4\pi}\mathrm{e}^{\mathrm{i}\mathbf{k}\cdot\mathbf{r}_{jm}}\hat{t}^{(l)}_j \tag{8.22}$$

$$\mathbf{V}_{sm'i}(\hat{k}) = \mathrm{e}^{-\mathrm{i}\mathbf{k}\cdot\mathbf{r}_{im'}}(\bar{\mathbf{I}} - \hat{k}\hat{k}) \cdot \overline{\boldsymbol{\beta}}_n w_i \tag{8.23}$$

are the receiving pattern and radiation pattern, respectively. In the above, $i = (n-1) + q$ represents the ith source point while $j = (m-1) + p$ represents jth observation or field point. Note that m is used in both the MLFMA and Nyström discretization but it has a different meaning in the two situations. Similarly, the matrix element related to $\mathcal{K}$ operator can be found

$$A^{(l)}_{ji} = \int \mathrm{d}^2\hat{k}\mathbf{V}^{(l)}_{fmj}(\hat{k}) \cdot \alpha_{mm'}(\mathbf{k}, \mathbf{r}_{mm'})\mathbf{V}_{sm'i}(\hat{k}) \tag{8.24}$$

where

$$\mathbf{V}^{(l)}_{fmj}(\hat{k}) = -\frac{\mathrm{i}k}{4\pi}\hat{k} \times \mathrm{e}^{\mathrm{i}\mathbf{k}\cdot\mathbf{r}_{jm}}\hat{t}^{(l)}_j \tag{8.25}$$

$$\mathbf{V}_{sm'i}(\hat{k}) = \mathrm{e}^{-\mathrm{i}\mathbf{k}\cdot\mathbf{r}_{im'}}\overline{\boldsymbol{\beta}}_n w_i. \tag{8.26}$$

If the CFIE is solved with the MLFMA acceleration, then the above decompositions of matrix elements related to the $\mathcal{L}$ operator and $\mathcal{K}$ operator should be combined together with the aforementioned manner.

8.3.3 Numerical Examples

We present two numerical examples to illustrate the Nyström-based MLFMA for solving electrically large problems with conducting objects. The first example is the EM scattering by a conducting sphere and its geometry is shown in Figure 8.2. The sphere is assumed to be a perfectly electric conductor (PEC) and has a radius $a = 10.0\lambda$. The incident plane wave with a frequency $f = 300$ MHz is propagating along the $-z$ direction and impinges upon the object to induce a current density on its surface in both vertical and horizontal polarizations. The sphere surface is discretized into 415628 triangular patches and we calculate the bistatic radar cross section (RCS) for the sphere. Figure 8.3 plots the solution when the incident wave is in vertical polarization while Figure 8.4 sketches the solution when the incident wave is in horizontal polarization. It can be seen that they are in good agreement with the corresponding exact or Mie-series solutions. Note that the former figure is obtained by solving the EFIE while the latter figure is obtained by solving the MFIE with MLFMA acceleration.

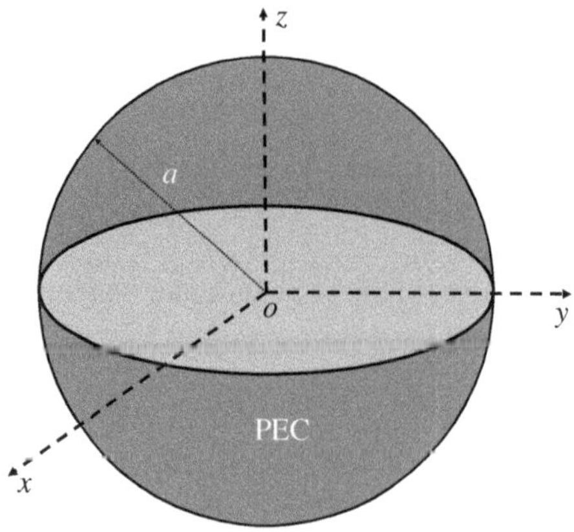

Figure 8.2 Geometry of a conducting sphere.

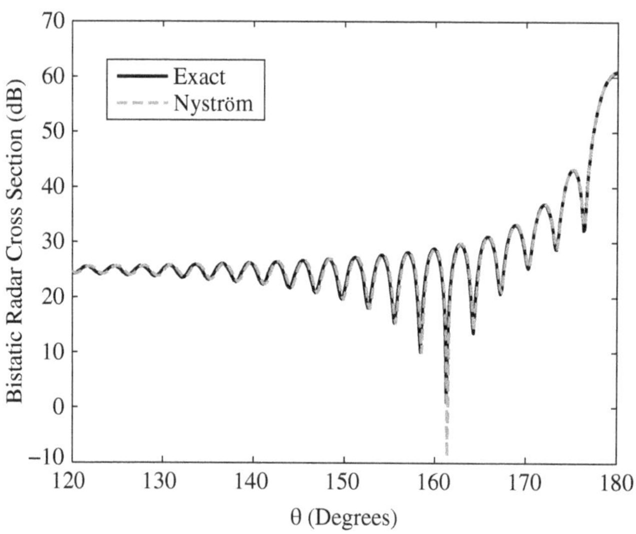

Figure 8.3 Bistatic RCS solutions of a conducting sphere with a radius $a = 10.0\lambda$. The incident plane wave is in vertical polarization (VV) and is propagating along the $-z$ direction.

The second example is the EM scattering by a real-world object, i.e. a conducting airplane wing whose geometry is shown in Figure 8.5. The profile of the wing is defined by a length $a = 2.541\lambda$, a longer width $b_1 = 1.723\lambda$, a shorter width $b_2 = 1.416\lambda$, and a height $c = 0.061\lambda$, and its surface is discretized into 21752. The incident plane wave with a frequency $f = 300$ MHz is propagating along the $+x$ direction (the x axis is the central line from the left side to right side) to impinge upon the wing when it has vertical polarization while it is propagating along the $-y$ direction (the y axis is the central line from the bottom side to top side) when it has horizontal polarization. Figure 8.6 illustrates its bistatic RCS solution when the incident wave is in vertical polarization while 8.7 depicts its bistatic RCS solution when the incident wave is in horizontal polarization. We use

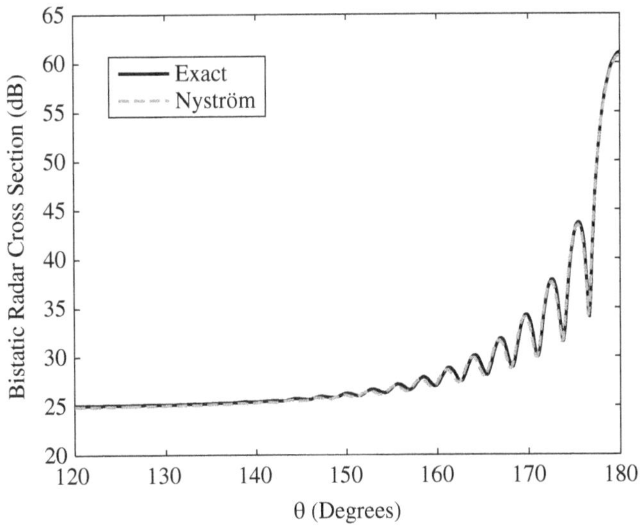

Figure 8.4 Bistatic RCS solutions of a conducting sphere with a radius $a = 10.0\lambda$. The incident plane wave is in horizontal polarization (HH) and is propagating along the $+z$ direction.

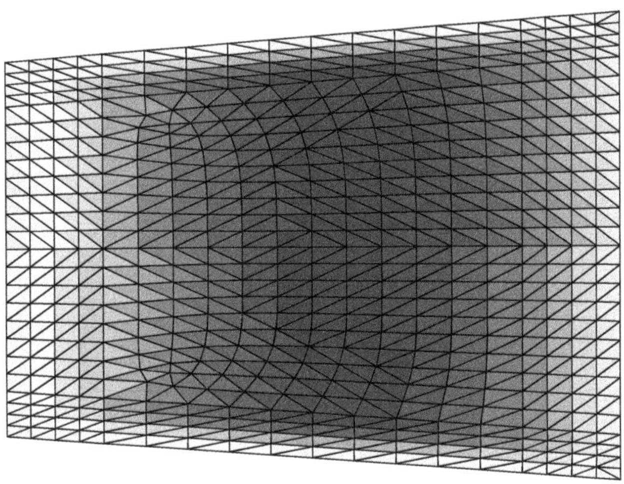

Figure 8.5 Geometry of a conducting airplane wing.

the CFIE to solve the problem with the MLFMA acceleration and the solutions are close to the corresponding results obtained from the MoM-based MLFMA.

8.4 Surface Integral Equations for Penetrable Objects

As we have seen in the preceding chapters, the SIEs are preferred for homogeneous and isotropic penetrable media although VIEs can also be used to formulate the same problems. The SIEs for penetrable media include EFIEs, MFIEs, CFIEs,

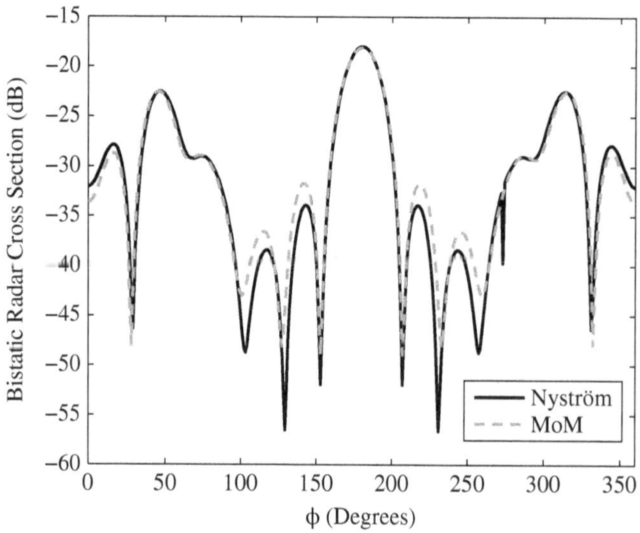

Figure 8.6 Bistatic RCS solutions of a conducting airplane wing with a length $a = 2.541\lambda$, a longer width $b_1 = 1.723\lambda$, a shorter width $b_2 = 1.416\lambda$, and a height $c = 0.061\lambda$. The incident plane wave is in a vertical polarization (VV) and is propagating along the $+x$ direction.

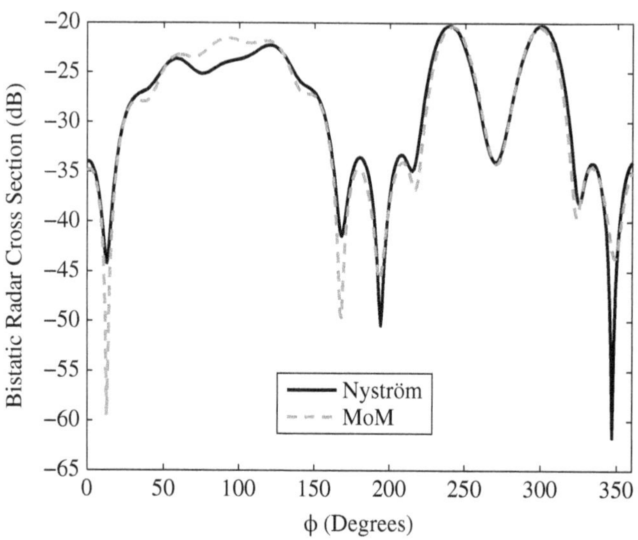

Figure 8.7 Bistatic RCS solutions of a conducting airplane wing with a length $a = 2.541\lambda$, a longer width $b_1 = 1.723\lambda$, a shorter width $b_2 = 1.416\lambda$, and a height $c = 0.061\lambda$. The incident plane wave is in vertical polarization (VV) and is propagating along the $-y$ direction.

the Poggio-Miller-Chang-Harrington-Wu-Tsai (PMCHWT) formulation, and the Müller formulation. Although the PMCHWT formulation is the most widely used, it may not be suitable for radiation problems because it requires an excitation of magnetic field [30] and also it may not be appropriate for highly conductive media because some entries are always dominant to others in its resulting system matrix [31]. In addition, the Müller formulation is only good for low-contrast media and has not been very widely used [1]. The EFIEs represent the simplest form of integral equations and also the MFIEs are fully dual to them for penetrable media, so we take the EFIEs as an example to illustrate the incorporation with the MLFMA [32]. The incorporation of MFIEs with the MLFMA and thus the incorporation of CFIEs or other combinations (PMCHWT formulation and Müller formulation) with the MLFMA can be easily extended. For the EFIEs, the traditional MoM will meet the embarrassment of selecting two basis functions for two current densities at the material interfaces and the Nyström method may be a particularly good alternative to the MoM in this scenario. Furthermore, the unique features of the Nyström method make it be much easier to incorporate with the MLFMA.

8.4.1 Integral Equations

EM scattering by penetrable objects can be described by VIEs or SIEs if the inhomogeneity of materials can be neglected. The SIEs are preferred whenever available because one only needs to discretize the interfaces of materials with much less unknowns when solving them. For the scattering by a single three-dimensional (3D) penetrable body as shown in Figure 8.8(a), the SIEs can be written as

$$\hat{n} \times \mathbf{E}_2^{S^+}(\mathbf{J}, \mathbf{M}) = -\hat{n} \times \mathbf{E}_{\text{inc}}^{S^+} \tag{8.27}$$

$$\hat{n} \times \mathbf{E}_1^{S^-}(\mathbf{J}, \mathbf{M}) = 0 \tag{8.28}$$

which is the EFIE or

$$\hat{n} \times \mathbf{H}_2^{S^+}(\mathbf{J}, \mathbf{M}) = -\hat{n} \times \mathbf{H}_{\text{inc}}^{S^+} \tag{8.29}$$

$$\hat{n} \times \mathbf{H}_1^{S^-}(\mathbf{J}, \mathbf{M}) = 0 \tag{8.30}$$

which is the MFIE. In the above, $\hat{n}$ is the unit normal vector on the surface, **E** and **H** are the electric and magnetic field, respectively, **J** and **M** are the equivalent electric and magnetic current densities on the boundary S, respectively, the subscript "*inc*" represents an incident field, and "+" and "−" on the interface S indicate its interior side and exterior side, respectively. We have used Region 1 and Region 2 to denote the interior and exterior of the object, respectively, and all variables with the subscript 1 or 2 are related to the corresponding regions. The SIEs are derived by decomposing the original problem into the equivalent external problem and equivalent internal problem as illustrated in Figure 8.8(b) and Figure 8.8(c), respectively. The electric field and magnetic field can be related to the current sources by a $\mathcal{L}$ or $\mathcal{K}$ operator, namely,

$$\begin{aligned}\mathbf{E}(\mathbf{J}, 0) &= -\mathrm{i}\omega\mu \int_S \left[\bar{\mathbf{I}} + \frac{\nabla\nabla}{k^2}\right] g(\mathbf{r}, \mathbf{r}') \cdot \mathbf{J}(\mathbf{r}')\, \mathrm{d}S' \\ &= \mathcal{L}(\mathbf{J})\end{aligned} \tag{8.31}$$

$$\begin{aligned}\mathbf{E}(0, \mathbf{M}) &= -\int_S \nabla g(\mathbf{r}, \mathbf{r}') \times \mathbf{M}(\mathbf{r}')\, \mathrm{d}S' \\ &= \mathcal{K}(\mathbf{M})\end{aligned} \tag{8.32}$$

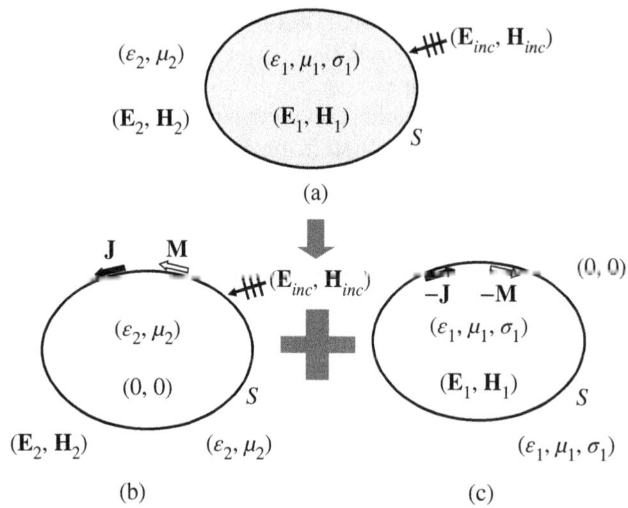

Figure 8.8 Scattering by a single penetrable body. (a) Original problem. (b) Equivalent external problem. (c) Equivalent internal problem.

$$\mathbf{H}(0, \mathbf{M}) = -\frac{ik}{\eta}\mathcal{L}(\mathbf{M}) \tag{8.33}$$

$$\mathbf{H}(\mathbf{J}, 0) = -\mathcal{K}(\mathbf{J}) \tag{8.34}$$

where $\bar{\mathbf{I}}$ is an identity dyad and $g(\mathbf{r}, \mathbf{r}') = e^{ikR}/(4\pi R)$ is the 3D scalar Green's function in which $R = |\mathbf{r} - \mathbf{r}'|$ is the distance between a field point $\mathbf{r}$ and a source point $\mathbf{r}'$. Also, ϵ, μ, η and k are the permittivity, permeability, wave impedance, and wavenumber of related medium, respectively.

For a more general case, i.e. a penetrable body with L-layer full penetrable coatings, we have $L + 1$ interfaces where the equivalent electric and magnetic current densities $(\mathbf{J}_\ell, \mathbf{M}_\ell)$ $(\ell = 1, 2, \cdots, L+1)$ reside. The corresponding EFIEs can be written as

$$\hat{n} \times \mathbf{E}_1^{S_1^-}(\mathbf{J}_1, \mathbf{M}_1) = 0 \tag{8.35}$$

$$\vdots$$

$$\hat{n} \times \left[\mathbf{E}_\ell^{S_{\ell-1}^+}(\mathbf{J}_{\ell-1}, \mathbf{M}_{\ell-1}) - \mathbf{E}_\ell^{S_{\ell-1}^+}(\mathbf{J}_\ell, \mathbf{M}_\ell)\right] = 0 \tag{8.36}$$

$$\hat{n} \times \left[\mathbf{E}_\ell^{S_\ell^-}(\mathbf{J}_{\ell-1}, \mathbf{M}_{\ell-1}) - \mathbf{E}_\ell^{S_\ell^-}(\mathbf{J}_\ell, \mathbf{M}_\ell)\right] = 0 \tag{8.37}$$

$$\vdots$$

$$\hat{n} \times \mathbf{E}_{L+2}^{S_{L+1}^+}(\mathbf{J}_{L+1}, \mathbf{M}_{L+1}) = -\hat{n} \times \mathbf{E}_{\text{inc}}^{S_{L+1}^+} \tag{8.38}$$

$$\ell = 2, 3, \cdots, L+1. \tag{8.39}$$

If the scatterer is a fully or partially coated (one layer) PEC, then the original fields are generated not only by the equivalent current densities on the coating surface but also by the conduction current density on the conductor surface. The corresponding EFIEs are modified as

$$\hat{n} \times \left[\mathbf{E}_1^{S_1^+}(\mathbf{J}_1, 0) - \mathbf{E}_1^{S_1^+}(\mathbf{J}_2, \mathbf{M}_2)\right] = 0 \tag{8.40}$$

$$\hat{n} \times \left[\mathbf{E}_1^{S_2^-}(\mathbf{J}_1, 0) - \mathbf{E}_1^{S_2^-}(\mathbf{J}_2, \mathbf{M}_2) \right] = 0 \tag{8.41}$$

$$\hat{n} \times \mathbf{E}_2^{S_2^+}(\mathbf{J}_2, \mathbf{M}_2) = -\hat{n} \times \mathbf{E}_{\text{inc}}^{S_2^+} \tag{8.42}$$

for a fully coated conductor, and

$$\hat{n} \times \left[\mathbf{E}_2^{S_1^+}(\mathbf{J}_1, 0) - \mathbf{E}_2^{S_1^+}(\mathbf{J}_2, \mathbf{M}_2) \right] = -\hat{n} \times \mathbf{E}_{\text{inc}}^{S_1^+} \tag{8.43}$$

$$\hat{n} \times \left[\mathbf{E}_2^{S_2^+}(\mathbf{J}_1, 0) - \hat{n} \times \mathbf{E}_2^{S_2^+}(\mathbf{J}_2, \mathbf{M}_2) \right] = -\hat{n} \times \mathbf{E}_{\text{inc}}^{S_2^+} \tag{8.44}$$

$$\hat{n} \times \mathbf{E}_1^{S_2^-}(\mathbf{J}_2, \mathbf{M}_2) = 0 \tag{8.45}$$

for a partially coated conductor, where $\mathbf{J}_2$ and $\mathbf{M}_2$ are the equivalent electric and magnetic current densities on the coating surface S_2, respectively, and $\mathbf{J}_1$ is the conduction current density on the conductor surface S_1. The corresponding MFIEs can be obtained by replacing $\mathbf{E}$ with $\mathbf{H}$ in the above. We can also obtain the CFIEs, PMCHWT equations, and Müller equations through different combinations of the EFIEs and MFIEs [1]. Since these equations are usually thought of as more stable and thus have been widely solved, we do not solve them here and we only solve the EFIEs and MFIEs, which are less frequently solved. The EFIEs and MFIEs are less expensive although they may have an internal resonance problem. Also, the solutions for these two equations could be a good test for the robustness of the used numerical approaches because they are less stable and also sensitive to the basis function in the MoM.

8.4.2 MLFMA Acceleration

For the previous EFIEs and MFIEs, the fields are generated through both $\mathcal{L}$ operator and $\mathcal{K}$ operator. After taking the tangential components by dot multiplying a unit tangential vector $\mathbf{t}(\mathbf{r})$ at a field point $\mathbf{r}$, the two operators can be written as

$$\mathbf{t}(\mathbf{r}) \cdot \mathcal{L}(\mathbf{C}) = \mathbf{t}(\mathbf{r}) \cdot \int_S \left[\bar{\mathbf{I}} + \frac{\nabla\nabla}{k^2} \right] g(\mathbf{r}, \mathbf{r}') \cdot \mathbf{C}(\mathbf{r}') \, \mathrm{d}S' \tag{8.46}$$

$$\begin{aligned} \mathbf{t}(\mathbf{r}) \cdot \mathcal{K}(\mathbf{C}) &= \mathbf{t}(\mathbf{r}) \cdot \int_S \nabla g(\mathbf{r}, \mathbf{r}') \times \mathbf{C}(\mathbf{r}') \, \mathrm{d}S' \\ &= -\nabla \times \mathbf{t}(\mathbf{r}) \cdot \int_S g(\mathbf{r}, \mathbf{r}') \mathbf{C}(\mathbf{r}') \end{aligned} \tag{8.47}$$

where $\mathbf{C} = \mathbf{J}$ or $\mathbf{M}$ represents a current density. Note that the integral in the $\mathcal{K}$ operator is of a principal-value sense and there is a $\frac{1}{2}\mathbf{t}(\mathbf{r}) \cdot \hat{n} \times \mathbf{C}(\mathbf{r})$ term in the self-interaction elements. This term, together with the near-interaction terms, will be treated traditionally because the MLFMA only acts on the far-interaction elements. Based on (8.7) and (8.46), the matrix element related to the $\mathcal{L}$ operator can be found by

$$A_{ji}^L = \int \mathrm{d}^2\hat{k} \mathbf{V}_{fmj}^L(\hat{k}) \cdot \alpha_{mm'}(\mathbf{k}, \mathbf{r}_{mm'}) \mathbf{V}_{sm'i}^L(\hat{k}) \tag{8.48}$$

where

$$\mathbf{V}_{fmj}^L(\hat{k}) = \frac{1}{4\pi} \mathrm{e}^{\mathrm{i}\mathbf{k}\cdot\mathbf{r}_{jm}} \mathbf{t}_j \tag{8.49}$$

$$\mathbf{V}_{sm'i}^L(\hat{k}) = \mathrm{e}^{-\mathrm{i}\mathbf{k}\cdot\mathbf{r}_{im'}} (\bar{\mathbf{I}} - \hat{k}\hat{k}) \cdot \mathbf{q}_n w_i. \tag{8.50}$$

Note that the above radiation pattern and receiving pattern look similar to those in MoM discretization, but the meaning is slightly different. Here, $\mathbf{t}_j = \mathbf{t}(\mathbf{r}_j)$ in the receiving pattern is a unit tangential vector at the field point $\mathbf{r}_j$ used to take a tangential component, as mentioned above, and it is like a testing function. However, $\mathbf{q}_n$ in the radiation pattern is not like a basis function, instead, it is a column vector of the transform matrix for the transform from a local to a global coordinate system for the current densities on the nth patch. Since the current densities are the two-dimensional (2D) surface current densities, they only have two independent components in a local coordinate system defined on the related triangular patch. We need a transform matrix

$$\overline{\mathbf{Q}}_n = \begin{bmatrix} q_x(n,1) & q_x(n,2) & q_x(n,3) \\ q_y(n,1) & q_y(n,2) & q_y(n,3) \\ q_z(n,1) & q_z(n,2) & q_z(n,3) \end{bmatrix} \tag{8.51}$$

to change the local representation into the global representation for the current densities on the nth patch because the MLFMA tree structure is usually created in a global coordinate system. Here, $\left[q_x(n,k), q_y(n,k), q_z(n,k)\right]$ $(k = 1, 2, 3)$ are the direction cosines of describing the kth axis of the local coordinate system (u_n, v_n, w_n) defined on the nth triangular patch in the global coordinate system (x, y, z). With the transform of coordinates, the current density on the ith point $\mathbf{C}_i = C_x^i\hat{x} + C_y^i\hat{y} + C_z^i\hat{z}$, which is expressed in the global coordinate system, can be written as

$$\begin{bmatrix} C_x^i \\ C_y^i \\ C_z^i \end{bmatrix} = \overline{\mathbf{Q}}_n \cdot \begin{bmatrix} C_u^i \\ C_v^i \\ 0 \end{bmatrix} = \begin{bmatrix} q_x(n,1) & q_x(n,2) \\ q_y(n,1) & q_y(n,2) \\ q_z(n,1) & q_z(n,2) \end{bmatrix} \begin{bmatrix} C_u^i \\ C_v^i \end{bmatrix} \tag{8.52}$$

where (C_u^i, C_v^i) are the two independent components of the current density expressed in the local coordinate system and $\mathbf{q}_n = \left[q_x(n,k), q_y(n,k), q_z(n,k)\right]^T$ $(k = 1, 2)$ is the vector used in the radiation pattern in (8.50).

Similarly, the matrix element related to the $\mathcal{K}$ operator can be found by

$$A_{ji}^{K} = \int \mathrm{d}^2\hat{k}\mathbf{V}_{fmj}^{K}(\hat{k}) \cdot \alpha_{mm'}(\mathbf{k}, \mathbf{r}_{mm'})\mathbf{V}_{sm'i}^{K}(\hat{k}) \tag{8.53}$$

where

$$\mathbf{V}_{fmj}^{K}(\hat{k}) = -\frac{ik}{4\pi}\hat{k} \times \mathrm{e}^{\mathrm{i}\mathbf{k}\cdot\mathbf{r}_{jm}}\mathbf{t}_j \tag{8.54}$$

$$\mathbf{V}_{sm'i}^{K}(\hat{k}) = \mathrm{e}^{-\mathrm{i}\mathbf{k}\cdot\mathbf{r}_{im'}}\mathbf{q}_n w_i. \tag{8.55}$$

Consequently, the total matrix element can be calculated by adding the contribution from the $\mathcal{L}$ operator to the contribution from the $\mathcal{K}$ operator. Note that only θ and ϕ components need to be stored for the radiation pattern in the $\mathcal{L}$ operator since $(\bar{\mathbf{I}} - \hat{k}\hat{k}) \cdot \mathbf{q} = q_\theta\hat{\theta} + q_\phi\hat{\phi}$. Also, only θ and ϕ components are needed in the radiation pattern in the $\mathcal{K}$ operator because the receiving pattern only has θ and ϕ components. These are similar to those in the MoM discretization [2].

For the multilevel implementation, the interpolation or anterpolation technique is used as for the conducting objects.

8.4.3 Numerical Examples

We use the proposed MLFMA to solve the EFIEs (the MFIEs are fully dual to the EFIEs) for EM scattering by large penetrable or composite scatterers. The one-point quadrature rule is used in the Nyström scheme because high-order quadrature rules will dramatically increase the number of unknowns. The scatterers are chosen as radially stratified spheres as sketched in Figure 8.9 so that the available analytical solutions (Mie-series solutions) can be used for validation. It is assumed that the incident wave has a frequency $f = 300$ MHz and is propagating along the $-z$ direction in free space in all examples. We then calculate the bistatic RCS [33] observed along the principal cut ($\phi = 0°$ and $\theta = 0°–180°$) for the scatterers in both vertical polarization (VV) and horizontal polarization (HH), and compare the solutions with corresponding analytical solutions. The first example is the scattering by a PEC sphere with one-layer full dielectric coating as shown in Figure 8.9(a). The inner radius is $a_1 = 12\lambda$ and the outer radius is $a_2 = 15\lambda$. The coating material is characterized by a relative permittivity $\epsilon_r = 3.0$ and the relative permeability is always assumed $\mu_r = 1.0$ except when stated otherwise in all examples. We discretize the inner and outer interfaces into 224618 and 317524 triangular patches, respectively, resulting in $N = 1719332$ unknowns in total. The RCS solutions in the vertical polarization and horizontal polarization are shown in Figures 8.10 and 8.11, respectively, and they are close to the analytical solutions. Note that only the solutions within $\theta = 120°–180°$ are shown in order to see the comparison clearly.

The second example illustrates the scattering by a dielectric sphere with one-layer full dielectric coating, as sketched in Figure 8.9(b). The inner and outer radii are the same as in the first example. The spherical core and coating are both dielectric and their relative permittivities are $\epsilon_{r1} = 4.0$ and $\epsilon_{r2} = 3.0$, respectively. We use the same meshes as in the first example, resulting in $N = 2168568$ unknowns in total. Figures 8.12 and 8.13

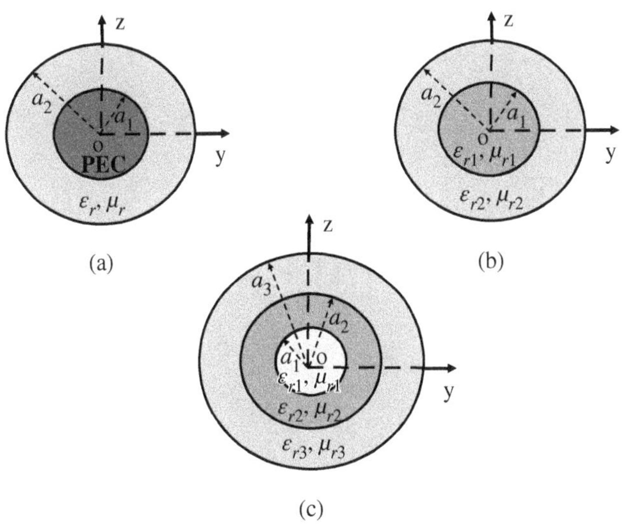

Figure 8.9 Geometries of radially stratified spherical scatterers. (a) A PEC sphere with one-layer full coating. (b) A dielectric sphere with one-layer full coating. (c) A dielectric sphere with two-layer full coating.

plot the RCS solutions within $\theta = 0°–60°$ for the vertical and horizontal polarizations, respectively, and they agree with the corresponding analytical solutions.

The third example considers the scattering by a dielectric sphere with two-layer full dielectric coating as shown in Figure 8.9(c). The radii of three interfaces are $a_1 = 5\lambda$, $a_2 = 8\lambda$, and $a_3 = 10\lambda$, respectively, and the materials are characterized by $\epsilon_{r1} = 2.0$, $\epsilon_{r2} = 4.0$, and $\epsilon_{r3} = 3.0$, respectively. We discretize the first, second and third interface into 75212, 137748 and 186352 triangular patches, respectively, resulting in

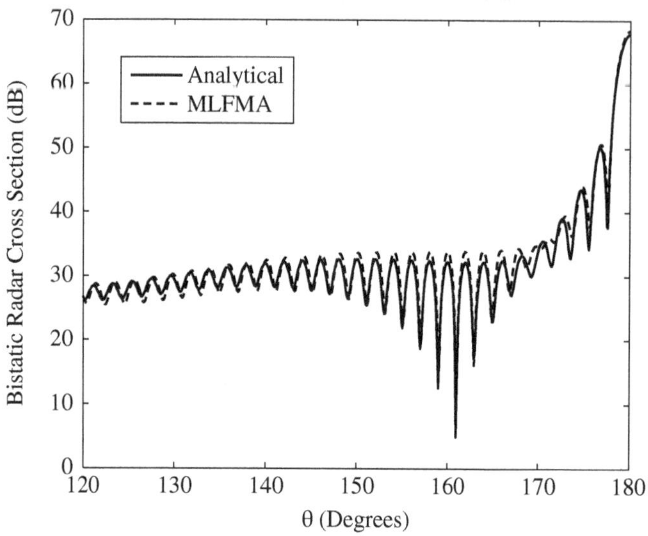

Figure 8.10 Bistatic RCS solutions for a PEC sphere with one-layer full dielectric coating in the vertical polarization.

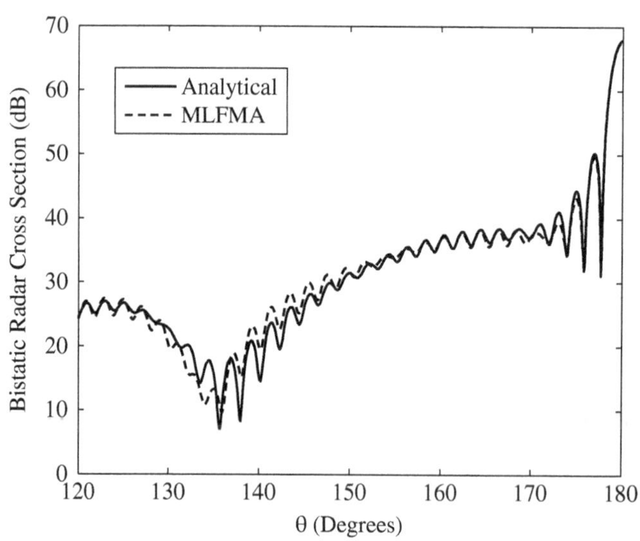

Figure 8.11 Bistatic RCS solutions for a PEC sphere with one-layer full dielectric coating in the horizontal polarization.

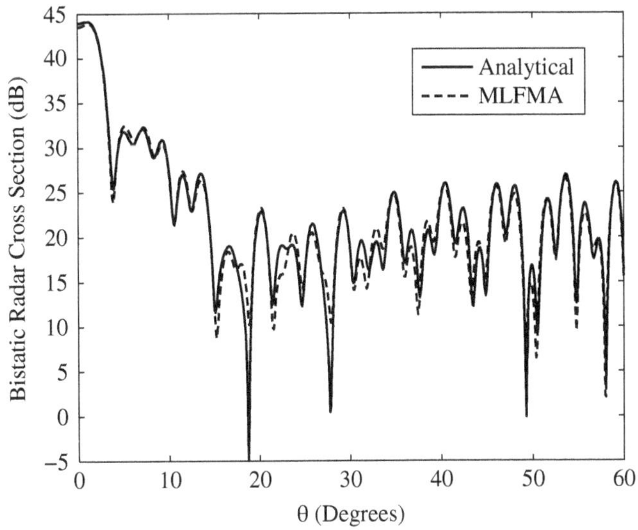

Figure 8.12 Bistatic RCS solutions for a dielectric sphere with one-layer full dielectric coating in the vertical polarization.

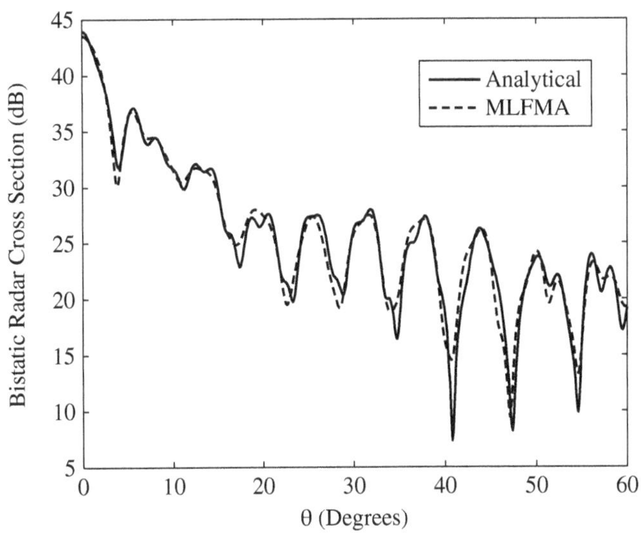

Figure 8.13 Bistatic RCS solutions for a dielectric sphere with one-layer full dielectric coating in the horizontal polarization.

$N = 1597248$ unknowns in total. The RCS solutions within the entire principal cut for vertical and horizontal polarizations are sketched in Figures 8.14 and 8.15, respectively, and again they are in good agreement with the analytical counterparts.

Finally, we summarize the CPU time T (Second) and memory usage M (GB) for these examples and they are shown in Table 8.1. To see the fast setup of the scheme, we distinguish the setup time T_s from the total solution time T_t in the table. All the calculations are performed on a Dell Precision 690 machine with two dual-core 3.0-GHz CPU and 16 GB RAM, but only one core is used.

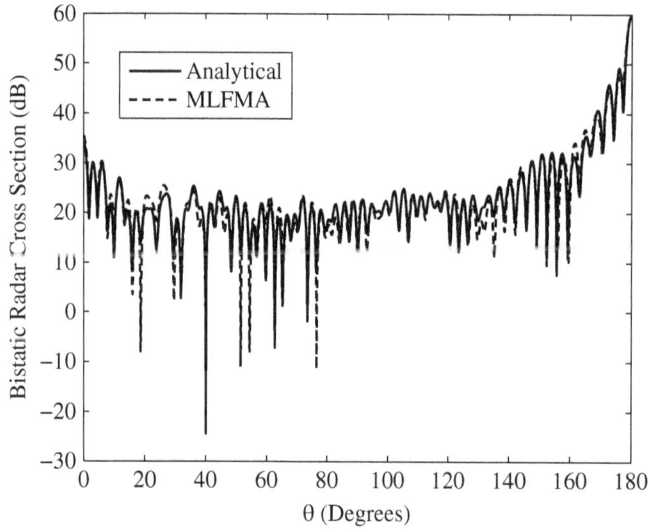

Figure 8.14 Bistatic RCS solutions for a dielectric sphere with two-layer full dielectric coating in the vertical polarization.

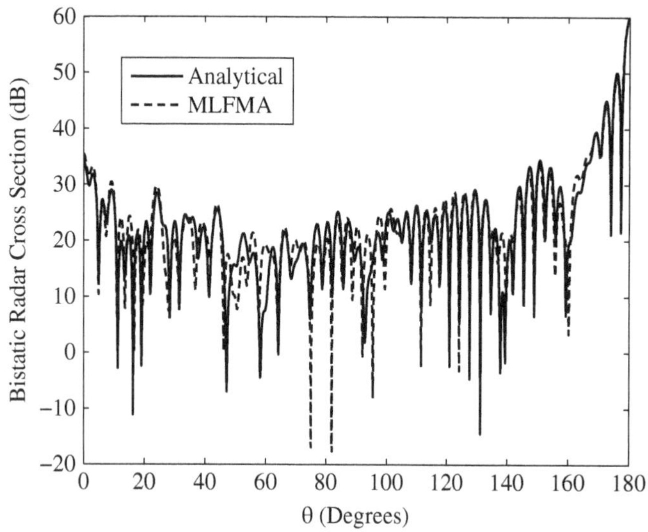

Figure 8.15 Bistatic RCS solutions for a dielectric sphere with two-layer full dielectric coating in the horizontal polarization.

Table 8.1 Summary of CPU time T (Second) and memory usage M (GB) for the three examples.

Examples	1	2	3
N (Number of unknowns)	1719332	2168568	1597248
T_s (Second)	23815	29371	18726
T_t (Second)	234892	263247	247958
M (GB)	10.6	11.9	9.7

8.5 Volume Integral Equations for Conductive Media

Solving EM problems with complex materials requires the formulation of VIEs in the integral equation approach [1]. The complex materials could be inhomogeneous, anisotropic, nonlinear, dispersive, magnetic, or lossily conducting and they are characterized by constitutive parameters. For conductive media or lossy conductors, accurate EM analysis requires considering the finite conductivity of the objects. The loss of the objects cannot be ignored when the frequency is low or the conductivity is small since the skin depth is large [34]. In the integral equation approach, one usually relies on SIEs with an approximate surface impedance when the skin depth is small [1]. For large skin depth caused by low frequency or small conductivity, one can treat the objects as dielectric-like objects and use a two-region scheme to formulate the SIEs [30, 31], which can be solved by the MoM with the RWG basis function [19]. However, the solutions may not be valid for a wide range of frequency and conductivity because the integral kernels are related to those parameters and the conditioning of SIEs is susceptible to them.

We can also treat the conductive objects as penetrable objects and use the VIEs to exactly formulate the problem. The VIEs that allow an inhomogeneity of materials are the second kind of integral equations, so they can result in well-conditioned system matrices in general [1]. Also, the VIEs only have one unknown function (electric current density) to be solved for dielectric (non-magnetic) objects, while two unknown functions (both electric and magnetic current densities) appear in the SIEs and two basis functions for representing them with a wisely chosen testing scheme are needed in the MoM solutions [35]. To solve the VIEs, we do not use the traditional MoM with the SWG basis function [20] but employ the Nyström scheme, which does not use any basis and testing functions and allows a geometric discretization of non-conforming meshes, resulting in much convenience in implementation.

Moreover, the scheme can choose current densities instead of flux densities as the unknowns to be solved so that the integral kernels of VIEs are free of the material parameters of objects (the material parameters are implied in the current densities). The integrals of matrix elements can thus be evaluated in an identical and regular way no matter how the material property varies, and the numerical solutions can endure a wide change of frequency and conductivity. For large conductive objects, we can also incorporate the MLFMA [2] to accelerate the solution. Since the material parameters do not appear in the integral kernels, the implementation of the MLFMA, which only acts on the far-interaction elements, is very convenient and is the same as that for regular dielectric media.

8.5.1 Integral Equations

When the conductive objects are treated as penetrable objects, we can use the VIEs to govern their interaction with EM waves. Consider the EM scattering by a 3D penetrable object in free space with a permittivity ϵ_0 and permeability μ_0. The VIEs can be written as [1]

$$\begin{aligned}\mathbf{E}(\mathbf{r}) = \mathbf{E}^{\rm inc}(\mathbf{r}) + {\rm i}\omega\mu_0 \int_V \overline{\mathbf{G}}(\mathbf{r},\mathbf{r}')\cdot\mathbf{J}_V(\mathbf{r}')d\mathbf{r}' \\ - \nabla\times\int_V \mathbf{G}(\mathbf{r},\mathbf{r}')\cdot\mathbf{M}_V(\mathbf{r}')d\mathbf{r}', \quad \mathbf{r}\in V\end{aligned} \tag{8.56}$$

$$\begin{aligned}\mathbf{H}(\mathbf{r}) = \mathbf{H}^{\rm inc}(\mathbf{r}) + {\rm i}\omega\epsilon_0 \int_V \overline{\mathbf{G}}(\mathbf{r},\mathbf{r}')\cdot\mathbf{M}_V(\mathbf{r}')d\mathbf{r}' \\ + \nabla\times\int_V \overline{\mathbf{G}}(\mathbf{r},\mathbf{r}')\cdot\mathbf{J}_V(\mathbf{r}')d\mathbf{r}', \quad \mathbf{r}\in V\end{aligned} \tag{8.57}$$

where $\mathbf{E}^{\rm inc}(\mathbf{r})$ and $\mathbf{H}^{\rm inc}(\mathbf{r})$ are the incident electric field and magnetic field, respectively, while $\mathbf{E}(\mathbf{r})$ and $\mathbf{H}(\mathbf{r})$ are the total electric field and magnetic field inside the object, respectively. The integral kernel is the dyadic Green's function given by

$$\overline{\mathbf{G}}(\mathbf{r},\mathbf{r}') = \left(\overline{\mathbf{I}} + \frac{\nabla\nabla}{k_0^2}\right)g(\mathbf{r},\mathbf{r}') \tag{8.58}$$

where $\overline{\mathbf{I}}$ is the identity dyad, $k_0 = \omega\sqrt{\epsilon_0\mu_0}$ is the free-space wavenumber (ω is the angular frequency), and $g(\mathbf{r},\mathbf{r}') = e^{ik_0R}/(4\pi R)$ is the scalar Green's function in which $R = |\mathbf{r} - \mathbf{r}'|$ is the distance between an observation point $\mathbf{r}$ and a source point $\mathbf{r}'$. The unknown functions to be solved are the volumetric electric current density and magnetic current density inside the object, which are related to the total electric field and magnetic field by

$$\mathbf{J}_V(\mathbf{r}') = {\rm i}\omega\left[\epsilon_0 - \epsilon(\mathbf{r}')\right]\mathbf{E}(\mathbf{r}') \tag{8.59}$$

$$\mathbf{M}_V(\mathbf{r}') = {\rm i}\omega\left[\mu_0 - \mu(\mathbf{r}')\right]\mathbf{H}(\mathbf{r}') \tag{8.60}$$

where $\epsilon(\mathbf{r}')$ and $\mu(\mathbf{r}')$ are the permittivity and permeability of the object, respectively. The above VIEs can be applicable to any penetrable objects including conductive objects although the property of resulting matrix equations could be quite different due to the significant difference in material parameters. If the conductive objects are non-magnetic or $\mu(\mathbf{r}') = \mu_0$, the magnetic current density vanishes and the VIEs can be reduced to

$$\frac{1}{{\rm i}\omega\left[\epsilon_0 - \epsilon(\mathbf{r}')\right]}\mathbf{J}_V(\mathbf{r}) - {\rm i}\omega\mu_0\int_V \overline{\mathbf{G}}(\mathbf{r},\mathbf{r}')\cdot\mathbf{J}_V(\mathbf{r}')d\mathbf{r}' = \mathbf{E}^{\rm inc}(\mathbf{r}), \quad \mathbf{r}\in V \tag{8.61}$$

where $\epsilon(\mathbf{r}') = \epsilon_r'\epsilon_0 + i\frac{\sigma}{\omega} = \epsilon_c$ for conductive objects in which ϵ_r' is the real part of relative permittivity and σ is the conductivity.

8.5.2 Nyström Discretization

The above VIEs for conductive objects can be discretized with the Nyström method. Since the current densities are chosen as the unknown functions to be solved, the integrands of the VIEs do not include the material parameters of the objects and the numerical implementation can be greatly facilitated. The VIEs are first expressed into a scalar form since we do not use any vector basis function to represent the unknown current densities. When an object is discretized into N small tetrahedral elements, the VIEs in Equations (8.56) and (8.57) can be changed into

$$\sum_{n=1}^{N}\int_{\Delta V_n} \mathrm{d}V' \left\{ \mathrm{i}\omega\mu_0 \overline{\mathbf{G}}(\mathbf{r}_m^{\mathrm{c}}, \mathbf{r}') \cdot \begin{bmatrix} J_V^x(\mathbf{r}'^{\mathrm{c}}_n) \\ J_V^y(\mathbf{r}'^{\mathrm{c}}_n) \\ J_V^z(\mathbf{r}'^{\mathrm{c}}_n) \end{bmatrix} - \begin{bmatrix} g_y(\mathbf{r}_m^{\mathrm{c}}, \mathbf{r}')M_V^z(\mathbf{r}'^{\mathrm{c}}_n) - g_z(\mathbf{r}_m^{\mathrm{c}}, \mathbf{r}')M_V^y(\mathbf{r}'^{\mathrm{c}}_n) \\ g_z(\mathbf{r}_m^{\mathrm{c}}, \mathbf{r}')M_V^x(\mathbf{r}'^{\mathrm{c}}_n) - g_x(\mathbf{r}_m^{\mathrm{c}}, \mathbf{r}')M_V^z(\mathbf{r}'^{\mathrm{c}}_n) \\ g_x(\mathbf{r}_m^{\mathrm{c}}, \mathbf{r}')M_V^y(\mathbf{r}'^{\mathrm{c}}_n) - g_y(\mathbf{r}_m^{\mathrm{c}}, \mathbf{r}')M_V^x(\mathbf{r}'^{\mathrm{c}}_n) \end{bmatrix} \right\}$$

$$- \frac{1}{\mathrm{i}\omega\left[\epsilon_0 - \epsilon(\mathbf{r}_m^{\mathrm{c}})\right]} \begin{bmatrix} J_V^x(\mathbf{r}_m^{\mathrm{c}}) \\ J_V^y(\mathbf{r}_m^{\mathrm{c}}) \\ J_V^z(\mathbf{r}_m^{\mathrm{c}}) \end{bmatrix} = - \begin{bmatrix} E_x^{\mathrm{inc}}(\mathbf{r}_m^{\mathrm{c}}) \\ E_y^{\mathrm{inc}}(\mathbf{r}_m^{\mathrm{c}}) \\ E_z^{\mathrm{inc}}(\mathbf{r}_m^{\mathrm{c}}) \end{bmatrix}, \quad m = 1, 2, \cdots, N \tag{8.62}$$

$$\sum_{n=1}^{N}\int_{\Delta V_n} \mathrm{d}V' \left\{ \mathrm{i}\omega\epsilon_0 \overline{\mathbf{G}}(\mathbf{r}_m^{\mathrm{c}}, \mathbf{r}') \cdot \begin{bmatrix} M_V^x(\mathbf{r}'^{\mathrm{c}}_n) \\ M_V^y(\mathbf{r}'^{\mathrm{c}}_n) \\ M_V^z(\mathbf{r}'^{\mathrm{c}}_n) \end{bmatrix} + \begin{bmatrix} g_y(\mathbf{r}_m^{\mathrm{c}}, \mathbf{r}')J_V^z(\mathbf{r}'^{\mathrm{c}}_n) - g_z(\mathbf{r}_m^{\mathrm{c}}, \mathbf{r}')J_V^y(\mathbf{r}'^{\mathrm{c}}_n) \\ g_z(\mathbf{r}_m^{\mathrm{c}}, \mathbf{r}')J_V^x(\mathbf{r}'^{\mathrm{c}}_n) - g_x(\mathbf{r}_m^{\mathrm{c}}, \mathbf{r}')J_V^z(\mathbf{r}'^{\mathrm{c}}_n) \\ g_x(\mathbf{r}_m^{\mathrm{c}}, \mathbf{r}')J_V^y(\mathbf{r}'^{\mathrm{c}}_n) - g_y(\mathbf{r}_m^{\mathrm{c}}, \mathbf{r}')J_V^x(\mathbf{r}'^{\mathrm{c}}_n) \end{bmatrix} \right\}$$

$$- \frac{1}{\mathrm{i}\omega\left[\mu_0 - \mu(\mathbf{r}_m^{\mathrm{c}})\right]} \begin{bmatrix} M_V^x(\mathbf{r}_m^{\mathrm{c}}) \\ M_V^y(\mathbf{r}_m^{\mathrm{c}}) \\ M_V^z(\mathbf{r}_m^{\mathrm{c}}) \end{bmatrix} = - \begin{bmatrix} H_x^{\mathrm{inc}}(\mathbf{r}_m^{\mathrm{c}}) \\ H_y^{\mathrm{inc}}(\mathbf{r}_m^{\mathrm{c}}) \\ H_z^{\mathrm{inc}}(\mathbf{r}_m^{\mathrm{c}}) \end{bmatrix}, \quad m = 1, 2, \cdots, N \tag{8.63}$$

where $\mathbf{r}_m^{\mathrm{c}}$ represents the center of the mth tetrahedron (observation element) and $\mathbf{r}'^{\mathrm{c}}_n$ denotes the center of the nth tetrahedron (source element). Also, $\{g_x(\mathbf{r}, \mathbf{r}'), g_y(\mathbf{r}, \mathbf{r}'), g_z(\mathbf{r}, \mathbf{r}')\}$ are the three components of the gradient of the scalar Green's function. The matrix elements in the above can be evaluated by integrating the components of the dyadic Green's function or the gradient of the scalar Green's function over each small tetrahedral element ΔV_n $(n = 1, 2, \cdots, N)$ for a different observation point $\mathbf{r}_m^{\mathrm{c}}$ $(m = 1, 2, \cdots, N)$. The dyadic Green's function has nine components but only six of them are independent due to its symmetry. When $m \neq n$, the integral kernels are regular and we can use a numerical quadrature rule like the Gauss-Legendre quadrature rule to evaluate the matrix elements. When $m = n$, however, the integral kernels are hypersingular and we need to specially treat them. Since the integral kernels do not include the material parameters, we can use the singularity treatment technique developed for regular dielectric media to evaluate the relevant matrix elements [28, 36].

8.5.3 Incorporation with the MLFMA

Considering the VIE in Equation (8.61) for a non-magnetic object, it can be discretized into N_t tetrahedrons by a point-matching scheme. In each tetrahedron, the electric current density is assumed to be constant and its value is sampled at the center of the tetrahedron, resulting in $N = 3N_t$ unknowns in total (the current density has three components at a point). The centers of tetrahedrons are also chosen as the observation points or field points and a collocation is performed over them to form a matrix equation. From Equation (8.4), we can see that the one component of field at the jth field point is related to the one component of electric current density J_V^i in the ith source tetrahedron by

$$F_{ji} = \left[\int d^2\hat{k}\,\mathcal{U}_{fmj}(\hat{k}) \cdot \alpha_{mm'}(\mathbf{k}_0, \mathbf{r}_{mm'})\mathcal{U}_{sm'i}(\hat{k})\right] J_V^i \tag{8.64}$$

where

$$\mathcal{U}_{fmj}(\hat{k}) = \frac{1}{4\pi} e^{i\mathbf{k}_0 \cdot \mathbf{r}_{jm}} \tag{8.65}$$

$$\mathcal{U}_{sm'i}(\hat{k}) = \int_{\Delta V_i} dV' e^{-i\mathbf{k}_0 \cdot \mathbf{r}_{im'}} f_c(\hat{k}). \tag{8.66}$$

Note that $\mathcal{U}_{fmj}(\hat{k})$ is the receiving pattern at the jth field point and $\mathcal{U}_{sm'i}(\hat{k})$ is the radiation pattern at the ith source point. The integration in $\mathcal{U}_{sm'i}(\hat{k})$ is about a source point $\mathbf{r}_s$ inside the ith tetrahedron ΔV_i and $f_c(\hat{k})$ is one of components resulting from the $\overline{\mathbf{f}}(\hat{k})$ operator, which represents a double gradient operation exerted on the scalar Green's function in the $\hat{k}$ domain. In the VIEs for non-magnetic objects where the magnetic current density vanishes, the integrals actually take a form of a $\mathcal{L}$ operator and $\overline{\mathbf{f}}(\hat{k}) = \overline{\mathbf{I}} - \hat{k}\hat{k}$. With the above relationship between one component of a field and one component of a current density, we can write the matrix element as

$$Z_{ji} = \int d^2\hat{k}\,\mathcal{U}_{fmj}(\hat{k}) \cdot \alpha_{mm'}(\mathbf{k}_0, \mathbf{r}_{mm'})\mathcal{U}_{sm'i}(\hat{k}) \tag{8.67}$$

while the unknown coefficient a_i is J_V^i in (8.1). The above formulas are used for a one-level implementation of the MLFMA and the multilevel scheme can be implemented as for the conducting objects in the previous section. Because the integral kernels are free of the material property, the above implementation can be unified for different materials except the calculation of self-interaction terms (with constitutive parameters) on which the MLFMA does not act.

8.5.4 Numerical Examples

We consider EM scattering by a conductive sphere as shown in Figure 8.16 to demonstrate the approach. The sphere is chosen because we can use its exact Mie-series solutions [33] to verify the numerical results. It is assumed that the incident wave with a frequency f is a plane wave and is propagating along the $+z$ direction. The sphere has a radius a and a conductivity σ. The relative permeability $\mu_r = 1.0$ is assumed (non-magnetic) and the real part of relative permittivity is chosen as $\epsilon_r' = 3.0$. We calculate the bistatic RCS observed along the principal cut ($\phi = 0°$ and $\theta = 0°–180°$) for the scatterer in both vertical polarization (VV) and horizontal polarization (HH).

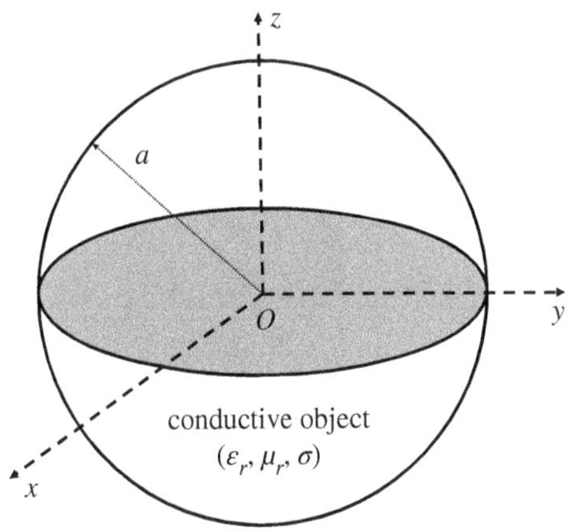

Figure 8.16 Geometry of a conductive sphere.

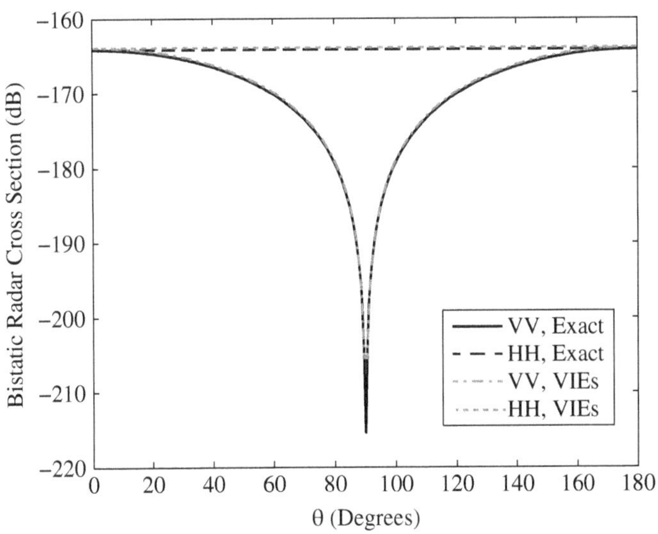

Figure 8.17 Bistatic RCS solutions for a conductive sphere with a radius $a = 1.0$ m and a conductivity $\sigma = 5.0\ \text{S m}^{-1}$ when the frequency is $f = 2$ KHz.

In the first example, we choose $a = 1.0$ m, $\sigma = 5.0\ \text{S m}^{-1}$, and $f = 2$ KHz, and the sphere is meshed into 1342 tetrahedrons. Figure 8.17 plots the bistatic RCS solutions and they agree with the corresponding exact solutions very well. In the second example, we choose $a = 1.0$ m, $\sigma = 1.0 \times 10^3\ \text{S m}^{-1}$, and $f = 100$ MHz, which are quite different from the parameters of the first example, and the sphere is meshed into 5318 tetrahedrons. Figure 8.18 sketches the bistatic RCS solutions for this example and they are also consistent with the corresponding exact solutions.

The scatterer in the above two examples is electrically small and we do not need to use the MLFMA for acceleration. If we select $a = 10.0$ m, $\sigma = 1.0\ \text{S m}^{-1}$, and $f = 300$ MHz, then $a = 10.0\lambda$, where λ is the wavelength in free space, and the scatterer can be viewed

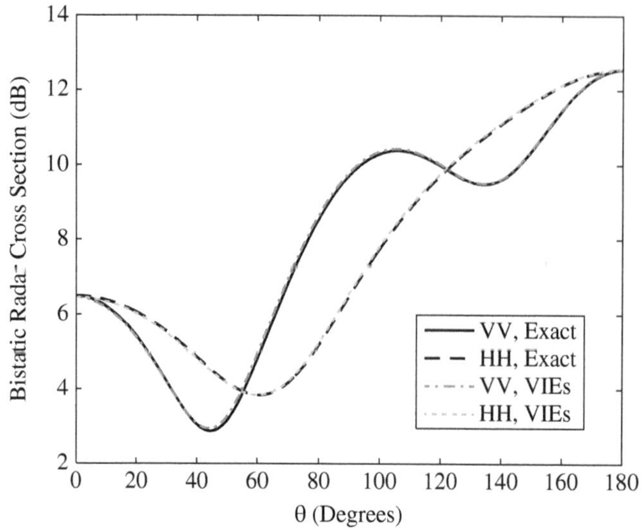

Figure 8.18 Bistatic RCS solutions for a conductive sphere with a radius $a = 1.0$ m and a conductivity $\sigma = 1.0 \times 10^3$ S m^{-1} when the frequency is $f = 100$ MHz.

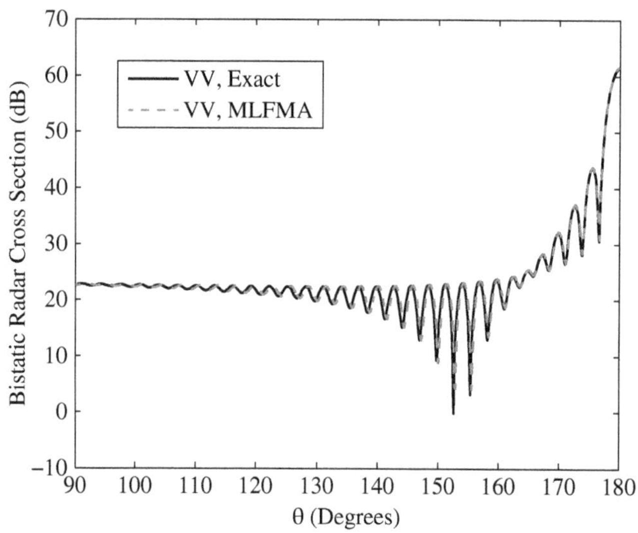

Figure 8.19 Bistatic RCS solutions (VV) for a conductive sphere with a radius $a = 10.0$ m and a conductivity $\sigma = 1.0$ S m^{-1} when the frequency is $f = 300$ MHz.

as electrically large. The scatterer is discretized into 154726 tetrahedrons, resulting in 464178 unknowns and we use the MLFMA to accelerate the solution. Figure 8.19 shows the bistatic RCS solutions in the vertical polarization while Figure 8.20 depicts the solutions in the horizontal polarization and it can be seen that both solutions are very close to the exact solutions.

We also consider the scattering by a conductive sphere, which is assumed to be magnetic, and its relative permeability is $\mu_r = 4.0$. The frequency of the incident plane

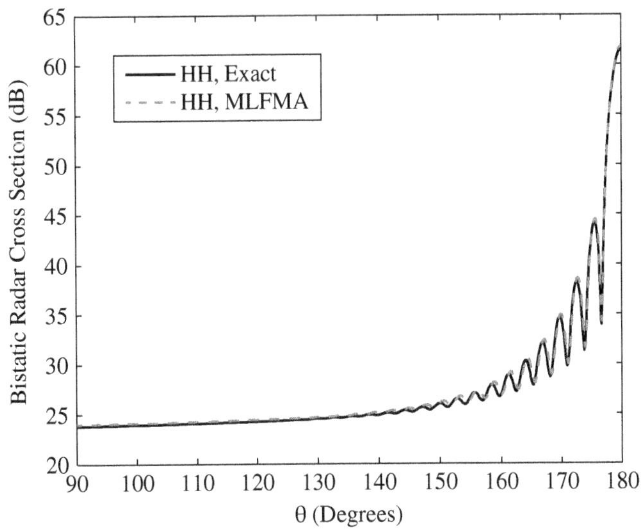

Figure 8.20 Bistatic RCS solutions (HH) for a conductive sphere with a radius $a = 10.0$ m and a conductivity $\sigma = 1.0\ \mathrm{S\,m^{-1}}$ when the frequency is $f = 300$ MHz.

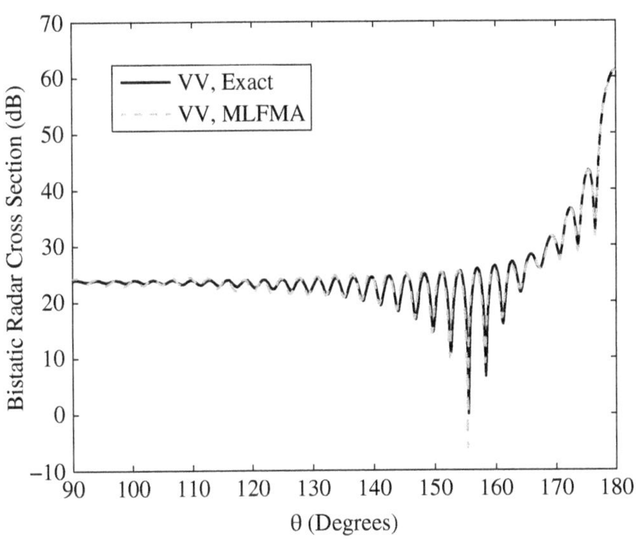

Figure 8.21 Bistatic RCS solutions with a vertical polarization for a lossy spherical conductor which is also assumed to be magnetic.

wave is 300 MHz and is propagating along the $+z$ direction. The sphere has a radius 10.0 m and a conductivity $\sigma = 1.0\ \mathrm{S\ m^{-1}}$. The solution is accelerated by the MLFMA and Figure 8.21 shows the bistatic RCS solutions in the vertical polarization while Figure 8.22 depicts the solutions in the horizontal polarization. It can seen that they are also in good agreement with the corresponding exact Mie-series solutions.

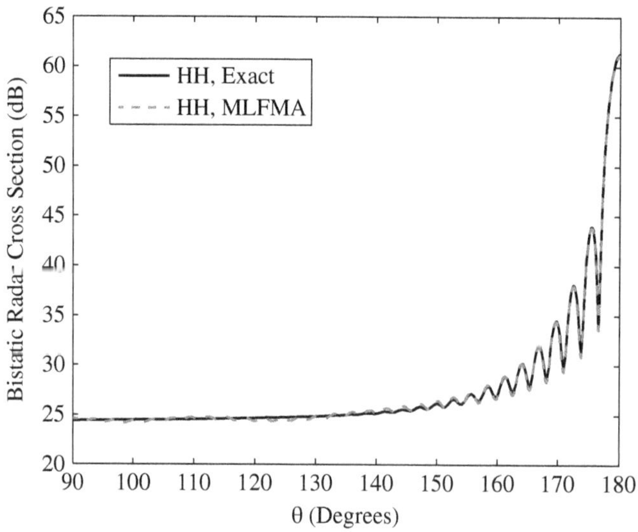

Figure 8.22 Bistatic RCS solutions with a horizontal polarization for a lossy spherical conductor which is also assumed to be magnetic.

8.6 Volume-Surface Integral Equations for Conducting-Anisotropic Media

In the real world, some devices or systems include inhomogeneous or anisotropic media in structures and the analysis relies on the accurate numerical solutions of VIEs in the integral equation approach [37]. The VIEs include an electric field VIE (EFVIE) and a magnetic field VIE (MFVIE) and they are solved simultaneously to determine volumetric electric and magnetic current densities for generalized media. If the structures also include conductors, the volume-surface integral equations (VSIEs) should then be solved [1].

Traditionally, the VIEs are solved by the MoM [18] with the divergence conforming SWG basis function [20] or curl conforming edge basis function [38]. These basis functions are defined over a pair of tetrahedral elements with a common face or common edge and require conforming meshes in geometric discretization, resulting in a higher preprocessing cost [39]. Also, the basis functions have to assume a homogeneous material in each tetrahedron and disregard the possible directional difference of material, so they may not be suitable for highly inhomogeneous and anisotropic structures. In contrast, the Nyström method may be more appropriate for such problems because of its different features [29]. The method represents unknown functions at discrete quadrature points with directional components and usually includes more degrees of freedoms than the MoM [40]. Furthermore, the method allows the use of the JM formulation [41] which does not explicitly include material property in the integral kernels in the VIEs. These characteristics indicate that it is very suitable for solving anisotropic problems and also very friendly to the incorporation of fast algorithms.

The MLFMA has also been applied to solve anisotropic EM problems based on the traditional MoM discretization of VIEs [42]. We present a different Nyström-based

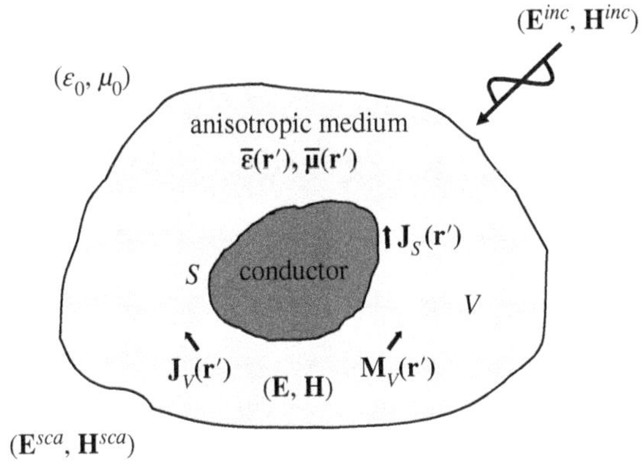

Figure 8.23 EM Scattering by a 3D conducting object coated with an anisotropic material characterized by a permittivity tensor $\bar{\bar{\epsilon}}(\mathbf{r}')$ and a permeability tensor $\bar{\bar{\mu}}(\mathbf{r}')$. A surface current density $\mathbf{J}_S(\mathbf{r}')$ is induced on the conducting surface S while a volumetric electric current density $\mathbf{J}_V(\mathbf{r}')$ and a volumetric magnetic current density $\mathbf{M}_V(\mathbf{r}')$ are induced inside the anisotropic material.

MLFMA by applying the Nyström discretization to the VIEs or VSIEs for rapidly solving large highly-anisotropic problems [43]. The use of the JM formulation and the feature of the Nyström method can greatly facilitate the implementation of MLFMA.

8.6.1 Integral Equations for Anisotropic Objects

In the integral equation approach, the EM scattering by anisotropic objects is described by VIEs. Consider a more general or typical case, namely, the scatterer is a 3D composite object including both conducting and anisotropic material, as shown in Figure 8.23. We can use the SIEs to describe the relationship between the fields and current source on the conducting surface and the VIEs to catch the EM feature for the anisotropic material, respectively. If the object only includes anisotropic materials, then the individual VIEs are enough. If the conducting material exists as well, then the SIEs and VIEs are coupled each other to form the VSIEs. Assuming that the object is embedded in free space whose permittivity and permeability are ϵ_0 and μ_0, respectively, we can write the EFIE for the conducting part which is assumed to be a PEC [1]

$$\hat{n} \times \mathbf{E}^{\mathrm{inc}}(\mathbf{r}) = -\hat{n} \times \mathrm{i}\omega\mu_0 \int_S \overline{\mathbf{G}}(\mathbf{r}, \mathbf{r}') \cdot \mathbf{J}_S(\mathbf{r}')\, \mathrm{d}S', \quad \mathbf{r} \in S \tag{8.68}$$

where $\mathbf{J}_S(\mathbf{r}')$ is the surface electric current density induced on the conducting surface, $\hat{n}$ is the unit normal vector of the conducting surface, and $\mathbf{E}^{\mathrm{inc}}(\mathbf{r})$ and $\mathbf{H}^{\mathrm{inc}}(\mathbf{r})$ are the incident electric field and magnetic field, respectively. Also, $\overline{\mathbf{G}}(\mathbf{r}, \mathbf{r}')$ is the 3D dyadic Green's function as defined previously.

For the anisotropic material in the object, the VIEs can be written as [1]

$$\mathbf{E}^{\mathrm{inc}}(\mathbf{r}) = \mathbf{E}(\mathbf{r}) + \int_V \nabla g(\mathbf{r}, \mathbf{r}') \times \mathbf{M}_V(\mathbf{r}')\mathrm{d}\mathbf{r}' - \mathrm{i}\omega\mu_0 \int_V \overline{\mathbf{G}}(\mathbf{r}, \mathbf{r}') \cdot \mathbf{J}_V(\mathbf{r}')\mathrm{d}\mathbf{r}', \quad \mathbf{r} \in V \tag{8.69}$$

$$\mathbf{H}^{\text{inc}}(\mathbf{r}) = \mathbf{H}(\mathbf{r}) - \int_V \nabla g(\mathbf{r},\mathbf{r}') \times \mathbf{J}_V(\mathbf{r}')\mathrm{d}\mathbf{r}' - \mathrm{i}\omega\epsilon_0 \int_V \overline{\mathbf{G}}(\mathbf{r},\mathbf{r}') \cdot \mathbf{M}_V(\mathbf{r}')\mathrm{d}\mathbf{r}', \quad \mathbf{r} \in V \tag{8.70}$$

where $\mathbf{E}(\mathbf{r})$ and $\mathbf{H}(\mathbf{r})$ are the total electric field and total magnetic field inside the material, respectively, while

$$\mathbf{J}_V(\mathbf{r}') = \mathrm{i}\omega \left[\epsilon_0 \overline{\mathbf{I}} - \overline{\epsilon}(\mathbf{r}')\right] \cdot \mathbf{E}(\mathbf{r}') \tag{8.71}$$

$$\mathbf{M}_V(\mathbf{r}') = \mathrm{i}\omega \left[\mu_0 \overline{\mathbf{I}} - \overline{\mu}(\mathbf{r}')\right] \cdot \mathbf{H}(\mathbf{r}') \tag{8.72}$$

are the induced volumetric electric current density and magnetic current density inside the material, respectively. Also, $\overline{\mu}_r(\mathbf{r}') = \overline{\mu}(\mathbf{r}')/\mu_0$ and $\overline{\epsilon}_r(\mathbf{r}') = \overline{\epsilon}(\mathbf{r}')/\epsilon_0$ are the relative permeability tensor and relative permittivity tensor, respectively.

When considering the coupling of fields generated from the current source at the conducting surface and the current sources inside the anisotropic material, we can form the following VSIEs

$$0 = \hat{n} \times \left[\mathbf{E}^{\text{inc}}(\mathbf{r}) + \mathrm{i}\omega\mu_0 \int_S \overline{\mathbf{G}}(\mathbf{r},\mathbf{r}') \cdot \mathbf{J}_S(\mathbf{r}')\,\mathrm{d}S' - \int_V \nabla g(\mathbf{r},\mathbf{r}') \times \mathbf{M}_V(\mathbf{r}')\mathrm{d}\mathbf{r}' \right.$$
$$\left. + \mathrm{i}\omega\mu_0 \int_V \overline{\mathbf{G}}(\mathbf{r},\mathbf{r}') \cdot \mathbf{J}_V(\mathbf{r}')\mathrm{d}\mathbf{r}' \right], \quad \mathbf{r} \in S \tag{8.73}$$

$$\mathbf{E}^{\text{inc}}(\mathbf{r}) = \mathbf{E}(\mathbf{r}) - \mathrm{i}\omega\mu_0 \int_S \overline{\mathbf{G}}(\mathbf{r},\mathbf{r}') \cdot \mathbf{J}_S(\mathbf{r}')\,\mathrm{d}S' + \int_V \nabla g(\mathbf{r},\mathbf{r}') \times \mathbf{M}_V(\mathbf{r}')\mathrm{d}\mathbf{r}'$$
$$- \mathrm{i}\omega\mu_0 \int_V \overline{\mathbf{G}}(\mathbf{r},\mathbf{r}') \cdot \mathbf{J}_V(\mathbf{r}')\mathrm{d}\mathbf{r}', \quad \mathbf{r} \in V \tag{8.74}$$

$$\mathbf{H}^{\text{inc}}(\mathbf{r}) = \mathbf{H}(\mathbf{r}) - \mathrm{i}\omega\mu_0 \int_S \overline{\mathbf{G}}(\mathbf{r},\mathbf{r}') \cdot \mathbf{J}_S(\mathbf{r}')\,\mathrm{d}S' - \int_V \nabla g(\mathbf{r},\mathbf{r}') \times \mathbf{J}_V(\mathbf{r}')\mathrm{d}\mathbf{r}'$$
$$- \mathrm{i}\omega\epsilon_0 \int_V \overline{\mathbf{G}}(\mathbf{r},\mathbf{r}') \cdot \mathbf{M}_V(\mathbf{r}')\mathrm{d}\mathbf{r}', \quad \mathbf{r} \in V \tag{8.75}$$

from which we can solve the unknown current densities.

8.6.2 Nyström Discretization

The above VIEs or VSIEs including anisotropic media can be discretized through the Nyström method. The method represents unknown functions at discrete quadrature points in the form of scalar components with more degrees of freedom and could account for the anisotropy of materials more appropriately. Moreover, the feature of the method allows the use of the JM formulation as defined in [41], which can greatly simplify the implementation of MLFMA for anisotropic problems.

To solve the VIEs or VSIEs, we choose the current densities $\mathbf{J}_V(\mathbf{r}')$ and $\mathbf{M}_V(\mathbf{r}')$ inside the anisotropic media as unknowns to be solved, leading to the use of the JM formulation. As discussed in [41], the choice is more suitable than the other two choices, i.e. the DB formulation using $\mathbf{D}(\mathbf{r}')$ and $\mathbf{B}(\mathbf{r}')$ as unknowns and the EH formulation using $\mathbf{E}(\mathbf{r}')$ and $\mathbf{H}(\mathbf{r}')$ as unknowns, for anisotropic problems, especially when the anisotropy is strong. Another advantage for such a choice is that we can avoid the involvement of material property, i.e. the permittivity tensor and permeability tensor inside the integrands, as can be seen from (8.73)–(8.75) since the current densities have accounted for the material property as shown in (8.71) and (8.72), leading to much convenience in numerical implementation for anisotropic media. The material property will appear in the total fields when they are expressed in terms of current densities through (8.71) and (8.72), namely,

$$\begin{bmatrix} E^x \\ E^y \\ E^z \end{bmatrix} = \mathbf{E}(\mathbf{r}') = \frac{1}{\mathrm{i}\omega}\left[\epsilon_0\bar{\mathbf{I}} - \bar{\epsilon}(\mathbf{r}')\right]^{-1} \cdot \mathbf{J}_V(\mathbf{r}')$$

$$= \frac{1}{\mathrm{i}\omega}\begin{bmatrix} \epsilon_0 - \epsilon_{11} & -\epsilon_{12} & -\epsilon_{13} \\ -\epsilon_{21} & \epsilon_0 - \epsilon_{22} & -\epsilon_{23} \\ -\epsilon_{31} & -\epsilon_{32} & \epsilon_0 - \epsilon_{33} \end{bmatrix}^{-1} \cdot \begin{bmatrix} J_V^x \\ J_V^y \\ J_V^z \end{bmatrix} \tag{8.76}$$

$$\begin{bmatrix} H^x \\ H^y \\ H^z \end{bmatrix} = \mathbf{H}(\mathbf{r}') = \frac{1}{\mathrm{i}\omega}\left[\mu_0\bar{\mathbf{I}} - \bar{\mu}(\mathbf{r}')\right]^{-1} \cdot \mathbf{M}_V(\mathbf{r}')$$

$$= \frac{1}{\mathrm{i}\omega}\begin{bmatrix} \mu_0 - \mu_{11} & -\mu_{12} & -\mu_{13} \\ -\mu_{21} & \mu_0 - \mu_{22} & -\mu_{23} \\ -\mu_{31} & -\mu_{32} & \mu_0 - \mu_{33} \end{bmatrix}^{-1} \cdot \begin{bmatrix} M_V^x \\ M_V^y \\ M_V^z \end{bmatrix}. \tag{8.77}$$

Note that we change the VIEs into a scalar or component form to implement in the Nyström method, but the SIE in (8.73) should be changed into a matrix equation by performing a testing-like procedure [1]. Although the permittivity tensor and permeability tensor appear in the VIEs or VSIEs, they are outside the integrals and only related to self-interaction terms or diagonal elements in the impedance matrix. This can greatly facilitate the implementation of the MLFMA since the MLFMA is designed for far-interaction terms and the self-interaction terms will be handled in a traditional way [37].

8.6.3 MLFMA Acceleration

The basic principle of MLFMA has been described in the previous section and we revise it here to accommodate the solution of VIEs or VSIEs with anisotropic materials.

When the VIEs are discretized with a Nyström scheme, each current density produces $N_c = PN_v$ unknowns where P is the number of quadrature points in each tetrahedral element and N_v is the number of tetrahedral elements. The quadrature points can be either field points (observation points) or source points, and i or $j = P(n-1) + p$ denotes the pth quadrature point in the nth tetrahedral element. From (8.4), we can see that the one component of field at the jth field point is related to the one component of current density in the nth source tetrahedron by

$$F_{jn} = \int \mathrm{d}^2\hat{k}\, U_{fmj}(\hat{k}) \cdot \alpha_{mm'}(\mathbf{k}_0, \mathbf{r}_{mm'}) U_{sm'n}(\hat{k}) \tag{8.78}$$

where

$$U_{fmj}(\hat{k}) = \frac{1}{4\pi}\mathrm{e}^{\mathrm{i}\mathbf{k}_0\cdot\mathbf{r}_{jm}} \tag{8.79}$$

$$\begin{aligned} U_{sm'n}(\hat{k}) &= \int_{\Delta V_n} \mathrm{d}V'\mathrm{e}^{-\mathrm{i}\mathbf{k}_0\cdot\mathbf{r}_{vm'}} f_c(\hat{k})C_V \\ &= \sum_{p=1}^{P} \mathrm{e}^{-\mathrm{i}\mathbf{k}_0\cdot\mathbf{r}_{im'}} f_c(\hat{k})w_i C_V^i \\ &= \sum_{p=1}^{P} U_{sm'i}(\hat{k})C_V^i. \end{aligned} \tag{8.80}$$

are the receiving pattern and radiation pattern, respectively. Note that the integration in $U_{sm'n}(\hat{k})$ is about a source point $\mathbf{r}_v$ inside the nth tetrahedron ΔV_n. Also, C_V could be J_V or M_V, which is a component of electric current density or magnetic current density on that source point. By using an appropriate quadrature rule, we have superseded the integration with a summation in which w_i is the weight while C_V^i is the value of current density at the ith quadrature point, respectively. In addition, $f_c(\hat{k})$ is one of components resulting from $\overline{\mathbf{f}}(\hat{k})$ or a $\mathbf{f}(\hat{k})$ operator which represents the gradient operation exerted on the scalar Green's function in the $\hat{k}$ domain. In the VIEs or VSIEs, the integrals actually take the form of a $\mathcal{L}$ operator or a $\mathcal{K}$ operator, and $\overline{\mathbf{f}}(\hat{k}) = (\overline{\mathbf{I}} - \hat{k}\hat{k})$ for a $\mathcal{L}$ operator while $\mathbf{f}(\hat{k}) = ik_0\hat{k}$ for a $\mathcal{K}$ operator. With the above relationship between one component of a field and one component of a current density, we can write the matrix element as

$$Z_{ji} = \int \mathrm{d}^2\hat{k}U_{fmj}(\hat{k})\cdot\alpha_{mm'}(\mathbf{k}_0, \mathbf{r}_{mm'})U_{sm'i}(\hat{k}) \tag{8.81}$$

while the unknown coefficient a_i is C_V^i in (8.1).

For VSIEs, the surface current density $\mathbf{J}_S$ on a conducting surface only has two independent components J_S^ξ and J_S^η, which are defined in a local coordinate system (ξ, η, ζ). The local coordinate system is established over the nth triangular patch where the current density reside. We need to use a transformation matrix to transform the representation of current density from the local coordinate system to a global coordinate system (x, y, z) because the MLFMA is implemented in the global coordinate system. The transformation can be expressed as

$$\begin{aligned} \begin{bmatrix} J_S^x \\ J_S^y \\ J_S^z \end{bmatrix} &= \overline{\mathbf{T}}_n \cdot \begin{bmatrix} J_S^\xi \\ J_S^\eta \\ 0 \end{bmatrix} \\ &= \begin{bmatrix} t_x(n,1) & t_x(n,2) & t_x(n,3) \\ t_y(n,1) & t_y(n,2) & t_y(n,3) \\ t_z(n,1) & t_z(n,2) & t_z(n,3) \end{bmatrix} \cdot \begin{bmatrix} J_S^\xi \\ J_S^\eta \\ 0 \end{bmatrix} \\ &= \begin{bmatrix} t_x(n,1) & t_x(n,2) \\ t_y(n,1) & t_y(n,2) \\ t_z(n,1) & t_z(n,2) \end{bmatrix} \cdot \begin{bmatrix} J_S^\xi \\ J_S^\eta \end{bmatrix} \end{aligned} \tag{8.82}$$

where $\left[t_x(n,l), t_y(n,l), t_z(n,l)\right]$ $(l = 1, 2, 3)$ represents the direction cosines characterizing the relationship of coordinate axes between the local coordinate system and the global coordinate system. Also, we need to take two tangential components of (8.73) in the VSIEs to transform it into a matrix equation and this can be done by dot multiplying the equation by two orthogonal unit tangential vectors $\mathbf{u}(\mathbf{r})$ at a field point $\mathbf{r}$ [28]. The corresponding receiving pattern and radiation pattern in the implementation of the MLFMA for this equation can be written as

$$\mathbf{U}^L_{fmj}(\hat{k}) = \frac{1}{4\pi} \mathrm{e}^{\mathrm{i}\mathbf{k}_0 \cdot \mathbf{r}_{jm}} \mathbf{u}_j \tag{8.83}$$

$$\mathbf{U}^L_{sm'i}(\hat{k}) = \mathrm{e}^{-\mathrm{i}\mathbf{k}_0 \cdot \mathbf{r}_{im'}} (\overline{\mathbf{I}} - \hat{k}\hat{k}) \cdot \mathbf{t}_n w_i \tag{8.84}$$

for a $\mathcal{L}$ operator, and

$$\mathbf{U}^K_{fmj}(\hat{k}) = -\frac{\mathrm{i}k_0}{4\pi} \hat{k} \times \mathrm{e}^{\mathrm{i}\mathbf{k}_0 \cdot \mathbf{r}_{jm}} \mathbf{u}_j \tag{8.85}$$

$$\mathbf{U}^K_{sm'i}(\hat{k}) = \mathrm{e}^{-\mathrm{i}\mathbf{k}_0 \cdot \mathbf{r}_{im'}} \mathbf{t}_n w_i \tag{8.86}$$

for a $\mathcal{K}$ operator, where $\mathbf{t}_n = \left[t_x(n,l), t_y(n,l), t_z(n,l)\right]^T$ $(l = 1, 2)$, which is from (8.82). The corresponding matrix element is then calculated by

$$Z^Q_{ji} = \int \mathrm{d}^2\hat{k} \mathbf{U}^Q_{fmj}(\hat{k}) \cdot \alpha_{mm'}(\mathbf{k}_0, \mathbf{r}_{mm'}) \mathbf{U}^Q_{sm'i}(\hat{k}) \tag{8.87}$$

where $Q = L$ or K.

The above formulas are used for a one-level implementation in the MLFMA and the multilevel scheme can be implemented by using the interpolation/anterpolation technique as described in the previous sections.

8.6.4 Numerical Examples

To demonstrate the proposed Nyström-based MLFMA for anisotropic objects, we consider the EM scattering by two large spherical scatterers with anisotropic material as shown in Figure 8.24. The spherical objects are chosen because they have exact analytical solutions for certain anisotropic materials, which can be used to verify the numerical solutions. The first object shown in Figure 8.24(a) is a uniaxially anisotropic sphere with a radius a and the second object sketched in Figure 8.24(b) is a PEC sphere coated with a uniaxially anisotropic material in which the radii of inner and outer interfaces are a_1 and a_2, respectively. The uniaxially anisotropic material in both objects is characterized by the following permittivity tensor and permeability tensor

$$\overline{\epsilon} = \epsilon_0 \begin{bmatrix} \epsilon_r & 0 & 0 \\ 0 & \epsilon_t & 0 \\ 0 & 0 & \epsilon_t \end{bmatrix} \cdot \overline{\mathbf{I}}, \quad \overline{\mu} = \mu_0 \begin{bmatrix} \mu_r & 0 & 0 \\ 0 & \mu_t & 0 \\ 0 & 0 & \mu_t \end{bmatrix} \cdot \overline{\mathbf{I}} \tag{8.88}$$

with $\overline{\mathbf{I}} = \hat{r}\hat{r} + \hat{\theta}\hat{\theta} + \hat{\phi}\hat{\phi}$ in a spherical coordinate system. Note that the permittivity tensor and permeability tensor will be full matrices (non-diagonal) in the Cartesian coordinate system after making a coordinate system transformation, so they can represent globally anisotropic materials. It is assumed that the incident wave is a plane wave with a frequency $f = 300$ MHz and is propagating along the $+z$ direction in free space. We calculate the bistatic RCS observed along the principal cut ($\phi = 0°$ and $\theta = 0°$–$180°$) for

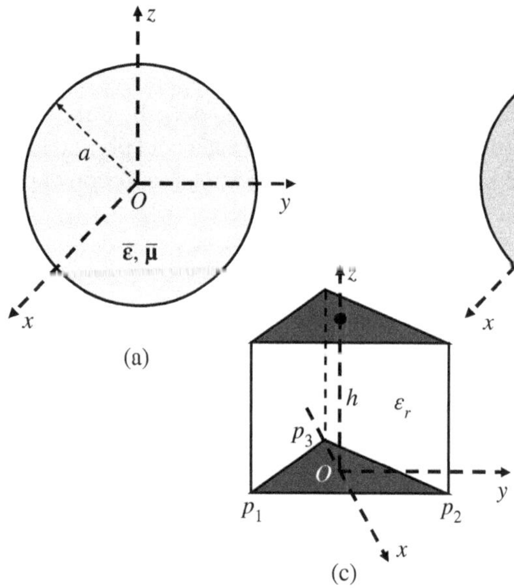

Figure 8.24 Geometries of anisotropic or isotropic scatterers. (a) An anisotropic sphere with a radius a. (b) A PEC sphere coated with a concentric anisotropic layer whose radii of inner and outer interfaces are a_1 and a_2, respectively. (c) An isotropic dielectric triangular cylinder with a base triangle $\Delta p_1 p_2 p_3$ and a height h.

the scatterers in the vertical polarization (VV) or horizontal polarization (HH) and then compare the solutions with corresponding analytical solutions or MoM-based MLFMA solutions with the SWG basis function.

In the first example, we select $a = 12.0\lambda$, $\epsilon_r = 10.0$, $\epsilon_t = 2.0$, $\mu_r = 6.0$, and $\mu_t = 3.0$ for the first object, which is discretized into 163512 tetrahedral elements, resulting in 981072 unknowns in the proposed approach (the number of unknowns per wavelength is 5.16). Note that the ratio of ϵ_r to ϵ_t is 5.0 and such an anisotropy can be thought of as high in practice [44]. The bistatic RCS solutions are plotted in Figure 8.25 for vertical polarization and Figure 8.26 for horizontal polarization, respectively, and we only show the figures in half of the observation range so that they can be seen more clearly. We use the generalized minimal residual (GMRES) iterative method to solve the problem and the plotted figure is a result when the number of iterations reaches 50, resulting in a relative error 2.5762×10^{-4}. It can be found that the Nyström-based MLFMA solutions are in good agreement with the corresponding analytical solutions, which can be obtained using the formulas from [45]. In the contrast, the MoM-based MLFMA solutions with the same discretization (952716 unknowns) are not very close to the analytical solutions due to the relatively high ratio of anisotropy of material.

In the second example, we select $a_1 = 6.0\lambda$ and $a_2 = 12.0\lambda$ for the second object, and the anisotropic material is characterized by $\epsilon_r = 2.0$, $\epsilon_t = 10.0$, $\mu_r = 3.0$, and $\mu_t = 6.0$. We discretize the object into 114726 tetrahedral elements for the anisotropic layer and 52948 triangular patches for the conducting surface, resulting in 794252 unknowns in the proposed approach (the number of unknowns per wavelength is 4.79). The bistatic RCS solutions after 50 iterations (with a relative error 2.1823×10^{-4}) are depicted

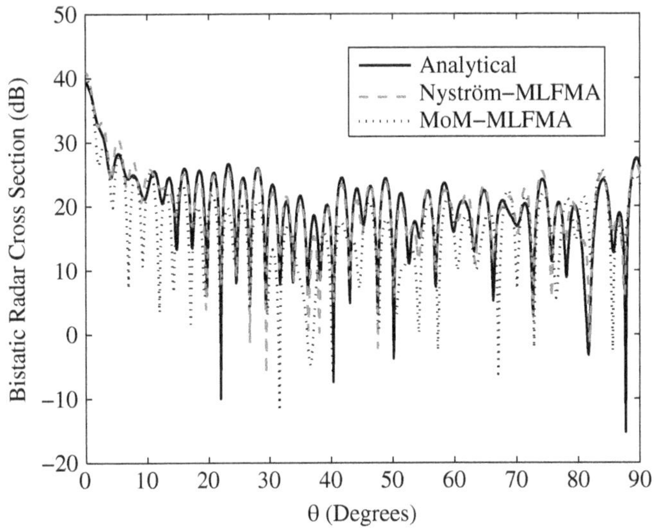

Figure 8.25 Bistatic RCS of a uniaxially anisotropic sphere in vertical polarization. The sphere has a radius $a = 12.0\lambda$ and its material is characterized by $\epsilon_r = 10.0$, $\epsilon_t = 2.0$, $\mu_r = 6.0$, and $\mu_t = 3.0$.

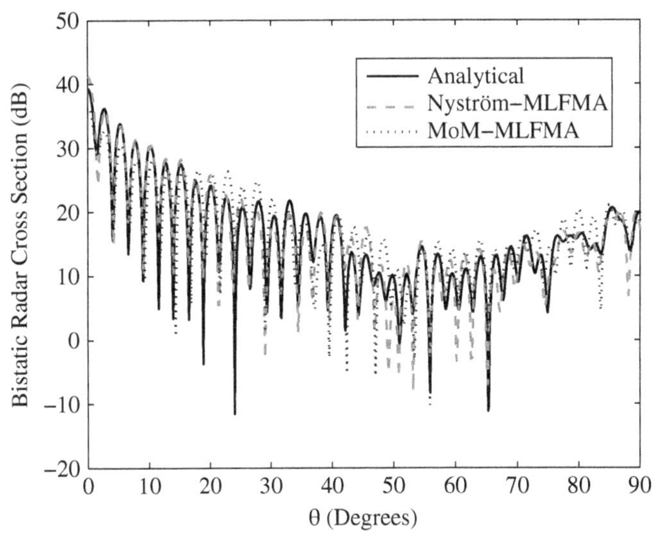

Figure 8.26 Bistatic RCS of a uniaxially anisotropic sphere in horizonal polarization. The sphere has a radius $a = 12.0\lambda$ and its material is characterized by $\epsilon_r = 10.0$, $\epsilon_t = 2.0$, $\mu_r = 6.0$, and $\mu_t = 3.0$.

in Figure 8.27 for vertical polarization and Figure 8.28 for horizontal polarization, respectively. It can be seen again that the Nyström-based MLFMA approach produces the solutions that well agree with the corresponding analytical solutions while the MoM-based MLFMA solutions with the same discretization (745168 unknowns) display larger numerical errors due to the high ratio of anisotropy of material. We use

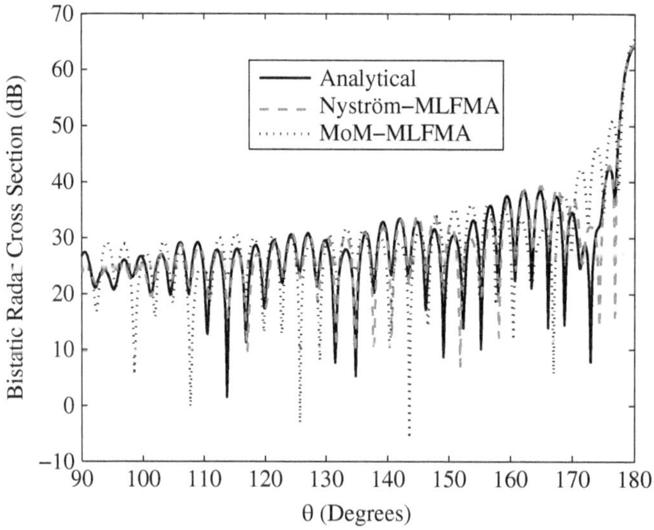

Figure 8.27 Bistatic RCS of a PEC sphere coated with a uniaxially anisotropic layer in vertical polarization. The scatterer has an inner radius $a_1 = 6.0\lambda$ and an outer radius $a_2 = 12.0\lambda$, and the coating material is characterized by $\epsilon_r = 2.0$, $\epsilon_t = 10.0$, $\mu_r = 3.0$, and $\mu_t = 6.0$.

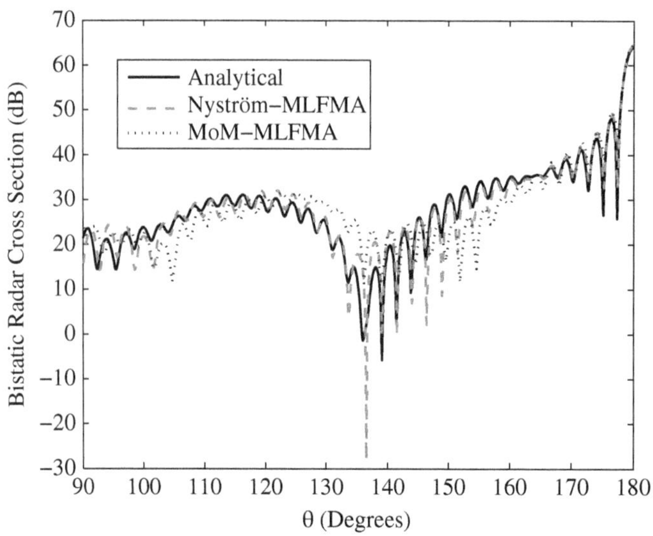

Figure 8.28 Bistatic RCS of a PEC sphere coated with a uniaxially anisotropic layer in horizonal polarization. The scatterer has an inner radius $a_1 = 6.0\lambda$ and an outer radius $a_2 = 12.0\lambda$, and the coating material is characterized by $\epsilon_r = 2.0$, $\epsilon_t = 10.0$, $\mu_r = 3.0$, and $\mu_t = 6.0$.

a one-point quadrature rule in the Nyström method for the two numerical examples so that its number of unknowns is similar to that of the MoM with the SWG basis function. The number of unknowns in the Nyström method is 3 times the number of tetrahedrons (for one unknown current density) in this case while the number of unknowns in the MoM is roughly 2.9 times the number of tetrahedrons, depending on

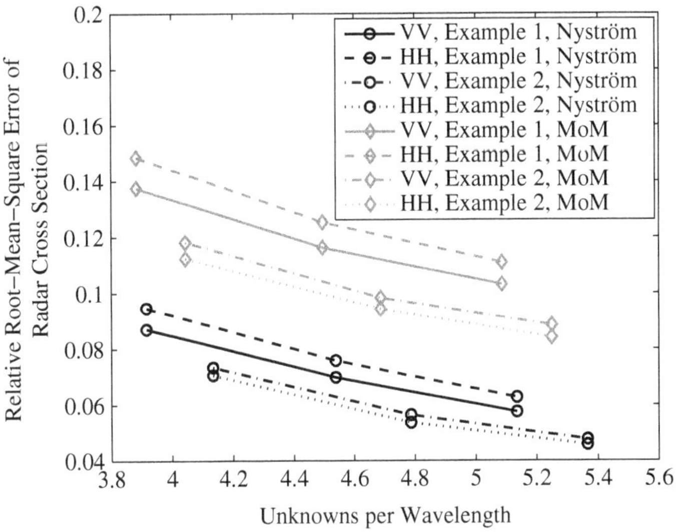

Figure 8.29 Convergence of RCS solutions based on the relative RMS error for the two examples.

the number of half-paired tetrahedrons on the object surface. Figure 8.29 illustrates the convergence of the relative root-mean-square (RMS) error versus the number of unknowns per wavelength for the RCS solutions and it is clear that the numerical error can be reduced by increasing the density of unknowns. The relative RMS error is defined as

$$e_{\mathrm{RMS}} = \frac{\sqrt{\frac{1}{N_0}\sum_{n=1}^{N_0}\left[\sigma_n^{\mathrm{numer}} - \sigma_n^{\mathrm{exact}}\right]^2}}{\sqrt{\frac{1}{N_0}\sum_{n=1}^{N_0}\left[\sigma_n^{\mathrm{exact}}\right]^2}} \tag{8.89}$$

where σ denotes the bistatic RCS and N_0 is the number of observation points in the RCS. The density of unknowns can be increased by using either finer meshes (h-refinement) or higher-order quadrature rules (p-refinement) and it has been demonstrated that the p-refinement has a higher convergence rate than the h-refinement when there is no geometrical approximation [22].

It should be noted that the numerical performance of the Nyström method is similar to the MoM for isotropic or weakly anisotropic structures. The Nyström method requires a little more unknowns (degree of freedoms) but its implementation is more flexible. As a demonstration, Figure 8.30 shows the bistatic RCS solutions of an isotropic dielectric triangular cylinder with a relative permittivity $\epsilon_r = 3.0$ and they are quite consistent even for such a non-smooth object. The object is sketched in Figure 8.24(c) and it has a base triangle defined with three vertices $p_1(0.1\lambda, -0.1\lambda, 0)$, $p_2(0.1\lambda, 0.1\lambda, 0)$, and $p_3(-0.1\lambda, 0, 0)$, and has a height $h = 0.2\lambda$. Finally, we summarize the computational costs, i.e. the CPU time T (Second) and memory usage M (GB), as shown in Table 8.2 for the first two examples and they are consumed on a T7500 Dell Precision workstation with 3.06-GHz CPU and 64-GB RAM. The stop criterion of GMRES iteration is that

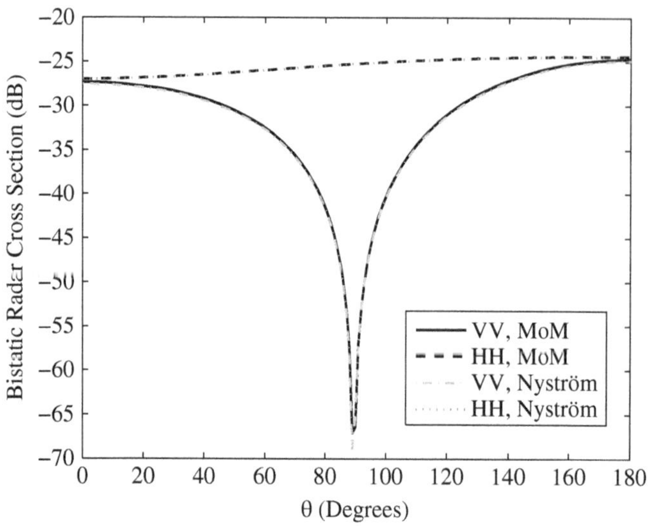

Figure 8.30 Bistatic RCS of an isotropic dielectric triangular cylinder with a relative permittivity $\epsilon_r = 3.0$. The scatterer has a base triangle defined with three vertices $p_1(0.1\lambda, -0.1\lambda, 0)$, $p_2(0.1\lambda, 0.1\lambda, 0)$, and $p_3(-0.1\lambda, 0, 0)$, and has a height $h = 0.2\lambda$.

Table 8.2 Comparison of computational costs between the Nyström-based MLFMA approach and the MoM-based MLFMA approach for the first two examples.

Costs		Example 1	Example 2
Nyström-based MLFMA	CPU time (Second)	52859	47662
	Memory usage (GB)	7.51	6.27
MoM-based MLFMA	CPU time (Second)	67245	65381
	Memory usage (GB)	5.23	4.46

the relative error (difference) between two neighboring iterations reaches 10^{-4} or the number of iterations reaches 50. We can see that the Nyström-based MLFMA approach requires more memory usage than the MoM-based MLFMA approach owing to its higher degree of freedoms or more unknowns, but less CPU time due to its simple mechanism of implementation and better GMRES convergence.

8.7 Summary

In this chapter, we have presented an efficient MLFMA based on the Nyström discretization of integral equations for EM scattering by large objects. Although the MoM-based MLFMA is quite robust whenever the RWG or SWG basis function is enough to represent unknown functions, it could have a problem when representing the magnetic current density in solving the EFIEs and MFIEs for composite structures. Using the Nyström discretization in the MLFMA could be a good solution for the problem because the Nyström scheme does not use any basis and testing functions. The benefits of the Nyström scheme also include the lower requirement on geometrical discretization and the simple mechanism of implementation. The defective meshes that are not allowed in the MoM can be used in the Nyström scheme. The ease of implementation is very desirable in the MLFMA, especially for penetrable or anisotropic objects, because it can easily be incorporated and can quickly complete the preprocessing stage. The only difficulty in the Nyström scheme is the need for local correction for self and near-interaction terms, but efficient local correction schemes have been developed for EM applications. The numerical examples for EM scattering by various large objects have demonstrated the effectiveness of the proposed MLFMA.

Bibliography

1 Chew, W.C., Tong, M.S., and Hu, B. (2008). *Integral Equation Methods for Electromagnetic and Elastic Waves*. San Rafael, CA: Morgan & Claypool.

2 Chew, W.C., Jin, J.M., Michielssen, E., and Song, J.M. (2001). *Fast and Efficient Algorithms in Computational Electromagnetics*. Boston: Artech House.

3 Sarkar, T.K., Arvas, E., and Rao, S.M. (1986). Application of FFT and the conjugate gradient method for the solution of electromagnetic radiation from electrically large and small conducting bodies. *IEEE Trans. Antennas Propagat.* AP-34 (5): 635–640.

4 Coifman, R., Rokhlin, V., and Wandzura, S. (1993). The fast multipole method for the wave equation: a pedestrian prescription. *IEEE Antennas Propagat. Mag.* 35: 7–12.

5 Bleszynski, E., Bleszynski, M., and Jaroszewicz, T. (1996). AIM: adaptive integral method for solving large-scale electromagnetic scattering and radiation problems. *Radio Sci.* 31: 1225–1251.

6 Phillips, J.R. and White, J.K. (1997). A precorrected-FFT method for electrostatic analysis of complicated 3-D structures. *IEEE Trans. Comput. Aided Des. Integr. Circuits Syst.* 16: 1059–1072.

7 Rokhlin, V. (1985). Rapid solution of integral equations of classical potential theory. *J. Comput. Phys.* 60 (2): 187–207.

8 Board, J. and Schulten, K. (2000). The fast multipole algorithm. *IEEE Comput. Sci. Eng.* 2 (1): 76–79.

9 Greengard, L. (1987). *The Rapid Evaluation of Potential Fields in Particle Systems*. Cambridge, MA: MIT Press.

10 Greengard, L. and Rokhlin, V. (1987). A fast algorithm for particle simulations. *J. Comp. Phys.* 73: 325–348.

11 Carrier, J., Greengard, L., and Rokhlin, V. (1988). A fast adaptive multipole algorithm for particle simulations. *SIAM J. Sci. Comput.* 9: 669–686.

12 Rokhlin, V. (1990). Rapid solution of integral equations of scattering theory in two dimensions. *J. Comp. Phys.* 86: 414–439.

13 Lu, C.C. and Chew, W.C. (1994). A multilevel algorithm for solving a boundary integral equation of wave scattering. *Microw. Opt. Technol. Lett.* 7 (10): 466–470.

14 Song, J.M. and Chew, W.C. (1995). Multilevel fast-multipole algorithm for solving combined field integral equations of electromagnetic scattering. *Microw. Opt. Technol. Lett.* 10: 14–19.

15 Dembart, B. and Yip, E. (1995). A 3D fast multipole method for electromagnetics with multiple levels. *Proceedings of 11th Annual Review of Progress in Applied Computational Electromagnetics*, Monterey, CA, vol. 1, pp. 621–628.
16 Song, J.M., Lu, C.C., and Chew, W.C. (1997). Multilevel fast multipole algorithm for electromagnetic scattering by large complex objects. *IEEE Trans. Antennas Propagat.* 45 (10): 1488–1493.
17 Ergül, Ö. and Gürel, L. (2013). Fast and accurate analysis of large-scale composite structures with the parallel multilevel fast multipole algorithm. *J. Opt. Soc. Am. A.* 30 (3): 509–517.
18 Harrington, R.F. (1993). *Field Computation by Moment Methods.* Piscataway, NJ: IEEE Press.
19 Rao, S.M., Wilton, D.R., and Glisson, A.W. (1982). Electromagnetic scattering by surfaces of arbitrary shape. *IEEE Trans. Antennas Propagat.* AP-30 (3): 409–418.
20 Schaubert, D.H., Wilton, D.R., and Glisson, A.W. (1984). A tetrahedral modeling method for electromagnetic scattering by arbitrary shaped inhomogeneous dielectric bodies. *IEEE Trans. Antennas Propagat.* AP-32 (1): 77–85.
21 Kot, J.S. (1990). Solution of thin-wire integral equations by Nyström methods. *Microw. Opt. Technol. Lett.* 3 (11): 393–396.
22 Canino, L.S., Ottusch, J.J., Stalzer, M.A. et al. (1998). Numerical solution of the Helmholtz equation in 2D and 3D using a high-order Nyström discretization. *J. Comput. Phys.* 146 (2): 627–663.
23 Gedney, S.D. (2003). On deriving a locally corrected Nyström scheme from a quadrature sampled moment method. *IEEE Trans. Antennas Propagat.* 51 (9): 2402–2412.
24 Burghignoli, P., Di Nallo, C., Frezza, F., and Galli, A. (2003). A simple Nyström approach for the analysis of 3D arbitrarily shaped conducting and dielectric bodies. *Int. J. Numer. Model.* 16 (2): 179–194.
25 Tong, M.S. and Chew, W.C. (2005). A higher-order Nyström scheme for electromagnetic scattering by arbitrarily shaped surfaces. *IEEE Antennas Wireless Propagat. Lett.* 4: 277–280.
26 Zhu, A., Gedney, S.D., and Visher, J.L. (2005). A study of combined field formulations for material scattering for a locally corrected Nyström discretization. *IEEE Trans. Antennas Propagat.* 53 (12): 4111–4120.
27 Strain, J. (1995). Locally corrected multidimensional quadrature rules for singular functions. *SIAM J. Sci. Comput.* 16 (4): 992–1017.
28 Tong, M.S. and Chew, W.C. (2007). Super-hyper singularity treatment for solving 3D electric field integral equations. *Microw. Opt. Technol. Lett.* 49 (6): 1383–1388.
29 Tong, M.S., Qian, Z.G., and Chew, W.C. (2010). Nyström method solution of volume integral equations for electromagnetic scattering by 3D penetrable objects. *IEEE Trans. Antennas Propagat.* 58 (5): 1645–1652.
30 Qian, Z.G. and Chew, W.C. (2007). Generalized impedance boundary condition for conductor modeling in surface integral equation. *IEEE Trans. Antennas Propagat.* 55 (11): 2354–2364.
31 Chu, Y.H. and Chew, W.C. (2005). A robust surface-integral-equation formulation for conductive media. *Microw. Opt. Technol. Lett.* 46 (2): 109–114.

32 Tong, M.S. and Chew, W.C. (2010). Multilevel fast multipole acceleration in the Nyström discretization of surface electromagnetic integral equations for composite objects. *IEEE Trans. Antennas Propagat.* 58 (10): 3411–3416.
33 Ruck, G.T., Barrick, D.E., Stuart, W.D., and Krichbaum, C.K. (1970). *Radar Cross Section Handbook*. New York: Plenum Press.
34 Coperich, K.M., Ruehli, A.E., and Cangellaris, A. (2000). Enhanced skin effect for partial-element equivalent-circuit (PEEC) models. *IEEE Trans. Microw. Theory Tech.* 48 (9): 1435–1442.
35 Sheng, X.Q., Jin, J.M., Song, J.M. et al. (1998). Solution of combined-field integral equation using multilevel fast multipole algorithm for scattering by homogeneous bodies. *IEEE Trans. Antennas Propagat.* 46 (11): 1718–1726.
36 Tong, M.S. and Chew, W.C. (2010). A novel approach for evaluating hypersingular and strongly singular surface integrals in electromagnetics. *IEEE Trans. Antennas Propagat.* 58 (11): 3593–3601.
37 Chew, W.C. (1995). *Waves and Fields in Inhomogeneous Media*. New York: IEEE Press.
38 Sun, L.E. and Chew, W.C. (2009). A novel formulation of the volume integral equation for electromagnetic scattering. *Wave. Random Complex.* 19 (1): 162–180.
39 Botha, M.M. (2007). Fully hierarchical divergence-conforming basis functions on tetrahedral cells, with applications. *Int. J. Numer. Meth. Eng.* 71 (2): 127–148.
40 Yang, K., Zhou, J.C., Sheng, W.T. et al. (2013). Efficient Nyström solutions of electromagnetic scattering by composite objects with inhomogeneous anisotropic media. *IEEE Trans. Antennas Propagat.* 61 (10): 5328–5332.
41 Markkanen, J., Ylä-Oijala, P., and Sihvola, A. (2012). Discretization of volume integral equation formulations for extremely anisotropic materials. *IEEE Trans. Antennas Propagat.* 60 (11): 5195–5202.
42 Kobidze, G. and Shanker, B. (2004). Integral equation based analysis of scattering from 3-D inhomogeneous anisotropic bodies. *IEEE Trans. Antennas Propagat.* 52 (10): 2650–2658.
43 Tong, M.S., Zhang, Y.Q., Chen, R.P., and Yang, C.X. (2014). Fast solutions of volume integral equations for electromagnetic scattering by large highly anisotropic objects. *IEEE Trans. Microw. Theory Tech.* 62 (7): 1429–1436.
44 Newnham, R.E. (2005). *Properties of Materials: Anisotropy, Symmetry, Structure*. New York: Oxford University Press.
45 Wong, K.-L. and Chen, H.-T. (1992). Electromagnetic scattering by a uniaxially anisotropic sphere. *IEEE Proc. H.* 139 (4): 314–318.

9

Application to Solve Multiphysics Problems

9.1 Introduction

Wave-type energetic illumination on materials or objects could cause multiple physics or multiphysics processes, depending on the properties of waves and materials. These processes could be electrodynamic, mechanical, thermodynamic, chemical or, even quantum-mechanical. Although there exist multiple physics or multiphysics, one or more of those processes may be dominant and others can be ignored, in general, to facilitate the analysis for involved mechanism. For example, in electromagnetic (EM) wave illumination on elastic media like piezoelectric material or biological tissues, the electrodynamic and mechanical processes are dominant when EM energy is limited and thermal effect can be ignored. The analysis for the processes is then based on the solution of corresponding governing equations that include the coupling of Maxwell's equations and the elastic wave equation. Although there are many types of multiphysics problems related to EM waves, such as EM-thermodynamic, EM-mechanical, EM-biological, EM-chemical, and EM-quantum-mechanical problems, we use the EM-elastodynamic problem as an example to illustrate the solution of multiphysics problems by using the Nyström method.

The study of the interaction of EM wave with elastic materials or piezoelectric materials was started by Kyame [1, 2] in 1949. He first investigated the propagation of an EM plane wave in unbounded elastic media with an analytical approach. Later on, many other researchers also studied the interaction of EM waves with various elastic media [3–9]. For example, Tseng and White studied the surface waves in hexagonal crystals [4], while Spaight and Koerber analyzed surface waves in lithium niobate [5]. Mindlin first studied EM radiation from a vibrating quartz plate in 1972. Sedov and Schmerr derived some exact solutions for the propagation of transient electroacoustic waves in piezoelectric media. However, these early investigations mainly focused on solving some simple problems by using analytical approaches. Numerical methods were developed for solving more complicated problems since the 1970s [10]. For instance, Mindlin derived a variational principle for piezoelectromagnetism in a compound continuum representing a diatomic material [11]. Lee gave a variational formulation for the fields inside and outside a body with continuity conditions at the interface between the body and free space [12]. Yang first derived a generalized variational principle with all mechanical and electromagnetic fields as independent variables [13] and his group conducted extensive studies on the interaction of EM waves with elastic waves in piezoelectric media [14–18]. In addition, the work in [19–21] was also very impressive. However, all solutions were

The Nyström Method in Electromagnetics, First Edition. Mei Song Tong and Weng Cho Chew.

based on the partial differential equation (PDE) form of the governing equations, i.e. the coupled Maxwell's equations and elastic wave equations. Therefore, only PDE solvers like the finite difference method (FDM) [22, 23] and the finite element method (FEM) [24, 25] can be used to find the solutions.

Although those differential equation solvers are robust to solve the problems in general, integral equation solvers (IESs) may be preferred in many applications, as demonstrated in previous chapters [26, 27]. We then develop an integral equation method for the interaction of an EM wave and elastic wave in piezoelectric-like elastic media with arbitrary three-dimensional (3D) shapes. The EM part of the integral equations is derived from the vector wave equation by using the equivalence principle and extinction theorem [28] and recognizing that the source current density includes the contribution from the varying displacement vector through strain tensors in Maxwell's equations. The integral equations take the electric field integral equation (EFIE) form although other forms like the magnetic field integral equation (MFIE) or the Poggio-Miller-Chang-Harrington-Wu-Tsai (PMCHWT) formulation [29] can be used. The elastic wave part of the integral equations is derived from the elastic wave equation in its PDE form and also uses the equivalence principle and extinction theorem [31] with the recognition that the source (stress tensor) of elastic wave includes the contribution from an electric field through the EM force. The two parts of integral equations are then fully coupled each other through the source. Using the developed integral equations that take the form of boundary integral equations (BIEs) or surface integral equations (SIEs), all numerical methods for the BIEs like the method of moments (MoM) [32], boundary element method (BEM) [33] or the Nyström method [34] can be used, and the elegance of IESs can be exhibited. We use the Nyström method to solve the coupled BIEs for scattering by 3D piezoelectric objects and the Nyström solution of multiphysics problems can then be illustrated.

The chapter begins with an introduction to elastic wave equations and then the illustration of the Nyström solution for elastic wave scattering by regular or small elastic objects. Since the involved singularity treatment is key and is different from that for EM integral equations, we address it individually in the first section. In the following section, we study the incorporation of the Nyström method with the multilevel fast multipole algorithm (MLFMA) based on the elastic wave equations for solving large elastic wave problems. Since an acoustic wave is a special case of an elastic wave and they share many common characteristics, we also investigate the acoustic wave equation and its Nyström solution with the MLFMA acceleration for solving acoustic wave scattering by very large objects. An investigation for acoustic waves may be very desirable because it is an important source of multiphysics processes and it can be found in many applications, such as thermoacoustic imaging or RFID antennas based on a surface acoustic wave induced through piezoelectric materials. Because of the close relation between elastic waves and acoustic waves, we develop unified integral equations for the two waves in the next section and show that the acoustic wave is a special case of an elastic wave when the shear modulus of the host medium vanishes. Finally, we develop coupled integral equations including both an EM wave and an elastic wave and present the Nyström solution for the equations to study the multiphysics interaction of the two waves in an elastic medium. We provide suitable numerical examples in all sections to demonstrate the Nyström solutions for different equations, either individual or coupled integral equations, and with or without an MLFMA acceleration.

9.2 Solution of Elastic Wave Problems

The study of the radiation and scattering of elastic wave forms the basis of solution for many physical problems in the realm of elastodynamics such as the dynamic stress concentration, non-destructive testing for materials and earth–structure interaction in an earthquake-induced wave environment. These studies usually require solving the elastic wave equations which can be in the form of PDEs or BIEs using numerical procedures. Although the direct solution for the PDEs has been a good approach due to its simplicity [35–38], the BIEs are more desirably solved in many applications and have received an extensive research attention [39–43]. The general BIEs for the elastic wave problems take the form of the Somigliana's formula and it is usually treated as the starting point for the solution procedure [42, 43].

The BEM may be the most widely used numerical tool in solving BIEs for elastic wave problems [44–47]. It is a well-established numerical method and is attractive because it reduces the dimensionality of problems if compared with the FEM or the FDM. The Nyström method belongs to the BEM which discretizes the boundary of a source domain and generates an algebraic matrix equation by the collocation method. The key problem in the Nyström method is the accurate numerical integration for the singular integral kernels. For elastic wave problems, the integral kernels in the Somigliana's formula are the Stokes' displacement tensor and traction tensor and they are singular over a singular element. The special singularity treatment technique known as the singularity regularization or singularity subtraction is needed to remove the singularities in the numerical integrations [45]. The technique usually requires reformulating the original Somigliana's equation into a non-singular or weakly-singular form and this will increase the complexity and computational costs of numerical implementation.

9.2.1 Boundary Integral Equations

Consider the elastic wave scattering problem as shown in Figure 9.1, where an elastic obstacle V_2 bounded with S is embedded in an infinite elastic medium V_1. The corresponding parameters of the media are $(\lambda_2, \mu_2, \rho_2)$ for V_2 and $(\lambda_1, \mu_1, \rho_1)$ for

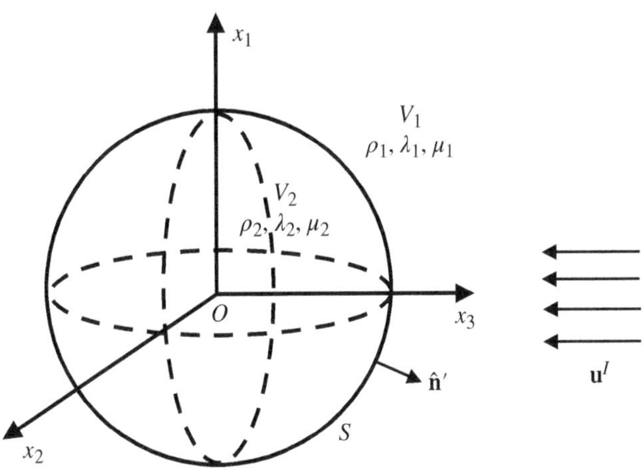

Figure 9.1 Elastic wave scattering by an obstacle embedded in an infinite elastic medium.

V_1 where λ and μ are Lamé constants and ρ is the mass density of the medium. The incident wave is a time-harmonic plane wave propagating along the $-x_3$ direction in V_1 and impinging upon the obstacle, yielding the scattered wave in V_1. The BIE for this problem can be derived from the governing PDE of wave propagation in a homogeneous medium [48]

$$(\lambda + \mu)\nabla\nabla \cdot \mathbf{u} + \mu\nabla^2\mathbf{u} + \omega^2\mathbf{u} = -\rho\mathbf{f} \tag{9.1}$$

where $\mathbf{u}$ is the displacement vector, $\mathbf{f}$ is the body force and ω is the angular frequency. Morse and Feshbach first derived a generalized BIE from the PDE of Helmholtz type by using Huygens' equivalence principle [49]. Pao and Varatharajulu formed the BIE for the elastic wave scattering from (9.1) in a similar way [48]

$$\int_S \left\{ \mathbf{t}(\mathbf{x}') \cdot \overline{\mathbf{G}}(\mathbf{x},\mathbf{x}') - \mathbf{u}(\mathbf{x}') \cdot \left[\hat{\mathbf{n}}' \cdot \overline{\overline{\boldsymbol{\Sigma}}}(\mathbf{x},\mathbf{x}') \right] \right\} \mathrm{d}S' + \mathbf{u}^I(\mathbf{x}) = \begin{cases} \mathbf{u}(\mathbf{x}), & \mathbf{x} \in V_2 \\ 0, & \mathbf{x} \notin V_2 \end{cases} \tag{9.2}$$

where $\mathbf{t}$ is the traction vector that can be related to $\mathbf{u}$ by Hooke's law, $\overline{\mathbf{G}}$ is the dyadic Green's function given by

$$\overline{\mathbf{G}} = \frac{1}{\mu}\left(\overline{\mathbf{I}} + \frac{\nabla\nabla}{\kappa_s^2}\right) g_s(\mathbf{x},\mathbf{x}') - \frac{1}{\gamma}\frac{\nabla\nabla}{\kappa_c^2} g_c(\mathbf{x},\mathbf{x}') \tag{9.3}$$

and $\overline{\overline{\boldsymbol{\Sigma}}}(\mathbf{x},\mathbf{x}')$ is a third-rank Green's tensor with $\Sigma_{\ell mn} = \lambda\delta_{\ell m}\partial_k G_{kn} + \mu(\partial_\ell G_{mn} + \partial_m G_{\ell n})$ in indicial notation. In (9.3), $g_s = \mathrm{e}^{\mathrm{i}\kappa_s r}/(4\pi r)$ and $g_c = \mathrm{e}^{\mathrm{i}\kappa_c r}/(4\pi r)$ are the scalar Green's functions in free space with $r = |\mathbf{x} - \mathbf{x}'|$ being the distance between the field point $\mathbf{x}$ and the source point $\mathbf{x}'$. The subscript s denotes the shear wave and c denotes the compressional wave. The corresponding wavenumbers are given by $\kappa_s^2 = \omega^2\rho/\mu$ and $\kappa_c^2 = \omega^2\rho/\gamma$ with $\gamma = \lambda + 2\mu$. In addition, the superscript I in (9.2) denotes an incident wave, single bar over a vector denotes a dyad, double bars over a vector denote a third-rank tensor and $\overline{\mathbf{I}}$ stands for the identity dyad in (9.2) and (9.3).

Equation (9.2) is actually the same as the Somigliana's identity [42]

$$\overline{\mathbf{C}}^T(\mathbf{x})\mathbf{u}(\mathbf{x}) = \int_S \left[\overline{\mathbf{U}}^T(\mathbf{x},\mathbf{x}')\mathbf{t}(\mathbf{x}') - \overline{\mathbf{T}}^T(\mathbf{x},\mathbf{x}')\mathbf{u}(\mathbf{x}') \right] \mathrm{d}S' + \mathbf{u}^{\mathrm{I}}(\mathbf{x}) \tag{9.4}$$

by identifying the Stokes' displacement and traction tensors $\overline{\mathbf{U}} = \overline{\mathbf{G}}$ and $\overline{\mathbf{T}} = \hat{\mathbf{n}}' \cdot \overline{\overline{\boldsymbol{\Sigma}}}$. In (9.4), the superscript T denotes the transpose and $\overline{\mathbf{C}}(\mathbf{x})$ is a tensor that takes the identity dyad for $\mathbf{x}$ in V_1, 0 for $\mathbf{x}$ in V_2, and a real function of the geometry of S in the vicinity of $\mathbf{x}$ for $\mathbf{x}$ on S. If the geometry is smooth at $\mathbf{x}$, then $\overline{\mathbf{C}}(\mathbf{x}) = \overline{\mathbf{I}}/2$. The Stokes' tensors can be expressed in an indicial notation [42]

$$\begin{aligned} U_{ij} &= \left(\overline{\mathbf{G}}\right)_{ij} \\ &= \frac{1}{4\pi\rho\omega^2 r^3}\left\{ \delta_{ij}\left[(\kappa_s r)^2 \mathrm{e}^{\mathrm{i}\kappa_s r} + D \right] + C\partial_i r \partial_j r \right\} \end{aligned} \tag{9.5}$$

$$\begin{aligned} T_{ij} &= \left(\hat{\mathbf{n}}' \cdot \overline{\overline{\boldsymbol{\Sigma}}}\right)_{ij} \\ &= \frac{1}{4\pi\rho\omega^2 r^4}\left\{ \lambda \mathrm{e}^{\mathrm{i}\kappa_c r}(\kappa_c r)^2(\mathrm{i}\kappa_c r - 1)n_i\partial_j r + \mu \mathrm{e}^{\mathrm{i}\kappa_s r}(\kappa_s r)^2(\mathrm{i}\kappa_s r - 1)\left(\delta_{ij}\frac{\partial r}{\partial n} + n_i\partial_j r \right) \right. \\ &\quad \left. + 2\mu\left[C\left(\delta_{ij}\frac{\partial r}{\partial n} + n_i\partial_j r + n_j\partial_i r \right) + F\partial_i r\partial_j r\frac{\partial r}{\partial n} \right] \right\} \end{aligned} \tag{9.6}$$

where δ_{ij} is the Kronecker delta, $\partial_i r = \partial r/\partial x_i$, $\partial_j r = \partial r/\partial x_j$, and

$$C = \Omega_s e^{i\kappa_s r} - \Omega_c e^{i\kappa_c r} \tag{9.7}$$

$$D = (i\kappa_s r - 1)e^{i\kappa_s r} - (i\kappa_c r - 1)e^{i\kappa_c r} \tag{9.8}$$

$$F = H_c e^{i\kappa_c r} - H_s e^{i\kappa_s r} \tag{9.9}$$

$$\Omega_s = 3 - 3i\kappa_s r - \kappa_s^2 r^2 \tag{9.10}$$

$$\Omega_c = 3 - 3i\kappa_c r - \kappa_c^2 r^2 \tag{9.11}$$

$$H_s = 15 - 15i\kappa_s r - 6\kappa_s^2 r^2 + i\kappa_s^3 r^3 \tag{9.12}$$

$$H_c = 15 - 15i\kappa_c r - 6\kappa_c^2 r^2 + i\kappa_c^3 r^3. \tag{9.13}$$

The Stokes' tensors are smooth for far interaction terms and the related integrals can be performed directly using numerical integrations. However, they possess fictitious $\mathcal{O}(1/r^3)$ and $\mathcal{O}(1/r^4)$ singularities on singular elements and the numerical integrations cannot be applied directly. If we expand the scalar Green's functions g_s and g_c in (9.3) into a series form

$$g(\mathbf{x}, \mathbf{x}') = \frac{e^{ikr}}{r} = \sum_{m=0}^{\infty} \frac{(i\kappa)^m r^{m-1}}{m!} \approx \sum_{m=0}^{M} \frac{(i\kappa)^m r^{m-1}}{m!} \tag{9.14}$$

then the fictitious singularities can be removed by cancelling the leading terms in $\overline{\mathbf{G}}$ and the Stokes' tensors become

$$U_{ij} \approx \frac{1}{4\pi\rho\omega^2}\left\{ \left[\frac{2}{r}(\kappa_s^2 - \kappa_c^2) - i(\kappa_s^3 - \kappa_c^3)\right] \partial_i r \partial_j r + \frac{\delta_{ij}}{r}(\kappa_c^2 - \kappa_s^2 + \kappa_s^2 e^{i\kappa_s r}) \right.$$
$$\left. + \sum_{m=2}^{M} \frac{i^m r^{m-3}}{m!} \left[(\kappa_s^m \Omega_s - \kappa_c^m \Omega_c)\partial_i r \partial_j r + \delta_{ij}\kappa_s^m (i\kappa_s r - 1) - \delta_{ij}\kappa_c^m (i\kappa_c r - 1)\right] \right\} \tag{9.15}$$

$$T_{ij} \approx \frac{1}{4\pi\rho\omega^2}\left[\lambda\kappa_c^2 n_i \partial_j r A + \mu\kappa_s^2 \left(\delta_{ij}\frac{\partial r}{\partial n} + n_j \partial_i r\right) B \right.$$
$$\left. + 2\mu\left(\delta_{ij}\frac{\partial r}{\partial n}P + n_i \partial_j r P + n_j \partial_i r P + \frac{\partial r}{\partial n}\partial_i r \partial_j r Q\right)\right] \tag{9.16}$$

where

$$A = \frac{(i\kappa_c r - 1)}{r^2} e^{i\kappa_c r}$$
$$= -\frac{1}{r^2} - \kappa_c^2 + \sum_{m=2}^{M} \frac{(i\kappa_c)^m}{m!}(i\kappa_c r - 1)r^{m-2} \tag{9.17}$$

$$B = \frac{(i\kappa_s r - 1)}{r^2} e^{i\kappa_s r}$$
$$= -\frac{1}{r^2} - \kappa_s^2 + \sum_{m=2}^{M} \frac{(i\kappa_s)^m}{m!}(i\kappa_s r - 1)r^{m-2} \tag{9.18}$$

$$P = \frac{1}{r^4}(\Omega_s e^{i\kappa_s r} - \Omega_c e^{i\kappa_c r})$$
$$= \frac{1}{2r^2}(\kappa_s^2 - \kappa_c^2) + \frac{ir}{6}(\kappa_s^5 - \kappa_c^5) + \sum_{m=2}^{M} \frac{i^m}{m!}(\kappa_s^m \Omega_s - \kappa_c^m \Omega_c)r^{m-4} \tag{9.19}$$

$$Q = \frac{1}{r^4}(H_c e^{i\kappa_c r} - H_s e^{i\kappa_s r})$$
$$= \frac{3}{2r^2}(\kappa_c^2 - \kappa_s^2) - \frac{1}{2}(\kappa_c^4 - \kappa_s^4) + \frac{i}{2}(\kappa_c^5 - \kappa_s^5)r + \frac{1}{6}(\kappa_c^6 - \kappa_s^6)r^2$$
$$+ \sum_{m=4}^{M} \frac{i^m}{m!}(\kappa_s^m \Omega_s - \kappa_c^m \Omega_c) r^{m-4}. \tag{9.20}$$

In the new form of the Stokes' tensors, the $\mathcal{O}(1/r)$ singularity in U_{ij} can be handled using the regularization technique in the Nyström method or other techniques like the Duffy's method [50] in electromagnetics. The $\mathcal{O}(1/r^2)$ singularity in T_{ij} is handled using the subtraction technique in the Nyström method, but we derive closed-form formulas in a Cauchy-principal-value (CPV) sense. Since the new form of the Stokes' tensors is only applied to the singular or near singular elements where r is small, the series converges fast and the number of truncated terms M is small.

If we incorporate boundary conditions that are the continuity of displacement and traction vectors $\mathbf{u}$ and $\mathbf{t}$ on S, the BIEs in indicial notation can be written as

$$\int_S \left[T_{ij}^{(1)}(\mathbf{x}, \mathbf{x}') u_j(\mathbf{x}') - U_{ij}^{(1)}(\mathbf{x}, \mathbf{x}') t_j(\mathbf{x}') \right] dS' + \frac{1}{2}\delta_{ij} u_i(\mathbf{x}) = u_i^I(\mathbf{x}), \quad \mathbf{x} \in S \tag{9.21}$$

$$\int_S \left[U_{ij}^{(2)}(\mathbf{x}, \mathbf{x}') t_j(\mathbf{x}') - T_{ij}^{(2)}(\mathbf{x}, \mathbf{x}') u_j(\mathbf{x}') \right] dS' + \frac{1}{2}\delta_{ij} u_i(\mathbf{x}) = 0, \quad \mathbf{x} \in S \tag{9.22}$$

on which the numerical procedure is based. The superscripts on the Stokes' tensors denote the media and we choose the field point $\mathbf{x}$ on a smooth surface leading to the coefficient of $\delta_{ij}/2$ in front of $u_i(\mathbf{x})$. If the obstacle is a traction-free cavity or a fixed rigid inclusion, the above BIEs can be simplified by enforcing $\mathbf{t} = 0$ or $\mathbf{u} = 0$ on S.

9.2.2 Singularity Treatment

As we have mentioned before, some singularity treatment techniques have not been addressed in Chapter 5 and we will present them as a supplementary in incorporation with physical applications, as appearing in this section. The Nyström method can also be used to transform the above BIEs into a matrix equation after discretizing the geometry of an object. We usually express the unknown function into interpolation functions with unknown coefficients and replacing the unknown function with such an expression leads to the integration of the kernels with the interpolation functions for each matrix entry. The interpolation functions are chosen as shape functions of an element in an isoparametric scheme and the unknown coefficients are the unknowns to be solved. The solution of the BIEs strongly relies on the accurate evaluation of singular matrix elements and we individually address the singularity treatment here because it is somewhat different from that for EM integral equations. Since the weakly-singular integrals with $\mathcal{O}(1/r)$ singularity can be handled by the regularization technique or Duffy's method, we will not discuss them here. We focus on those strongly-singular integrals with $\mathcal{O}(1/r^2)$ singularity, which are handled by using the singularity subtraction technique and we derive closed-form expressions here. If we combine the interpolation function with the integral kernel T_{ij} in (9.16), the strongly-singular integrals have the following terms in T_{ij}

$$f_1 = \frac{1}{r^2}\frac{\partial r}{\partial x_i} \tag{9.23}$$

$$f_2 = \frac{1}{r^2}\frac{\partial r}{\partial x_i}\frac{\partial r}{\partial x_j}\frac{\partial r}{\partial n}. \tag{9.24}$$

The resultant integrals take the following forms in a global coordinate systems (x_1, x_2, x_3)

$$I_1 = \int_{\Delta S} \frac{(x_1 - x_1')}{r^3}\,\mathrm{d}S'; \qquad I_2 = \int_{\Delta S} \frac{(x_2 - x_2')}{r^3}\,\mathrm{d}S'; \tag{9.25}$$

$$I_3 = \int_{\Delta S} \frac{(x_3 - x_3')}{r^3}\,\mathrm{d}S'; \qquad I_4 = \int_{\Delta S} \frac{(x_1 - x_1')^3}{r^5}\,\mathrm{d}S'; \tag{9.26}$$

$$I_5 = \int_{\Delta S} \frac{(x_2 - x_2')^3}{r^5}\,\mathrm{d}S'; \qquad I_6 = \int_{\Delta S} \frac{(x_3 - x_3')^3}{r^5}\,\mathrm{d}S'; \tag{9.27}$$

$$I_7 = \int_{\Delta S} \frac{(x_1 - x_1')^2(x_2 - x_2')}{r^5}\,\mathrm{d}S'; \quad I_8 = \int_{\Delta S} \frac{(x_1 - x_1')(x_2 - x_2')^2}{r^5}\,\mathrm{d}S'; \tag{9.28}$$

$$I_9 = \int_{\Delta S} \frac{(x_1 - x_1')^2(x_3 - x_3')}{r^5}\,\mathrm{d}S'; \quad I_{10} = \int_{\Delta S} \frac{(x_1 - x_1')^2(x_3 - x_3')}{r^5}\,\mathrm{d}S'; \tag{9.29}$$

$$I_{11} = \int_{\Delta S} \frac{(x_2 - x_2')^2(x_3 - x_3')}{r^5}\,\mathrm{d}S'; \quad I_{12} = \int_{\Delta S} \frac{(x_2 - x_2')(x_3 - x_3')^2}{r^5}\,\mathrm{d}S'; \tag{9.30}$$

$$I_{13} = \int_{\Delta S} \frac{(x_1 - x_1')(x_2 - x_2')(x_3 - x_3')}{r^5}\,\mathrm{d}S'. \tag{9.31}$$

We create a local coordinate system (u, v, w) over the singular element, as shown in Figure 9.2. The transformation relation between the global and local coordinate systems is

$$\begin{bmatrix} x_1 \\ x_2 \\ x_3 \end{bmatrix} = \begin{bmatrix} \ell_1 & \ell_2 & \ell_3 \\ m_1 & m_2 & m_3 \\ n_1 & n_2 & n_3 \end{bmatrix} \begin{bmatrix} u \\ v \\ w \end{bmatrix} + \begin{bmatrix} u_0 \\ v_0 \\ w_0 \end{bmatrix} \tag{9.32}$$

where (ℓ_1, m_1, n_1), (ℓ_2, m_2, n_2), and (ℓ_3, m_3, n_3) are the direction cosines of three local coordinate axes in the global coordinate system and (u_0, v_0, w_0) are the coordinates of the local coordinate system's origin in the global coordinate system. In this local coordinate system, those strongly-singular integrals can be expressed as

$$I_k = a_k^{(1)} I_u^{(3)} + a_k^{(2)} I_v^{(3)} \quad \text{for } k = 1, 2, 3; \tag{9.33}$$

$$I_k = a_k^{(3)} I_u^{(5)} + a_k^{(4)} I_{uv}^{(5)} + a_k^{(5)} I_{vu}^{(5)} + b_k^{(6)} I_v^{(5)} \quad \text{for } k = 4, \cdots, 13 \tag{9.34}$$

where

$$I_u^{(3)} = \int_{\Delta S} \frac{u'}{r^3}\,\mathrm{d}S'; \quad I_v^{(3)} = \int_{\Delta S} \frac{v'}{r^3}\,\mathrm{d}S'; \tag{9.35}$$

$$I_u^{(5)} = \int_{\Delta S} \frac{u'^3}{r^5}\,\mathrm{d}S'; \quad I_{uv}^{(5)} = \int_{\Delta S} \frac{u'^2 v'}{r^5}\,\mathrm{d}S'; \tag{9.36}$$

$$I_{vu}^{(5)} = \int_{\Delta S} \frac{u' v'^2}{r^5}\,\mathrm{d}S'; \quad I_v^{(5)} = \int_{\Delta S} \frac{v'^3}{r^5}\,\mathrm{d}S' \tag{9.37}$$

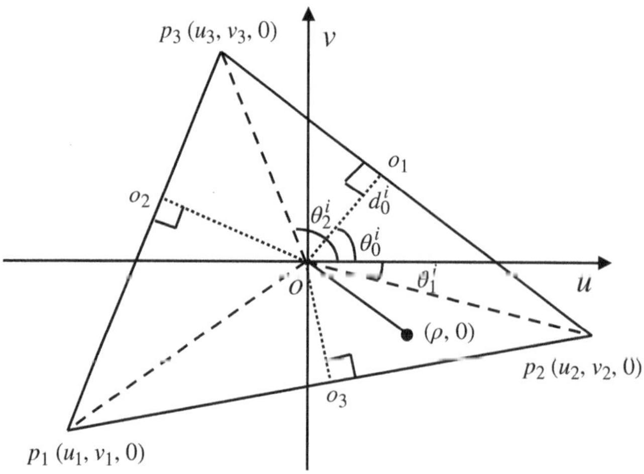

Figure 9.2 Local coordinate system over a singular element. The field point is initially located at $(0, 0, w)$ and approaches the element with $w \to 0$.

and $a_k^{(i)}$ $(i = 1, \cdots, 6)$ are the constants related to the direction cosines. The integrals in (9.35)–(9.37) are strongly singular and the accurate evaluation for them is the key part in the Nyström method. We evaluate these integrals under a CPV sense, i.e. we assume that the field point is initially off the singular element with a vertical distance of w from the singular element plane. After deriving the analytical solution for those integrals with $w \neq 0$ under the polar coordinate system as shown in Figure 9.2, we take the limit of $w \to 0$ to find the exact solutions for the integrals in (9.35)–(9.37). The final results can be found in Appendix 9.1.

9.2.3 Numerical Examples

Several numerical examples are presented to demonstrate the validity of the Nyström method. The one-point quadrature rule that corresponds to a constant approximation for the unknown function over an element is used in all cases. The higher-order quadrature rules corresponding to a higher-order interpolation for the unknown function can be used in a straightforward way. This is because the strongest singularity comes from the constant term of the interpolation function. The higher-order terms of interpolation functions will weaken the degree of singularity when exerting on the kernels and the resultant integrals can all be handled numerically. The tangential component in the following denotes the θ component in all cases.

We first consider the scattering by a fixed rigid sphere with a radius of $a = 1.0$. The surrounding medium has Poisson's ratio $\nu = 0.25$ and mass density $\rho = 1.0$. The incident wave is a time-harmonic plane dilatational wave with a unit circular frequency ($\omega = 1.0$) and normalized wavenumber of $\kappa_c a = 0.125, 0.913$, and π, respectively. We use 360, 960, and 960 triangle elements, respectively, in the discretization of the geometry for these wavenumbers. The meshes are over fine for the $\kappa_c a = 0.125$ and 0.913 cases (small and middle sizes) because we want to show more points on the curves. Figures 9.3–9.5 show the total radial and tangential components of traction along the principal cut ($\phi = 0°$ and

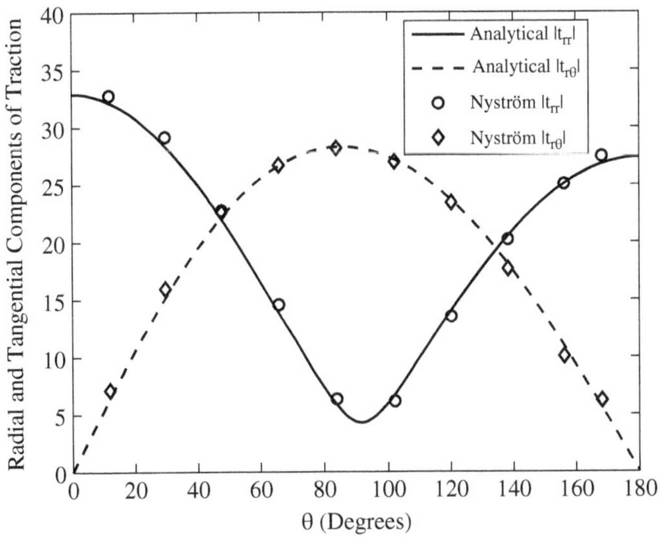

Figure 9.3 Total radial and tangential components of traction along the elevated cut at the surface of a rigid sphere, $k_c a = 0.125$.

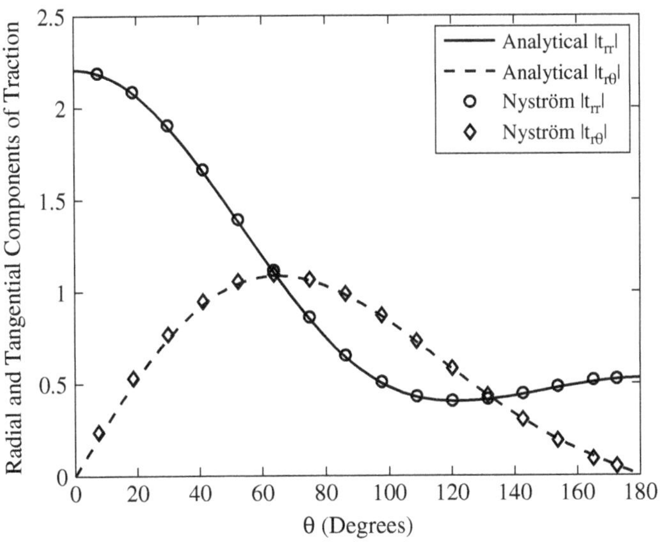

Figure 9.4 Total radial and tangential components of traction along the elevated cut at the surface of a rigid sphere, $k_c a = 0.913$.

$\theta = 0° - 180°$) at the surface. It can be seen that the solutions agree with the analytical solutions very well. The analytical solutions can be found in [51–53].

We next consider the scattering by a traction-free spherical cavity surrounded by an infinite elastic medium. The cavity also has a radius of $a = 1.0$ and the surrounding medium is characterized by Poisson's ratio $\nu = 1/3$, Young's modulus $E = 2/3$ and mass density $\rho = 1.0$. The incident wave is the same as before but the normalized

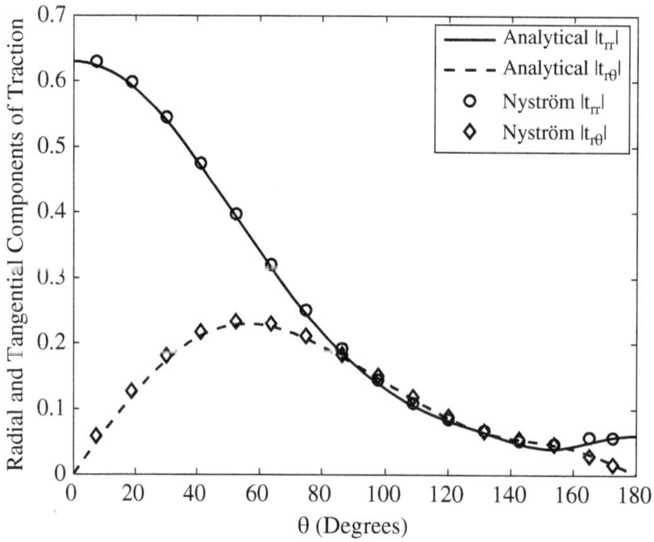

Figure 9.5 Total radial and tangential components of traction along the elevated cut at the surface of a rigid sphere, $k_c a = \pi$.

wavenumbers are chosen as $\kappa_c a = 0.09$ and 0.913 corresponding to 360 and 960 triangle meshes in the discretization, respectively. Figures 9.6 and 9.7 show the scattered radial and tangential displacements along the principal cut at $r = 5a$. The solutions are also compared with the analytical solutions and they are very close to each other.

We finally consider the generalized case with both the surrounding medium and obstacle being elastic. We select $\lambda_1 = 0.53486$, $\mu_1 = 0.23077$, and $\rho_1 = 1.0$ for the

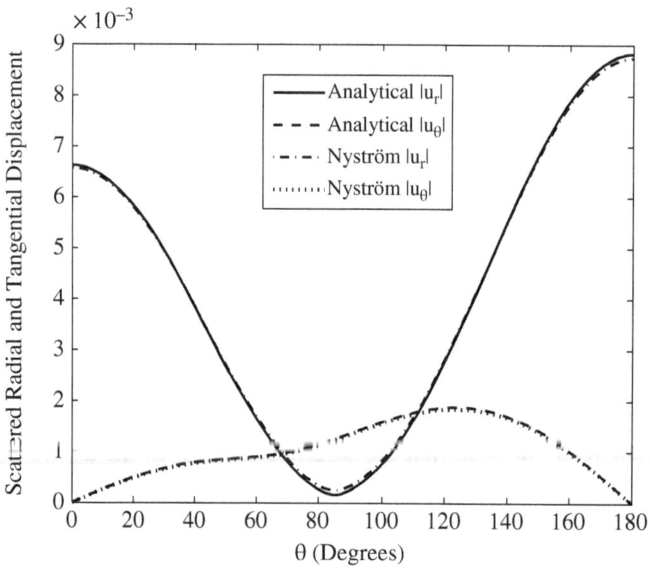

Figure 9.6 Scattered radial and tangential displacement of a spherical cavity at $r = 5a$, $k_c a = 0.09$.

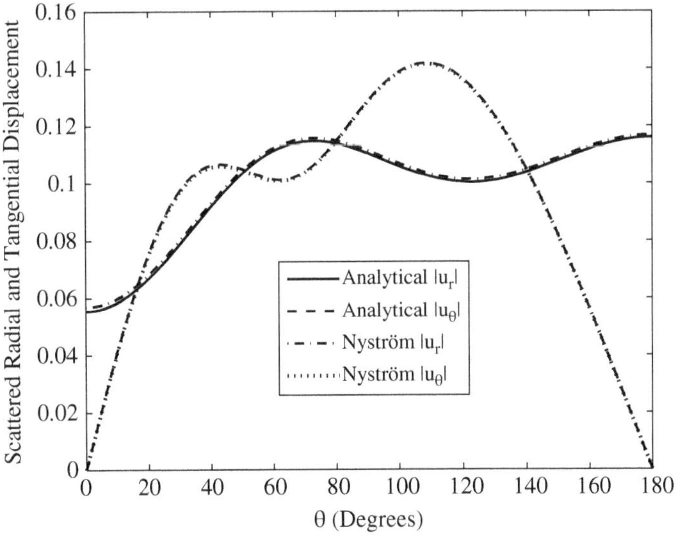

Figure 9.7 Scattered radial and tangential displacement of a spherical cavity at $r = 5a$, $k_c a = 0.913$.

surrounding medium, and $\lambda_2 = 0.23716$, $\mu_2 = 0.52641$, and $\rho_2 = 1.9852$ for the elastic spherical obstacle with a unit radius. The incident wave is also the same as before but the normalized wavenumbers are chosen as $\kappa_c a = 0.125$, 2.0 and 5.0, respectively. Figures 9.8–9.10 plot the total radial components of displacement and traction at the surface along the principal cut. These results, obtained using 360, 960, and 960 elements, respectively, are again in excellent agreement with the analytical solutions. Note that $\kappa_c a = 4.4934$ is a wavenumber close to the fictitious eigenfrequency in this

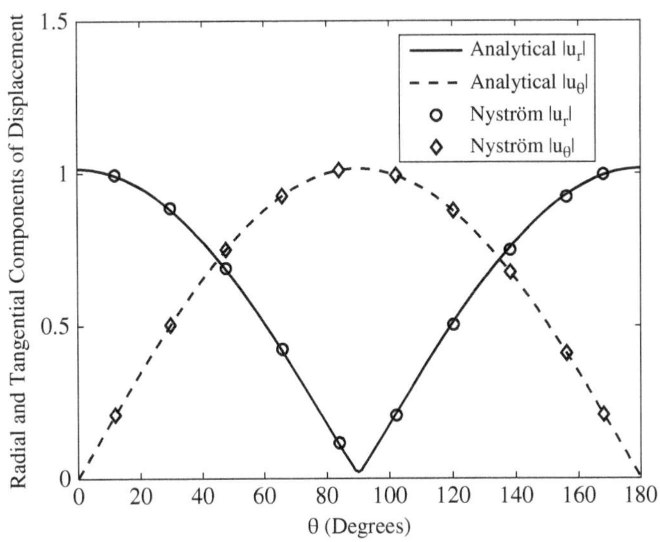

Figure 9.8 Total radial and tangential components of displacement along the elevated cut at the surface of an elastic sphere, $k_c a = 0.125$.

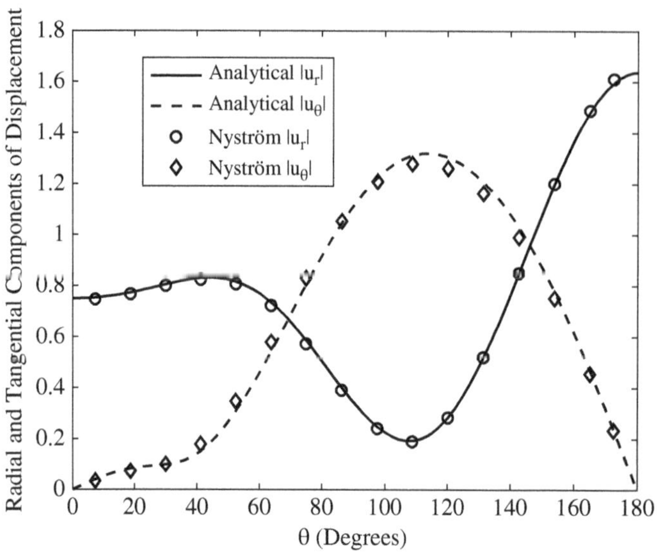

Figure 9.9 Total radial and tangential components of displacement along the elevated cut at the surface of an elastic sphere, $k_c a = 2.0$.

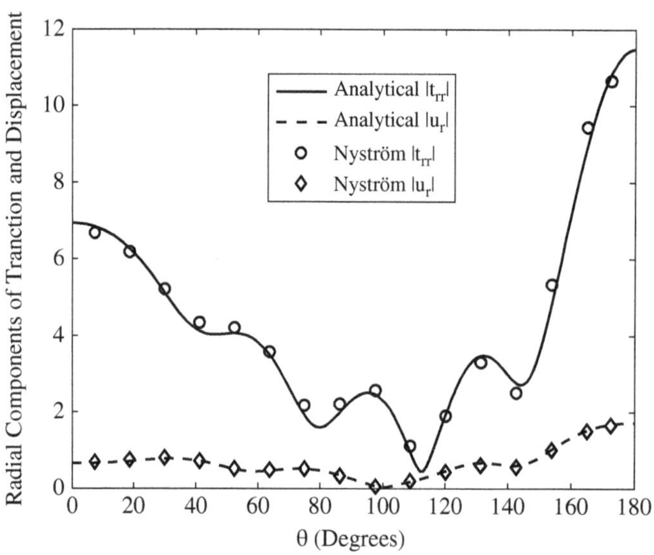

Figure 9.10 Total radial components of traction and displacement along the elevated cut at the surface of an elastic sphere, $k_c a = 5.0$.

case and it is hard to obtain a good result using the traditional BEM [54]. To get rid of the resonance problem, one usually needs to reformulate the integral equation and a higher cost is required in the solution process. However, like the method of fundamental solution [55], The Nyström method solutions do not sharply worsen near the fictitious eigenfrequency when solving the original equation. The results

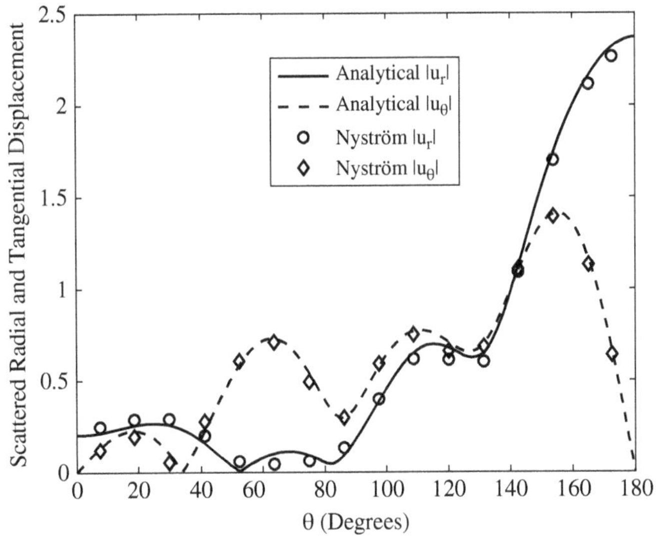

Figure 9.11 Scattered radial and tangential displacement of an elastic sphere along the principal cut at the surface, $k_c a = 4.4934$.

using 960 elements are still good enough, as shown in Figure 9.11. The insensitivity of the eigenfrequency may allow the method to be a good choice in seeking numerical solutions. Therefore, this method does not remove the internal resonance problem, but makes the solution robust close to the resonance frequency.

9.3 MLFMA Acceleration for Solve Large Elastic Wave Problems

The MLFMA and its application to rapidly solve large electromagnetic problems were introduced in the previous chapter. As pointed out in a review article for the fast multipole method (FMM) [56], the most impressive development for the method is in electrical engineering where workers use the method to solve very large problems by extending the method to MLFMA [57–59]. A typical application for this method is to calculate electromagnetic scattering by aircraft in which millions of unknowns are involved in the discretization of integral equations [60]. The solutions for the problems with 10–20 million unknowns were reported several years ago [61–63], and most recently the problem with more than 85 million unknowns was also solved in a parallel scheme [64, 65].

Compared to the rapid development of the FMA in electrical engineering, the BIE community in applied mechanics seems to remain less active with respect to the FMM [56], even in recent years. In particular, very few publications can be found for elastodynamic applications of the method. In those papers [66–71] mentioned by the review article [56], the first three only applied the FMA to solve two-dimensional (2D) problems for elastic wave scattering by a rough interface or by many cavities and cracks or in the time domain. The other publications dealt with 3D problems but the maximum number of unknowns was $N = 24576$ [71]. There was no significant advance in recent years, i.e. after the review paper in terms of literature search. Some recent works also mainly focused on the 2D case, the elastostatic case or the time domain approach for 3D problems [72–75]. We intend to enhance the application of the FMA in elastodynamics by presenting this work in which the MLFMA is developed for solving elastic wave scattering by large 3D objects in the frequency domain.

Unlike the acoustic wave and electromagnetic wave BIEs in which only one wave exists, the elastic wave BIEs include multiple waves and more complicated kernels. Hence the implementation of the FMA is quite different. Since the FMA is based on the addition theorem for the Green's function and the kernel for one medium involves two different scalar Green's functions, i.e. one for the compressional wave and the other for the shear wave, we need to create two different FMA trees for one medium. If the Nyström method is used as a matrix equation generator underlying the FMA and wisely formulate the radiation and receiving patterns, there will be nine radiation and receiving patterns and nine matrix-vector multiplications are needed in each iteration.

Since multiple trees are used, the difference in wavenumbers in different trees leads to the different definitions for well-separated groups by which the near terms and far terms are defined. The far terms are handled by the MLFMA but the near terms are still generated with the based Nyström method. In the original Nyström method implementation without MLFMA acceleration [76], we evaluate the near terms by combining the two scalar Green's functions in the kernels, viz. the compressional wave Green's function and the shear wave Green's function. This combination will cancel the hypersingularity in the leading terms of the series expansion of the scalar Green's functions, resulting in weakly-singular $1/R$ and strongly-singular $1/R^2$ kernels. However, the multiple-tree structure in the MLFMA will destroy the combination and require the evaluation of both near terms and far terms tree by tree. Fortunately, we can recombine those near terms from different trees by searching the common near patches and evaluate them using the scheme in the original Nyström method with appropriate reformulation.

We have implemented the proposed MLFMA as demonstrated in numerical examples. Due to the high cost of the multiple tree structure and the complexity of integral kernels, we can only solve the elastic wave scattering problems with 0.3–0.4 million unknowns currently on our Dell Precision 690 workstation. This machine has two dual core 3.0 GHz processors and 16 GB RAM, but we only use one core in the series implementation. Hence, the memory usage is intentionally limited to within 12 GB.

9.3.1 Formulations

Consider the typical elastic wave scattering by a 3D elastic object embedded in a homogeneous elastic medium, as shown in Figure 9.12. The elasticity of a medium is characterized by (ρ, λ, μ), where ρ is the mass density, λ is the bulk modulus, and μ is the shear modulus. The subscript 1 denotes the surrounding medium and 2 denotes the object. The governing BIEs for the problem can be found such that [42, 48]

$$\frac{1}{2}\mathbf{u}(\mathbf{x}) + \int_S \left[\overline{\mathbf{T}}_1^T(\mathbf{x},\mathbf{x}')\cdot\mathbf{u}(\mathbf{x}') - \overline{\mathbf{G}}_1^T(\mathbf{x},\mathbf{x}')\cdot\mathbf{t}(\mathbf{x}')\right] \mathrm{d}S' = \mathbf{u}^{\mathrm{I}}(\mathbf{x}), \quad \mathbf{x}\in S \tag{9.38}$$

$$\frac{1}{2}\mathbf{u}(\mathbf{x}) + \int_S \left[\overline{\mathbf{G}}_2^T(\mathbf{x},\mathbf{x}')\cdot\mathbf{t}(\mathbf{x}') - \overline{\mathbf{T}}_2^T(\mathbf{x},\mathbf{x}')\cdot\mathbf{u}(\mathbf{x}')\right] \mathrm{d}S' = 0, \quad \mathbf{x}\in S \tag{9.39}$$

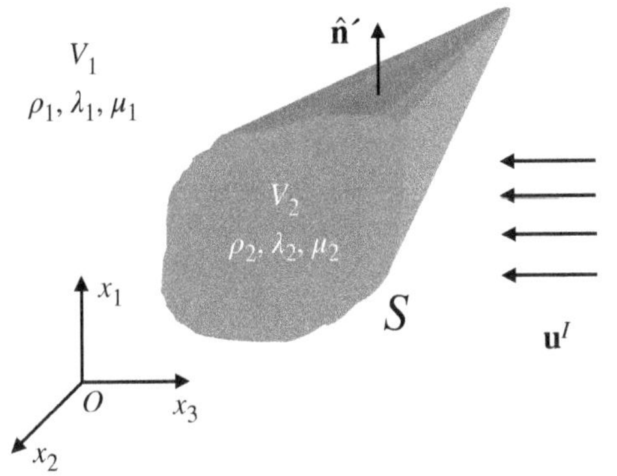

Figure 9.12 Elastic wave scattering by a 3D elastic object embedded in an elastic medium.

where **u** and **t**, the unknowns to be solved, are the total displacement and traction vectors at the surface S of the object, $\overline{\mathbf{G}}$ is the tensor Green's function given by

$$\overline{\mathbf{G}} = \frac{1}{\mu}\left(\overline{\mathbf{I}} + \frac{\nabla\nabla}{k_s^2}\right) g_s(\mathbf{x}, \mathbf{x}') - \frac{1}{\gamma}\frac{\nabla\nabla}{k_c^2} g_c(\mathbf{x}, \mathbf{x}') \tag{9.40}$$

and $\overline{\mathbf{T}} = \hat{n}' \cdot \overline{\overline{\boldsymbol{\Sigma}}}(\mathbf{x}, \mathbf{x}')$ where $\overline{\overline{\boldsymbol{\Sigma}}}(\mathbf{x}, \mathbf{x}') = \lambda \overline{\mathbf{I}} \nabla \cdot \overline{\mathbf{G}} + \mu\left(\nabla \overline{\mathbf{G}} + \overline{\mathbf{G}} \nabla\right)$ is a third-rank Green's tensor. $\overline{\mathbf{G}}$ and $\overline{\mathbf{T}}$ are also known as the Stokes' displacement and traction tensors, respectively, and the superscript T on them denotes the transpose. In (9.40), $g_s = e^{ik_s r}/(4\pi r)$ and $g_c = e^{ik_c r}/(4\pi r)$ are the scalar Green's functions in free space, and $r = |\mathbf{x} - \mathbf{x}'|$ is the distance between a field point $\mathbf{x}$ (without a prime) and a source point $\mathbf{x}'$ (with a prime). The subscript "s" denotes the shear wave and "c" denotes the compressional wave. The corresponding wavenumbers are given by $k_s^2 = \omega^2 \rho/\mu$ and $k_c^2 = \omega^2 \rho/\gamma$ with $\gamma = \lambda + 2\mu$, and ω is the angular frequency here. Also, the superscript I in (9.38) denotes an incident wave, the hat on a variable indicates a unit vector, the boldface of a variable implies a vector, a single bar over a vector denotes a dyad, double bars over a vector denote a third-rank tensor, and $\overline{\mathbf{I}}$ stands for the identity dyad in (9.40). If the object is a rigid inclusion or traction-free cavity, the above equations can be simplified and the individual MLFMA formula can be developed for each case.

Note that the multiple trees are needed in the implementation because two waves with different wavenumbers exist in each medium. The shear wave has a wavenumber k_s and the compressional wave has a wavenumber k_c. The ratio of these two wavenumbers will be

$$\frac{k_s}{k_c} = \sqrt{\frac{\gamma}{\mu}} = \sqrt{\frac{\lambda}{\mu} + 2}. \tag{9.41}$$

Since both λ and μ are positive, the above ratio will be greater than $\sqrt{2}$. If the object is elastic, then there are also two wavenumbers inside the object that are different from the two wavenumbers in the surrounding medium.

Due to the use of a multi-tree structure, the computational costs are quite different from those in EM problems. For elastic objects, we have four FMA trees and each tree has an independent aggregation–translation–disaggregation process. Therefore, compared with the single-tree EM problems, the costs will be at least four times.

Also, there are nine radiation and receiving patterns in total for each medium (two trees). This is because the unknown vectors in the elastic BIEs are three-dimensional (three components) over the boundary (surface) and we use the Nyström method to solve it. In the contrast, the unknown vectors (electric and magnetic current densities) in the EM SIEs are two-dimensional, namely, they are surface vectors. If we use the solving process for the EM SIEs to solve the elastic BIEs, namely use the MoM with the Rao-Wilton-Glisson (RWG) basis function to expand the surface component and pulse basis function to expand the normal component of unknown vectors in the elastic BIEs, the number of patterns will be four. The other contribution to the difference of the "CONSTANT" in the computational complexity expression [27] comes from the big difference in formulations between these two types of problems. From the next subsection, i.e. Section 9.3.2, we will see that the kernels in the elastic BIEs are much more complex than those in the EM SIEs, so the implementation costs are also much higher.

Scattering by a Rigid Object

If the object is rigid, the unknown displacement vector on the surface vanishes and the above BIEs can be reduced to

$$\int_S \overline{\mathbf{G}}_1^T(\mathbf{x},\mathbf{x}')\cdot\mathbf{t}(\mathbf{x}')\,\mathrm{d}S' = -\mathbf{u}^{\mathrm{I}}(\mathbf{x}),\quad \mathbf{x}\in S. \tag{9.42}$$

From (9.40), we can see that the above kernel $\overline{\mathbf{G}}$ includes two parts. One is related to the shear wave Green's function g_s and the other, to the compressional wave Green's function g_c. We have to create an individual FMA tree for each of the two Green's functions so that the far interaction terms between the ith source point and jth field point can be calculated by

$$A_{ji}^w = \int \mathrm{d}^2\hat{k}_w V_{fmj}^w(\hat{k}_w)\cdot\alpha_{mm'}^w(\mathbf{k}_w,\mathbf{r}_{mm'})V_{sm'i}^w(\hat{k}_w),\qquad w = s \text{ or } c \tag{9.43}$$

where $V_{sm'i}^w$ and V_{fmj}^w are radiation pattern and receiving pattern, respectively, and $\alpha_{mm'}^w$ is the translator. They are defined for the shear wave when $w = s$ and the compressional wave when $w = c$, respectively. From the above kernel and the addition theorem for the scalar Green's function as shown in the preceding chapter, we can find the corresponding receiving pattern and radiation pattern as follows

$$V_{fmj}^s(\hat{k}_s) = \frac{\mathrm{e}^{\mathrm{i}\mathbf{k}_s\cdot\mathbf{r}_{jm}}}{4\pi\mu} \tag{9.44}$$

$$V_{fmj}^c(\hat{k}_c) = \frac{\mathrm{e}^{\mathrm{i}\mathbf{k}_c\cdot\mathbf{r}_{jm}}}{4\pi\gamma} \tag{9.45}$$

$$\overline{\mathbf{V}}_{sm'i}^s(\hat{k}_s) = \int_{\Delta S_i}\mathrm{d}S'\overline{\mathbf{E}}_s\mathrm{e}^{-\mathrm{i}\mathbf{k}_s\cdot\mathbf{r}_{im'}} \tag{9.46}$$

$$\overline{\mathbf{V}}_{sm'i}^c(\hat{k}_c) = \int_{\Delta S_i}\mathrm{d}S'\overline{\mathbf{E}}_c\mathrm{e}^{-\mathrm{i}\mathbf{k}_c\cdot\mathbf{r}_{im'}} \tag{9.47}$$

where

$$\overline{\mathbf{E}}_s = \sum_{i=1}^{3}\sum_{j=1}^{3} E_{ij}^s\hat{x}_i\hat{x}_j \tag{9.48}$$

$$\overline{\mathbf{E}}_c = \sum_{i=1}^{3}\sum_{j=1}^{3} E_{ij}^c \hat{x}_i \hat{x}_j \tag{9.49}$$

with

$$E_{ij}^s = \delta_{ij} - k_i^s k_j^s \tag{9.50}$$

$$E_{ij}^c = k_i^c k_j^c. \tag{9.51}$$

In the above, we have used the indicial notation in the Cartesian coordinate system and defined $\mathbf{k}_w = k_w \hat{k}_w$ with $\hat{k}_w = k_1^w \hat{x}_1 + k_2^w \hat{x}_2 + k_3^w \hat{x}_3$ ($w = s$ or c). Also, ΔS_i represents a small triangle patch in the discretization of the object surface S and δ_{ij} is the Kronecker delta.

Scattering by a Traction-Free Cavity

If the object is a traction-free cavity such as a bubble in elastic material, the unknown traction vector at the surface vanishes and the BIEs are reduced to

$$\frac{1}{2}\mathbf{u}(\mathbf{x}) + \int_S \overline{\mathbf{T}}_1^T(\mathbf{x}, \mathbf{x}') \cdot \mathbf{u}(\mathbf{x}')\, \mathrm{d}S' = \mathbf{u}^{\mathrm{I}}(\mathbf{x}), \quad \mathbf{x} \in S. \tag{9.52}$$

The integral kernel can be written into two parts, i.e.

$$\begin{aligned}\overline{\mathbf{T}} &= \hat{n}' \cdot \overline{\overline{\boldsymbol{\Sigma}}}(\mathbf{x}, \mathbf{x}') \\ &= \lambda \hat{n}' \nabla \cdot \overline{\mathbf{G}} + \mu \hat{n}' \cdot \left(\nabla \overline{\mathbf{G}} + \overline{\mathbf{G}} \nabla\right) \\ &= \overline{\mathbf{A}} + \overline{\mathbf{B}}\end{aligned} \tag{9.53}$$

where

$$\begin{aligned}\overline{\mathbf{A}} &= \lambda \hat{n}' \nabla \cdot \overline{\mathbf{G}} \\ &= \frac{\lambda}{\gamma} \hat{n}' \nabla g_c\end{aligned} \tag{9.54}$$

$$\overline{\mathbf{B}} = \mu \hat{n}' \cdot \left(\nabla \overline{\mathbf{G}} + \overline{\mathbf{G}} \nabla\right). \tag{9.55}$$

The first part is only related to the compressional wave and its expansion in $\hat{k}$ domain can be found as

$$\overline{\mathbf{A}} = \int \mathrm{d}^2 \hat{k}_c \mathrm{e}^{\mathrm{i}\mathbf{k}_c \cdot \mathbf{r}_{jm}} \cdot \alpha_{mm'}(\mathbf{k}_c, \mathbf{r}_{mm'}) \overline{\mathbf{Q}}_c \mathrm{e}^{-\mathrm{i}\mathbf{k}_c \cdot \mathbf{r}_{im'}} \tag{9.56}$$

where

$$\overline{\mathbf{Q}}_c = c_3 \sum_{i=1}^{3}\sum_{j=1}^{3} q_{ij}^c \hat{x}_i \hat{x}_j \tag{9.57}$$

with $c_3 = i\lambda k_c/\gamma$ and $q_{ij}^c = n_i' k_j^c$. Here n_i' is a component of the unit normal vector on the object surface $\hat{n}'$, i.e. $\hat{n}' = n_1' \hat{x}_1 + n_2' \hat{x}_2 + n_3' \hat{x}_3$.

The second part includes both shear wave and compressional wave and we can separate these two waves in the $\hat{k}$ domain, i.e.

$$\begin{aligned}\overline{\mathbf{B}} = &\int \mathrm{d}^2 \hat{k}_s \mathrm{e}^{\mathrm{i}\mathbf{k}_s \cdot \mathbf{r}_{jm}} \cdot \alpha_{mm'}(\mathbf{k}_s, \mathbf{r}_{mm'}) \overline{\mathbf{P}}_s \mathrm{e}^{-\mathrm{i}\mathbf{k}_s \cdot \mathbf{r}_{im'}} \\ &- \int \mathrm{d}^2 \hat{k}_c \mathrm{e}^{\mathrm{i}\mathbf{k}_c \cdot \mathbf{r}_{jm}} \cdot \alpha_{mm'}(\mathbf{k}_c, \mathbf{r}_{mm'}) \overline{\mathbf{P}}_c \mathrm{e}^{-\mathrm{i}\mathbf{k}_c \cdot \mathbf{r}_{im'}}\end{aligned} \tag{9.58}$$

where

$$\overline{\mathbf{P}}_s = c_1 \sum_{i=1}^{3} \sum_{j=1}^{3} p_{ij}^s \hat{x}_i \hat{x}_j \tag{9.59}$$

$$\overline{\mathbf{P}}_c = c_2 \sum_{i=1}^{3} \sum_{j=1}^{3} p_{ij}^c \hat{x}_i \hat{x}_j \tag{9.60}$$

and

$$c_1 = -\frac{2i\mu k_s^3}{4\pi\omega^2\rho} \tag{9.61}$$

$$c_2 = -\frac{2i\mu k_c^3}{4\pi\omega^2\rho} \tag{9.62}$$

$$p_{ij}^s = k_i^s k_j^s (n_1' k_1^s + n_2' k_2^s + n_3' k_3^s) \tag{9.63}$$

$$p_{ij}^c = k_i^c k_j^c (n_1' k_1^c + n_2' k_2^c + n_3' k_3^c). \tag{9.64}$$

After combining these two parts together, the corresponding receiving pattern and radiation pattern for the kernel can be written as

$$V_{fmj}^s(\hat{k}_s) = \mathrm{e}^{\mathrm{i}\mathbf{k}_s\cdot\mathbf{r}_{jm}} \tag{9.65}$$

$$\overline{\mathbf{V}}_{sm'i}^s(\hat{k}_s) = \sum_{i=1}^{3} \sum_{j=1}^{3} (c_1 p_{ij}^s + c_4 q_{ij}^s) \int_{\Delta S_i} \mathrm{d}S' \mathrm{e}^{-\mathrm{i}\mathbf{k}_s\cdot\mathbf{r}_{im'}} \tag{9.66}$$

$$V_{fmj}^c(\hat{k}_c) = \mathrm{e}^{\mathrm{i}\mathbf{k}_c\cdot\mathbf{r}_{jm}} \tag{9.67}$$

$$\overline{\mathbf{V}}_{sm'i}^c(\hat{k}_c) = \sum_{i=1}^{3} \sum_{j=1}^{3} (c_3 q_{ij}^c - c_2 p_{ij}^c) \int_{\Delta S_i} \mathrm{d}S' \mathrm{e}^{-\mathrm{i}\mathbf{k}_c\cdot\mathbf{r}_{im'}} \tag{9.68}$$

where

$$q_{ij}^s = n_i' k_j^s + \delta_{ij}(n_1' k_1^s + n_2' k_2^s + n_3' k_3^s) \tag{9.69}$$

and $c_4 = c_1/2$. Note that we do not include the identity or constant term (1/2) of the BIE in the receiving pattern or radiation pattern because it belongs to near terms (diagonal terms) that the MLFMA will not work on.

Scattering by an Elastic Object

If both the object and the surrounding medium are elastic, we have two BIEs corresponding to the waves in the exterior medium and interior medium of the object. In each medium, there are two types of waves, i.e. a shear wave and a compressional wave, so we have four wavenumbers in total and need four FMA trees. In an indicial notation, the BIEs in (9.38)–(9.39) can be written as

$$\frac{1}{2}u_i + \int_S \sum_{j=1}^{3} \left[-G_{ij}^{(1)} t_j' + T_{ij}^{(1)} u_j'\right] \mathrm{d}S' = u_i^I \tag{9.70}$$

$$\frac{1}{2}u_i + \int_S \sum_{j=1}^{3} \left[G_{ij}^{(2)} t_j' - T_{ij}^{(2)} u_j'\right] \mathrm{d}S' = 0 \tag{9.71}$$

where $i = 1, 2, 3$, and in a matrix form

$$\begin{bmatrix} -G_{11}^{(1)} & -G_{21}^{(1)} & -G_{31}^{(1)} & T_{11}^{(1)} & T_{21}^{(1)} & T_{31}^{(1)} \\ -G_{12}^{(1)} & -G_{22}^{(1)} & -G_{32}^{(1)} & T_{12}^{(1)} & T_{22}^{(1)} & T_{32}^{(1)} \\ -G_{13}^{(1)} & -G_{23}^{(1)} & -G_{33}^{(1)} & T_{13}^{(1)} & T_{23}^{(1)} & T_{33}^{(1)} \\ G_{11}^{(2)} & G_{21}^{(2)} & G_{31}^{(2)} & -T_{11}^{(2)} & -T_{21}^{(2)} & -T_{31}^{(2)} \\ G_{12}^{(2)} & G_{22}^{(2)} & G_{32}^{(2)} & -T_{12}^{(2)} & -T_{22}^{(2)} & -T_{32}^{(2)} \\ G_{13}^{(2)} & G_{23}^{(2)} & G_{33}^{(2)} & -T_{13}^{(2)} & -T_{23}^{(2)} & -T_{33}^{(2)} \end{bmatrix} \begin{bmatrix} t_1' \\ t_2' \\ t_3' \\ u_1' \\ u_2' \\ u_3' \end{bmatrix} = \begin{bmatrix} u_1^I \\ u_2^I \\ u_3^I \\ 0 \\ 0 \\ 0 \end{bmatrix}. \tag{9.72}$$

In the matrix form, we again omit the identity or constant terms (1/2) of the BIEs because the MLFMA only acts on far interaction terms. Each matrix element above includes subkernels related to shear wave and compressional wave, respectively. These subkernels can be calculated for far interactions following the MLFMA formulas developed in the previous subsections for the rigid object and traction-free cavity.

9.3.2 Reformulation of Near Terms

The near terms in the original Nyström method without the MLFMA acceleration are calculated in the integrated form of the kernels [76]. Now we have to separate the compressional wave part from the shear wave part in the kernels because they belong to different FMA trees and reformulate the expressions for common near patches after cancelling the strongest singularities. The kernels in indicial notation have been shown in Subsection 9.2.1 and we repeat them again here [76]

$$\begin{aligned} G_{ij} &= \left(\overline{\mathbf{G}}\right)_{ij} \\ &= \frac{c_0}{r^3}\left\{\delta_{ij}\left[(k_s r)^2 e^{ik_s r} + D\right] + C\partial_i r \partial_j r\right\} \end{aligned} \tag{9.73}$$

$$\begin{aligned} T_{ij} &= \left(\hat{n}' \cdot \overline{\overline{\Sigma}}\right)_{ij} \\ &= \frac{c_0}{r^4}\left\{ \lambda e^{ik_c r}(k_c r)^2(ik_c r - 1)n_i \partial_j r + \mu e^{ik_s r}(k_s r)^2(ik_s r - 1)\left(\delta_{ij}\frac{\partial r}{\partial n} + n_i \partial_j r\right)\right. \\ &\quad \left. + 2\mu\left[C\left(\delta_{ij}\frac{\partial r}{\partial n} + n_i\partial_j r + n_j \partial_i r\right) + F\partial_i r \partial_j r \frac{\partial r}{\partial n}\right]\right\} \end{aligned} \tag{9.74}$$

where $\partial_i r = \partial r/\partial x_i$, $\partial_j r = \partial r/\partial x_j$, and

$$c_0 = \frac{1}{4\pi\rho\omega^2} \tag{9.75}$$

$$C = \Omega_s e^{ik_s r} - \Omega_c e^{ik_c r} \tag{9.76}$$

$$D = (ik_s r - 1)e^{ik_s r} - (ik_c r - 1)e^{ik_c r} \tag{9.77}$$

$$F = H_c e^{ik_c r} - H_s e^{ik_s r} \tag{9.78}$$

$$\Omega_s = 3 - 3ik_s r - k_s^2 r^2 \tag{9.79}$$

$$\Omega_c = 3 - 3ik_c r - k_c^2 r^2 \tag{9.80}$$

$$H_s = 15 - 15ik_s r - 6k_s^2 r^2 + ik_s^3 r^3 \tag{9.81}$$

$$H_c = 15 - 15ik_c r - 6k_c^2 r^2 + ik_c^3 r^3. \tag{9.82}$$

If we use the series expansion of the scalar Green's function

$$\frac{\mathrm{e}^{\mathrm{i}kr}}{r} = \sum_{m=0}^{\infty} \frac{(\mathrm{i}k)^m r^{m-1}}{m!} \approx \sum_{m=0}^{M} \frac{(\mathrm{i}k)^m r^{m-1}}{m!} \tag{9.83}$$

The leading terms in the kernels, which include the strongest singularities, will be cancelled. Since the expansion is only applied to the near terms with a small r, the series converges very fast. The typical M is chosen as 10 and the accuracy of the series can reach 10^{-8} when $r = 0.1$ and $k = 2\pi$ (unit wavelength).

With the use of the series expansion and the cancellation of the strongest singularities in the kernels, we obtain

$$\begin{aligned} G_{ij} = c_0 \Bigg\{ & \left[\frac{2}{r}(k_\mathrm{s}^2 - k_\mathrm{c}^2) - \mathrm{i}(k_\mathrm{s}^3 - k_\mathrm{c}^3)\right] \partial_i r \partial_j r + \frac{\delta_{ij}}{r}(k_\mathrm{c}^2 - k_\mathrm{s}^2 + k_\mathrm{s}^2 \mathrm{e}^{\mathrm{i}k_\mathrm{s} r}) \\ & + \sum_{m=2}^{M} \frac{i^m r^{m-3}}{m!} \left[(k_\mathrm{s}^m \Omega_s - k_\mathrm{c}^m \Omega_c)\partial_i r \partial_j r + \delta_{ij} k_\mathrm{s}^m (\mathrm{i}k_\mathrm{s} r - 1) - \delta_{ij} k_\mathrm{c}^m (\mathrm{i}k_\mathrm{c} r - 1)\right] \Bigg\} \end{aligned} \tag{9.84}$$

$$\begin{aligned} T_{ij} = c_0 \Bigg[& \lambda k_\mathrm{c}^2 n_i \partial_j r A + \mu k_\mathrm{s}^2 \left(\delta_{ij} \frac{\partial r}{\partial n} + n_j \partial_i r\right) B \\ & + 2\mu \left(\delta_{ij} \frac{\partial r}{\partial n} P + n_i \partial_j r P + n_j \partial_i r P + \frac{\partial r}{\partial n} \partial_i r \partial_j r Q\right)\Bigg] \end{aligned} \tag{9.85}$$

where

$$\begin{aligned} A &= \frac{(\mathrm{i}k_\mathrm{c} r - 1)}{r^2} \mathrm{e}^{\mathrm{i}k_\mathrm{c} r} \\ &= -\frac{1}{r^2} - k_\mathrm{c}^2 + \sum_{m=2}^{M} \frac{(\mathrm{i}k_\mathrm{c})^m}{m!} (\mathrm{i}k_\mathrm{c} r - 1) r^{m-2} \end{aligned} \tag{9.86}$$

$$\begin{aligned} B &= \frac{(\mathrm{i}k_\mathrm{s} r - 1)}{r^2} \mathrm{e}^{\mathrm{i}k_\mathrm{s} r} \\ &= -\frac{1}{r^2} - k_\mathrm{s}^2 + \sum_{m=2}^{M} \frac{(\mathrm{i}k_\mathrm{s})^m}{m!} (\mathrm{i}k_\mathrm{s} r - 1) r^{m-2} \end{aligned} \tag{9.87}$$

$$\begin{aligned} P &= \frac{1}{r^4}(\Omega_s \mathrm{e}^{\mathrm{i}k_\mathrm{s} r} - \Omega_c \mathrm{e}^{\mathrm{i}k_\mathrm{c} r}) \\ &= \frac{1}{2r^2}(k_\mathrm{s}^2 - k_\mathrm{c}^2) + \frac{\mathrm{i}r}{6}(k_\mathrm{s}^5 - k_\mathrm{c}^5) + \sum_{m=2}^{M} \frac{i^m}{m!}(k_\mathrm{s}^m \Omega_s - k_\mathrm{c}^m \Omega_c) r^{m-4} \end{aligned} \tag{9.88}$$

$$\begin{aligned} Q &= \frac{1}{r^4}(H_c \mathrm{e}^{\mathrm{i}k_\mathrm{c} r} - H_s \mathrm{e}^{\mathrm{i}k_\mathrm{s} r}) \\ &= \frac{3}{2r^2}(k_\mathrm{c}^2 - k_\mathrm{s}^2) - \frac{1}{2}(k_\mathrm{c}^4 - k_\mathrm{s}^4) + \frac{\mathrm{i}}{2}(k_\mathrm{c}^5 - k_\mathrm{s}^5) r + \frac{1}{6}(k_\mathrm{c}^6 - k_\mathrm{s}^6) r^2 \\ &\quad + \sum_{m=4}^{M} \frac{i^m}{m!}(k_\mathrm{s}^m \Omega_s - k_\mathrm{c}^m \Omega_c) r^{m-4}. \end{aligned} \tag{9.89}$$

The FMA trees require calculating the two parts related to two different wavenumbers in the kernels individually, so the above two kernels can be rewritten as

$$G_{ij} = G_{ij}^s + G_{ij}^c \tag{9.90}$$

$$T_{ij} = T^s_{ij} + T^c_{ij} \tag{9.91}$$

where

$$\begin{aligned} G^s_{ij} = c_0 \Bigg\{ & \left[\frac{k_s^2}{2r} + \frac{rk_s^4}{8} + \frac{ir^2k_s^5}{24} - \frac{r^3k_s^6}{24} + \sum_{m=5}^{M} \frac{(ik_s)^m}{m!} \Omega_s r^{m-5} \right] \partial_i r \partial_j r \\ & + \delta_{ij} \left[-\frac{k_s^2}{2r} - \frac{ik_s^3}{2} + \frac{k_s^2 e^{ik_s r}}{r} + \sum_{m=3}^{M} \frac{(ik_s)^m}{m!} (ik_s r - 1) r^{m-3} \right] \Bigg\} \end{aligned} \tag{9.92}$$

$$\begin{aligned} G^c_{ij} = c_0 \Bigg\{ & \left[-\frac{k_c^2}{2r} - \frac{rk_c^4}{8} - \frac{ir^2k_c^5}{24} + \frac{r^3k_c^6}{24} - \sum_{m=5}^{M} \frac{(ik_c)^m}{m!} \Omega_c r^{m-5} \right] \partial_i r \partial_j r \\ & + \delta_{ij} \left[\frac{k_c^2}{2r} + \frac{ik_c^3}{2} - \sum_{m=3}^{M} \frac{(ik_c)^m}{m!} (ik_c r - 1) r^{m-3} \right] \Bigg\} \end{aligned} \tag{9.93}$$

$$\begin{aligned} T^s_{ij} = c_0 \Bigg[& 2\mu \left(\delta_{ij} \frac{\partial r}{\partial n} P_s + n_i \partial_j r P_s + n_j \partial_i r P_s + \frac{\partial r}{\partial n} \partial_i r \partial_j r Q_s \right) \\ & + \mu k_s^2 \left(\delta_{ij} \frac{\partial r}{\partial n} + n_j \partial_i r \right) B \Bigg] \end{aligned} \tag{9.94}$$

$$T^c_{ij} = c_0 \left[\lambda k_c^2 n_i \partial_j r A + 2\mu \left(\delta_{ij} \frac{\partial r}{\partial n} P_c + n_i \partial_j r P_c + n_j \partial_i r P_c + \frac{\partial r}{\partial n} \partial_i r \partial_j r Q_c \right) \right] \tag{9.95}$$

with

$$P_w = \pm \left[\frac{k_w^2}{2r^2} + \frac{irk_w^5}{6} + \sum_{m=2}^{M} \frac{(ik_w)^m}{m!} \Omega_w r^{m-4} \right] \tag{9.96}$$

$$Q_w = \mp \left[\frac{3k_w^2}{2r^2} - \frac{k_w^4}{2} + \frac{irk_w^5}{2} + \frac{r^2k_w^6}{6} - \sum_{m=4}^{M} \frac{(ik_w)^m}{m!} \Omega_w r^{m-4} \right]. \tag{9.97}$$

In the above, $w = s$ or c, and the upper sign is taken when $w = s$ whereas the lower sign is taken when $w = c$. We can see that the above kernels only include the weak $1/R$ and strong $1/R^2$ singularities and they can be easily handled. We have derived closed form solutions for these singularities and the treatment technique can be found in [76]. The $1/R^2$ singularity is a strong singularity compared with the $1/R$ singularity but it is weaker and much easier to handle compared to the $1/R^3$ hypersingularity. After the two parts are evaluated in the two different FMA tree frames, they will be combined together for common near patches.

9.3.3 Reduction of Number of Patterns

We have showed an implementation scheme for the MLFMA in the preceding section. This scheme is suitable for BEM workers in elastodynamics because we fully follow their conventions to formulate. These conventions include changing the vector BIEs into a scalar form with the use of indicial notations in denoting the components of kernels, and using a collocation procedure to form a matrix equation without basis and testing functions involved. Although the implementation looks very simple, it is not efficient. If we introduce a MoM-like procedure in the Nyström method, the efficiency of implementation can be improved by reducing the number of patterns. Consider the

BIE in (9.42)

$$\int_S \overline{\mathbf{G}}(\mathbf{x},\mathbf{x}')\cdot\mathbf{t}(\mathbf{x}')\,\mathrm{d}S' = -\mathbf{u}^{\mathrm{I}}(\mathbf{x}),\quad \mathbf{x}\in S \tag{9.98}$$

where we have omitted the transpose on $\overline{\mathbf{G}}$ due to its symmetry and the subscript for simplicity. We can decompose the unknown traction vector $\mathbf{t}$ into three orthogonal components, i.e.

$$\mathbf{t}(\mathbf{x}') = \xi(\mathbf{x}')\hat{\xi}(\mathbf{x}') + \eta(\mathbf{x}')\hat{\eta}(\mathbf{x}') + \zeta(\mathbf{x}')\hat{\zeta}(\mathbf{x}') \tag{9.99}$$

where $(\hat{\xi},\hat{\eta},\hat{\zeta})$ are three orthogonal unit vectors and $\left[\xi(\mathbf{x}'),\eta(\mathbf{x}'),\zeta(\mathbf{x}')\right]$ are three unknown components to be solved. We may choose $\hat{\xi}$ and $\hat{\eta}$ as two unit tangential vectors to the object surface at the point $\mathbf{x}'$ and $\hat{\zeta}$ as the outward normal unit vector at the same point. When discretizing the object surface into N small triangle patches ΔS_i $(i = 1,2,\cdots,N)$ and applying a quadrature rule to the integral over a patch, we can rewrite the BIE in (9.98) as

$$\begin{aligned}&\sum_{i=1}^{N}\sum_{j=1}^{Q}\overline{\mathbf{G}}(\mathbf{x},\mathbf{x}'_{ij})\cdot\left[\xi(\mathbf{x}'_{ij})\hat{\xi}(\mathbf{x}'_{ij}) + \eta(\mathbf{x}'_{ij})\hat{\eta}(\mathbf{x}'_{ij}) + \zeta(\mathbf{x}'_{ij})\hat{\zeta}(\mathbf{x}'_{ij})\right]w_{ij}\\ &= -\mathbf{u}^{\mathrm{I}}(\mathbf{x}),\quad \mathbf{x}\in S\end{aligned} \tag{9.100}$$

where $\mathbf{x}'_{ij}$ represents the jth quadrature point within the ith patch, w_{ij} is the corresponding weight at the same point, and Q is the total number of quadrature points over a patch. Testing the above discretized equation with $\hat{\xi}$, $\hat{\eta}$, and $\hat{\zeta}$ at an observation point $\mathbf{x}_{pq}$, which is the qth quadrature point in the pth patch, we can obtain a matrix equation

$$\begin{aligned}&\sum_{i=1}^{N}\sum_{j=1}^{Q}\Big\{\left[\hat{\alpha}(\mathbf{x}_{pq})\cdot\overline{\mathbf{G}}(\mathbf{x}_{pq},\mathbf{x}'_{ij})\cdot\hat{\xi}(\mathbf{x}'_{ij})\right]w_{ij}\xi(\mathbf{x}'_{ij}) + \left[\hat{\alpha}(\mathbf{x}_{pq})\cdot\overline{\mathbf{G}}(\mathbf{x}_{pq},\mathbf{x}'_{ij})\cdot\hat{\eta}(\mathbf{x}'_{ij})\right]w_{ij}\eta(\mathbf{x}'_{ij})\\ &\quad+\left[\hat{\alpha}(\mathbf{x}_{pq})\cdot\overline{\mathbf{G}}(\mathbf{x}_{pq},\mathbf{x}'_{ij})\cdot\hat{\zeta}(\mathbf{x}'_{ij})\right]w_{ij}\zeta(\mathbf{x}'_{ij})\Big\}\\ &= -\hat{\alpha}(\mathbf{x}_{pq})\cdot\mathbf{u}^{\mathrm{I}}(\mathbf{x}_{pq}),\quad \mathbf{x}_{pq}\in S\end{aligned} \tag{9.101}$$

where $\hat{\alpha}$ represents $\hat{\xi}$, $\hat{\eta}$ or $\hat{\zeta}$, and $p = 1,2,\cdots,N$, $q = 1,2,\cdots,Q$. We have two scalar Green's functions in the kernel that are related to the shear wave and compressional wave, respectively, and we need to handle them individually. For the shear wave part, we can write the kernel as

$$\begin{aligned}\overline{\mathbf{G}}_s(\mathbf{x}_{pq},\mathbf{x}'_{ij}) &= \frac{1}{\mu}\left(\overline{\mathbf{I}} + \frac{\nabla\nabla}{k_s^2}\right)g_s(\mathbf{x}_{pq},\mathbf{x}'_{ij})\\ &= \int \mathrm{d}^2\hat{k}_s \mathrm{e}^{\mathrm{i}\mathbf{k}_s\cdot(\mathbf{r}_{bm}-\mathbf{r}_{am'})}\alpha_{mm'}(\mathbf{k}_s,\mathbf{r}_{mm'})(\overline{\mathbf{I}} - \hat{k}_s\hat{k}_s)\end{aligned} \tag{9.102}$$

where the subscript $a = ij$ represents a source point and $b = pq$ denotes an observation point. The terms in the square brackets in the matrix equation (9.101) are then

$$\begin{aligned}\hat{\alpha}(\mathbf{x}_{pq})\cdot\overline{\mathbf{G}}_s(\mathbf{x}_{pq},\mathbf{x}'_{ij})\cdot\hat{\beta}(\mathbf{x}'_{ij}) &= \hat{\alpha}(\mathbf{x}_{pq})\cdot\frac{1}{\mu}\int \mathrm{d}^2\hat{k}_s \mathrm{e}^{\mathrm{i}\mathbf{k}_s\cdot\mathbf{r}_{bm}}\alpha_{mm'}(\mathbf{k}_s,\mathbf{r}_{mm'})\cdot\\ &\quad\left[\mathrm{e}^{-\mathrm{i}\mathbf{k}_s\cdot\mathbf{r}_{am'}}\cdot(\overline{\mathbf{I}} - \hat{k}_s\hat{k}_s)\cdot\hat{\beta}(\mathbf{x}'_{ij})\right]\end{aligned} \tag{9.103}$$

where $\hat{\beta}$ also represents $\hat{\xi}$, $\hat{\eta}$ or $\hat{\zeta}$. Since $\bar{\mathbf{I}} - \hat{k}_s\hat{k}_s = \hat{\theta}^s_k\hat{\theta}^s_k + \hat{\phi}^s_k\hat{\phi}^s_k$, we can express the radiation pattern in (9.103) as

$$(\bar{\mathbf{I}} - \hat{k}_s\hat{k}_s) \cdot \hat{\beta}(\mathbf{x}'_{ij}) = \hat{\theta}^s_k \left[\hat{\theta}^s_k \cdot \hat{\beta}(\mathbf{x}'_{ij})\right] + \hat{\phi}^s_k \left[\hat{\phi}^s_k \cdot \hat{\beta}(\mathbf{x}'_{ij})\right]. \tag{9.104}$$

From the above expression, we can see that only two components or two radiation patterns need to be kept for each unknown component at a source point in the matrix equation.

For the compressional wave part, the related kernel can be written as

$$\begin{aligned}\bar{\mathbf{G}}_c(\mathbf{x}_{pq}, \mathbf{x}'_{ij}) &= -\frac{1}{\gamma}\frac{\nabla\nabla}{k_c^2} g_c(\mathbf{x}_{pq}, \mathbf{x}'_{ij}) \\ &= \frac{1}{\gamma}\int \mathrm{d}^2\hat{k}_c \mathrm{e}^{\mathrm{i}\mathbf{k}_c\cdot(\mathbf{r}_{bm}-\mathbf{r}_{am'})}\alpha_{mm'}(\mathbf{k}_c, \mathbf{r}_{mm'})\hat{k}_c\hat{k}_c\end{aligned} \tag{9.105}$$

and the terms in the square brackets in the matrix equation (9.101) are then

$$\begin{aligned}\hat{\alpha}(\mathbf{x}_{pq}) \cdot \bar{\mathbf{G}}_c(\mathbf{x}_{pq}, \mathbf{x}'_{ij}) \cdot \hat{\beta}(\mathbf{x}'_{ij}) = \hat{\alpha}(\mathbf{x}_{pq}) \cdot \frac{1}{\gamma}\int \mathrm{d}^2\hat{k}_c \mathrm{e}^{\mathrm{i}\mathbf{k}_c\cdot\mathbf{r}_{bm}}\alpha_{mm'}(\mathbf{k}_c, \mathbf{r}_{mm'}) \cdot \\ \left\{\mathrm{e}^{-\mathrm{i}\mathbf{k}_c\cdot\mathbf{r}_{am'}} \cdot \hat{k}_c\left[\hat{k}_c \cdot \hat{\beta}(\mathbf{x}'_{ij})\right]\right\}.\end{aligned} \tag{9.106}$$

It can be seen that only one radiation pattern needs to be kept for each unknown component in this case. Therefore, there are three radiation patterns (the shear wave part has $\hat{\theta}_k$ and $\hat{\phi}_k$ components and the compressional wave part has a k component) in total for each unknown component. Since we have three unknown components for each unknown vector, we have nine radiation patterns in total now and a half number of radiation patterns is reduced if compared with the original scheme. The key step in the pattern reduction scheme is the use of spherical coordinates in the $\hat{k}$ space where $\hat{k}$ corresponds to $\hat{r}$ in a spherical coordinate system. The radiation patterns are the $\hat{\theta}_k$, $\hat{\phi}_k$ and $\hat{k}$ components in this space. When we make a translation in $\hat{r}$ space, these patterns remain invariant because they belong to $\hat{k}$ space.

9.3.4 Numerical Examples

We demonstrate the proposed MLFMA by solving several relatively large problems with spherical geometries and comparing the solutions with the corresponding analytical solutions. In the implementation, the object geometries are discretized into triangular meshes and the Nyström method with a one-point quadrature rule is used. This is because the cost will dramatically increase if we use a higher-order quadrature rule (the Nyström method with higher-order quadrature rules is usually used for small problems). Also, the Nyström method with the one-point quadrature rule is equivalent to the BEM in computational costs.

We first consider scattering by a fixed rigid sphere with a normalized radius of $k_c a = 16.0$, where k_c is the wavenumber of the incident wave and a is the radius of the sphere. The surrounding medium has Poisson's ratio $\nu = 0.1$ and mass density $\rho = 1.0$. The incident wave is a time-harmonic plane compressional wave with a unit circular frequency ($\omega = 1.0$). Figure 9.13 illustrates the solution for the total traction field along the principal cut ($\phi = 0°$ and $\theta = 0° - 180°$) at the sphere surface and the solution is close to the

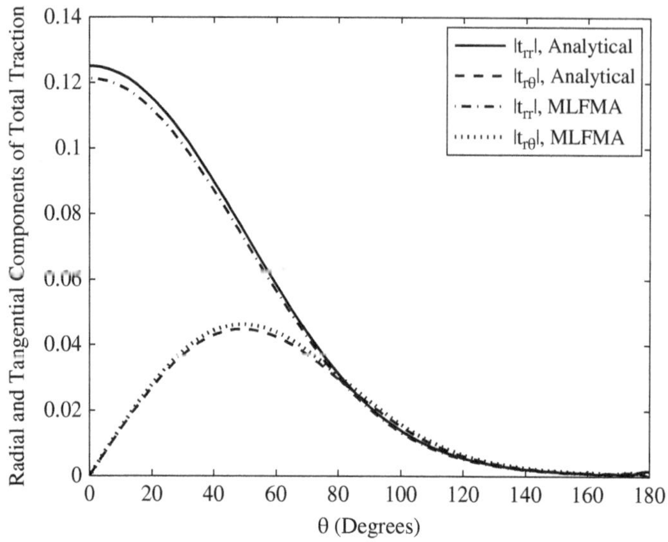

Figure 9.13 Radial and tangential (elevated) components of total traction field along the principal cut at the surface of a rigid sphere, $k_c a = 16.0$.

analytical counterpart. The used number of levels is $L = 5$ and the number of unknowns reaches $K = 305280$. The consumed CPU time is $T = 219167$ seconds and the memory usage is $M = 12.0$ GB. Note that although the same number of levels is used for different trees here, the definition for well-separation groups and near terms is different in different trees due to the difference of the wavenumbers and they cannot share the same trees. We also use the pattern reduction scheme to calculate the scattered displacement field along the principal cut at the $r = 5a$ observation surface and Figure 9.14 shows the solution with a comparison to the analytical counterpart. The computational conditions are the same as before except that $k_c a = 20.0$, $K = 317196$, and $T = 182347$ now. The proposed MLFMA can of course be applied to arbitrary 3D geometries. Figure 9.15 illustrates the scattering solution observed along the principal cut at the $r = 10a$ surface for a rigid cube and no exact solutions can be used to compare with. The computational conditions are the same as those in the first case except that $K = 343224$, $T = 230672$ seconds, and the side length of the cube is $2a$ with $k_c a = 20.0$ now.

We then consider the scattering by a traction-free spherical cavity embedded in an elastic medium. The cavity also has a normalized radius of $k_c a = 16.0$ and the surrounding medium is characterized by Poisson's ratio $\nu = 0.1$, Young's modulus $E = 2/3$ and mass density $\rho = 1.0$. The incident wave is the same as that for the rigid sphere. Figure 9.16 plots the solution for the scattered displacement field along the principal cut at the $r = 5a$ observation surface. The number of levels and number of unknowns are the same as those for the rigid sphere. The consumed CPU time is $T = 245986$ seconds and the memory usage is $M = 12.0$ GB.

We finally consider the generalized case, i.e. both the object and surrounding media are elastic. We select $\lambda_1 = 0.1$, $\mu_1 = 0.4$, and $\rho_1 = 1.0$ for the surrounding medium, and $\lambda_2 = 0.2$, $\mu_2 = 0.5$, and $\rho_2 = 2.0$ for the elastic spherical object with a normalized

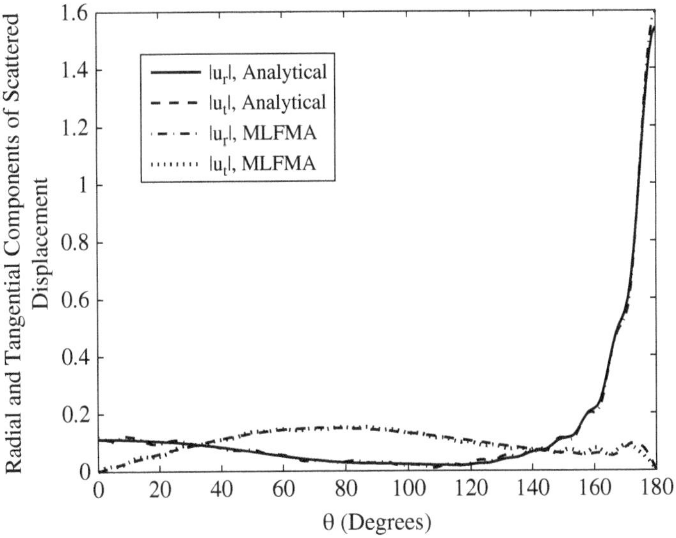

Figure 9.14 Radial and tangential (elevated) components of scattered displacement field along the principal cut at the $r = 5a$ surface for a spherical cavity, $k_c a = 20.0$.

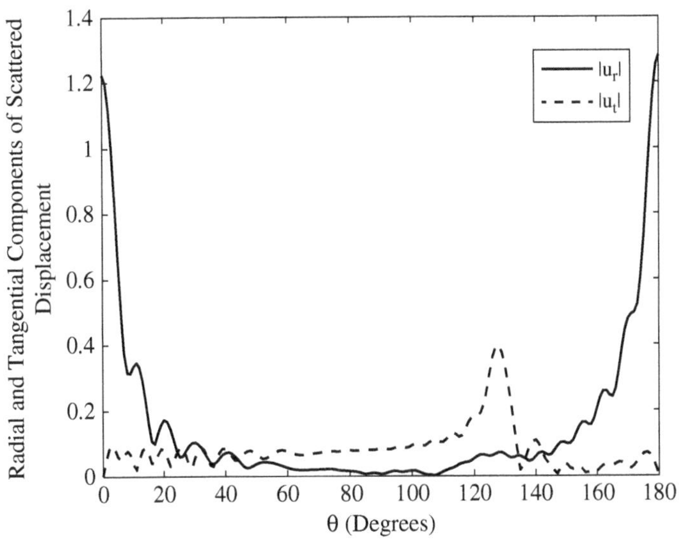

Figure 9.15 Radial and tangential (elevated) components of scattered displacement field along the principal cut at the $r = 10a$ surface for a rigid cube with a side length of $2a$.

radius of $k_c a = 8.0$. The incident wave is also the same as before. Figure 9.17 shows the solution for the total displacement field along the principal cut at the sphere surface. The number of levels is $L = 4$ and the number of unknowns is $K = 237600$.

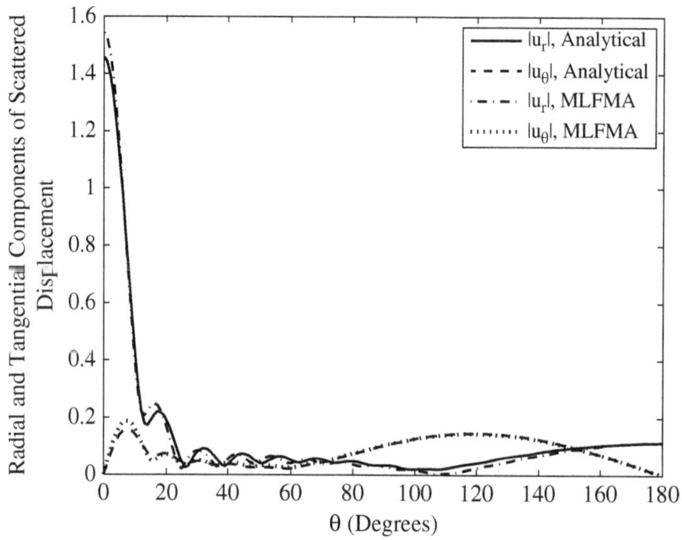

Figure 9.16 Radial and tangential (elevated) components of scattered displacement field along the principal cut at the $r = 5a$ surface for a spherical cavity, $k_c a = 16.0$.

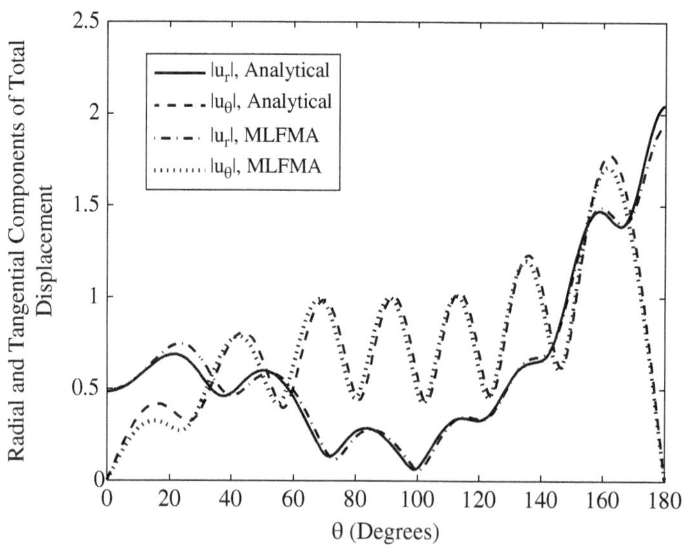

Figure 9.17 Radial and tangential (elevated) components of total displacement field along the principal cut at the surface of an elastic sphere, $k_c a = 8.0$.

This problem is relatively small because there are four FMA trees in this case and the computational cost is much higher than those for rigid sphere and cavity scattering. The consumed CPU time and memory usage in this case are $T = 201129$ seconds and $M = 12.0$ GB, respectively. From these figures, we can see that the MLFMA solutions are in good agreement with corresponding analytical solutions, which can be found in [51].

9.4 Solution of Acoustic Wave Problems with MLFMA Acceleration

As described in the preceding chapter, the MLFMA is an efficient accelerator for solving large matrix equations. The MLFMA has been widely used in solving EM integral equations [27] and very large problems with 10 – 30 million unknowns have been solved with the help of supercomputers [58, 62, 63, 77]. Although the FMM dealt with acoustic wave problems very early [78], the development of the method in acoustics has been relatively slow [79–83] compared with that in the EM community. Recently, there has been a growing interest in the acoustic applications of the MLFMA [84–89]. In references [84] and [85], the authors considered scattering by multiple spheres based on the spherical harmonic expansion for the scattered fields. Since semi-analytical formulas instead of BIE are used, the generation of matrix elements is simple. The scheme is not suitable for solving the scattering by arbitrarily shaped objects, although it has been demonstrated to solve problems involving up to 10^4 spheres (about one million unknowns). In fact, the scheme is different from the BIE-based MLFMA that we developed in this work. In references [86–89], however, the MLFMA together with the BIE is applied to solve for acoustic scattering by 3D arbitrarily shaped scatterers, but the implementation is based on the BEM.

We modify the MLFMA for solving acoustic wave scattering by very large objects with 3D arbitrary shapes. In the studies for mitigating gun blast noise at proving grounds, we design a trench structure in the ground to block the propagation of sound wave. The simulation for acoustic wave scattering by such a structure requires solving a very large matrix equation resulting from the acoustic BIE. The MLFMA is then used to accelerate the solution process and fast solutions for problems with millions of unknowns can be obtained on workstations without involving parallel algorithms. Unlike the work in references [85–89], the MLFMA is based on the Nyström method to form matrix elements. This is also different from the MLFMA implementation in EM, which is mainly based on the MoM with the RWG basis function [90] to create matrix entries.

For the near interaction terms of the matrix in the MLFMA implementation, the original Nyström method is recalled and we derive very simple and efficient formulas for the singular or near-singular integrations of the acoustic BIE kernel. This technique is different from the regularization of singularity in the BEM. Numerical examples from the simulation for mitigating gun blast noise show that the proposed MLFMA has a robust performance in solving for acoustic wave scattering by very large and complex structures. A typical problem with one million unknowns can be solved within two days on a Dell Precision 690 machine using 5-GB memory. This machine has two dual-core 3.0-GHz CPUs with 16-GB RAM, but we only employ one of CPUs since the series algorithm is utilized.

9.4.1 Implementation of the MLFMA for the Acoustic BIE

The MLFMA is described in [27], which factorizes the Green's function, and hence the matrix element that describes the interaction between a source point and a field point.

In this section, we will describe the use of the MLFMA for expediting the solution of the acoustic BIE.

Acoustic BIE

For EM wave scattering, the governing equations are vector integral equations with the vector current densities as the unknown functions, and the MoM with the RWG basis function may be the appropriate method incorporating the MLFMA. In contrast, acoustic wave scattering is represented by a scalar BIE [96], i.e.

$$p^{\mathrm{i}}(\mathbf{r}) + \int_S \left[\frac{\partial g(\mathbf{r},\mathbf{r}')}{\partial n'} p(\mathbf{r}') - g(\mathbf{r},\mathbf{r}') \frac{\partial p(\mathbf{r}')}{\partial n'} \right] \mathrm{d}S' = \frac{1}{2} p(\mathbf{r}), \quad \mathbf{r} \in S \tag{9.107}$$

where p is the pressure field, $g(\mathbf{r},\mathbf{r}') = \mathrm{e}^{\mathrm{i}kR}/(4\pi R)$ is the 3D Green's function, which is the same as that in EM, and $\hat{n}'$ is the outward unit normal vector on the surface S of a scatterer. The superscript "i" on p denotes the incident wave and $R = |\mathbf{r} - \mathbf{r}'|$ represents the distance between a field point (or observation point) $\mathbf{r}$ and a source point $\mathbf{r}'$. If we incorporate the impedance boundary condition [97]

$$\frac{\partial p(\mathbf{r}')}{\partial n'} - \eta p(\mathbf{r}') = 0 \tag{9.108}$$

where $\eta = \mathrm{i}\omega\rho/z_s$ and z_s is the surface impedance, the above BIE becomes

$$\frac{1}{2} p(\mathbf{r}) - \int_S G_0(\mathbf{r},\mathbf{r}') p(\mathbf{r}')\, \mathrm{d}S' = p^{\mathrm{i}}(\mathbf{r}), \quad \mathbf{r} \in S. \tag{9.109}$$

where $G_0 = (\hat{n}' \cdot \nabla' - \eta) g(\mathbf{r},\mathbf{r}')$. Note that the above BIE may possibly have the internal resonance problem, which refers to the non-uniqueness of the solution when the frequency of incident wave coincides with the resonance frequency of a cavity with the same shape as that of the scatterer [98]. This problem can only be removed completely by using different BIEs, such as the combined Helmholtz integral equation formulation [99] or the Burton and Miller formulation [100]. We do not meet the internal resonance problem even using the BIE (9.107) in the solution because the iteration process converges normally. If the resonant frequency is approached or touched, the system matrix will be near singular or singular and the iteration, which is very sensitive to the conditioning of the matrix, will tend to be divergent.

Radiation and Receiving Patterns

Due to the scalar property of the acoustic BIE above, the BEM or Nyström method may be more appropriate than the MoM for being combined with the MLFMA in order to accelerate the solution. If the Nyström method with a one-point quadrature rule is used (the higher-order quadrature rules can be used, but the number of unknowns will dramatically increases), the corresponding terms in the matrix equation in (8.1) will be

$$A_{ji} = - \int_{\Delta S_i} \mathrm{d}S' G_0(\mathbf{r}_j, \mathbf{r}') \tag{9.110}$$

$$b_j = p^{\mathrm{i}}(\mathbf{r}_j) \tag{9.111}$$

and a_i is the unknown pressure field at the centroid of the ith triangle patch ΔS_i. We do not include the identity or constant term (1/2) of the BIE in A_{ij} here because it belongs to the near terms (diagonal terms) and the matrix element here only represents the far interaction between a field point and a source point, which the MLFMA works on. Using the factorization of the Green's function as shown in (8.2), the integral kernel can be written as

$$G_0(\mathbf{r}_j - \mathbf{r}_i) = \frac{1}{4\pi} \int \mathrm{d}^2\hat{k}(-\hat{n}' \cdot \mathrm{i}\mathbf{k} - \eta)\mathrm{e}^{\mathrm{i}\mathbf{k}\cdot(\mathbf{r}_{jm} - \mathbf{r}_{im'})} \cdot \alpha_{mm'}(\mathbf{k}, \mathbf{r}_{mm'}). \tag{9.112}$$

With the above kernel, the radiation and receiving patterns are

$$V_{sm'i}(\hat{k}) = \int_{\Delta S_i} \mathrm{d}S'\mathrm{e}^{-\mathrm{i}\mathbf{k}\cdot(\mathbf{r}' - \mathbf{r}_{m'})}(-\hat{n}' \cdot \mathrm{i}\mathbf{k} - \eta), \quad \mathbf{r}' \in \Delta S_i \tag{9.113}$$

$$V_{fmj}(\hat{k}) = \frac{1}{4\pi}\mathrm{e}^{\mathrm{i}\mathbf{k}\cdot\mathbf{r}_{jm}} \tag{9.114}$$

and the translator is the same as that for EM integral equations. We will implement the MLFMA based on the above radiation and receiving patterns for far interactions and the solution process will be accelerated dramatically. The far interaction or well-separated interaction is defined as the interaction between a field point $\mathbf{r}_j$ and a source point $\mathbf{r}_i$ whose separation is at least one box at the same level [27].

The integral over $\hat{k}$ above is evaluated by the Gaussian quadrature rule with $Q = 2L^2$ points, where L is the truncated number of the summation series in the translator. The use of the quadrature rule results in

$$\begin{aligned} A_{ji} &= -\sum_{q=1}^{Q} V_{fjm}(\hat{k}_q) \cdot \alpha_{mm'}(\mathbf{k}_q, \mathbf{r}_{mm'}) \cdot V_{sm'i}(\hat{k}_q) \\ &= \mathbf{V}_{jm} \cdot \overline{\alpha}_{mm'} \cdot \mathbf{V}_{m'i} \end{aligned} \tag{9.115}$$

where

$$\mathbf{V}_{jm} = \left[w_1 V_{fjm}(\hat{k}_1), w_2 V_{fjm}(\hat{k}_2), \cdots, w_Q V_{fjm}(\hat{k}_Q)\right] \tag{9.116}$$

$$\overline{\alpha}_{mm'} = \mathrm{diag}\left\{\alpha_{mm'}(\mathbf{k}_1, \mathbf{r}_{mm'}), \alpha_{mm'}(\mathbf{k}_2, \mathbf{r}_{mm'}), \cdots, \alpha_{mm'}(\mathbf{k}_Q, \mathbf{r}_{mm'})\right\} \tag{9.117}$$

$$\mathbf{V}_{m'i} = \left[V_{sm'i}(\hat{k}_1), V_{sm'i}(\hat{k}_2), \cdots, V_{sm'i}(\hat{k}_Q)\right]^T \tag{9.118}$$

and $w_1, w_2, \cdots, w_Q$ are the weights of the quadrature rule. We will store the above vectors and matrix with an efficient manner in the preprocessing stage and they are used in each matrix-vector multiply in the iterative process.

Near Terms

The near terms in the MLFMA are calculated by the original numerical methods on which the solution of the BIE is based without MLFMA acceleration. One usually employs the MoM with the RWG basis function in EM and utilizes BEM in acoustics. Here we use the Nyström method to generate the near terms. The near terms include the self interaction terms in which the source point and field point reside in the same triangle patch, and the near interaction terms in which the source point and field point are located at different but nearby triangle patches. The acoustic BIE includes two

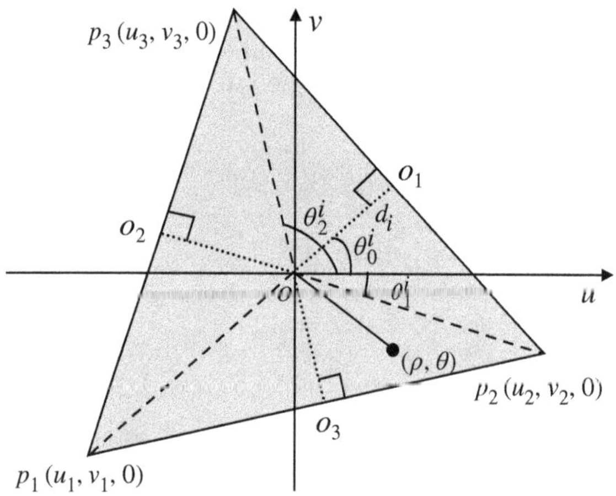

Figure 9.18 Local coordinate system on a triangular patch plane.

kernels, which are the Green's function itself and its gradient. The integral related to the Green's function itself has a week singularity, which can be easily handled [50, 95] and we will not deal with it here. The integral with the gradient of the Green's function has a $1/R^2$ singularity on which the special care should be taken. In a Cartesian coordinate system, the gradient of the Green function becomes

$$\begin{aligned}\nabla g(\mathbf{r}, \mathbf{r}') &= g_x\hat{x} + g_y\hat{y} + g_z\hat{z} \\ &= c_0(x - x')\hat{x} + c_0(y - y')\hat{y} + c_0(z - z')\hat{z}\end{aligned} \tag{9.119}$$

where

$$c_0 = \frac{(ikR - 1)}{4\pi R^3}e^{ikR}. \tag{9.120}$$

In the above, (x, y, z) and (x', y', z') are the coordinates of the field point and the source point on a triangular patch, respectively. If we create a local coordinate system on the source triangle patch plane whose origin is the projection of the field point on the plane, as shown in Figure 9.18, then the three components of the gradient can be written as

$$g_u = c_0(-u); \quad g_v = c_0(-v); \quad g_w = c_0 w_0 \tag{9.121}$$

where $R = \sqrt{u^2 + v^2 + w_0^2}$. The source point is located at $(u, v, 0)$ on the triangle patch and the field point is located at $(0,0, w_0)$ in the local coordinate system now. If we expand the exponential term in the Green's function into a series form

$$e^{ikR} = \sum_{m=0}^{\infty} \frac{(ik)^m}{m!} R^m \approx \sum_{m=0}^{M} \frac{(ik)^m}{m!} R^m \tag{9.122}$$

which converges very rapidly (M is small) for near and self interaction terms since R is small, then the most singular integrals including those three components of gradient will be

$$I_1 = \int_{\Delta S} \frac{u}{R^3}\,\mathrm{d}S;\quad I_2 = \int_{\Delta S} \frac{v}{R^3}\,\mathrm{d}S;\quad I_3 = \int_{\Delta S} \frac{w_0}{R^3}\,\mathrm{d}S. \tag{9.123}$$

Other integrals are either regular or with a weak singular kernel of $1/R$, which can be easily handled. By referring to Figure 9.18, which includes a polar coordinate system (ρ, θ) for a source point on the triangle patch and $R = \sqrt{\rho^2 + w_0^2}$, the first integral above can be written as

$$\begin{aligned}
I_1 &= \int_{\Delta S} \frac{u}{R^3}\,\mathrm{d}S = \sum_{i=1}^{3} \int_{\theta_1^i}^{\theta_2^i} \mathrm{d}\theta \cos\theta \int_0^{d_\theta^i} \frac{\rho^2\,\mathrm{d}\rho}{(\rho^2 + w_0^2)^{3/2}} \\
&= \sum_{i=1}^{3} \left[\int_{\theta_{10}^i}^{\theta_{20}^i} \mathrm{d}\phi \frac{-d_i \cos(\phi + \theta_0^i)}{\sqrt{d_i^2 + w_0^2 \cos^2\phi}} + \int_{\theta_{10}^i}^{\theta_{20}^i} \mathrm{d}\phi \cos(\phi + \theta_0^i) \ln\left(d_i + \sqrt{d_i^2 + w_0^2\cos^2\phi}\right) \right.\\
&\quad \left. - \int_{\theta_{10}^i}^{\theta_{20}^i} \mathrm{d}\phi \cos(\phi + \theta_0^i) \ln(\cos\phi) \right] \\
&= \sum_{i=1}^{3} (I_{1a}^i + I_{1b}^i + I_{1c}^i)
\end{aligned} \tag{9.124}$$

where $d_\theta^i = d_i / \cos(\theta - \theta_0^i)$, $\theta_{10}^i = \theta_1^i - \theta_0^i$, and $\theta_{20}^i = \theta_2^i - \theta_0^i$. The first and third integrals above can be found analytically [102], i.e.

$$\begin{aligned}
I_{1a}^i = -\frac{d_i}{w_0} \Big\{ &\cos\theta_0^i \left[\arcsin(s_0 a_2^i) - \arcsin(s_0 a_1^i)\right] + \sin\theta_0^i \left[\ln\left(w_0 b_2^i + \sqrt{d_i^2 + (w_0 b_2^i)^2} \right) \right.\\
&\left.\left. - \ln\left(w_0 b_1^i + \sqrt{d_i^2 + (w_0 b_1^i)^2} \right) \right] \right\}
\end{aligned} \tag{9.125}$$

$$\begin{aligned}
I_{1c}^i = t_0 &\left[(1 + a_2^i)\ln(1 + a_2^i) - (1 + a_1^i)\ln(1 + a_1^i) - (1 - a_2^i)\ln(1 - a_2^i) - 2(a_2^i - a_1^i) \right.\\
&\left. + (1 - a_1^i)\ln(1 - a_1^i) \right] + \sin\theta_0^i \left[b_2^i \ln(b_2^i) - b_1^i \ln(b_1^i) - b_2^i + b_1^i \right]
\end{aligned} \tag{9.126}$$

where $t_0 = 0.5\cos\theta_0^i$, $s_0 = w_0\left(d_i^2 + w_0^2\right)^{-\frac{1}{2}}$, $a_1^i = \sin\theta_{10}^i$, $a_2^i = \sin\theta_{20}^i$, $b_1^i = \cos\theta_{10}^i$ and $b_2^i = \cos\theta_{20}^i$. The second integral I_{1b}^i is never singular no matter how the field point is close to the triangle patch, so it can be integrated numerically. We can use a similar way to handle I_2. For I_3, it is easy to show that

$$\begin{aligned}
I_3 &= \int_{\Delta S} \frac{w_0}{R^3}\,\mathrm{d}S = w_0 \sum_{i=1}^{3} \int_{\theta_1^i}^{\theta_2^i} \mathrm{d}\theta \int_0^{d_\theta^i} \frac{\rho\,\mathrm{d}\rho}{(\rho^2 + w_0^2)^{3/2}} \\
&= t_1 \sum_{i=1}^{3} \left[\theta_2^i - \theta_1^i - \arcsin(s_0 a_2^i) + \arcsin(s_0 a_1^i) \right]
\end{aligned} \tag{9.127}$$

where $t_1 = \text{sign}(w_0)$. When the field point is inside the triangle patch, we can take a limit of $w_0 \to 0$ to obtain

$$I_1 = \sum_{i=1}^{3} \left[(\sin\theta_2^i - \sin\theta_1^i)\ln(d_i) - I_A \cos\theta_0^i + I_B \sin\theta_0^i \right] \tag{9.128}$$

$$I_2 = \sum_{i=1}^{3} \left[(\cos\theta_1^i - \cos\theta_2^i)\ln(d_i) - I_B \cos\theta_0^i - I_A \sin\theta_0^i \right] \tag{9.129}$$

$$I_3 = 0 \tag{9.130}$$

where

$$I_A = 0.5 \left\{ (1 + a_2^i) \left[\ln(1 + a_2^i) - 1\right] - (1 + a_1^i) \left[\ln(1 + a_1^i) - 1\right] - (1 - a_2^i) \left[\ln(1 - a_2^i) - 1\right] + (1 - a_1^i) \left[\ln(1 - a_1^i) - 1\right] \right\} \tag{9.131}$$

$$I_B = b_2^i \left[1 - \ln(b_2^i)\right] - b_1^i \left[1 - \ln(b_1^i)\right]. \tag{9.132}$$

The formulas in (9.128)–(9.130) may not be necessary since the dot product of the unit normal vector with the gradient of the Green's function on a self patch will vanish. We present these formulas for completeness, and also they may be useful for underwater acoustics in which a different BIE is employed.

9.4.2 Numerical Examples

We use the developed MLFMA with the generalized minimal residual method as the iterative solver to solve for the acoustic wave scattering by very large scatterers. We first consider the scattering by a sphere so that the solutions can be verified by comparing with the corresponding analytical solutions. The sphere has a large radius $a = 20\lambda$, 25λ, and 30λ, respectively, where λ is the wavelength in free space, and is meshed with flat triangle patches, as illustrated in Figure 9.19. The surface status of the sphere is characterized by the value of η and the incident wave is generated by a normalized point source

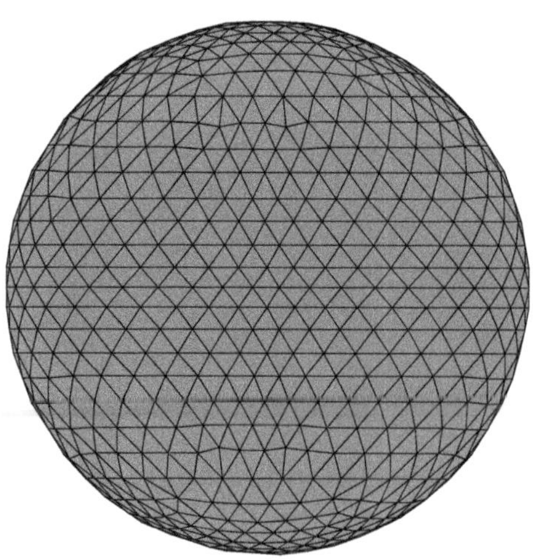

Figure 9.19 A meshed sphere surface.

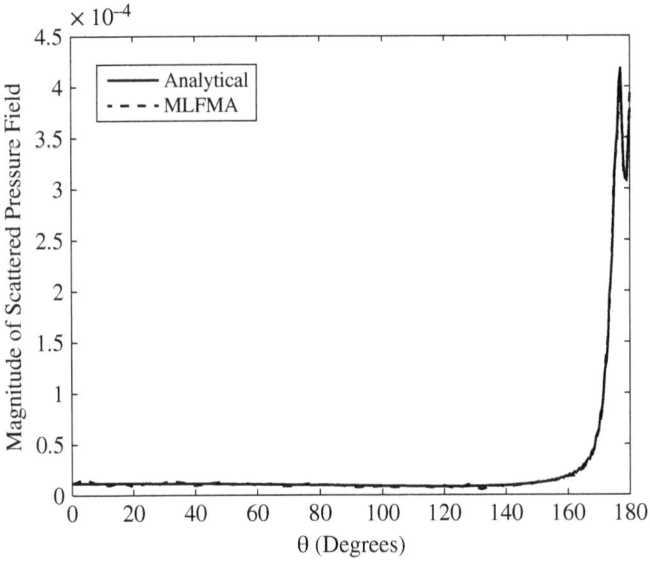

Figure 9.20 Magnitude of the scattered pressure field at the principal cut on the $r = 50a$ spherical surface, $a = 20\lambda$ and $\eta = 10–10i$.

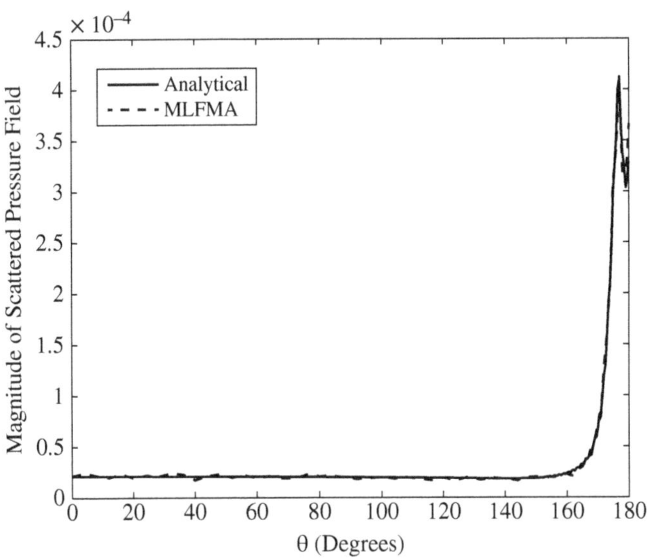

Figure 9.21 Magnitude of the scattered pressure field at the principal cut on the $r = 50a$ spherical surface, $a = 20\lambda$ and $\eta = 10$.

located at $(0,0,10a)$ with a frequency $f = 1700$ Hz. The propagation medium is the air with a standard acoustic velocity $c = 340$ m s^{-1}.

Figures 9.20–9.23 show the scattered pressure fields along the principal cut ($\phi = 0°$ and $\theta = 0° - 180°$) at the $r = 50a$ surface. Figure 9.24 is a zoom-in plot for the details between $\theta = 160°$ and $180°$ for the $a = 30\lambda$ case. We can see that the solutions are very

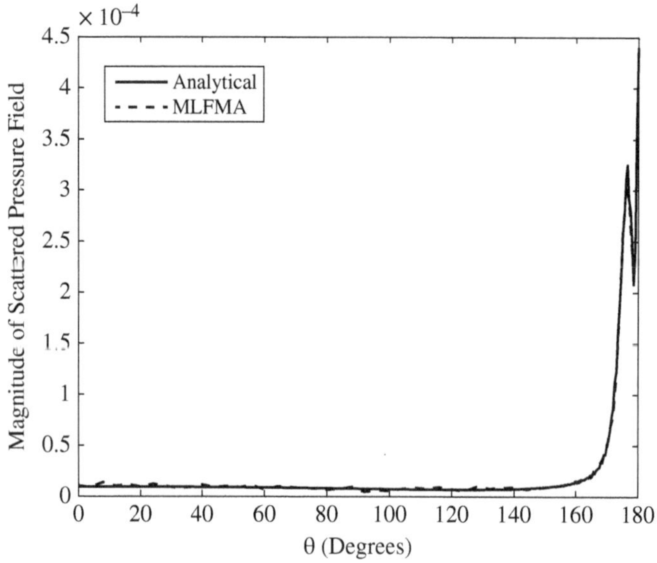

Figure 9.22 Magnitude of the scattered pressure field at the principal cut on the $r = 50a$ spherical surface, $a = 25\lambda$ and $\eta = 10$–$10i$.

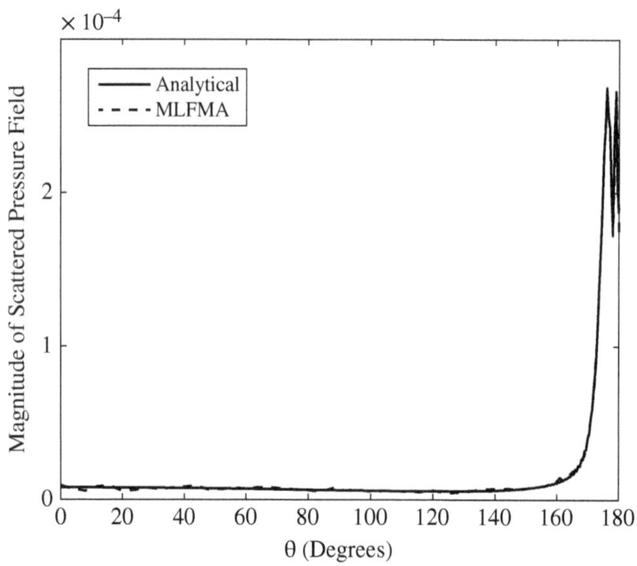

Figure 9.23 Magnitude of the scattered pressure field at the principal cut on the $r = 50a$ spherical surface, $a = 30\lambda$ and $\eta = 10$–$10i$.

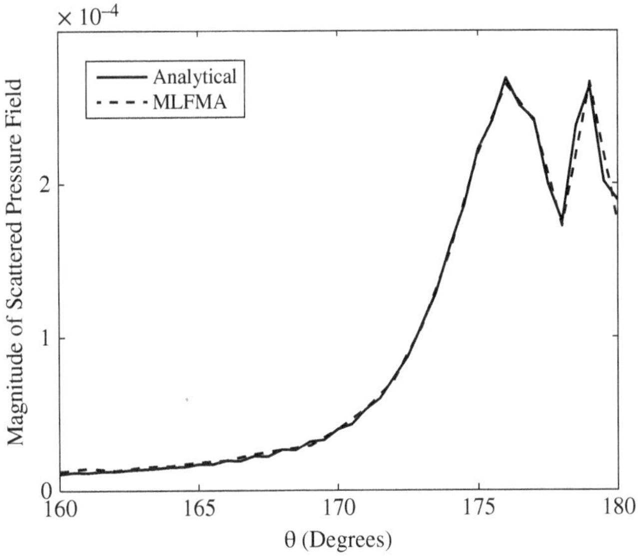

Figure 9.24 Zoom-in plot for the magnitude of the scattered pressure field at the principal cut on the $r = 50a$ spherical surface, $a = 30\lambda$ and $\eta = 10\text{–}10i$.

Table 9.1 CPU time T (Second) and memory usage M (GB) for scattering by a large sphere.

a	20	20	25	30
η	10–10i	10	10–10i	10–10i
N	1046528	1046528	1635840	2556800
T	54267	58728	107037	145887
M	4.8	4.8	9.2	13.0

close to the corresponding analytical solutions. Table 9.1 summarizes the CPU time T (Second) and memory usage M (GB) for solving these large problems, where N is the number of unknowns. It is seen from the table that solving a one million unknown problem only consumes about 16 hours when the stopping error of the iteration is set up 10^{-5}.

After validating the code, we apply it to solve for the acoustic wave scattering by large trench structures that are needed in the design for mitigating gun blast noise. Figure 9.25 illustrates a meshed trench structure embedded in a circularly truncated ground (top view) and Figure 9.26 defines the geometrical sizes for each trench. We assume that the ground and trench surfaces are rigid, so $\eta = 0$ here. The trenches are concentrically shaped and the entire structure is a body of revolution (BOR), but we treat it as a full 3D geometry and the solutions are compared with those obtained from BOR formulas we developed previously. This ensures the correctness of the solutions when lacking analytical solutions to compare with. In the calculations, we choose the frequency of

Figure 9.25 A meshed trench structure embedded in truncated ground (top view).

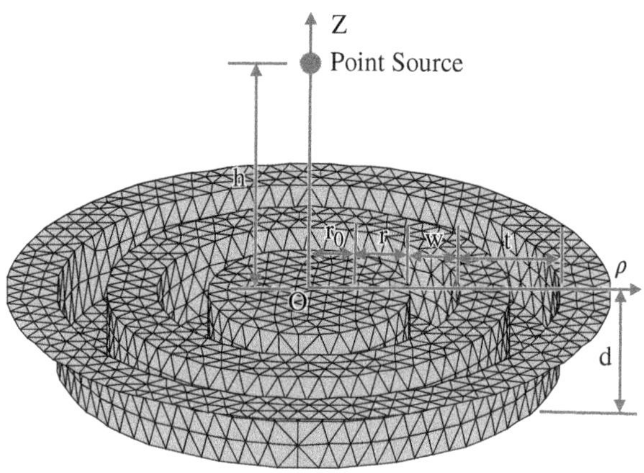

Figure 9.26 Geometrical parameters of the trench structure, in which r_0 is the distance between the ground center and the inner wall of the first trench, w and d are the width and depth of each trench, respectively, and t is the periodicity of trenches. The ground width between two neighboring trenches is $t - w$.

the incident wave generated by a point source as $f = 20$ Hz, which is close to the central frequency of a large-calibre gun blast noise, and the wave velocity as $c = 350$ m s^{-1} [103]. The point source is located at a location above the ground center with a height of $h = 0.1\lambda$. The truncated radius of the ground is chosen as $r_g = 20\lambda$, which is large enough for the specific point source above [104]. The trench geometry has a width of $w = 0.1\lambda$, a periodicity of $t = 0.2\lambda$ and a depth of $d = 0.25\lambda$. The distance between the ground center and the inner wall of the first trench is $r_0 = 0.2\lambda$.

Figures 9.27–9.30 depict the radiation patterns (RP) or excess attenuations (EA) for the trench structures when the number of concentric trenches is $K = 1$, 5, 10 and 50,

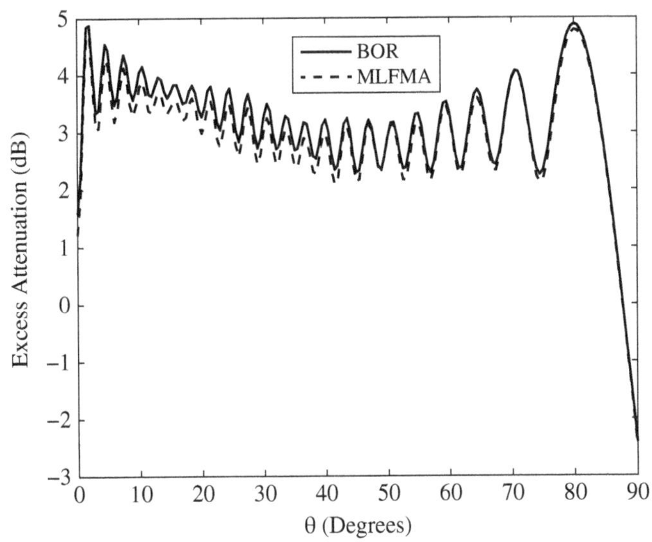

Figure 9.27 Radiation pattern for the truncated ground with 1 trench.

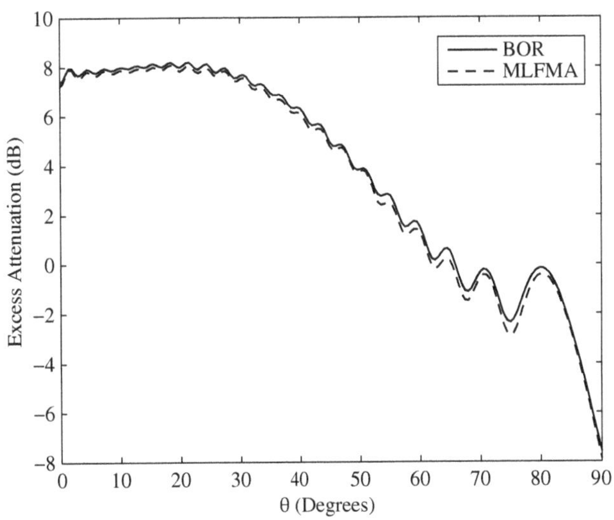

Figure 9.28 Radiation pattern for the truncated ground with 5 trenches.

respectively. RP or EA is defined as [105]

$$\mathrm{RP} = \mathrm{EA} = 20\log_{10}\left|\frac{p(\mathbf{r})}{p_0(\mathbf{r})}\right| \ \mathrm{dB} \tag{9.133}$$

where p is the total pressure field (with scatterer) and p_0 is the free-space pressure field (without scatterer) at an observation point $\mathbf{r}$. The observation points are located at the principal cut defined by $\theta = 0^\circ - 90^\circ$ and $\phi = 0^\circ$ and 180° on the sphere surface of $|\mathbf{r}| = 2000$ m. It can be seen that the solutions based on the MLFMA are close to those

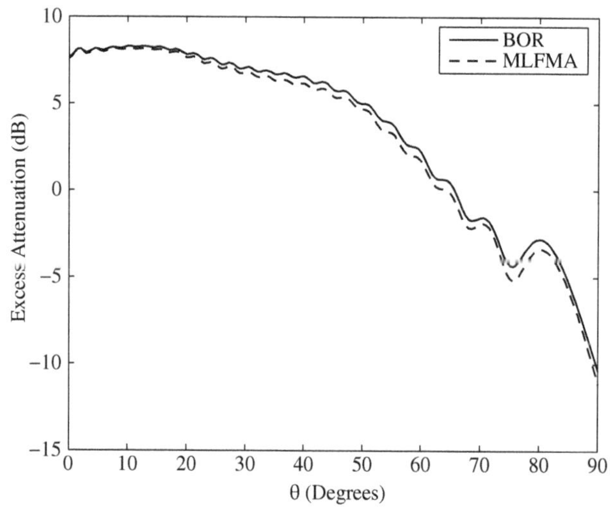

Figure 9.29 Radiation pattern for the truncated ground with 10 trenches.

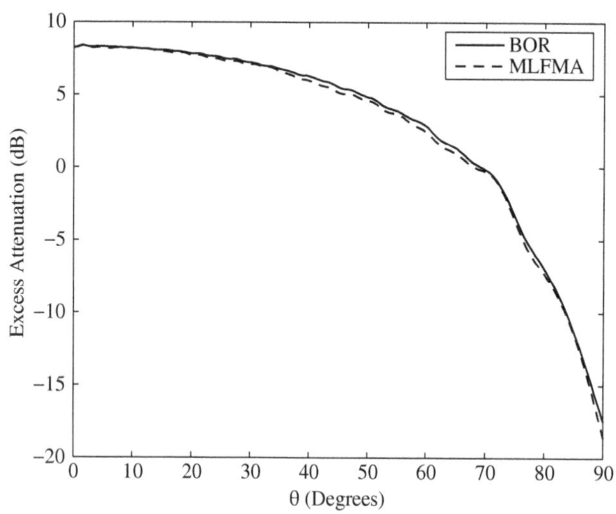

Figure 9.30 Radiation pattern for the truncated ground with 50 trenches.

obtained from BOR formulas and also the propagation of acoustic wave is suppressed along the ground surface as the number of trenches increases. Figure 9.31 presents the comparison of radiation patterns when the larger truncated sizes of ground $r_g = 30\lambda$ and 40λ with 50 trenches are chosen and it is clear that $r_g = 20\lambda$ is large enough. Table 9.2 gives the CPU times and memory usage for running these cases, where K is the number of trenches. For the cases with more than 2 million unknowns in the two tables, we believe that they are the largest cases in acoustics that numerical solvers have solved without resorting to parallel algorithms and supercomputers. It should be noted that the commercial software IDEAS we employed to generate meshes in this work usually cannot generate more than one million triangular meshes. If more than one million triangular meshes are needed for a geometry, we have to generate coarse

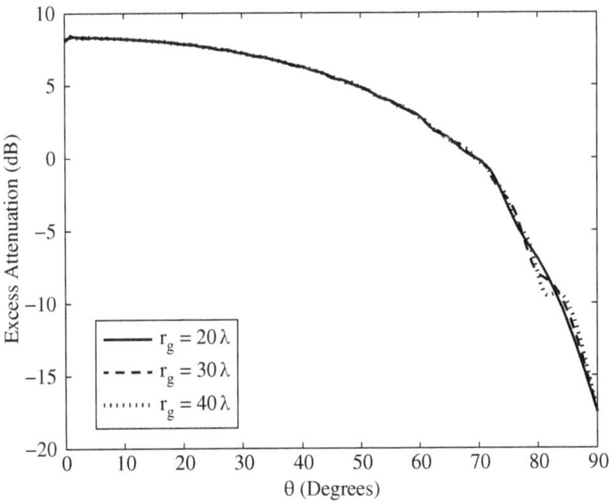

Figure 9.31 Comparison of radiation patterns for different truncated sizes of ground with 50 trenches.

Table 9.2 CPU time T (Second) and memory usage M (GB) for scattering by truncated ground with trenches.

r_g	20	20	20	20	30	40
K	1	5	10	50	50	50
N	801116	804718	822476	991943	1736988	2441136
T	63202	66547	71281	108259	167642	248357
M	3.6	3.6	3.9	4.7	9.5	12.0

meshes using IDEAS first, and then refine these meshes by dividing a triangle into three smaller triangles. The three smaller triangles are formed by connecting three vertices with the centroid of the larger triangle. Since such a refinement is based on the coarse IDEAS meshes, the small difference between three curves in Figure 9.31 may be caused by the disagreement of geometrical approximation of the coarse IDEAS meshes between three cases.

9.5 Unified Boundary Integral Equations for Elastic Wave and Acoustic Wave

The study of elastic wave or acoustic wave behavior requires solving the corresponding wave equations. These equations can be in the PDE form or the BIE form. It is very clear that the acoustic wave equation is a special case of the elastic wave equation and they can be solved in a unified form in PDE [35–38]. Conventionally, however, the elastic wave BIEs and the acoustic wave BIE are treated differently [106, 107]. This is because the acoustic wave BIE has a simpler form and it is unnecessary to resort to solving the full fledge elastic wave BIEs in most cases. In this section, we unify the acoustic wave BIE and elastic wave BIEs and show that the acoustic wave case can be obtained as a special case of the full elastic wave BIEs. The unified BIEs can provide a new approach to solve those problems in a more versatile manner. This is accomplished by introducing a shielding loss for a small shear modulus in the elastic wave BIEs and solving them as an elastic wave problem. This shielding loss attenuates the shear wave in the medium. The advantage of this approach is that it requires the maintenance of only one numerical code that can account for both elastic wave physics and acoustic wave physics. This is especially important for today's computational engineering where numerical codes for complex structures often require high maintenance due to the complexity of the codes.

This work fills a void in the numerical elastic scattering solutions for integral equation solvers where the acoustic wave scattering is never derived from a full-fledge code for elastic wave scattering, even though it has been done in differential equation solvers. Furthermore, we also derive the asymptotic Green's function and the BIEs for the $\mu \to 0$ case, where μ is the shear modulus of the host medium, and solve the corresponding vector BIEs. The new approach has the same complexity as the traditional acoustic wave BIE, but can provide extra information about the displacement field in addition to the potential in its solution.

In acoustics, one usually works with the velocity potential or pressure as the unknowns leading to a simpler scalar BIE. If the obstacle is soft or elastic, surface impedance concept is used to account for the property of the obstacle. However, for fluid or underwater acoustics problems, the traditional scalar acoustic wave BIE may not be appropriate; hence one has to resort to the vector elastic wave BIEs with displacement and stress vectors as unknowns [108–110].

Although we address the Nyström method in this book, we choose the MoM as a numerical tool to solve the unified BIEs in this section. The purpose is that we want to compare the Nyström method solutions with the MoM solutions because we intentionally solve the same problems as in the previous section before the elastic wave part and acoustic wave part are united. The MoM is a robust numerical method that has been widely used in electromagnetics and has also been introduced to solve for the scalar acoustic BIE [111–116], so we cannot ignore its existence, and any non-traditional numerical methods like the Nyström method should be compared to it when available. Although the MoM or Galerkin's method has been used to solve the scalar acoustic BIE [111–120], it has not been implemented yet for the vector elastic wave BIEs.

Since the elastic wave BIEs have 3D unknown vectors over a boundary surface, the implementation of the MoM is not straightforward compared with EM BIEs. We have to separate the unknown vectors into tangential components and normal components along the surface and expand them in terms of basis functions individually. We choose the RWG basis function [90] to represent the tangential components and pulse basis function to represent the normal components. These basis functions are also used as the weighting functions to test the BIEs, resulting in a Galerkin's process.

In the numerical process for solving the unified BIEs, we will encounter the integral kernels with $1/R^3$ singularity, where R is the distance between a field point and a source point. This singularity is generated from the double gradient of the scalar compressional-wave Green's function. In the elastic wave scattering, the shear wave exists and this singularity is cancelled by the same term in the double gradient of the scalar shear-wave Green's function, leading to an easily-handled $1/R^2$ singularity. In the acoustic wave problem, the shear wave vanishes and the degree of singularity increases to $1/R^3$. This may have been a challenging problem before, but we have developed a technique based on the Cauchy principal value (CPV) to treat this kind of singular integral in electromagnetics [121], and the same procedure can be followed here.

9.5.1 Elastic Wave BIEs

The elastic wave BIEs have been shown in the previous sections, but we present them again with minor changes for facilitating the late unification. Consider elastic wave scattering by a homogeneous obstacle V_0 with boundary S, as shown in Figure 9.32. The obstacle, whose properties are characterized by the mass density ρ_0 and Lamé constants λ_0 and μ_0, is embedded in an infinite 3D isotropic elastic medium V. Here we use the subscript 0 to indicate the parameters of the obstacle. Similarly, the host medium is characterized by ρ, λ and μ with no subscript for its parameters. The coordinate system is labelled in indicial notation (x_1, x_2, x_3) corresponding to (x, y, z) for convenience.

The incident wave is a time-harmonic compressional (longitudinal) plane wave propagating along the $-x_3$ direction and impinging upon the obstacle. If we incorporate boundary conditions, which are the continuity of displacement and traction vectors, the BIEs can be written as [31]

$$\frac{1}{2}\mathbf{u}(\mathbf{x}) + \int_S \left[\overline{\mathbf{R}}^T(\mathbf{x},\mathbf{x}') \cdot \mathbf{u}(\mathbf{x}') - \overline{\mathbf{G}}^T(\mathbf{x},\mathbf{x}') \cdot \mathbf{t}(\mathbf{x}')\right] \mathrm{d}S' = \mathbf{u}^{\mathrm{I}}(\mathbf{x}), \quad \mathbf{x} \in S \tag{9.134}$$

$$\frac{1}{2}\mathbf{u}(\mathbf{x}) + \int_S \left[\overline{\mathbf{G}}_0^T(\mathbf{x},\mathbf{x}') \cdot \mathbf{t}(\mathbf{x}') - \overline{\mathbf{R}}_0^T(\mathbf{x},\mathbf{x}') \cdot \mathbf{u}(\mathbf{x}')\right] \mathrm{d}S' = 0, \quad \mathbf{x} \in S \tag{9.135}$$

where $\mathbf{u}$ and $\mathbf{t}$, the unknowns to be solved, are the total displacement and traction vectors at the surface of the obstacle. $\overline{\mathbf{G}}$ is the dyadic Green's function given by (9.40) and $\overline{\mathbf{R}} = \hat{\mathbf{n}}' \cdot \overline{\overline{\boldsymbol{\Sigma}}}(\mathbf{x},\mathbf{x}')$ where $\overline{\overline{\boldsymbol{\Sigma}}}(\mathbf{x},\mathbf{x}') = \lambda\overline{\mathbf{I}}\nabla \cdot \overline{\mathbf{G}} + \mu\left(\nabla\overline{\mathbf{G}} + \overline{\mathbf{G}}\nabla\right)$ is a third-rank Green's tensor. All other variables have been defined in Section 9.3.1 and the superscript T on $\overline{\mathbf{G}}$ and $\overline{\mathbf{R}}$ is the transpose which can be removed if their positions with the vector $\mathbf{u}$ or $\mathbf{t}$ are exchanged. The constant $\frac{1}{2}$ in front of $\mathbf{u}$ follows from the assumption that the observation point is on a locally-smooth surface on S. Note that the integrals in all BIEs are defined in the CPV sense.

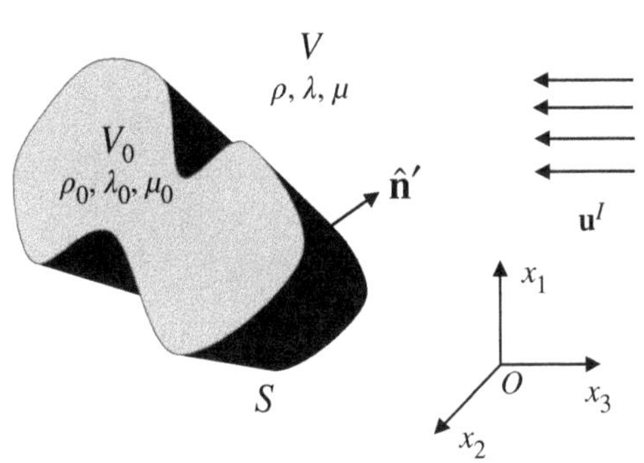

Figure 9.32 Elastic wave scattering by an arbitrarily shaped homogeneous obstacle embedded in an infinite 3D isotropic elastic medium.

If the obstacle is a traction-free cavity, then the total traction on the surface vanishes and the above equations reduce to

$$\frac{1}{2}\mathbf{u}(\mathbf{x}) + \int_S \overline{\mathbf{R}}^T(\mathbf{x},\mathbf{x}') \cdot \mathbf{u}(\mathbf{x}')\, \mathrm{d}S' = \mathbf{u}^{\mathrm{I}}(\mathbf{x}), \quad \mathbf{x} \in S. \tag{9.136}$$

If the obstacle is a fixed rigid inclusion, then the total displacement on the surface vanishes and the above equations reduce to

$$\int_S \overline{\mathbf{G}}^T(\mathbf{x},\mathbf{x}') \cdot \mathbf{t}(\mathbf{x}')\, \mathrm{d}S' = -\mathbf{u}^{\mathrm{I}}(\mathbf{x}), \quad \mathbf{x} \in S. \tag{9.137}$$

As we have mentioned before, there is a numerical difficulty in solving for the BIEs when $\mu \to 0$ in the host medium. However, if we introduce a loss for the host medium that causes the shear wave to decay quickly, then we can solve the pertinent BIEs. This can be accomplished by choosing a negative small value for μ to approximate the original positive value of μ. After doing so, the wavenumber $\kappa_s = \omega\sqrt{\rho/\mu}$ will have a large imaginary part with a chosen plus sign. The Green's function now decays exponentially like an evanescent wave with no oscillatory behavior. Then the integration with $e^{i\kappa_s R}$ in the kernel is no longer difficult. The approach will give us a good approximate solution for the small value of μ as demonstrated in the numerical examples.

9.5.2 Limit of Dyadic Green's Function

For acoustic waves, there is no shear wave propagating in the surrounding non-viscous medium and $\mu = 0$. The limit of the dyadic Green's function $\overline{\mathbf{G}}$ when $\mu \to 0$ in the host medium may not be derived from its explicit expression. When $\mu \to 0$, $\kappa_s \to \infty$ and the scalar shear Green's function g_s becomes highly oscillatory. This high oscillation makes numerical computation with this Green's function difficult in this limit. Hence, there is no direct way of calculating the asymptotic dyadic Green's function. However, a highly-oscillatory integrand contributes a small value to the integral eventually due to the cancellation within the integrand. Therefore, we can derive the limit by inspecting the derivation of the dyadic Green's function using the Fourier-Laplace transform [122].

The elastic wave equation in PDE form is

$$(\lambda + \mu)\nabla\nabla \cdot \mathbf{u} + \mu\nabla^2\mathbf{u} - \rho\ddot{\mathbf{u}} = -\mathbf{f}. \tag{9.138}$$

Applying the Fourier-Laplace transform

$$\mathbf{u}(\mathbf{x}, t) = \frac{1}{(2\pi)^4}\int_{-\infty}^{\infty} \mathrm{d}\omega e^{-i\omega t} \int_{-\infty}^{\infty} \mathrm{d}\mathbf{k} e^{i\mathbf{k}\cdot\mathbf{x}} \tilde{\mathbf{u}}(\mathbf{k}, \omega) \tag{9.139}$$

we obtain

$$\begin{aligned}\tilde{\mathbf{u}}(\mathbf{k}, \omega) &= \left[(\lambda + \mu)\mathbf{k}\mathbf{k} + \mu\kappa^2\overline{\mathbf{I}} - \omega^2\rho\overline{\mathbf{I}}\right]^{-1} \cdot \tilde{\mathbf{f}}(\mathbf{k}, \omega) \\ &= (\alpha\overline{\mathbf{I}} + \beta\mathbf{k}\mathbf{k}) \cdot \tilde{\mathbf{f}}(\mathbf{k}, \omega) \\ &= \left(\overline{\mathbf{I}} - \frac{\mathbf{k}\mathbf{k}}{\kappa_s^2}\right) \cdot \frac{\tilde{\mathbf{f}}(\mathbf{k}, \omega)}{\mu(\kappa^2 - \kappa_s^2)} + \frac{\mathbf{k}\mathbf{k} \cdot \tilde{\mathbf{f}}(\mathbf{k}, \omega)}{\kappa_c^2\gamma(\kappa^2 - \kappa_c^2)}\end{aligned} \tag{9.140}$$

where

$$\alpha = \frac{1}{\mu\kappa^2 - \omega^2\rho} \tag{9.141}$$

$$\beta = -\frac{1}{\kappa_s^2\mu(\kappa^2 - \kappa_s^2)} + \frac{1}{\kappa_c^2\gamma(\kappa^2 - \kappa_c^2)}. \tag{9.142}$$

If $\mu \neq 0$, which is the elastic wave case, it can be shown that

$$\int_{-\infty}^{\infty} d\mathbf{k}e^{i\mathbf{k}\cdot\mathbf{x}} \left(\bar{\mathbf{I}} - \frac{\mathbf{kk}}{\kappa_s^2}\right) \frac{1}{\mu(\kappa^2 - \kappa_s^2)} = \left(\bar{\mathbf{I}} + \frac{\nabla\nabla}{\kappa_s^2}\right) \frac{g_s}{\mu} \tag{9.143}$$

$$\int_{-\infty}^{\infty} d\mathbf{k}e^{i\mathbf{k}\cdot\mathbf{x}} \frac{\mathbf{kk}}{\kappa_c^2\gamma(\kappa^2 - \kappa_c^2)} = -\frac{\nabla\nabla}{\kappa_c^2} \frac{g_c}{\gamma} \tag{9.144}$$

yielding the dyadic Green's function, as shown in (9.40). However, the equation (9.143) is not valid if $\mu = 0$, which is the acoustic wave case. This is because the identity used in deriving that equation

$$\int_{-\infty}^{\infty} \frac{d\mathbf{k}e^{i\mathbf{k}\cdot\mathbf{x}}}{\kappa^2 - \kappa_s^2} = -g_s \tag{9.145}$$

is not well defined as in the lossless medium case in electromagnetics [28]. In fact, when $\mu = 0$, Equations (9.141) and (9.142) are reduced to

$$\alpha = -\frac{1}{\omega^2\rho} \tag{9.146}$$

$$\beta = \frac{1}{\kappa_c^2\lambda(\kappa^2 - \kappa_c^2)} \tag{9.147}$$

and Equation (9.143) becomes

$$\begin{aligned} \lim_{\mu\to 0} \int_{-\infty}^{\infty} d\mathbf{k}e^{i\mathbf{k}\cdot\mathbf{x}} \left(\bar{\mathbf{I}} - \frac{\mathbf{kk}}{\kappa_s^2}\right) \frac{1}{\mu(\kappa^2 - \kappa_s^2)} &= -\frac{\bar{\mathbf{I}}}{\omega^2\rho} \int_{-\infty}^{\infty} d\mathbf{k}e^{i\mathbf{k}\cdot\mathbf{x}} \\ &= -\frac{\delta(\mathbf{x} - \mathbf{x}')}{\omega^2\rho} \cdot \bar{\mathbf{I}} \end{aligned} \tag{9.148}$$

where $\delta(\mathbf{x} - \mathbf{x}')$ is the Dirac delta function. Hence, the limit of the dyadic Green's function for $\mu = 0$ is

$$\overline{\mathbf{G}}(\mathbf{x}, \mathbf{x}') = -\frac{1}{\omega^2\rho} \left[\delta(\mathbf{x} - \mathbf{x}')\bar{\mathbf{I}} + \nabla\nabla g_c(\mathbf{x}, \mathbf{x}')\right]. \tag{9.149}$$

9.5.3 Vector BIE for Acoustic Wave

It is of interest to obtain the limit of the dyadic Green's function for acoustic waves when $\mu \to 0$. Having the asymptotic Green's function in (9.149), the other related quantities can be simplified with $\mu = 0$ as follows

$$\begin{aligned} \mathbf{t}(\mathbf{x}') &= \hat{\mathbf{n}}' \cdot \left\{\lambda\bar{\mathbf{I}}\nabla' \cdot \mathbf{u}(\mathbf{x}') + \mu\left[\nabla'\mathbf{u}(\mathbf{x}') + \mathbf{u}(\mathbf{x}')\nabla'\right]\right\} \\ &= \hat{\mathbf{n}}'\left[\lambda\nabla' \cdot \mathbf{u}(\mathbf{x}')\right] \end{aligned} \tag{9.150}$$

$$\begin{aligned} \mathbf{R}(\mathbf{x}, \mathbf{x}') &= \hat{\mathbf{n}}' \cdot \left\{\lambda\bar{\mathbf{I}}\nabla \cdot \overline{\mathbf{G}}(\mathbf{x}, \mathbf{x}') + \mu\left[\nabla\overline{\mathbf{G}}(\mathbf{x}, \mathbf{x}') + \overline{\mathbf{G}}(\mathbf{x}, \mathbf{x}')\nabla\right]\right\} \\ &= \hat{\mathbf{n}}'\nabla g_c(\mathbf{x}, \mathbf{x}') \end{aligned} \tag{9.151}$$

and the equation (9.134) becomes

$$\int_S \left\{ \left[-\lambda\nabla' \cdot \mathbf{u}(\mathbf{x}')\right] \hat{\mathbf{n}}' \cdot \overline{\mathbf{G}}(\mathbf{x},\mathbf{x}') + \mathbf{u}(\mathbf{x}') \cdot \left[\hat{\mathbf{n}}'\nabla g_c(\mathbf{x},\mathbf{x}')\right] \right\} \mathrm{d}S'$$
$$= -\frac{1}{2}\mathbf{u}(\mathbf{x}) + \mathbf{u}^{\mathrm{I}}(\mathbf{x}), \quad \mathbf{x} \in S \tag{9.152}$$

where we have removed the transpose on the kernels by exchanging their order with **u** or **t** for convenience. This is a vector BIE for acoustic waves. This vector BIE is a limiting case of the elastic wave BIEs and can be solved together with the equation (9.135).

Notice that the boundary conditions here may be different from before and can be expressed as [108]

$$\hat{\mathbf{n}} \cdot \mathbf{u}_+ = \hat{\mathbf{n}} \cdot \mathbf{u}_- \tag{9.153}$$

$$\hat{\mathbf{n}} \cdot \mathbf{t}_+ = \hat{\mathbf{n}} \cdot \mathbf{t}_- \tag{9.154}$$

$$\hat{\mathbf{n}} \times \mathbf{t}_- = 0 \tag{9.155}$$

where + and − imply that the observation points approach the boundary surface from the exterior and interior of the obstacle, respectively. These conditions state that the normal components of displacement and traction vectors are continuous and the tangential component of the traction vector vanishes (also is continuous), but the tangential component of displacement may not be continuous at the surface due to the assumption that the host medium may be non-viscous [123].

In fact, the above vector BIE is equivalent to the conventional scalar acoustic wave BIE. If we take a divergence on the vector BIE and define the potential $\Phi = \nabla \cdot \mathbf{u}$, we have

$$\int_S \left\{ -\lambda\Phi(\mathbf{x}')\nabla \cdot \left[\hat{\mathbf{n}}' \cdot \overline{\mathbf{G}}(\mathbf{x},\mathbf{x}')\right] + \mathbf{u}(\mathbf{x}') \cdot \hat{\mathbf{n}}'\nabla^2 g_c(\mathbf{x},\mathbf{x}') \right\} \mathrm{d}S'$$
$$= -\frac{1}{2}\Phi(\mathbf{x}) + \Phi^I(\mathbf{x}), \quad \mathbf{x} \in S. \tag{9.156}$$

Since

$$\begin{aligned}\nabla \cdot \left[\hat{\mathbf{n}}' \cdot \overline{\mathbf{G}}(\mathbf{x},\mathbf{x}')\right] &= \nabla \cdot \overline{\mathbf{G}}(\mathbf{x},\mathbf{x}') \cdot \hat{\mathbf{n}}' \\ &= \frac{1}{\lambda}\nabla' g_c(\mathbf{x},\mathbf{x}') \cdot \hat{\mathbf{n}}'\end{aligned} \tag{9.157}$$

and

$$\nabla^2 g_c(\mathbf{x},\mathbf{x}') = -\kappa_c^2 g_c(\mathbf{x},\mathbf{x}') - \delta(\mathbf{x}-\mathbf{x}') \tag{9.158}$$

with $\delta(\mathbf{x}-\mathbf{x}') = 0$ due to the CPV definition, we can write (9.156) as

$$\int_S \left[-\Phi(\mathbf{x}')\hat{\mathbf{n}}' \cdot \nabla' g_c(\mathbf{x},\mathbf{x}') - \mathbf{u}(\mathbf{x}') \cdot \hat{\mathbf{n}}'\kappa_c^2 g_c(\mathbf{x},\mathbf{x}')\right] \mathrm{d}S'$$
$$= -\frac{1}{2}\Phi(\mathbf{x}) + \Phi^i(\mathbf{x}), \quad \mathbf{x} \in S. \tag{9.159}$$

Also, when $\mu = 0$, the elastic wave equation in the PDE form reduces to

$$\nabla'\nabla' \cdot \mathbf{u}(\mathbf{x}') + \kappa_c^2\mathbf{u}(\mathbf{x}') = 0 \tag{9.160}$$

or

$$\hat{\mathbf{n}}' \cdot \mathbf{u}(\mathbf{x}') = -\frac{1}{\kappa_c^2}\hat{\mathbf{n}}' \cdot \nabla'\nabla' \cdot \mathbf{u}(\mathbf{x}')$$
$$= -\frac{1}{\kappa_c^2}\hat{\mathbf{n}}' \cdot \nabla'\Phi(\mathbf{x}'). \tag{9.161}$$

Substituting (9.161) to (9.159), we obtain

$$\int_S \left[-\Phi(\mathbf{x}')\hat{\mathbf{n}}' \cdot \nabla' g_c(\mathbf{x},\mathbf{x}') + \hat{\mathbf{n}}' \cdot \nabla'\Phi(\mathbf{x}') g_c(\mathbf{x},\mathbf{x}')\right] \mathrm{d}S'$$
$$= -\frac{1}{2}\Phi(\mathbf{x}) + \Phi^i(\mathbf{x}), \quad \mathbf{x} \in S \tag{9.162}$$

which is consistent with the conventional scalar acoustic wave BIE.

The derivation above indicates that we can use the vector BIE (9.152) to solve for acoustic wave problems. The vector BIE is a special case of the elastic wave BIEs and the solution for it will provide additional information about the displacement in addition to the potential if compared with the conventional scalar BIE. The potential is included in the traction vector in this case ($\mathbf{t} = \hat{\mathbf{n}}'[\lambda\nabla' \cdot \mathbf{u}] = \hat{\mathbf{n}}'\lambda\Phi$).

9.5.4 Method of Moments (MoM) Solutions

The MoM is used in this section because we want to compare its solutions with the Nyström method solutions of the same problems, as presented in previous sections, or the problems before the elastic wave part and acoustic wave part are united. Although the MoM has been extensively studied and used in electromagnetics, it is rarely used in acoustics and has not been used in elastodynamics yet. To illustrate the MoM for solving the elastic wave or acoustic wave scattering problems governed by the vector BIEs, we consider the BIE in (9.137) describing the scattering by a rigid obstacle. The general case is a natural extension of the example shown here.

As the first step in the MoM, we separate the unknown traction vector into tangential and normal components, i.e.

$$\mathbf{t}(\mathbf{x}') = \mathbf{t}_t(\mathbf{x}') + \mathbf{t}_n(\mathbf{x}'). \tag{9.163}$$

We then expand the two components using the RWG and pulse basis functions, respectively,

$$\mathbf{t}_t(\mathbf{x}') = \sum_{n=1}^{N_t} \alpha_n \mathbf{f}_n(\mathbf{x}') \tag{9.164}$$

$$\mathbf{t}_n(\mathbf{x}') = \sum_{n=1}^{N_n} \beta_n \hat{\mathbf{n}}_n(\mathbf{x}'). \tag{9.165}$$

In the above, $\mathbf{f}_n(\mathbf{x}')$ is the RWG basis function, as shown in Figure 9.33, $\hat{\mathbf{n}}_n(\mathbf{x}')$ is the unit normal vector of the nth triangle patch used as a pulse basis function, and α_n and β_n represent the unknown expansion coefficients to be solved. The RWG basis function is

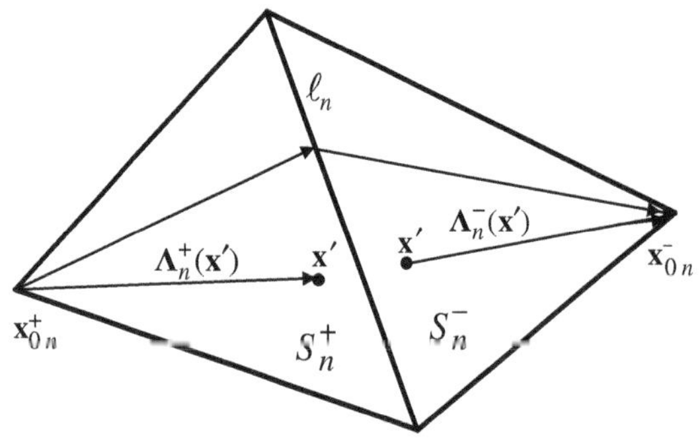

Figure 9.33 RWG basis function $\mathbf{f}_n$ defined in two neighboring triangles S_n^+ and S_n^-. These two triangles share the nth non-boundary edge whose length is ℓ_n. S_n^+ and S_n^- also denote the corresponding areas of the two triangles.

defined as [90]

$$\mathbf{f}_n(\mathbf{x}') = \begin{cases} \frac{\ell_n}{2S_n^+}\Lambda_n^+(\mathbf{x}'), & \mathbf{x}' \in S_n^+ \\ \frac{\ell_n}{2S_n^-}\Lambda_n^-(\mathbf{x}'), & \mathbf{x}' \in S_n^- \\ 0, & \text{otherwise} \end{cases} \tag{9.166}$$

where ℓ_n is the length of the common edge of two neighboring triangles, S_n^+ and S_n^- are the areas of the two triangles, and $\Lambda_n^+(\mathbf{x}')$ and $\Lambda_n^-(\mathbf{x}')$ are the distance vectors as indicated in Figure 9.33. We have in total N_t non-boundary edges connecting two neighboring triangles in which the RWG basis functions are defined and N_n triangles in which the pulse basis functions are defined.

After using the above expansion, the BIE can be written as

$$\sum_{n=1}^{N_t} \alpha_n \int_{S_n} \overline{\mathbf{G}}(\mathbf{x},\mathbf{x}') \cdot \mathbf{f}_n(\mathbf{x}')\, dS' + \sum_{n=1}^{N_n} \beta_n \int_{S_n} \overline{\mathbf{G}}(\mathbf{x},\mathbf{x}') \cdot \hat{\mathbf{n}}_n(\mathbf{x}')\, dS' = -\mathbf{u}^{\mathrm{I}}(\mathbf{x}) \tag{9.167}$$

where we have omitted the transpose T on $\overline{\mathbf{G}}(\mathbf{x},\mathbf{x}')$ without changing the order with $\mathbf{t}$. This is because $\overline{\mathbf{G}}(\mathbf{x},\mathbf{x}')$ is symmetrical [42].

The next step is using the basis functions as the weighting or testing functions to test the equation. By doing so, the following matrix equations are formed

$$\sum_{n=1}^{N_t} \alpha_n A_{mn} + \sum_{n=1}^{N_n} \beta_n B_{mn} = E_{mn}, \quad m = 1, \cdots, N_t \tag{9.168}$$

$$\sum_{n=1}^{N_t} \alpha_n C_{mn} + \sum_{n=1}^{N_n} \beta_n D_{mn} = F_{mn}, \quad m = 1, \cdots, N_n \tag{9.169}$$

where

$$\begin{aligned} A_{mn} &= \left\langle \mathbf{f}_m(\mathbf{x}), \overline{\mathbf{G}}(\mathbf{x},\mathbf{x}'), \mathbf{f}_n(\mathbf{x}') \right\rangle \\ &= \frac{\ell_m \ell_n}{4S_m^+ S_n^+} \left\langle \Lambda_m^+(\mathbf{x}), \overline{\mathbf{G}}(\mathbf{x},\mathbf{x}'), \Lambda_n^+(\mathbf{x}') \right\rangle + \frac{\ell_m \ell_n}{4S_m^+ S_n^-} \left\langle \Lambda_m^+(\mathbf{x}), \overline{\mathbf{G}}(\mathbf{x},\mathbf{x}'), \Lambda_n^-(\mathbf{x}') \right\rangle \end{aligned}$$

$$+\frac{\ell_m \ell_n}{4S_m^- S_n^+}\left\langle \boldsymbol{\Lambda}_m^-(\mathbf{x}), \overline{\mathbf{G}}(\mathbf{x},\mathbf{x}'), \boldsymbol{\Lambda}_n^+(\mathbf{x}')\right\rangle + \frac{\ell_m \ell_n}{4S_m^- S_n^-}\left\langle \boldsymbol{\Lambda}_m^-(\mathbf{x}), \overline{\mathbf{G}}(\mathbf{x},\mathbf{x}'), \boldsymbol{\Lambda}_n^-(\mathbf{x}')\right\rangle \tag{9.170}$$

$$B_{mn} = \left\langle \mathbf{f}_m(\mathbf{x}), \overline{\mathbf{G}}(\mathbf{x},\mathbf{x}'), \hat{\mathbf{n}}_n(\mathbf{x}')\right\rangle$$

$$= \frac{\ell_m}{2S_m^+}\left\langle \boldsymbol{\Lambda}_m^+(\mathbf{x}), \overline{\mathbf{G}}(\mathbf{x},\mathbf{x}'), \hat{\mathbf{n}}_n(\mathbf{x}')\right\rangle + \frac{\ell_m}{2S_m^-}\left\langle \boldsymbol{\Lambda}_m^-(\mathbf{x}), \overline{\mathbf{G}}(\mathbf{x},\mathbf{x}'), \hat{\mathbf{n}}_n(\mathbf{x}')\right\rangle \tag{9.171}$$

$$C_{mn} = \left\langle \hat{\mathbf{n}}_m(\mathbf{x}), \overline{\mathbf{G}}(\mathbf{x},\mathbf{x}'), \mathbf{f}_n(\mathbf{x}')\right\rangle$$

$$= \frac{\ell_n}{2S_n^+}\left\langle \hat{\mathbf{n}}_m(\mathbf{x}), \overline{\mathbf{G}}(\mathbf{x},\mathbf{x}'), \boldsymbol{\Lambda}_n^+(\mathbf{x}')\right\rangle + \frac{\ell_n}{2S_n^-}\left\langle \hat{\mathbf{n}}_m(\mathbf{x}), \overline{\mathbf{G}}(\mathbf{x},\mathbf{x}'), \boldsymbol{\Lambda}_n^-(\mathbf{x}')\right\rangle \tag{9.172}$$

$$D_{mn} = \left\langle \hat{\mathbf{n}}_m(\mathbf{x}), \overline{\mathbf{G}}(\mathbf{x},\mathbf{x}'), \hat{\mathbf{n}}_n(\mathbf{x}')\right\rangle \tag{9.173}$$

$$E_{mn} = -\left\langle \mathbf{f}_m(\mathbf{x}), \mathbf{u}^{\mathrm{I}}(\mathbf{x})\right\rangle$$

$$= -\frac{\ell_m}{2S_m^+}\left\langle \boldsymbol{\Lambda}_m^+(\mathbf{x}), \mathbf{u}^{\mathrm{I}}(\mathbf{x})\right\rangle - \frac{\ell_m}{2S_m^-}\left\langle \boldsymbol{\Lambda}_m^-(\mathbf{x}), \mathbf{u}^{\mathrm{I}}(\mathbf{x})\right\rangle \tag{9.174}$$

$$F_{mn} = -\left\langle \hat{\mathbf{n}}_m(\mathbf{x}), \mathbf{u}^{\mathrm{I}}(\mathbf{x})\right\rangle \tag{9.175}$$

with

$$\boldsymbol{\Lambda}_n^+(\mathbf{x}') = \mathbf{x}' - \mathbf{x}_{0n}^+ \tag{9.176}$$

$$\boldsymbol{\Lambda}_n^-(\mathbf{x}') = \mathbf{x}_{0n}^- - \mathbf{x}'. \tag{9.177}$$

The involved inner products above are defined, for example, as

$$\left\langle \mathbf{f}_m(\mathbf{x}), \overline{\mathbf{G}}(\mathbf{x},\mathbf{x}'), \mathbf{f}_n(\mathbf{x}')\right\rangle = \int_{S_m} \mathrm{d}S\mathbf{f}_m(\mathbf{x}) \cdot \int_{S_n} \overline{\mathbf{G}}(\mathbf{x},\mathbf{x}') \cdot \mathbf{f}_n(\mathbf{x}')\,\mathrm{d}S' \tag{9.178}$$

$$\left\langle \mathbf{f}_m(\mathbf{x}), \mathbf{u}^{\mathrm{I}}(\mathbf{x})\right\rangle = \int_{S_m} \mathbf{f}_m(\mathbf{x}) \cdot \mathbf{u}^{\mathrm{I}}(\mathbf{x})\,\mathrm{d}S. \tag{9.179}$$

The main work in the MoM is the evaluation of those two-fold integrals to generate the system matrix. For the elastic wave scattering, the integral kernels have only $1/R$ and $1/R^2$ singularities for singular elements and they can be easily handled. For acoustic wave scattering, however, there is a $1/R^3$ singularity in the evaluation of integrals for singular elements and we use the strategy developed in electromagnetics to perform such a kind of singular integrations [121]. To reduce the overhead of numerical integrations, the one-point quadrature rule can be used for the outer integrals in the testing procedure. Since the unknown coefficients in the RWG basis expansions are counted in terms of the non-boundary edges connecting two neighboring triangles, the MoM uses fewer unknowns compared with the collocation-based methods in the same mesh size [26].

9.5.5 Numerical Examples

Elastic Wave Scattering

To demonstrate the use of MoM in elastodynamics, we first consider the scattering by a fixed rigid sphere with a radius of $a = 1.0$. The host medium has a Poisson's ratio $\nu = 0.25$ and mass density $\rho = 1.0$. The incident wave has a unit circular frequency ($\omega = 1.0$) and normalized wavenumber of $\kappa_c a = 0.125$, 0.913, and π, respectively. Figures 9.34–9.36 show the radial and tangential (elevated) components of total traction along the principal cut ($\phi = 0°$ and $\theta = 0° - 180°$) at the surface. The surface is discretized into 960 flat triangles in all cases and may be overly meshed for the spheres of small sizes. It can be seen that the solutions agree with the analytical solutions very well. Note that the analytical solutions for general elastic wave scattering by a sphere can be found in [123], but the asymptotic solutions with $\mu \to 0$ for acoustic wave scattering are not available and we have derived them in Appendix 9.2.

We then consider the scattering in an elastic medium by a traction-free spherical cavity (void). The cavity has a radius of $a = 1.0$ and the host medium is characterized by a Poisson's ratio $v = 1/3$, Young's modulus $E = 2/3$ and mass density $\rho = 1.0$. The incident wave has a unit circular frequency ($\omega = 1.0$) with normalized wavenumber of $\kappa_c a = 0.125$ and 0.913, respectively. Figures 9.37 and 9.38 show the total displacement components along the principal cut at the surface. The solutions are also very close to the analytical solutions.

For the generalized case with both the host medium and obstacle being elastic, we select $\lambda = 0.53486$, $\mu = 0.23077$ and $\rho = 1.0$ for the host medium, and $\lambda_0 = 0.23716$, $\mu_0 = 0.52641$, and $\rho_0 = 1.9852$ for the elastic spherical inclusion with a unit radius. Figures 9.39 and 9.40 plot the tangential or radial components of total traction and displacement at the surface along the principal cut for $\kappa_c a = 0.125$ and 0.913, respectively. These results are also in excellent agreement with the analytical solutions.

We choose the spherical objects as scatterers because there are available analytical solutions to compare with in numerical solutions for scattering by such objects. The MoM can of course be used to solve the scattering by arbitrarily shaped objects and Figure 9.41 is an illustration for the solutions of scattering by an elastic cube with

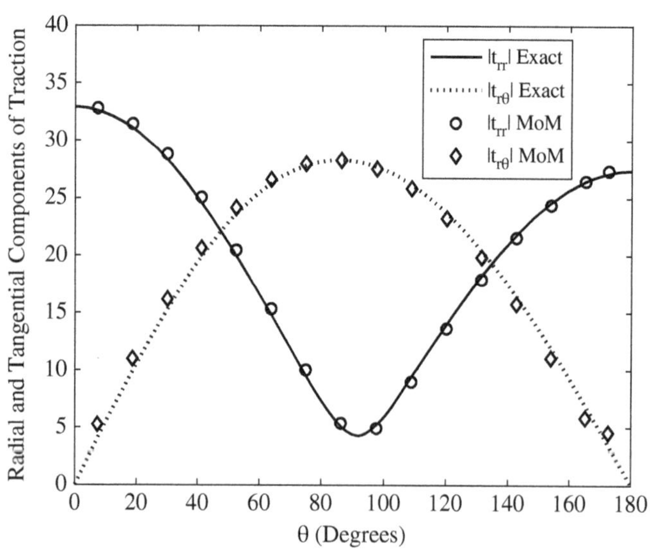

Figure 9.34 Radial and tangential components of total traction along the principal cut at the surface of a rigid sphere, $k_c a = 0.125$.

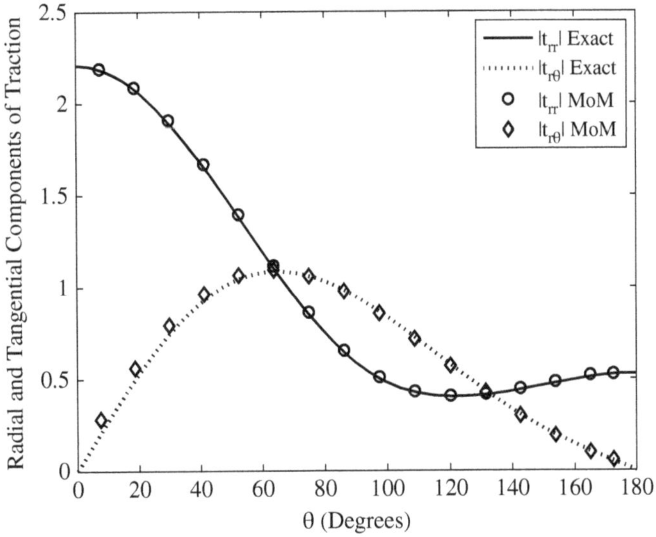

Figure 9.35 Radial and tangential components of total traction along the principal cut at the surface of a rigid sphere, $k_c a = 0.913$.

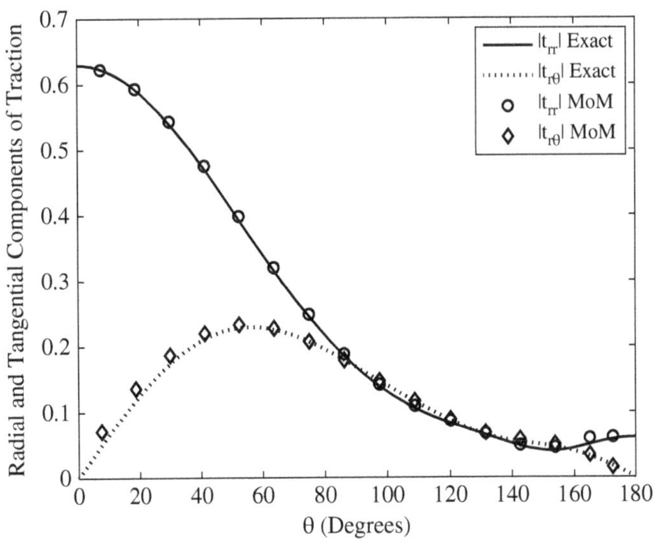

Figure 9.36 Radial and tangential components of total traction along the principal cut at the surface of a rigid sphere, $k_c a = \pi$.

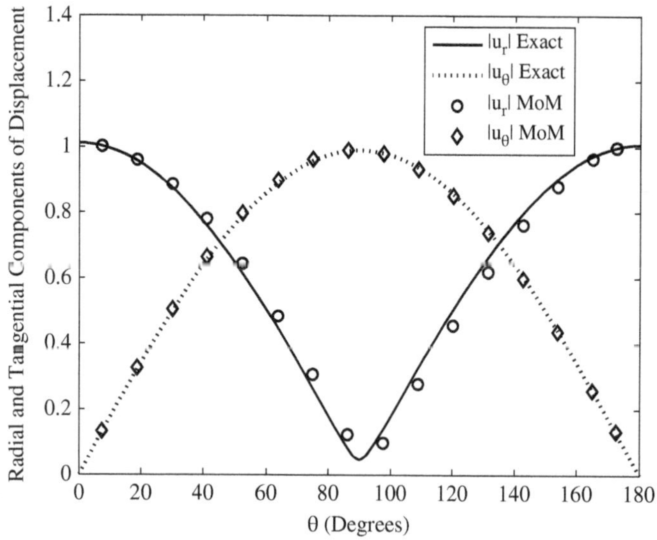

Figure 9.37 Radial and tangential components of total displacement along the principal cut at the surface of a cavity, $k_c a = 0.125$.

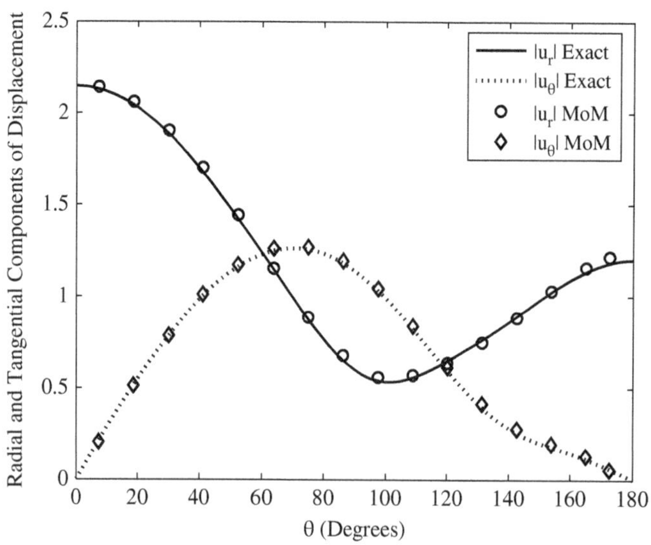

Figure 9.38 Radial and tangential components of total displacement along the principal cut at the surface of a cavity, $k_c a = 0.913$.

a side length of $2a$. We calculate the scattered displacement field at $r = 5a$ surface along the principal cut under the same material properties and incident wave as in the generalized case.

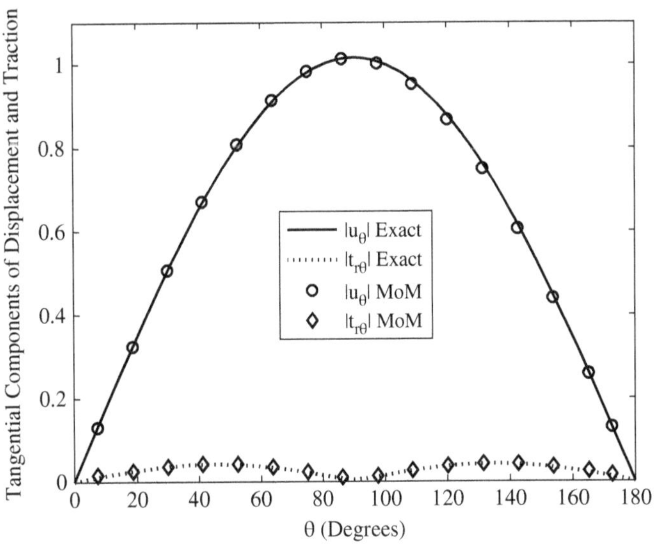

Figure 9.39 Tangential components of total displacement and traction along the principal cut at the surface of an elastic sphere, $k_c a = 0.125$.

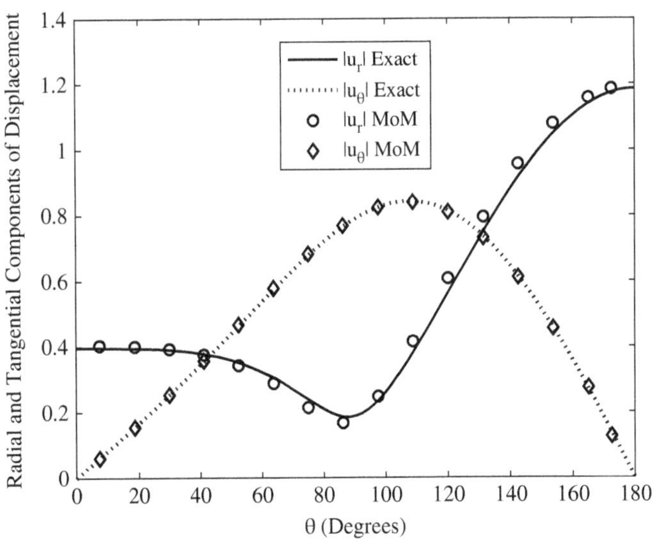

Figure 9.40 Radial and tangential components of total displacement along the principal cut at the surface of an elastic sphere, $k_c a = 0.913$.

Acoustic Wave Scattering

In principle, the acoustic wave equation can be derived from the elastic wave equations by letting $\mu \rightarrow 0$, as shown in the previous subsection. The shear wave part of the dyadic Green's function becomes a Dirac delta distribution as shown by Equation (9.149). However, from the numerical experimentation, it is difficult to achieve this numerically

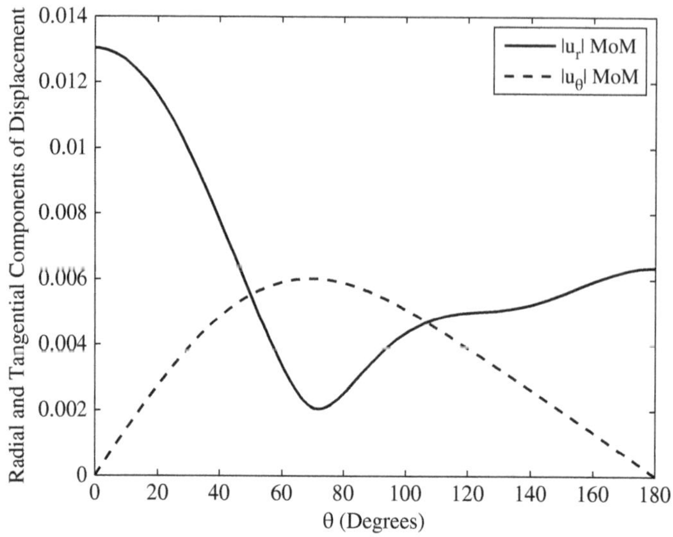

Figure 9.41 Radial and tangential components of scattered displacement along the principal cut at the $r = 5a$ surface by an elastic cube, $k_c a = 0.125$.

by setting $\mu \to 0$, as the Green's function for the shear wave becomes highly oscillatory. This oscillation makes the numerical evaluation of the matrix elements difficult. We have also experimented with $\mu = -i\delta$ where δ is a small number. This is analogous to the highly conductive medium case in electromagnetics, where the shear wave will be highly attenuated. However, an oscillatory factor of the Green's function still persists with this choice, making its accurate numerical evaluation difficult. We find that the best approach is to let $\mu = -\delta$ where δ is a small number. This is analogous to a plasma medium in electromagnetics where waves become evanescent with no oscillatory component, or are shielded in a short length-scale.

In the following, the acoustic wave scattering solution can be derived numerically using the original elastic wave BIE with a negative small value of μ. This approach allows one to solve for the scattering of both waves in a unified manner using one numerical code. Figure 9.42 shows the solutions of scattering by a rigid sphere with $a = 1.0$, $\rho = 1.0$, $\omega = 1.0$, and $\kappa_c a = 1.0$. We take $\mu = -10^{-3}$ for this case and the solution is very close to the analytical counterpart. It can be seen that the $\mu = -10^{-3}$ solution approximates the $\mu = 0$ solution. On the other hand, the $\mu = 10^{-3}$ solution approximates the $\mu = 0$ case with a difficulty and less accuracy due to the highly oscillatory integral kernels.

Figure 9.43 illustrates the scattering by an elastic sphere with the same parameters as the rigid sphere case except the penetrable property of the sphere. The elasticity of the sphere is the same as the third case for the elastic wave scattering. The numerical solution is also in good agreement with the analytical one for the smaller value of μ and also the approximation between negative μ and $\mu = 0$ gets better if μ is closer to 0.

If μ is exactly equal to 0, we need to use the vector BIE (9.152) to solve for acoustic wave scattering. Similar to the elastic wave scattering, we consider the scattering by a rigid sphere, a spherical cavity and elastic sphere, all with a unit radius, respectively. For the rigid sphere case, the host medium is characterized by $\rho = 1.0$, $\mu = 0$ and λ, which is

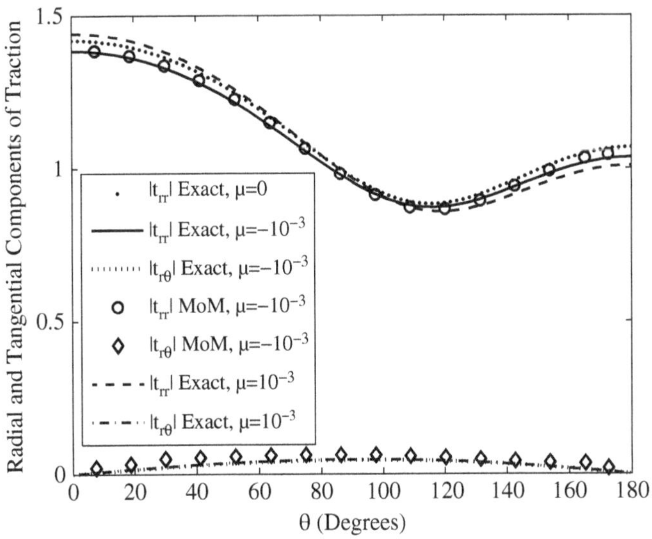

Figure 9.42 Radial and tangential components of total traction along the principal cut at the surface of a rigid sphere with a small value of μ.

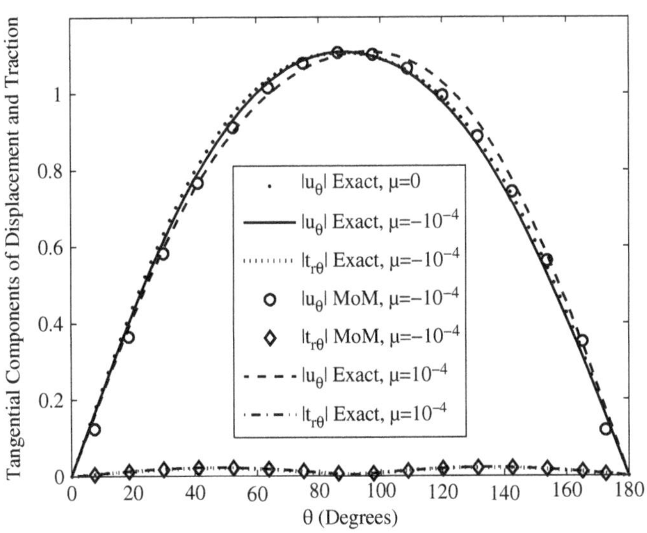

Figure 9.43 Tangential components of total displacement and traction along the principal cut at the surface of an elastic sphere with a small value of μ.

determined by other parameters. The incident wave has a unit circular frequency ($\omega = 1.0$) and normalized wavenumber of $\kappa_c a = 0.5$ and π, respectively. Figures 9.44 and 9.45 show the radial components of total traction along the principal cut at the surface of the rigid sphere. Since there is no shear wave, the tangential component of traction is zero. It can be seen that the MoM solutions are very close to the analytical solutions.

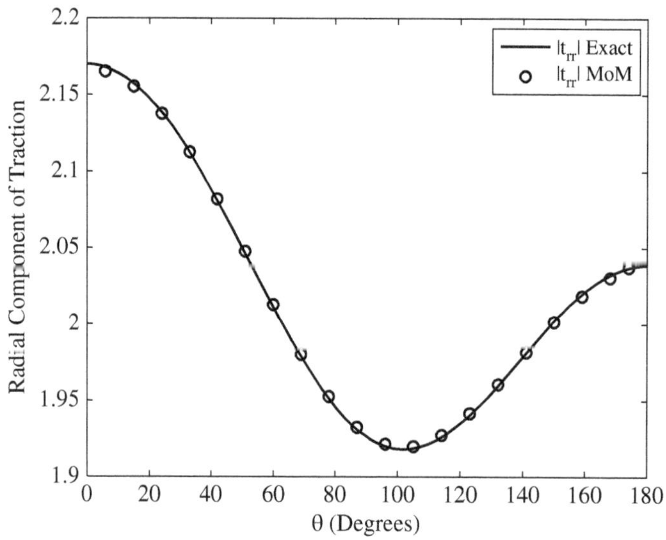

Figure 9.44 Radial components of total traction along the principal cut at the surface of a rigid sphere, $k_c a = 0.5$.

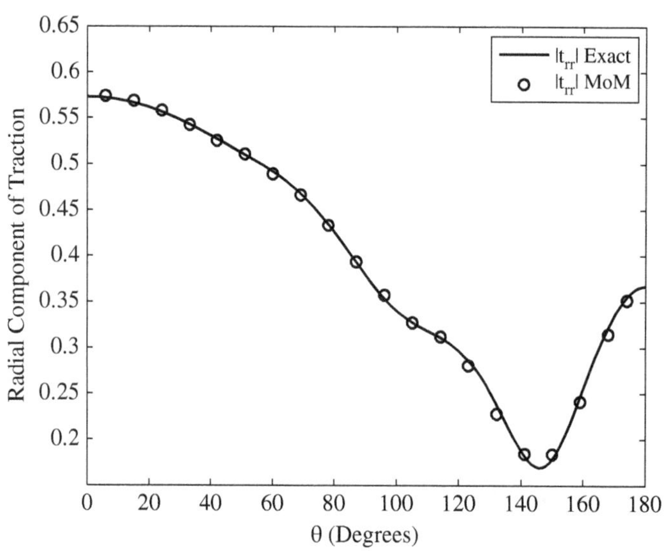

Figure 9.45 Radial components of total traction along the principal cut at the surface of a rigid sphere, $k_c a = \pi$.

For the spherical cavity case, the host medium and incident wave are the same as before except $\kappa_c a = 0.913$ now. The numerical and analytical solutions are shown in Figure 9.46 and a good agreement is seen again. Finally, for the elastic inclusion case, we choose the same parameters as in the preceding elastic wave scattering, except $\mu = 0$ for the host medium now. Figure 9.47 plots the solutions for $\kappa_c a = 0.913$ and they are also consistent with the analytical results.

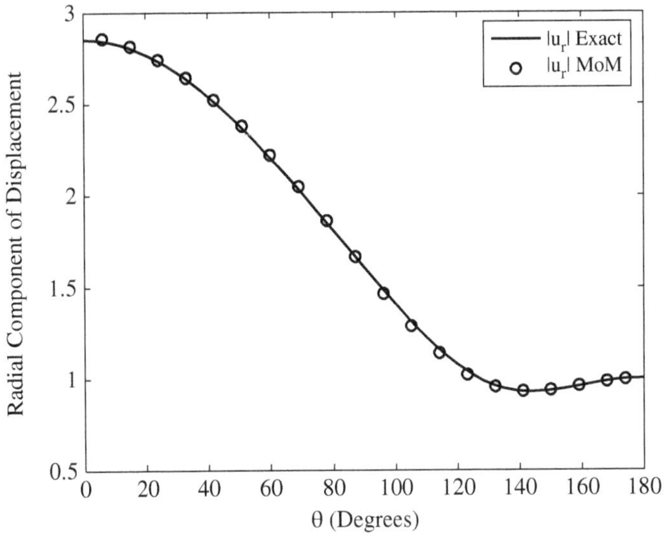

Figure 9.46 Radial components of total displacement along the principal cut at the surface of a spherical cavity, $k_c a = 0.913$.

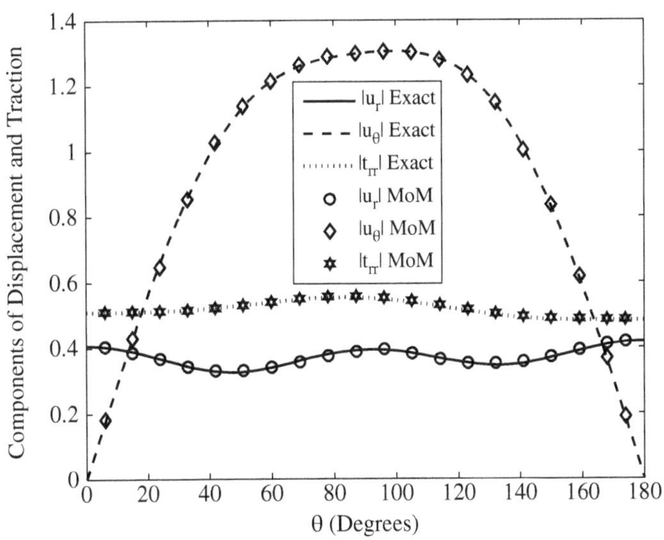

Figure 9.47 Radial and tangential components of total displacement and traction along the principal cut at the surface of an elastic sphere, $k_c a = 0.913$.

9.6 Coupled Integral Equations for Electromagnetic Wave and Elastic Wave

The interaction of EM wave with elastic materials represents one of the important multiphysics problems and could be encountered in many applications. When an EM wave illuminates an elastic media, electrodynamic and elastodynamic processes will occur simultaneously and coupled Maxwell's equations and elastic wave equations should be solved together in order to fully understand the involved physics. Although the coupled equations have been derived earlier, they were only in a PDE form and thus only PDE numerical solvers can be employed in seeking solutions. We first develop coupled BIEs for the problem by using Huygens' equivalence principle and extinction theorem. The coupling is reflected in the excitations or right-hand sides in both EM wave part and elastic wave parts of the integral equations. The EM force is the new excitation for elastic wave and the current density due to the strain tensor is the new excitation for EM wave in the coupled BIEs. Using the BIEs, the solution process can take all advantages of IESs like the MoM, BEM and Nyström methods. We use the Nyström method with an iterative process to solve the coupled BIEs and some numerical results for the interaction of EM wave with piezoelectric materials are illustrated. The developed approach could also find applications in many other areas. For example, one has tried to use the illumination of microwave and the measurement of induced elastic wave to reconstruct objects in geophysical exploration and medical imaging because the microwave illumination can result in a high contrast while the measurement of an elastic wave can yield high resolution in the imaging. Using the coupled integral equations could greatly facilitate the analysis for such problems.

9.6.1 EM Wave Integral Equations

The EM wave integral equations for scattering by a dielectric object, when ignoring the elasticity of the object and surrounding medium, can be derived from the vector wave equations

$$\nabla \times \nabla \times \mathbf{E}_2(\mathbf{r}) - \omega^2 \mu_2 \epsilon_2 \mathbf{E}_2(\mathbf{r}) = \mathrm{i}\omega\mu_2 \mathbf{J}_e(\mathbf{r}), \quad \mathbf{r} \in V_2 \tag{9.180}$$

$$\nabla \times \nabla \times \mathbf{E}_1(\mathbf{r}) - \omega^2 \mu_1 \epsilon_1 \mathbf{E}_1(\mathbf{r}) = 0, \quad \mathbf{r} \in V_1 \tag{9.181}$$

where we designate the interior V_1 and exterior V_2 of the object as Region 1 and Region 2, respectively, and assume that there is a current density source $\mathbf{J}_e$ in region 2 to excite the incident wave upon the object. Also, ϵ_i and μ_i $(i = 1{,}2)$ are the permittivity and permeability of materials, and $\mathbf{E}_i$ $(i = 1{,}2)$ is the electric field in the two regions, respectively. By introducing the dyadic Green's function

$$\overline{\mathbf{G}}(\mathbf{r}, \mathbf{r}') = \left(\overline{\mathbf{I}} + \frac{\nabla\nabla}{k^2}\right) g(\mathbf{r}, \mathbf{r}') \tag{9.182}$$

where k is the wavenumber, $\overline{\mathbf{I}}$ is an identity dyad, and $g(\mathbf{r}, \mathbf{r}') = \mathrm{e}^{\mathrm{i}kR}/(4\pi R)$ is the 3D scalar Green's function with $R = |\mathbf{r} - \mathbf{r}'|$ being the distance between a field point $\mathbf{r}$ and a source point, we can derive the electric field integral equation [28]

$$\int_S \left\{ \mathrm{i}\omega\mu_2 \overline{\mathbf{G}}_2(\mathbf{r}, \mathbf{r}') \cdot \mathbf{J}(\mathbf{r}') + \left[\nabla \times \overline{\mathbf{G}}_2(\mathbf{r}, \mathbf{r}')\right] \cdot \mathbf{M}(\mathbf{r}') \right\} \mathrm{d}S' = -\mathbf{E}_{\mathrm{inc}}(\mathbf{r}), \quad \mathbf{r} \in S \tag{9.183}$$

$$\int_S \left\{ \mathrm{i}\omega\mu_1 \overline{\mathbf{G}}_1(\mathbf{r}, \mathbf{r}') \cdot \mathbf{J}(\mathbf{r}') + \left[\nabla \times \overline{\mathbf{G}}_1(\mathbf{r}, \mathbf{r}')\right] \cdot \mathbf{M}(\mathbf{r}') \right\} \mathrm{d}S' = 0, \quad \mathbf{r} \in S \tag{9.184}$$

where S is the object surface, $\mathbf{J}(\mathbf{r}')$ and $\mathbf{M}(\mathbf{r}')$ are the unknown equivalent electric current density and magnetic current density on the surface, respectively, and

$$\mathbf{E}_{\text{inc}}(\mathbf{r}) = \mathrm{i}\omega\mu_2 \int_{V_2} \overline{\mathbf{G}}(\mathbf{r},\mathbf{r}') \cdot \mathbf{J}_e(\mathbf{r}')\, \mathrm{d}V' \tag{9.185}$$

is the incident field.

Alternatively, the integral equations can be directly derived by the equivalent electric and magnetic current method with the help of vector and scalar potentials relating the fields to their sources. Consider the scattering by a 3D dielectric body, as shown in Figure 9.48(a). The original problem can be decomposed into an equivalent external problem shown in 9.48(b) and an equivalent internal problem shown in Figure 9.48(c). In the equivalent external problem, the original electric and magnetic fields exist outside the object surface S and no fields exist inside S. To support such fields, there must exist equivalent electric and magnetic surface current densities $\mathbf{J}$ and $\mathbf{M}$ on the boundary S, which are unknowns to be solved. Since the tangential components of fields are continuous at the boundary, we can write the integral equations as follows

$$\hat{n} \times \mathbf{E}_2^{S^+}(\mathbf{J},\mathbf{M}) = -\hat{n} \times \mathbf{E}_{\text{inc}}^{S^+} \tag{9.186}$$

$$\hat{n} \times \mathbf{H}_2^{S^+}(\mathbf{J},\mathbf{M}) = -\hat{n} \times \mathbf{H}_{\text{inc}}^{S^+} \tag{9.187}$$

where $\hat{n}$ is the unit normal vector on the surface, and $\mathbf{E}$ and $\mathbf{H}$ are the electric and magnetic field, respectively. The subscript "inc" represents an incident field and the superscript "S^+" denotes the interface where observation points are located. We use "+" and "−" on it to indicate the interior side and exterior side of the surface S, respectively. In the equivalent internal problem described in Figure 9.48(c), the original fields exist inside the object surface and no fields exist outside the surface. Hence, there must exist

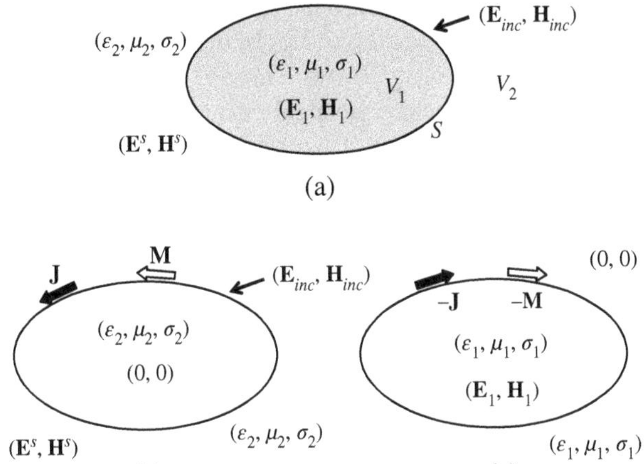

Figure 9.48 Scattering by a dielectric object. (a) Original problem. (b) Equivalent external problem. (c) Equivalent internal problem.

an equivalent electric current density $-\mathbf{J}$ and magnetic current density $-\mathbf{M}$ on the surface S to produce the original fields inside the object. The boundary condition requires

$$\hat{n} \times \mathbf{E}_1^{S^-}(\mathbf{J}, \mathbf{M}) = 0 \tag{9.188}$$

$$\hat{n} \times \mathbf{H}_1^{S^-}(\mathbf{J}, \mathbf{M}) = 0 \tag{9.189}$$

which is another group of integral equations. Equations (9.186) and (9.188) about the electric field are known as the electric field integral equations (EFIEs) or E-field formulations, and Equations (9.187) and (9.189) about the magnetic field are known as the magnetic field integral equations (MFIEs) or H-field formulations. The electric field and magnetic field can be expressed in terms of the equivalent current densities on the surface, $\mathbf{J}$ and $\mathbf{M}$ by recalling the vector and scalar potentials, i.e.

$$\begin{aligned}\mathbf{E}(\mathbf{J}, 0) &= -\mathrm{i}\omega \mathbf{A}(\mathbf{J}) - \nabla \Phi(\mathbf{J}) \\ &= -\mathrm{i}\omega\mu \int_S \overline{\mathbf{G}}(\mathbf{r}, \mathbf{r}') \cdot \mathbf{J}(\mathbf{r}')\, \mathrm{d}S' \\ &= -ik\eta \mathcal{L}(\mathbf{J})\end{aligned} \tag{9.190}$$

$$\begin{aligned}\mathbf{E}(0, \mathbf{M}) &= -\frac{1}{\epsilon} \nabla \times \mathbf{F}(\mathbf{M}) \\ &= \int_S \mathbf{M}(\mathbf{r}') \times \nabla g(\mathbf{r}, \mathbf{r}')\, \mathrm{d}S' \\ &= \mathcal{K}(\mathbf{M})\end{aligned} \tag{9.191}$$

$$\begin{aligned}\mathbf{H}(0, \mathbf{M}) &= -\mathrm{i}\omega \mathbf{F}(\mathbf{M}) - \nabla \Psi(\mathbf{M}) \\ &= -\frac{ik}{\eta} \mathcal{L}(\mathbf{M})\end{aligned} \tag{9.192}$$

$$\mathbf{H}(\mathbf{J}, 0) = \frac{1}{\mu} \nabla \times \mathbf{A}(\mathbf{J}) = -\mathcal{K}(\mathbf{J}) \tag{9.193}$$

where $\mathbf{A}$ and $\mathbf{F}$ are the magnetic and electric vector potentials, and Φ and Ψ are the electric and magnetic scalar potentials, respectively. Also, η and k are the wave impedance and wavenumber of the relevant media, respectively. We have used $\mathcal{L}$ and $\mathcal{K}$ operators to represent the relationship between the fields and equivalent current densities in the above.

If we add the internal field equations to the external field equations by a weighted-sum method [26], then we can obtain PMCHWT equations or Müller formulations [30], but we only use the EFIEs here, which takes the following specific form

$$\hat{n} \times \int_S \left[\mathrm{i}\omega\mu_2 g_2 \mathbf{J} - \frac{1}{\mathrm{i}\omega\epsilon_2} \nabla g_2 (\nabla' \cdot \mathbf{J}) - \mathbf{M} \times \nabla g_2 \right] \mathrm{d}S' = \hat{n} \times \mathbf{E}_{\mathrm{inc}}, \quad \mathbf{r} \in S \tag{9.194}$$

$$\hat{n} \times \int_S \left[\mathrm{i}\omega\mu_1 g_1 \mathbf{J} - \frac{1}{\mathrm{i}\omega\epsilon_1} \nabla g_1 (\nabla' \cdot \mathbf{J}) - \mathbf{M} \times \nabla g_1 \right] \mathrm{d}S' = 0, \quad \mathbf{r} \in S \tag{9.195}$$

where g_1 and g_2 are the 3D scalar Green's function related to the medium in Region 1 and Region 2, respectively. The above EM integral equations will be revised when considering the elasticity of the object and the surrounding medium.

9.6.2 Elastic Wave Integral Equations

The elastic wave scattering from an elastic object is governed by the following PDE [126]

$$(\lambda+\mu)\nabla\nabla\cdot\mathbf{u}+\mu\nabla^2\mathbf{u}+\omega^2\rho\mathbf{u}=-\mathbf{f} \tag{9.196}$$

or

$$\gamma\nabla\nabla\cdot\mathbf{u}-\mu\nabla\times\nabla\times\mathbf{u}+\omega^2\rho\mathbf{u}=-\mathbf{f} \tag{9.197}$$

where $\gamma=\lambda+2\mu$. These two equations are equivalent because $\nabla^2\mathbf{u}=\nabla\nabla\cdot\mathbf{u}-\nabla\times\nabla\times\mathbf{u}$. In the above, the elasticities of the elastic object and the surrounding medium are characterized by (λ_1,μ_1,ρ_1) and (λ_2,μ_2,ρ_2), respectively, where λ and μ are Lamé constants and ρ is the mass density of the object or medium. Also, $\mathbf{u}$ stands for the displacement vector, $\mathbf{f}$ denotes the body force per unit mass and ω is the angular frequency of a time-harmonic excitation. From the above PDE, and using the equivalence principle and extinction theorem, we can derive the corresponding integral equations. First, Equation (9.196) can be solved using either Helmholtz decomposition or the Fourier-Laplace transform [26] for a homogeneous medium as

$$\mathbf{u}(\mathbf{r})=\int d\mathbf{r}'\left[\frac{1}{\mu}\left(\bar{\mathbf{I}}+\frac{\nabla\nabla}{k_s^2}\right)g_s(\mathbf{r}-\mathbf{r}')-\frac{1}{\gamma}\frac{\nabla\nabla}{k_c^2}g_c(\mathbf{r}-\mathbf{r}')\right]\cdot\mathbf{f}(\mathbf{r}') \tag{9.198}$$

where

$$g_s=\frac{e^{ik_sR}}{4\pi R},\quad g_c=\frac{e^{ik_cR}}{4\pi R} \tag{9.199}$$

are the scalar Green's functions for a shear wave with $k_s=\omega\sqrt{\rho/\mu}$ and a compressional wave with $k_c=\omega\sqrt{\rho/\gamma}$, respectively. Here we have two kinds of wave in each elastic media, i.e. shear wave with a subscript "s" and compressional wave with a subscript "c". We then identify that the dyadic Green's function is

$$\bar{\mathbf{G}}(\mathbf{r},\mathbf{r}')=\frac{1}{\mu}\left(\bar{\mathbf{I}}+\frac{\nabla\nabla}{k_s^2}\right)g_s(\mathbf{r},\mathbf{r}')-\frac{1}{\gamma}\frac{\nabla\nabla}{k_c^2}g_c(\mathbf{r},\mathbf{r}') \tag{9.200}$$

which satisfies

$$\gamma\nabla\nabla\cdot\bar{\mathbf{G}}(\mathbf{r},\mathbf{r}')-\mu\nabla\times\nabla\times\bar{\mathbf{G}}(\mathbf{r},\mathbf{r}')+\omega^2\rho\bar{\mathbf{G}}(\mathbf{r},\mathbf{r}')=\bar{\mathbf{I}}\delta(\mathbf{r}-\mathbf{r}'). \tag{9.201}$$

Right multiplying (9.197) by $\bar{\mathbf{G}}(\mathbf{r},\mathbf{r}')$, and left multiplying (9.201) by $\mathbf{u}(\mathbf{r})$ and subtracting the result, we have

$$\begin{aligned}&\gamma\left\{\nabla\left[\nabla\cdot\mathbf{u}(\mathbf{r})\right]\cdot\bar{\mathbf{G}}(\mathbf{r},\mathbf{r}')-\mathbf{u}(\mathbf{r})\cdot\nabla\nabla\cdot\bar{\mathbf{G}}(\mathbf{r},\mathbf{r}')\right\}-\mu\left[\nabla\times\nabla\times\mathbf{u}(\mathbf{r})\cdot\bar{\mathbf{G}}(\mathbf{r},\mathbf{r}')\right.\\&\left.-\mathbf{u}(\mathbf{r})\cdot\nabla\times\nabla\times\bar{\mathbf{G}}(\mathbf{r},\mathbf{r}')\right]=-\mathbf{f}(\mathbf{r})\cdot\bar{\mathbf{G}}(\mathbf{r},\mathbf{r}')-\delta(\mathbf{r}-\mathbf{r}')\mathbf{u}(\mathbf{r}').\end{aligned} \tag{9.202}$$

Using the vector identities

$$\nabla\cdot\left[(\nabla\cdot\mathbf{u})\mathbf{B}-\mathbf{u}\nabla\cdot\mathbf{B}\right]=\nabla(\nabla\cdot\mathbf{u})\cdot\mathbf{B}-\mathbf{u}\cdot(\nabla\nabla\cdot\mathbf{B}) \tag{9.203}$$

$$\nabla\cdot\left[(\nabla\times\mathbf{u})\times\mathbf{B}+\mathbf{u}\times\nabla\times\mathbf{B}\right]=(\nabla\times\nabla\times\mathbf{u})\cdot\mathbf{B}-\mathbf{u}\cdot\nabla\times\nabla\times\mathbf{B} \tag{9.204}$$

we can integrate (9.202) over a volume V surrounded by S to get

$$\gamma \int_S \hat{n} \cdot \left\{ \left[\nabla \cdot \mathbf{u}(\mathbf{r})\right] \overline{\mathbf{G}}(\mathbf{r}, \mathbf{r}') - \mathbf{u}(\mathbf{r}) \nabla \cdot \overline{\mathbf{G}}(\mathbf{r}, \mathbf{r}') \right\} \, \mathrm{d}S - \mu \int_S \hat{n} \cdot \left\{ \left[\nabla \times \mathbf{u}(\mathbf{r})\right] \times \overline{\mathbf{G}}(\mathbf{r}, \mathbf{r}') \right.$$
$$\left. + \mathbf{u}(\mathbf{r}) \times \nabla \times \overline{\mathbf{G}}(\mathbf{r}, \mathbf{r}') \right\} \, \mathrm{d}S + \int_V \mathbf{f}(\mathbf{r}) \cdot \overline{\mathbf{G}}(\mathbf{r}, \mathbf{r}') \, \mathrm{d}\mathbf{r} = \begin{cases} -\mathbf{u}(\mathbf{r}'), & \mathbf{r}' \in V \\ 0, & \mathbf{r}' \notin V \end{cases} . \quad (9.205)$$

By changing the primed into the unprimed and the unprimed into the primed, the above equation can be rewritten as

$$\mathbf{u}_{\mathrm{inc}}(\mathbf{r}) + \mu \int_S \left\{ \hat{n}' \times \left[\nabla' \times \mathbf{u}(\mathbf{r}')\right] \cdot \overline{\mathbf{G}}(\mathbf{r}', \mathbf{r}) + \left[\hat{n}' \times \mathbf{u}(\mathbf{r}')\right] \cdot \nabla' \times \overline{\mathbf{G}}(\mathbf{r}', \mathbf{r}) \right\} \, \mathrm{d}S'$$
$$- \gamma \int_S \left[\nabla' \cdot \mathbf{u}(\mathbf{r}') \hat{n}' \cdot \overline{\mathbf{G}}(\mathbf{r}', \mathbf{r}) - \hat{n}' \cdot \mathbf{u}(\mathbf{r}') \nabla' \cdot \overline{\mathbf{G}}(\mathbf{r}', \mathbf{r}) \right] \, \mathrm{d}S' = \begin{cases} \mathbf{u}(\mathbf{r}), & \mathbf{r} \in V \\ 0, & \mathbf{r} \notin V \end{cases} \quad (9.206)$$

where

$$\mathbf{u}_{\mathrm{inc}}(\mathbf{r}) = \int_V \mathbf{f}(\mathbf{r}') \cdot \overline{\mathbf{G}}(\mathbf{r}', \mathbf{r}) \, \mathrm{d}\mathbf{r}' \quad (9.207)$$

represents the incident displacement field. Also, the extinction theorem is embedded in the above equation. The above equation includes $\nabla' \cdot \mathbf{u}$, $\hat{n}' \cdot \mathbf{u}$, $\nabla' \times \mathbf{u}$, and $\hat{n}' \times \mathbf{u}$ as unknown functions but they are not solvable since the boundary conditions, i.e. the continuity of the displacement vector **u** and traction vector **t** across a boundary, cannot be applied to. The traction vector **t** is related to the stress tensor $\overline{\mathbf{T}}$ or the displacement vector **u** by Hooke's law, i.e.

$$\begin{aligned} \mathbf{t} &= \hat{n} \cdot \overline{\mathbf{T}} \\ &= \hat{n} \cdot \left[\lambda \overline{\mathbf{I}} \nabla \cdot \mathbf{u} + \mu (\nabla \mathbf{u} + \mathbf{u} \nabla) \right] . \end{aligned} \quad (9.208)$$

To incorporate the boundary conditions, we must reformulate the integrals in Equation (9.206), namely (we omit **r** and **r**′ for simplicity),

$$\begin{aligned} I_1 &= -\gamma \int_S \left[(\nabla' \cdot \mathbf{u}) \left(\hat{n}' \cdot \overline{\mathbf{G}} \right) - (\hat{n}' \cdot \mathbf{u}) \left(\nabla' \cdot \overline{\mathbf{G}} \right) \right] \, \mathrm{d}S' \\ &\quad + \mu \int_S \left[\hat{n}' \times (\nabla' \times \mathbf{u}) \cdot \overline{\mathbf{G}} + (\hat{n}' \times \mathbf{u}) \cdot \left(\nabla' \times \overline{\mathbf{G}} \right) \right] \, \mathrm{d}S' \\ &= -\gamma \int_S \left[(\nabla' \cdot \mathbf{u}) \left(\hat{n}' \cdot \overline{\mathbf{G}} \right) - (\hat{n}' \cdot \mathbf{u}) \left(\nabla' \cdot \overline{\mathbf{G}} \right) \right] \, \mathrm{d}S' \\ &\quad + \mu \int_S \left[\hat{n}' \times (\nabla' \times \mathbf{u}) \cdot \overline{\mathbf{G}} - \mathbf{u} \cdot \hat{n}' \times \left(\nabla' \times \overline{\mathbf{G}} \right) \right] \, \mathrm{d}S' \\ &= \int_S \left[-(\lambda + 2\mu)(\nabla' \cdot \mathbf{u}) \left(\hat{n}' \cdot \overline{\mathbf{G}} \right) + (\lambda + 2\mu)(\hat{n}' \cdot \mathbf{u}) \left(\nabla' \cdot \overline{\mathbf{G}} \right) \right. \\ &\quad \left. + \mu \hat{n}' \cdot (\nabla' \mathbf{u} - \mathbf{u} \nabla') \cdot \overline{\mathbf{G}} + \mu \mathbf{u} \cdot \hat{n}' \cdot \left(\overline{\mathbf{G}} \nabla' - \nabla' \overline{\mathbf{G}} \right) \right] \, \mathrm{d}S' \end{aligned} \quad (9.209)$$

where we have used the identity $(\mathbf{A} \times \mathbf{B}) \cdot \overline{\mathbf{C}} = -\mathbf{B} \cdot \left(\mathbf{A} \times \overline{\mathbf{C}}\right)$ and $\hat{n} \times (\nabla \times \mathbf{A}) = \hat{n} \cdot (\mathbf{A}\nabla - \nabla\mathbf{A})$. If we add the following null quantity to the above I_1 [31]

$$
\begin{aligned}
I' &= 2\mu \int_V \nabla' \cdot \left\{ \nabla' \times \left[\mathbf{u}(\mathbf{r}') \times \overline{\mathbf{G}}(\mathbf{r}', \mathbf{r}) \right] \right\} \, \mathrm{d}\mathbf{r}' \\
&= 2\mu \int_S \hat{n}' \cdot \left\{ \nabla' \times \left[\mathbf{u}(\mathbf{r}') \times \overline{\mathbf{G}}(\mathbf{r}', \mathbf{r}) \right] \right\} \, \mathrm{d}S' \\
&= 0
\end{aligned}
\tag{9.210}
$$

and apply the identity

$$
\nabla' \times \left(\mathbf{u} \times \overline{\mathbf{G}} \right) = \mathbf{u} \left(\nabla' \cdot \overline{\mathbf{G}} \right) - (\nabla' \cdot \mathbf{u})\overline{\mathbf{G}} + (\nabla'\mathbf{u})^T \cdot \overline{\mathbf{G}} - \mathbf{u} \cdot \left(\nabla'\overline{\mathbf{G}} \right)
\tag{9.211}
$$

with $(\nabla'\mathbf{u})^T = \mathbf{u}\nabla'$, where T denotes a transpose, then we have

$$
\begin{aligned}
I_2 &= \int_S \left[-\lambda(\nabla' \cdot \mathbf{u}) \left(\hat{n}' \cdot \overline{\mathbf{G}} \right) - \mu\hat{n}' \cdot (\nabla'\mathbf{u} + \mathbf{u}\nabla') \cdot \overline{\mathbf{G}} \right. \\
&\qquad \left. + \lambda\mathbf{u}\hat{n}' \left(\nabla' \cdot \overline{\mathbf{G}} \right) + \mu\mathbf{u} \cdot \hat{n}' \cdot \left(\nabla'\overline{\mathbf{G}} + \overline{\mathbf{G}}\nabla' \right) \right] \, \mathrm{d}S' \\
&= \int_S \left\{ -\hat{n}' \cdot \left[\lambda\bar{\mathbf{I}}(\nabla' \cdot \mathbf{u}) + \mu(\nabla'\mathbf{u} + \mathbf{u}\nabla') \right] \cdot \overline{\mathbf{G}} \right. \\
&\qquad \left. + \mathbf{u} \cdot \hat{n}' \cdot \left[\lambda\bar{\mathbf{I}}\nabla' \cdot \overline{\mathbf{G}} + \mu \left(\nabla'\overline{\mathbf{G}} + \overline{\mathbf{G}}\nabla' \right) \right] \right\} \, \mathrm{d}S' \\
&= \int_S \left[-\mathbf{t} \cdot \overline{\mathbf{G}} + \mathbf{u} \cdot \left(\hat{n}' \cdot \overline{\overline{\mathbf{W}}} \right) \right] \, \mathrm{d}S'.
\end{aligned}
\tag{9.212}
$$

where

$$
\overline{\overline{\mathbf{W}}} = \lambda\bar{\mathbf{I}}\nabla \cdot \overline{\mathbf{G}} + \mu \left(\nabla\overline{\mathbf{G}} + \overline{\mathbf{G}}\nabla \right)
\tag{9.213}
$$

is a third-rank Green's tensor and $\bar{\mathbf{I}}$ is the identity dyad. The above I_2 is for the exterior problem, namely, V is the complementary space of the object. For the interior problem, namely V is the space occupied by the object, then

$$
I_2 = \int_S \left[\mathbf{t} \cdot \overline{\mathbf{G}} - \mathbf{u} \cdot \left(\hat{n}' \cdot \overline{\overline{\mathbf{W}}} \right) \right] \, \mathrm{d}S'.
\tag{9.214}
$$

Substituting the above I_2 into Equation (9.206), considering that the observation point is approaching the boundary S from the interior of the object (exterior problem) and from the exterior of the object (interior problem), respectively, and taking a transpose on the resultant equations, we can obtain

$$
\frac{1}{2}\mathbf{u}(\mathbf{r}) + \int_S \left[\overline{\mathbf{Q}}_2^T(\mathbf{r}, \mathbf{r}') \cdot \mathbf{u}(\mathbf{r}') - \overline{\mathbf{G}}_2^T(\mathbf{r}, \mathbf{r}') \cdot \mathbf{t}(\mathbf{r}') \right] \, \mathrm{d}S' = \mathbf{u}_{\mathrm{inc}}(\mathbf{r}), \quad \mathbf{r} \in S
\tag{9.215}
$$

$$
\frac{1}{2}\mathbf{u}(\mathbf{r}) + \int_S \left[\overline{\mathbf{G}}_1^T(\mathbf{r}, \mathbf{r}') \cdot \mathbf{t}(\mathbf{r}') - \overline{\mathbf{Q}}_1^T(\mathbf{r}, \mathbf{r}') \cdot \mathbf{u}(\mathbf{r}') \right] \, \mathrm{d}S' = 0, \quad \mathbf{r} \in S
\tag{9.216}
$$

where $\overline{\mathbf{Q}} = \hat{n}' \cdot \overline{\overline{\mathbf{W}}}(\mathbf{r}, \mathbf{r}')$, the subscript 1 or 2 is the region index, and $\mathbf{u}$ and $\mathbf{t}$ are the total displacement and traction vectors at the object surface S, which are the unknowns to be solved. Equations (9.215) and (9.216) are the BIEs for elastic wave scattering by an object and we have incorporated the boundary conditions in it, so it is solvable now. Note that the BIEs are consistent with the one in publications [127], but the derivation here is new.

9.6.3 Coupled Integral Equations

When considering the interaction of EM wave with elastic wave in elastic media, the governing equations are the coupled electrodynamic and elastodynamic equations. In a PDE form, they are [10]

$$\epsilon_{ijk}E_{k,j} = -\dot{B}_i, \quad \epsilon_{ijk}H_{k,j} = \dot{D}_i, \quad B_{i,i} = 0, \quad D_{i,i} = 0 \tag{9.217}$$

which are the Maxwell's equations and

$$T_{ji,j} + \rho f_i = \rho \ddot{u}_i \tag{9.218}$$

which is the elastic wave equation equivalent to (9.196). Note that one uses **x** instead of **r** to denote a position vector in space in elastodynamics so that indicial notation can be used conveniently (in the indicial notation, x_1, x_2, and x_3 correspond to x, y, and z, respectively, and the components of a vector or tensor are denoted with indices). In the above, ϵ_{ijk} is the permutation tensor, B_i and D_i are the components of the magnetic induction (magnetic flux density) vector **B** and electric displacement (electric flux density) vector **D**, respectively, and T_{ij} are the components of stress tensor $\overline{\mathbf{T}}$. A dot and two dots over a variable imply the first-order and second-order derivatives with respect to time, respectively, and a comma followed by an index denotes the partial differentiation with respect to the coordinate associated with the index. The coupling is reflected in the constitutive relations [10]

$$T_{ij} = c_{ijkl}S_{kl} - e_{kij}E_k \tag{9.219}$$

$$D_i = e_{ijk}S_{jk} + \epsilon_{ij}E_j \tag{9.220}$$

where $S_{ij} = \frac{1}{2}(u_{j,i} + u_{i,j})$ are the components of the strain tensor $\overline{\mathbf{S}}$, $c_{ijkl} = \lambda\delta_{ij}\delta_{kl} + \mu(\delta_{jk}\delta_{il} + \delta_{il}\delta_{ik})$ is the elastic stiffness, ϵ_{ij} is the permittivity, and e_{ijk} is the stress constant. If we consider the symmetries in the strain and stress tensors, which usually hold, and re-order the indices 11, 22, 33, 23 or 32, 31 or 13, and 12 or 21 in ij and kl as 1, 2, 3, 4, 5, and 6 in p and q, respectively, then the components of those tensors can be re-indexed as

$$c_{ijkl} \to c_{pq}, \quad e_{ikl} \to e_{ip}, \quad T_{ij} \to T_p \tag{9.221}$$

$$S_1 = S_{11}, \quad S_2 = S_{22}, \quad S_3 = S_{33}, \tag{9.222}$$

$$S_4 = 2S_{23}, \quad S_5 = 2S_{31}, \quad S_6 = 2S_{12}. \tag{9.223}$$

With the new indices, the above constitutive relations can be written as

$$T_p = \sum_{q=1}^{6} c_{pq}S_q - \sum_{q=1}^{3} e_{qp}E_q, \quad p = 1, \cdots, 6 \tag{9.224}$$

$$D_p = \sum_{q=1}^{6} e_{pq}S_q + \sum_{q=1}^{3} \epsilon_{pq}E_q, \quad p = 1,2,3. \tag{9.225}$$

To derive the coupled integral equations, we need to consider the EM wave part and elastic wave part separately. For the elastic wave part, we recall the derivation of the elastic wave equation in the PDE form in (9.196), which is a force balance on the elastic body, namely [126],

$$\int_V \rho \frac{\partial^2 \mathbf{u}}{\partial t^2}\, \mathrm{d}V' = \int_V \nabla \cdot \overline{\mathbf{T}}\, \mathrm{d}V' + \int_V \rho \mathbf{f}\, \mathrm{d}V' \tag{9.226}$$

where V is an arbitrary volume in the object. The integrand on the left-hand side above is actually from the Newton's second law of motion. Since (9.226) holds true for an arbitrary volume, we have

$$\rho \frac{\partial^2 \mathbf{u}}{\partial t^2} = \nabla \cdot \overline{\mathbf{T}} + \rho \mathbf{f}. \tag{9.227}$$

The divergence of stress tensor above can be found [128]

$$\nabla \cdot \overline{\mathbf{T}} = (\lambda + \mu)\nabla\nabla \cdot \mathbf{u} + \mu \nabla^2 \mathbf{u}. \tag{9.228}$$

Substituting (9.228) into (9.227) in a time harmonic case, we can obtain (9.196).

When an EM wave is incident on the elastic body and there is no other external force, the force in the above equations is the EM force, i.e.

$$\mathbf{f} = \mathbf{f}^{\mathrm{E}}(\mathbf{r}') = \nabla' \cdot \overline{\mathbf{T}}^{\mathrm{E}} \tag{9.229}$$

where the superscript "E" implies that the variable is EM-related. From the equation (9.224), it is clear that the EM force is generated by the second sum in the stress tensor expression which is contributed by electric field, i.e.

$$T_p^{\mathrm{E}} = \sum_{q=1}^{3} e_{qp} E_q, \quad p = 1, \cdots, 6. \tag{9.230}$$

Therefore, the EM force can be found as

$$f_1^{\mathrm{E}} = T_{1,1}^{\mathrm{E}} + T_{6,2}^{\mathrm{E}} + T_{5,3}^{\mathrm{E}} \tag{9.231}$$

$$f_2^{\mathrm{E}} = T_{6,1}^{\mathrm{E}} + T_{2,2}^{\mathrm{E}} + T_{4,3}^{\mathrm{E}} \tag{9.232}$$

$$f_3^{\mathrm{E}} = T_{5,1}^{\mathrm{E}} + T_{4,2}^{\mathrm{E}} + T_{3,3}^{\mathrm{E}}. \tag{9.233}$$

By treating the EM force as the external force in the derivation of integral equations in the proceeding section, the integral equations for elastic wave are the same as (9.215) and (9.216), except that the incident displacement field $\mathbf{u}_{\mathrm{inc}} = 0$ in the equation (9.215) and the right-hand side in the equation (9.216) should be

$$\mathbf{u}_1^{\mathrm{ex}} = \int_{V_1} \overline{\mathbf{G}}_1(\mathbf{r}, \mathbf{r}') \cdot \mathbf{f}_1^{\mathrm{E}}(\mathbf{r}')\, \mathrm{d}\mathbf{r}' \tag{9.234}$$

where V_1 is the volume of the object and $\mathbf{f}_1^{\mathrm{E}}(\mathbf{r}')$ is the EM force in V_1. When the EM force exists outside the object, the excitation of the elastic wave, namely the right-hand side in the equation (9.215) is

$$\mathbf{u}_2^{\mathrm{ex}} = \int_{V_2} \overline{\mathbf{G}}_2(\mathbf{r}, \mathbf{r}') \cdot \mathbf{f}_2^{\mathrm{E}}(\mathbf{r}')\, \mathrm{d}\mathbf{r}' \tag{9.235}$$

where V_2 is the volume of the surrounding medium that can be truncated in calculations and $\mathbf{f}_2^{\mathrm{E}}(\mathbf{r}')$ is the EM force in V_2.

For the coupled integral equations for the EM wave part, we can use the same strategy as for the elastic wave part to derive them. The integral equations are the same as

(9.194) and (9.195) except that the excitation on the right-hand side should include the contribution of the source from strain tensor. From Maxwell's equations, we have

$$\epsilon_{ijk}H_{k,j} = \dot{D}_i = e_{ijk}\dot{S}_{jk} + \epsilon_{ij}\dot{E}_j. \tag{9.236}$$

We can treat the first term above as an effective current density or new excitation current density and classify it into $\mathbf{J}_e$ producing the incident field in Region 2. The extra incident field by the effective current density in Region 2 is

$$\mathbf{E}_2^{ex} = \int_{V_2} \overline{\mathbf{G}}_2(\mathbf{r}, \mathbf{r}') \cdot \mathbf{J}_2^u(\mathbf{r}') \, \mathrm{d}V' \tag{9.237}$$

where $\mathbf{J}_2^u(\mathbf{r}')$ is the effective current density in V_2 and the superscript u implies that it is related to the displacement field. From (9.225), the current density has the following three components

$$J_p^u = -\mathrm{i}\omega \sum_{q=1}^{6} e_{pq}S_q, \quad p = 1,2,3 \tag{9.238}$$

for time-harmonic fields and S_q is determined by $S_{ij} = \frac{1}{2}(u_{j,i} + u_{i,j})$. In Region 1, there is no incident field and the excitation is only from the strain tensor, namely

$$\mathbf{E}_1^{ex} = \int_{V_1} \overline{\mathbf{G}}_1(\mathbf{r}, \mathbf{r}') \cdot \mathbf{J}_1^u(\mathbf{r}') \, \mathrm{d}V' \tag{9.239}$$

where $\mathbf{J}_1^u(\mathbf{r}')$ is the effective current density in V_1. Using the excitation in (9.237) and (9.239) as the right-hand side in Equations (9.194) and (9.195), respectively, we can solve it to obtain the extra electric fields in the two regions. These fields will create EM force by (9.231)–(9.233) and become the new excitations for the elastic wave integral equations in (9.215)–(9.216).

9.6.4 Solving Method

The developed integral equations can be solved by any IESs, such as the MoM, the BEM or the Nyström method. We use the Nyström method with a different local correction scheme to solve the equations [34, 95]. However, since the EM wave part and elastic wave part are coupled with each other by excitation, a special solving strategy is needed. One choice is to use an iterative method, namely solving the EM wave integral equations first by assuming no contribution from the elastic wave in the excitation, and substituting the obtained electric field to the elastic wave integral equations as an excitation so that the elastic wave integral equations can be solved. When the displacement field is obtained, we substitute it back into the EM wave integral equations as a new excitation or a perturbation and solve the resultant EM wave integral equations to obtain new electric field. This new electric field is a new excitation for the elastic wave integral equations and we go through this iterative process until the solutions converge. In the implementation, we use the central difference of electric field or displacement field to approximate the differential of the fields at a point because the partial derivatives of fields are required in calculating the coupled excitations. The medium bodies are discretized into small cubes or tetrahedrons in which the partial derivatives of fields are represented by the values at the centers of cells in an average sense. Also, the surrounding medium needs to be truncated in terms of the attenuation of fields in the calculations of coupled excitations. The elastic stiffness c_{pq} and stress constant e_{pq} needed in calculations can be found from [10] for some typical elastic piezoelectric materials. For example, langasite has

$$\left[c_{pq}\right] = \begin{bmatrix} 1.89 & 1.05 & 0.96 & -0.14 & 0 & 0 \\ 1.05 & 1.89 & 0.96 & 0.14 & 0 & 0 \\ 0.96 & 0.96 & 2.61 & 0 & 0 & 0 \\ -0.14 & 0.14 & 0 & 0.54 & 0 & 0 \\ 0 & 0 & 0 & 0 & 0.54 & -0.14 \\ 0 & 0 & 0 & 0 & -0.14 & 0.42 \end{bmatrix} \tag{9.240}$$

$$\left[e_{pq}\right] = \begin{bmatrix} -0.44 & 0.44 & 0 & -0.08 & 0 & 0 \\ 0 & 0 & 0 & 0 & 0.08 & 0.44 \\ 0 & 0 & 0 & 0 & 0 & 0 \end{bmatrix} \tag{9.241}$$

and lithium niobate has

$$\left[c_{pq}\right] = \begin{bmatrix} 2.03 & 0.53 & 0.75 & 0.09 & 0 & 0 \\ 0.53 & 2.03 & 0.75 & -0.09 & 0 & 0 \\ 0.75 & 0.75 & 2.45 & 0 & 0 & 0 \\ 0.09 & -0.09 & 0 & 0.60 & 0 & 0 \\ 0 & 0 & 0 & 0 & 0.60 & 0.09 \\ 0 & 0 & 0 & 0 & 0.09 & 0.75 \end{bmatrix} \tag{9.242}$$

$$\left[e_{pq}\right] = \begin{bmatrix} 0 & 0 & 0 & 0 & 3.70 & -2.50 \\ -2.50 & 2.50 & 0 & 3.70 & 0 & 0 \\ 0.20 & 0.20 & 1.30 & 0 & 0 & 0 \end{bmatrix} \tag{9.243}$$

where the unit is $\times 10^{11}$ N m^{-2} for c_{pq} and C m^{-2} for e_{pq}. We will use these parameters in the numerical examples.

9.6.5 Numerical Examples

We use the Nyström method to solve the coupled integral equations for an elastic cube illuminated by a time harmonic EM plane wave along the $-z$ direction and in vertical polarization. The cube has a side length $s = 0.2\lambda$, where λ is the wavelength in free space, and its elasticity is characterized by lithium niobate with $\lambda_1 = 0.53486$, $\mu_1 = 0.23077$ and $\rho_1 = 0.5743$. Also, the cube has an isotropic relative permittivity $\epsilon_{r1} = 2.0$ and relative permeability $\mu_{r1} = 1.0$. The surrounding medium is the same as the langasite in elasticity with $\lambda_2 = 0.23716$, $\mu_2 = 0.52641$, and $\rho_2 = 0.47$, and the same as free space in electric properties.

We first ignore the elasticities of both the scatterer and surrounding medium and solve the EM wave part of the integral equations. Figures 9.49 and 9.50 show the bistatic radar cross section (RCS) of the cube and a dielectric sphere with a radius $a = 0.2\lambda$ and the same material as the cube. This is to verify the solutions of the EM wave part of the integral equations by comparing with the MoM solutions or analytical solutions.

We then ignore the electric properties of both the scatterer and surrounding medium and solve the elastic wave part of the integral equations. Figure 9.51 sketches the normalized total displacement fields on the surface of an elastic sphere with a radius $a = 0.2\lambda$

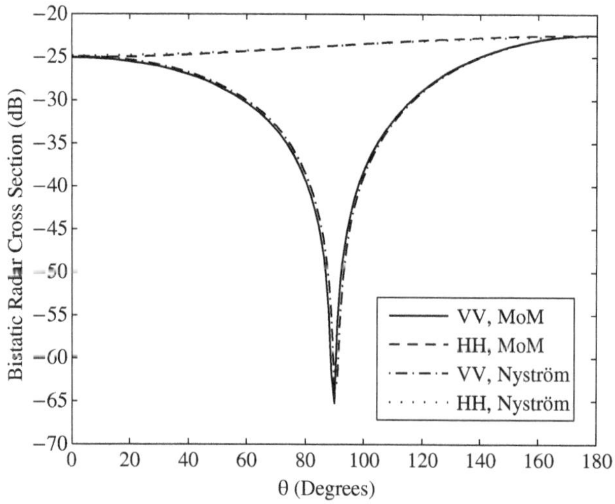

Figure 9.49 Bistatic RCS solutions for a dielectric cube with a side length $s = 0.2\lambda$.

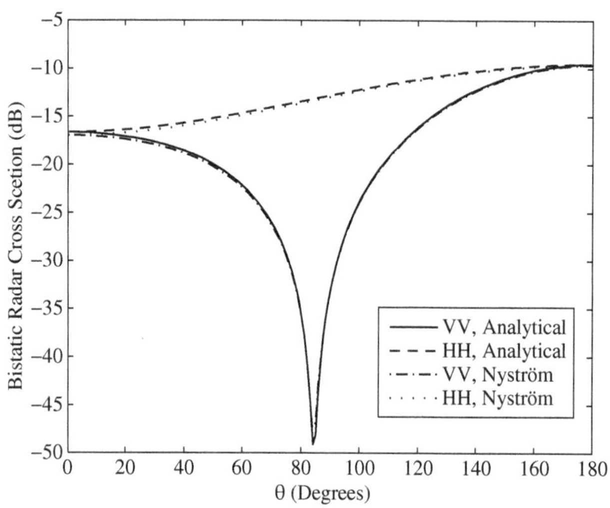

Figure 9.50 Bistatic RCS solutions for a dielectric sphere with a radius $a = 0.2\lambda$.

and the same material as the cube when an elastic plane wave illuminates in the elastic surrounding medium defined as before. The fields are observed along the principal cut ($\phi = 0°$ and $\theta = 0°–180°$) and compared with corresponding analytical solutions. This is to verify the elastic wave part of the integral equations.

For solving the coupled integral equations, we first consider a simplified case, i.e. the dielectric and elastic cube is put into a vacuum-like free space, so there is no excited elastic wave outside the cube. Figure 9.52 plots the normalized displacement fields of the elastic wave inside the cube excited by the EM wave. The fields are observed along the central line from the center of the cube or the origin to the end facet in the $+x$ direction. We then consider another simplified case which is opposite to the previous case,

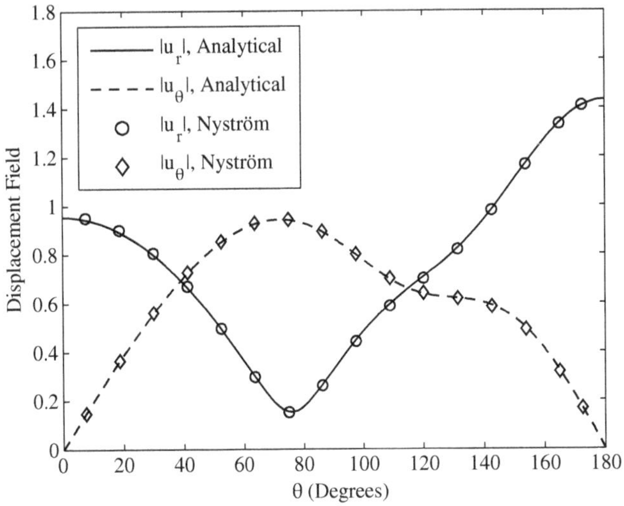

Figure 9.51 Normalized displacement field scattered by an elastic sphere with a radius $a = 0.2\lambda$ when an elastic plane wave illuminates.

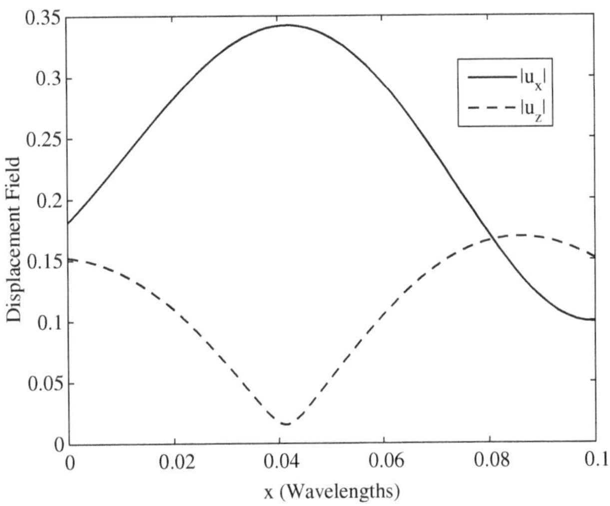

Figure 9.52 Normalized displacement field inside an elastic cube illuminated by an EM wave in free space.

i.e. the cube is replaced by a traction-free and bubble-like cavity and the surrounding medium takes the material of the cube. There is no excited elastic wave inside the cubic cavity in this case and it only exists in the surrounding medium. Figure 9.53 illustrates the normalized displacement field in the surrounding medium that is observed along the principal cut at the $r = 5s$ surface. Finally, we consider a generalized case, i.e. both the object and surrounding medium are elastic. The properties of object and surrounding medium are those described in the first paragraph. Figure 9.54 depicts the normalized

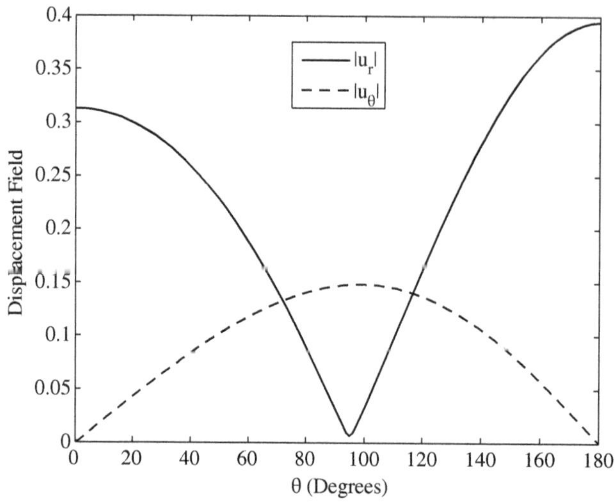

Figure 9.53 Normalized displacement field from a traction-free cubic cavity illuminated by an EM wave in an elastic medium.

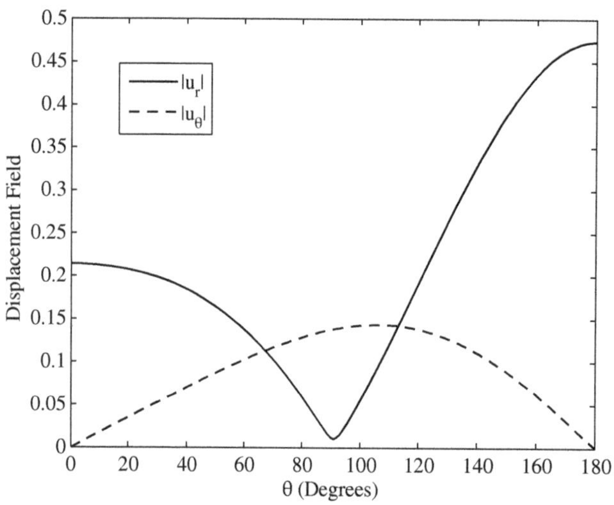

Figure 9.54 Normalized displacement field from an elastic cube illuminated by EM wave in elastic medium.

displacement field of the elastic wave in the surrounding medium that is observed along the principal cut at the $r = 5s$ surface.

Table 9.3 summarizes the computational costs, i.e. CPU time T (Second) and memory usage M (MB) for all the cases above. The cube and sphere are discretized into 768 and 1520 triangular meshes, respectively, and all calculations are performed on a Dell Precision 690 workstation with 3.0-GHz CPU and 16-GB RAM. Note that the used Nyström method is slightly more efficient than the widely-used MoM because of its simpler mechanism of implementation. The results in Figure 9.49 take 176 seconds in CPU time

Table 9.3 Summary of CPU time T (Second) and memory usage M (MB) for solving all examples.

Examples	Figure 9.49	Figure 9.50	Figure 9.51	Figure 9.52	Figure 9.53	Figure 9.54
T (Second)	162	321	446	237	229	1068
M (MB)	115	286	342	181	175	418

and 116 MB in memory usage if the MoM is used. Since there exists a magnetic current density in addition to the electric current density on the object surface, we use the RWG basis function [90] to expand the electric current density and the dual basis function [129] to expand the magnetic current density, respectively, in the MoM, because the system matrix will be very ill-conditioned if using the RWG basis function to represent both the electric and magnetic current densities. The dual basis function is complicated in construction and requires high costs in implementation. Also, the MoM is not suitable for solving the elastic wave part of the integral equations because the unknown **u** and **t** vectors on the object surface are 3D vectors instead of surface vectors and the Nyström method is more convenient in implementation.

Although we only consider the simple geometry in the above, the coupled integral equations are developed based on 3D arbitrary shapes and they can be applied to more complicated objects with an increase of computational effort. Also, the simple cubic shape is not just for illustration and actually it can find many applications in practice because many piezoelectric sensors take this simple geometry. The developed approach is amenable to implementation as in other IESs because we only need to discretize the object surface or source region no matter how large the field space is. Usually the discretization for the source region is much easier than for the field space. The limitations of the approach could be the internal resonance problem, as found in both EM and elastic wave integral equations, and the low-frequency breakdown problem, as appearing in the EM integral equations. The relevant remedy techniques in the EM and elastodynamics can be incorporated in the future.

9.7 Summary

Although there are many types of multiphysics problems related to an EM wave, we select the coupled EM–elastodynamic problem as an example to illustrate the solution by the Nyström method. We start with the introduction of elastic wave equations, with which EM researchers may not be familiar, and their numerical solutions by the Nyström method. We then consider the acoustic wave equation and its unification with the elastic wave equations. These equations are solved by the Nyström method and accelerated by the MLFMA. Although the MLFMA acceleration for the acoustic wave equation could be simpler than for EM equations, it is not the case for the elastic wave equations. Since the governing BIEs include multiple wavenumbers, a multiple-tree structure is used in the implementation of the MLFMA. However, the multiple-tree frame will result in a non-uniform definition for well-separated groups and the cancellation of the strongest singularities in the series representation of kernels for near terms may be destroyed due to the disagreement of near-patch definition. We overcome this drawback by searching common near patches in different trees and reformulating the expressions of the near terms so that the cancellation of the strongest singularities is kept and the resulting singular integrals can be handled conveniently.

With the basic knowledge of elastic waves and acoustic waves and their special relationship, we then address the coupled EM–elastodynamic problem. We consider the EM wave illumination on elastic media, which includes both electrodynamic and elastodynamic processes, and coupled Maxwell's equations and elastic wave equations should be solved simultaneously to understand the involved physics. Although the coupled equations have been derived earlier, they were only in a PDE form and thus only PDE numerical solvers can be employed in seeking solutions. We first develop coupled BIEs for the problem by using Huygens' equivalence principle and extinction theorem. The coupling is reflected in the excitations or right-hand sides in both the EM wave part and the elastic wave part of the integral equations. The EM force is the new excitation for elastic wave and the current density due to the strain tensor is the new excitation for EM wave in the coupled BIEs. Using the BIEs, the solution process can use all advantages of IESs like the MoM, the BEM and the Nyström method. We use the Nyström method with an iterative process to solve the coupled BIEs and some numerical results for the interaction of an EM wave with piezoelectric materials are illustrated. The developed approach could also find applications in many other areas. For example, we have tried to use the illumination of microwave and measurement of an induced elastic wave to reconstruct objects in geophysical exploration and medical imaging because the microwave illumination can result in high contrast while the measurement of elastic wave can yield a high resolution in the imaging. Using the coupled integral equations could greatly facilitate the analysis for such problems.

Appendix 9.1 Analytical Solutions for the Strongly-Singular Integrals in (9.35)–(9.37)

In the polar coordinate system as shown in Figure 9.2, we divide the triangle element into three subtriangles ΔS_i $(i = 1,2,3)$ by connecting the origin to the three vertices of the triangle element. Each subtriangle is specified by four parameters d_0^i, θ_0^i, θ_1^i and θ_2^i. Here, d_0^i is the vertical distance $\overline{OO_i}$ between the origin and the ith side of the triangle element, θ_0^i is the angle of $\overline{OO_i}$, and θ_k^i $(k = 1,2)$ are the angles of connecting lines from the origin to the two end points of the ith side, respectively. With the above definitions, those integrals in (9.35)–(9.37) can be derived in the following forms

$$
\begin{aligned}
I_u^{(3)} &= \int_{\Delta S} \frac{u'}{r^3}\,\mathrm{d}S' \\
&= \sum_{i=1}^{3} \left[(\sin\theta_2^i - \sin\theta_1^i)\ln(d_0^i) - I_a\cos\theta_0^i + I_b\sin\theta_0^i\right]
\end{aligned}
\tag{9.244}
$$

$$
\begin{aligned}
I_v^{(3)} &= \int_{\Delta S} \frac{v'}{r^3}\,\mathrm{d}S' \\
&= \sum_{i=1}^{3} \left[(\cos\theta_1^i - \cos\theta_2^i)\ln(d_0^i) - I_b\cos\theta_0^i - I_a\sin\theta_0^i\right]
\end{aligned}
\tag{9.245}
$$

$$I_u^{(5)} = \int_{\Delta S} \frac{u'^3}{r^5}\,\mathrm{d}S'$$
$$= \sum_{i=1}^{3} \big[(\sin\theta_2^i - \sin\theta_1^i + t_1)\ln(d_0^i) - I_A\cos^3\theta_0^i + 1.5\sin 2\theta_0^i(I_B\cos\theta_0^i - I_C\sin\theta_0^i)$$
$$+I_D\sin^3\theta_0^i\big] \tag{9.246}$$

$$I_{uv}^{(5)} = \int_{\Delta S} \frac{u'^2 v'}{r^5}\,\mathrm{d}S'$$
$$= \sum_{i=1}^{3} \big[t_2\ln(d_0^i) - I_A\sin\theta_0^i\cos^2\theta_0^i - I_D\cos\theta_0^i\sin^2\theta_0^i + I_B\cos\theta_0^i(3\sin^2\theta_0^i - 1)$$
$$+I_C\sin\theta_0^i(3\cos^2\theta_0^i - 1)\big] \tag{9.247}$$

$$I_{vu}^{(5)} = \int_{\Delta S} \frac{u' v'^2}{r^5}\,\mathrm{d}S'$$
$$= \sum_{i=1}^{3} \big[I_D\sin\theta_0^i\cos^2\theta_0^i - I_A\cos\theta_0^i\sin^2\theta_0^i - t_1\ln(d_0^i) - I_B\sin\theta_0^i(3\cos^2\theta_0^i - 1)$$
$$+I_C\cos\theta_0^i(3\sin^2\theta_0^i - 1)\big] \tag{9.248}$$

$$I_v^{(5)} = \int_{\Delta S} \frac{v'^3}{r^5}\,\mathrm{d}S'$$
$$= \sum_{i=1}^{3} \big[(\cos\theta_1^i - \cos\theta_2^i - t_2)\ln(d_0^i) - I_A\sin^3\theta_0^i - 1.5\sin 2\theta_0^i(I_B\sin\theta_0^i - I_C\cos\theta_0^i)$$
$$-I_D\cos^3\theta_0^i\big] \tag{9.249}$$

where

$$I_a = \frac{1}{2}\big[b_1(\ln b_1 - 1) - a_1(\ln a_1 - 1) - b_2(\ln b_2 - 1) + a_2(\ln a_2 - 1)\big] \tag{9.250}$$
$$I_b = b_0(1 - \ln b_0) - a_0(1 - \ln a_0) \tag{9.251}$$
$$I_A = 0.5(I_a - I_p + I_q + I_r - I_s) \tag{9.252}$$
$$I_B = \frac{a_0^3}{3}\left(\ln a_0 - \frac{1}{3}\right) - \frac{b_0^3}{3}\left(\ln b_0 - \frac{1}{3}\right) \tag{9.253}$$
$$I_C = 0.5(I_p - I_q - I_r + I_s) \tag{9.254}$$
$$I_D = I_b - I_B \tag{9.255}$$
$$I_p = \frac{b_1^3}{3}\left(\ln b_1 - \frac{1}{3}\right) - b_1^2\left(\ln b_1 - \frac{1}{2}\right) + b_1(\ln b_1 - 1) \tag{9.256}$$
$$I_q = \frac{a_1^3}{3}\left(\ln a_1 - \frac{1}{3}\right) - a_1^2\left(\ln a_1 - \frac{1}{2}\right) + a_1(\ln a_1 - 1) \tag{9.257}$$
$$I_r = \frac{b_2^3}{3}\left(\ln b_2 - \frac{1}{3}\right) - b_2^2\left(\ln b_2 - \frac{1}{2}\right) + b_2(\ln b_2 - 1) \tag{9.258}$$
$$I_s = \frac{a_2^3}{3}\left(\ln a_2 - \frac{1}{3}\right) - a_2^2\left(\ln a_2 - \frac{1}{2}\right) + a_2(\ln a_2 - 1) \tag{9.259}$$

$$a_0 = \cos(\theta_1^i - \theta_0^i); \qquad b_0 = \cos(\theta_2^i - \theta_0^i) \tag{9.260}$$

$$a_1 = 1 + \sin(\theta_1^i - \theta_0^i); \qquad b_1 = 1 + \sin(\theta_2^i - \theta_0^i) \tag{9.261}$$

$$a_2 = 1 - \sin(\theta_1^i - \theta_0^i); \qquad b_2 = 1 - \sin(\theta_2^i - \theta_0^i) \tag{9.262}$$

$$t_1 = \frac{1}{3}(\sin^3\theta_1^i - \sin^3\theta_2^i); \quad t_2 = \frac{1}{3}(\cos^3\theta_1^i - \cos^3\theta_2^i). \tag{9.263}$$

Appendix 9.2 Asymptotic Solutions of Elastic Wave Scattering by Spherical Obstacles with $\mu \to 0$

Both elastic wave and acoustic wave scattering by a spherical obstacle have analytical solutions and they are usually used as a benchmark to verify numerical results. Although Pao and Mow provided the general solutions for elastic wave scattering [123], the asymptotic solutions for acoustic wave scattering are not available; hence we present the results here. For a rigid sphere, the solution for the total traction field is

$$\tau_{rr} = \frac{2\mu}{r^2}\sum_{n=0}^{\infty}(-\Phi_0\mathcal{E}_3 + A_n\mathcal{E}_{31} + B_n\mathcal{E}_{32})P_n(\cos\theta) \tag{9.264}$$

$$\tau_{r\theta} = \frac{2\mu}{r^2}\sum_{n=0}^{\infty}(-\Phi_0\mathcal{E}_4 + A_n\mathcal{E}_{41} + B_n\mathcal{E}_{42})\frac{P_n(\cos\theta)}{d\theta} \tag{9.265}$$

where we have used the notations in [123]. Here, $\boldsymbol{\tau}$ is the traction vector, Φ_0 is the amplitude of the incident plane compressional wave travelling in the x_3 direction, P_n is the Legendre polynomial of order n, A_n and B_n are the coefficients to be determined, and all $\mathcal{E}$'s are defined in (18) in [123]. In the limit of $\mu \to 0$, we have

$$\lim_{\mu\to 0}\mu\mathcal{E}_3 = \frac{1}{2}i^{-n}(2n+1)\omega^2\rho r^2 j_n(\alpha r) = \mathcal{E}_3^0 \tag{9.266}$$

$$\lim_{\mu\to 0}\mu\mathcal{E}_{31} = -\frac{1}{2}\omega^2\rho r^2 h_n(\alpha r) = \mathcal{E}_{31}^0 \tag{9.267}$$

and all other terms incorporating μ will be zeros. Here j_n is the spherical Bessel function of order n, h_n is the first kind of spherical Hankel function of order n and α is the wavenumber of the compressional wave in the host medium. Thus the solution reduces to

$$\tau_{rr} = \frac{2}{r^2}\sum_{n=0}^{\infty}(-\Phi_0\mathcal{E}_3^0 + A_n\mathcal{E}_{31}^0)P_n(\cos\theta) \tag{9.268}$$

$$\tau_{r\theta} = 0 \tag{9.269}$$

with

$$A_n = \Phi_0\frac{E_1}{E_{11}} = -\Phi_0 i^{-n}(2n+1)\frac{nj_n(\alpha a) - \alpha a j_{n+1}(\alpha a)}{nh_n(\alpha a) - \alpha a h_{n+1}(\alpha a)} \tag{9.270}$$

where E is the value of $\mathcal{E}$ at $r = a$ and a is the radius of the sphere. This solution, when only considering the scattered part, is the same as the conventional analytical solution for the scattered potential by a rigid sphere in acoustics [124]

$$\Phi_s(r,\theta) = \Phi_i\sum_{m=0}^{\infty} i^{m+1}(2m+1)P_m(\cos\theta)\sin\delta_m(\kappa a)h_m(\kappa r)e^{i\delta_m(\kappa a)} \tag{9.271}$$

if we note that $\boldsymbol{\tau} = \hat{\mathbf{n}}\lambda\nabla \cdot \mathbf{u} = \hat{\mathbf{n}}\lambda\Phi$. The notations in (9.271) can be found in [125] and $\delta_m(z)$ is the phase of $\frac{\mathrm{d}}{\mathrm{d}z}h_m(z)$ minus $\frac{\pi}{2}$.

For an elastic inclusion, we can derive the following equations from the equation (20) in [123] by taking the $\mu \to 0$ limit

$$E_{11}A_n + E_{13}C_n + E_{14}D_n = \Phi_0 E_1 \tag{9.272}$$

$$E_{31}^0 A_n + E_{33}C_n + E_{34}D_n = \Phi_0 E_3^0 \tag{9.273}$$

$$E_{43}C_n + E_{44}D_n = 0 \tag{9.274}$$

where E_3^0 and E_{31}^0 are $\mathcal{E}_3^0$ and $\mathcal{E}_{31}^0$ with $r = a$ in (9.266)–(9.267), respectively. With the solved coefficients from (9.272)–(9.274), the analytical solutions are

$$u_r = \frac{1}{r}\sum_{n=0}^{\infty}(-\Phi_0\mathcal{E}_1 + A_n\mathcal{E}_{11})P_n(\cos\theta) \tag{9.275}$$

$$u_\theta = \frac{1}{r}\sum_{n=0}^{\infty}(-\Phi_0\mathcal{E}_2 + A_n\mathcal{E}_{21})\frac{dP_n(\cos\theta)}{d\theta} \tag{9.276}$$

$$\tau_{rr} = \frac{2}{r^2}\sum_{n=0}^{\infty}(-\Phi_0\mathcal{E}_3^0 + A_n\mathcal{E}_{31}^0)P_n(\cos\theta) \tag{9.277}$$

$$\tau_{r\theta} = 0. \tag{9.278}$$

Similarly, we can find the following analytical solution for the acoustic scattering by a traction-free cavity

$$u_r = \frac{1}{r}\sum_{n=0}^{\infty}(-\Phi_0\mathcal{E}_1 + A_n\mathcal{E}_{11})P_n(\cos\theta) \tag{9.279}$$

$$u_\theta = 0 \tag{9.280}$$

where

$$A_n = \Phi_0\frac{E_3^0}{E_{31}^0} = -\Phi_0 i^{-n}(2n+1)\frac{j_n(\alpha a)}{h_n(\alpha a)}. \tag{9.281}$$

Bibliography

1 Kyame, J.J. (1949). Wave propagation in piezoelectric crystals. *J. Acoust. Soc. Am.* 21 (3): 159–167.

2 Kyame, J.J. (1953). Conductivity and viscosity effects on wave propagation in piezoelectric crystals. *J. Acoust. Soc. Am.* 26 (6): 990–993.

3 Hruska, H. (1966). The rate of propagation of ultrasonic waves in ADP and in Voigt's theory. *Czech. J. Phys.* B16: 446–453.

4 Tseng, C.C. and White, P.M. (1967). Propagation of piezoelectric and elastic surface waves on the basal plane of hexagonal piezoelectric crystals. *J. Appl. Phys.* 38: 4274–4280.

5 Spaight, R.N. and Koerber, G.G. (1971). Piezoelectric surface waves on $LiNbO_3$. *IEEE Trans. Sonics Ultrason.* 18: 237–238.

6 Mindlin, R.D. (1972). Electromagnetic radiation from a vibrating quartz plate. *Int. J. Solids Struct.* 9: 697–702.
7 Sedov, A. and Schmerr, L.W. Jr. (1986). Some exact solutions for the propagation of transient electroacoustic waves I: piezoelectric half-space. *Int. J. Eng. Sci.* 24: 557–568.
8 Schmerr, L.W. Jr, and Sedov, A. (1986). Some exact solutions for the propagation of transient electroacoustic waves II: plane interface between two piezoelectric media. *Int. J. Eng. Sci.* 24: 921–932.
9 Lee, P.C.Y. (1989). Electromagnetic radiation from an AT-cut quartz plate under lateral-field excitation. *J. Appl. Phys.* 65: 1395–1399.
10 Yang, J.S. (2005). *An Introduction to the Theory of Piezoelectricity*. New York: Springer.
11 Mindlin, R.D. (1978). A variational principle for the equations of piezoelectromagnetism in a compound medium. In: *Complex Variable Analysis and Its Applications*, 379–400. Moscow, USSR: Academy of Sciences.
12 Lee, P.C.Y. (1991). A variational principle for the equations of piezoelectromagnetism in elastic dielectric crystals. *J. Appl. Phys.* 69: 7470–7473.
13 Yang, J.S. (1991). A generalized variational principle for piezoelectromagnetism in an elastic medium. *Arch. Mech.* 43: 795–798.
14 Yang, J.S. (1993). Variational principles for the vibration of an elastic dielectric. *Arch. Mech.* 45: 279–284.
15 Yang, J.S. and Wu, X.Y. (1995). The vibration of an elastic dielectric with piezoelectromagnetism. *Q. Appl. Math.* 53: 753–760.
16 Yang, J.S. (2000). Bleustein–Gulyaev waves in piezoelectromagnetic materials. *Int. J. Appl. Electromagn. Mech.* 12: 235–240.
17 Yang, J.S. (2004). Piezoelectromagnetic waves in a ceramic plate. *IEEE Trans. Ultrason. Ferroelectr. Freq. Control.* 51: 1035–1039.
18 Li, X.F. and Yang, J.S. (2005). Electromagnetoelastic behavior induced by a crack under antiplane mechanical and inplane electric impacts. *Int. J. Fract.* 132: 49–65.
19 Lee, P.C.Y., Kim, Y.G., and Prevost, J.H. (1990). Electromagnetic radiation from doubly rotated piezoelectric crystal plates vibrating at thickness frequencies. *J. Appl. Phys.* 67: 6633–6642.
20 Campbell, C.F. and Weber, R.J. (1993). Calculation of radiated electromagnetic power from bulk acoustic wave resonators. *IEEE International Frequency Control Symposium, Salt Lake City, UT*, pp. 472–475.
21 Li, S. (1996). The electromagneto-acoustic surface wave in piezoelectric medium: the Bleustein–Gulyaev mode. *J. Appl. Phys.* 80: 5264–5269.
22 Mitchell, A.R. and Griffiths, D.F. (1980). *The Finite Difference Method in Partial Differential Equations*. New York: Wiley.
23 Taflove, A. and Hagness, S.C. (2005). *Computational Electrodynamics: The Finite-Difference Time-Domain Method*, 3e. Norwood, MA: Artech House.
24 Huebner, K.H. and Thornton, E.A. (1982). *The Finite Element Method for Engineers*, 2e. New York: Wiley.
25 Jin, J.M. (2002). *The Finite Element Method in Electromagnetics*, 2e. New York: Wiley.
26 Chew, W.C., Tong, M.S., and Hu, B. (2008). *Integral Equation Methods for Electromagnetic and Elastic Waves*. San Rafael, CA: Morgan & Claypool.

27 Chew, W.C., Jin, J.M., Michielssen, E., and Song, J.M. (eds.) (2001). *Fast and Efficient Algorithms in Computational Electromagnetics*. Boston, MA: Artech House.
28 Chew, W.C. (1990). *Waves and Fields in Inhomogeneous Media*. New York: Van Nostrand Reinhold.
29 Poggio, A.J. and Miller, E.K. (1973). Integral equation solutions of three-dimensional scattering problems, Chapter 4. In: *Computer Techniques for Electromagnetics* (ed. R. Mittra). Oxford: Pergamon Press.
30 Müller, C. (1969). *Foundations of the Mathematical Theory of Electromagnetic Waves*. Berlin: Springer-Verlag.
31 Pao, Y.H. and Varatharajulu, V. (1976). Huygens' principle, radiation conditions, and integral formulas for the scattering of elastic waves. *J. Acoust. Soc. Am.* 59 (6): 1361–1371.
32 Harrington, R.F. (1993). *Field Computation by Moment Methods*. Piscataway, NJ: IEEE Press.
33 Aliabadi, M.H. (2002). *The Boundary Element Method*, vol. 2. New York: John Wiley & Sons.
34 Tong, M.S. and Chew, W.C. (2005). A higher-order Nyström scheme for electromagnetic scattering by arbitrarily shaped surfaces. *IEEE Antennas Wireless Propagat. Lett.* 4: 277–280.
35 Chen, Y.-H., Chew, W.C., and Liu, Q.H. (1998). A three-dimensional finite difference code for the modeling of sonic logging tools. *J. Acoust. Soc. Am.* 93: 702–712.
36 Chew, W.C. and Liu, Q.H. (1996). Perfectly matched layers for elastodynamics: a new absorbing boundary condition. *J. Comput. Acoust.* 4: 341–359.
37 Randall, C.J. (1989). Absorbing boundary condition for the elastic wave equation: velocity-stress formulation. *Geophysics.* 54: 1141–1152.
38 Liu, Q.-H., Schoen, E., Daube, F. et al. (1996). A three-dimensional finite difference simulation of sonic logging. *J. Acoust. Soc. Am.* 100: 72–79.
39 Gonsalves, I.R., Shippy, D.J., and Rizzo, F.J. (1990). Direct boundary integral equations for elastodynamics in three-dimensional half-spaces. *Comput. Mech.* 6: 279–292.
40 Rizzo, F.J. and Shippy, D.J. (1977). An advanced boundary integral equation method for three-dimensional thermoelasticity. *Int. J. Numer. Meth. Eng.* 11: 1753–1768.
41 Cruse, T.A. (1974). An improved boundary-integral equation method for three dimensional elastic stress analysis. *Comput. Struct.* 4: 741–754.
42 Rizzo, F.J., Shippy, D.J., and Rezayat, M. (1985). A boundary integral equation method for radiation and scattering of elastic waves in three dimensions. *Int. J. Numer. Meth. Eng.* 21: 115–129.
43 Liu, Y. and Rizzo, F.J. (1993). Hypersingular boundary integral equations for radiation and scattering of elastic waves in three dimensions. *Comput. Meth. Appl. Meth. Eng.* 107: 131–144.
44 Hall, W.S. (1994). *The Boundary Element Method*. London: Kluwer Academic Publishers.
45 Manolis, G.D. and Beskos, D.E. (1988). *Boundary Element Methods in Elastodynamics*. London: Unwin Hyman.
46 Schafbuch, P.J., Thompson, R.B., and Rizzo, F.J. (1990). Application of the boundary element method to the elastic wave scattering by irregular defects. *J. Nondestruc. Eval.* 2: 113–127.

47 Kondapalli, P.S. (1995). Frequency response of elastic bodies of revolution by the boundary element method. *J. Acoust. Soc. Am.* 98: 1558–1564.

48 Pao, Y.-H. and Varatharajulu, V. (1976). Huygens' principle, radiation conditions, and integral formulas for the scattering of elastic waves. *J. Acoust. Soc. Am.* 59: 1361–1371.

49 Morse, P.M. and Feshbach, H. (1953). *Methods of Theoretical Physics*. New York: McGraw-Hill.

50 Duffy, M.G. (1982). Quadrature over a pyramid or cube of integrands with a singularity at a vertex. *SIAM J. Numer. Anal.* 19: 1260–1262.

51 Pao, Y.-H. and Mow, C.C. (1963). Scattering of plane compressional waves by a spherical obstacle. *J. Appl. Phys.* 34: 493–499.

52 Moon, F.C. and Pao, Y.-H. (1967). The influence of the curvature of spherical waves on dynamic stress concentration. *ASME J. Appl. Mech.* 34: 373–379.

53 Ying, C.F. and Truell, R. (1956). Scattering of a plane longitudinal wave by a spherical obstacle in an isotropically elastic solid. *J. Appl. Phys.* 27: 1086–1097.

54 Gonsalves, I.R., Shippy, D.J., and Rizzo, F.J. (1991). The direct boundary integral equation method for the three-dimensional elastodynamic transmission problem. *Math. Comput. Modell.* 15: 155–164.

55 Kondapalli, P.S., Shippy, D.J., and Fairweather, G. (1992). The method of fundamental solutions for transmission and scattering of elastic waves. *Comput. Meth. Appl. Meth. Eng.* 96: 255–269.

56 Nishimura, N. (2002). Fast multipole accelerated boundary integral equation methods. *ASME Trans. Appl. Mech. Rev.* 55: 299–324.

57 Lu, C.C. and Chew, W.C. (1994). A multilevel algorithm for solving a boundary integral equation of wave scattering. *Microw. Opt. Technol. Lett.* 7: 466–470.

58 Song, J.M. and Chew, W.C. (1995). Multilevel fast-multipole algorithm for solving combined field integral equations of electromagnetic scattering. *Microw. Opt. Technol. Lett.* 10: 14–19.

59 Dembart, B. and Yip, E. (1995). A 3D fast multipole method for electromagnetics with multiple levels. *Proceedings of the 11th Annual Review of Progress in Applied Computational Electromagnetics, Monterey, CA*, vol. 1, pp. 621–618.

60 Song, J.M. and Chew, W.C. (1998). The fast Illinois solver code: requirements and scaling properties. *IEEE Comput. Sci. Eng.* 5: 19–23.

61 Song, J.M. and Chew, W.C. (2000). Large scale computations using FISC. *IEEE Antennas Propagat. Int. Symp.* 4: 1856–1859.

62 Velamparambil, S., Chew, W.C., and Song, J.M. (2003). 10 Million unknowns: is it that big? *IEEE Antennas Propagat. Mag.* 45: 43–58.

63 Hastriter, M.L. (2003). A study of MLFMA for large scale scattering problems. Ph.D. Dissertation. ECE Department, University of Illinois, Urbana.

64 Gürel, L. and Ergül, Ö. (2007). Fast and accurate solutions of extremely large integral-equation problems discretized with tens of millions of unknowns. *Electron. Lett.* 43: 499–500.

65 Ergül, Ö. and Gürel, L. (2008). Efficient parallelization of the multilevel fast multipole algorithm for the solution of large-scale scattering problems. *IEEE Trans. Antennas Propagat.* 56: 2335–2345.

66 Chen, Y.-H. and Chew, W.C. (1997). Fast multipole method as an efficient solver for 2D elastic wave surface integral equations. *Comput. Mech.* 20: 495–506.

67 Fukui, T. and Inoue, K. (1998). Fast multipole boundary element method in 2D elastodynamics. *J. Appl. Mech. JSCE.* 1: 373–380. (in Japanese).
68 Fujiwara, H. (1998). The fast multipole method for integral equations of seismic scattering problems. *Geophys. J. Int.* 133: 773–782.
69 Takahashi, T., Nishimura, N., and Kobayashi, S. (2001). Fast boundary integral equation method for elastodynamic problems in 2D in time domain. *Trans. JSMS A.* 67: 1409–1416. (in Japanese).
70 Yoshida, K., Nishimura, N., and Kobayashi, S. (2000). Analysis of three dimensional scattering of elastic waves by crack with fast multipole boundary integral equation method. *J. Appl. Mech. JSCE.* 3: 143–150. (in Japanese).
71 Fujiwara, H. (2000). The fast multipole method for solving integral equations of three-dimensional topography and basin problems. *Geophys. J. Int.* 140: 198–210.
72 Takahashi, T., Nishimura, N., and Kobayashi, S. (2003). A fast BIEM for three-dimensional elastodynamics in time domain. *Eng. Anal. Boundary Elem.* 27: 491–506.
73 Otani, Y. and Nishimura, N. (2006). A fast multipole boundary integral equation method for period boundary value problems in three dimensional elastostatics and its application to homogenisation. *Int. J. Multiscale Comput. Eng.* 4: 487–500.
74 Saitoh, T., Hirose, S., and Fukui, T. (2007). Application of fast multipole boundary element method to multiple scattering analysis of acoustic and elastic waves. *AIP Conference Proc.* 894: 79–86.
75 Otani, Y. and Nishimura, N. (2008). An FMM for periodic boundary value problems for cracks for Helmhotz' equation in 2D. *Int. J. Numer. Meth. Eng.* 73: 381–406.
76 Tong, M.S. and Chew, W.C. (2007). Nyström method for elastic wave scattering by three-dimensional obstacles. *J. Comput. Phys.* 226: 1845–1858.
77 Gürel, L. and Ergül, Ö. (2007). Fast and accurate solutions of extremely large integral-equation problems discretized with tens of millions of unknowns. *Electron. Lett.* 43: 499–500.
78 Rokhlin, V. (1990). Rapid solution of integral equations of scattering theory in two dimensions. *J. Comput. Phys.* 86: 414–439.
79 Coifman, R., Rokhlin, V., and Wandzura, S. (1993). The fast multipole method for the wave equation: a pedestrian prescription. *IEEE Antennas Propagat. Mag.* 35: 7–12.
80 Rokhlin, V. (1993). Diagonal forms of translation operators for the Helmholtz equation in three dimensions. *Appl. Comput. Harmon. Anal.* 1: 82–93.
81 Epton, M.A. and Dembart, B. (1995). Multipole translation theory for the three-dimensional laplace and Helmholtz equations. *SIAM J. Sci. Comput.* 16: 865–897.
82 Gyure, M.F. and Stalzer, M.A. (1998). A prescription for the multilevel Helmholtz FMM. *IEEE Comput. Sci. Eng.* 5: 39–47.
83 Koc, S. and Chew, W.C. (1998). Calculation of acoustical scattering from a cluster of scatterers. *J. Acoust. Soc. Am.* 103: 721–734.
84 Gumerov, N.A. and Duraiswami, R. (2004). *Fast Multipole Methods for the Helmholtz Equation in Three Dimensions.* Oxford: Elsevier.
85 Gumerov, N.A. and Duraiswami, R. (2005). Computation of scattering from clusters of spheres using the fast multipole method. *J. Acoust. Soc. Am.* 117: 1744–1761.

86 Sakuma, T. and Yasuda, Y. (2002). Fast multipole boundary element method for large-scale steady-state sound field analysis. *Part I: Setup and validation. Acta. Acust. Acust.* 88: 513–525.

87 Yasuda, Y. and Sakuma, T. (2003). Fast multipole boundary element method for large-scale steady-state sound field analysis. Part II: Examination of numerical items. *Acta. Acust. Acust.* 89: 28–38.

88 Yasuda, Y. and Sakuma, T. (2005). An effective setting of hierarchical cell structure for the fast multipole boundary element method. *J. Comput. Acoust.* 13: 47–70.

89 Yasuda, Y. and Sakuma, T. (2005). A technique for plane-symmetric sound field analysis in the fast multipole boundary element method. *J. Comput. Acoust.* 13: 71–85.

90 Rao, S.M., Wilton, D.R., and Glisson, A.W. (1982). Electromagnetic scattering by surfaces of arbitrary shape. *IEEE Trans. Antennas Propagat.* 30: 409–418.

91 Kot, J.S. (1990). Solution of thin-wire integral equations by Nyström methods. *Microw. Opt. Technol. Lett.* 3: 393–396.

92 Canino, L.S., Ottusch, J.J., Stalzer, M.A. et al. (1998). Numerical solution of the Helmholtz equation in 2D and 3D using a highorder Nyström discretization. *J. Comput. Phys.* 146: 627–663.

93 Gedney, S.D. (2003). On deriving a locally corrected Nyström scheme from a quadrature sampled moment method. *IEEE Trans. Antennas Propagat.* 51: 2402–2412.

94 Gedney, S.D. (2000). Application of the high-order Nyström scheme to the integral equation solution of electromagnetic interaction problems. *IEEE International Symposium on Electromagnetic Compatibility*, Washington, DC, vol. 1, pp. 289–294.

95 Tong, M.S. and Chew, W.C. (2007). Super-hyper singularity treatment for solving 3D electrical field integration equations. *Microw. Opt. Technol. Lett.* 49: 1383–1388.

96 Lam, Y.W. (1998). A boundary integral formulation for the prediction of acoustic scattering from periodic structures. *J. Acoust. Soc. Am.* 105: 762–769.

97 Ylä-Oijala, P. and Järvenpää, S. (2006). Iterative solution of high-order boundary element method for acoustic impedance boundary value problems. *J. Sound Vib.* 291: 824–843.

98 Chandrasekhar, B. and Rao, S.M. (2004). Elimination of internal resonance problem associated with acoustic scattering by three-dimensional rigid body. *J. Acoust. Soc. Am.* 115: 2731–2737.

99 Schenck, H.A. (1968). Improved integral formulation for acoustic radiation problems. *J. Acoust. Soc. Am.* 44: 41–48.

100 Burton, A.J. and Miller, G.F. (1971). The application of integral equation methods to the numerical solution of some exterior boundary value problems. *Proc. R. Soc. A London.* 323: 201–210.

101 Stroud, A.H. (1971). *Approximate Calculation of Multiple Integrals.* Englewood Cliffs: Prentice Hall.

102 Dwight, H.B. (1961). Tables of Integrals and Other Mathematical Data, *4e*. New York: Macmillan.

103 Swenson, G.W.J., Sandeen, E.R., Pater, L.L., and Zhuang, H.C. (1992). The potential for mitigation of gun blast noise through sheltering of the source. *USACERL Tech. Rep., No. N-92/09*, pp. 1–26.

104 Tong, M.S., Chew, W.C., and White, M.J. (2007). Surface impedance design with ground corrugation for mitigation of large-caliber gun blast noise. *CCEML Tech. Rep., No. 4-07*.

105 Duhamel, D. (1996). Efficient calculation of the three-dimensional sound pressure field around a noise barrier. *J. Sound Vib.* 197: 547–571.

106 Visscher, W.M. (1980). A new way to calculate scattering of acoustic and elastic waves I. Theory illustrated for scalar waves. *J. Appl. Phys.* 51: 825–834.

107 Kondapalli, P.S. and Shippy, D.J. (1992). Analysis of acoustic scattering in fluids and solids by the method of fundamental solutions. *J. Acoust. Soc. Am.* 91: 1844–1854.

108 Boström, A. (1980). Scattering of stationary acoustic waves by an elastic obstacle immersed in a fluid. *J. Acoust. Soc. Am.* 67: 390–398.

109 Boström, A. (1984). Scattering of acoustic waves by a layered elastic obstacle in a fluid–an improved null field approach. *J. Acoust. Soc. Am.* 76: 588–593.

110 Lim, R. (1998). Acoustic scattering by a partially buried three-dimensional elastic obstacle. *J. Acoust. Soc. Am.* 104: 769–782.

111 Boag, A., Leviatan, Y., and Boag, A. (1989). Analysis of two-dimensional acoustic scattering from periodic structures using a hybrid source model. *J. Acoust. Soc. Am.* 86: 387–394.

112 Balanis, C.A. (1989). *Advanced Engineering Electromagnetics*. New York: John Wiley & Sons.

113 Shi, Y. and Shirahatti, U.S. (1996). The use of the method of moments with finite difference operator in acoustics. *J. Acoust. Soc. Am.* 99: 2667–2673.

114 Rao, S.M. and Sridhara, B.S. (1992). Acoustic scattering from arbitrarily shaped multiple bodies in half-space: method of moments solution. *J. Acoust. Soc. Am.* 91: 652–657.

115 Rao, S.M. and Sridhara, B.S. (1991). Application of the method of moments to acoustic scattering from arbitrary shaped rigid bodies coated with lossless, shearless materials of arbitrary thickness. *J. Acoust. Soc. Am.* 90: 1601–1607.

116 Rao, S.M. and Raju, P.K. (1989). Application of the method of moments to acoustic scattering from multiple bodies of arbitrary shape. *J. Acoust. Soc. Am.* 86: 1143–1148.

117 Fenlon, F.H. (1969). Calculation of the acoustic radiation field at the surface of a finite cylinder by the method of weighted residuals. *Proc. IEEE.* 57: 291–306.

118 De Lautour, N.J. (1999). A Galerkin method for the numerical analysis of diffraction by a rectangular screen. *J. Acoust. Soc. Am.* 106: 3072–3080.

119 Erickson, R.R. and Zinn, B.T. (2003). Modeling of finite amplitude acoustic waves in closed cavities using the Galerkin method. *J. Acoust. Soc. Am.* 113: 1863–1870.

120 Ding, Y., Forestier, A., and Duong, T.H. (1989). A Galerkin scheme for the time domain integral equation of acoustic scattering from a hard surface. *J. Acoust. Soc. Am.* 86: 1566–1572.

121 Tong, M.S. and Chew, W.C. (2006). A direct approach for solving 3D EFIE with double gradient of the Green function. *Proc. IEEE Antennas Propagat. Soc. Int. Symp.* 4: 2855–2858.

122 Chew, W.C. (1991). *Elastic Wave Lecture Notes, ECE471*. Illinois: University of Illinois at Urbana-Champaign.

123 Pao, Y.-H. and Mow, C.C. (1963). Scattering of plane compressional waves by a spherical obstacle. *J. Appl. Phys.* 34: 493–499.

124 Seybert, A.F., Soenarko, B., Rizzo, F.J., and Shippy, D.J. (1985). An advanced computational method for radiation and scattering of acoustic waves in three dimensions. *J. Acoust. Soc. Am.* 77: 362–368.

125 Skudrzyk, E. (1971). *The Foundations of Acoustics*. New York: Springer-Verlag.

126 Hudson, J.A. (1980). *The Excitation and Propagation of Elastic Waves*. Combridge: Combridge University Press.

127 Chen, Y.H., Chew, W.C., and Zeroug, S. (1997). Fast multipole method as an efficient solver for 2D elastic wave surface Integral equations. *Comput. Mech.* 20: 495–506.

128 Achenbach, J.D. (1973). *Wave Propagation in Elastic Solids*. Amsterdam: North Holland.

129 Chen, Q.L. (1990). Electromagnetic modeling of three-dimensional piecewise homogeneous material bodies of arbitrary composition and geometry. Ph.D. Dissertation. Department of Electrical Engineering, University of Houston.

10

Application to Solve Time Domain Integral Equations

10.1 Introduction

Accurate and stable solution for transient electromagnetic (EM) problems is essential in many applications such as the design of radar cross section for aircrafts and EM-compatibility design for various electronic devices. The problems include non-linear, broadband, and time-varying characteristics and direct solution of time domain governing equations is preferred although converting frequency-domain solutions into time domain counterparts is also a choice [1]. In the integral equation approach, the time domain form of the integral equations presented in the preceding chapters should be formulated and they are the time domain electric field integral equation (TDEFIE), the time domain magnetic field integral equation (TDMFIE), and the time domain combined field integral equation (TDCFIE) for both conducting and penetrable media. For penetrable media, there are also the time domain Poggio-Miller-Chang-Harrington-Wu-Tsai (PMCHWT) formulation, the time domain Müller formulation, and time domain volume integral equations (TDVIEs) [2]. If both conducting and penetrable media exist in a structure, then time domain volume-surface integral equations (TDVSIEs) could be needed in some applications [3].

The time domain integral equations (TDIEs) are usually solved by combining the method of moments (MoM) in space domain and the march-on-in-time (MOT) scheme in the time domain [4]. As we have seen in the preceding chapters, the space domain MoM usually requires conforming meshes in geometric discretization. Furthermore, for surface integral equations (SIEs) of penetrable materials, two basis functions are required to represent two current densities on material boundaries. The well-known Rao-Wilton-Glisson (RWG) basis function [5] is a natural choice to represent electric current density, but one usually uses the $\hat{n}\times$RWG basis function, where $\hat{n}$ is a unit normal vector on a material interface, to represent magnetic current density [6, 7]. The $\hat{n}\times$RWG basis function may not be a good choice because it is not divergence conforming or curl conforming [8] and it will produce a fictitious charge deposit on mesh edges when a divergence is taken in some integral equations. Also, the MOT scheme has a well-known late-time instability problem [9–13] and it will significantly aggravate in time domain SIEs (TDSIEs) with penetrable materials [14]. Although the combined form integral equations, in particular the CFIEs and PMCHWT formulation, are widely used in the frequency domain because of their unique merits [2], they may not be so in the time domain scenario. It has been shown that all combined form integral equations could be unstable in the traditional time-marching schemes in the time domain [15, 16] and the interior resonance of scatterers could affect the stability, but it is not the unique source of instability [17]. It is said that the instability is mainly caused by inappropriate sampling and accumulation of round-off or truncation errors and cannot be fully eliminated in a simple and low-cost way [17].

We propose a different hybrid approach to solve the TDIEs in which the Nyström method is used to discretize the space domain while a Galerkin method (GM) with Laguerre basis and testing functions is employed to discretize the time domain. Although the Nyström method has been widely used in the frequency domain as shown in the previous chapters [18–24], it only has very limited applications in the time domain [25–27]. The merits of the Nyström method could be specially desirable for solving TDIEs because it can greatly simplify numerical implementation. In the time domain, we use the GM with Laguerre basis and testing functions to discretize the TDIEs. The method was proposed by Jung *et al.* [28] and has received attention in recent years [29–31]. The Laguerre function has several merits, i.e. going to zero as the time increases, leading to analytical formulations, and fully eliminating the late-time instability problem. The first merit allows a natural enforcement for the causality, which is essential for transient problems and the second merit makes the implementation more convenient. The third merit is due to the GM procedure, which fully removes the time dependence by replacing the MOT scheme with a march-on-in-degree (MOD) scheme in solved matrix equations [28]. Although the proposed approach may not be more efficient than the traditional MOT scheme [32], the instability problem of the MOT scheme can be thoroughly solved with reasonable complexity. In the proposed approach, we actually switch the strategies used in the spatial domain and temporal domain, respectively, in the traditional approach, i.e. using a finite-difference-time domain-like (FDTD-like) strategy [33] in the spacial domain while employing a finite-element-method-like (FEM-like) strategy [34] in the temporal domain. The switch allows us to acquire the benefits of both the FDTD method and FEM of differential equation methods (DEMs) or the benefits of both the Nyström method and the MoM of integral equation methods [35].

In this chapter, we first address the derivation of time domain versions of various integral equations and then show the proposed hybrid approach applied to solve them. We start with the simplest TDEFIE and followed by the TDMFIE for conducting media, and then illustrate the TDSIEs for penetrable media. We also investigate the TDVIEs, which are indispensable for inhomogeneous and anisotropic media, and study more complicated TDCFIEs and TDVSIEs for mixed media, which include both conducting and penetrable components. For each kind of integral equation we present the details of numerical implementation and provide several numerical examples to demonstrate the effectiveness and robustness of proposed hybrid approach.

10.2 Time Domain Surface Integral Equations for Conducting Media

10.2.1 Time Domain Electric Field Integral Equation

When the transient EM wave interacts with conducting media or objects, the problem can be formulated by the TDEFIE, which is the simplest form of TDIEs. As mentioned before, the TDEFIE is traditionally solved by combining the MoM in the spatial domain and the MOT scheme in the temporal domain. The MoM employs the divergence conforming RWG basis function [5], which requires conforming triangular meshes in geometric discretization and double surface integrations in evaluating matrix elements [2]. On the other hand, the MOT scheme in its early stage might be very inefficient and unstable due to the low- and high-frequency modes creeping into the solution [28]. Also, the MOT scheme suffers from a late-time instability that cannot be thoroughly removed, although many improved versions have been developed [36–38].

We propose a different hybrid approach to accurately solve the TDEFIE. The Nyström method instead of the traditional MoM is used in the spatial domain while the GM with Laguerre basis and testing functions instead of the conventional MOT is employed in the temporal domain. Although the Nyström method has been widely used in the frequency domain, it is almost ignored in the time domain and to the best of our knowledge there are very few papers that have used it to deal with three-dimensional (3D) problems [25–27]. Note that the technique of treating singularities [39–45] presented in the previous chapters should be revised slightly for the spatial domain hypersingular integrals of the TDIEs since the integral kernels do not have an exponential function coming from the Green's function now. On the other hand, the MOD scheme could be a good alternative to the MOT scheme due to its desirable features, particularly the full elimination of late-time instability. Even if the MOD may not be more efficient than the MOT, it is simpler in implementation than the MOT due to the derived elegant analytical formulations and also the removal of line integrals over surface patches when piecewise polynomial basis functions are used in the MOT [46, 47].

Formulations

Consider the scattering of a transient EM wave by a conducting object with a surface S in free space. The problem can be formulated by the TDEFIE, i.e.

$$\left[\mathbf{E}^{\text{inc}}(\mathbf{r}, t) + \mathbf{E}^{\text{sca}}(\mathbf{r}, t)\right]_{\tan} = 0, \quad \mathbf{r} \in S \tag{10.1}$$

where $\mathbf{E}^{\text{inc}}(\mathbf{r}, t)$ is the incident electric field and $\mathbf{E}^{\text{sca}}(\mathbf{r}, t)$ is the scattered electric field that is related to the magnetic vector potential $\mathbf{A}(\mathbf{r}, t)$ and electric scalar potential $\Phi(\mathbf{r}, t)$ by

$$\begin{aligned}\mathbf{E}^{\text{sca}}(\mathbf{r}, t) &= -\frac{\partial}{\partial t}\mathbf{A}(\mathbf{r}, t) - \nabla\Phi(\mathbf{r}, t) \\ &= -\frac{\mu_0}{4\pi}\int_S \frac{1}{R}\frac{\partial}{\partial t}\mathbf{J}(\mathbf{r}', \tau)\, \mathrm{d}S' - \frac{1}{4\pi\epsilon_0}\nabla\int_S \frac{1}{R} q(\mathbf{r}', \tau)\, \mathrm{d}S'.\end{aligned} \tag{10.2}$$

In the above, ϵ_0 and μ_0 are the permittivity and permeability of free space, respectively, $\mathbf{J}(\mathbf{r}', t)$ and $q(\mathbf{r}', t)$ are the current density and charge density induced on the conductor surface, respectively, $R = |\mathbf{r} - \mathbf{r}'|$ is the distance between an observation point $\mathbf{r}$ and a source point $\mathbf{r}'$, and $\tau = t - R/c$ is the retarded time with c being the speed of light in free space. The charge density is related to the current density by the continuity equation, i.e.

$$\nabla \cdot \mathbf{J}(\mathbf{r}, t) = -\frac{\partial}{\partial t} q(\mathbf{r}, t). \tag{10.3}$$

By introducing a new source vector $\mathbf{e}(\mathbf{r}, t)$, which is related to the current density by [7]

$$\mathbf{J}(\mathbf{r}, t) = \frac{\partial}{\partial t}\mathbf{e}(\mathbf{r}, t) \tag{10.4}$$

we have

$$q(\mathbf{r}, t) = -\nabla \cdot \mathbf{e}(\mathbf{r}, t). \tag{10.5}$$

By using the above relations, the TDEFIE can be written as

$$\left[\mu_0 \int_S \frac{1}{R} \frac{\partial^2 \mathbf{e}(\mathbf{r}', \tau)}{\partial t^2} \, \mathrm{d}S' - \frac{1}{\epsilon_0} \nabla \int_S \frac{1}{R} \nabla' \cdot \mathbf{e}(\mathbf{r}', \tau) \, \mathrm{d}S' \right]_{\tan} = 4\pi \mathbf{E}^{\mathrm{inc}}(\mathbf{r}, t)\Big|_{\tan}, \quad \mathbf{r} \in S. \tag{10.6}$$

Since the Nyström method does not use any basis function to represent the unknown function $\mathbf{e}(\mathbf{r}', \tau)$, the divergence operation on it should be removed by a transformation. Recalling the vector identity and Gauss' theorem, we have the following transformation

$$\begin{aligned}
\int_S \frac{1}{R} \nabla' \cdot \mathbf{e}(\mathbf{r}', \tau) \, \mathrm{d}S' &= \int_S \left\{ \nabla' \cdot \left[\frac{1}{R} \mathbf{e}(\mathbf{r}', \tau)\right] - \left[\nabla' \left(\frac{1}{R}\right) \cdot \mathbf{e}(\mathbf{r}', \tau)\right] \right\} \mathrm{d}S' \\
&= \int_S \left[-\nabla' \left(\frac{1}{R}\right) \cdot \mathbf{e}(\mathbf{r}', \tau)\right] \mathrm{d}S' \\
&= \int_S \nabla \left(\frac{1}{R}\right) \cdot \mathbf{e}(\mathbf{r}', \tau) \, \mathrm{d}S'
\end{aligned} \tag{10.7}$$

because we have

$$\int_S \nabla' \cdot \left[\frac{1}{R} \mathbf{e}(\mathbf{r}', \tau)\right] \mathrm{d}S' = \oint_C \frac{1}{R} \mathbf{e}(\mathbf{r}', \tau) \cdot \hat{n}' \, \mathrm{d}l' = 0 \tag{10.8}$$

where C is the boundary of S and $\hat{n}'$ is the unit normal vector on S. Applying the above transformation, the TDEFIE can written as

$$\left[\mu_0 \int_S \frac{1}{R} \frac{\partial^2 \mathbf{e}(\mathbf{r}', \tau)}{\partial t^2} \, \mathrm{d}S' + \frac{1}{\epsilon_0} \int_S \nabla\nabla \left(\frac{1}{R}\right) \cdot \mathbf{e}(\mathbf{r}', \tau) \, \mathrm{d}S' \right]_{\tan} = 4\pi \mathbf{E}^{\mathrm{inc}}(\mathbf{r}, t)\Big|_{\tan}, \quad \mathbf{r} \in S \tag{10.9}$$

from which we can solve for the unknown function.

Numerical Solution

The conductor surface is meshed into N small triangular patches and ΔS_n is the area of the nth patch. The unknown source vector $\mathbf{e}(\mathbf{r}', t)$ is separated into the time domain part and space domain part in each patch, i.e.

$$\mathbf{e}(\mathbf{r}', t) = \sum_{n=1}^{N} e_n(t) \mathbf{e}_n(\mathbf{r}'). \tag{10.10}$$

With the above expansion, the TDEFIE can be written as

$$\begin{aligned}
&\left[\mu_0 \sum_{n=1}^{N} \int_{\Delta S_n} \frac{1}{R} \frac{\partial^2 e_n(t)}{\partial t^2} \mathbf{e}_n(\mathbf{r}') \, \mathrm{d}S' + \frac{1}{\epsilon_0} \sum_{n=1}^{N} \int_{\Delta S_n} \nabla\nabla \left(\frac{1}{R}\right) \cdot \mathbf{e}_n(\mathbf{r}') e_n(\tau) \, \mathrm{d}S' \right]_{\tan} \\
&= 4\pi \mathbf{E}^{\mathrm{inc}}(\mathbf{r}, t)\Big|_{\tan}, \quad \mathbf{r} \in S.
\end{aligned} \tag{10.11}$$

We apply the Nyström method in the spatial domain, i.e. the integration over a small triangular patch can be replaced with a summation under a quadrature rule provided that the integrand is regular. By doing so, we have

$$\left[\mu_0 \sum_{n=1}^{N}\sum_{q=1}^{Q} \frac{w_{nq}}{R}\frac{\partial^2 e_n(\tau)}{\partial t^2}\mathbf{e}(\mathbf{r}'_{nq}) + \frac{1}{\epsilon_0}\sum_{n=1}^{N}\sum_{q=1}^{Q} w_{nq}\nabla\nabla\left(\frac{1}{R}\right)\cdot\mathbf{e}(\mathbf{r}'_{nq})e_n(\tau)\right]_{\tan}$$

$$= 4\pi\mathbf{E}^{\text{inc}}(\mathbf{r},t)\bigg|_{\tan}, \quad \mathbf{r}\in S. \tag{10.12}$$

where $\mathbf{e}(\mathbf{r}'_{nq}) = \mathbf{e}_{nq} = e^u_{nq}\hat{u} + e^v_{nq}\hat{v}$ is the value of unknown source vector and w_{nq} is the weight of a quadrature rule at the qth quadrature point in the nth patch. The unknown source vector has two independent components e^u_{nq} and e^v_{nq} in a local coordinate system $\{u,v,w\}$ established over the patch plane. If we choose two orthogonal unit tangential vectors $\mathbf{t}^{(l)}(\mathbf{r}_{mp}) = \mathbf{t}^{(l)}_{mp} = t^{(l)}_{ump}\hat{u} + t^{(l)}_{vmp}\hat{v} + t^{(l)}_{wmp}\hat{w}$ at the pth quadrature point in the mth observation patch ($l = 1,2$) to test the above equation, then we have

$$\mu_0 \sum_{n=1}^{N}\sum_{q=1}^{Q} \frac{w_{nq}}{R}\frac{\partial^2 e_n(\tau)}{\partial t^2}\left[\mathbf{t}^{(l)}_{mp}\cdot\mathbf{e}_{nq}\right] + \frac{1}{\epsilon_0}\sum_{n=1}^{N}\sum_{q=1}^{Q} w_{nq}\mathbf{t}^{(l)}_{mp}\cdot\nabla\nabla\left(\frac{1}{R}\right)\cdot\mathbf{e}_{nq}e_n(\tau)$$

$$= 4\pi\mathbf{t}^{(l)}_{mp}\cdot\mathbf{E}^{\text{inc}}(\mathbf{r}_{mp},t), \quad m = 1,\cdots,N; \quad p = 1,\cdots,Q. \tag{10.13}$$

The above procedure could have a singularity problem that occurs when $m = n$ but we have developed a robust treatment technique [43] and it can also be used here.

In the time domain, we use the following Laguerre function [7]

$$\phi_j(t) = \mathrm{e}^{-t/2}L_j(t) \tag{10.14}$$

as a basis function to expand $e_n(t)$, i.e.

$$e_n(t) = \sum_{j=0}^{\infty} e_{nj}\phi_j(st) \tag{10.15}$$

where s is the scaling factor that can control the support of expansion. Substituting this representation into Equation (10.13) and using $\phi_i(st)$ as a testing function to test it, we obtain

$$\sum_{n=1}^{N}\sum_{q=1}^{Q}\sum_{j=0}^{\infty}\left[\left(0.25s^2a^{(l)}_{mpnq} + b^{(l)}_{mpnq}\right)e_{nj} + s^2a^{(l)}_{mpnq}\sum_{k=0}^{j-1}(j-k)e_{nk}\right]I_{ij}\left(\frac{s}{c}R_{mpnq}\right) = 4\pi V_{mi},$$

$$m = 1,\cdots,N; \quad p = 1,\cdots,Q \tag{10.16}$$

where

$$a^{(l)}_{mpnq} = \frac{w_{nq}\mu_0}{R_{mpnq}}\mathbf{t}^{(l)}_{mp}\cdot\mathbf{e}_{nq} \tag{10.17}$$

$$b^{(l)}_{mpnq} = \frac{w_{nq}}{\epsilon_0}\mathbf{t}^{(l)}_{mp}\cdot\nabla\nabla\left(\frac{1}{R}\right)\cdot\mathbf{e}_{nq} \tag{10.18}$$

$$I_{ij}\left(\frac{s}{c}R_{mpnq}\right) = \int_0^\infty \phi_i(st)\phi_j\left(st - \frac{s}{c}R_{mpnq}\right)\,\mathrm{d}(st) \tag{10.19}$$

$$V_{mpi} = \int_0^\infty \phi_i(st)\mathbf{E}^{\mathrm{inc}}(\mathbf{r}_{mp}, t)\,\mathrm{d}(st). \tag{10.20}$$

By using the revelent formulations in the appendix of [7], we arrive at

$$\begin{aligned}
&\sum_{n=1}^{N}\sum_{q=1}^{Q}\left(0.25s^2a_{mpnq}^{(l)} + b_{mpnq}^{(l)}\right)e_{ni}I_{ii}\left(\frac{s}{c}R_{mpnq}\right) = V_{mpi} - \\
&\sum_{n=1}^{N}\sum_{q=1}^{Q}\sum_{j=0}^{i-1}\left(0.25s^2a_{mpnq}^{(l)} + b_{mpnq}^{(l)}\right)e_{nj}I_{ij}\left(\frac{s}{c}R_{mpnq}\right) - \\
&\sum_{n=1}^{N}\sum_{q=1}^{Q}\sum_{j=0}^{i}s^2a_{mpnq}^{(l)}\sum_{k=0}^{j-1}(j-k)e_{nk}I_{ij}\left(\frac{s}{c}R_{mpnq}\right), \\
&m = 1,\cdots,N;\quad p = 1,\cdots,Q
\end{aligned} \tag{10.21}$$

from which the unknown expansion coefficients can be solved in an MOD manner ($i = 0, 1, 2, \cdots, \infty$). Note that there are two space domain coefficients in $a_{mpnq}^{(l)}$ and $b_{mpnq}^{(l)}$, which are the two independent components e_{nq}^{u} and e_{nq}^{v} of $\mathbf{e}_{nq}$ and they should be combined with the time domain coefficient e_{ni} together to form two new unknown coefficients that can be determined by two groups of Equation (10.21) when $l = 1, 2$.

Numerical Examples

We consider two numerical examples to demonstrate the approach. The first example is transient EM scattering by a conducting sphere while the second example is transient EM scattering by a conducting cylinder. The geometry of the scatterers centered in a rectangular coordinate system is sketched in Figure 10.1. A Gaussian plane wave is illuminating the scatterers and the electric field and magnetic field are defined by

$$\mathbf{E}^{\mathrm{inc}}(\mathbf{r}, t) = \mathbf{E}_0\frac{4\mathrm{e}^{-\gamma^2}}{\sqrt{\pi}T} \tag{10.22}$$

$$\mathbf{H}^{\mathrm{inc}}(\mathbf{r}, t) = \frac{1}{\eta}\hat{k}\times\mathbf{E}^{\mathrm{inc}}(\mathbf{r}', t) \tag{10.23}$$

$$\gamma = \frac{4}{T}(ct - ct_0 - \mathbf{r}\cdot\hat{k}) \tag{10.24}$$

where $\hat{k}$ is the unit vector representing the direction of the incident wave, T is the width of Gaussian impulse, and t_0 is the time delay denoting the time when the pulse peaks at the origin. It is assumed that the incident wave has a direction $\hat{k} = -\hat{z}$ and is polarized with $\mathbf{E}_0 = \hat{x}$. We choose a Gaussian pulse with $T = 8.0$ lm (light meter) and $ct_0 = 12.0$ lm as an incident wave.

In the first example, the sphere is centered at the origin of the coordinate system and has a radius $a = 0.5$ m. Its surface is discretized into 722 triangular patches by commercial software ANSYS and the discretization is a one-time job without the need of

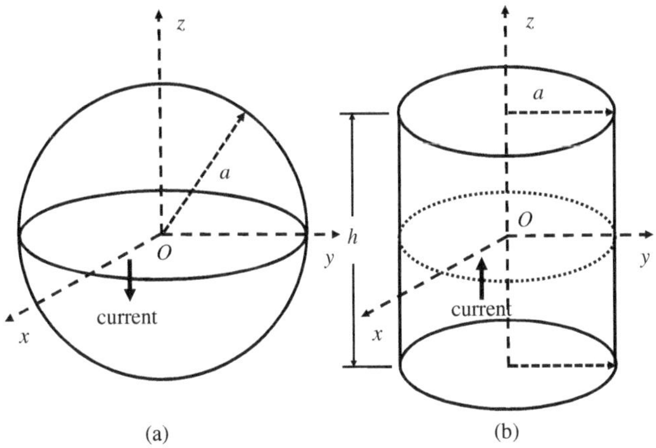

Figure 10.1 Geometry of two scatterers sitting in a rectangular coordinate system. (a) A conducting sphere. (b) A conducting cylinder.

examining the conformity of meshes. Figures 10.2–10.4 show the solution of transient θ-directed current density on the surface at $\theta = 90^\circ$, the solution of the θ-component of normalized far-zone scattered electric field observed along the backward direction, and the solution of the monostatic radar cross section (RCS), respectively. In the second example, the cylinder is also centered at the origin of the coordinate system and has a radius $a = 0.5$ m in its cross section and a height $h = 1.0$ m. Its surface is discretized

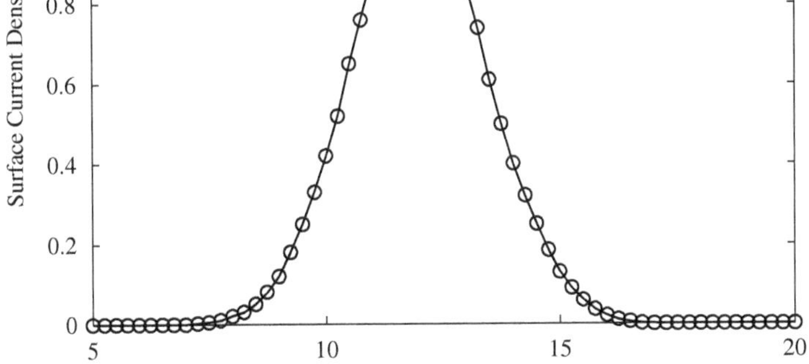

Figure 10.2 θ-directed transient current density at $\theta = 90^\circ$ on the surface of a conducting sphere.

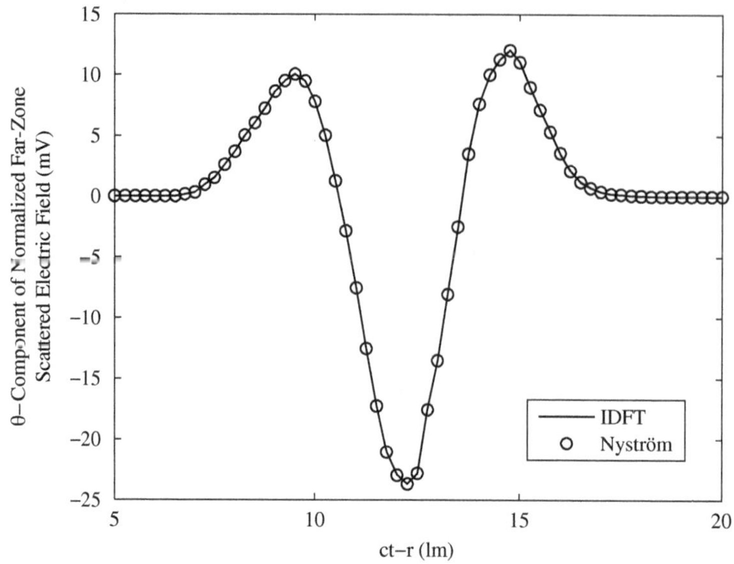

Figure 10.3 θ-component of normalized far-zone scattered electric field observed along the backward direction for a conducting sphere.

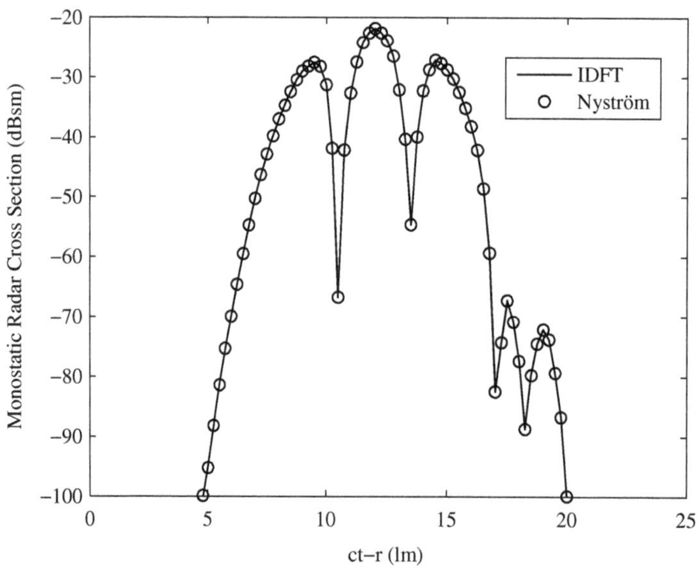

Figure 10.4 Monostatic radar cross section of a conducting sphere.

into 938 triangular patches. Figures 10.5–10.7 show the solution of same parameters as in the first example (but the current density is z-directed now). We select the scaling factor $s = 10^9$ and the number of temporal basis functions $M = 80$ in the two examples. The algorithm is implemented with FORTRAN codes and runs on a T7500 Dell Precision workstation with 3.06-GHz CPU and 64-GB RAM. All the results are compared with

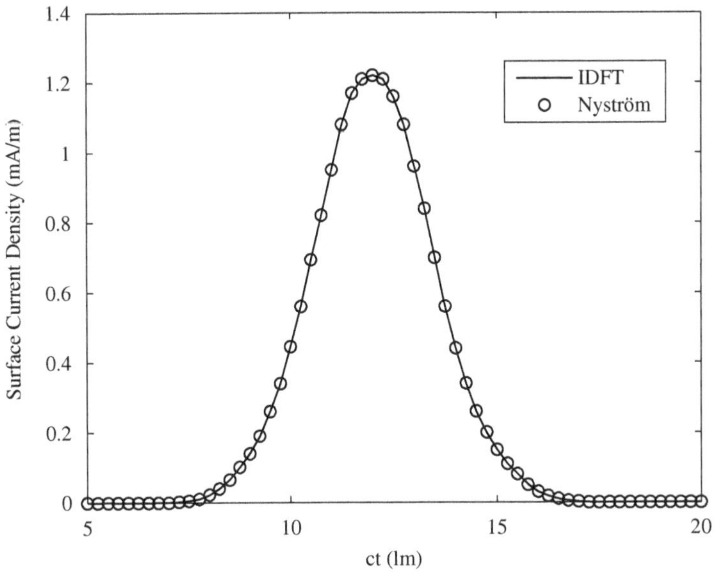

Figure 10.5 z-directed transient current density at $\theta = 90°$ on the surface of a conducting cylinder.

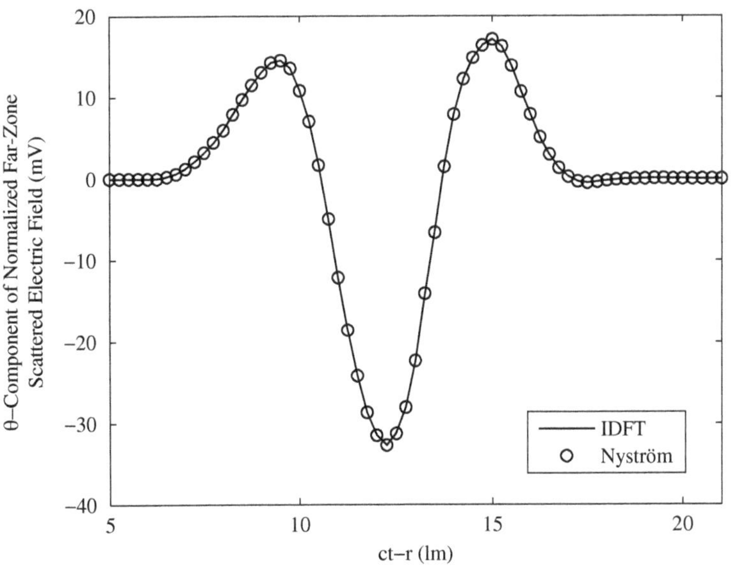

Figure 10.6 θ-component of normalized far-zone scattered electric field observed along the backward direction for a conducting cylinder.

the solutions obtained from the inverse discrete Fourier transform (IDFT) of frequency domain solutions and it is found that they are in good agreement each other.

Note that the low-order Nyström method may require higher computational costs than the traditional MoM for solving SIEs with the same geometric discretization. This is because the number of unknowns is 2 times the number of triangles for a one-point

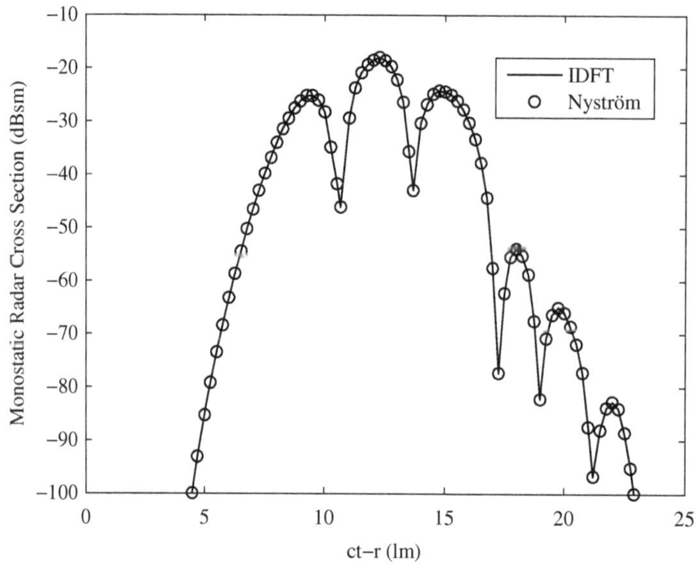

Figure 10.7 Monostatic radar cross section of a conducting cylinder.

quadrature rule in the Nyström method while it is 1.5 times in the MoM. However, the Nyström method may achieve a higher accuracy than the MoM with a similar number of unknowns by using higher-order quadrature rules with coarse meshes because it has been demonstrated that p-refinement has a higher convergence rate than h-refinement [19]. The Nyström method with higher-order quadrature rules has been implemented with a comparison to the MoM and the above conclusion can be observed [24, 30]. Also, the Nyström method can be applied to more complicated structures such as the surface-wire structure in the monopole antenna on a plate [59]. The solution does not rely on any special basis function because enforcing the normal continuity of current density is not necessary and non-conforming meshes with a low quality can be used in the Nyström method.

10.2.2 Time Domain Magnetic Field Integral Equation

The transient EM interaction with conducting media or objects can also be formulated by the TDMFIE. Although the TDMFIE is less widely used in practice than the TDEFIE, its importance cannot be underestimated because the TDCFIE includes its operator and also it appears in the TDIEs of penetrable media as an operator. Therefore, investigating the effective approach of solving the TDMFIE is necessary. The problem of using the traditional approach to solving the TDMFIE will be the same as solving the TDEFIE and could be more serious due to the nature of integral kernel. In the time domain, although the MOT scheme has been significantly improved [36–38], the instability problem still puzzles us and acquiring good solutions strongly depends upon the choice of time step [26].

In the hybrid approach we have proposed, the TDMFIE may be more difficult to solve than the TDEFIE because of the nature of its integral kernel. Note that the Nyström method we have used is different from the traditional Nyström method with which several papers have dealt by combining the traditional MOT scheme in the time domain [25–27] because its singularity treatment is different from those of the traditional Nyström method [39–44]. We use the strategy of transforming surface integrals into boundary integrals and the resulting formulations are more accurate and convenient in implementation [45]. In the time domain, the GM with Laguerre basis and testing functions may be more desirable because it can more significantly simplify the numerical implementation than in the TDEFIE.

Formulations

Consider the transient EM scattering by a perfectly electric conductor (PEC) with a surface S in free space. The problem can be formulated by the TDMFIE, i.e.

$$\hat{n} \times \left[\mathbf{H}^{\text{inc}}(\mathbf{r}, t) + \mathbf{H}^{\text{sca}}(\mathbf{r}, t)\right] = \mathbf{J}(\mathbf{r}, t), \quad \mathbf{r} \in S \tag{10.25}$$

where $\mathbf{H}^{\text{inc}}(\mathbf{r}, t)$ and $\mathbf{H}^{\text{sca}}(\mathbf{r}, t)$ are the incident magnetic field and scattered magnetic field, respectively, $\hat{n}$ is the unit normal vector at an observation point $\mathbf{r}$, and $\mathbf{J}(\mathbf{r}, t)$ is the induced electric current density on the conductor surface. The scattered magnetic field is related to the vector potential $\mathbf{A}(\mathbf{r}, t)$ by

$$\mathbf{H}^{\text{sca}}(\mathbf{r}, t) = \frac{1}{\mu_0}\nabla \times \mathbf{A}(\mathbf{r}, t) \tag{10.26}$$

where μ_0 is the permeability of free space and the tangential component of the magnetic field can be written as

$$\begin{aligned} \hat{n} \times \mathbf{H}^{\text{sca}}(\mathbf{r}, t) &= \hat{n} \times \frac{1}{4\pi}\nabla \times \int_S \frac{\mathbf{J}(\mathbf{r}', \tau)}{R}\, dS' \\ &= \frac{1}{2}\mathbf{J}(\mathbf{r}, t) + \hat{n} \times \frac{1}{4\pi} \fint_S \nabla \times \frac{\mathbf{J}(\mathbf{r}', \tau)}{R}\, dS' \end{aligned} \tag{10.27}$$

where $R = |\mathbf{r} - \mathbf{r}'|$ is the distance between an observation point $\mathbf{r}$ and a source point $\mathbf{r}'$ and $\tau = t - R/c$ is the retarded time in which c is the speed of wave propagation in free space. Also, the integral with a dash indicates that it is defined in a Cauchy-principal-value (CPV) sense. With the above derivation, the TDMFIE can be formulated by

$$\frac{1}{2}\mathbf{J}(\mathbf{r}, t) - \hat{n} \times \frac{1}{4\pi} \fint_S \nabla \times \frac{\mathbf{J}(\mathbf{r}', \tau)}{R}\, dS' = \hat{n} \times \mathbf{H}^{\text{inc}}(\mathbf{r}, t) \tag{10.28}$$

from which the unknown current density can be solved. Note that the following transformation for the integral kernel will be used in the implementation [48]

$$\nabla \times \frac{\mathbf{J}(\mathbf{r}', \tau)}{R} = \frac{1}{c}\frac{\partial}{\partial t}\mathbf{J}(\mathbf{r}', \tau) \times \frac{\mathbf{R}}{R^2} + \mathbf{J}(\mathbf{r}', \tau) \times \frac{\mathbf{R}}{R^3} \tag{10.29}$$

where $\mathbf{R} = \mathbf{r} - \mathbf{r}'$.

Numerical Solution

The conducting surface is discretized into N small triangular patches and ΔS_n is the area of the nth patch. We introduce the unknown source vector $\mathbf{e}(\mathbf{r}', t)$, which is related to the unknown current density by

$$\mathbf{J}(\mathbf{r}', t) = \frac{\partial}{\partial t}\xi(\mathbf{r}', t) \tag{10.30}$$

and it can be expanded as

$$\xi(\mathbf{r}', t) = \sum_{n=1}^{N} \xi_n(t)\mathbf{J}_n(\mathbf{r}'). \tag{10.31}$$

The unknown current density can then be expanded as

$$\mathbf{J}(\mathbf{r}', t) = \sum_{n=1}^{N} \mathbf{J}_n(\mathbf{r}')\frac{\partial}{\partial t}\xi_n(t). \tag{10.32}$$

If we substitute the expansion of current density and apply a Nyström scheme in the spatial domain, i.e. the integration over a small triangular patch is replaced with a summation under a quadrature rule provided that the integrand is regular, then the above TDMFIE becomes

$$\frac{1}{2}\frac{\partial}{\partial t}\xi_m(t)\hat{t}_{mp}^{(l)} \cdot \mathbf{J}_{mp} - \frac{1}{4\pi}\hat{t}_{mp}^{(l)} \cdot \hat{n}_{mp} \times \sum_{n=1}^{N}\sum_{q=1}^{Q} w_q \left[\frac{1}{c}\frac{\partial^2 \xi_n(\tau_{mpnq})}{\partial t^2}\mathbf{J}_{nq} \times \frac{\mathbf{R}_{mpnq}}{R_{mpnq}^2} + \frac{\partial \xi_n(\tau_{mpnq})}{\partial t}\mathbf{J}_{nq} \times \frac{\mathbf{R}_{mpnq}}{R_{mpnq}^3}\right] = V_{mp}^{(l)}(t), \quad m = 1, 2, \cdots, N; \;\; p = 1, 2, \cdots, Q \tag{10.33}$$

where $\mathbf{J}_{nq} = \mathbf{J}(\mathbf{r}'_{nq}) = J_{nq}^{u}\hat{u} + J_{nq}^{v}\hat{v}$ is the spatial vector of unknown current density at the source point $\mathbf{r}'_{nq}$, which is the qth quadrature point in the nth patch and it only has two independent components J_{nq}^{u} and J_{nq}^{v} in a local coordinate system $\{u, v, w\}$ established over the patch plane because it is a surface current density. Also, $\mathbf{J}_{mp} = \mathbf{J}(\mathbf{r}_{mp}) = J_{mp}^{u}\hat{u} + J_{mp}^{v}\hat{v}$ has an analogous meaning but for the observation point $\mathbf{r}_{mp}$ located at the pth quadrature point in the mth patch. In addition, $\mathbf{R}_{mpnq} = \mathbf{r}_{mp} - \mathbf{r}'_{nq}$ and $R_{mpnq} = |\mathbf{R}_{mpnq}|$. These spatial components of unknown current density should be transformed into the corresponding components of the global coordinate system $\{x, y, z\}$ in the implementation. We have chosen two orthogonal unit tangential vectors $\hat{t}_{mp}^{(l)} = \hat{t}^{(l)}(\mathbf{r}_{mp}) = t_{mp}^{x(l)}\hat{x} + t_{mp}^{y(l)}\hat{y} + t_{mp}^{z(l)}\hat{z}$ at the observation point $(l = 1, 2)$ to test the above equation so that the TDMFIE can be transformed into a matrix equation in the space domain. The right-hand side of the equation is given by

$$V_{mp}^{(l)}(t) = \hat{t}_{mp}^{(l)} \cdot \hat{n}_{mp} \times \mathbf{H}^{\mathrm{inc}}(\mathbf{r}_{mp}, t) \tag{10.34}$$

where $\hat{n}_{mp}$ is the unit normal vector at the observation point. The above procedure can only be used for a far interaction between the observation point and a source patch or $m \neq n$, which allows the direct application of a quadrature rule. If the observation point is inside the source patch or $m = n$, there is a singularity problem and we have developed a singularity treatment technique, which is different from those traditional techniques [39–44], to handle it.

In the temporal domain, we use the Laguerre function $\alpha_j(t) = \mathrm{e}^{-t/2}L_j(t)$ as a basis function to expand $\xi_n(t)$ [28]

$$\xi_n(t) = \sum_{j=0}^{\infty} \xi_{nj}\alpha_j(st) \tag{10.35}$$

where s is the scaling factor to control the support of expansion. Substituting this representation into Equation (10.33) and using $\alpha_i(st)$ as a testing function to test Equation (10.33), we can obtain

$$\hat{t}_{mp}^{(l)} \cdot \frac{s}{2}\sum_{j=0}^{i}\left(\frac{1}{2}\xi_{mj} + \sum_{k=0}^{j-1}\xi_{mk}\right) - \sum_{n=1}^{N}\sum_{q=1}^{Q}\left\{\frac{s^2}{c}I_{1nq}^{(l)}\sum_{j=0}^{i}\left[\frac{1}{4}\xi_{nj} + \sum_{k=0}^{j-1}(j-k)\xi_{nk}\right]I_{ij}(\Delta t_{mpnq})\right.$$
$$\left. + sI_{2nq}^{(l)}\sum_{j=0}^{i}\left(\frac{1}{2}\xi_{nj} + \sum_{k=0}^{j-1}\xi_{nk}\right)I_{ij}(\Delta t_{mpnq})\right\} = V_{mpi}^{(l)}, \quad i = 0, 1, \cdots, \infty \tag{10.36}$$

where $\Delta t_{mpnq} = sR_{mpnq}/c$ and

$$I_{1nq}^{(l)} = \frac{w_q}{4\pi}\hat{t}_{mp}^{(l)} \cdot \hat{n}_{mp} \times \mathbf{J}_{nq} \times \frac{\mathbf{R}_{mpnq}}{R_{mpnq}^2} \tag{10.37}$$

$$I_{2nq}^{(l)} = \frac{I_{1nq}^{(l)}}{R_{mpnq}} \tag{10.38}$$

$$I_{ij}(\Delta t_{mpnq}) = \int_0^\infty \alpha_i(st)\alpha_j(st - \Delta t_{mpnq})\, \mathrm{d}(st) \tag{10.39}$$

$$V_{mpi}^{(l)} = \int_0^\infty \alpha_i(st)V_{mp}^{(l)}(t)\, \mathrm{d}(st). \tag{10.40}$$

The integral $I_{ij}(\Delta t_{mpnq})$ in the above can be calculated according to the closed form formulations in the appendix in [28] while $V_{mpi}^{(l)}$ is evaluated numerically by replacing the integral interval with the real duration of a signal. Note that there are two spatial domain coefficients in $I_{1nq}^{(l)}$ and $I_{2nq}^{(l)}$, which are the two independent components J_{nq}^u and J_{nq}^v of $\mathbf{J}_{nq}$, and they should be combined with the time domain coefficient ξ_{ni} to form two new unknown coefficients J_{nqi}^u and J_{nqi}^v ($i = 0, 1, \cdots, M$, where M is the number of Laguerre basis functions in the temporal domain). The two new coefficients can be determined by two groups of Equation (10.36) when taking $l = 1, 2$ for a given i.

Numerical Examples

We consider the transient EM scattering by a conducting sphere, a conducting cube, and a conducting dumbbell-like object, respectively, to demonstrate the approach. The objects are centered at the origin of a rectangular coordinate system, as shown in Figure 10.8. The sphere has a radius $a = 0.5$ m, the cube has a side length $s = 1.0$ m, and the dumbbell-like object is defined by $h_1 = 0.35$ m, $h_2 = 0.3$ m, $h_3 = 0.35$ m, $a_1 = 0.5$ m, $a_2 = 0.3$ m, and $a_3 = 0.5$ m. Note that the dumbbell-like object is concave and this kind of object is seldom used in numerical demonstration. The concave objects are much easier to trigger a late-time instability because they can produce multiple scattering [1]. The surfaces of the first and third objects are discretized into 746 and 1156 triangular patches, respectively, by using the commercial software HFSS. As a sample, the meshed conducting dumbbell-like object is shown in Figure 10.8(d). The second object is meshed manually because we need low-quality meshes to demonstrate the approach.

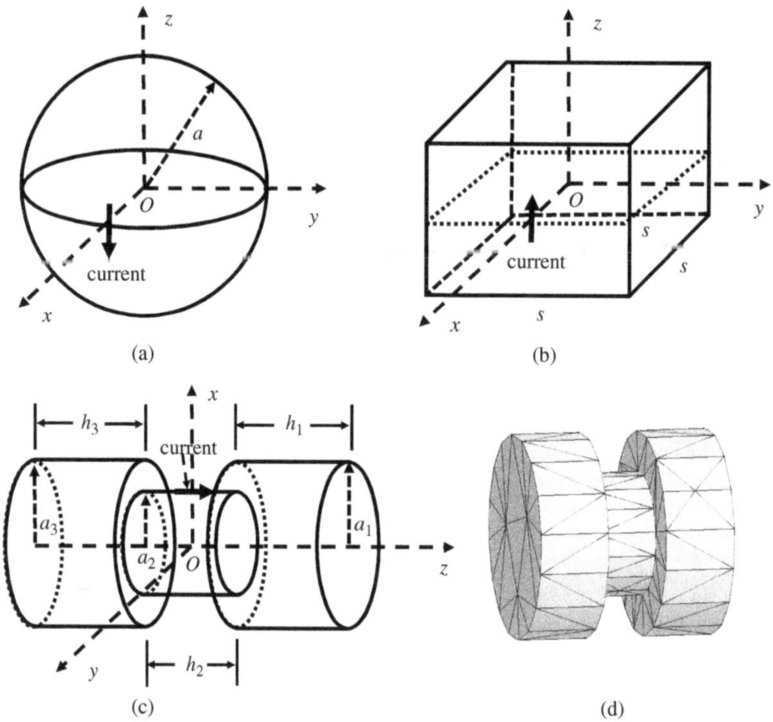

Figure 10.8 Geometry of the scatterers sited in a rectangular coordinate system. (a) A conducting sphere. (b) A conducting cube. (c) A conducting dumbbell-like object. (d) Meshes of the conducting dumbbell-like object.

A Gaussian plane wave is illuminating the scatterers along the $-z$ direction and the electric field and magnetic field are defined by

$$\mathbf{E}^{\mathrm{inc}}(\mathbf{r}, t) = \hat{x}\frac{4e^{-\gamma^2}}{\sqrt{\pi}T} \tag{10.41}$$

$$\mathbf{H}^{\mathrm{inc}}(\mathbf{r}, t) = -\frac{1}{\eta}\hat{z} \times \mathbf{E}^{\mathrm{inc}}(\mathbf{r}', t) \tag{10.42}$$

where $\gamma = 4(ct - ct_0 + \mathbf{r} \cdot \hat{z})/T$, T is the width of Gaussian impulse, and t_0 is the time delay for the pulse to arrive at the origin. We choose a Gaussian pulse with $T = 8.0$ lm and $ct_0 = 12.0$ lm as an incident wave. We select $s = 1.0 \times 10^9$ and $M = 80$ in all examples and the method of selecting them can be found in [46]. Figures 10.9 and 10.10 show the solutions of θ-directed surface current density at $z = 0$ and the θ-component of a normalized far-zone scattered electric field observed along the backward direction for the sphere, respectively. To examine the late-time stability, we show the results in a wide range of time and compare the result from the proposed approach with that from the traditional MOT approach. It can be seen that the former is still stable and agrees well with the result from the IDFT of corresponding frequency domain solution, while the latter becomes unstable or oscillating in late time. The computational performance of the proposed approach is related to the number of temporal basis functions M and

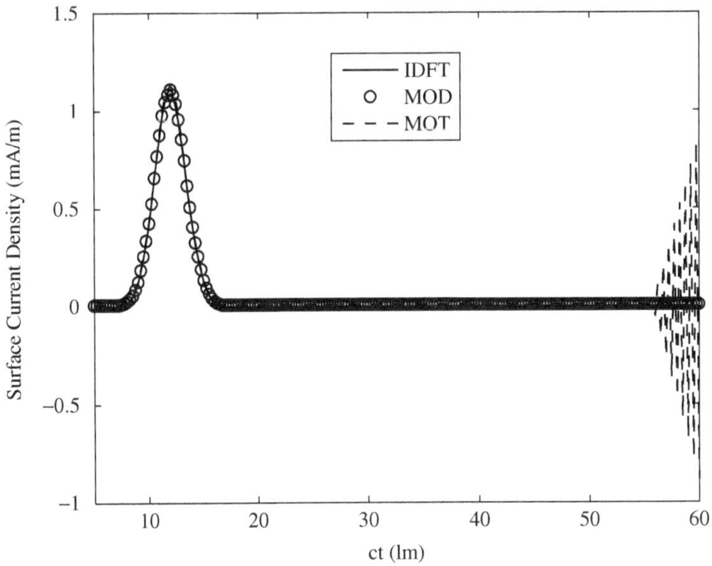

Figure 10.9 θ-directed transient current density at $z = 0$ on the surface of a conducting sphere.

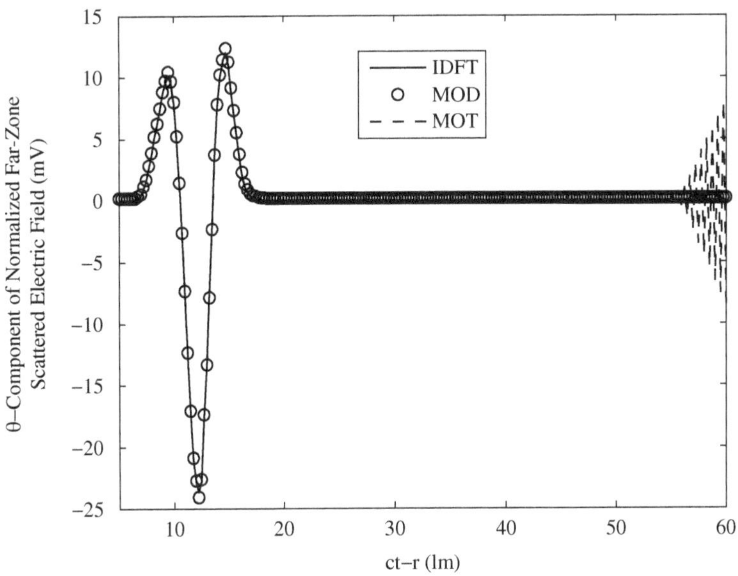

Figure 10.10 θ-component of normalized far-zone scattered electric field observed along the backward direction for a conducting sphere.

Figure 10.11 demonstrates that the numerical error of the solution compared to the IDFT result will rapidly decrease as M increases.

Figures 10.12 and 10.13 plot the solutions of the same parameters as before for the cube (but the surface current density is z-directed). We use this object as a sample to demonstrate that the proposed approach can still work well even for low-quality meshes.

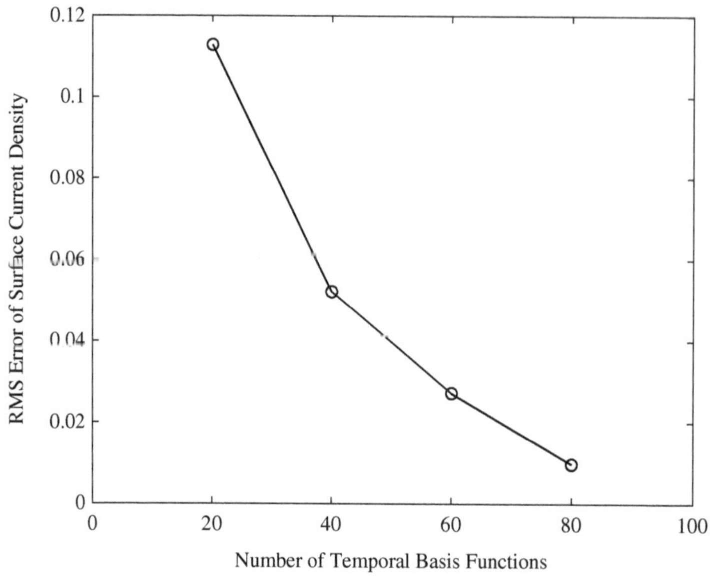

Figure 10.11 Root-mean-square (RMS) error of surface current density compared to the IDFT result versus the number of temporal basis function for a conducting sphere.

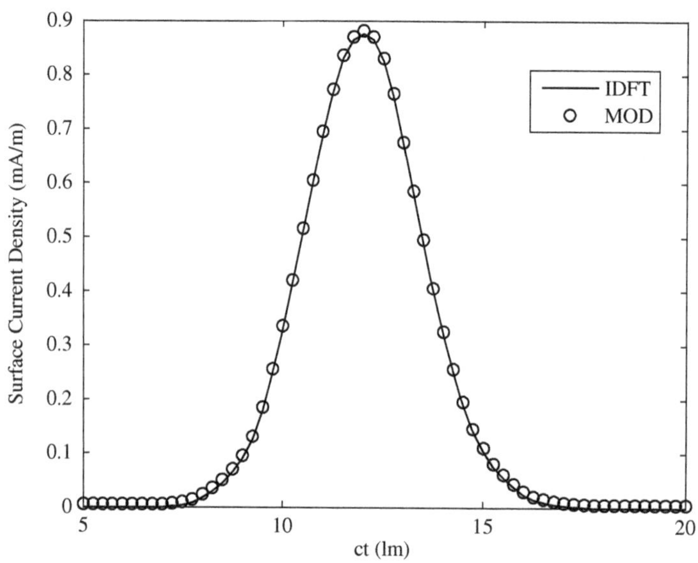

Figure 10.12 z-directed transient current density at $z = 0$ on the surface of a conducting cube.

We manually discretize the surface of the object by dividing the length into 25 segments and the width into 4 segments for a face, resulting in 200 triangular meshes for each face and 1200 triangular meshes in total. The quality factor of each triangular mesh is QF = 0.2909, which is quite low (QF= 1.0 for an equilateral triangle), but the solutions

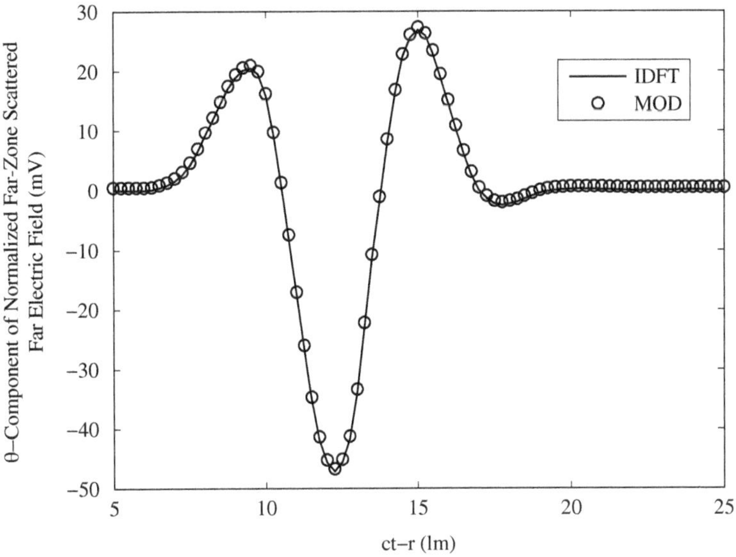

Figure 10.13 θ-component of a normalized far-zone scattered electric field observed along the backward direction for a conducting cube.

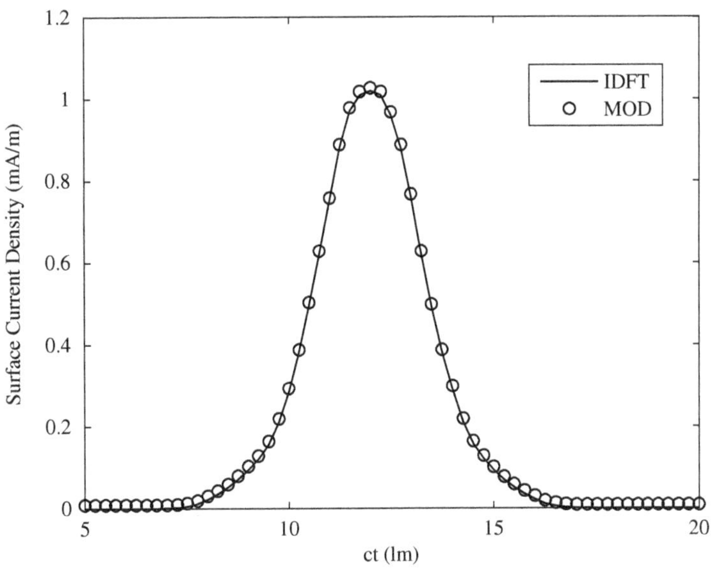

Figure 10.14 z-directed transient current density at $z = 0$ on the surface of a conducting dumbbell-like object.

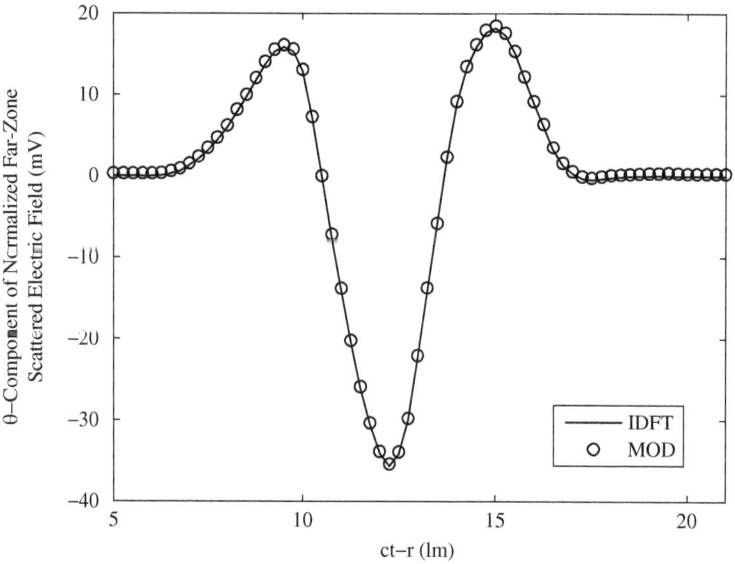

Figure 10.15 θ-component of a normalized far-zone scattered electric field observed along the backward direction for a conducting dumbbell-like object.

are still good compared with those from the IDFT. Figures 10.14 and 10.15 sketch the solutions of the same parameters as before for the dumbbell-like object (the surface current density is z-directed) and they agree well with those from the IDFT.

10.3 Time Domain Surface Integral Equations for Penetrable Media

Transient EM interaction with penetrable media can be formulated by TDSIEs or TDVIEs, but TDSIEs are preferred when the media are homogeneous. The TDSIEs include the time domain version of the PMCHWT formulation [49] and Müller formulation [50] in addition to the regular TDEFIEs, TDMFIEs, and TDCFIEs. Although the combined-form integral equations, in particular the CFIEs and the PMCHWT formulation, are widely used in frequency domain because of their unique merits [2], they may not be so in the time domain. It has been shown that all combined-form integral equations could be unstable in the traditional time-marching schemes in the time domain [15, 16] and the interior resonance of scatterers could affect the stability but it is not the unique source of instability [17]. We study the TDEFIEs as an example in this section because they represent the simplest form of TDIEs and the TDMFIEs are purely dual to the TDEFIEs for penetrable media. The TDEFIEs could be more desirable when those combined-form integral equations do not show an obvious advantage in the time domain [6].

In the traditional approach to solve the TDEFIEs, the space domain MoM requires two basis functions to represent two current densities at material boundaries and the $\hat{n}\times$RWG basis function, where $\hat{n}$ is a unit normal vector at a material interface, to represent the magnetic current density [7] may not be a good choice since it is not divergence conforming or curl conforming. In the frequency domain, one usually solves the problem by using the dual basis function [51], which possesses similar properties to the RWG basis function and is approximately orthogonal to it, but it is very complicated and inconvenient in implementation [52]. Another choice is to use the RWG basis function to represent both the electric and magnetic current densities but the resultant system matrix will be extremely ill-conditioned. Sheng *et al.* proposed a complicated testing scheme by including the $\hat{n}\times$RWG in the testing function to improve the conditioning of system matrix [53], but the testing scheme does not allow the movement of the Del operator onto the testing function. On the other hand, the late-time instability problem of the MOT scheme in the time domain will significantly aggravate in the TDSIEs with penetrable materials [54]. The proposed hybrid approach could be more desirable for solving TDSIEs of penetrable media because the embarrassment of selecting two basis functions can be avoided and the serious late-time instability of the MOT scheme can be fully removed. Note that the proposed approach is quite different from those in [26, 55] that either use the traditional locally corrected Nyström method in [19] or use the traditional MOT scheme. Also, the GM procedure in the time domain can produce many analytical formulations to make the implementation more convenient and avoid the line integrals over surface patches when piecewise polynomial basis functions are used in the MOT [7].

10.3.1 Formulations

Consider transient EM scattering by a dielectric object that is assumed to be homogeneous and embedded in free space. We use region 0 to denote the free space with permittivity ϵ_0 and μ_0, and region 1 to indicate the dielectric object with permittivity ϵ_1 and μ_1. The dielectric object has a surface or boundary S with an outward unit normal vector $\hat{n}$. A transient EM wave with an electric field $\mathbf{E}^{\text{inc}}$ is incident upon the object and there are equivalent electric current density $\mathbf{J}$ and equivalent magnetic current density $\mathbf{M}$ on S to account for the scattered fields from the dielectric object based on the equivalence principle. The TDEFIEs are formulated according to the boundary condition, i.e. the tangential component of total electric field vanishes on S

$$\left[\mathbf{E}_0^{\text{sca}}(\mathbf{J},\mathbf{M}) + \mathbf{E}^{\text{inc}}\right]_{\text{tan}} = 0, \quad \mathbf{r} \in S^- \tag{10.43}$$

$$\left[\mathbf{E}_1^{\text{sca}}(-\mathbf{J},-\mathbf{M})\right]_{\text{tan}} = 0, \quad \mathbf{r} \in S^+ \tag{10.44}$$

where $\mathbf{E}^{\text{sca}}$ is the scattered electric field in the corresponding region specified by its subscripts. Also, "tan" represents a tangential component and "+" and "−" on S denotes its interior side and exterior side, respectively. The scattered electric field can be related to the vector potential and scalar potential by

$$\mathbf{E}_j^{\text{sca}}\left[\mathbf{J}(\mathbf{r},t)\right] = -\frac{\partial}{\partial t}\mathbf{A}_j(\mathbf{r},t) - \nabla\Phi_j(\mathbf{r},t) \tag{10.45}$$

$$\mathbf{E}_j^{\text{sca}}\left[\mathbf{M}(\mathbf{r},t)\right] = -\frac{1}{\epsilon_j}\nabla\times\mathbf{F}_j(\mathbf{r},t) \tag{10.46}$$

where $\mathbf{A}_j(\mathbf{r},t)$ and $\mathbf{F}_j(\mathbf{r},t)$ are the magnetic and electric vector potentials, respectively, while $\Phi_j(\mathbf{r},t)$ is the electric scalar potential [31]. The subscripts $j = 0, 1$ on the potentials and fields indicate the corresponding regions in which the potentials and fields reside. We introduce new source vectors $\mathbf{e}(\mathbf{r}', t)$ and $\mathbf{h}(\mathbf{r}', t)$ that are related to $\mathbf{J}(\mathbf{r}', t)$ and $\mathbf{M}(\mathbf{r}', t)$, respectively, by

$$\mathbf{J}(\mathbf{r}', t) = \frac{\partial}{\partial t}\mathbf{e}(\mathbf{r}', t), \quad \mathbf{M}(\mathbf{r}', t) = \frac{\partial}{\partial t}\mathbf{h}(\mathbf{r}', t) \tag{10.47}$$

so that the unknown source vectors in the TDEFIEs can be unified. With the new source vectors, the TDEFIEs can be written as

$$\begin{aligned}\hat{t}(\mathbf{r}) \cdot &\left\{ \mu_0 \int_S \frac{\partial^2 \mathbf{e}(\mathbf{r}', \tau_0)}{R\partial t^2}\, dS' - \frac{1}{\epsilon_0} \int_S \overline{\mathbf{G}} \cdot \mathbf{e}(\mathbf{r}', \tau_0)\, dS' \right. \\ &\left. +2\pi\hat{n}(\mathbf{r}) \times \frac{\partial}{\partial t}\mathbf{h}(\mathbf{r}, \tau_0) + \fint_S \nabla \times \left[\frac{\partial}{R\partial t}\mathbf{h}(\mathbf{r}', \tau_0)\right] dS' \right\} \\ &= 4\pi\hat{t}(\mathbf{r}) \cdot \mathbf{E}^{\text{inc}}(\mathbf{r}, t), \quad \mathbf{r} \in S \end{aligned} \tag{10.48}$$

$$\begin{aligned}\hat{t}(\mathbf{r}) \cdot &\left\{ \mu_1 \int_S \frac{\partial^2 \mathbf{e}(\mathbf{r}', \tau_1)}{R\partial t^2}\, dS' - \frac{1}{\epsilon_1} \int_S \overline{\mathbf{G}} \cdot \mathbf{e}(\mathbf{r}', \tau_1)\, dS' \right. \\ &\left. -2\pi\hat{n}(\mathbf{r}) \times \frac{\partial}{\partial t}\mathbf{h}(\mathbf{r}, \tau_1) + \fint_S \nabla \times \left[\frac{\partial}{R\partial t}\mathbf{h}(\mathbf{r}', \tau_1)\right] dS' \right\} \\ &= 0, \quad \mathbf{r} \in S \end{aligned} \tag{10.49}$$

where $R = |\mathbf{r} - \mathbf{r}'|$ is the distance between an observation point $\mathbf{r}$ and a source point $\mathbf{r}'$ and $\tau_j = t - R/c_j$ with $c_j = 1/\sqrt{\epsilon_j \mu_j}$ being the velocity of wave propagation in the medium of the jth region. Also, $\hat{t}(\mathbf{r})$ is a tangential unit vector at the observation point $\mathbf{r}$ and $\overline{\mathbf{G}} = \overline{\mathbf{G}}(\mathbf{r}, \mathbf{r}') = \nabla\nabla(1/R)$ can be viewed as a dyadic Green's function. The integral with a dash in the above means that it is defined in a CPV sense.

10.3.2 Numerical Solution

Space Domain Discretization

We discretize the object surface S into N small triangular patches with ΔS_n being the area of the nth patch. The unknown source vectors on the surface can be expanded as

$$\mathbf{e}(\mathbf{r}', t) = \sum_{n=1}^{N} e^n(t)\mathbf{J}^n(\mathbf{r}'), \quad \mathbf{h}(\mathbf{r}', t) = \sum_{n=1}^{N} h^n(t)\mathbf{M}^n(\mathbf{r}'). \tag{10.50}$$

Substituting the above expansions and applying a Nyström scheme in the space domain, i.e. replacing the integration over a small triangular patch ΔS_n with a summation under a quadrature rule provided that the integrand is regular, we can discretize the TDEFIEs into a summation form, as shown in [31]. We establish a local coordinate system $\{u, v, w\}$ over a source patch plane and the unknown source vectors only have two independent $\hat{u}$ and $\hat{v}$ components in the coordinate system because they are surface vectors (surface current densities). We define $\mathbf{J}^n_{nq}$ and $\mathbf{M}^n_{nq}$ as the spatial parts of two unknown source vectors, respectively, at the source point $\mathbf{r}'_{nq}$, which is the qth quadrature point in the nth patch (there are Q quadrature points in total). Also, $\mathbf{J}^m_{mp}$ is the counterpart of $\mathbf{J}^n_{nq}$ and $\mathbf{M}^m_{mp}$ is the counterpart of $\mathbf{M}^n_{nq}$ at an observation point $\mathbf{r}_{mp}$, which is the pth quadrature point in the mth patch. In addition, w_{nq} is the weight of a quadrature rule at the qth quadrature point in the nth patch. If we use two orthogonal tangential unit vectors at the pth quadrature point in the mth observation patch, i.e.

$\hat{t}_{mp}^{(l)} = \hat{t}^{(l)}(\mathbf{r}_{mp}) = t_{mp}^{u(l)}\hat{u} + t_{mp}^{v(l)}\hat{v} + t_{mp}^{w(l)}\hat{w}$ $(l = 1, 2)$ to test the discretized TDEFIEs, then we can obtain the following matrix equation

$$\begin{aligned}
&\sum_{n=1}^{N}\sum_{q=1}^{Q}\left[\frac{\mu_0 w_{nq}}{R_{mpnq}}\frac{\mathrm{d}^2\mathrm{e}^n(\tau_0)}{\mathrm{d}t^2}A_{mpnq}^{(l)} - \frac{w_{nq}}{\epsilon_0}\mathrm{e}^n(\tau_0)B_{mpnq}^{(l)}\right]\\
&+\sum_{n=1}^{N}\sum_{q=1}^{Q}\left[\left(\frac{w_{nq}}{c_0 R_{mpnq}}\frac{\mathrm{d}}{\mathrm{d}t} + \frac{w_{nq}}{R_{mpnq}^2}\right)\frac{\mathrm{d}h^n(\tau_0)}{\mathrm{d}t}C_{mpnq}^{(l)}\right]\\
&+2\pi D_{mp}^{(l)}\frac{\mathrm{d}h^m(\tau_0)}{\mathrm{d}t} = 4\pi\hat{t}_{mp}^{(l)}\cdot\mathbf{E}^{\mathrm{inc}}(\mathbf{r}_{mp}, t),\ \ \mathbf{r}_{mp}\in S,\\
&m = 1, 2, \cdots, N;\ \ q = 1, 2, \cdots, Q
\end{aligned} \tag{10.51}$$

$$\begin{aligned}
&\sum_{n=1}^{N}\sum_{q=1}^{Q}\left[\frac{\mu_1 w_{nq}}{R_{mpnq}}\frac{\mathrm{d}^2\mathrm{e}^n(\tau_1)}{\mathrm{d}t^2}A_{mpnq}^{(l)} - \frac{w_{nq}}{\epsilon_1}\mathrm{e}^n(\tau_1)B_{mpnq}^{(l)}\right]\\
&+\sum_{n=1}^{N}\sum_{q=1}^{Q}\left[\left(\frac{w_{nq}}{c_1 R_{mpnq}}\frac{\mathrm{d}}{\mathrm{d}t} + \frac{w_{nq}}{R_{mpnq}^2}\right)\frac{\mathrm{d}h^n(\tau_1)}{\mathrm{d}t}C_{mpnq}^{(l)}\right]\\
&-2\pi D_{mp}^{(l)}\frac{\mathrm{d}h^m(\tau_1)}{\mathrm{d}t} = 0,\ \ \mathbf{r}_{mp}\in S,\\
&m = 1, 2, \cdots, N;\ \ q = 1, 2, \cdots, Q
\end{aligned} \tag{10.52}$$

where

$$\mathbf{J}_{nq}^n = \mathbf{J}^n(\mathbf{r}'_{nq}) = J_{nq}^{nu}\hat{u} + J_{nq}^{nv}\hat{v} \tag{10.53}$$

$$\mathbf{M}_{nq}^n = \mathbf{M}^n(\mathbf{r}'_{nq}) = M_{nq}^{nu}\hat{u} + M_{nq}^{nv}\hat{v} \tag{10.54}$$

$$A_{mpnq}^{(l)} = \hat{t}_{mp}^{(l)}\cdot\mathbf{J}_{nq}^n \tag{10.55}$$

$$B_{mpnq}^{(l)} = \hat{t}_{mp}^{(l)}\cdot\overline{\mathbf{G}}_{mpnq}\cdot\mathbf{J}_{nq}^n \tag{10.56}$$

$$C_{mpnq}^{(l)} = \hat{t}_{mp}^{(l)}\cdot\mathbf{M}_{nq}^n\times\hat{R}_{mpnq} \tag{10.57}$$

$$D_{mp}^{(l)} = \hat{t}_{mp}^{(l)}\cdot(\hat{n}_{mp}\times\mathbf{M}_{mp}^m). \tag{10.58}$$

Also, $R_{mpnq} = |\mathbf{r}_{mp} - \mathbf{r}'_{nq}|$, $\hat{R}_{mpnq} = (\mathbf{r}_{mp} - \mathbf{r}'_{nq})/R_{mpnq}$, and $\overline{\mathbf{G}}_{mpnq} = \overline{\mathbf{G}}(\mathbf{r}_{mp}, \mathbf{r}'_{nq})$. Note that the above matrix equation includes singular integrals which occur when $m = n$ and we have developed a robust technique to evaluate them as shown in Chapter 5 or in [45].

Time Domain Discretization

In the time domain, we use the Galerkin method to discretize the TDEFIEs. The Laguerre function is used as a basis function to expand the temporal parts of source vectors [7]

$$\mathrm{e}^n(t) = \sum_{j=0}^{\infty}\mathrm{e}_j^n\phi_j(st),\ \ h^n(t) = \sum_{j=0}^{\infty}h_j^n\phi_j(st) \tag{10.59}$$

where $\phi_j(t) = \mathrm{e}^{-t/2}L_j(t)$ is the Laguerre function and $L_j(t)$ is the Laguerre polynomial in the jth degree. Also, s is the scaling factor used to change the support of expansion. Substituting the above expansions to the TDEFIEs which have been discretized in space domain as shown in Equation (10.51), and using the Laguerre function $\phi_i(st)$ as a testing function to test it, we can remove the time dependence in Equation (10.51). After separating known terms from unknown terms in the equations, we can obtain the following recursive equations

$$
\begin{aligned}
&\sum_{n=1}^{N}\sum_{q=1}^{Q}\left\{\left[\frac{s^2\mu_0 w_{nq}}{4R_{mpnq}}A_{mpnq}^{(l)}-\frac{w_{nq}}{\epsilon_0}B_{mpnq}^{(l)}\right]\mathrm{e}_i^n\right\}I_{ii}^{(0)}+\\
&\sum_{n=1}^{N}\sum_{q=1}^{Q}\left\{\left(\frac{s^2 w_{nq}}{4c_0R_{mpnq}}+\frac{sw_{nq}}{2R_{mpnq}^2}\right)C_{mpnq}^{(l)}h_i^n\right\}I_{ii}^{(0)}\\
&+\pi sD_{mp}^{(l)}h_i^m I_{ii}^{(0)}=4\pi V_{mpi}^{(l)}-\\
&\sum_{n=1}^{N}\sum_{q=1}^{Q}\sum_{j=0}^{i-1}\left\{\left[\frac{s^2\mu_0 w_{nq}}{4R_{mpnq}}A_{mpnq}^{(l)}-\frac{w_{nq}}{\epsilon_0}B_{mpnq}^{(l)}\right]\mathrm{e}_j^n\right\}I_{ij}^{(0)}-\\
&\sum_{n=1}^{N}\sum_{q=1}^{Q}\sum_{j=0}^{i-1}\left\{\left(\frac{s^2 w_{nq}}{4c_0R_{mpnq}}+\frac{sw_{nq}}{2R_{mpnq}^2}\right)C_{mpnq}^{(l)}h_j^n\right\}I_{ij}^{(0)}-\\
&\pi sD_{mp}^{(l)}\sum_{j=0}^{i-1}h_j^m I_{ij}^{(0)}-2\pi sD_{mp}^{(l)}\sum_{j=0}^{i}\sum_{k=0}^{j-1}h_k^m I_{ij}^{(0)}-\\
&\sum_{n=1}^{N}\sum_{q=1}^{Q}\sum_{j=0}^{i}\left[\frac{s^2\mu_0 w_{nq}}{R_{mpnq}}A_{mpnq}^{(l)}\sum_{k=0}^{j-1}(j-k)\mathrm{e}_k^n\right]I_{ij}^{(0)}-\\
&\sum_{n=1}^{N}\sum_{q=1}^{Q}\sum_{j=0}^{i}\left[\frac{s^2 w_{nq}}{c_0R_{mpnq}^2}C_{mpnq}^{(l)}\sum_{k=0}^{j-1}(j-k)h_k^n\right]I_{ij}^{(0)}-\\
&\sum_{n=1}^{N}\sum_{q=1}^{Q}\sum_{j=0}^{i}\left[\frac{sw_{nq}}{R_{mpnq}^2}C_{mpnq}^{(l)}\sum_{k=0}^{j-1}h_k^n\right]I_{ij}^{(0)}\\
&m=1,\cdots,N;\ \ p=1,\cdots,Q
\end{aligned}
\tag{10.60}
$$

$$
\begin{aligned}
&\sum_{n=1}^{N}\sum_{q=1}^{Q}\left\{\left[\frac{s^2\mu_1 w_{nq}}{4R_{mpnq}}A_{mpnq}^{(l)}-\frac{w_{nq}}{\epsilon_1}B_{mpnq}^{(l)}\right]\mathrm{e}_i^n\right\}I_{ii}^{(1)}+\\
&\sum_{n=1}^{N}\sum_{q=1}^{Q}\left\{\left(\frac{s^2 w_{nq}}{4c_1R_{mpnq}}+\frac{sw_{nq}}{2R_{mpnq}^2}\right)C_{mpnq}^{(l)}h_i^n\right\}I_{ii}^{(1)}\\
&-\pi sD_{mp}^{(l)}h_i^m I_{ii}^{(1)}=4\pi V_{mpi}^{(l)}-\\
&\sum_{n=1}^{N}\sum_{q=1}^{Q}\sum_{j=0}^{i-1}\left\{\left[\frac{s^2\mu_1 w_{nq}}{4R_{mpnq}}A_{mpnq}^{(l)}-\frac{w_{nq}}{c_1}B_{mpnq}^{(l)}\right]\mathrm{e}_j^n\right\}I_{ij}^{(1)}-\\
&\sum_{n=1}^{N}\sum_{q=1}^{Q}\sum_{j=0}^{i-1}\left\{\left(\frac{s^2 w_{nq}}{4c_1R_{mpnq}}+\frac{sw_{nq}}{2R_{mpnq}^2}\right)C_{mpnq}^{(l)}h_j^n\right\}I_{ij}^{(1)}+\\
&\pi sD_{mp}^{(l)}\sum_{j=0}^{i-1}h_j^m I_{ij}^{(1)}+2\pi sD_{mp}^{(l)}\sum_{j=0}^{i}\sum_{k=0}^{j-1}h_k^m I_{ij}^{(1)}-
\end{aligned}
$$

$$\sum_{n=1}^{N}\sum_{q=1}^{Q}\sum_{j=0}^{i}\left[\frac{s^2\mu_1 w_{nq}}{R_{mpnq}}A_{mpnq}^{(l)}\sum_{k=0}^{j-1}(j-k)e_k^n\right]I_{ij}^{(1)}-$$
$$\sum_{n=1}^{N}\sum_{q=1}^{Q}\sum_{j=0}^{i}\left[\frac{s^2 w_{nq}}{c_1 R_{mpnq}^2}C_{mpnq}^{(l)}\sum_{k=0}^{j-1}(j-k)h_k^n\right]I_{ij}^{(1)}-$$
$$\sum_{n=1}^{N}\sum_{q=1}^{Q}\sum_{j=0}^{i}\left[\frac{s w_{nq}}{R_{mpnq}^2}C_{mpnq}^{(l)}\sum_{k=0}^{j-1}h_k^n\right]I_{ij}^{(1)}$$
$$m=1,\cdots,N;\quad p=1,\cdots,Q \tag{10.61}$$

where the definitions of $I_{ij}^{(0)}$, $I_{ij}^{(0)}$, and $V_{mpi}^{(l)}$ can be found in [31]. In the above equations, the left-hand side only involves the unknown time domain coefficients in the current ith degree while the right-hand side includes all previous degree time domain coefficients that have been known. Also, the space domain unknown coefficients and time domain unknown coefficients should be unified together in the solution process.

10.3.3 Numerical Examples

We present transient EM scattering by two typical objects, i.e. a dielectric sphere and a dielectric cube, to demonstrate the approach and the geometries of scatterers are sketched in Figure 10.16. The sphere has a radius $a = 0.5$ m and the cube has a side length $s = 1.0$ m. The relative permittivity of both objects is $\epsilon_r = 2.0$ (their relative permeability $\mu_r = 1.0$ is assumed). We discretize the surface of the sphere into 812 triangular patches and the surface of the cube into 1068 triangular patches. Both scatterers are centered at the origin of a rectangular coordinate system $\{x, y, z\}$ and a Gaussian plane wave is illuminating the scatterers along the $-z$ direction. The incident electric field is defined by

$$\mathbf{E}^{\text{inc}}(\mathbf{r},t)=\hat{x}\frac{4e^{-\gamma^2}}{\sqrt{\pi}T},\quad \gamma=\frac{4}{T}(ct-ct_0+\mathbf{r}\cdot\hat{z}) \tag{10.62}$$

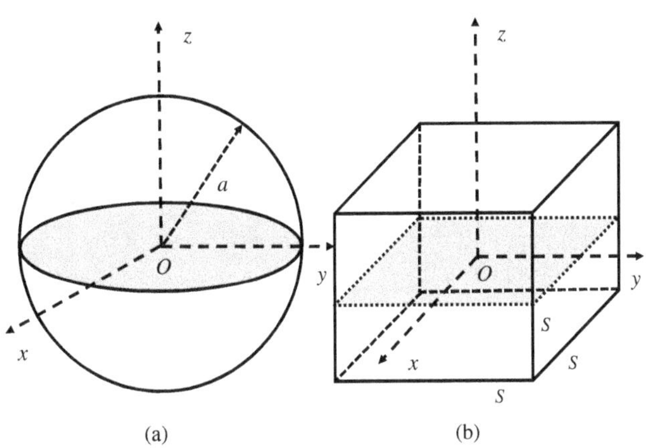

Figure 10.16 Geometries of dielectric scatterers centered at the origin of a rectangular coordinate system. (a) A dielectric sphere. (b) A dielectric cube.

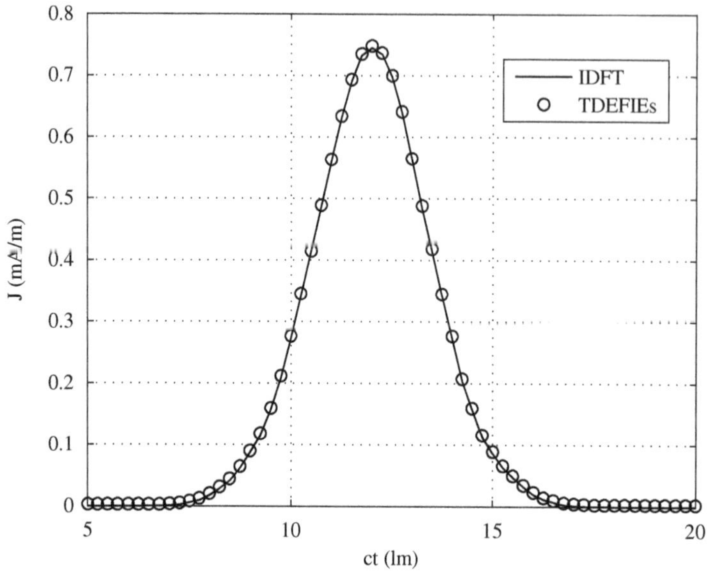

Figure 10.17 θ-directed electric current density at the centroid of a triangular patch close to $\theta = 90°$ and $\phi = 0°$ on the surface of a dielectric sphere.

where T is the width of Gaussian impulse, and t_0 is the time delay denoting the time when the pulse peaks at the origin. We choose a Gaussian pulse with $T = 8.0$ lm and $ct_0 = 12.0$ lm as an incident wave. Figures 10.17 and 10.18 plot the solutions of θ-directed electric current density and ϕ-directed magnetic current density, respectively, which are sampled at the centroid of a triangular patch close to $\theta = 90°$ and $\phi = 0°$ on the surface of the sphere, while Figure 10.19 depicts the solution of the θ-component of normalized far-zone scattered electric field observed along the backward direction for the sphere. Similarly, Figures 10.20 and 10.21 illustrate the solutions of z-directed electric current density and y-directed magnetic current density, respectively, which are sampled at the centroid of a triangular patch close to $x = 0.5s$, $y = 0$, and $z = 0$ on the surface of the cube, while Figure10.22 depicts the solution of the θ-component of normalized far-zone scattered electric field observed along the backward direction for the cube. To see if there is a late-time instability, we extend the time duration up to 30 lm for these two figures and the solutions are still stable. All the results are compared with those obtained from the IDFT of the corresponding frequency domain solutions and they are in good agreement.

We also examine the numerical errors and their convergence in terms of the refinement of spatial meshes and temporal basis functions. First, we choose $M = 80$ Laguerre basis functions in the time domain and discretize the surfaces of the two objects into different numbers of triangular meshes in the space domain. Figure 10.23 shows the root-mean-square (RMS) errors versus the number of meshes for the electric current density and normalized far-zone scattered electric field compared with the IDFT results and it is clear that the errors converges with respect to the refinement of meshes. Second, we fix the geometric discretization of the two objects into 1068 triangular meshes in

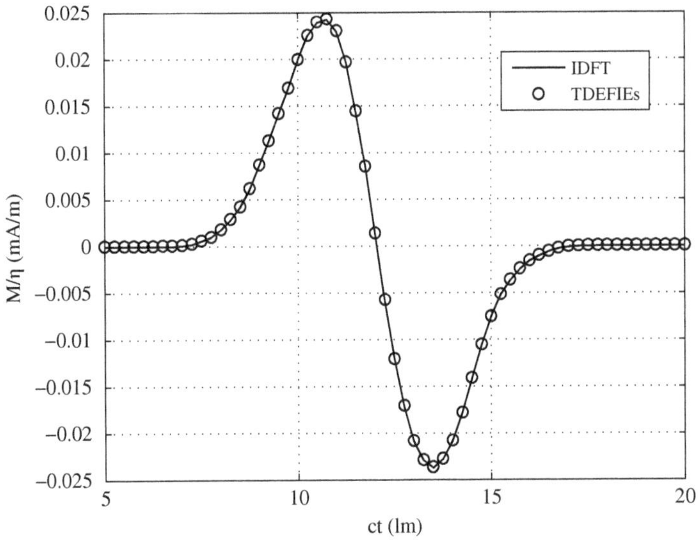

Figure 10.18 ϕ-directed magnetic current density at the centroid of a triangular patch close to $\theta = 90°$ and $\phi = 0°$ on the surface of a dielectric sphere.

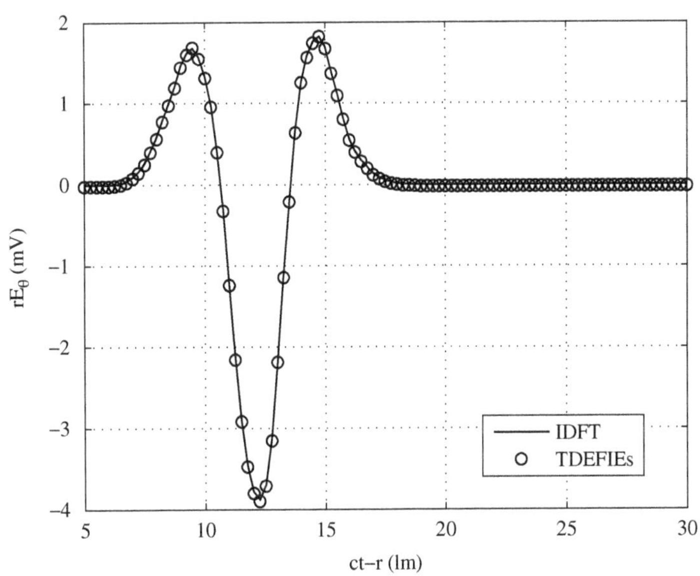

Figure 10.19 θ-component of normalized far-zone scattered electric field observed along the backward direction for a dielectric sphere.

space domain and select different numbers of Laguerre basis functions in time domain. Figure 10.24 sketches the RMS errors versus the number of Laguerre basis functions for the electric current density and normalized far-zone scattered electric field compared with the IDFT results and it is also apparent that the errors converges with respect to

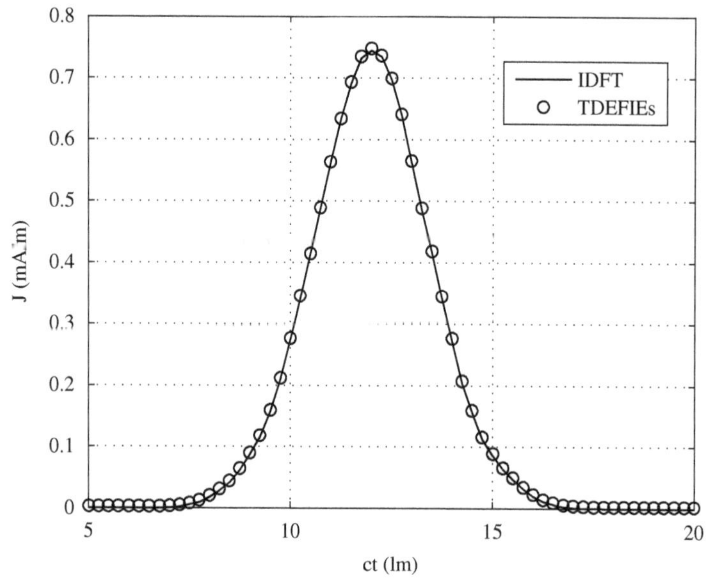

Figure 10.20 z-directed electric current density at the centroid of a triangular patch close to $x = 0.5s$, $y = 0$, and $z = 0$ on the surface of a dielectric cube.

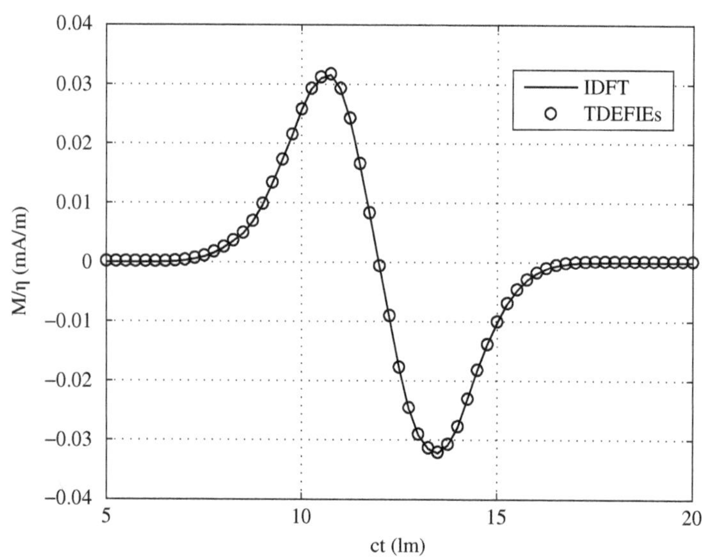

Figure 10.21 y-directed magnetic current density at the centroid of a triangular patch close to $x = 0.5s, y = 0$, and $z = 0$ on the surface of a dielectric cube.

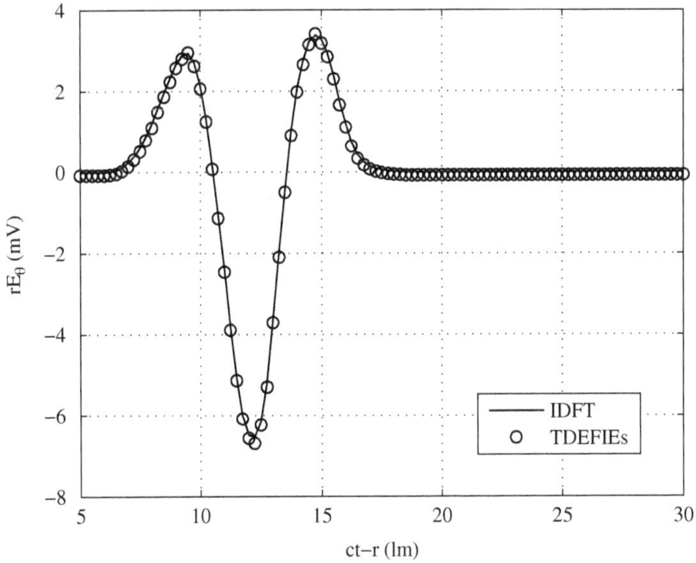

Figure 10.22 θ-component of normalized far-zone scattered electric field observed along the backward direction for a dielectric cube.

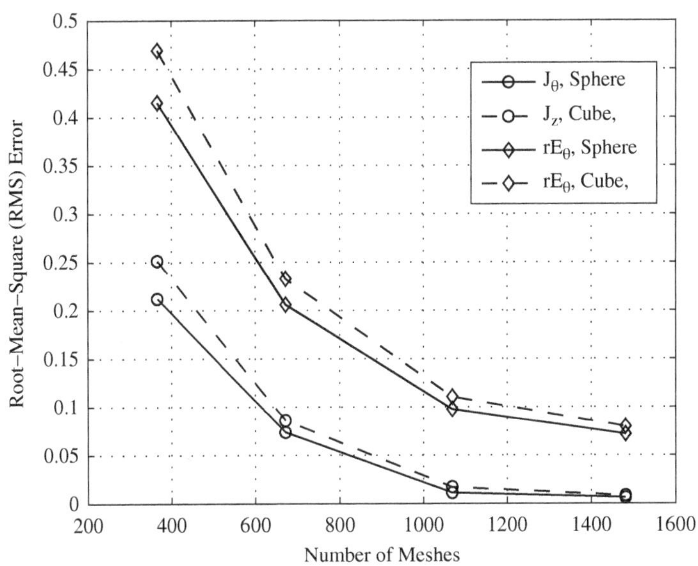

Figure 10.23 RMS errors versus number of meshes for electric current density and normalized far-zone scattered electric field compared with the IDFT results.

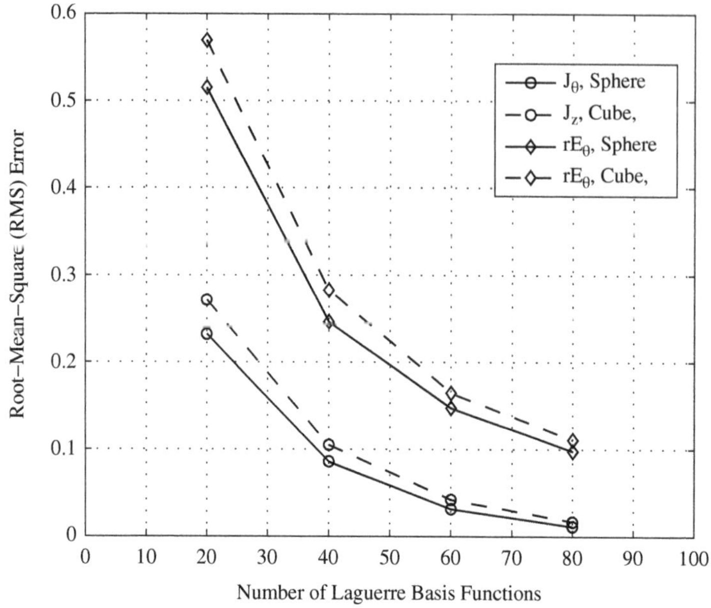

Figure 10.24 RMS errors versus number of Laguerre basis functions for electric current density and normalized far-zone scattered electric field compared with the IDFT results.

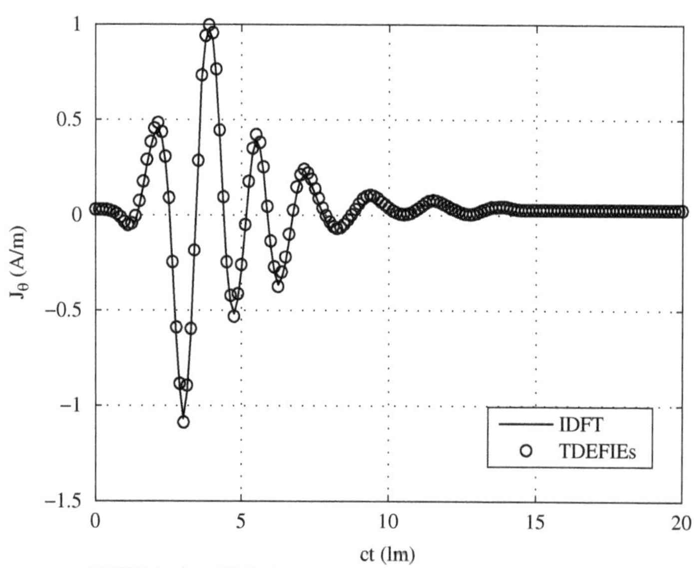

Figure 10.25 θ-directed electric current density at the centroid of a triangular patch close to $\theta = 90°$ and $\phi = 0°$ on the surface of a dielectric sphere for modulated Gaussian pulse scattering.

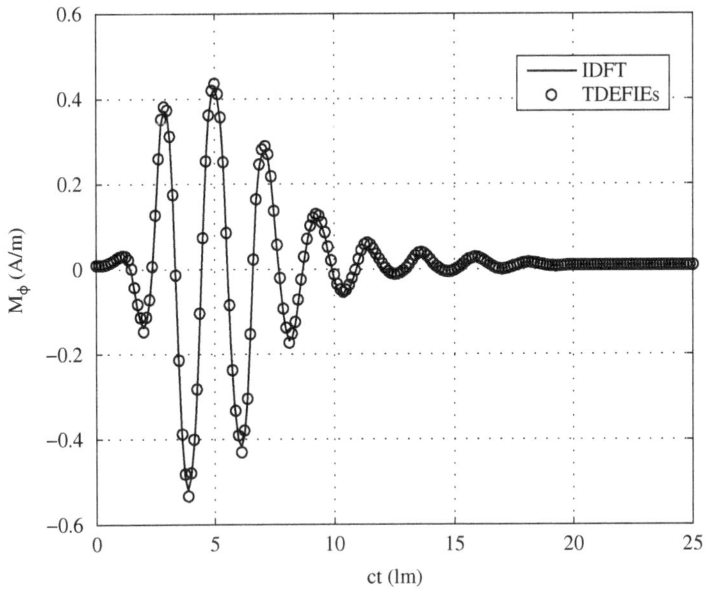

Figure 10.26 ϕ-directed magnetic current density at the centroid of a triangular patch close to $\theta = 90°$ and $\phi = 0°$ on the surface of a dielectric sphere for modulated Gaussian pulse scattering.

the refinement of basis functions. The above examples only consider the scattering of an unmodulated Gaussian pulse. We can of course use a modulated Gaussian pulse as an incident field to demonstrate the approach. Suppose the modulated Gaussian pulse is defined by

$$\mathbf{E}^{\text{inc}}(\mathbf{r}, t) = \hat{x}\cos(2\pi f_0 \tau)e^{-\left(\frac{\tau-\Delta t}{\sqrt{2}\sigma}\right)^2} \tag{10.63}$$

where f_0 is the center frequency, $\tau = t - \mathbf{r} \cdot \hat{k}/c$, $\sigma = 6/(2\pi f_b)$ is the pulse width (f_b is the bandwidth), and Δt is the temporal offset of the pulse. The pulse is propagating along the $\hat{k} = -\hat{z}$ direction and scattered by the first scatterer (dielectric sphere) whose relative permittivity is $\epsilon_r = 4.0$. We choose $f_0 = 150$ MHz, $f_b = 300$ MHz, and $\Delta t = 8\sigma$ in the numerical simulations. Figures 10.25 and 10.26 plot the solutions of θ-directed electric current density and ϕ-directed magnetic current density, respectively, sampled at the centroid of a triangular patch close to $\theta = 90°$ and $\phi = 0°$ on the surface of scatterer, and they are consistent with the corresponding IDFT solutions. We have intentionally extended the time duration for the figures and the solutions are also stable in the late-time period.

10.4 Time Domain Volume Integral Equations for Penetrable Media

As discussed in previous chapters, SIEs are preferred for penetrable media when they are homogeneous and isotropic. However, VIEs are indispensable for formulating inhomogeneous or anisotropic penetrable structures in an integral equation approach. Also, VIEs could be selected even for homogeneous and isotropic objects in some applications because the VIEs belong to the second kind of integral equations, which will result in a better conditioning in general [2]. In addition, for non-magnetic media that are very common, there is only one unknown current density to be solved in the VIEs while there are two unknown current densities to be solved in the SIEs. When a transient EM interaction is described with the VIEs, the time domain version of the VIEs (TDVIEs) should be applied.

In the traditional scheme to solve the TDVIEs, the MoM with divergence conforming Schaubert-Wilton-Glisson (SWG) basis function [56] or curl conforming edge basis function [8] is used to discretize the spatial domain while a MOT scheme is employed to discretize the temporal domain. The MoM requires conforming meshes in geometric discretization and also needs to perform two-fold integrations in evaluating matrix elements, resulting in certain inconvenience. Also, the basis function defined over a pair of tetrahedral elements is used to expand the unknown flux density and the homogeneity of material in each element has to be assumed [56]. This assumption may result in extra errors for inhomogeneous materials and produce fictitious surface charges at the common faces of paired tetrahedrons. On the other hand, in addition to the well-known instability problem of the MOT scheme due to its spurious oscillation [13], it will greatly complicate the correct evaluation of volume integrals in matrix entries when taking into account the intersections of the phase front spheres with the tetrahedral elements and surface integrals over the surfaces of tetrahedrons may be necessary if piecewise time domain basis functions are applied [35]. It is found that the stability of the MOT scheme is strongly related to the accurate evaluation of matrix elements [57–61], so the correct evaluation of volume integrals is very essential to stable solutions.

As mentioned before, the Nyström method has a very simple mechanism of implementation and its complexity is analogous to that of the FDTD method [33] in the DEMs while the MoM is more like the FEM [62] in the DEMs for its complexity because both of them strongly rely on the basis functions of representing unknown functions. The Nyström method discretizes integral equations by using the summation of a quadrature rule to directly replace a non-singular integral and this idea is quite analogous to the FDTD method, which uses a difference to directly replace a differential for discretizing a differential equation. Also, the use of the Nyström method, which is a point-matching scheme, allows the selection of volumetric current densities as unknowns to be solved, making integral kernels free of material parameters and facilitating an incorporation with fast algorithms [63].

10.4.1 Formulations

Consider the scattering of transient EM wave by a dielectric object with a volume V in free space with permittivity ϵ_0 and permeability μ_0. The problem can be formulated by the TDVIEs which are reduced to a single electric field TDVIE (EF-TDVIE) when the object is non-magnetic. The EF-TDVIE states that the total electric field is equal to the summation of incident electric field and scattered electric field, i.e.

$$\mathbf{E}(\mathbf{r},t) = \mathbf{E}^{\mathrm{inc}}(\mathbf{r},t) + \mathbf{E}^{\mathrm{sca}}(\mathbf{r},t) \tag{10.64}$$

where $\mathbf{E}(\mathbf{r},t)$ is the total electric field, $\mathbf{E}^{\mathrm{inc}}(\mathbf{r},t)$ is the incident electric field, and $\mathbf{E}^{\mathrm{sca}}(\mathbf{r},t)$ is the scattered electric field, which can be related to the magnetic vector potential $\mathbf{A}(\mathbf{r},t)$

and the electric scalar potential $\Phi(\mathbf{r}, t)$ by

$$\mathbf{E}^{\text{sca}}(\mathbf{r}, t) = -\frac{\partial}{\partial t}\mathbf{A}(\mathbf{r}, t) - \nabla\Phi(\mathbf{r}, t)$$
$$= -\frac{\mu_0}{4\pi}\int_V \frac{1}{R}\frac{\partial}{\partial t}\mathbf{J}(\mathbf{r}', \tau)\,\mathrm{d}V' - \frac{1}{4\pi\epsilon_0}\nabla\int_V \frac{1}{R}q(\mathbf{r}', \tau)\,\mathrm{d}V'. \tag{10.65}$$

In the above, $\mathbf{J}(\mathbf{r}, t)$ and $q(\mathbf{r}', t)$ are the volume current density and volume charge density in the object, respectively, $R = |\mathbf{r} - \mathbf{r}'|$ is the distance between an observation point $\mathbf{r}$ and a source point $\mathbf{r}'$, and $\tau = t - R/c$ is the retarded time with c being the speed of light in free space. By using the continuity equation

$$\nabla \cdot \mathbf{J}(\mathbf{r}, t) = -\frac{\partial}{\partial t}q(\mathbf{r}, t) \tag{10.66}$$

and introducing a new source vector $\mathbf{e}(\mathbf{r}, t)$, which is related to the current density by

$$\mathbf{J}(\mathbf{r}, t) = \frac{\partial}{\partial t}\mathbf{e}(\mathbf{r}, t) \tag{10.67}$$

we have

$$q(\mathbf{r}, t) = -\nabla \cdot \mathbf{e}(\mathbf{r}, t). \tag{10.68}$$

Also, since the volume current density is related to the total electric field by

$$\mathbf{J}(\mathbf{r}, t) = [\epsilon(\mathbf{r}) - \epsilon_0]\frac{\partial}{\partial t}\mathbf{E}(\mathbf{r}, t)$$
$$= \frac{\partial}{\partial t}\left\{[\epsilon(\mathbf{r}) - \epsilon_0]\mathbf{E}(\mathbf{r}, t)\right\} \tag{10.69}$$

we can obtain

$$\mathbf{e}(\mathbf{r}, t) = [\epsilon(\mathbf{r}) - \epsilon_0]\mathbf{E}(\mathbf{r}, t). \tag{10.70}$$

Based on the above changes, the EF-TDVIE can be written as

$$\frac{\mu_0}{4\pi}\int_V \frac{1}{R}\frac{\partial^2\mathbf{e}(\mathbf{r}', \tau)}{\partial t^2}\,\mathrm{d}V' + \frac{1}{4\pi\epsilon_0}\int_V \nabla\nabla\left(\frac{1}{R}\right)\cdot\mathbf{e}(\mathbf{r}', \tau)\,\mathrm{d}V'$$
$$= \frac{\mathbf{e}(\mathbf{r}, t)}{[\epsilon_0 - \epsilon(\mathbf{r})]} + \mathbf{E}^{\text{inc}}(\mathbf{r}, t), \quad \mathbf{r} \in V. \tag{10.71}$$

10.4.2 Numerical Solution

The dielectric object is discretized into N small tetrahedral elements and ΔV_n is the volume of the nth tetrahedron. The unknown source vector $\mathbf{e}(\mathbf{r}', t)$ can then be expanded as

$$\mathbf{e}(\mathbf{r}', t) = \sum_{n=1}^{N}\xi_n(t)\mathbf{f}_n(\mathbf{r}'). \tag{10.72}$$

If we apply the Nyström method in the spatial domain, i.e. the integration over a small tetrahedron is replaced with a summation under a quadrature rule provided that the integrand is regular, then we have

$$\frac{\mathbf{e}(\mathbf{r}, t)}{[\epsilon_0 - \epsilon(\mathbf{r})]} + \mathbf{E}^{\text{inc}}(\mathbf{r}, t) = \frac{\mu_0}{4\pi}\sum_{n=1}^{N}\sum_{q=1}^{Q}\frac{w_{nq}}{R}\frac{\partial^2\xi_n(\tau)}{\partial t^2}\mathbf{f}_n(\mathbf{r}'_{nq})$$
$$+ \frac{1}{4\pi\epsilon_0}\sum_{n=1}^{N}\sum_{q=1}^{Q}w_{nq}\nabla\nabla\left(\frac{1}{R}\right)\cdot\mathbf{f}_n(\mathbf{r}'_{nq})\xi_n(\tau) \tag{10.73}$$

where $\mathbf{f}_n(\mathbf{r}'_{nq}) = \mathbf{f}_{nq}$ is the space part value of an unknown source vector at the qth quadrature point in the nth tetrahedron and w_{nq} is the weight at that point under a quadrature rule. If we choose the pth quadrature point in the mth tetrahedron as an observation point to perform a collocation procedure, then we have the following space domain matrix equation

$$\frac{\mathbf{f}_{mp}\xi_m(t)}{\left[\epsilon(\mathbf{r}_{mp}) - \epsilon_0\right]} + \sum_{n=1}^{N}\sum_{q=1}^{Q}\left[\frac{\partial^2 \xi_n(\tau)}{\partial t^2}\mathbf{a}_{mpnq} + \xi_n(\tau)\mathbf{b}_{mpnq}\right] = \mathbf{E}^{\text{inc}}(\mathbf{r}_{mp}, t),$$
$$m = 1, \cdots, N; \quad p = 1, \cdots, Q \tag{10.74}$$

where

$$\mathbf{a}_{mpnq} = \frac{w_{nq}\mu_0}{4\pi R_{mpnq}}\mathbf{f}_{nq} \tag{10.75}$$

$$\mathbf{b}_{mpnq} = \frac{w_{nq}}{4\pi\epsilon_0}\left[\nabla\nabla\left(\frac{1}{R}\right)\right]_{mpnq} \cdot \mathbf{f}_{nq} \tag{10.76}$$

and $\mathbf{f}_{mp} = \mathbf{f}_m(\mathbf{r}_{mp})$ is the space part value of an unknown source vector at the observation point or the pth quadrature point in the mth tetrahedron. The above procedure could have a singularity problem that occurs when $m = n$ but we have developed a robust treatment technique [2].

In the temporal domain, we use the Laguerre function

$$\phi_j(t) = \mathrm{e}^{-t/2}L_j(t) \tag{10.77}$$

as a basis function to expand $\xi_n(t)$ [7], where $L_j(t)$ is the Laguerre polynomial of jth degree. Multiplying the time variable by a scaling factor s, we can control the support of expansion and $\xi_n(t)$ can be expanded as

$$\xi_n(t) = \sum_{j=0}^{\infty} \xi_{nj}\phi_j(st). \tag{10.78}$$

Substituting this expansion to Equation (10.74), we have

$$\sum_{n=1}^{N}\sum_{q=1}^{Q}\sum_{j=0}^{\infty}\left\{s^2\left[0.25\xi_{nj} + \sum_{k=0}^{j-1}(j-k)\xi_{nk}\right]\mathbf{a}_{mpnq} + \xi_{nj}\mathbf{b}_{mpnq}\right\}\phi_j(st - \Delta t_{mpnq})$$
$$= \frac{\mathbf{f}_{mp}}{\left[\epsilon_0 - \epsilon(\mathbf{r}_{mp})\right]}\sum_{j=0}^{\infty}\xi_{mj}\phi_j(st) + \mathbf{E}^{\text{inc}}(\mathbf{r}_{mp}, t) \tag{10.79}$$

where $\Delta t_{mpnq} = sR_{mpnq}/c$. Furthermore, using Galerkin's testing procedure, i.e. using $\phi_i(st)$ as a testing function to test the equation, we obtain

$$\sum_{n=1}^{N}\sum_{q=1}^{Q}\left(0.25s^2\mathbf{a}_{mpnq} + \mathbf{b}_{mpnq}\right)\xi_{ni}I_{ii}(\Delta t_{mpnq}) +$$
$$\sum_{n=1}^{N}\sum_{q=1}^{Q}\sum_{j=0}^{i-1}\left(0.25s^2\mathbf{a}_{mpnq} + \mathbf{b}_{mpnq}\right)\xi_{nj}I_{ij}(\Delta t_{mpnq}) +$$

$$\sum_{n=1}^{N}\sum_{q=1}^{Q}\sum_{j=0}^{i}\sum_{k=0}^{j-1}(j-k)s^2\mathbf{a}_{mpnq}\xi_{nk}I_{ij}(\Delta t_{mpnq}) = \frac{\mathbf{f}_{mp}\xi_{mi}}{\left[\epsilon_0-\epsilon(\mathbf{r}_{mp})\right]} + \mathbf{V}_{mpi} \tag{10.80}$$

where

$$I_{ij}(\Delta t_{mpnq}) = \int_0^\infty \phi_i(st)\phi_j(st-\Delta t_{mpnq})\,\mathrm{d}(st) \tag{10.81}$$

can be calculated according to the appendix in [7] and

$$\mathbf{V}_{mpi} = \int_0^\infty \phi_i(st)\mathbf{E}^{\mathrm{inc}}(\mathbf{r}_{mp},t)\,\mathrm{d}(st) \tag{10.82}$$

is calculated numerically by replacing the upper limit with sT_d, where T_d is the real duration of a signal. Also, we have used the following result

$$\begin{aligned}\sum_{j=0}^{\infty}\xi_{mj}\int_0^\infty \phi_i(st)\phi_j(st)\,\mathrm{d}(st) &= \sum_{j=0}^{i}\xi_{mj}\int_0^\infty \mathrm{e}^{-st}L_i(st)L_j(st)\,\mathrm{d}(st)\\ &= \sum_{j=0}^{i}\xi_{mj}\delta_{ij} = \xi_{mi}\end{aligned} \tag{10.83}$$

in the above testing procedure. Thus, Equation (10.80) can be written as

$$\begin{aligned}&\frac{\mathbf{f}_{mp}\xi_{mi}}{\left[\epsilon_0-\epsilon(\mathbf{r}_{mp})\right]} - \sum_{n=1}^{N}\sum_{q=1}^{Q}\left(0.25s^2\mathbf{a}_{mpnq}+\mathbf{b}_{mpnq}\right)\xi_{ni}I_{ii}(\Delta t_{mpnq})\\ &= -\mathbf{V}_{mpi} + \sum_{n=1}^{N}\sum_{q=1}^{Q}\sum_{j=0}^{i-1}\left(0.25s^2\mathbf{a}_{mpnq}+\mathbf{b}_{mpnq}\right)\xi_{nj}I_{ij}(\Delta t_{mpnq})\\ &\quad+\sum_{n=1}^{N}\sum_{q=1}^{Q}\sum_{j=0}^{i}\sum_{k=0}^{j-1}(j-k)s^2\mathbf{a}_{mpnq}\xi_{nk}I_{ij}(\Delta t_{mpnq})\end{aligned} \tag{10.84}$$

i.e. the present expansion coefficient ξ_{ni} is determined based on all the previous expansion coefficients ξ_{nj} $(j=0,1,\cdots,i-1)$. Note that the space domain coefficients in $\mathbf{a}_{mpnq}$ and $\mathbf{b}_{mpnq}$, i.e. $\mathbf{f}_{nq}$, should be combined with the time domain expansion coefficients ξ_{ni} together to form new coefficients $\mathbf{f}_{nqi}=\mathbf{f}_{nq}\xi_{ni}$. With this combination, the above equation can be written as a matrix form

$$\begin{aligned}&\alpha_{mp}\mathbf{f}_{mpi} - \sum_{n=1}^{N}\sum_{q=1}^{Q}\overline{\mathbf{Z}}_{mpnq}\cdot\mathbf{f}_{nqi} = -\mathbf{V}_{mpi}+\mathbf{P}_{mpi}\\ &m=1,\cdots,N;\;\; p=1,\cdots,Q\end{aligned} \tag{10.85}$$

where

$$\alpha_{mp} = \frac{1}{\left[\epsilon_0-\epsilon(\mathbf{r}_{mp})\right]} \tag{10.86}$$

$$\overline{\mathbf{Z}}_{mpnq} = \left\{\frac{s^2\mu_0}{4R_{mpnq}}\overline{\mathbf{I}} + \frac{1}{\epsilon_0}\left[\nabla\nabla\left(\frac{1}{R}\right)\right]_{mpnq}\right\}\frac{w_{nq}}{4\pi}I_{ii}(\Delta t_{mpnq}) \tag{10.87}$$

$$\mathbf{P}_{mpi} = \sum_{n=1}^{N}\sum_{q=1}^{Q}\sum_{j=0}^{i-1}\left(0.25s^2\mathbf{a}_{mpnq} + \mathbf{b}_{mpnq}\right)\xi_{nj}I_{ij}(\Delta t_{mpnq})$$
$$+\sum_{n=1}^{N}\sum_{q=1}^{Q}\sum_{j=0}^{i}\sum_{k=0}^{j-1}(j-k)s^2\mathbf{a}_{mpnq}\xi_{nk}I_{ij}(\Delta t_{mpnq}). \quad (10.88)$$

In the above, the matrix element $\overline{\mathbf{Z}}_{mpnq}$ is a dyad that includes an identity dyad $\overline{\mathbf{I}}$ and can be determined by the space domain variables. The matrix equation is written into a scalar or component form and then solved with an MOD or recursive manner ($i = 0, 1, 2, \ldots, \infty$). Once the combined coefficients have been solved, the current density at the qth quadrature point in the nth tetrahedron at the time t can be calculated by

$$\begin{aligned}
\mathbf{J}(\mathbf{r}'_{nq}, t) &= \frac{\partial}{\partial t}\mathbf{e}(\mathbf{r}'_{nq}, t) \\
&= \mathbf{f}_{nq}\frac{\partial}{\partial t}\xi_n(t) \\
&= \mathbf{f}_{nq}s\sum_{j=0}^{M-1}\left(0.5\xi_{nj} + \sum_{k=0}^{j-1}\xi_{nk}\right)\phi_j(st) \\
&= s\sum_{j=0}^{M-1}\left(0.5\mathbf{f}_{nqj} + \sum_{k=0}^{j-1}\mathbf{f}_{nqk}\right)\phi_j(st)
\end{aligned} \quad (10.89)$$

where M is the number of Laguerre basis functions in the temporal domain. On the other hand, the scattered electric field $\mathbf{E}(\mathbf{r}_0, t)$ which is at a point $\mathbf{r}_0$ in the space and at the time t can be computed by

$$\begin{aligned}
\mathbf{E}(\mathbf{r}_0, t) = \sum_{n=1}^{N}\sum_{q=1}^{Q}\sum_{j=0}^{M-1}&\left\{s^2\left[0.25\mathbf{f}_{nqj} + \sum_{k=0}^{j-1}(j-k)\mathbf{f}_{nqk}\right]u_{r_0nq}\right. \\
&\left. + \overline{\mathbf{v}}_{r_0nq}\cdot\mathbf{f}_{nqj}\right\}\phi_j(st - \Delta t_{r_0nq})
\end{aligned} \quad (10.90)$$

where

$$u_{r_0nq} = \frac{w_{nq}\mu_0}{4\pi R_{r_0nq}} \quad (10.91)$$

$$\overline{\mathbf{v}}_{r_0nq} = \frac{w_{nq}}{4\pi\epsilon_0}\left[\nabla\nabla\left(\frac{1}{R}\right)\right]_{r_0nq} \quad (10.92)$$

and R_{r_0nq} is the distance between the point $\mathbf{r}_0$ in the space and the qth quadrature point in the nth tetrahedron, and $\Delta t_{r_0nq} = sR_{r_0nq}/c$.

10.4.3 Numerical Examples

We present transient EM scattering by typical dielectric objects to demonstrate the approach. The geometries of dielectric objects are shown in Figure 10.27 and they are a spherical, cubic, cylindrical, and double-cone object, respectively. The spherical

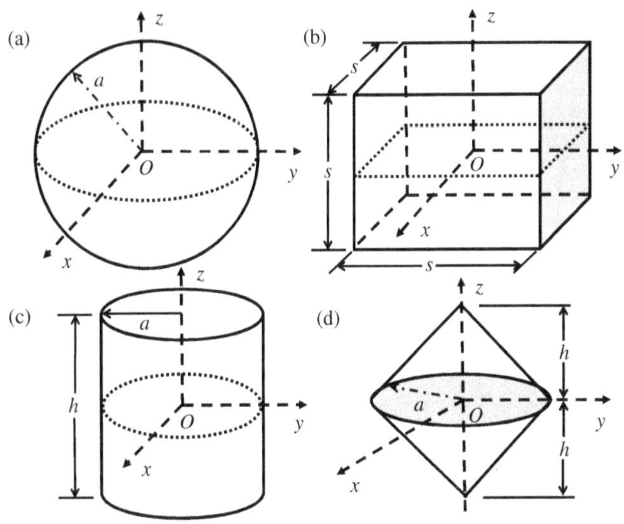

Figure 10.27 Geometries of dielectric scatterers centered at the origin of a rectangular coordinate system. (a) A dielectric sphere. (b) A dielectric cube. (c) A dielectric cylinder. (d) A dielectric double-cone object.

object has a radius $a = 0.5$ m and the cubic object has a side length $s = 1.0$ m. Also, the cylindrical object has a radius $a = 0.5$ m at its cross section and a height $h = 1.0$ m while the double-cone object has a maximum radius $a = 0.5$ m at its cross section and a height $h = 1.0$ m for each cone. The relative permittivity of these objects is $\epsilon_r = 2.0$ and their relative permeability $\mu_r = 1.0$ is assumed. The objects are centered at the origin of a rectangular coordinate system $\{x, y, z\}$ and we discretize them into 726, 942, 916, and 1478 tetrahedrons, respectively. A Gaussian plane wave is illuminating these scatterers along the $-z$ direction and the electric field is defined by

$$\mathbf{E}^{\text{inc}}(\mathbf{r}, t) = \hat{x}\frac{4\mathrm{e}^{-\gamma^2}}{\sqrt{\pi}T}, \quad \gamma = \frac{4}{T}(ct - ct_0 + \mathbf{r}\cdot\hat{z}) \tag{10.93}$$

where T is the width of Gaussian impulse, and t_0 is the time delay denoting the time when the pulse peaks at the origin. We choose a Gaussian pulse with $T = 4.0$ lm and $ct_0 = 6.0$ lm as an incident wave. Also, we choose $s = 10^9$, $M = 80$, and $sT_\text{d} = 40$ lm in the calculations. Figures 10.28–10.30 plots the solutions of the θ-component of normalized far-zone scattered electric field observed along the backward direction for the first three scatterers. In Figure 10.31, we depict the x-component of near-zone scattered electric field observed at $x = y = 0$ and $z = 1.0$ m for the double-cone scatterer. All the results are in good agreement with those obtained from the IDFT of corresponding frequency domain solutions. Note that the sharp corners of the double-cone object are usually thought of as challenging in numerical modeling and simulations, thus the object can serve as a good tester for numerical solvers, although its geometry is not very complicated.

Although the problem with high relative permittivity is less addressed for time domain integral equations, we choose a relatively high relative permittivity $\epsilon_r = 4.5$ for the spherical object to further demonstrate the approach. The incident wave is the same as before

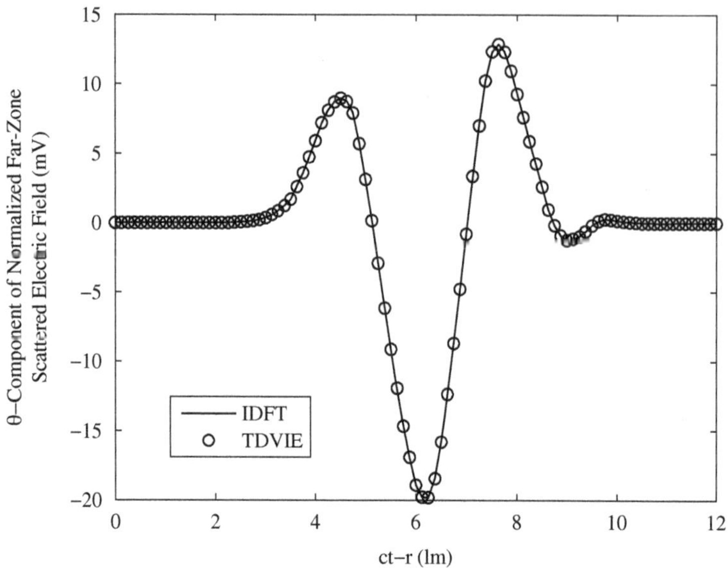

Figure 10.28 θ-component of normalized far-zone scattered electric field observed along the backward direction for a dielectric sphere. The sphere has a radius $a = 0.5$ m and a relative permittivity $\epsilon_r = 2.0$.

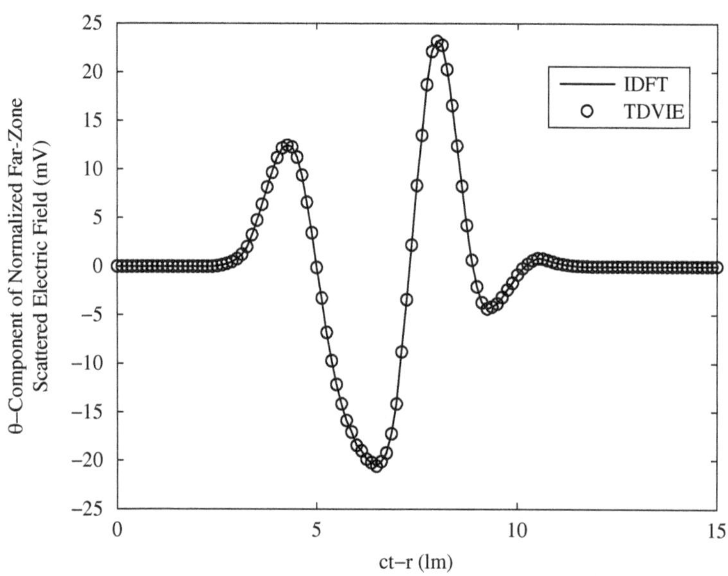

Figure 10.29 θ-component of normalized far-zone scattered electric field observed along the backward direction for a dielectric cube. The cube has a side length $s = 1.0$ m and a relative permittivity $\epsilon_r = 2.0$.

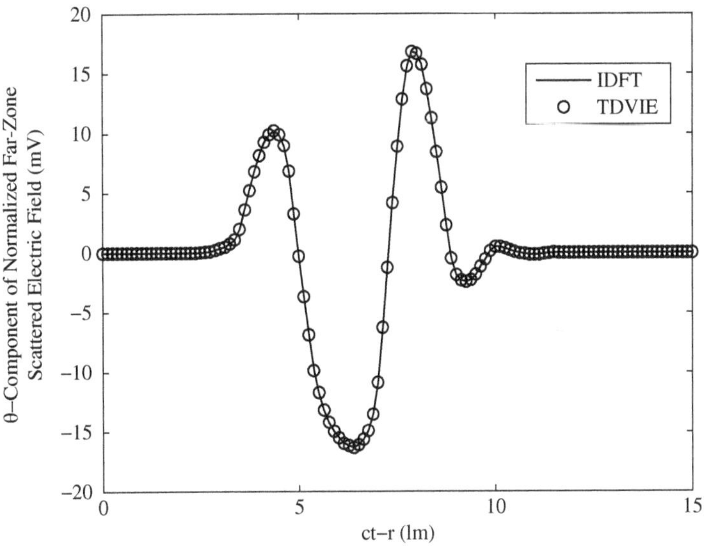

Figure 10.30 θ-component of normalized far-zone scattered electric field observed along the backward direction for a dielectric cylinder. The cylinder has a radius $a = 0.5$ m, a height $h = 1.0$ m, and a relative permittivity $\epsilon_r = 2.0$.

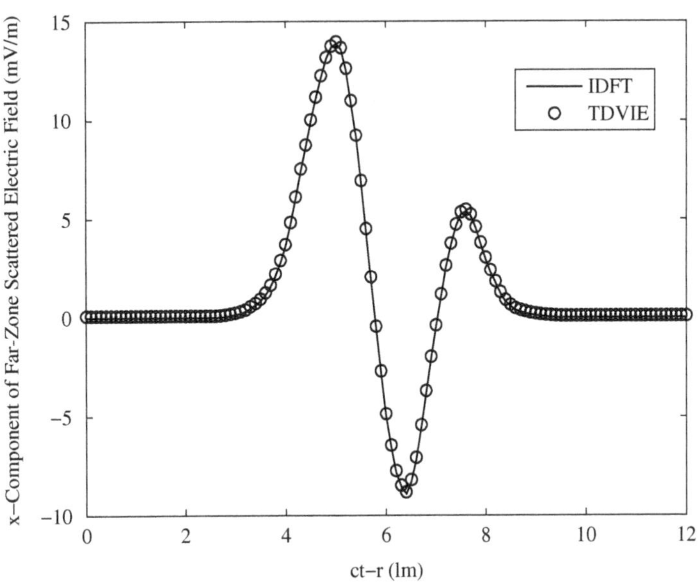

Figure 10.31 x-component of near-zone scattered electric field observed at $x = y = 0$ and $z = 1.0$ m for a dielectric double-cone object. The object has a radius $a = 0.5$ m, a height $h = 0.5$ m, and a relative permittivity $\epsilon_r = 2.0$.

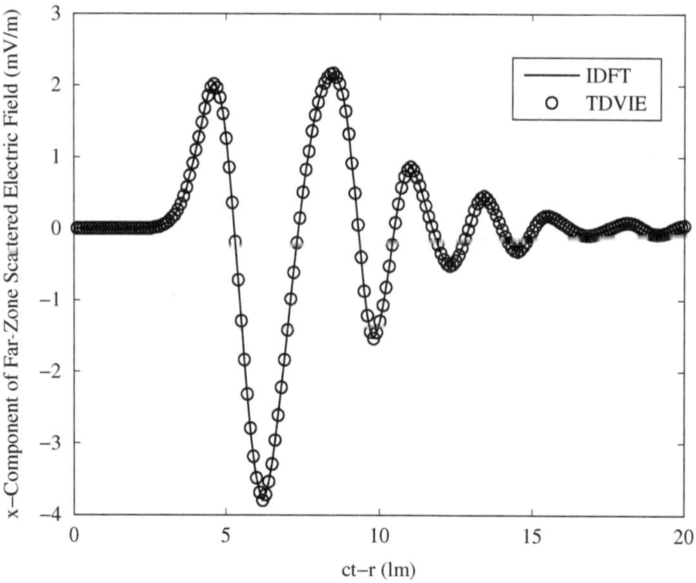

Figure 10.32 x-component of far-zone scattered electric field at $x = y = 0$ and $z = 10.0$ m for a dielectric sphere. The sphere has a radius $a = 0.5$ m and a relative permittivity $\epsilon_r = 4.5$.

and Figure 10.32 plots the x-component of scattered electric field observed at $x = y = 0$ and $z = 10.0$ m for the scatterer. Moreover, we change the radius into $a = 1.0$ m and the relative permittivity into $\epsilon_r = 5.0$ for the spherical object, and then calculate its bistatic RCS so that the solution of scattered field in more directions can be seen. The incident wave is a Gaussian pulse defined as

$$\mathbf{E}^{\text{inc}}(\mathbf{r}, t) = \hat{x}\cos(2\pi f_0 \tau)e^{-\gamma^2} \tag{10.94}$$

$$\gamma = \frac{\tau - t_0}{2\sigma^2}, \quad \tau = t - \mathbf{r} \cdot \hat{k}/c, \quad t_0 = 8\sigma, \quad \sigma = \frac{6}{2\pi f_{bw}} \tag{10.95}$$

where $\hat{k} = \hat{z}$ is the direction of propagation, f_0 is the central frequency, and f_{bw} is the bandwidth. Figure 10.33 shows the RCS solution at $f = 30$ MHz when we choose $f_0 = 20$ MHz and $f_{bw} = 10$ MHz and it agrees well with the corresponding Mie-series solution. It is well known that the Nyström method can easily have a higher-order convergence of solutions and we demonstrate this feature by using this example because the Mie-series solutions can be used as exact solutions to calculate numerical errors. We discretize the sphere into different densities of tetrahedral meshes and apply the Nyström method with a one-point quadrature rule and four-point quadrature rule, respectively, in the space domain and calculate the relative RMS errors for the RCS solutions by comparing them to the exact solutions. Figure 10.34 illustrates the convergence rates of solutions under the two quadrature rules and it is clear that the higher-order quadrature rule gives a higher rate of convergence.

For the computational costs, it has been found that the MOD scheme may not be more efficient than the traditional MOT scheme [28]. It is hard to make a fair comparison between these two methods because the MOT scheme will usually result in a late-time instability or inaccurate solution. However, we use the fourth example (double-cone

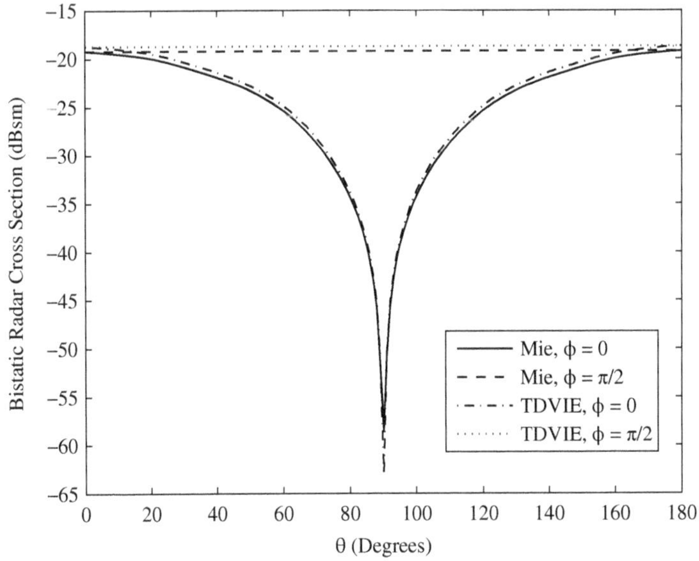

Figure 10.33 Bistatic RCS solutions at the frequency $f = 30$ MHz for a dielectric sphere. The sphere has a radius $a = 1.0$ m and a relative permittivity $\epsilon_r = 5.0$.

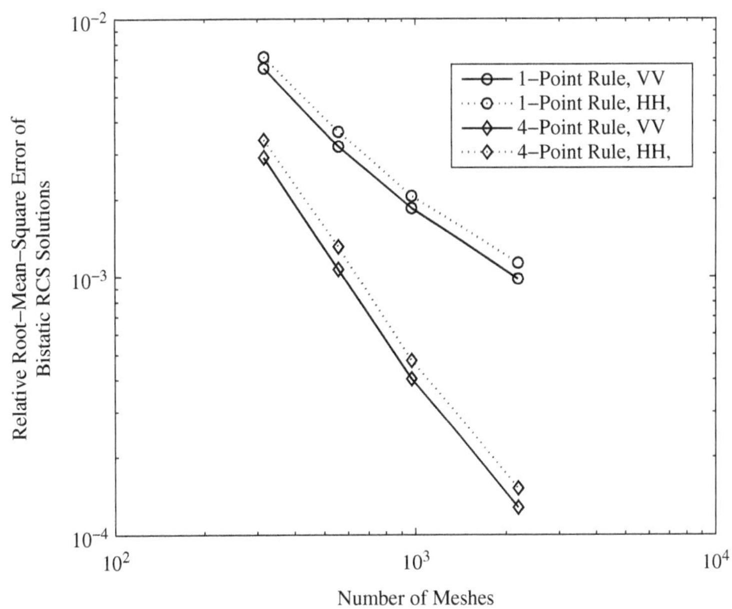

Figure 10.34 Convergence rates of bistatic RCS solutions at the frequency $f = 30$ MHz for a dielectric sphere. The sphere has a radius $a = 1.0$ m and a relative permittivity $\epsilon_r = 5.0$.

object) to make a relatively fair comparison between the proposed approach and the method used in [26], i.e. the traditional MoM based on the SWG basis function in the space domain plus the MOD in the time domain. It is found that the proposed approach consumes 52.3 seconds in CPU time and 738 MB in memory while the method in [26] consumes 57.6 seconds in CPU time and 722 MB in memory. The algorithms are executed on a T7500 Dell Precision workstation with 3.06-GHz CPU and 64-GB RAM under the same or similar conditions. The comparison indicates that the two methods require very similar computational costs because the Nyström method usually has a similar number of unknowns to the traditional MoM in the space domain when same geometric (volumetric) discretization is used.

10.5 Time Domain Combined Field Integral Equations for Mixed Media

Real world objects are seldom purely conducting or penetrable and the co-existence of different materials is more popular. Transient EM interaction with such objects can be found in many practical applications, so it is very desirable to investigate a way of accurately and efficiently modeling and simulating such a phenomenon. In the integral equation approach, we use the TDSIEs to formulate the problem when the involved penetrable materials are homogeneous and isotropic because they require a fewer number of unknowns compared to TDVIEs for penetrable materials and are also more flexible in geometric discretization [2]. Although there are several choices for the TDSIEs, we select the TDCFIEs since they can fully eliminate the interior resonance of frequency domain, which has an impact on the time domain stability, and also they are rarely addressed in time domain [31].

The use of the Nyström method in the space domain can avoid the embarrassment of selecting two basis functions as required in the MoM. When the RWG basis function is used to represent the electric current density, the $\hat{n}\times$RWG basis function, where $\hat{n}$ is a unit normal vector on the material interface, may not well represent the magnetic current density because it is not divergence conforming or curl conforming and will produce a fictitious charge deposit on mesh edges when a divergence is taken in the TDCFIEs [52]. Also, using non-conforming meshes in the Nyström method can dramatically lower the requirement on mesh quality and this is desirable in particular for composite structures since it is hard to have conforming meshes at the interfaces of two kinds of very different materials [52]. In addition, the late-time instability is more serious in the TDSIEs with penetrable materials, especially with mixed media [6], so the unconditionally stable scheme like the time domain GM with Laguerre basis and testing functions is particularly valuable.

10.5.1 Formulations

Consider the transient EM scattering by a composite structure including both conducting and dielectric materials, as shown in Figure 10.35. The structure is embedded in free space and the conducting material is assumed to be a PEC while the dielectric material is assumed to be homogeneous. There are three regions with different media, i.e. region 0, which is the free space with a permittivity ϵ_0 and μ_0, region 1, which is the conducting

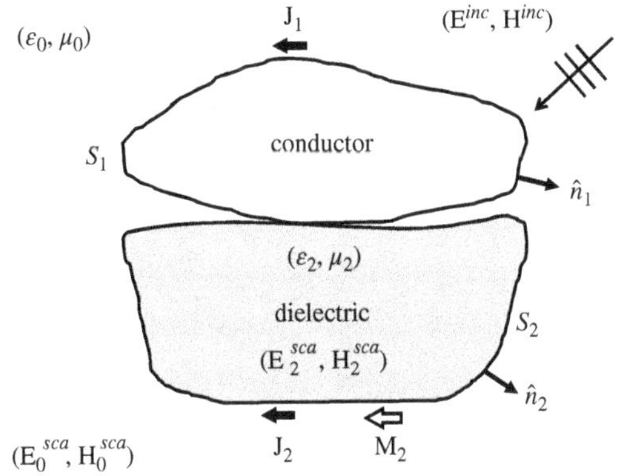

Figure 10.35 Illustration of transient EM scattering by a composite scatterer including both conducting and dielectric materials.

material, and region 2, which is the dielectric material with a permittivity ϵ_2 and μ_2. The conducting part has a surface or boundary S_1 with an outward unit normal vector $\hat{n}_1$ while the dielectric part has a surface S_2 with an outward unit normal vector $\hat{n}_2$. A transient EM wave with electric field $\mathbf{E}^{\rm inc}$ and magnetic field $\mathbf{H}^{\rm inc}$ is incident upon the structure and an electric current density $\mathbf{J}_1$ is induced on S_1. Also, there are equivalent electric current density $\mathbf{J}_2$ and equivalent magnetic current density $\mathbf{M}_2$ on S_2 to account for the scattered fields from the dielectric part based on the equivalence principle. The problem can be formulated by TDEFIEs

$$\left[\mathbf{E}_0^{\rm sca}(\mathbf{J}_1, 0) + \mathbf{E}_0^{\rm sca}(\mathbf{J}_2, \mathbf{M}_2) + \mathbf{E}^{\rm inc}\right]_{\rm tan} = 0, \quad \mathbf{r} \in S_1 \tag{10.96}$$

$$\left[\mathbf{E}_0^{\rm sca}(\mathbf{J}_1, 0) + \mathbf{E}_0^{\rm sca}(\mathbf{J}_2, \mathbf{M}_2) + \mathbf{E}^{\rm inc}\right]_{\rm tan} = 0, \quad \mathbf{r} \in S_2^- \tag{10.97}$$

$$\left[\mathbf{E}_2^{\rm sca}(-\mathbf{J}_2, -\mathbf{M}_2)\right]_{\rm tan} = 0, \quad \mathbf{r} \in S_2^+ \tag{10.98}$$

or by the TDMFIEs

$$\left[\mathbf{H}_0^{\rm sca}(\mathbf{J}_1, 0) + \mathbf{H}_0^{\rm sca}(\mathbf{J}_2, \mathbf{M}_2) + \mathbf{H}^{\rm inc}\right]_{\rm tan} = \mathbf{J}_1, \quad \mathbf{r} \in S_1 \tag{10.99}$$

$$\left[\mathbf{H}_0^{\rm sca}(\mathbf{J}_1, 0) + \mathbf{H}_0^{\rm sca}(\mathbf{J}_2, \mathbf{M}_2) + \mathbf{H}^{\rm inc}\right]_{\rm tan} = 0, \quad \mathbf{r} \in S_2^- \tag{10.100}$$

$$\left[\mathbf{H}_2^{\rm sca}(-\mathbf{J}_2, -\mathbf{M}_2)\right]_{\rm tan} = 0, \quad \mathbf{r} \in S_2^+ \tag{10.101}$$

where $\mathbf{E}^{\rm sca}$ and $\mathbf{H}^{\rm sca}$ are the scattered electric field and scattered magnetic field, respectively, in the corresponding region specified by their subscripts. Also, "tan" represents a tangential component and "+" and "−" on S_2 denote its interior side and exterior side, respectively. The scattered electric field and magnetic field can be related to the vector potentials and scalar potentials by

$$\mathbf{E}_j^{\rm sca}\left[\mathbf{J}_\ell(\mathbf{r}, t)\right] = -\frac{\partial}{\partial t}\mathbf{A}_j(\mathbf{r}, t) - \nabla\Phi_j(\mathbf{r}, t) \tag{10.102}$$

$$\mathbf{E}_j^{\rm sca}\left[\mathbf{M}_\ell(\mathbf{r}, t)\right] = -\frac{1}{\epsilon_j}\nabla \times \mathbf{F}_j(\mathbf{r}, t) \tag{10.103}$$

$$\mathbf{H}_j^{\rm sca}\left[\mathbf{J}_\ell(\mathbf{r}, t)\right] = \frac{1}{\mu_j}\nabla \times \mathbf{A}_j(\mathbf{r}, t) \tag{10.104}$$

$$\mathbf{H}_j^{\rm sca}\left[\mathbf{M}_\ell(\mathbf{r}, t)\right] = -\frac{\partial}{\partial t}\mathbf{F}_j(\mathbf{r}, t) - \nabla\Psi_j(\mathbf{r}, t) \tag{10.105}$$

where $\mathbf{A}_j(\mathbf{r}, t)$ and $\mathbf{F}_j(\mathbf{r}, t)$ are the magnetic and electric vector potentials, respectively, while $\Phi_j(\mathbf{r}, t)$ and $\Psi_j(\mathbf{r}, t)$ are the electric and magnetic scalar potentials, respectively. The subscript $j = 0, 2$ indicates the corresponding region where the potentials and fields are defined and the subscript $\ell = 1, 2$ represents the corresponding surface where the current densities reside. The potentials are related to the current densities by

$$\mathbf{A}_j(\mathbf{r}, t) = \frac{\mu_j}{4\pi} \int_{S_\ell} \frac{\mathbf{J}_\ell(\mathbf{r}', \tau_j)}{R} \, \mathrm{d}S' \tag{10.106}$$

$$\mathbf{F}_j(\mathbf{r}, t) = \frac{\epsilon_j}{4\pi} \int_{S_\ell} \frac{\mathbf{M}_\ell(\mathbf{r}', \tau_j)}{R} \, \mathrm{d}S' \tag{10.107}$$

$$\Phi_j(\mathbf{r}, t) = \frac{1}{4\pi\epsilon_j} \int_{S_\ell} \frac{q_\ell^e(\mathbf{r}', \tau_j)}{R} \, \mathrm{d}S' \tag{10.108}$$

$$\Psi_j(\mathbf{r}, t) = \frac{1}{4\pi\mu_j} \int_{S_\ell} \frac{q_\ell^m(\mathbf{r}', \tau_j)}{R} \, \mathrm{d}S' \tag{10.109}$$

where $R = |\mathbf{r} - \mathbf{r}'|$ is the distance between an observation point $\mathbf{r}$ and a source point $\mathbf{r}'$ and $\tau_j = t - R/c_j$ with $c_j = 1/\sqrt{\epsilon_j\mu_j}$ being the velocity of wave propagation in the medium of region j. Also, q_ℓ^e and q_ℓ^m are the electric and magnetic charge densities at the ℓth surface (but $q_1^m = 0$ since there is no magnetic current density at S_1) and they are related to the current densities by the continuity equation

$$\nabla' \cdot \mathbf{J}_\ell(\mathbf{r}', t) = -\frac{\partial}{\partial t} q_\ell^e(\mathbf{r}', t), \quad \nabla' \cdot \mathbf{M}_\ell(\mathbf{r}', t) = -\frac{\partial}{\partial t} q_\ell^m(\mathbf{r}', t). \tag{10.110}$$

We introduce new source vectors $\mathbf{e}_\ell(\mathbf{r}', t)$ and $\mathbf{h}_\ell(\mathbf{r}', t)$, which are related to $\mathbf{J}_\ell(\mathbf{r}', t)$ and $\mathbf{M}_\ell(\mathbf{r}', t)$, respectively, by

$$\mathbf{J}_\ell(\mathbf{r}', t) = \frac{\partial}{\partial t} \mathbf{e}_\ell(\mathbf{r}', t), \quad \mathbf{M}_\ell(\mathbf{r}', t) = \frac{\partial}{\partial t} \mathbf{h}_\ell(\mathbf{r}', t) \tag{10.111}$$

so that the unknown source vectors in the integral equations can be unified. With the new source vectors, the TDEFIEs can be written as

$$\begin{aligned}
&\hat{t}(\mathbf{r}) \cdot \left\{ \mu_0 \int_{S_1} \frac{\partial^2 \mathbf{e}_1(\mathbf{r}', \tau_0)}{R \partial t^2} \, \mathrm{d}S' - \frac{1}{\epsilon_0} \int_{S_1} \overline{\mathbf{G}} \cdot \mathbf{e}_1(\mathbf{r}', \tau_0) \, \mathrm{d}S' \right. \\
&+ \mu_0 \int_{S_2} \frac{\partial^2 \mathbf{e}_2(\mathbf{r}', \tau_0)}{R \partial t^2} \, \mathrm{d}S' - \frac{1}{\epsilon_0} \int_{S_2} \overline{\mathbf{G}} \cdot \mathbf{e}_2(\mathbf{r}', \tau_0) \, \mathrm{d}S' \\
&\left. + 2\pi \hat{n}(\mathbf{r}) \times \frac{\partial}{\partial t} \mathbf{h}_2(\mathbf{r}', \tau_0) + \fint_{S_2} \nabla \times \left[\frac{\partial}{R \partial t} \mathbf{h}_2(\mathbf{r}', \tau_0) \right] \mathrm{d}S' \right\} \\
&= 4\pi \hat{t}(\mathbf{r}) \cdot \mathbf{E}^{\mathrm{inc}}(\mathbf{r}, t), \quad \mathbf{r} \in S_1
\end{aligned} \tag{10.112}$$

$$\begin{aligned}
&\hat{t}(\mathbf{r}) \cdot \left\{ \mu_0 \int_{S_1} \frac{\partial^2 \mathbf{e}_1(\mathbf{r}', \tau_0)}{R \partial t^2} \, \mathrm{d}S' - \frac{1}{\epsilon_0} \int_{S_1} \overline{\mathbf{G}} \cdot \mathbf{e}_1(\mathbf{r}', \tau_0) \, \mathrm{d}S' \right. \\
&+ \mu_0 \int_{S_2} \frac{\partial^2 \mathbf{e}_2(\mathbf{r}', \tau_0)}{R \partial t^2} \, \mathrm{d}S' - \frac{1}{\epsilon_0} \int_{S_2} \overline{\mathbf{G}} \cdot \mathbf{e}_2(\mathbf{r}', \tau_0) \, \mathrm{d}S' \\
&\left. + 2\pi \hat{n}(\mathbf{r}) \times \frac{\partial}{\partial t} \mathbf{h}_2(\mathbf{r}', \tau_0) + \fint_{S_2} \nabla \times \left[\frac{\partial}{R \partial t} \mathbf{h}_2(\mathbf{r}', \tau_0) \right] \mathrm{d}S' \right\} \\
&= 4\pi \hat{t}(\mathbf{r}) \cdot \mathbf{E}^{\mathrm{inc}}(\mathbf{r}, t), \quad \mathbf{r} \in S_2
\end{aligned} \tag{10.113}$$

$$\begin{aligned}
&\hat{t}(\mathbf{r}) \cdot \left\{ \mu_2 \int_{S_2} \frac{\partial^2 \mathbf{e}_2(\mathbf{r}', \tau_0)}{R \partial t^2} \, dS' - \frac{1}{\epsilon_2} \int_{S_2} \overline{\mathbf{G}} \cdot \mathbf{e}_2(\mathbf{r}', \tau_0) \, dS' \right. \\
&\left. -2\pi \hat{n}(\mathbf{r}) \times \frac{\partial}{\partial t} \mathbf{h}_2(\mathbf{r}', \tau_0) + \fint_{S_2} \nabla \times \left[\frac{\partial}{R \partial t} \mathbf{h}_2(\mathbf{r}', \tau_0) \right] \, dS' \right\} \\
&= 0, \quad \mathbf{r} \in S_2 \qquad (10.114)
\end{aligned}$$

where $\hat{t}(\mathbf{r})$ is a tangential unit vector at the observation point $\mathbf{r}$ and $\overline{\mathbf{G}} = \overline{\mathbf{G}}(\mathbf{r}, \mathbf{r}') = \nabla\nabla(1/R)$ can be viewed as a dyadic Green's function which is obtained from a transformation [30]. Also, the integral with a dash means that it is defined in a CPV sense. Similarly, the TDMFIEs can be written as

$$\begin{aligned}
&\hat{t}(\mathbf{r}) \cdot \left\{ 2\pi \hat{n}(\mathbf{r}) \times \frac{\partial}{\partial t} \mathbf{e}_1(\mathbf{r}', \tau_0) + \fint_{S_1} \nabla \times \left[\frac{\partial}{R \partial t} \mathbf{e}_1(\mathbf{r}', \tau_0) \right] \, dS' \right. \\
&- \epsilon_0 \int_{S_2} \frac{\partial^2 \mathbf{h}_2(\mathbf{r}', \tau_0)}{R \partial t^2} \, dS' - \frac{1}{\mu_0} \int_{S_2} \overline{\mathbf{G}} \cdot \mathbf{h}_2(\mathbf{r}', \tau_0) \, dS' \\
&\left. +2\pi \hat{n}(\mathbf{r}) \times \frac{\partial}{\partial t} \mathbf{e}_2(\mathbf{r}', \tau_0) + \fint_{S_2} \nabla \times \left[\frac{\partial}{R \partial t} \mathbf{e}_2(\mathbf{r}', \tau_0) \right] \, dS' \right\} \\
&= -4\pi \hat{t}(\mathbf{r}) \cdot \mathbf{H}^{\mathrm{inc}}(\mathbf{r}, t), \quad \mathbf{r} \in S_1 \qquad (10.115)
\end{aligned}$$

$$\begin{aligned}
&\hat{t}(\mathbf{r}) \cdot \left\{ 2\pi \hat{n}(\mathbf{r}) \times \frac{\partial}{\partial t} \mathbf{e}_1(\mathbf{r}', \tau_0) + \fint_{S_1} \nabla \times \left[\frac{\partial}{R \partial t} \mathbf{e}_1(\mathbf{r}', \tau_0) \right] \, dS' \right. \\
&- \epsilon_0 \int_{S_2} \frac{\partial^2 \mathbf{h}_2(\mathbf{r}', \tau_0)}{R \partial t^2} \, dS' - \frac{1}{\mu_0} \int_{S_2} \overline{\mathbf{G}} \cdot \mathbf{h}_2(\mathbf{r}', \tau_0) \, dS' \\
&\left. +2\pi \hat{n}(\mathbf{r}) \times \frac{\partial}{\partial t} \mathbf{e}_2(\mathbf{r}', \tau_0) + \fint_{S_2} \nabla \times \left[\frac{\partial}{R \partial t} \mathbf{e}_2(\mathbf{r}', \tau_0) \right] \, dS' \right\} \\
&= -4\pi \hat{t}(\mathbf{r}) \cdot \mathbf{H}^{\mathrm{inc}}(\mathbf{r}, t), \quad \mathbf{r} \in S_2 \qquad (10.116)
\end{aligned}$$

$$\begin{aligned}
&\hat{t}(\mathbf{r}) \cdot \left\{ \epsilon_2 \int_{S_2} \frac{\partial^2 \mathbf{h}_2(\mathbf{r}', \tau_2)}{R \partial t^2} \, dS' + \frac{1}{\mu_2} \int_{S_2} \overline{\mathbf{G}} \cdot \mathbf{h}_2(\mathbf{r}', \tau_2) \, dS' \right. \\
&\left. +2\pi \hat{n}(\mathbf{r}) \times \frac{\partial}{\partial t} \mathbf{e}_2(\mathbf{r}, \tau_2) - \fint_{S_2} \nabla \times \left[\frac{\partial}{R \partial t} \mathbf{e}_2(\mathbf{r}', \tau_2) \right] \, dS' \right\} \\
&= 0, \quad \mathbf{r} \in S_2. \qquad (10.117)
\end{aligned}$$

Note that the integrands of the CPV integrals in the above equations can be changed into a different form to facilitate the implementation [48]. The TDCFIEs are formed by combining the corresponding equations in the TDEFIEs and TDMFIEs according to the well-known manner [2].

10.5.2 Numerical Solution

Referring to the geometry in Figure 10.35, we discretize the conducting surface S_1 into N_1 small triangular patches and the dielectric surface S_2 into N_2 small triangular patches, respectively, with ΔS_n being the area of nth patch. The unknown source vectors on both surfaces can be expanded as

$$\mathbf{e}_\xi(\mathbf{r}',t) = \sum_{n=1}^{N_\xi} e_\xi^n(t)\mathbf{J}_\xi^n(\mathbf{r}'), \quad \mathbf{h}_2(\mathbf{r}',t) = \sum_{n=1}^{N_2} h_2^n(t)\mathbf{M}_2^n(\mathbf{r}') \tag{10.118}$$

where $\xi = 1, 2$. Substituting the above expansions and applying a Nyström scheme in the space domain, i.e. replacing the integration over a small triangular patch ΔS_n with a summation under a quadrature rule provided that the integrand is regular, we can discretize the TDEFIEs. If we use two orthogonal tangential unit vectors at the pth quadrature point in the mth observation patch, i.e. $\hat{t}_{mp}^{(l)} = \hat{t}^{(l)}(\mathbf{r}_{mp}) = t_{mp}^{u(l)}\hat{u} + t_{mp}^{v(l)}\hat{v} + t_{mp}^{w(l)}\hat{w}$ $(l = 1, 2)$ to test the discretized TDEFIEs, then we can obtain the following matrix equation

$$\begin{aligned}
&\sum_{n=1}^{N_1}\sum_{q=1}^{Q}\left[\frac{\mu_0 w_{nq}}{R_{mpnq}}\frac{d^2 e_1^n(\tau_0)}{dt^2}A_{1mpnq}^{J(l)} - \frac{w_{nq}}{\epsilon_0}e_1^n(\tau_0)B_{1mpnq}^{J(l)}\right]\\
&+\sum_{n=1}^{N_2}\sum_{q=1}^{Q}\left[\frac{\mu_0 w_{nq}}{R_{mpnq}}\frac{d^2 e_2^n(\tau_0)}{dt^2}A_{2mpnq}^{J(l)} - \frac{w_{nq}}{\epsilon_0}e_2^n(\tau_0)B_{2mpnq}^{J(l)}\right]\\
&+\sum_{n=1}^{N_2}\sum_{q=1}^{Q}\left[\left(\frac{w_{nq}}{c_0 R_{mpnq}}\frac{d}{dt} + \frac{w_{nq}}{R_{mpnq}^2}\right)\frac{dh_2^n(\tau_0)}{dt}C_{2mpnq}^{M(l)}\right]\\
&= 4\pi\hat{t}_{mp}^{(l)}\cdot\mathbf{E}^{\mathrm{inc}}(\mathbf{r}_{mp},t), \quad \mathbf{r}_{mp}\in S_1,\\
&m = 1, 2, \cdots, N_1; \quad q = 1, 2, \cdots, Q
\end{aligned} \tag{10.119}$$

$$\begin{aligned}
&\sum_{n=1}^{N_1}\sum_{q=1}^{Q}\left[\frac{\mu_0 w_{nq}}{R_{mpnq}}\frac{d^2 e_1^n(\tau_0)}{dt^2}A_{1mpnq}^{J(l)} - \frac{w_{nq}}{\epsilon_0}e_1^n(\tau_0)B_{1mpnq}^{J(l)}\right]\\
&+\sum_{n=1}^{N_2}\sum_{q=1}^{Q}\left[\frac{\mu_0 w_{nq}}{R_{mpnq}}\frac{d^2 e_2^n(\tau_0)}{dt^2}A_{2mpnq}^{J(l)} - \frac{w_{nq}}{\epsilon_2}e_2^n(\tau_0)B_{2mpnq}^{J(l)}\right]\\
&+\sum_{n=1}^{N_2}\sum_{q=1}^{Q}\left[\left(\frac{w_{nq}}{c_0 R_{mpnq}}\frac{d}{dt} + \frac{w_{nq}}{R_{mpnq}^2}\right)\frac{dh_2^n(\tau_0)}{dt}C_{2mpnq}^{M(l)}\right]\\
&+2\pi D_{2mp}^{M(l)}\frac{dh_2^m(\tau_0)}{dt} = 4\pi\hat{t}_{mp}^{(l)}\cdot\mathbf{E}^{\mathrm{inc}}(\mathbf{r}_{mp},t), \quad \mathbf{r}_{mp}\in S_2,\\
&m = 1, 2, \cdots, N_2; \quad q = 1, 2, \cdots, Q
\end{aligned} \tag{10.120}$$

$$\begin{aligned}
&\sum_{n=1}^{N_2}\sum_{q=1}^{Q}\left[\frac{\mu_2 w_{nq}}{R_{mpnq}}\frac{d^2 e_2^n(\tau_2)}{dt^2}A_{2mpnq}^{J(l)} - \frac{w_{nq}}{\epsilon_2}e_2^n(\tau_2)B_{2mpnq}^{J(l)}\right]\\
&+\sum_{n=1}^{N_2}\sum_{q=1}^{Q}\left[\left(\frac{w_{nq}}{c_2 R_{mpnq}}\frac{d}{dt} + \frac{w_{nq}}{R_{mpnq}^2}\right)\frac{dh_2^n(\tau_2)}{dt}C_{2mpnq}^{M(l)}\right]\\
&-2\pi D_{2mp}^{M(l)}\frac{dh_2^m(\tau_2)}{dt} = 0, \quad \mathbf{r}_{mp}\in S_2,\\
&m = 1, 2, \cdots, N_2; \quad q = 1, 2, \cdots, Q
\end{aligned} \tag{10.121}$$

where

$$\mathbf{J}_{1nq}^n = \mathbf{J}_1^n(\mathbf{r}_{nq}') = J_{1nq}^{nu}\hat{u} + J_{1nq}^{nv}\hat{v} \tag{10.122}$$

$$\mathbf{J}^n_{2nq} = \mathbf{J}^n_2(\mathbf{r}'_{nq}) = J^{nu}_{2nq}\hat{u} + J^{nv}_{2nq}\hat{v} \tag{10.123}$$

$$\mathbf{M}^n_{2nq} = \mathbf{M}^n_2(\mathbf{r}'_{nq}) = M^{nu}_{2nq}\hat{u} + M^{nv}_{2nq}\hat{v} \tag{10.124}$$

are the spatial parts of unknown source vectors at the source point $\mathbf{r}'_{nq}$, which is the qth quadrature point in the nth patch (there are Q quadrature points in total). Since the unknown source vectors are surface vectors (the current densities are surface vectors), each source vector only has two independent components in a local coordinate system $\{u, v, w\}$ where the patch plane is chosen as the u–v plane. Also, w_{nq} is the weight of a quadrature rule at the qth quadrature point in the nth patch, $R_{nq} = |\mathbf{r} - \mathbf{r}'_{nq}|$, $\hat{R}_{nq} = (\mathbf{r} - \mathbf{r}'_{nq})/R_{nq}$, and $\overline{\mathbf{G}}_{nq} = \overline{\mathbf{G}}(\mathbf{r}, \mathbf{r}'_{nq})$. The coefficients in the equations are defined by

$$A^{J(l)}_{1mpnq} = \hat{t}^{(l)}_{mp} \cdot \mathbf{J}^n_{1nq} \tag{10.125}$$

$$A^{J(l)}_{2mpnq} = \hat{t}^{(l)}_{mp} \cdot \mathbf{J}^n_{2nq} \tag{10.126}$$

$$B^{J(l)}_{1mpnq} = \hat{t}^{(l)}_{mp} \cdot \overline{\mathbf{G}}_{mpnq} \cdot \mathbf{J}^n_{1nq} \tag{10.127}$$

$$B^{J(l)}_{2mpnq} = \hat{t}^{(l)}_{mp} \cdot \overline{\mathbf{G}}_{mpnq} \cdot \mathbf{J}^n_{2nq} \tag{10.128}$$

$$C^{M(l)}_{2mpnq} = \hat{t}^{(l)}_{mp} \cdot \mathbf{M}^n_{2nq} \times \hat{R}_{mpnq} \tag{10.129}$$

$$D^{M(l)}_{2mp} = \hat{t}^{(l)}_{mp} \cdot (\hat{n}_{mp} \times \mathbf{M}^m_{2mp}) \tag{10.130}$$

where $R_{mpnq} = |\mathbf{r}_{mp} - \mathbf{r}'_{nq}|$, $\hat{R}_{mpnq} = (\mathbf{r}_{mp} - \mathbf{r}'_{nq})/R_{mpnq}$, and $\overline{\mathbf{G}}_{mpnq} = \overline{\mathbf{G}}(\mathbf{r}_{mp}, \mathbf{r}'_{nq})$. In a similar way, the TDMFIEs can be transformed into a matrix equation but we omit them to save space because they are purely dual to the TDEFIEs. Note that the above procedure can only be used for a far interaction between the observation point and a source patch or $m \neq n$, which allows the direct application of a quadrature rule. If the observation point is inside the source patch or $m = n$, there is a singularity problem and we can make use of the singularity treatment or local correction technique developed in [43] to handle it.

We use the Galerkin method to discretize the TDEFIEs and TDMFIEs in the time domain. The Laguerre function is used as a basis function to expand the temporal parts of unknown source vectors [28]

$$e^n_\xi(t) = \sum_{j=0}^{\infty} \mathrm{e}^n_{\xi j}\phi_j(st), \quad h^n_2(t) = \sum_{j=0}^{\infty} h^n_{2j}\phi_j(st) \tag{10.131}$$

where $\xi = 1, 2$ and $\phi_j(t) = \mathrm{e}^{-t/2}L_j(t)$ is the Laguerre function in which $L_j(t)$ is the Laguerre polynomial of jth degree. Also, s is the scaling factor used to change the support of expansion. Substituting the above expansions into the TDEFIEs, which have been discretized in space domain as shown in Equations (10.119)–(10.121), and using the Laguerre function $\phi_i(st)$ as a testing function to test, we can obtain

$$\sum_{n=1}^{N_1}\sum_{q=1}^{Q}\left\{\left[\frac{s^2\mu_0 w_{nq}}{4R_{mpnq}}A^{J(l)}_{1mpnq} - \frac{w_{nq}}{\epsilon_0}B^{J(l)}_{1mpnq}\right]\mathrm{e}^n_{1i}\right\}I^{(0)}_{ii} +$$

$$
\begin{aligned}
&\sum_{n=1}^{N_2}\sum_{q=1}^{Q}\left\{\left[\frac{s^2\mu_0 w_{nq}}{4R_{mpnq}}A_{2mpnq}^{J(l)}-\frac{w_{nq}}{\epsilon_0}B_{2mpnq}^{J(l)}\right]\mathrm{e}_{2i}^{n}\right\}I_{ii}^{(0)}+\\
&\sum_{n=1}^{N_2}\sum_{q=1}^{Q}\left\{\left(\frac{s^2 w_{nq}}{4c_0R_{mpnq}}+\frac{sw_{nq}}{2R_{mpnq}^2}\right)C_{2mpnq}^{M(l)}h_{2i}^{n}\right\}I_{ii}^{(0)}\\
&=4\pi V_{mpi}^{E(l)}-\\
&\sum_{n=1}^{N_1}\sum_{q=1}^{Q}\sum_{j=0}^{i-1}\left\{\left[\frac{s^2\mu_0 w_{nq}}{4R_{mpnq}}A_{1mpnq}^{J(l)}-\frac{w_{nq}}{\epsilon_0}B_{1mpnq}^{J(l)}\right]\mathrm{e}_{1j}^{n}\right\}I_{ij}^{(0)}-\\
&\sum_{n=1}^{N_2}\sum_{q=1}^{Q}\sum_{j=0}^{i-1}\left\{\left[\frac{s^2\mu_0 w_{nq}}{4R_{mpnq}}A_{2mpnq}^{J(l)}-\frac{w_{nq}}{\epsilon_0}B_{2mpnq}^{J(l)}\right]\mathrm{e}_{2j}^{n}\right\}I_{ij}^{(0)}-\\
&\sum_{n=1}^{N_2}\sum_{q=1}^{Q}\sum_{j=0}^{i-1}\left\{\left(\frac{s^2 w_{nq}}{4c_0R_{mpnq}}+\frac{sw_{nq}}{2R_{mpnq}^2}\right)C_{2mpnq}^{M(l)}h_{2j}^{n}\right\}I_{ij}^{(0)}-\\
&\sum_{n=1}^{N_1}\sum_{q=1}^{Q}\sum_{j=0}^{i}\left[\frac{s^2\mu_0 w_{nq}}{R_{mpnq}}A_{1mpnq}^{J(l)}\sum_{k=0}^{j-1}(j-k)\mathrm{e}_{1k}^{n}\right]I_{ij}^{(0)}-\\
&\sum_{n=1}^{N_2}\sum_{q=1}^{Q}\sum_{j=0}^{i}\left[\frac{s^2\mu_0 w_{nq}}{R_{mpnq}}A_{2mpnq}^{J(l)}\sum_{k=0}^{j-1}(j-k)\mathrm{e}_{2k}^{n}\right]I_{ij}^{(0)}-\\
&\sum_{n=1}^{N_2}\sum_{q=1}^{Q}\sum_{j=0}^{i}\left[\frac{s^2 w_{nq}}{c_0R_{mpnq}}C_{2mpnq}^{M(l)}\sum_{k=0}^{j-1}(j-k)h_{2k}^{n}\right]I_{ij}^{(0)}-\\
&\sum_{n=1}^{N_2}\sum_{q=1}^{Q}\sum_{j=0}^{i}\left[\frac{sw_{nq}}{R_{mpnq}^2}C_{2mpnq}^{M(l)}\sum_{k=0}^{j-1}h_{2k}^{n}\right]I_{ij}^{(0)}\\
&m=1,\cdots,N_1;\quad p=1,\cdots,Q
\end{aligned}
\tag{10.132}
$$

$$
\begin{aligned}
&\sum_{n=1}^{N_1}\sum_{q=1}^{Q}\left\{\left[\frac{s^2\mu_0 w_{nq}}{4R_{mpnq}}A_{1mpnq}^{J(l)}-\frac{w_{nq}}{\epsilon_0}B_{1mpnq}^{J(l)}\right]\mathrm{e}_{1i}^{n}\right\}I_{ii}^{(0)}+\\
&\sum_{n=1}^{N_2}\sum_{q=1}^{Q}\left\{\left[\frac{s^2\mu_0 w_{nq}}{4R_{mpnq}}A_{2mpnq}^{J(l)}-\frac{w_{nq}}{\epsilon_0}B_{2mpnq}^{J(l)}\right]\mathrm{e}_{2i}^{n}\right\}I_{ii}^{(0)}+\\
&\sum_{n=1}^{N_2}\sum_{q=1}^{Q}\left\{\left(\frac{s^2 w_{nq}}{4c_0R_{mpnq}}+\frac{sw_{nq}}{2R_{mpnq}^2}\right)C_{2mpnq}^{M(l)}h_{2i}^{n}\right\}I_{ii}^{(0)}\\
&+\pi sD_{2mp}^{M(l)}h_{2i}^{m}I_{ii}^{(0)}=4\pi V_{mpi}^{E(l)}-\\
&\sum_{n=1}^{N_1}\sum_{q=1}^{Q}\sum_{j=0}^{i-1}\left\{\left[\frac{s^2\mu_0 w_{nq}}{4R_{mpnq}}A_{1mpnq}^{J(l)}-\frac{w_{nq}}{\epsilon_0}B_{1mpnq}^{J(l)}\right]\mathrm{e}_{1j}^{n}\right\}I_{ij}^{(0)}-\\
&\sum_{n=1}^{N_2}\sum_{q=1}^{Q}\sum_{j=0}^{i-1}\left\{\left[\frac{s^2\mu_0 w_{nq}}{4R_{mpnq}}A_{2mpnq}^{J(l)}-\frac{w_{nq}}{\epsilon_0}B_{2mpnq}^{J(l)}\right]\mathrm{e}_{2j}^{n}\right\}I_{ij}^{(0)}-
\end{aligned}
$$

$$\sum_{n=1}^{N_2}\sum_{q=1}^{Q}\sum_{j=0}^{i-1}\left\{\left(\frac{s^2 w_{nq}}{4c_0 R_{mpnq}}+\frac{s w_{nq}}{2R_{mpnq}^2}\right)C_{2mpnq}^{M(l)} h_{2j}^n\right\} I_{ij}^{(0)}-$$
$$\pi s D_{2mp}^{M(l)} \sum_{j=0}^{i-1} h_{2j}^m I_{ij}^{(0)} - 2\pi s D_{2mp}^{M(l)} \sum_{j=0}^{i}\sum_{k=0}^{j-1} h_{2k}^m I_{ij}^{(0)}-$$
$$\sum_{n=1}^{N_1}\sum_{q=1}^{Q}\sum_{j=0}^{i}\left[\frac{s^2\mu_0 w_{nq}}{R_{mpnq}} A_{1mpnq}^{J(l)} \sum_{k=0}^{j-1}(j-k)\mathrm{e}_{1k}^n\right] I_{ij}^{(0)}-$$
$$\sum_{n=1}^{N_2}\sum_{q=1}^{Q}\sum_{j=0}^{i}\left[\frac{s^2\mu_0 w_{nq}}{R_{mpnq}} A_{2mpnq}^{J(l)} \sum_{k=0}^{j-1}(j-k)\mathrm{e}_{2k}^n\right] I_{ij}^{(0)}-$$
$$\sum_{n=1}^{N_2}\sum_{q=1}^{Q}\sum_{j=0}^{i}\left[\frac{s^2 w_{nq}}{c_0 R_{mpnq}^2} C_{2mpnq}^{M(l)} \sum_{k=0}^{j-1}(j-k)h_{2k}^n\right] I_{ij}^{(0)}-$$
$$\sum_{n=1}^{N_2}\sum_{q=1}^{Q}\sum_{j=0}^{i}\left[\frac{s w_{nq}}{R_{mpnq}^2} C_{2mpnq}^{M(l)} \sum_{k=0}^{j-1} h_{2k}^n\right] I_{ij}^{(0)}$$
$$m=1,\cdots,N_2;\quad p=1,\cdots,Q \tag{10.133}$$

$$\sum_{n=1}^{N_2}\sum_{q=1}^{Q}\left\{\left[\frac{s^2\mu_2 w_{nq}}{4R_{mpnq}} A_{2mpnq}^{J(l)} - \frac{w_{nq}}{\epsilon_2} B_{2mpnq}^{J(l)}\right]\mathrm{e}_{2i}^n\right\} I_{ii}^{(2)}+$$
$$\sum_{n=1}^{N_2}\sum_{q=1}^{Q}\left\{\left(\frac{s^2 w_{nq}}{4c_2 R_{mpnq}}+\frac{s w_{nq}}{2R_{mpnq}^2}\right)C_{2mpnq}^{M(l)} h_{2i}^n\right\} I_{ii}^{(2)}$$
$$-\pi s D_{2mp}^{M(l)} h_{2i}^m I_{ii}^{(2)} = 4\pi V_{mpi}^{E(l)}-$$
$$\sum_{n=1}^{N_2}\sum_{q=1}^{Q}\sum_{j=0}^{i-1}\left\{\left[\frac{s^2\mu_2 w_{nq}}{4R_{mpnq}} A_{2mpnq}^{J(l)} - \frac{w_{nq}}{\epsilon_2} B_{2mpnq}^{J(l)}\right]\mathrm{e}_{2j}^n\right\} I_{ij}^{(2)}-$$
$$\sum_{n=1}^{N_2}\sum_{q=1}^{Q}\sum_{j=0}^{i-1}\left\{\left(\frac{s^2 w_{nq}}{4c_2 R_{mpnq}}+\frac{s w_{nq}}{2R_{mpnq}^2}\right)C_{2mpnq}^{M(l)} h_{2j}^n\right\} I_{ij}^{(2)}+$$
$$\pi s D_{2mp}^{M(l)} \sum_{j=0}^{i-1} h_{2j}^m I_{ij}^{(2)} + 2\pi s D_{2mp}^{M(l)} \sum_{j=0}^{i}\sum_{k=0}^{j-1} h_{2k}^m I_{ij}^{(2)}-$$
$$\sum_{n=1}^{N_2}\sum_{q=1}^{Q}\sum_{j=0}^{i}\left[\frac{s^2\mu_2 w_{nq}}{R_{mpnq}} A_{2mpnq}^{J(l)} \sum_{k=0}^{j-1}(j-k)\mathrm{e}_{2k}^n\right] I_{ij}^{(2)}-$$
$$\sum_{n=1}^{N_2}\sum_{q=1}^{Q}\sum_{j=0}^{i}\left[\frac{s^2 w_{nq}}{c_2 R_{mpnq}^2} C_{2mpnq}^{M(l)} \sum_{k=0}^{j-1}(j-k)h_{2k}^n\right] I_{ij}^{(2)}-$$
$$\sum_{n=1}^{N_2}\sum_{q=1}^{Q}\sum_{j=0}^{i}\left[\frac{s w_{nq}}{R_{mpnq}^2} C_{2mpnq}^{M(l)} \sum_{k=0}^{j-1} h_{2k}^n\right] I_{ij}^{(2)}$$
$$m=1,\cdots,N_2;\quad p=1,\cdots,Q \tag{10.134}$$

where

$$I_{ij}^{(0)} = \int_0^{\infty} \phi_i(st)\phi_j\left(st - \frac{s}{c_0}R_{mpnq}\right) \mathrm{d}(st) \tag{10.135}$$

$$I_{ij}^{(2)} = \int_0^{\infty} \phi_i(st)\phi_j\left(st - \frac{s}{c_2}R_{mpnq}\right) \mathrm{d}(st) \tag{10.136}$$

$$V_{mpi}^{E(l)} = \int_0^{\infty} \phi_i(st)\hat{t}_{mp}^{(l)} \cdot \mathbf{E}^{\mathrm{inc}}(\mathbf{r}_{mp}, t)\, \mathrm{d}(st). \tag{10.137}$$

The integrals $I_{ij}^{(0)}$ and $I_{ij}^{(2)}$ in the above can be calculated using the closed form formulations in the appendix in [28] while $V_{mpi}^{E(l)}$ is evaluated numerically by replacing the integral interval with the real duration of a signal. Also, the left-hand side of each equation in the above only involves the unknown time domain coefficients in the current ith degree while the right-hand side of each equation includes all previous degree time domain coefficients that have been known. With the same procedure, we can obtain the recursive formulations for the TDMFIEs, which are omitted due to their duality to those of TDEFIEs, and the corresponding recursive matrix equation for the TDCFIEs can be formed by the aforementioned combination [2]. We have two space domain coefficients and one time domain coefficient at a quadrature point for each current density (i.e. $\mathbf{J}_{1nq}^n$, $\mathbf{J}_{2nq}^n$, and $\mathbf{M}_{2nq}^n$). The space domain coefficients and time domain coefficient should be combined together to form new unknown coefficients (for instance, $J_{1nq}^{nu} e_{1i}^n = J_{1nqi}^{nu}$ and $J_{1nq}^{nv} e_{1i}^n = J_{1nqi}^{nv}$ for $\mathbf{J}_{1nq}^n$) so that they can be solved by two groups of equations corresponding to $l = 1, 2$ for a given i.

10.5.3 Numerical Examples

To demonstrate the proposed approach, we present several numerical examples regarding the transient EM scattering by composite structures. The geometries of scatterers are sketched in Figure 10.36 and they are a dielectric cone backed with a conducting thin circular disk, a conducting cone attached to a dielectric hemisphere, and a conducting hemisphere joint to a dielectric hemisphere, respectively. The conducting parts are assumed to be PECs and the dielectric parts have a relative permittivity $\epsilon_r = 3.0$ (the relative permeability $\mu_r = 1.0$). The first object as shown in Figure 10.36(a) has a radius $a = 0.25$ m and a thickness $d = 0.01$ m for its disk while a radius $a = 0.25$ m and a height $h = 0.25$ m for its cone. The second object as shown in Figure 10.36(b) has a radius $a = 0.2$ m for its hemisphere and cone while a height $a = 0.5$ m for its cone. The third object as shown in Figure 10.36(c) has a radius $a = 0.5$ m for both conducting and dielectric hemispheres. The surfaces of the three objects are discretized into 2362, 986, and 1758 triangular patches, respectively, and we discretize the conducting and dielectric parts independently without considering the conformity in the contacting area of two parts. A rectangular coordinate system $\{x, y, z\}$ is established as shown in the figure and a Gaussian plane wave is illuminating the scatterers along the $-z$ direction. For the first two objects, the incident electric field and magnetic field are defined by

$$\mathbf{E}^{\mathrm{inc}}(\mathbf{r}, t) = 120\pi\hat{x}\frac{4\mathrm{e}^{-\chi^2}}{\sqrt{\pi}T}, \quad \mathbf{H}^{\mathrm{inc}}(\mathbf{r}, t) = -\frac{\hat{z}}{\eta} \times \mathbf{E}^{\mathrm{inc}}(\mathbf{r}, t) \tag{10.138}$$

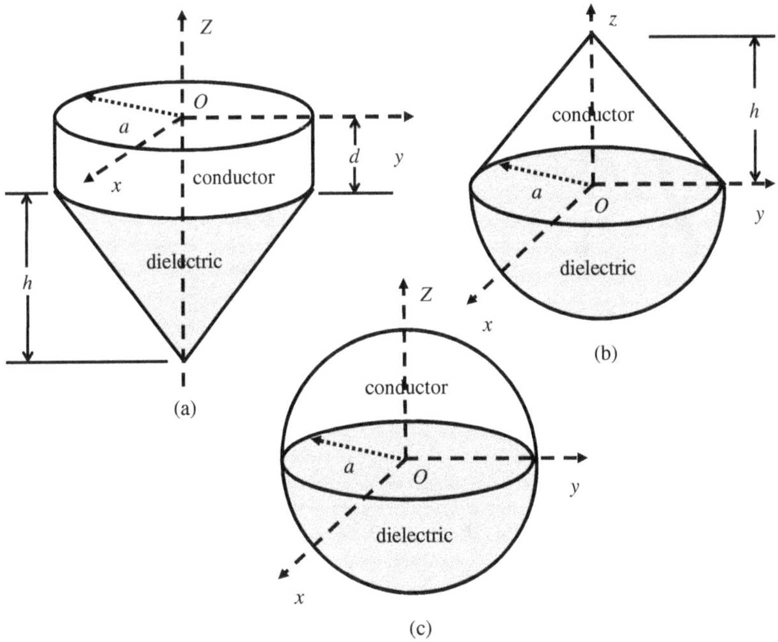

Figure 10.36 Geometry of composite scatterers. (a) A dielectric cone backed with a conducting circular disk. (b) A conducting cone attached to a dielectric hemisphere. (c) A conducting hemisphere joint to a dielectric hemisphere.

respectively, where $\chi = 4(ct - c\Delta t + \mathbf{r} \cdot \hat{z})/T$, T is the duration of Gaussian pulse and Δt is the needed time for the pulse to arrive at the origin. We choose a Gaussian pulse with $T = 4.0$ lm and $c\Delta t = 6.0$ lm as an incident wave. For the third object, we use a modulated Gaussian pulse as an incident wave which is defined by

$$\mathbf{E}^{\text{inc}}(\mathbf{r}, t) = \hat{x}e^{-\left(\frac{\tau-\Delta t}{\sqrt{2}\sigma}\right)^2} \cos(2\pi f_0 \tau) \tag{10.139}$$

where f_0 is the center frequency and $\tau = t + \mathbf{r} \cdot \hat{z}/c$. We choose $f_0 = 500$ MHz while $\Delta t = 8\sigma$ and $\sigma = 6/(2\pi f_b)$ with $f_b = 1$ GHz being the nominal bandwidth.

Figure 10.37 shows the solution for the x-component of surface current density (J_x) sampled at $x = z = 0$ and $y = 0.125$ m on the surface of conducting disk for the first object while Figure 10.38 displays the θ-component of surface current density (J_θ) sampled at $z = 0.2$ m and $\phi = 180°$ on the surface of conducting cone for the second object. Figure 10.39 depicts the x-component of surface current density (J_x) sampled at the north pole on the surface of conducting hemisphere for the third object. All results are compared with the solutions obtained from the IDFT of corresponding frequency domain solutions and they agree well each other. These results are obtained by selecting $s = 10^9$ and $M = 80$, where M is the largest number or degree of temporal Laguerre basis functions. The accuracy of numerical solutions is related to the choice of M when geometric discretization is fixed. Figure 10.40 plots the RMS errors of the solutions compared to the IDFT counterparts when $M = 20, 40, 60$, and 80, respectively, for the three examples and they are clearly convergent as M increases. For the computational complexity of the proposed method, the Nyström scheme is equivalent to the MoM when

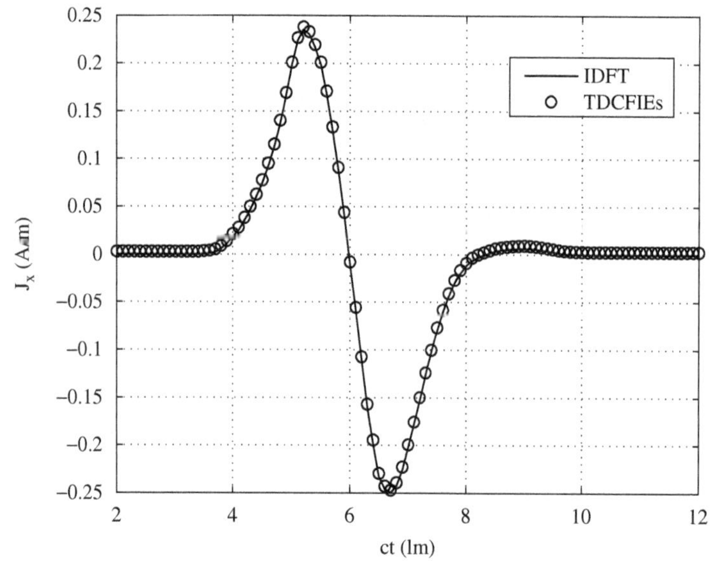

Figure 10.37 x-component of transient current density sampled at $x = z = 0$ and $y = 0.125$ m on the surface of a conducting circular disk attached to a dielectric cone.

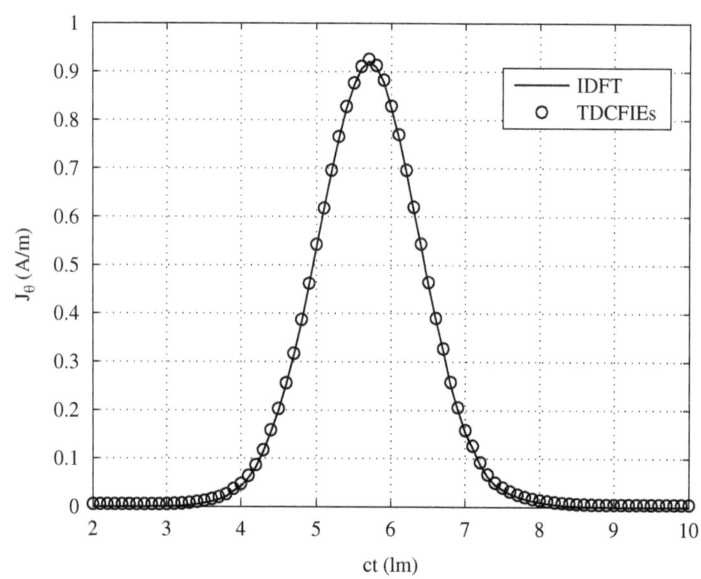

Figure 10.38 θ-component of transient current density sampled at $z = 0.2$ m and $\phi = 180°$ on the surface of a conducting cone attached to a dielectric hemisphere.

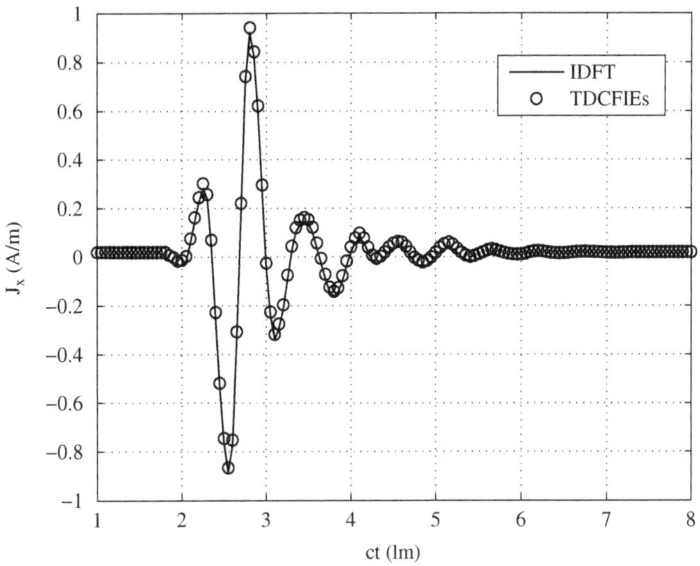

Figure 10.39 *x*-component of transient current density sampled at the north pole on the surface of a conducting hemisphere joint to a dielectric hemisphere.

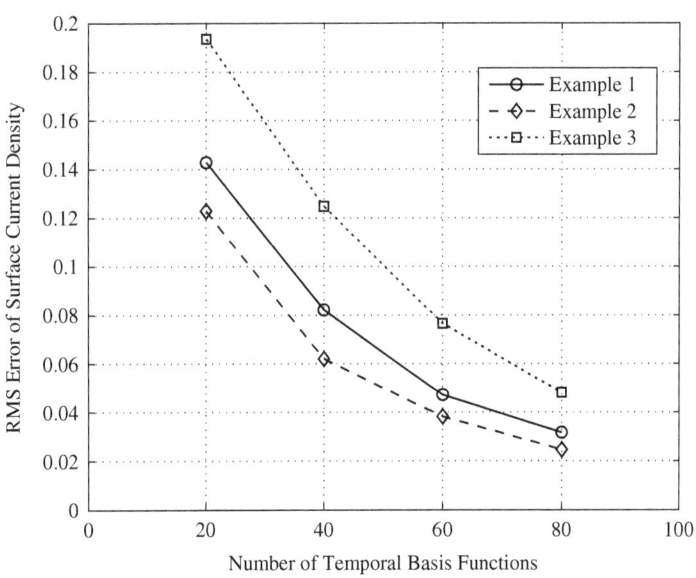

Figure 10.40 RMS error of surface current density versus the number of temporal basis functions for the three examples.

one-point quadrature rules are used in space domain as demonstrated in frequency domain [24] while the comparison between the Galerkin method with Laguerre function and traditional MOT in the time domain is similar to that between the MoM and the finite difference method for time domain discretization.

10.6 Summary

In this chapter, we consider the transient EM interaction with conducting, penetrable, or mixed media, respectively, through the time domain version of various integral equations. We do not use the traditional approach to solve the TDIEs, i.e. incorporating the space domain MoM with a time domain MOT scheme. The space domain MoM requires conforming meshes in geometric discretization, resulting in higher preprocessing cost. Also, two basis functions are required to represent two current densities on material interfaces for the TDSIEs of penetrable media, and the choice of a $\hat{n}\times$RWG basis function for representing the magnetic current density may be problematic for some SIEs. Furthermore, the space domain MoM may be more inconvenient to implement in the TDVIEs or TDVSIEs. On the other hand, the MOT scheme suffers from a well-known late-time instability problem that will become worse in the TDSIEs with penetrable media. The instability is mainly caused by inappropriate sampling in the time domain, just like in the FDTD method, and cannot be easily eliminated in general. We propose a different hybrid approach to solve the TDIEs by combining the Nyström method in the space domain and the GM with the Laguerre basis and testing functions in the time domain. The Nyström method has several notable advantages, as described in the preceding chapters, and can overcome the inconvenience of the MoM. Also, the time domain GM with the Laguerre function possesses some desirable merits, in particular, the full elimination of late-time instability. Numerical examples have been provided to illustrate the proposed approach for each case and good results have been observed.

Bibliography

1 Rao, S.M. (1999). *Time Domain Electromagnetics*. San Diego: Academic Press.

2 Chew, W.C., Tong, M.S., and Hu, B. (2008). *Integral Equation Methods for Electromagnetic and Elastic Waves*. San Rafael, CA: Morgan & Claypool.

3 Sarkar, T.K., Lee, W., and Rao, S.M. (2000). Analysis of transient scattering from composite arbitrarily shaped complex structures. *IEEE Trans. Antennas Propagat.* 48: 1625–1634.

4 Rao, S.M. and Wilton, D.R. (1991). Transient scattering by conducting surfaces of arbitrary shape. *IEEE Trans. Antennas Propagat.* 39 (1): 56–61.

5 Rao, S.M., Wilton, D.R., and Glisson, A.W. (1982). Electromagnetic scattering by surfaces of arbitrary shape. *IEEE Trans. Antennas Propagat.* AP-30 (3): 409–418.

6 Rao, S.M. and Sarkar, T.K. (2002). Numerical solution of time domain integral equations for arbitrarily shaped conductor/dielectric composite bodies. *IEEE Trans. Antennas Propagat.* 50 (12): 1831–1837.

7 Jung, B.H., Sarkar, T.K., and Salazar-Palma, M. (2004). Time domain EFIE and MFIE formulations for analysis of transient electromagnetic scattering from 3-D dielectric objects. *Prog. Electromagn. Res.* 49: 113–142.
8 Sun, L.E. and Chew, W.C. (2009). A novel formulation of the volume integral equation for electromagnetic scattering. *Wave. Random Complex.* 19 (1): 162–180.
9 Manara, G., Monorchio, A., and Reggiannini, R. (1997). A space-time discretization criterion for a stable time-marching solution of the electric field integral equation. *IEEE Trans. Antennas Propagat.* 45: 527–532.
10 Dodson, S., Walker, S.P., and Bluck, M.J. (1998). Implicitness and stability of time domain integral equation scattering analysis. *Appl. Comput. Electromagn. Soc. J.* 13: 291–301.
11 Shanker, B., Ergin, A.A., and Aygün, K. (2000). Analysis of transient electromagnetic scattering phenomena using a two-level plane wave time domain algorithm. *IEEE Trans. Antennas Propagat.* 48: 510–523.
12 Yilmaz, A., Weile, D.S., Michielssen, E., and Jin, J.-M. (2002). A hierarchical FFT algorithm (HIL-FFT) for the fast analysis of transient electromagnetic scattering phenomena. *IEEE Trans. Antennas Propagat.* 51: 971–982.
13 Weile, D.S., Pisharody, G., Chen, N.-W. et al. (2004). A novel scheme for the solution of the time domain integral equations of electromagnetics. *IEEE Trans. Antennas Propagat.* 52 (1): 283–294.
14 Shanker, B., Lu, M., Yuan, J., and Michielssen, E. (2009). Time domain integral equation analysis of scattering from composite bodies via exact evaluation of radiation fields. *IEEE Trans. Antennas Propagat.* 57 (5): 1506–1520.
15 Rynne, B.P. and Smith, P.D. (1990). Stability of time marching algorithm for the electric field integral equation. *J. Electromagn. Waves Appl.* 4: 1181–1205.
16 Smith, P.D. (1990). Instability in time marching methods for scattering: cause and rectification. *Electromagnetics* 10 (4): 439–451.
17 Vechinski, D.A. and Rao, S.M. (1992). A stable procedure to calculate the transient scattering by conducting surfaces of arbitrary shape. *IEEE Trans. Antennas Propagat.* 40 (6): 661–665.
18 Kot, J.S. (1990). Solution of thin-wire integral equations by Nyström methods. *Microw. Opt. Technol. Lett.* 3: 393–396.
19 Canino, L.S., Ottusch, J.J., Stalzer, M.A. et al. (1998). Numerical solution of the Helmholtz equation in 2D and 3D using a high-order Nyström discretization. *J. Comput. Phys.* 146: 627–663.
20 Burghignoli, P., Di Nallo, C., Frezza, F., and Galli, A. (2003). A simple Nyström approach for the analysis of 3D arbitrarily shaped conducting and dielectric bodies. *Int. J. Numer. Model.* 16: 179–194.
21 Gedney, S.D. (2003). On deriving a locally corrected Nyström scheme from a quadrature sampled moment method. *IEEE Trans. Antennas Propagat.* 51: 2402–2412.
22 Wildman, R.A. and Weile, D.S. (2006). Mixed-order testing functions on triangular patches for the locally corrected Nyström method. *IEEE Antennas Wireless Propagat. Lett.* 5: 370–372.
23 Tong, M.S. and Chew, W.C. (2010). E-Field, H-Field, and combined-field based Nyström method analysis for electromagnetic scattering by complex-material bodies. *IEEE Trans. Electromagn. Compat.* 52 (3): 620–628.

24 Tong, M.S., Qian, Z.G., and Chew, W.C. (2010). Nyström method solution of volume integral equations for electromagnetic scattering by 3D penetrable objects. *IEEE Trans. Antennas Propagat.* 58 (5): 1645–1652.

25 Wildman, R.A. and Weile, D.S. (2005). Two-dimensional transverse-magnetic time domain scattering using the Nyström method and bandlimited extrapolation. *IEEE Trans. Antennas Propagat.* 53 (7): 2259–2266.

26 Shi, Y. and Jin, J.-M. (2011). A higher-order Nyström scheme for a march-on-in-degree solution of the magnetic field integral equation. *IEEE Antennas Wireless Propagat. Lett.* 10: 1059–1062.

27 Cao, J., Chen, R.S., Hu, Y., and Tao, S. (2015). A higher order Nyström scheme for a marching-on-in-time solution of time domain integral equation. *IEEE Trans. Antennas Propagat.* 63 (6): 2762–2767.

28 Jung, B.H., Chung, Y.S., and Sarkar, T.K. (2003). Time domain EFIE, MFIE, and CFIE formulations using Laguerre polynomials as temporal basis functions for the analysis of transient scattering from arbitrary shaped conducting structures. *Prog. Electromagn. Res.* 39: 1–45.

29 Shi, Y. and Jin, J.-M. (2011). A time domain volume integral equation and its marching-on-in-degree solution for analysis of dispersive dielectric objects. *IEEE Trans. Antennas Propagat.* 59 (3): 969–978.

30 Tong, M.S. and Chen, W.J. (2015). A hybrid scheme with Nyström discretization for solving transient electromagnetic scattering by conducting objects. *IEEE Trans. Antennas Propagat.* 63 (9): 4219–4224.

31 Tong, M.S. and Wang, P.C. (2016). Stable solution of time domain combined field integral equations for transient electromagnetic scattering by composite structures based on Nyström scheme and Laguerre function. *IEEE Trans. Antennas Propagat.* 64 (7): 3239–3244.

32 Jung, B.H., Ji, Z., Sarkar, T.K. et al. (2007). A comparison of marching-on in time method with marching-on in degree method for the TDIE solver. *Prog. Electromagn. Res.* 70: 281–296.

33 Taflove, A. and Hagness, S.C. (2005). *Computational Electrodynamics: The Finite-Difference Time Domain Method,* 3e. Boston, MA: Artech House.

34 Lee, J.-F., Lee, R., and Cangellaris, A. (1997). Time domain finite-element methods. *IEEE Trans. Antennas Propagat.* 45 (3): 430–442.

35 Tong, M.S. and Zhang, J. (2016). Efficient solution of time domain volume integral equations for transient electromagnetic scattering by dielectric objects. *IEEE Trans. Antennas Propagat.* 64 (10): 4246–4251.

36 Bluck, M.J. and Walker, S.P. (1997). Time domain BIE analysis of large three dimensional electromagnetic scattering problems. *IEEE Trans. Antennas Propagat.* 45: 894–901.

37 Hu, J.-L. and Chan, C.H. (1999). Improved temporal basis functions using for time domain electric field integral equation method. *Electron. Lett.* 35: 883–885.

38 Hu, J.-L., Chan, C.H., and Xu, Y. (2001). A new temporal basis function for the time domain integral equation method. *IEEE Microw. Wireless Comp. Lett.* 11: 465–466.

39 Gray, L.J., Glaeser, J.M., and Kaplan, T. (2004). Direct evaluation of hypersingular Galerkin surface integrals. *SIAM J. Sci. Comput.* 25 (5): 1534–1556.

40 Khayat, M.A. and Wilton, D.R. (2005). Numerical evaluation of singular and near-singular potential integrals. *IEEE Trans. Antennas Propagat.* 53: 3180–3190.

41 Järvenpää, S., Taskinen, M., and Ylä-Oijala, P. (2006). Singularity subtraction technique for high-order polynomial vector basis functions on planar triangles. *IEEE Trans. Antennas Propagat.* 54 (1): 42–49.

42 Tang, W. H. and Gedney, S.D. (2006). An efficient evaluation of near singular surface integrals via the Khayat-Wilton transformation. *Microw. Opt. Technol. Lett.* 48 (8): 1583–1586.

43 Tong, M.S. and Chew, W.C. (2007). Super-hyper singularity treatment for solving 3D electric field integral equations. *Microw. Opt. Technol. Lett.* 49: 1383–1388.

44 Graglia, R.D. and Lombardi, G. (2008). Machine precision evaluation of singular and nearly singular potential integrals by use of Gauss quadrature formulas for rational functions. *IEEE Trans. Antennas Propagat.* 56: 981–998.

45 Tong, M.S. and Chew, W.C. (2010). A novel approach for evaluating hypersingular and strongly singular surface integrals in electromagnetics. *IEEE Trans. Antennas Propagat.* 58 (11): 3593–3601.

46 Chung, Y.S., Sarkar, T.K., Jung, B.H. et al. (2004). Solution of time domain electric field integral equation using the Laguerre polynomials. *IEEE Trans. Antennas Propagat.* 52: 2319–2328.

47 Ji, Z., Sarkar, T.K., Jung, B.H. et al. (2006). Solving time domain electric field integral equation without the time variable. *IEEE Trans. Antennas Propagat.* 54: 258–262.

48 Van Bladel, J. (1964). *Electromagnetic Fields*. New York: McGraw-Hill.

49 Poggio, A.J. and Miller, E.K. (1973). Integral equation solutions of three-dimensional scattering problems, Chapter 4. In: *Computer Techniques for Electromagnetics* (ed. R. Mittra). Oxford: Pergamon Press.

50 Müller, C. (1969). *Foundations of the Mathematical Theory of Electromagnetic Waves*. Berlin: Springer-Verlag.

51 Chen, Q.L. (1990). Electromagnetic modeling of three-dimensional piecewise homogeneous material bodies of arbitrary composition and geometry. Ph.D. Dissertation. Department of Electrical Engineering, University of Houston, Houston, TX.

52 Tong, M.S., Chew, W.C., Rubin, B.J. et al. (2009). On the dual basis for solving electromagnetic surface integral equations. *IEEE Trans. Antennas Propagat.* 57 (10): 3136–3146.

53 Sheng, X.Q., Jin, J.M., Song, J.M. et al. (1998). Solution of combined-field integral equation using multilevel fast multipole algorithm for scattering by homogeneous bodies. *IEEE Trans. Antennas Propagat.* 46: 1718–1726.

54 Shanker, B., Ergin, A.A., and Michielssen, E. (2002). Plane-wave-time domain-enhanced marching-on-in-time scheme for analyzing scattering from homogeneous dielectric structures. *J. Opt. Soc. Am. A.* 19: 716–726.

55 Wildman, R.A. and Weile, D.S. (2005). Numerical solution of time domain integral equations using the Nyström method. *IEEE Antennas and Propagation Society International Symposium, Washington, DC (July 2005)*, 3A: 179–182.

56 Schaubert, D.H., Wilton, D.R., and Glisson, A.W. (1984). A tetrahedral modeling method for electromagnetic scattering by arbitrary shaped inhomogeneous dielectric bodies. *IEEE Trans. Antennas Propagat.* AP-32 (1): 77–85.

57 Lu, M. and Michielssen, E. (2002). Closed form evaluation of time domain fields due to Rao-Wilton-Glisson sources for use in marching-on-in-time based EFIE solvers. *IEEE Antennas and Propagation Society International Symposium, San Antonio, TX (June 2002)*, 1: 74–77.

58 Pingenot, J., Chakraborty, S., and Jandhyala, V. (2006). Polar integration for exact space-time quadrature in time domain integral equations. *IEEE Trans. Antennas Propagat.* 54 (10): 3037–3042.

59 Zhang, G.H., Xia, M.Y., and Jiang, X.M. (2009). Transient analysis of wire structures using time domain integral equation method with exact elements. *Prog. Electromagn. Res.* 92: 281–298.

60 Guo, X.Y., Xia, M.Y., and Chan, C.H. (2014). Stable TDIE–MOT solver for transient scattering by two-dimensional conducting structures. *IEEE Trans. Antennas Propagat.* 62 (4): 2149–2157.

61 Shi, Y.F., Xia, M.Y., Chen, R.S. et al. (2011). Stable electric field TDIE solvers via quasi-exact evaluation of MOT matrix elements. *IEEE Trans. Antennas Propagat.* 59 (2): 574–584.

62 Jin, J.-M. (2002). *The Finite Element Method in Electromagnetics*, 2e. New York, NY: John Wiley & Sons.

63 Chew, W.C., Jin, J.M., Michielssen, E., and Song, J.M. (2001). *Fast and Efficient Algorithms in Computational Electromagnetics.* Boston: Artech House.

Index

The Nyström Method in Electromagnetics, First Edition. Mei Song Tong and Weng Cho Chew.

d

e

f

www.ingramcontent.com/pod-product-compliance
Lightning Source LLC
LaVergne TN
LVHW062203200926
848484LV00006B/1305
* 9 7 8 1 1 1 9 2 8 4 8 4 0 *